AF790883

Amphibians in the Caribbean

The Conservation and Biogeography of **Amphibians in the Caribbean**

Edited by

Neftalí Ríos-López
and
Harold Heatwole

Pelagic Publishing

Published by Pelagic Publishing
20–22 Wenlock Road
London N1 7GU
United Kingdom

www.pelagicpublishing.com

The Conservation and Biogeography of Amphibians in the Caribbean

Copyright © 2023 Neftalí Ríos-López, Harold Heatwole and respective contributors to the various chapters

ISBN 978-1-78427-267-8 Hbk
ISBN 978-1-78427-268-5 ePub
ISBN 978-1-78427-269-2 PDF

This book should be quoted as Ríos-López, N. and Heatwole, H. (eds) (2023) *The Conservation and Biogeography of Amphibians in the Caribbean*. London, Pelagic Publishing.

It is part of the series 'Amphibian Biology', comprising Volume 9: Status of Decline of Amphibians: Western Hemisphere, Part 5: The Caribbean.

All rights reserved. No part of this document may be produced, stored in a retrieval system, or transmitted in any form or by any means, electronic, mechanical, photocopying, recording or otherwise without prior permission of the publisher.

While every effort has been made in the preparation of this book to ensure the accuracy of the information presented, the information contained in this book is sold without warranty, either express or implied. Neither the author, nor Pelagic Publishing, its agents or distributors will be held liable for any damage or loss caused or alleged to be caused indirectly by this book.

British Library Cataloguing in Publication Data
A catalogue record for this book is available from the British Library.

https://doi.org/10.53061/HUCG2445

Cover image: *Eleutherodactylus patriciae*, a species endemic to the Cordillera Central, Dominican Republic. Photo © Luis M. Díaz (taken in 2011 at Valle Nuevo Scientific Reserve, Constanza).

Book frontispiece: Adult female *Eleutherodactylus portoricensis*, a species listed as Endangered by the International Union for Conservation of Nature (IUCN). Taken at Cerro La Tabla (805 m asl; WGS84, 18° 3' 17.31" N, 66° 7' 23.35" W), Guayama, southeastern Puerto Rico, 2 November 2019. Photo © Neftalí Ríos-López

Typeset by Deanta Global Publishing Services, Chennai, India

Dedication

Stephen Blair Hedges (in his late 30s) presumably at the parking lot of the Greyhound Bus Station in University Park campus, Pennsylvania State University (Pennsylvania, USA) in April 1994. Cuban herpetologist Alberto R. Estrada took this picture as part of a one-week working session, between Blair's house and his facility at Müller Laboratory (University Park), with material collected by Estrada in 1993 from Cuba. This work resulted in the descriptions of *Eleutherodactylus glamirus*, *E. iberia*, and *E. tetajulia*, which were published elsewhere in separate papers. S. Blair Hedges has contributed to the taxonomy, phylobiogeography, evolution, and conservation of the Caribbean amphibians in profound and significant ways: by the time of this volume, Blair's legacy includes 298 publications (10 of these books and monographs) in 43 years of publishing, and continuing; he has authored and coauthored the description of 36 species of amphibians in which 21 are from the Caribbean islands. His work – an ongoing endeavour – has motivated younger generations of herpetologists into applying the methodic and rigorous approach to the study of amphibians and their conservation in the Caribbean.

Calling male *Eleutherodactylus auriculatus*. Cupeyal del Norte, Guantánamo, Cuba. Photo © Luis M. Díaz.

Contents

Contributors

EDITORS

RÍOS-LÓPEZ, Neftalí, University of Puerto Rico–Humacao Campus, Call Box 860, Humacao, Puerto Rico, 00792.
neftali.rios@upr.edu
HEATWOLE, Harold, Department of Zoology, University of New England, Armidale, NSW 2351, Australia.
harold_heatwole@ncsu.edu

AUTHORS

ANGELI, Nicole F., Division of Fish and Wildlife, Department of Planning and Natural Resources, Government of the U.S. Virgin Islands, USA.
nicoleangeli1@gmail.com
AUGUSTE, Renoir J., Department of Life Sciences, The University of the West Indies, St. Augustine, Trinidad and Tobago.
renguste@gmail.com
BIGNOTTE-GIRÓ, Irelis, Centro Oriental de Ecosistemas y Biodiversidad (BIOECO), Museo de Historia Natural 'Tomás Romay', Enramadas # 601, Santiago de Cuba 90100, Cuba.
irelis@bioeco.cu
CÁDIZ, Antonio, Department of Biology, University of Miami, 215 Cox Science Center, 1301 Memorial Drive, Coral Gables, FL, 33146, USA.
tonycadiz@miami.edu
COSSIO, Ricardo, University of Veterinary Medicine, Institute of Zoology, Bünteweg 17, 30559 Hannover, Lower Saxony, Germany.
calcalifer2@gmail.com
DÁVILA-CASANOVA, Daniel, Puerto Rico Biodiversity Conservation Office, Inter American University of Puerto Rico, Metropolitan Campus, P.O. Box 19597, San Juan, Puerto Rico, 00919; Department of Environmental Sciences, University of Puerto Rico–Río Piedras Campus, P.O. Box 70377, San Juan, Puerto Rico, 00936.
coquidanny@gmail.com
DÍAZ, Luis M., Museo Nacional de Historia Natural de Cuba. Obispo #61 Esquina Oficios, Plaza de Armas, Habana Vieja, CP 10100, Cuba.
luistromcuba@yahoo.es
DOWNIE, J. Roger, School of Life Sciences, Graham Kerr Building, University of Glasgow, Glasgow G12 8QQ, United Kingdom.
Roger.Downie@glasgow.ac.uk
ESTRADA, Alberto R., 13355 SW 9th Ct APT 415 H, Pembroke Pines, FL 33027, USA.
peleuth@gmail.com
FERNÁNDEZ de ARCILA, Rolando, Centro Nacional de Áreas Protegidas. Calle 18A, #4114 e/ 43 y 47, Playa, La Habana, Cuba.
roland@snap.cu

FONG, Ansel, Centro Oriental de Ecosistemas y Biodiversidad (BIOECO), Museo de Historia Natural 'Tomás Romay', Enramadas # 601, Santiago de Cuba 90100, Cuba.
ansel@bioeco.cu

FREEBORN, Layla, Department of Biological Sciences, Kenneth P. Dietrich School of Arts and Sciences, University of Pittsburgh, 4249 Fifth Avenue, Pittsburgh, PA 15260, USA.
LRF18@pitt.edu

GALEANO, Sandra P., Instituto de Investigación de Recursos Biológicos Alexander von Humboldt, Bogotá, Colombia and Grupo Herpetológico de Antioquia, Instituto de Biología, Universidad de Antioquia, Medellín, Colombia.
spgalean@gmail.com

GIERY, Sean T., Department of Ecology and Evolutionary Biology, University of Connecticut, Storrs, CT 06269, USA.
stgiery@gmail.com

GONZÁLEZ, Amnerys, Centro Nacional de Áreas Protegidas. Calle 18A #4114 e/ 43 y 47, Playa, La Habana, Cuba.
amnerys@snap.cu

HENDERSON, Robert W., Vertebrate Zoology, Milwaukee Public Museum, 800 W. Wells St., Milwaukee, WI 53233, USA.
rh@mpm.edu

INCHÁUSTEGUI, Sixto J., Grupo Jaragua, El Vergel 33, Santo Domingo, Republica Dominicana.
sixtojinchaustegui@yahoo.com

JOWERS, Michael J., CIBIO, Centro de Investigação em Biodiversidade e Recursos Genéticos, InBIO Laboratório Associado, Campus de Vairão, Universidade do Porto, 4485–661 Vairão, Portugal. Departamento de Zoología, Facultad de Ciencias, Universidad de Granada, 18071 Granada, Spain.
michaeljowers@hotmail.com

LEHTINEN, Richard M., Department of Biology, The College of Wooster, 931 College Mall, Wooster, Ohio 44691, USA.
rlehtinen@wooster.edu

MOHAMMED, Ryan, Biological Department, Auburn University, Alabama, USA.
ryansmohammed2020@gmail.com

MURPHY, John C., Science and Education, Field Museum of Natural History, 1400 Lake Shore Drive, Chicago, IL 60616, USA.
serpentresearch@gmail.com

POWELL, Robert, Department of Biology, Avila University, 11901 Wornall Road, Kansas City, MO 64145, USA.
robert.powell@avila.edu

PUENTE-ROLÓN, Alberto R., Department of Biology, University of Puerto Rico–Mayagüez Campus, Carr. 108 Barrio Miradero Km 1.3, Mayagüez, Puerto Rico, 00681-9012.
alberto.puente@upr.edu

REYNOLDS, R. Graham, Department of Biology, University of North Carolina Asheville, One University Heights, Asheville, NC 28804, USA.
greynold@unca.edu

STEPHENSON, Kimberly, Department of Life Sciences, Faculty of Science and Technology, University of the West Indies, Mona, Kingston 7, Jamaica.
kimberly.stephenson02@uwimona.edu.jm

SUNYER, Javier, Departamento de Biología, Facultad de Ciencias y Technología, Universidad Nacional Autónoma de Nicaragua-León, León, Nicaragua.
jsunyermaclennan@gmail.com
van BUURT, Gerard, Kaya Oy Sprock 18, Curaçao, Dutch West Indies.
gvanbuurt@gmail.com
VEGA-CASTILLO, Sondra I., Department of Biology, University of Puerto Rico–Arecibo Campus, Carreter. 653 Km. 0.8 Sector Las Dunas, Arecibo, Puerto Rico, 00614-4010.
sondra.vega@upr.edu
WILSON, Byron, Department of Life Sciences, University of the West Indies, Mona, Kingston 7, Jamaica.
byron.wilson@uwimona.edu.jm

Preface

Amphibians are notoriously poor colonizers of islands, for several reasons: (1) Although a few exceptional species are somewhat resistant, most amphibians osmotically lose their body water rapidly and survive only short periods of exposure to seawater (only hours versus the survival of lizards for days in similar situations; Heatwole, unpub. data), thereby negating the likelihood of their being dispersed as waifs either directly in seawater or in marine-transported flotsam. (2) Some islands, especially small sand cays, lack standing water for breeding and do not provide shelter from heat, thereby preventing establishment even if amphibians do reach them. (3) The peripheral habitats on islands (beaches and strands), where dispersing amphibians would most likely make first contact, usually are dry, and such waifs would have to traverse that inhospitable barrier to reach any favourable, moist habitats inland. Contrariwise, frogs sometimes are introduced onto islands, either deliberately by humans, or inadvertently in plants or other materials. The various chapters in the present volume need to be interpreted against this background.

The region covered in this book reaches somewhat beyond the Caribbean proper; the Bahamas and the Turks & Caicos islands probably should be considered as Western Atlantic, rather than Caribbean, and some of the small nearshore islands of coastal northern South America and eastern Central America perhaps should be seen as extensions of the continental landmass. However, some of the latter islands are considered here in order to illustrate the range of types of Caribbean islands.

Like volumes 8 and 10 and the various Parts of volumes 9 and 11 already published in the series *Amphibian Biology*, the present book not only (1) looks backward in assessing how the amphibian fauna of particular Caribbean regions came to be what they are, and (2) deals with the present by describing the current status of those faunas and the effectiveness of mitigations now in place; it (3) also looks forward in suggesting measures to ameliorate the worst-case scenario of the unprecedented, catastrophic declines and extinctions originating from diseases and from anthropogenic pollution, destruction of habitat, introduction of invasive species, and climatic change, which are now in progress.

Amphibians are perhaps the most important taxon for such study because their thin, permeable skins make them especially vulnerable to environmental insults and they are undergoing declines and extinctions at a greater rate than most other taxa. Hence, they may be harbingers of the fate of the rest of us who share their planet and who ourselves will increasingly succumb to the worsening environmental pressures we inflict on them – unless we heed their timely warning.

The taxonomic nomenclature of amphibians is in a state of flux and can be controversial. In this book, we use familial and generic names that could change as relationships and the associated nomenclature are resolved.

Similar to how the resolution of relationships can affect taxonomic nomenclature, information regarding emerging and ongoing threats and current conditions can alter assessments of the conservation status of species. Of the 222 assessments of Caribbean amphibians on the IUCN Red List of Threatened Species prior to 2020, 100 were published as the result of the Global Amphibian Assessment (GAA), a major initiative in 2004 to assess all known amphibian species. As of 2014, all of those assessments were designated in need of updating. Since the GAA, one assessment was published in 2006, 18 in 2008, one in 2009, 74 in 2010, 13 in 2013, seven in 2014, two in 2015, three in 2017, nine in 2020, 45 in 2021, and 144 in 2022. Some of the later assessments are updates and several address continental species with island populations established via human mediation. Hence a few of the assessments cited in the chapters of this book are out of date, and the IUCN SSC Amphibian Specialist Group continues the process of re-evaluating the conservation status of the region's species as new threats emerge and populations and entire species respond to global climatic changes and human-mediated alterations of their habitats. In addition, the conservation status of a number of newly described species from the Caribbean region have yet to be assessed.

Amphibian Biology thus serves as a time capsule, providing a perspective by present-day experts on the conservation of amphibians and the challenges faced by each taxon in the region. Because many species are declining and some have become extinct or are on the verge of extinction while only a select few benefit from human activities, in one sense the present treatise will be out of date by the time it is published. In another sense, however, it is timeless because one can turn to it for information on status at a specific time as a means of assessing subsequent extents and rates of change.

Harold Heatwole
Guyra NSW, Australia
December 2022

Eleutherodactylus glamyrus male vocalizing among understorey vegetation, as they frequently do at low heights above ground. Taken on 19 July 2016, by Luis M. Díaz at Aguada de Joaquín, Sierra Maestra, Cuba. This species is listed as Endangered by the International Union for Conservation of Nature; it is restricted to areas of rainforest over 800 m asl in the Sierra Maestra mountain range, eastern Cuba.

1 Theories of Insular Biogeography and Ecology

Harold Heatwole and Neftalí Ríos-López

I. Background

The Caribbean archipelagos contain islands of various sizes, topographies, geologic origins, vegetation formations, and history of exploitation by humans, all of which have affected their biotas. Accordingly, the story of the decline and conservation of their amphibians is intimately entwined with the insular biogeography and the ecology of the region, and it is germane to provide a brief review of these fields as a context for the specific treatment of individual Caribbean archipelagos. This introductory chapter is not intended to be a comprehensive review of all the literature pertaining to insular biogeography and ecology; such a treatise would be longer than this entire volume. Accordingly, we selectively emphasize seminal papers and reviews, when possible, rather than the contained primary references.

The early Greek philosophers believed that some organisms were spontaneously generated from forces residing in air, water, fire, or earth (mud, slime) (Aristotle, not dated; Peck 1963) and this view enjoyed wide acceptance for centuries and even still had some adherents as late as the 19th century (Nordenskiöld 1928). Thus, the occurrence of some animals on islands was easily explained: they could generate spontaneously there as easily as on a mainland. Even the later Judeo-Christian concept of creation via Jehovah's direct verbal order was not always deemed to be in conflict with the theory of spontaneous generation; the waters brought forth birds (Moses, not dated a) and the earth brought forth terrestrial animals (Moses, not dated b). Saint Augustine (413–426) apparently was not completely satisfied with spontaneous generation being

Harold Heatwole and Neftalí Ríos-López, 'Chapter 1: Theories of Insular Biogeography and Ecology' in: *The Conservation and Biogeography of Amphibians in the Caribbean*. Pelagic Publishing (2023). © Harold Heatwole and Neftalí Ríos-López. DOI: 10.53061/USVO8651 responsible for the occurrence of insular faunas although he considered it probably accounted for frogs. He suggested that in addition to the spontaneous generation of lower life, sexually reproducing animals reached islands via swimming, transport by humans, dispersal by angels under divine decree, or by local creation *in situ* directly by God. The first really scientific approach, and indeed probably the first exposition of evolution, was by a Benedictine monk, Don Calmut (White 1895), who realized that the ark could not possibly hold all of the species of animals that were then being discovered around the world. He suggested that the ark only held a prototype for each general kind of animal and that after liberation from the ark, each prototypic species then diverged into a number of different, but related, species.

The role of isolation of species on islands was a major impetus in the independent formulation of the concept of evolution by Charles R. Darwin and Alfred R. Wallace, and they can reasonably be viewed as the originators of the modern field of insular biogeography (Darwin and Wallace; read publicly in 1858 and published in 1859). For an account of the development of ideas regarding the species–area curve by naturalists in the late eighteenth century to its more mathematical description in the 1920s, and the more sophisticated modelling of the present time, see the review by Tjørve *et al.* (2020).

There is a vast emerging literature that extends the study of insular biogeography backwards in time to examine its evolutionary implications (e.g. adaptive shift; divergence of isolated populations into endemic species); we only perfunctorily touch upon this topic herein; for an entree into this broader subject the reader is referred to papers by Wilson (1959, 1961), Clegg (2010), Grant and Grant (2010), Ricklefs (2010), Vellend and Orrock (2010), Chen *et al.* (2011), Walker and Bellingham (2011), Warren *et al.* 2015; Whittaker *et al.* (2017), Ali and Meiri (2019), Gravel and Massol (2020), Heimus and Behm (2020), Triantis and Matthews (2020), and Valente *et al.* (2020). Similarly, the relevance of the theories of insular biogeography to practical matters such as restoration, assessing fragmentation of habitats, and planning the effective sizes of refuges for purposes of conservation, is not reviewed here as they are treated, where applicable, in the following chapters. The publications of Laurance (2010) and Temperton *et al.* (2013) provide the reader with some theoretical background for this topic. Finally, the theory arising from insular biogeography has increasingly had a profound influence on the development of biogeography and community ecology generally, including that of mainlands and 'habitat islands' (habitats surrounded not by water but by other habitats of a different type (see Matthews 2021); this trend is briefly treated in section II. C. *The legacy of the equilibrium theory*). A good general background on these topics and biogeography, in general, is available in the textbook *Biogeography* (Lomolino *et al.* 2010). For one of the most comprehensive and important, but heretofore overlooked, treatments of all aspects of the dynamics of insular biogeography for any archipelago, see Abbott and Wills (2016).

Three modern theories of insular biogeography, the Equilibrium theory of MacArthur and Wilson (1963, 1967), the Trophic Structure Stability theory of Heatwole and Levins (1972), and the Assembly Rules of Diamond (1975), are treated herein. All were promulgated within slightly more than a decade of each other. They have now been scrutinized and tested for about half a century and variously have been supported, modified, or considered invalid by later researchers, sometimes in emotive, confrontational language, rather than the cool objectivism generally considered as the foundation of scientific debate. This range of views currently persists, although a fair assessment seems to be that all three theories are valid general descriptors of nature, but not applicable to all situations and in need of modification or expansion in a number of ways.

The basis of much of the controversy about all three theories was whether insular communities were structured randomly or by more deterministic processes. Early naïve, biased null hypotheses that were irrefutable, whether true or not, gave way eventually to more refined and realistic null hypotheses of improved validity. In the following pages, these three theories are treated *seriatim*.

II. The equilibrium theory of insular biogeography

MacArthur and Wilson (1963, 1967) proposed that the rate of immigration of species new to an island should be higher for islands closer to the mainland than for those more distant from the source and for larger islands than for smaller ones, everything else being equal. Extirpations[1] were presumed to occur more frequently on small islands than on larger ones, again, everything else being equal. It follows, therefore, that the intersection between the curves for immigration and for extirpation would represent the number of species on an island at equilibrium (their graph occurs in their two publications cited above, as well as in a great number of the papers cited in this and following sections and is not reproduced herein). They acknowledged that the intervention of 'stepping-stone' islands between a given island and its mainland source of biota would affect the dynamics of immigration.

There are several implications of this model: (1) The equilibrium number of species on an island should be relatively stable, but the species-composition should vary with the vagaries of dispersal and extirpation; the species-composition may have a stochastic component because it may partly be a matter of chance which of the available species in the pool of possible immigrants happens to successfully disperse to the island at any given time. Some species, however, may be better dispersers and reach the island more often than others. (2) There should be a measurable turnover of species on islands. (3) The number of species on an island does not necessarily reflect the total number of species that an island could accommodate at 'saturation'; the number of resident species could change via the introduction of additional ones by humans (an increase in the immigration rate). (4) On a geologic scale, the equilibrium number for an island could change because of alterations in an island's size or its distance from the source, e.g. by rises or falls in sea level or by volcanic action. Note, however, that (5) introduction by humans could affect extirpation as well as immigration if an introduced species eliminated a resident one, rather than occupying a vacant niche.

A. Preludes to the theory

The prelude to the development of the MacArthur–Wilson equilibrium theory was abundant descriptive evidence that the number of resident insular species increases with area of island (species–area curve) and that insular habitats support fewer species than do equivalent areas on a mainland (see review by Johnson and Raven 1970; Allison 2002).

Levins and Heatwole (1963) reasoned that the probability of a species being found on an island depends on the probability of it reaching the island and the probability of its survival once there. They let D = the average duration of a species on the island and M = the average waiting time for new immigration to occur. Thus, since all species eventually are eliminated at one time or other:

$$\frac{D}{M + D}$$

---

1 Note that the early literature used the term 'local extinction' or merely 'extinction' to indicate the disappearance of a species from an island. Extinction, however, refers to the complete disappearance of a taxon, with no living representatives left anywhere in the world and hence 'local extinction' is an oxymoron. The correct term now generally accepted by many biologists for the complete disappearance of a taxon from only a part of its range, such as an island, but leaving extant populations elsewhere, is 'extirpation'. This term should not be confused with 'extermination', which implies a deliberate attempt to eradicate an individual, population, or taxon.

indicates the average time that a species will occur on the island, and for N potential inhabitants

$$\frac{ND}{M+D}$$

indicates the average number of species present on the island, roughly equivalent to equilibrium number although the approach is from an inverse perspective, i.e. as the expected time to extirpation and the expected time to recolonization, rather than as the rates of extirpation and colonization (see Hanski 2010).

N depends on the size of the pool of species at the source of migration, D on the insular environment and the species' ecological requirements, and M on the dispersal capabilities of the species and its tolerance to the conditions of transport. Levins and Heatwole further discussed four specific processes that influence M and D: (1) *Random Extirpation*, which in turn is influenced by population size; populations that are very small, either because few immigrants have just arrived and have not yet built up a population, or because the island is tiny, are more vulnerable to this effect. (2) *Environmental Tolerance* depends both on the environmental diversity of the island and the breadth of habitats it offers, as well as on the physiological and ecological tolerances of the immigrants, eurytopic species being more likely than stenotopic ones to find suitable habitat among the array present. (3) *Genetic Drift* in small populations works against natural selection and reduces the ability of a population to respond adaptively to environmental change. (4) *Community Effects*: the mix of other species may affect the chances of a species persisting on an island. Thus, D may be reduced in the presence of a competing species, and species-pairs may not be distributed independently among islands but rather exhibit mutual exclusion. For information about the interaction of Heatwole and Levins with MacArthur and Wilson, see section VII. *Acknowledgements*.

More directly related to the work of MacArthur and Wilson was an early (1948) thesis by Eugene G. Monroe that captured most of the essential elements of the equilibrium theory independently but never being otherwise published, failed to receive public attention or acclaim (see Brown and Lomolino 1989).

B. Tests of the equilibrium theory and its modifications

The publication of the equilibrium theory stimulated a flurry of investigations designed to test its validity. For example, Farrow (1984) demonstrated measurable rates of immigrations of aerially dispersed insects even to a distant, small island and Heatwole *et al.* (1981a) found repeated immigrations, extinctions, and species-turnover on a coral cay even over a short period of time. An early experimental study by Simberloff (1976a) and Simberloff and Wilson (1969, 1970) traced the recolonization by arthropods of some Floridian mangrove islands after the cays had been enclosed in tents and the fauna extirpated by fumigation.

Not all investigators accepted the validity of the theory, however. For instance, Gilbert (1980) concluded that 'Studies purporting to validate the [equilibrium] theory are critically examined, and it is concluded that there is little support for its tenets.' Few ecologists accepted such an extreme view, but many found that modifications were necessary to round out what was an overly simplistic generalization (for reviews of insular dynamics and the range of viewpoints on the equilibrium theory, see Whittaker and Fernández-Palacios 2007; Lomolino 2000; Losos and Ricklefs 2010; and Chase *et al.* 2019). A few examples are summarized below to illustrate the types of modifications that have been made.

1. Overshoot

The first modification recognized was by Simberloff and Wilson (1969, 1970), whose results supported the equilibrium theory but added an important new observation, *viz.*, that during the colonization of an island devoid of any biota, there is first an overshoot in the number of species, followed by a relaxation to the lower, more stable equilibrium number. An obvious interpretation of this phenomenon is that during the early stages of colonization, or in this case recolonization, the numbers of individuals of invading species are low, and populations have not yet reached a critical size at which biological interactions, such as interspecific competition or predation, may be sufficient to eliminate some of the species, and/or prevent newly arriving ones from becoming established.

2. *Size of island may not be as important as once thought*

The area of an island may not be the proximate factor at all, but rather a surrogate for habitat diversity (see Fox and Fox 2000) or topography (Irl *et al.* 2015). Heatwole (1991a), for example, showed that the number of species of insular plants was more precisely related to the height of an island than it was to the basal area. A large, sandy island, or a large mangrove island, is merely more of the same kind of habitat occurring on a small island of the same type, whereas a higher island may have sandy habitat or mangroves around the periphery, with other habitats added farther inland and along environmental (elevational) gradients (see section III. B. g. *The small-island effect and minimum insular area*). Accordingly, the number of habitats – as an indicator of ecological diversity and number of niches – might be a more direct determinant of equilibrium number than mere physical area. Other important attributes for which size of island may be a surrogate are productivity (Terborgh 2010) and length of food chain (Holt 1996). Whittaker and Fernández-Palacios (2007) reviewed the use of available energy, or a surrogate for it, instead of – or incorporated with – area in the species–area curve. At least in some cases, the component of energy explained more of the variation in number of species than that obtained by the species–area curve. Taxa, however, differed in the extent of this effect.

3. *Species are not all equivalent*

Many modifications centred around the recognition that species are not interchangeable cyphers in insular dynamics but that interspecific differences such as autecology need to be considered. Heatwole *et al.* (1981a) found that some taxa (soil microarthropods) had low rates of immigration, extirpation, and species-turnover compared to other taxa (epigeal insects) on the same island. Low species-turnover may be maintained by high immigration rates via sea dispersal and low extirpation rates because of hardiness or habitat-related factors (e.g. Ríos-López and Aide 2007). Franzén *et al.* (2012) found that reproductive capacity, abundance, size of range, trend of the population, temperature, preferred habitat, breadth of diet, and taxonomic category modified the species–area relationship by altering the slope of the species–area curve. Ohyama *et al.* (2021) found that for ants, there was a negative relationship between the slope of the species–area curve and precipitation. Clearly, the equilibrium theory is tempered by the autecology of component species and of their interactions with each other in different environments. Especially important are the trophic relations of individual species and their incorporation into food webs (e.g. see Holt 1996; Holt *et al.* 1999, 2010, 2021; Cirtwill and Stouffer 2016; Guzman *et al.* 2019).

In the equilibrium theory, local food webs are random samples of the regional one, and the theory makes no prediction of the outcome of the association between the breadth of the diet of component species to insular areas. An important paper on this topic is that of Gravel *et al.* (2011), who extended the equilibrium theory by including the trophic relations of the species involved to form a model termed the 'Trophic Theory of Island Biogeography'. In this model, the breadth of diet was included (e.g. the number of prey species in the diet of a predator), thereby differentiating between roles of stenotrophic (dietary specialists) and eurytrophic (dietary generalists) species' roles in the dynamics of insular communities. Their model predicted a species-specific relationship between diet-breadth and occupancy; for example, there was a much higher number of links per food web and a higher connectance than under the hypothesis of random sampling. On small islands, species with a broader breadth of diet are more likely to occur than are more specialized consumers. This prediction suggests that selective establishment of species may be based on their trophic status, and thus it is relevant to the theory of trophic structure stability (see section III. *The theory of trophic structure stability*).

Massol *et al.* (2017) found that food-web connectance and size on the mainland increased species diversity and that more heavily tailed degree distributions on the mainland food web correlated with less frequent but potentially more important extirpation cascades on islands.

An extensive literature on how individual trophic categories exert influence on the structure of meta-communities on mainlands can only be treated here briefly to indicate some of the main features that may have relevance to islands. Ryberg and Chase (2012) produced a model describing how generalist and specialist predators may alter the diversity of prey. Donohue *et al.* (2017) showed that even the loss of a single predatory species could help trigger rapid secondary cascades of extirpation in natural communities. Grainger *et al.* (2017) found that predator-resistant species showed the predicted increase in rates of occupancy with increasing size of patch. For species that were susceptible to predators, however, predators consistently altered the impact of biogeographic constraints, rather than acting independently.

4. *Natural or anthropogenic episodes may alter the equilibrium number (Helmus and Behm 2018, 2020)*

Temporary departures from equilibrium may be caused by episodic events such as drought (Heatwole *et al.* 1981a), wash-over by waves (Morrison 1997), disturbances by humans (Cody 2006; Schoener 2010), or introduction of invasive species by humans (Krause 2002; Graham *et al.* 2018) that may upset the equilibrium of the number of species on an island. Piovia-Scott *et al.* (2017) found that experimental introduction of a predator (lizard) reduced the various taxa of resident prey on islands to different extents and differently on small versus larger islands.

5. *Rescue effect*

Extirpation of an insular species may be prevented or occur less frequently than would otherwise be the case if its population is continually replenished via a sufficiently high rate of dispersal of individuals from the source area. The rescue effect has received empirical validation (e.g. Brown and Kodric-Brown 1977; Foufopoulos and Mayer 2007). It should be noted, however, that the calculation of species-turnover is sensitive to the frequency of sampling; if the interval between samplings is long, a species may become extirpated and subsequently re-immigrate multiple times during that interval, undetected, thereby leading to a considerable underestimation of turnover (Heatwole 1984).

6. *Geologic factors need to be considered*

Coral cays form, change in size, and shift location on their reefs under the influence of wind, waves, and storms or may disappear altogether (Flood and Heatwole 1986; Heatwole 1994). Furthermore, oceanic volcanic islands arise *de novo* from the deeps and disappear again beneath the waves (Darwin 1842) and rising and falling sea levels alternately alter the sizes of islands and their distances from others, alternately isolating and re-connecting them with each other, or at highwater may even submerge some completely. On a long-term basis, isolation of islands can lead to speciation of their resident populations. Thus, an equilibrium number of species on an island may be relatively stable in the short-term but unstable on the geological scale.

Partial inundation of the Puerto Rican Bank during rising sea levels toward the end of the last glaciation provides an example of how the configuration of land and sea influences gene flow between populations on islands of different sizes. Rising sea levels flooded the lower elevations of this once continuous landmass and dissected it into the Puerto Rican/Virgin Island archipelago composed of islands of varying sizes and distances from each other and the main island of Puerto Rico. This new configuration has remained relatively stable for the past 6,000–7,000 years and during that time has adjusted not only species-richnesses in accordance with the equilibrium theory but also faunal similarity related to the relative time of isolation between the various pairs of islands (Heatwole and MacKenzie 1967; Heatwole *et al.* 1981b; Pregill and Olson 1981).

7. The small-island effect and minimum insular area

Empirical data demonstrate that often the equilibrium number of species of a given taxon does not seem to increase with increasing size of island over the lower range of islands' sizes, i.e. the species–area curve is flat at its lower end. This phenomenon is known as the 'small-island effect' (Triantis *et al.* 2006). Morrison (1997) suggested that part of the small-island effect might reflect the fact that small islands are more susceptible to natural events causing area-independent extirpation, such as wash-over by waves. Another explanation could be that over a range of smaller sizes of islands, the number of habitats does not increase with larger insular area, but rather, there is merely a greater extent of the same kind of habitat (see section II. B. 2. *Size of island may not be as important as once thought*). This explanation applies particularly to small sand cays that have only pioneer, salt-tolerant, arid-adapted vegetation that can thrive on nutrient-poor soils and are adapted to shifting sand driven by wind (Heatwole 1991b). Somewhat larger islands still have that kind of habitat, and the increased diversity of habitats is possible only when the sand cays are large enough to retain rainfall as a lens of fresh water in the soil, floating on the lateral seepage of seawater. Sandy islands need to have at least a minimum width of about 350 feet (107 metres) before they can develop a freshwater lens (Whitehead and Jones 1969). The number of species on very small islands increases only slightly with increasing size of island, probably because (1) they are mainly salt-tolerant species that are sea-dispersed and (2) these are available in limited number in the pool of species from the mainland source. This relationship accounts for the small-island effect, above which there is an inflexion in the species–area curve as a freshwater lens develops and provides a suitable environment on larger islands for salt-intolerant species dispersed by birds or wind. Finally, a second-inflexion point is superimposed on the still-larger islands inhabited by humans who introduce additional species, either deliberately or inadvertently. Mechanisms leading to diversity, however, vary with the scale of observation (Levin 1992: 1943) (also see Fig. 6.21 in Hubbell 2001), which requires 'the study of how pattern and variability change with the scale of description, and the development of laws for simplification, aggregation, and scaling.' As such, spatially structured models would provide the coupling of dispersal and speciation for a broader understanding of geographic patterns in diversity (Holyoak and Loreau 2006).

C. The legacy of the equilibrium theory

In addition to individual papers dealing with various aspects of the equilibrium theory (see above), several comprehensive, critical anthologies of it by a number of the most prominent researchers in the field have been published from a modern perspective. The first was edited by Lomolino (2000) and consisted of a series of papers occupying a complete issue of *Global Ecology and Biogeography*. The second arose from a symposium held in 2007 to celebrate the 40th anniversary of the publication of the MacArthur–Wilson theory and to assess how well it had withstood the test of time. This collection of papers was published three years later as the book *The Theory of Island Biogeography Revisited* (edited by Losos and Ricklefs 2010). Several critiques were published about 50 years after the promulgation of the equilibrium theory, here collectively called the Golden Jubilee series. Finally, *The Species–Area Relationship, Theory and Application* (edited by Matthews *et al.* 2021) appeared as a book of 19 chapters by different authors. The following synopsis of the conclusions emanating from these anthologies of papers provides a capstone to the discussions in the previous sections.

1. The critique of 2000

Lomolino (2000) introduced this series of seven papers by recognizing that the equilibrium theory dominated thinking in the field of insular biogeography for several decades, but that more recently, many ecologists have questioned whether it remains a useful paradigm. The reasons for this were elaborated in the six other papers in the series. Lomolino (2000) also presented a diagrammatic tripartite model illustrating the extra effect speciation added to that exerted by immigration and extirpation/extinction on islands of different sizes and degrees of isolation. Brown and Lomolino (2000) summarized the entire group of papers by recognizing that the equilibrium theory had revolutionized the field of evolutionary biology but that it had not kept pace with 'relevant theory and our growing appreciation of the complexity of nature, especially with empirical findings that species diversity on many islands: (1) is not in equilibrium; (2) is influenced by differences in speciation, colonization, and extinction among taxa; and (3) is influenced by differences among islands in characteristics other than area and isolation.'

2. The critique of 2010

In the preface to the book, Losos and Ricklefs (2010) set the tone by making a more positive assessment of the status of the equilibrium theory than portrayed in the critique of 2000. In their words:

> the legacy of the *Theory of Island Biogeography* is alive and thriving. When we first envisioned this book, we expected most contributions to be retrospective, reviewing the views laid forth in that book and assessing how they had fared. By contrast, a glance at this book will indicate that many of the chapters are looking primarily forward, rather than backward. Some of the most exciting areas in ecology and evolutionary biology—metapopulation theory, the neutral theory of biological diversity, trophic cascade theory, the synthesis of ecological and phylogenetic evolutionary approaches, to name a few—were inspired by, or are being integrated with island biogeography. Much of this work was at most only hinted at by MacArthur and Wilson, yet exciting developments today have a clear intellectual thread leading back to that work.

The published proceedings reviewed the antecedent research that led to the inception of the theory, recognized the theory's massive heuristic insight, and, while acknowledging its broad generality, noted the ways subsequent research assessed its limitations, enlarged its scope, and expanded the range of contributory parameters from merely the sizes of islands and their distances from mainlands to include a host of biological attributes of species and the ecological attributes of islands. With few exceptions, the tenor of the papers was in the best tradition of science by valuing theories as stepping-stones toward greater understanding obtained through modification based on empirical testing.

3. The Golden Jubilee critiques

There were at least four critiques of the equilibrium theory about 50 years following its promulgation. The first was a treatise by Warren *et al.* (2015) that emphasized how the attributes of islands, as propounded by MacArthur and Wilson, combined with modern advances in other disciplines to develop a more comprehensive understanding of the formation of ecological networks, the distributions of species' abundances, and of community ecology generally, with implications far beyond merely islands, and hence beyond the confines of the present chapter.

One of their emphases was on the relative continuum from ecological processes initially to a relative increase of evolutionary ones over a longer time frame. The main point to stress here is that as important as the stimulus of the MacArthur–Wilson was for island biogeography, its influence has now permeated far beyond that field in its stimulatory influence.

The second jubilee issue was a report (Patiño *et al.* 2017) of an amazing online canvasing of 400 attendees of the Biology Conference held in 2014 in Honolulu, 407 attendees of the Island Biology conference in 2016 held in the Azores, members of seven national ecological societies (France, Great Britain, Greece, New Zealand, Portugal, Spain, and the United States) and several international ones, and persons on lists related to island biology that the survey's coordinators solicited, i.e. practically everyone in the world with scientific expertise in insular biogeography and ecology, to identify at least 15 questions that they viewed as of fundamental interest within the various subject areas related to the MacArthur–Wilson equilibrium theory. After winnowing the responses for duplications, 187 questions remained. These were further pared by various means (such as several rounds of voting) to obtain 50 key questions deemed to be of greatest importance (see Appendix 1.I). These were grouped into topics and subtopics, and for each, there was a comprehensive review of the literature relating to each subtopic so that the reader could gain a thorough appreciation of the current state of comprehension of that subject, be brought abreast of controversial issues, and informed of fruitful lines for further investigation. For the purposes of the present chapter, this unique approach not only highlights again the vast array of disciplines whose roots were nourished by the equilibrium theory but also reveals its ongoing stimulus. The emphasis of the equilibrium theory and their hope for its continuance as a source of inspiration was summarized by (Patiño *et al.* 2017) as:

> Despite the long and critical influence of islands on ecological and evolutionary theories, the focus of efforts has typically remained limited to the scale of individual islands or single archipelagos. In the coming years, the analysis of biogeographical dynamics performed through the comparative study of multiple archipelagos may provide us with a better understanding of the regulation of biodiversity at higher levels of spatial organization [...] The 50 fundamental questions identified [...] the potential for island biology to inspire and guide empirical, theoretical and applied research questions related to ecological, evolutionary and conservation science. We hope that this first list of questions compiled under the legacy of MacArthur and Wilson's *Theory of Island Biogeography* provides a source of inspiration for constructive discussions about the future agenda of island research and a fruitful arena for the coming generations of island biologists.

The third critique of the equilibrium theory after five decades or so was a paper by Whittaker *et al.* (2017), who reviewed the projection of an expanded temporal dimension onto the short-term dynamics of the equilibrium theory (see section II. B. 6. *Geologic factors need to be considered*). The emphasis was on the changes in the biota during the cycle of the emergence of remote volcanic islands from the sea throughout their cycle of changes in topography, elevation, and environment, before eventual erosion and/or submergence into the sea. Quaternary or even earlier changes in eustatic changes in sea level also can influence the insular cycle. Whittaker *et al.* (2017) hypothesized that the initial equilibrium number was adjusted upward through community and evolutionary processes (e.g. *in situ* diversification into endemic species) followed by extinction upon submergence of the island. With the movement of tectonic plates, 'hotspots' of islands in different stages can form and exchange biota and may even export species to a mainland. Any single-island endemic species that fails to disperse from its island of origin eventually becomes extinct with the disappearance of its home island. Wilson's (1961) taxon cycle, a model of speciation on islands believed by some to have been superseded by the equilibrium theory, was incorporated by Whittaker *et al.* (2017) into their own model.

The concluding assessment by Whittaker *et al.* (2017) of the legacy of the equilibrium theory was that 'Fifty years on from the landmark Theory of Island Biogeography, the subject is in a vigorous period of development'.

Finally, the fourth critique was a paper by Warren *et al.* (2019), who reviewed CESAB's collective perspective on how key attributes of islands foster an understanding of the ecology and evolution of biological communities in general. They regarded the core of the equilibrium theory as (1) a highly deterministic perspective on species richness based on the assumption that the geographic context of islands determines the rates of immigration and extirpation, and (2) that MacArthur and Wilson recognized that this not only represented a great number of islands worldwide but that there also were links stretching far beyond real islands to other situations. In time, the theory became central to debates regarding practical issues in conservation and 'a highly diversified field of fragmentation research can be traced back to Core IBT [Island Biogeography Theory]', e.g. fragmented crops, intertidal habitats, and guidelines for establishing reserves. These differed from true islands in that the intervening media through which dispersal took place were different, and some had limited value, but such studies did force decision-makers to think about population turnover within habitat fragments, as well as about the influence of the size and isolation of fragments on colonization and extirpation. The core theory also had direct relevance to studies of metapopulations connected by movements of individuals (i.e. source–sink dynamics).

4. The 2021 critique

The 2021 critique (Matthews *et al.* 2021) recognized the species–area curve as a valid generality, but as with most generalities, there are exceptions that prove the rule, and that when predicted relationships between insular area and species-richness did not occur, such departures were attributable to variables other than area of island exerting an overriding influence on diversity. It was further recognized that factors that influence species-richness vary across spatial scales and that distinction should be made between nested and non-nested sampling. Many different functions can be used to model the species–area relationship, and when speciation-dominated systems are compared to those characterized by the dynamics of the MacArthur–Wilson theory, different slopes of the species–area curve may be expected. Increasing emphasis was placed on functional traits (those that relate to fitness) rather than merely the identity of species, and thus ecological attributes, especially trophic relations, assumed greater attention. The distinction was made between nested and non-nested sampling and between phenomena that are unified by theory and those that are idiosyncratic. The geometry of the archipelago (spatial arrangement of islands) was identified as an important influence upon the interconnectivity of islands, their colonization, their loss of habitat, and the stability of their equilibrium.

A number of chapters were concerned with the details of mathematical modelling of the species–area curve and the factors determining its various shapes. Others related how the data on the species-relationship is affected by humans or are related to conservation. An appeal was made to consider species–time curves (the increase in number of species reported at a site as the duration of observation increases) as well as species–area curves.

5. Summary

The initial response to the equilibrium theory was to hail it as a major explanation of the dynamics of the biota on islands, which changed to disillusionment by some because it was not all-inclusive and answered all questions, and finally to celebrating its pivotal heuristic role in inspiring numerous ecological theories and schools of thought and spearheading a more comprehensive understanding not only of islands but of other isolates and of theoretical and practical community ecology in general. In the words of Patiño *et al.* (2017):

> MacArthur and Wilson's (1967) book remains one of the most influential texts on ecology and evolution, with continued debate over its strengths and limitations. It has been, and will continue to be, a springboard for research on the origin and maintenance of biological communities, with particular reference to marine island systems, but also extending to other island-like systems.

In the above critiques, there has been a progressive shift in emphasis from assessment of the validity, or not, of the equilibrium theory to islands (small parcels of land surrounded by water) to its relevance to other kinds of ecological isolates or partial isolates on mainlands (e.g. mountain tops, peninsulas, particular vegetational zones, and meta-populations).

III. The theory of trophic structure stability

Heatwole and Levins (1972) considered the possibility that the trophic structure of communities on newly emerging islands or nascent communities on defaunated islands may be important in the dynamics of stepdown from the initial overshoot in species richness. They further analysed Simberloff's and Wilson's (1969) data and showed that the communities in succession on defaunated mangrove islands reached a stable trophic structure even before the equilibrium number of species had been attained. This result led to the theory of trophic structure stability, i.e. that a stable trophic structure of insular communities develops even before the equilibrium number stabilizes. From this theory, a mechanism for a deterministic effect of trophic structure on the dynamics of establishment of newly arriving species was suggested. For instance, if there are proportionally fewer species of predators than characteristic of the stable trophic structure, new species of predators may become established, but if there are already the requisite proportion of predators present, newly arriving species of predators may find it difficult to become established, or else they displace ones already present. Similar arguments apply to other trophic categories.

Initially, Simberloff and Wilson (1969: 278) opined: 'We believe the curves are produced by colonization involving little if any interaction, *then a gradual decline as interaction becomes important*, and finally, a *lasting dynamic interaction*' [italics ours] and Simberloff (1974: 166) accepted the theory of trophic structure stability and stated that:

> Heatwole & Levins have re-examined the colonization data for the defaunated islands and have shown that, if the arthropod colonists are divided into broad niche types (detritus-feeders, herbivores, wood-borers, etc.), the defaunated islands not only gradually achieved their original number of species, but also their original broad trophic structure, or distribution of species into the different niche types. This observation is exciting, for it implies that the groups on these islands are true interacting communities, and not just haphazard assemblages that arrived and were extinguished fortuitously. Although the arboreal red mangrove system does not appear to be one in which species interactions are particularly pronounced, it is possible that evolution has gradually molded the source fauna so that the mangrove resource is utilized in some canonical, nearly optimal way, with the result that random subsets of this source biota are likely to be rather close to the canonical structure.

Later, however, Simberloff (1976b: 395) reversed that position and concluded that the concept of Heatwole and Levins 'conflicts with Simberloff and Wilson's (1969) view of these processes on the mangrove islands as largely abiotic, mediated primarily by physical factors, and with interactions playing a subsidiary role'. Perhaps their change in viewpoint was related to their failure to falsify null hypotheses of no difference between observed patterns of community structure and those expected by random assembly only. Accordingly, they considered community structure to depend on stochastic processes. Resolution of such fundamental differences in viewpoint depends on testing counter theories (Holyoak and Loreau 2006; Whittaker *et al.* 2010; Sax and Gaines 2011). Some decades now have passed, and such testing has taken place. An assessment can now be made from a more distant perspective.

Giller (1984: 137), in a review, recognized that Simberloff's (1976b) null hypothesis was faulty because 'it assumes that all species are available for colonization at the same time', which, of course, is not correct, and concluded that the results of Heatwole and Levins (1972) 'suggest a remarkably stable trophic structure determined by species interaction'. Other support came from Moran and Southwood (1982), who found an impressive uniformity in the proportional representation of species in three guilds of insects (phytophages, epiphyte feeders, and predators) on trees. Similarly, Briand and Cohen (1984) found that the proportion of top predators, intermediate predators and prey, and bottom prey remained independent of the total number of species.

Putnam (1994) reviewed the concept of trophic structure stability and reported that such striking similarities in the trophic organization both between islands and within an island both before and after recolonization are strongly suggestive of some 'ideal optimal structure—' but went on to conclude from Simberloff's (1976b) failure to reject the null hypothesis of no difference from purely random assembly of trophic categories that although the data supporting the theory of trophic structure stability 'seem to tell a most convincing story [...] the case for such adaptation is not proven'. The fact that Simberloff's analysis was invalidated because it tended to accept the null hypothesis, whether or not it was true, softens Putnam's hesitation in giving the trophic theory credence. The theory of trophic structure stability may have broader application than just to islands as Fetcher *et al.* (2015) indicated that their results from a study of the reclamation on a severely contaminated mainland site were similar to those of Heatwole's and Levins' (1972) assessment of the recovery of the biota on mangrove islands.

Cash *et al.* (2012) found that local assemblages (on mainland sites) tended to contain a relatively constant representation of guilds, especially in the most disturbed areas; they concluded that 'there is only limited truth in the frequent assumption that community structure is stronger in undisturbed, equilibrium communities.'

Terborgh (2010) noted that it is not always a superior competitor replacing a less competitive one from a similar niche. When a particular niche becomes vacant because of extirpation of its occupying species (for whatever reason), a less competitive immigrant species might then be able to successfully establish and fill that vacancy, thereby restoring the trophic structure. Subsequent re-immigration by a better competitor might then eliminate the less competitive species, again maintaining the trophic structure. In this way, there could be a succession of various species, all with similar niches, stochastically replacing each other in some cases or by deterministic competitive exclusion in others.

Gravel *et al.* (2011) did not stop at the production of a predictive model but extended their study to testing it against empirical data of natural communities, one of which was the recolonization of defaunated mangrove islands studied by Simberloff and Wilson (1969), coincidentally the same dataset that constituted the basis for formulation of the theory of trophic structure stability by Heatwole and Levins (1972). Gravel *et al.* (2011) found that their trophic model fits the empirical data better than did the equilibrium model of MacArthur and Wilson (1963, 1967) (also see section III. B. *Modelling of trophic relations within the context of insular ecosystems*).

Several advances have been made in the analysis of trophic structure, and modifications of the theory have been made. One is the distinction between food chains (straight-line transfer of material and energy from one species to another) and food webs that have varying degrees of complexity and trophic specialization, including some generalized feeders involved in more than one link (see reviews by Levine 1980; Holt 1996; Verhoef and Olff 2010; Morin 2011). Another is a methodological advance. Martínez *et al.* (1999) found that the re-analysis of the defaunation study of Simberloff and Wilson (1969) by Heatwole and Levins (1972) foreshadowed their result but did not answer their question. Their study suggested that 'comparative analyses of connectance among trophic-species webs constructed with varying degrees of effort are generally robust' but that 'the fractions of species appear to require corrections for variable sample sizes [...] These corrections may be easier for trophic-species webs because of their relatively simple changes with sampling effort.'

Finally, third is the recognition that trophic relations can change ontogenetically and that an animal may undergo extensive shifts in diet and preferred habitat as well as face a different suite of predators and competitors as it grows (Werner and Gilliam 1984). Consequently, the species is not necessarily the fundamental unit for trophic analysis (e.g. Hubbell 2001; Rosindell *et al.* 2011), especially for amphibians and holometabolous insects. For example, a species of frog, which as a tadpole is aquatic and consumes algae but as a terrestrial adult feeds on insects, not only occupies different trophic levels in its food web but is a member of two entirely different communities depending on which life-history stage is considered. In this example, the species is not a single functional unit and cannot be used, as such, for any analysis of trophic structure or for the construction of food webs. Instead, the tadpole stage and the adult have to be treated as separate functional entities belonging to different guilds. Heatwole (1989: 18) coined the term *'econe'* for such an infraspecific functional unit and defined it as 'a species or some component of a species (such as a life-history stage, age-class, morph, or sex) whose members share common patterns of resource utilization and niche characteristics but differ from other such components or species' or put more succinctly, but less precisely, as an 'inhabitant of a niche'. Putnam (1994) noted that omnivorous species might belong to more than one trophic category. In cases of more than one econe per species, the assignment of trophic categories becomes even more imprecise.

Piechnik *et al.* (2008) also re-examined the data from Simberloff's and Wilson's (1969) defaunation study and found that the proportion of specialist species increased and the proportion of generalists decreased during re-assembly of the food web. However, late-colonizing generalists maintained constant connectance at later dates. They concluded that the relationship between success in colonization and trophic breadth helps explain the patterns of food webs and corroborates assertions that the assembly of communities is influenced by species' trophic breadths.

Aggressive invasive species that are not generally within the pool of natural immigrants to islands, but are introduced by humans, can disrupt insular trophic structures by inserting new trophic levels (e.g. cats and rats as top predators), thereby replacing trophic constituents and increasing trophic efficiency; synergisms among invasives can also enhance their transforming power (Krause 2002; Sax and Gaines 2011). Curiously, despite the fact that most insular herbivores are insects, they do not seem to have the devastating effect of herbivorous mammals when they reach islands (also see review by Lugo *et al.* 2012). Terborgh (2010) proposed that this may be because most herbivorous insects are more restricted in their range of suitable food plants than are most mammals. Because insects and plants disperse independently, many insects may fail to find their preferred food plants on an island and consequently fail to become established or be forced to subsist on less preferred plants on which their larvae develop slowly or in reduced numbers.

In addition to the above considerations, detailed empirical studies have supported the theory of trophic structure stability. For instance, Heatwole *et al.* (1981a) found that during recovery from a drought on One-Tree Island, the number both of individuals and of species of epigeal insects on the island changed greatly (Fig. 1.1 Upper), but the trophic structure remained remarkably stable (Fig. 1.1 Lower). Moreover, Mikkelson (1993) compared the trophic structures of patches fragmented from formerly continuous habitat. He found no tendency for any of the trophic categories to lose species proportionally faster than any others, nor did different trophic categories become extirpated or colonize independently. Rather, extirpations and colonizations in different trophic categories corresponded in such a way that the same trophic structure persisted in the various fragments as their number of species declined during isolation from the once-contiguous habitat from which they originated. Ríos-López and Aide (2007) examined the herpetofaunal dynamics during the early stages of secondary succession following reforestation. They found that species richness and abundance changed rapidly during the early stages of succession whereas the trophic structure changed slowly; they also predicted similar trophic structures, albeit different herpetofaunal compositions, throughout time. Results from these studies are parallel to those of Heatwole *et al.* (1981a) in providing strong support for the theory of trophic structure stability; in these cases, the trophic structure remained relatively stable in the face of fluctuating numbers of species. All of these studies, however, differed from the analysis by Heatwole and Levins (1972) of the data of Simberloff and Wilson (1969) on the recolonization of defaunated mangrove islands. The mangrove islands had lost their complete assemblage of animals through fumigation but retained their plants and hence subsequent colonization was more analogous to the initial colonization of a newly forming island previously without animals. Disregarding marine input supporting a scavenger-based food chain (see section III. A. *Trophic-structure depends on the type of resource-base of an island*), it is obvious that there will be changes in trophic structure of initial colonization of bare islands. The initial species to become established would need to be a species of plant (trophic structure 100% autotrophs); then, say one species of herbivore immigrated and became established (50% autotroph:50% herbivore), followed by another species of herbivore (33% autotroph:67% herbivore) and then by another (second) species of plant (back to 50% autotroph:50% herbivore), and so on. Clearly, trophic structure is very labile in the early colonization of a bare island, and a defaunated one is similar except that the autotrophic base already is fully established. The fundamental prediction of the theory of trophic structure stability is that well before the equilibrium number of species is reached, the trophic structure becomes stable and once above such a threshold, it remains stable even in the face of marked fluctuations in the total number of species and individuals involved. A challenge both for modellers of ecosystems (see review by Verhoef and Olff 2010) and for empirical studies of specific communities would seem to lie in determining where such thresholds are, whether they differ among communities with different resource-bases (see following section), and how they relate to the overshoot of numbers of species during colonization or recovery, either from primary or secondary succession.

A. Trophic structure depends on the type of resource-base of an island

One of the main modifications of the theory of trophic structure stability is the recognition that it operates over a limited time. Clearly, as islands develop and disappear over geologic time, or perhaps even shorter periods, the basic ecology of the island may change in important ways that affect trophic structure. Heatwole *et al.* (1981a) and Heatwole (2018a) showed that all kinds of

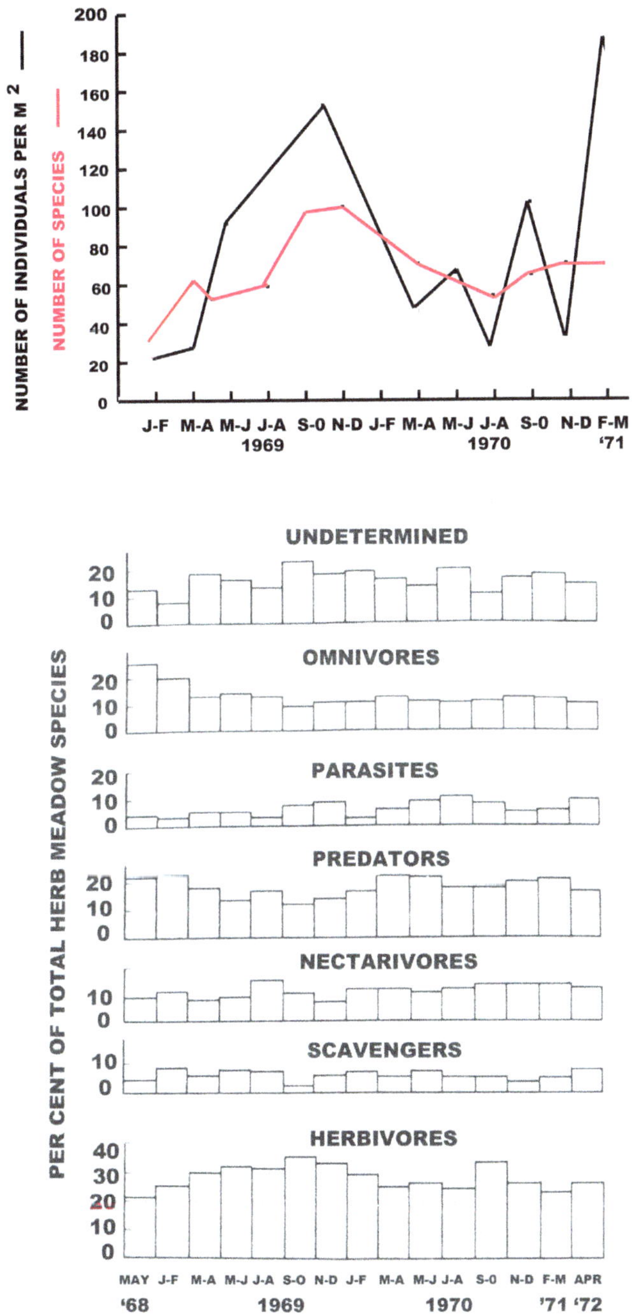

Figure 1.1. Upper: Temporal change in the number of species (red line) and population density (black line) of all epigean arthropods in the herb meadow of One-Tree Island during the period of intensive sampling. Lower: Temporal change in the proportional representation of species from the herb meadow of One-Tree Island in different trophic categories. The first visit to the island (May 1968; pre-drought) was a preliminary one to test the feasibility of One-Tree for the projected study and no quantitative sampling was carried out on the epigean arthropods at that time; only voucher specimens were collected for identification of resident species. Note that the trophic structure of even that incomplete collection was similar to the quantitative estimates made throughout the rest of the investigation. (Redrawn from Heatwole *et al.* 1981a).

islands do not have the same trophic structure. Rather, different types of islands have different bases of their food webs that impose a particular trophic structure.

Islands are not permanent entities. They form anew from volcanic action, intense storms piling up coral rubble wrested from the reef, or by deposition of sand onto sand bars that gradually accrete and become emergent under the action of water and wind. At first, these are devoid of vegetation, but sea-dispersed, salt-tolerant plants arrive (Flood and Heatwole 1986; Heatwole 1991b). These mostly are creeping forms that stabilize the sand or shrubs that form a peripheral ring around the island, thereby protecting the interior from salt spray (Charley and Heatwole, in preparation). With a less-harsh environment, a greater variety of ground-nesting seabirds are attracted, and these add nutrients such as nitrogen and phosphorus to the soil. When a sufficient size of island is reached, a freshwater lens forms in the soil. These ameliorating changes provide a suitable environment for the establishment of plants other than pioneer species, and the interiors of islands develop a lush growth of low vegetation, and eventually shrubs and even trees, on some islands culminating in forests and in hordes of tree-nesting birds whose guano further enriches the soil.

There are not only temporal trends toward increasing complexity of biotic communities on islands, but there are also deteriorative changes that revert islands to previous stages of insular development (Flood and Heatwole 1986). These include erosion that can decrease the size of an island or, by erosion on one side and prograding on the other, cause it to 'walk' across a reef or even disappear altogether over the edge into deeper water. In addition, nesting sea turtles alter the developing cover of vegetation and may return an island to pioneer vegetation that is more resistant to mechanical disruption because of the creeping tendencies of some plants; wash-over during severe storms may also transform the vegetation and otherwise alter islands and, as noted above, droughts and other episodic events can exert a significant influence.

It is in the context of this dynamic scenario that trophic structure stability must be considered. In the sense that virtually nothing in the universe or in geologic time is entirely permanent, except the operation of natural laws, trophic structure may be stable, but only as part of a certain stage of an island's development, or more accurately in terms of the island's base of resources. Take the example of sandy cays on the Great Barrier Reef that have no vegetation and yet sustain an assemblage of animals whose food web is based upon the transfer of energy and nutrients from the marine ecosystem by seabirds (excreta, food scraps, and cadavers), by intertidal organisms, and by washup of marine carrion and driftwood (Heatwole 1971). The terrestrial food web consists of scavengers on marine carrion (beetles, carrion flies) preyed upon by predators. There being no plants, herbivores are also absent except for transients and will continue to be so as long as no plants colonize the island to provide a food source. Small islands from other archipelagos also have been shown to feature higher trophic levels supported by marine input (Polis and Hurd 1995a,b, 1996).

Evidently, a bare island with no vegetation will have a different trophic structure than will one vegetated by forest. We propose that what drives a particular stable trophic structure is the resource-base of the island and that as this base changes, such as during an island's ontogeny (Whittaker *et al.* 2010), the trophic structure adopts a new configuration. Thus, trophic structure stability is maintained within a given resource-base but not between different resource-bases throughout spatio-temporal scales. This hypothesis can be tested empirically. If it is true, then different islands with similar resource-bases should have similar trophic structures, even if the component taxa are not the same. This offers a stimulus for future research and a better understanding of insular dynamics.

A corollary hypothesis that is subject to empirical analysis is that an island's trophic structure is related to its size. The amount of vegetation on an island *ceteris paribus* is related to the area of the island and thus to the number of trophic resources for herbivores, while the perimeter of an island is related to the islands' interface with the marine environment and hence to the transfer of marine resources from the sea to scavengers in the terrestrial habitat (Heatwole 1971; Hodkinson *et al.* 2002; Ellis *et al.* 2006). The smaller the island, the greater is its perimeter-to-surface-area ratio, and consequently, it would be expected to have proportionately more scavengers relative to herbivores than would a larger island. As the size of an island increases even more, one would expect proportionally more herbivores than scavengers. This relationship between the scavenger:herbivore ratio and insular size is empirically testable. One needs to realize, however, that there is an upper limit to both the number of species of scavengers and the number of species of herbivores, both of which depend on the size of the pool of species of those types available for dispersal from the mainland source.

There has been some headway in examining these hypotheses. Table 1.1 compares the trophic structures of a series of islands of different sizes and attributes. All of the islands without vegetation, such as bare sand cays, or those with scant vegetation such as very small cays or volcanic islands whose denuded surface is beginning to be recolonized, have a similar trophic structure with marine-derived materials servicing a high proportion of scavengers at the base of the food web. The trophic structure of the islands that are forested or otherwise have a larger proportional area of vegetation have convergently shifted their trophic base to plants as the prime resource base. Data from many more islands of various ecologies, sizes, and area-to-perimeter ratios are needed to expand the base of comparison and refine their relationship to trophic structure.

An important advance in the understanding of the interaction of marine and insular food webs is the finding by Piovia-Scott *et al.* (2017) that predatory lizards experimentally introduced

Table 1.1. Comparison of the trophic structures on islands with different bases of resources.

NAME & TYPE OF ISLAND	N	LOCATION	PERCENT OF SPECIES IN CATEGORY:							SOURCE/NOTES
			H	S	P	PA	O	E	?	
Bare sand cays	10	Coral Sea, Australia, and Puerto Rico	2*	**63**	7	0	0	27	0	Heatwole (1971); Heatwole and Levins (1973)
Motmot Island; volcanic Island	1	New Guinea	12	**34**	25	0	30	0	0	Ball and Glucksman (1975); recolonizing 3 and 4 years after eruption.
Cayo Ahogado; sand cay with low vegetation	1	Puerto Rico	23	**34**	24	0	17	0	+	Heatwole and Levins (1973)
One-Tree Island; coral rubble cay	1	Great Barrier Reef, Australia								Heatwole *et al.* (1981a)
Herb meadow			**45**	6	18	11	17	0	2	13 samplings
Pisonia foliage			**51**	4	21	7	10	0	7	8 samplings
Heliotropium foliage			**61**	0	19	2	11	0	/	8 samplings
Mangrove keys	6	Florida, USA	**39**	13	20	7	19	0	+	Heatwole and Levins (1972, 1973), based on Simberloff and Wilson (1969)

N = number of islands in the group. Dominant (largest) category in **boldface**; + indicates present in negligibly small proportions; H = herbivores (sucking and chewing herbivores; granivores; woodeaters, and nectarivores; S = scavengers, dung eaters, and detritivores; P = predators; Pa = entomophagous parasites; O = ants and other omnivores; ? = trophic category unknown; * signifies a single individual transient. From Heatwole *et al.* 1981a).

onto islands reduced the abundances of their prey and that the nature of the effect was modified by the area of the islands, the taxon of prey, and the input of seaweed from the marine environment. For salticid spiders, the effect of the lizards was less on the larger island than on the smaller ones, whereas the effect on wasps was the reverse, and area of island did not affect the degree of effect of lizards on orb-weaving spiders. In addition, there was a weak effect of 'marine subsidy' (the amount of seaweed washed onto islands from the sea on small islands); greater deposition resulted in a lessening of the effect of lizards. These authors did not speculate on the mechanism of this effect, but the amphipods and isopods commonly inhabiting such wash-up may have served as supplementary prey for the lizards, thereby reducing predation-pressure on the spiders and other resident taxa.

Roslin *et al.* (2014) found that the length of food chains decreased from large to small islands, with the species–area slope steepening from plants through herbivores to primary parasitoids to secondary parasitoids. Since the ratio between generalists and specialists was not affected by island size, either of herbivores or parasitoids, they concluded that the observed patterns reflected changes in the overall number of nodes and levels in local food webs than in their linkage structure.

Two islands of the same size and distance from the source of their biotas, but with different numbers of habitats within that given area may differ in their dynamic (see above), with the number of different habitats not only influencing the biodiversity, but the sizes of each of those habitats being differentially important for different trophic categories. For instance, Van Noordwijk *et al.* (2015) found that the species–area relationship in mainland meta-communities was modulated not only by dispersal ability and affinity for specific habitats but also by trophic rank and stated that in small habitats surrounded by a hostile matrix, food chains might be disrupted. In general, studies of meta-communities on mainlands have found that trophic relations have a bearing on aspects of community structure that may perhaps also apply to islands.

B. Modelling of trophic relations within the context of insular ecosystems

Major heuristic advances have been made in relating trophic relations to insular biogeography in a realistic meld of its deterministic and stochastic components, thereby providing functional links between the MacArthur–Wilson theory of insular biogeography and the Heatwole–Levins theory of trophic structure stability. Holt (1996) analysed the relationship between the incidence of different trophic levels in relation to the area of islands and their distance from a mainland source of biota and found the following trends:

(1) the incidence of predators increases in concert with the lower extinction rates found on larger islands and with the increasing colonization rates characteristic of larger islands or those close to the mainland source.

(2) given biogeographic donor control (extirpation of a prey species leading to the extirpation of its predator, but without the predator altering the rate of extirpation of its prey), the incidence of prey is as predicted without predation.

(3) food chains are longer on larger or closer islands, but counterintuitively, with a large enough destabilizing effect of predation, the reverse may occur.

(4) a prey species may be less frequently encountered on larger islands because predators persist there long enough to be able to eliminate their food base – but more frequently found on distant islands where colonization of predators is reduced.

Furthermore, he and colleagues (Holt *et al.* 1999) modelled the effect of trophic rank on the species–area relationship. Using simple linear food chains (each plant species supporting its own specialist herbivore, which in turn supports its own specialist predator), he predicted that the slope (z) of the species–area curve should increase with trophic rank. Some, but not all, empirical examples fit this prediction, and even some more complex food webs involving generalists may do so. However, generalizations were hard to reach, and the effect of trophic rank can be weakened or even reversed if (1) top-down interactions lead to extirpations, (2) communities are open to recurrent immigration (especially at the higher trophic levels), (3) consumers are facultatively able to utilize a wide range of resource species, or (4) systems are not in equilibrium.

Gravel *et al.* (2011) found that the MacArthur–Wilson theory of island biogeography did not entirely account for the highly complex dynamics of insular biotic communities. Therefore, they added a trophic dimension into their theory of insular biogeography that improved predictions across trophic levels and scales compared to the original equilibrium theory (for more details, see the first part of section III. *The theory of trophic structure stability*).

Pokarzhevskii *et al.* (2000) reviewed the relationship of soil microbes to the trophic structure of ecosystems and found that they can control the migration of some species and affect food chains. This remains a relatively unexplored aspect of insular biogeography.

Vandermeer *et al.* (1980) reported that previous workers had found that locally unstable communities could be stabilized by allowing migrations among habitats, but that the opposite also occurred, with instability arising from bringing together competitive species that were previously isolated. Their own work with communities in temperate leaf litter showed that migration influenced trophic guilds in different ways. Whereas there was no difference in biodiversity of non-predators between controls and assemblages subjected to migration, there were fewer predatory species in comparison to controls. Hajian-Forooshani *et al.* (2020) conducted similar experiments with communities in tropical leaf litter with similar results. Although these experiments were not conducted in an insular context, they are nonetheless germane to the present topic. In terms of the equilibrium theory, large islands near the source have higher immigration rates than small islands distant from that source, and thus, the consequences of that dynamic would be that the former islands would have a greater immigration-effect on some guilds than on others. Heatwole (1971, 2018a) found that the resource-base of islands differ with island size (see section III. A. *Trophic structure depends on the type of resource-base of an island*), and thus, if immigration affects different guilds in disparate ways, then this phenomenon may be partly responsible for the difference in trophic structure on islands with different resource bases. Curiously, however, Hajian-Forooshani *et al.* (2020) found that although the number of species of predators was affected by migration, the entire community of predators plus non-predators was not, presumably because the number of predatory species was small enough that their change in number was insufficient to be detectable when the community as a whole was examined.

IV. Incidence function and assembly rules

Diamond's (1975, 1979) concept of the dynamics and structure of insular biotic communities is that they are assembled through selection of colonists, adjustment of their abundances, and compression of their niches in ways that match their combined consumption of resources with the production of resources on the island. He developed this theory from a body of empirical data primarily from guilds of birds on the islands of the New Guinean and Pacific archipelagos (Diamond 1970a,b, 1972, 1973, 1975). There are two ingredients in his approach: (1) incidence analysis and (2) assembly rules.

A. Incidence (J) analysis

The incidence (indicated by 'J') is the fraction of the islands in a given archipelago that is occupied by a designated species plotted against the size of island. The shapes of incidence curves vary from species to species. For example, some species have a nearly bell-shaped curve at the lower end of the scale of the number of equilibrium species, thus occurring only on the smallest, or most remote, species-poor islands in the archipelago (Fig. 1.2); these species Diamond called 'Supertramps'. The opposite extreme consists of the species whose incidence curve is sigmoidal in shape and with the species occurring primarily on large, close, species-rich islands. These he called 'High-S' species. A range of intermediate types of curves was identified and the species

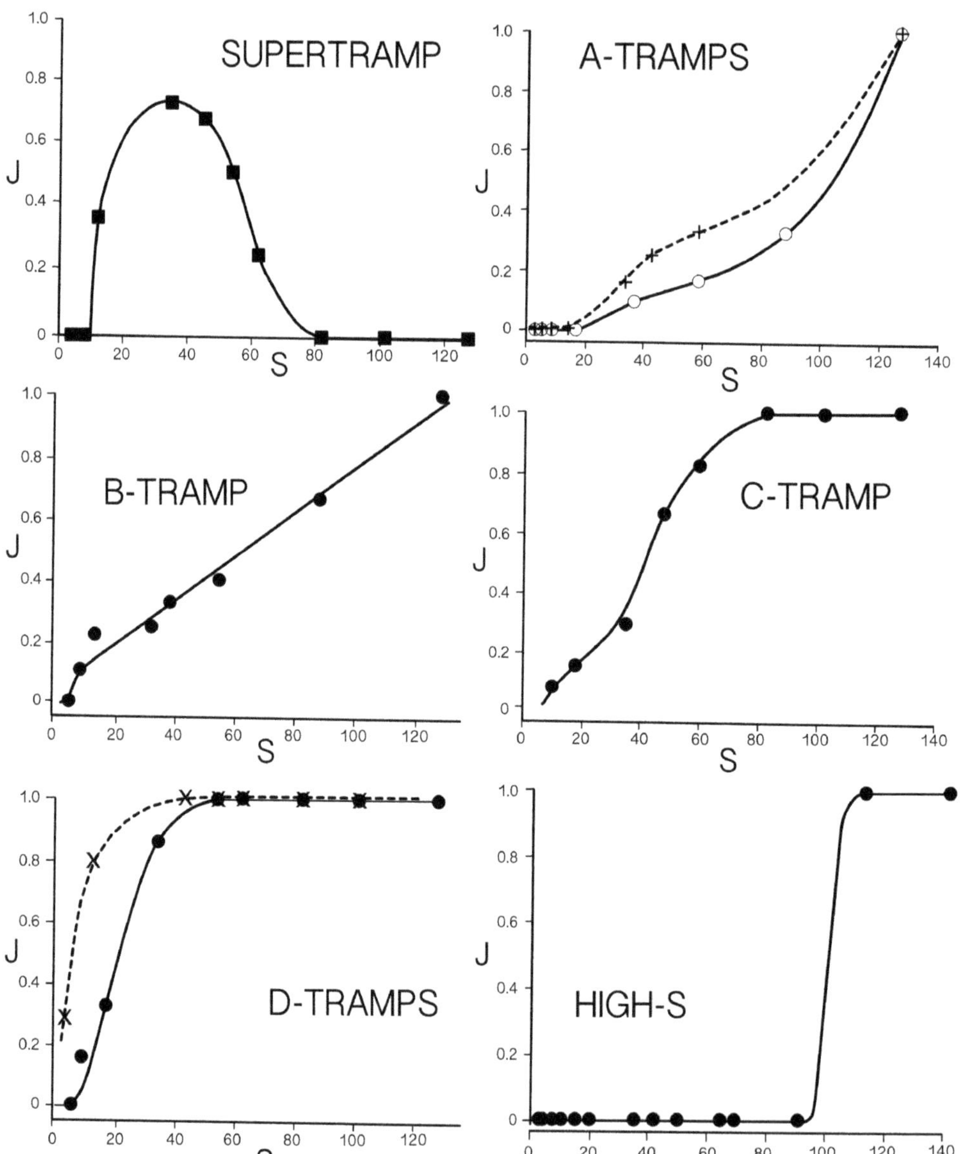

Figure 1.2. Shapes of J curves (see text) characteristic of types of species named by Diamond (1975). S = the number of species on islands and J = the fraction of the islands in a given archipelago that are occupied by a designated species. Excerpted and modified from Diamond (1975).

characterizing each of them dubbed progressively from relatively species-poor islands to relatively species-rich islands as 'A-tramps', 'B-tramps', 'C-tramps', and 'D-tramps'.

The beauty of incidence analysis is that it can be related to the autecology of the species on islands with different properties to provide insight into the dynamics of community function. Certain species may be restricted to the larger, more ecologically diverse islands because (1) some of their required resources, such as a sufficient area for a minimum territory, are lacking on the smaller islands, or because (2) the harsh physical conditions of high salinity and lack of standing water characteristic of small islands lie beyond their tolerance limits.

B. Assembly rules

Diamond (1975) posited competition as one of the determinants of the composition of insular communities and that there were some combinations of species that were possible and others that were not; for instance, a Super-tramp and a High-S species would be unlikely co-inhabitants on any island. His assembly rules addressed the principles governing the construction of observed insular assemblages (Diamond 1975: 423):

Permissible combinations resist invaders that would transform them into forbidden combinations.

A combination that is stable on a large island or species-rich island may be unstable on a small or species-poor island.

On a small or species-poor island, a combination may resist invaders that would be incorporated on a larger or more species-rich island.

Some pairs of species never coexist, either by themselves or as part of a larger combination.

Some pairs of species that form an unstable combination by themselves may form part of a stable larger combination.

Conversely, some combinations that are composed entirely of stable sub-combinations are themselves unstable.

These combinations can be visualized by comparing the curve for the production of a resource along some axis of one of its characteristics (e.g. size of food item, elevation of habitat above sea level) against the curve for the minimum utilization of that resource required by various species. Only those combinations of species whose utilization curves fit under the production curve along the same axis, without undue overlap, are permissible. There are, of course, multiple axes and exclusion of a species could occur along any one of them.

These rules can thus serve as heuristic hypotheses for empirical testing of particular species-groups to ascertain whether their interaction affects coexistence and thus biodiversity. Diamond (1975) concluded that (1) the use of resources must be a major factor in the origin of assembly rules, (2) permitted combinations may leave fewer resources unutilized than forbidden ones with the same number of species, (3) species of a permitted combination must have roughly comparable abilities to tolerate lowered levels of resources ('companions in starvation'), (4) on smaller, more remote islands, the ability to arrive frequently and maintain a large population becomes a significant consideration in determining which species will occur, and (5) some combinations are difficult to assemble if their component species colonize infrequently. He also indicated that unsolved problems were (a) constructing mathematical models of how size of territory, population density, birth and death rates, properties of competing species, and area and isolation of islands relate to incidence functions, (b) relating incidence functions to non-insular communities, and (c) elucidating the relative roles of chance and ecological processes in influencing the species-composition of islands.

In a review of the subject, Beleya and Lancaster (1999) proposed that assembly rules are deterministic and mechanistic and identified three principal determinants: dispersal-constraints, which define the available pool of potential colonists; environmental constraints, which govern establishment; and internal processes, which control a species' responses to the external determinants. They also pointed out that even at the equilibrium number of species, environmental constraints influence the outcome of assembly rules.

C. Checkerboard distributions

Diamond (1975) used 'checkerboards' to visualize the utilization of resources of a particular species on a given island. Shown in Figure 1.3 are four such checkerboards in which columns represent progressively larger sizes of fruits eaten by pigeons in the Bismarck Archipelago, and rows represent different types of habitats. Each square is coded according to the extent of the frequency of use of fruits of different sizes by the species of bird in question. Species 1 mostly used the smaller fruits more often in all habitats on one island but was absent from the other two islands (slight intra-island variation in selection of fruit and occupancy of all habitats on one island but with greater inter-island differences). Species 4 had a nearly opposite pattern of diet and tended to select the heavier fruits more often in all the habitats occupied (A–C) and was restricted to the same three habitats on all three islands with only slight variation in the intensity with which different sizes of fruit were eaten among them. Species 2 and 3 were intermediate between these extremes; both showed a predilection for the larger fruits on all islands and in all habitats occupied; in this respect, they resembled species 4 but there was less inter-island consistency in the habitats used from island to island.

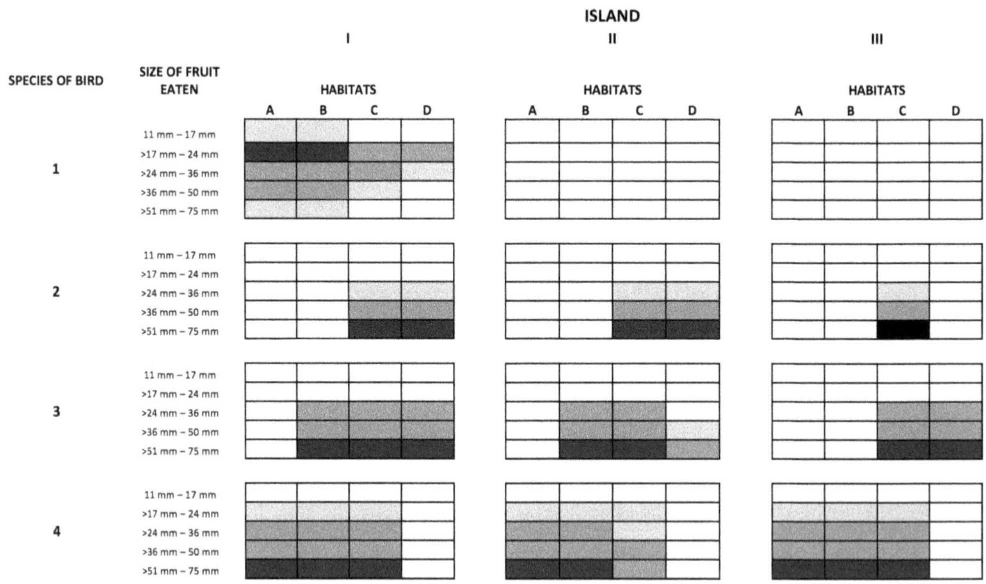

Figure 1.3. A selection of a few exemplary checkerboards based on Diamond (1975) to illustrate the informative nature of the checkerboard technique for four species in the fruit-pigeon guild: (1) *Ptilinopus insolitus*; (2) *Ducula melanochroa*; (3) *Ducula finschii*; (4) *Ptilinopus rivoli* in four types of habitat: (A) second growth; (B) lowland forest; (C) hill forest; (D) mountain forest, on Fundamental Island, Bismarck Islands. The black rectangles indicate the frequency with which fruits of different sizes were selected; black rectangles indicate high use of fruits of a given size, dark grey rectangles, moderate use; light grey rectangles, light use; and blank squares indicate absence of the bird from the habitat and hence no use.

D. Current status of assembly rules

About 25 years after Diamond's (1975) promulgation of the concept of assembly rules, a symposium *'Assembly Rules'* resulted in a 418-page book entitled *Ecological Assembly Rules, Perspectives, Advances, Retreats* (Weiher and Keddy 2004) that assessed the status of assembly rules as of 1999. Herein we cite the individual chapters of this book only by author(s) and title of article but do not include them in the list of references at the end of the present chapter unless we refer to them elsewhere in a different context. Because of the necessity of severely abstracting such a voluminous tome, we recommend that readers consult the original chapters for a fuller exposition of the supporting data.

The motive for producing their book, according to Keddy and Weiher (*Introduction: The scope and goals of research on assembly rules*), was to address the problem of how to get from the pool of potential donor species to the community. They identified two paradigms, (1) the island paradigm because it deals with mainlands, islands, immigration, and coexistence, and (2) the trait–environment paradigm because it begins with pools, habitats as filters, and convergence. The book was divided into two parts: I. *The Search for Meaningful Patterns in Species Assemblages*, and II. *Other Perspectives on Community Assembly*. The first section is the one most relevant to islands, although reference also is made to some chapters in the second section.

One of the problems with studying the assembly of insular communities is that the definition of terms varies among researchers. Booth and Larson (ch. 7: 'Impact of language, history, and choice of system on the study of assembly rules') noted three features of the discussion about assembly rules that lessen their impact on ecologists: (1) Different researchers define assembly rules in different ways; (2) insufficient account is given to how earlier ecologists addressed the concept prior to Diamond's introduction of the term 'Assembly Rules'; and (3) 'the primary evidence used to develop and test ideas on assembly rules is often derived from complex ecosystems that do not necessarily follow simple trajectories'. Rather, ecologists should start with simple natural systems where assembly rules are easier to analyse and later expand their investigations to more complex ones. In this regard, it is instructive to compare the way different authors defined the assembly of biotic communities (see Appendix 1.II).

In addition to these three difficulties, we can add attempts by some investigators to perpetuate demonstrably invalid but cherished ideas by trying to discredit opposing views rather than by seeking genuine understanding. For example, Keddy and Weiher (2004) rather humorously characterized this confrontational attitude as (1) 'Mine is bigger than yours', (2) 'I know more natural history than you', and (3) 'My friends are more important than yours' but on a more serious note opined that 'self-indulgent work aimed at building up egos may not cause irreparable harm to science, but it certainly is no substitute for thoughtful goals, collegiality, care, respect and social responsibility' and further 'when an ecologist builds a pointless model or publishes a bad paper, it is possible for it to persist and cause unnecessary debates for decades'. Clearly, the need to bring clarity and objectivity to the discourse by countering emotive attacks, squanders time, energy, and financial resources that otherwise could be used more profitably in rationally pursuing the resolution of unsolved problems.

Despite the above impediments, there does seem to be some conceptual agreement among authors of most chapters, albeit in some cases somewhat grudgingly, that attention to empirical autecological studies could bring important information to bear and that environmental and biotic factors both are involved in the construction of insular biotic communities. For instance, Simberloff, Stone, and Dayan (ch. 2: 'Ruling out a community assembly rule: the method of favored states') indicated that '[with] sufficient knowledge of the basic biology of each species in a local species pool, including far more information than is usually brought to bear on the precise habitat preferences and requirements, it would probably be possible, in a narrowly defined region, to predict which species will be found at which site and which species will not coexist' and Wilson (ch. 5: 'Assembly rules in plant communities') identified two kinds of patterns in community ecology: 'Environmentally mediated patterns, i.e. correlations between species due to their shared responses to the physical environment' and 'Assembly rules, i.e. patterns due to interactions between species such as competition, allelopathy, facilitation, mutualism, and all the other biotic interactions that we know about in theory'. Fox (ch. 1: 'The genesis and development of guild assembly rules') opined: 'The guild assembly rule should be seen as a combination of stochastic and deterministic components.'

Nevertheless, on most issues, little consensus has been reached on the value or validity of assembly rules or the role of competition (and other interactions) (see section IV. *Incidence function and assembly rules*) in determining the structure of biotic communities. Much the same stance, as five decades previously, was maintained and some of the arguments seem much the same, only using more recent datasets or more refined null hypotheses. The deep division in this debate can be illustrated by the contrasting statements by Fox (ch. 1) that 'the assembly rule concept has survived and prospered to move on to a further stage' with that by Lockwood, Moulton, and Ballent (ch. 4: 'Introduced avifaunas as natural experiments in community assembly') that '[Diamond's] work sparked a heated debate and the concept of "assembly rules" was largely put to rest'.

Much of the past and present controversy centres on methodological procedures and statistical methods, whereby two investigators examining the same datasets come to opposite conclusions, e.g. Nevada test site as reviewed by Fox (ch. 1) versus Simberloff, Stone and Dayan (ch. 2). In some cases, such impasses seem to arise from proponents of one view striving to maintain a previous stance and seeking arguments to justify it and discredit the opposite view rather than seeking genuine understanding. A more charitable interpretation is that it is exceptionally challenging to construct a realistic, unbiased hypothesis whose testing will produce unambiguous results, and since the way in which to proceed still is being contested, perhaps neither approach alone is up to the task. Although there has been progress in formulating more precise and reliable null hypotheses (see section III. B. *Modelling of trophic relations within the context of insular ecosystems*), the question arises as to how one decides when an adequate null hypothesis, capable of analysing highly complex systems, has been reached. The test of the validity of a hypothesis should not be whether or not it confirms one's cherished ideas but how well it explains empirically based data.

In addition to prolonged attacks over the use of null hypotheses and statistical procedures and the need to employ defensive measures against attacks, there have been some positive suggestions for improvement of the study of the assembly of biotic communities. Lockwood, Moulton, and Balent (ch. 4) pointed out that to understand an assembly, one needs to define the species-pool relevant to the target community accurately and to take account of priority effects. Their conclusion was that this has seldom been achieved but that the available data suggest that interspecific competition is involved in the structuring of insular biotic communities. They also indicated that '[t]o some degree, all species that strive for membership in a community face a novel situation. The species they encounter will likely not be the same as those present in the community from which they originated. The variability in abiotic conditions will also not be the same.'

We suggest that a fruitful way to extend such work is to conduct detailed studies of the arrival, establishment, and extirpation of species on an island, such as conducted by Heatwole *et al.* (1981a) but over a much longer period. Because of species-turnover, an immigrant may face a very different set of resident species, and hence different challenges and opportunities on one occasion than at a later re-immigration after an intervening extirpation. Study of sufficient duration that multiple cycles of immigration, establishment (or failure to do so), extirpation, and re-immigration would be required even to begin to understand the degree of complexity of the dynamics of assembly. When one considers the multiplication of the number of potential vectors of interaction that occurs with an increase in the size of the species-pool of available immigrants, the difficulty in devising realistic null hypotheses becomes evident. In fact, it may be impossible to do so comprehensively, and it may mean that realistic null hypotheses are possible only for a reduced set of interactions. Most refined null hypotheses have been more restrictive, e.g. confined to a specific guild (see section V. *Stochasticism versus determinism*). Booth and Larson (ch. 7) pointed out that 'the decision about what null model to test patterns against requires more information than we have about the structure and function of ecosystems' and that 'Few statistical tests would be able to discriminate among any of the individual species distributional patterns'.

There seems to have been increased awareness of the separate roles of the physical environment and the biotic one as determinants of the structure of insular communities. A species may be excluded from an island either because of the island's inhospitable climatic conditions or because of the species' inability to cope with competing residents; conversely, its probability of establishment and persistence could be enhanced by a benign environment and/or lack of competitors. These may occur in sequence. Wilson (ch. 5: 'Assembly rules in plant communities') noted that '[t]his combination of processes [physical environment and biotic interactions] can be seen as the initial exclusion of species that are unable to tolerate the physical environment (i.e. environmental filtering), followed by the operation of assembly rules (i.e. biotic filtering)'.

Diamond's original concept of assembly rules as relating to patterns of interactions among particular species has broadened over the years to include processes as well as patterns, including environmentally mediated ones, and to be applied more generally than on the species-level. Wilson (ch. 5) remarked: 'In ecology terms are often used, for effect, so far outside their original intent that they lose any meaning. If this happens to "assembly rule" with the inclusion of the environmental sieve [...] another term will have to be invented to cover real assembly rules, based only on species interactions. That would be a nuisance.'

Much of the discussion in *Ecological Assembly Rules, Perspectives, Advances, Retreats* centred around assembly rules in communities generally, rather than specifically related to islands, and hence were somewhat peripheral to the present treatise. However, some of the content has a direct bearing on insular scenarios. For example, Wilson (ch. 5) pointed out that environmental filtering is followed by biotic filtering (assembly rules) and that their interaction produces the final pattern of the community and that much of the difficulty in framing assembly rules arises from choosing inappropriate statistics or null models. He recommended three safeguards: (1) the method should be tested on random datasets; (2) the null model should exclude every feature of the observed data except the one it is intended to test; and (3) should represent an explicit ecological process by which the community could conceivably have been assembled. An important question to address is whether there is 'limiting similarity', i.e. do pairs of species that co-occur have less overlap than expected at random?

Most of the original formulation of assembly rules was based on the study of birds, but some later studies have tested their taxonomic generality. For instance, Kelt and Brown (ch. 3: 'Community structure and assembly rules: confronting conceptual and statistical issues with data on desert rodents') reviewed their applicability to desert rodents. They found that:

> The vast majority of assembly rules indicate that the species that coexist are more different than expected by chance; species in local communities are non-random samples from the regional species pool, and some species of desert rodents in the regional pools are non-random samples from the larger pool of species inhabiting

> the entire southwestern [United States] desert region [...] And the experimental evidence for interspecific competition is overwhelming,

but went on to say: 'Requiring further investigation, however, is the relationship between phylogenetic, biogeographic, and ecological relationships' and that 'many of the assembly rules describe probabilistic, rather than absolute, patterns of co-occurrence'.

They provided an example in which two species of closely related and morphologically and functionally similar species co-occurred for at least 19 years and suggested that even strong competitors could coexist indefinitely if the presence of each were reinforced either by immigration or if environmental shifts alternately favoured opposite members of the species-pair. They finally concluded that if there truly are general patterns and processes that characterize the organization of communities, these do not appear to be reflected in any very direct way in the well-documented assembly rules that describe the structure of North American desert-rodent assemblages.

Cody (ch. 6: 'Assembly rules at different scales in plant and bird communities') emphasized the concept of nestedness (the degree to which biotas of small islands are subsets of those of larger ones) because of its relevance to whether insular communities are dispersal-driven or extirpation-driven. Nestedness signals Type A assembly rules (those that govern which species are added as the size of the community increases). He also noted: 'Species that are particularly close, ecologically, may coexist only in certain years (e.g. high productivity years) or in certain sites or that permit their divergence, but otherwise they segregate by habitat use.'

Drake, Zimmerman, Purucker, and Rojo (ch. 8: 'On the nature of the assembly trajectory') emphasized the multiplicity of control of the assembly process and that these vary in space and time, with invasions of new species causing changes in the trajectory 'as the community exhibits sensitivity to a new set of initial conditions.' They noted that both environmental and stochastic events modify the mechanism of assembly and that 'Communities likely exist on a continuum from what are essentially random assemblages up to highly integrated self-organized structures' and may experience high rates of disrupting invasions and never settle onto a self-organized trajectory.

Weiher and Keddy (ch. 9: 'Assembly rules as general constraints on community composition') addressed the use of null models and concluded that they led to several important ideas about the structure of biotic communities, especially when focussing on the dispersion of traits rather than on species, but that they had provided little information about the nature of assembly rules. They concluded that 'null models can best be used to increase our understanding of community structure, but they rarely provide us with defined assembly rules' and that 'If assembly rules are going to be useful, then attempts must be made to use them to predict composition'.

Lomolino (ch. 10: 'A species-based, hierarchical model of island biogeography') presented a model of insular biogeography based on the assumptions that (a) species are common on those islands where their duration on the island exceeds the time between immigrations, (b) duration of the focal species should increase with island-area, and (c) the time between successive immigrations should increase with an island's isolation. He linked this to community structure and found that the model predicted two assembly rules: (1) When better dispersers tend to require fewer resources, regardless of the biogeographic space sampled, communities should accumulate species in order of their increasing demand upon resources and decreasing ability to immigrate and (2) when better dispersers tend to require more resources, the sequence of accumulation will vary depending on the range in area and isolation sampled (e.g. first A, then B, then C on the near islands; first C, then B, then A on the large islands). He concluded: 'Only when the results of these studies are combined, does the ambiguity clear to reveal the combined importance of immigration and extinction at the community level. Assembly "rules" should differ, but in a manner consistent with the resource requirements and immigration abilities of the pool of focal species.'

Strange and Foin (ch. 11: 'Interaction of physical and biological processes in the assembly of stream fish communities'), Díaz, Cabido, and Casanoves (ch. 12: 'Functional implications of trait–environment linkages in plant communities'), and Lockwood and Pimm (ch. 13: 'When does restoration succeed?') deal either with non-insular systems or with the applications of community ecology and hence are extralimital for most topics covered in the present chapter.

In summing up the book, Keddy (ch. 14: 'Epilogue: from global exploration to community assembly') stated that '[t]he intractable number of components still remains a central problem in the causal analysis of ecological communities' and that '[t]he finer the scale becomes, the more difficult the explanation seems to become.'

Götzenberger *et al.* (2012), some years after the 50-year celebration of the concept of assembly rules, noted that non-random assemblages had been demonstrated only infrequently and recognized there still were many possible biases and shortcomings in the methodology. They defined the concept in a more general framework in which the co-occurrence of species is a product of chance, historical patterns of speciation and migration, dispersal, abiotic environmental factors and biotic interactions, with none of these processes being mutually exclusive. They suggested including new comparative studies that simultaneously address not only different categories of biotic assembly rules (e.g. co-occurrence, niche-limitation, guild proportionality, limiting similarity), but also different types of assembly rules (abiotic, biotic, dispersal, even phylogeographic) as well as applying consistent sets of methods to different datasets from different types of ecosystems; different traits and functional types of organisms; and studying patterns throughout a range of scales from local to landscape to biogeographic; and to include different trophic levels. They concluded that non-random assemblages are infrequently demonstrated but noted that biases and methodological shortcomings mean that the evidence is not conclusive.

The study of assembly rules of communities in the years following the 50-year celebration of the concept has moved on to greater emphasis on their applicability to communities on mainlands rather than solely to insular ones. Consequently, the body of this voluminous literature deals with topics beyond the scope of the present review. However, a few relevant comparisons between the islands and mainlands should be made: (1) Mainland communities are spatially contiguous and do not have the same barriers to immigration from one to another that apply to islands whose dynamics are constrained by the exigencies of overwater dispersal. It is uncertain to what extent one can carry over generalizations from mainland communities to insular ones or vice versa. (2) The study of mainland communities has not contributed significantly to resolving opposing opinions on the validity of assembly rules as illustrated by the exchange between Grime (2006), who considered the assembly-rules model to lack empirical support and Wilson's (2007) reply to Grime in which he stated: 'Grime declares that the assembly-rules model [...] is "not supported by empirical study of plant communities".' This is, he [Grime] says, 'the agony of community ecology. I show that there is now abundant evidence for assembly rules, and no agony.' (3) Study of both situations has converged in the realization that the ecology of the species involved needs to be given greater consideration. (4) The study of assembly of mainland communities have involved plant assemblages more often than did the earlier literature on islands. (5) The study of mainland communities treated a wider range of topics than insular ones.

Simberloff and Collins (2010), in a symposium on the equilibrium theory, also discussed assembly rules and reinterpreted some of Diamond's (1975) conclusions:

> Specifically at the subset of species pairs in which competition would be most expected, we found that no exclusively distributed pairs quite conformed to the checkerboard model and that the exclusive patterns might be explained by a combination of colonization history and timing, behavioral traits (especially propensity to fly over water), and habitat preferences. [...] Some [Supertramps] may be only on small islands because they are excluded elsewhere by competition. Other explanations are possible, however. They may prefer habitats disproportionately present on small islands [...] it is also possible that some checkerboards have been created by undocumented anthropogenic extinction. Additionally, the possibility of incomplete censuses [...] should be borne in mind; some absences may be artifacts and rectifying them would be more likely to obliterate checkerboards than to generate them.

By contrast, Fox and Fox (2000) posited that 'the complement of species established on islands, or in insular habitats, may be governed by assembly rules. These contributions may assist in taking a renewed theory into the new millennium.' Colorado-Zuluaga (2015) concluded, in a review of how ecological communities are assembled, that 'a growing body of theoretical and empirical studies [...] seems to provide evidence to support the idea that assembly rules can govern certain communities' but went on to add that 'patterns may not be the result of contemporary ecological processes but of events that occurred sometime during community assembly'.

V. Stochasticism versus determinism

The controversies over stochasticism versus determinism have been discussed in relation to the theory of trophic structure stability (see section III. B. *Modelling of trophic relations within the context of insular ecosystems*) and assembly rules (see section IV. D. *Current status of assembly rules*). Here, these two approaches are addressed in a more general way.

Clearly, the physical environment has a decisive role in screening out species that are maladapted either to environmental rigours of oversea dispersal to islands or to survival under the insular environments they meet upon arrival. In terms of dispersal, one species could be superior to another species either by being hardy and able to survive salinity or the other physical rigours of oversea dispersal or by producing a greater number of propagules, thereby improving the chances that at least some reach an island. In this sense, the competition has already taken place in the source area, whereas competition for resources needed for establishment and persistence take place after arrival. Indeed, Hubbell proposed a 'neutral' theory that integrates rate of speciation, species' abundances (in a zero-sum ecological drift leading to saturation of space or resources), and the assumption of equivalency among species (Hubbell 2001, 2010; also see Rosindell *et al.* 2011). In Hubbell's theory, either individuals or species have an equal chance of immigrating and becoming extirpated, and the outcome is determined mainly by their relative abundance at the source area. Others echoed elements of Hubbell's neutral theory: Weiher *et al.* (2011) suggested that neutral processes may explain a significant source of variation in community assembly, whereas Morin (2011) highlighted the importance of including species' abundances in such models. Therefore, Hubbell (2001) generated a dispersal-assembly model that also 'has the potential to have a strong, multidecadal trajectory, not unlike that of the ETIB' [the Equilibrium Theory of Island Biogeography] (Sax and Gaines 2011: 232; also see Vellend 2010).

Most debate, therefore, has centred around two schools of thought, one espousing stochasticism in influencing the structure of insular assemblages and the other championing determinism. Diamond and his colleagues were the leading advocates for determinism, while Simberloff and co-workers favoured the stochastic view – even listing it as the default position in constructing null hypotheses. The debate continued at rather high intensity over a number of decades. This controversy was too prolonged and detailed to treat on a paper-by-paper basis in such a brief review as the present one, and only a few of the most important ones are discussed herein. We merely list a sample of the papers from the stochastic side (Simberloff 1971, 1976a,b, 1978; Connor and Simberloff 1979a,b; Connor and McCoy 1979; Strong *et al.* 1979, 1984) and from the deterministic side (Diamond 1975, 1979; Diamond and Gilpin 1980, 1982; Grant and Abbott 1980; Colwell and Winkler 1984; Gilpin and Diamond 1982, 1984), some of which are treated in more detail below. For more recent examples, compare Connor *et al.* (2013) with Diamond *et al.* (2015). Gotelli and McCabe (2002) took a more middle-of-the-road position by concluding: 'Although these [meta-analyses] do not confirm the mechanisms of Diamond's controversial assembly rules model, they do establish that observed co-occurrence in most natural communities is usually less than expected by chance. These results contrast with previous analyses of species co-occurrence patterns and bridge the apparent gap between experimental and correlative studies in community ecology.'

Islands often have simplified habitats and are ecologically less diverse relative to the mainland supplying the pool of immigrants. The question arises as to whether this situation means that potential competitors respond to each other differently on islands than they do on mainlands. Grant (1966a,b) addressed this by comparing the population densities of congeneric birds on islands compared to those on adjacent mainlands and concluded that there was a different pattern of interaction among species on the islands than at their source. Simberloff (1971) criticized these results and pointed out some errors in the data. However, when these were corrected (Grant 1971), Grant's conclusions were more strongly supported than before. Simberloff also criticized Grant for the omission of occurrences of rare birds in samples (which probably were unestablished waifs and not part of the breeding population on the islands) and the exclusion of irrelevant species, thereby missing the point of the critical question Grant asked, which was 'What happens to species in a move from mainland to island?' To answer that, only the species occurring in both places were pertinent.

Cirtwill and Stouffer (2015) found that incorporating information on the trophic structure of insular communities, particularly the species-specific availability of resources, substantially alters the probability of extirpation, whereas the probability of immigration was largely stochastic. Colorado-Zuluaga (2015) concluded that random and non-random processes influence community structure and that both types of hypotheses should be tested in order to obtain an accurate portrayal of the processes involved.

VI. Amphibians as island hoppers

Most species of frogs contain at least two *'econes'* (see section III. *The theory of trophic structure stability*). However, many Caribbean species have direct development in which the larval stage occurs within the egg and is nourished by the mother's deposition of yolk, and the young only participate directly in the trophic activities of the community upon hatching as miniatures of the adult; in this case, the species may consist of only one econe. Another example is *Eleutherodactylus jasperi* from Puerto Rico, the only Neotropical example of a frog with internal fertilization and development of the young within the mother, and now possibly extinct (Burrowes and Joglar 1991; Moreno 1991). In this case, as well as in direct-development, the energetic cost of the development of the embryos is encapsulated within the nutritional requirements and trophic relations of the mother, with the larvae or juveniles exerting their independent trophic impact only after hatching or birth. Moreover, amphibians are notoriously poor dispersers across barriers posed by seawater (Ali and Meiri 2019). Although there are a few exceptional species that are somewhat resistant to seawater, most amphibians osmotically lose their body water rapidly and survive only short periods of time when exposed to the sea (only hours versus the survival of lizards for days in similar situations; Heatwole, unpublished data), thereby negating the likelihood of their being dispersed as waifs either directly in seawater or marine-transported flotsam. Thus, they are much less commonly represented in insular faunas than are lizards.

The minimum habitable insular area (MIA) of amphibians on sandy cays may be larger than for many other taxa because rain seeps into sandy soils without leaving surface water for aquatically breeding species. Consequently, they may only be able to inhabit islands that are larger and have habitats other than a sandy strand. A limitation of the database used in studies of the insular biogeography of amphibians generally is that 'blank' surveys of small marine cays often are not published because only 'negative' information is obtained. The fact that a species occurs on a particular island is deemed more noteworthy than that it does not occur there, so documentation of absences frequently is not provided. Serendipitously, herpetological surveys usually include both reptiles and amphibians, and since reptiles are not so directly tied to water for reproduction and are more resistant to evaporative loss than amphibians, they are found on small islands that are climatically unsuitable for amphibians. Thus, islands where the complete herpetofauna has been surveyed, and with reptiles found to be present, but not amphibians, can provide an estimate of the minimum size of island suitable for amphibians; comparative data for amphibians from islands in marine situations are badly needed, as are minimum insular areas of islands over a geographic range of archipelagos. In this regard, it would be interesting to note whether the minimum size of island inhabitable by amphibians differs between sandy cays and small rocky islands as the latter are more likely to have small temporary puddles – and humid microhabitats – that may facilitate breeding.

In summary, most amphibians have three strikes against them in terms of their representation on islands compared with reptiles. Specifically: (1) their vulnerability to high salinity limits their ability to disperse to islands; (2) lack of standing water for breeding, and shelter from the heat on small islands, especially sandy cays, prevents their establishment; and (3) the peripheral habitats (beaches and strands) where they most likely first land, even on more benign islands, are not habitats favourable for most amphibians, but which they have to cross to find somewhere suitable for survival and establishment.

VII. Acknowledgements

The roots of the present chapter go back to the early 1960s. I (HH) had just finished my PhD dissertation on frogs (Heatwole 1959) and was launching into a combined career of research in herpetology and insular biogeography at the University of Puerto Rico, the latter in collaboration with the late Richard Levins. The late Robert MacArthur spent brief periods in our laboratory in those years when he and E. O. Wilson were thinking deeply about island biogeography at a time when Levins and I were conducting surveys of insular taxa in order to draw biogeographical conclusions. These two pairs of scientists were similar in that each was a productive partnership of a mathematical ecologist (MacArthur; Levins) with a naturalist (Wilson; Heatwole). Eventually, the two groups concentrated on different aspects of the topic, with MacArthur and Wilson pursuing the study of the number of species on islands and Levins and Heatwole the distribution of various taxa among islands (see Warren *et al.* 2019); inevitably there were overlaps. The discussions with MacArthur and my sustained collaboration with Levins over the years were influential in determining the trajectory of my subsequent career and to both I am grateful. I previously paid tribute to Levins on the occasion of the Festschrift honouring him on his 85th birthday, shortly before his death (Heatwole 2018b), but I felt that my debt to him merited repeating here.

In the early 1960s, Levins and I had an NSF grant to conduct a survey of various taxa on the Puerto Rican/Virgin Island archipelago as it related to island biogeography. We had a friendly relationship with both Wilson and MacArthur; Wilson identified our ants, and MacArthur visited our laboratory in either 1961 or 1962 and Levins and I discussed island biogeography with him at length. Since Levins' and my paper came out the same year as the MacArthur-Wilson one, I have been asked several times: "Who got the idea first?" They were not quite the same idea, only very similar approaches to the same problem, with the original question leading to slightly different formulations (see section II. A. *Preludes to the theory*). All four of us probably could trace some of our insights into island biogeography to heuristic input from each of the others. Perhaps an insight into these discussions is best expressed in MacArthur's (1972) own words in his book *Geographical Ecology*, that these shared theoretical ideas were 'on such a give-and-take basis that I could not possibly trace the history of many'.

Neither Levins nor I ever felt territorial or had any competitive interaction with MacArthur/Wilson about the ideas we discussed. When asked which paper came out first? I had to admit I hadn't any idea as it had never occurred to me as an important question. I don't recall now whether we sent them a copy of our paper for comment before we submitted it, or whether we knew about theirs before they published. At the time of our discussions about these matters, none of us had fully crystallized our thinking about these matters and later, when we had, there was no way of identifying the impetus for a particular idea. In fact, similar ideas had been published even before any of our papers came out, so the origins of the theory of island biogeography antedated 1963 (see section II. A. *Preludes to the theory*).

Whatever such contributions from Levins or myself to MacArthur and Wilson might have been should not detract from the fact that they developed the theory far beyond the scope of the paper Levins and I wrote, and theirs had a correspondingly far greater impact on the further development of theory in that field. I certainly learned more from our discussions than I contributed. The high reputation they received for their theory was well deserved and should not be minimized.

I am grateful to my late wife, Audry Yoder Heatwole, and to the faculty members, many undergraduate students, graduate students, and research assistants at the University of Puerto Rico, especially Frank Torres, Sheila Clements (née Blasini), the late Faustino MacKenzie, the late Francis Rolle, and the late Manuel Vélez for participating in the field surveys. It is sad that so many of the persons acknowledged above had to be designated as 'late'. I am grateful to my present wife, Sue King, who kindly rendered technical assistance in the preparation of this manuscript and offered constructive criticism.

I (NR-L) am grateful to several colleagues at the University of Puerto Rico who influenced me as a naturalist and herpetofaunal ecologist: Rafael L. Joglar initiated me in the study of amphibians in Puerto Rico on my way to my MS thesis (Ríos-López 1999); T. Michell Aide is responsible for my development as a community ecologist using the Puerto Rican herpetofauna as a study model, while J. P. Richard Thomas expanded my interest and knowledge about the herpetofauna at large for my PhD dissertation (Ríos-López 2007). Ariel E. Lugo has been instrumental for a better understanding of the influence of natural and human phenomena on ecosystem processes that maintain and generate biodiversity. I am also grateful to my undergraduate students for their extraordinary enthusiasm and support in the field and laboratory at the University of Puerto Rico in Humacao.

I am most grateful to Elsie Rivera for her unconditional support as a colleague and, more so, as my beloved wife. To our daughters, Laura and Sofía, for their love, patience, and understanding of the many long nights that I had to go out 'counting coquí frogs and lizards'. Finally, my most sincere gratitude goes to Harold Heatwole for allowing me to contribute to a treatise that expands his extraordinary legacy to herpetology and ecology in particular and science in general; it is the greatest of honours.

VIII. References

Abbott, I. and Wills, A., 2016. Review and synthesis of knowledge of insular ecology, with emphasis on islands of Western Australia. *Conservation Science Western Australia* **11**: 1–209.

Ali, J. R. and Meiri, S., 2019. Biodiversity growth on the volcanic ocean islands and the roles of *in situ* cladogenesis and immigration: case with the reptiles. *Ecography* **42**: 989–999. https://doi.org/10.1111/ecog.04024

Allison, A., 2002. Patterns of geographic distribution: animals. Ch. 1.6.4.4 in *Oceanic Islands*, volume 1.6.4., edited by H. Heatwole in 5. *Natural Resource System Challenge: Oceans and Aquatic Ecosystems*, edited by E. Wolanski, Encyclopedia of Life Support Systems. Developed under the auspices of the UNESCO. EOLSS Publishers, Oxford, UK [http://www.eolss.net]

Aristotle., Περι Ζωιων Γενεσεως. *Generation of Animals;* see Peck (1963).

Ball, E. and Glucksman, J., 1975. Biological colonization of Motmot, a recently-created tropical island. *Proceedings of the Royal Society, London* **190**: 421–442. https://doi.org/10.1098/rspb.1975.0104

Belyea, L. R. and Lancaster, J., 1999. Assembly rules within a contingent ecology. *Oikos* **86**: 402–416. https://doi.org/10.2307/3546646

Briand, F. and Cohen, J., 1984. Community food webs have scale invariant structure. *Nature* **307**: 264–267. https://doi.org/10.1038/307264a0

Brown, J. H. and Kodric-Brown, K., 1977. Turnover rates in insular biogeography: Effect of immigration on extinction. *Ecology* **58**: 445–449. https://doi.org/10.2307/1935620

Brown, J. H. and Lomolino, M. V., 1989. Independent discovery of the equilibrium theory of island biogeography. *Ecology* **70**: 1954–1957. https://doi.org/10.2307/1938125

Brown, J. H. and Lomolino, M. V., 2000. Concluding remarks: historical perspective and the future of island biogeography theory. *Global Ecology and Biogeography* **9**: 87–92. https://doi.org/10.1046/j.1365-2699.2000.00186.x

Burrowes, P. A. and Joglar, R. L., 1991. A survey of the population status and ecological evaluation of three Puerto Rican frogs. *Publicación Cientifica Miscelánea Departamento de Recursos Naturales de Puerto Rico* **1**: 42–46.

Cash, F. B. J., Conn, A., Coutts, S., Stephen, M., Mason, N. W. H., Anderson, B. J., and Wilson, B., 2012. Assembly rules operate only in equilibrium communities: is it true? *Austral Ecology* **37**: 903–914. https://doi.org/10.1111/j.1442-9993.2011.02349.x

Chase, J. M., Gooriah, L., May, F., Ryberg, W. A., Schuler, M. S., Craven, D., and Knight, T. M., 2019. A Framework for disentangling ecological mechanisms underlying the island species–area relationship. *Frontiers of Biogeography* **11**: e40844. https://doi.org/10.21425/F5FBG40844

Chen, X. Y., Jaio, J., and Tong, H., 2011. A generalized model of island biogeography *Science China (Life sciences)* **54**: 1055–1061. https://doi.org/10.1007/s11427-011-4238-y

Cirtwill, A. R. and Stouffer, D. B., 2016. Knowledge of predator-prey interactions improves predictions of immigration and extinction in island biogeography. *Global Ecology and Biogeography* **25**: 900–911. https://doi.org/10.1111/geb.12332

Clegg, S., 2010. Evolutionary changes following island colonization in birds, empirical insights into roles of microevolutionary processes. Pp. 293–325 in *The Theory of Island Biogeography Revisited*, edited by J. B. Losos and R. E. Ricklefs. Princeton University Press, Princeton, New Jersey. https://doi.org/10.1515/9781400831920.293

Cody, M. L., 2006. *Plants on Islands: Diversity and Dynamics on a Continental Archipelago.* University of California Press, Berkeley, California. https://doi.org/10.1525/9780520932722

Colorado-Zuluaga, G. J., 2015. How ecological communities are structured: a review on ecological assembly rules. *Revista Escuela de Ingenieria Antioquia* (in Colombia) **12**: 27–53.

Colwell, R. K. and Winkler, D. W., 1984. A null model for null models in biogeography. Ch. 20, pp. 344–359 in *Ecological Communities: Conceptual Issues and the Evidence*, edited by D. R. Strong, Jr., D. Simberloff, L. G. Abele, and A. B. Thistle. Princeton University Press, Princeton, New Jersey. https://doi.org/10.1515/9781400857081.344

Connor, E. F. and McCoy, E. D., 1979. The statistics and biology of the species–area relationship. *The American Naturalist* **113**: 791–833. https://doi.org/10.1086/283438

Connor, E. F. and Simberloff, D., 1979a. The assembly of species communities: chance or competition? *Ecology* **60**: 1132–1140. https://doi.org/10.2307/1936961

Connor, E. F. and Simberloff, D., 1979b. You can't falsify ecological hypotheses without data. *Bulletin of the Ecological Society of America* **60**: 154–155.

Connor, E. F., Collins, M. D., and Simberloff, D., 2013. The checkered history of checkerboard distributions. *Ecology* **94**: 2403–2414. https://doi.org/10.1890/12-1471.1

Darwin, C., 1842. *The Structure and Distribution of Coral Reefs. Being the first part of the geology of the voyage of the Beagle, under the command of Capt. Fitzroy, R.N. during the years 1832 to 1836*. Smith Elder and

Company, London, England, UK. https://doi.org/10.2307/1797986

Darwin, C. and Wallace, A., 1859. On the tendency of species to form varieties: and on the perpetuation of varieties and species by natural means of selection. *Journal of the Linnean Society* **3**: 45. https://doi.org/10.1111/j.1096-3642.1858.tb02500.x

Diamond, J. M., 1970a. Ecological consequences of island colonization by southwest Pacific birds. I. Types of niche shifts. *Proceedings of the National Academy of Sciences of the United States of America* **67**: 529–536. https://doi.org/10.1073/pnas.67.2.529

Diamond, J. M., 1970b. Ecological consequences of island colonization by southwest Pacific birds. II. The effect of species diversity on total population density. *Proceedings of the National Academy of Sciences of the United States of America* **67**: 1715–1721. https://doi.org/10.1073/pnas.67.4.1715

Diamond, J. M., 1972. Biogeographic kinetics: estimation of relaxation times of avifaunas of southwest Pacific islands. *Proceedings of the National Academy of Sciences of the United States of America* **69**: 3199–3203. https://doi.org/10.1073/pnas.69.11.3199

Diamond, J. M., 1973. Distributional ecology of New Guinea birds. *Science* **179**: 759–769. https://doi.org/10.1126/science.179.4075.759

Diamond, J. M., 1975. Assembly of species communities. Pp. 342–444 in *Ecology and Evolution of Communities*, edited by M. L. Cody and J. M. Diamond. Harvard University Press, Cambridge, England, UK.

Diamond, J. M., 1979. Community structure: is it random or is it shaped by species-differences and competition? Pp. 165–181 in *Population Dynamics*, edited by R. M. Anderson, B. D. Turner, and L. R. Taylor. Blackwell Scientific Publications, Oxford, England, UK.

Diamond, J. M. and Gilpin, M. E., 1980. Turnover noise: Contribution to variance in species number and prediction from immigration and extinction curves. *The American Naturalist* **115**: 884–889. https://doi.org/10.1086/283607

Diamond, J. M. and Gilpin M. E., 1982. Examination of the null model of Connor and Simberloff for species co-occurrences on islands. *Oecologia* **52**: 64–74. https://doi.org/10.1007/BF00349013

Diamond, J. M., Pimm, S. L., and Sanderson, J. G., 2015. The checkered history of checkerboard distributions: Comment. *Ecology* **965**: 3386–3388. https://doi.org/10.1890/14-1848.1

Donohue, I., Petchey, O. L., Kéfi, S., Génin, A., Jackson, A. L., Yang, Q., and O'Connor, N. E. 2017. Loss of predator species, not intermediate consumers, triggers rapid and dramatic extinction cascades. *Global Change Biology* **23**: 2962–2972. https://doi.org/10.1111/gcb.13703

Ellis, J. C., Farina, J. M., and Witman, J. D., 2006. Nutrient transfer from land to sea: the case of gulls and cormorants in the Gulf of Maine. *Journal of Animal Ecology* **75**: 565–574. https://doi.org/10.1111/j.1365-2656.2006.01077.x

Farrow, R. A., 1984. Detection of transoceanic migration of insects to a remote island in the Coral Sea, Willis Island. *Australian Journal of Ecology* **9**: 253–272. https://doi.org/10.1111/j.1442-9993.1984.tb01362.x

Fetcher, N., Agosta, S. J., Moore, J. C., J. A. Stratford, J. A. and Steele, M. A., 2015. The food web of severely contaminated site following reclamation with warm season grasses. Restoration Ecology **23**: 421–429. https://doi.org/10.1111/rec.12202

Flood, P. G. and Heatwole, H., 1986. Coral cay instability and species-turnover of plants at Swain Reefs, southern Great Barrier Reef, Australia. *Journal of Coastal Research* **2**: 479–496.

Foufopoulos, J. and Mayer, G. C., 2007. Turnover of passerine birds on islands in the Aegean Sea (Greece). *Journal of Biogeography* **34**: 1113–1123. https://doi.org/10.1111/j.1365-2699.2007.01695.x

Fox, B. J. and Fox, M. D., 2000. Factors determining mammal species richness on habitat islands and isolates: habitat diversity disturbance, species interactions and guild assembly rules. *Global Ecology and Biogeography* **9**: 19–37. https://doi.org/10.1046/j.1365-2699.2000.00184.x

Franzén, M., Schweiger, O., and Betzholtz, P.-E. 2012. Species–area relationships are controlled by species traits. *PLoS ONE* **7**: e37359. https://doi.org/10.1371/journal.pone.0037359

Gilbert, F. S., 1980. The equilibrium theory of island biogeography: fact of fiction? *Journal of Biogeography* **7**: 209–235. https://doi.org/10.2307/2844629

Giller, P. S., 1984. *Community Structure and the Niche.* Chapman and Hall, London, England, UK. https://doi.org/10.1007/978-94-009-5558-5

Gilpin, M. E. and Diamond J. M., 1982. Factors contributing to non-randomness in species co-occurrences on islands. *Oecologia* **52**: 75–84. https://doi.org/10.1007/BF00349014

Gilpin, M. E. and Diamond, J. M., 1984. Are species co-occurrences on islands non-random, and are null hypotheses useful in community ecology? Ch. 17, pp. 297–315 in *Ecological Communities: Conceptual Issues and the Evidence*, edited by D. R. Strong Jr., D. Simberloff, L. G. Abele, and A. B. Thistle. Princeton University Press, Princeton, New Jersey. https://doi.org/10.1515/9781400857081.297

Gotelli, N. J. and McCabe, D. J., 2002. Species co-occurrence: a meta-analysis of J. M. Diamond's assembly rules model. *Ecology* **83**: 2091–2096. https://doi.org/10.1890/0012-9658(2002)083[2091:SCOAMA]2.0.CO;2

Götzenberger, L., de Bello, F., Brathen, K. A., Davison, J., Dubuis, A., Guisan, A., Leps, J., Lindborg, R., Moora, M., Pärtel, M., Pellissier, L., Pottier, J., Vittoz, P., Zobel, K., and Zobel, M., 2012. Ecological assembly rules in plant communities—approaches, patterns and prospects. *Biological Reviews* **87**: 111–127. https://doi.org/10.1111/j.1469-185X.2011.00187.x

Graham, N. A. J., Wilson, S. K., Carr, P., Hoey, A. S., Jennings, S., and McNeil, M. A., 2018. Seabirds enhance coral reef productivity and functioning in the absence of invasive rats. *Nature* **559**: 250–253. https://doi.org/10.1038/s41586-018-0202-3

Grainger, T. N., Germain, R. M., Jones, N. T., and Gilbert, B., 2017. Predators modify biogeographic constraints on species distributions in an insect metacommunity. *Ecology* **98**: 851–860. https://doi.org/10.1002/ecy.1712

Grant, P. R., 1966a. The density of land birds on the Tres Marias Islands in Mexico. II. Distribution of abundances in the community. *Canadian Journal of Zoology* **44**: 1023–1030. https://doi.org/10.1139/z66-108

Grant, P. R., 1966b. Ecological compatibility of bird species on islands. *The American Naturalist* **100**: 451–462. https://doi.org/10.1086/282438

Grant, P. R., 1971. Comment on Simberloff's letter. *The American Naturalist* **105**: 194–197. https://doi.org/10.1086/282716

Grant, P. R. and Abbott, I., 1980. Interspecific competition, island biogeography and null hypotheses. *Evolution* **34**: 332–341. https://doi.org/10.1111/j.1558-5646.1980.tb04822.x

Grant, P. R. and Grant, B. R., 2010. Sympatric speciation, immigration, and hybridization in island birds. Pp. 326–357 in *The Theory of Island Biogeography Revisited*, edited by J. B. Losos and R. E. Ricklefs. Princeton University Press, Princeton. https://doi.org/10.1515/9781400831920.326

Gravel, D. and Massol, F., 2020. Toward a general theory of metacommunity ecology. Ch. 12, pp. 195–220 in *Theoretical Ecology: Concepts and Applications*, edited by K. S. McCann and G. Gellner. Oxford University Press, Oxford, UK. https://doi.org/10.1093/oso/9780198824282.003.0012

Gravel, D., Massol, F., Canard, E., Mouillot, D., and Mouquet, N., 2011. Trophic theory of island biogeography. *Ecology Letters* **14**: 1010–1016. https://doi.org/10.1111/j.1461-0248.2011.01667.x

Grime, J. P., 2006. Trait convergence and trait divergence in herbaceous plant communities: Mechanisms and consequences. *Journal of Vegetation Science* **17**: 255–260. https://doi.org/10.1111/j.1654-1103.2006.tb02444.x

Guzman, L. M., Germain, R. M, Forbes, C., Strauss, S., O'Connor, M. I., Gravel, D., Srivastava, D. S., and Thompson, P. L., 2018. Towards a multi-trophic extension of metacommunity ecology. Ecology Letters 22: 19–33. https://doi.org/10.1111/ele.13162

Hajian-Forooshani, Z., Schmitt, L., Medina, N., and Vandermeer, J., 2020. Trophic-specific responses to migration in empirical metacommunities. *Oikos* **129**: 413–419. https://doi.org/10.1111/oik.06342

Hall, C. M., 2010. An island biographical approach to island tourism and biodiversity: an exploratory study of the Caribbean and Pacific islands. *Asia Pacific Journal of Tourism Research* **15**: 383–399. https://doi.org/10.1080/10941665.2010.503628

Hanski, I., 2010. The theories of island biogeography and metapopulation dynamics. Pp. 186–213 in *The Theory of Island Biogeography Revisited*, edited by J. B. Losos and R. E. Ricklefs. Princeton University Press, Princeton, New Jersey. https://doi.org/10.1515/9781400831920.186

Heatwole, H. F., 1959. *The Forest Floor and its Utilization by Selected Species of Terrestrial Amphibians*. Ph.D. Dissertation, University of Michigan, Ann Arbor, Michigan.

Heatwole, H., 1971. Marine-dependent terrestrial biotic communities on some cays in the Coral Sea. *Ecology* **52**: 563–566. https://doi.org/10.2307/1934598

Heatwole, H., 1984. Terrestrial vegetation of the coral cays, Capricornia Section, Great Barrier Reef Marine Park. *Royal Society of Queensland Symposium Capricornia Section, Great Barrier Reef*, pp. 87–139.

Heatwole, H., 1989. The concept of the econe, a fundamental ecological unit. *Tropical Ecology* **30**: 13–19.

Heatwole, H., 1991a. Factors affecting the number of species of plants on islands of the Great Barrier Reef, Australia. *Journal of Biogeography* **18**: 213–221. https://doi.org/10.2307/2845294

Heatwole, H., 1991b. The colonization of coral islands. Pp. 146–153 in *The Encyclopedia of the Earth, Oceans and Islands*, edited by F. H. Talbot and R. E. Stevenson. Weldon Owen Pty. Ltd., Sydney, AU and Weldon Owen, Inc., San Francisco, California.

Heatwole, H., 1994. Colonization of coral islands. Ch. 5, pp. 56–65 in *Islands, the Illustrated Library of the Earth*, edited by R. E. Stevenson and F. H. Talbot. Readers Digest Press, Surrey Hills (Sydney), and Weldon Owen Pty. Ltd., Sydney.

Heatwole, H., 2018a. Trophic structure stability in insular biotic communities. Pp. 220–243 in *Essays in Honor of Richard Levins, The Truth is the Whole*, edited by T. Awerbuch, M. S. Clark, and P. J. Taylor. The Pumping Station, Arlington, Virginia.

Heatwole, H., 2018b. Rescue at Sandy Cay. Pp. 260–262 in *The Truth is the Whole: Essays in Honor of Richard Levins*, edited by T. Awerbuch, M. S., Clark, and P. J. Taylor. The Pumping Station, Arlington, Virginia.

Heatwole, H. and R. Levins., 1972. Trophic structure stability and faunal change during recolonization. *Ecology* **53**: 531–534. https://doi.org/10.2307/1934248

Heatwole, H. and R. Levins., 1973. Biogeography of the Puerto Rican Bank: Species-turnover on a small cay, Cayo Ahogado. *Ecology* **54**: 1042–1055. https://doi.org/10.2307/1935570

Heatwole, H. and MacKenzie, F., 1967. Herpetogeography of Puerto Rico. IV. Paleogeography, faunal similarity and endemism. *Evolution* **21**: 429–438. https://doi.org/10.1111/j.1558-5646.1967.tb03400.x

Heatwole, H., Done, T., and Cameron, E., 1981a. *Community Ecology of a Coral Cay, a Study of One-Tree Island, Great Barrier Reef, Australia. Monographiae Biologicae* **43**: 1–379. Dr W. Junk Publishers, The Hague, Netherlands.

Heatwole, H., Levins, R., and Byer, M. D., 1981b. Biogeography of the Puerto Rican Bank. *Atoll Research Bulletin* **251**: 1–63. https://doi.org/10.5479/si.00775630.251.1

Helmus, M. R. and Behm, J. E., 2018. Anthropocene: Island Biogeography. Pp. 37–43 in *Encyclopedia of the Anthropocene* **3**: 37–43. https://doi.org/10.1016/B978-0-12-809665-9.09856-6

Helmus, M. R. and Behm, J. E., 2020. Island geography revisited. *Encyclopedia of the World's Biomes* **1**: 51–56. https://doi.org/10.1016/B978-0-12-409548-9.11945-1

Hodkinson, I. D., Webb, N. R., and Coulson, S. J., 2002. Primary community assembly on land – the missing stages: why are the heterotrophic organisms always there first? *Journal of Ecology* **90**: 569–577. https://doi.org/10.1046/j.1365-2745.2002.00696.x

Holt, R. D., 1996. Food webs in space: an island biogeographic perspective. Ch. 29, pp. 313–323 in *Food Webs: Contemporary Perspectives*, edited by G. Polis and K. Winemiller, Chapman and Hall, London, England, UK. https://doi.org/10.1007/978-1-4615-7007-3_30

Holt, R. D., 2010. Toward a trophic island biogeography: Reflections on the interface of island biogeography and food web ecology. Pp 143–185 in *The Theory of Island Biogeography Revisited*, edited by J. B. Losos and R. E. Ricklefs. Princeton University Press, Princeton. https://doi.org/10.1515/9781400831920.143

Holt, R. D., Lawton, J. H., Polis, G. A., and Martínez, N. D., 1999. Trophic rank and the species area relationship. *Ecology* **80**: 1495–1504. https://doi.org/10.1890/0012-9658(1999)080[1495:TRATSA]2.0.CO;2

Holt, R. D., Gravel, D., Stier, A., and Rosindell, J. 2021. On the interface of food webs and spatial ecology: The trophic dimension of species–area relationships. Ch. 12, pp. 289–318 in *The Species–Area Relationship: Theory and Application*, edited by T. Matthews, K. Triantis, and R. Whittaker. Cambridge University Press, Cambridge, England, UK. https://doi.org/10.1017/9781108569422.017

Holyoak, M. and Loreau, M., 2006. Reconciling empirical ecology with neutral community models. *Ecology* **87**: 1370–1377. https://doi.org/10.1890/0012-9658(2006)87[1370:REEWNC]2.0.CO;2

Hubbell, S. P., 2001. *The Unified Neutral Theory of Biodiversity and Biogeography. Monographs in Population Biology* **32**, Princeton University Press, Princeton, New Jersey.

Hubbell, S. P., 2010. Neutral theory and the theory of island biogeography. Pp. 263–292 in *The Theory of Island Biogeography Revisited*, edited by J. B. Losos and R. E. Ricklefs. Princeton University Press, Princeton, New Jersey. https://doi.org/10.1515/9781400831920.264

Irl, S. D. H., Harter, D. E. V., Steinbauer, M. J., Puyol, D. G., Fernández-Palacios, J. M., Jentsch, A., and Beierkuhnlein, C., 2015. Climate vs. topography – spatial patterns of plant species diversity and endemism on a high-elevation island. *Journal of Ecology* **103**: 1621–1633. https://doi.org/10.1111/1365-2745.12463

Johnson, M. P. and Raven, P. H., 1970. Natural regulation of plant species diversity. *Evolutionary Biology* **4**: 127–162.

Kalmar, A. and Currie, D. J., 2006. A global model of island biogeography. *Global Ecology and Biogeography* **15**: 72–81. https://doi.org/10.1111/j.1466-822X.2006.00205.x

Keddy, P. and Weiher, E., 2004. Introduction: The scope and goals of research on assembly rules. Pp. 1–20 in *The Theory of Island Biogeography Revisited*, edited by J. B. Losos and R. E. Ricklefs. Princeton University Press, Princeton, New Jersey. https://doi.org/10.1017/CBO9780511542237.001

Keil, P. and Chase, J. M., 2019. Global patterns and drivers of tree diversity integrated across a continuum of spatial grains. *Nature Ecology & Evolution* 3: 390–399. https://doi.org/10.1101/367961

Krause, F., 2002. Introduced species and their control. Ch. 1.6.4.5 in *Oceanic Islands*, volume 1.6.4., edited by H. Heatwole, in *Encyclopedia of Life Support Systems (EOLSS)*, Developed under the auspices of the UNESCO, EOLSS Publishers, Oxford. [http://www.eolss.net]

Laurance, W. F., 2010. Understanding biogeography theory: Understanding habitat fragmentation Pp. 214–236 in *The Theory of Island Biogeography Revisited*, edited by J. B. Losos and R. E. Ricklefs. Princeton University Press, Princeton, New Jersey. https://doi.org/10.1515/9781400831920.214

Levin, S. A., 1992. The problem of pattern and scale in ecology: The Robert H. MacArthur Award

Lecture. *Ecology* **73**: 1943–1967. https://doi.org/10.2307/1941447

Levine, S., 1980. Several Measures of trophic structure applicable to complex food webs. *Journal of Theoretical Biology* **83**: 195–207. https://doi.org/10.1016/0022-5193(80)90288-X

Levins, R. and Heatwole, H., 1963. On the distribution of organisms on islands. *Caribbean Journal of Science* **3**: 173–177.

Liu, J., Matthews, T. J., Zhong, L., Liu, J., Wu, J., and Yu, M., 2019. Environmental filtering underpins the island species–area relationship in a subtropical anthropogenic archipelago. *Journal of Ecology* **108**: 424–432. https://doi.org/10.1111/1365-2745.13272

Lomolino, M. V., 2000. A call for a new paradigm of island biogeography. *Global Ecology and Biogeography* **9**:1–6. https://doi.org/10.1046/j.1365-2699.2000.00185.x

Lomolino, M. V., Brown, J. H., and Sax, D. F., 2010. Island biogeography theory. Pp. 13–51 in *The Theory of Island Biogeography Revisited*, edited by J. B. Losos and R. E. Ricklefs. Princeton University Press, Princeton, New Jersey. https://doi.org/10.1515/9781400831920.13

Lomolino, M. V., Riddle, B. R., Whittaker, R. J., and Brown, J. H., 2010. *Biogeography* (Fourth Edition). Sinauer, Sunderland, UK.

Losos, J. B. and Ricklefs, R. E., 2010. *The Theory of Island Biogeography Revisited*, edited by J. B. Losos and R. E. Ricklefs. Princeton University Press, Princeton, New Jersey.

Lugo, A. E., Carlo, T. A., and Wunderle, J. M. Jr., 2012. Natural mixing of species: novel plant-animal communities on Caribbean Islands. *Animal Conservation* **15**: 233–241. https://doi.org/10.1111/j.1469-1795.2012.00523.x

MacArthur, R. H., 1972. *Geographical Ecology: Patterns in the Distribution of Species*. Princeton University Press, Princeton.

MacArthur, R. H. and Wilson, E. O., 1963. An equilibrium theory of insular zoogeography. *-Evolution* **17**: 373–387. https://doi.org/10.1111/j.1558-5646.1963.tb03295.x

MacArthur, R. H. and Wilson, E. O., 1967. *The Theory of Island Biogeography*. Princeton University Press, Princeton, New Jersey.

Martínez, N. D., Hawkins, B. A., Dawah, H. A., and Feifarek, B. P., 1999. Effects of sampling effort on characterization of food-web structure. *Ecology* **80**: 1044–1055. https://doi.org/10.1890/0012-9658(1999)080[1044:EOSEOC]2.0.CO;2

Massol, F., Dubart, M., Calcagno, V., Cazelles, K., Jacquet, C., Kéfi, S., and Gravel, D., 2017. Island biogeography of food webs. *Advances in Ecological Research* **56**: 183–262. https://doi.org/10.1016/bs.aecr.2016.10.004

Matthews, T., 2021. On thew biogeography of habitat islands: The Importance of matrix effects, noncore species, and source-sink dynamics. *The Quarterly Review of Biology* **96**: 730–104. https://doi.org/10.1086/714482

Matthews, T., Triantis, K., and Whittaker, R. (eds), 2021. *The Species–Area Relationship: Theory and Application*. Cambridge University Press, Cambridge, England, UK. https://doi.org/10.1017/9781108569422

Mikkelson, G. M., 1993. How do food webs fall apart? A study of changes in trophic structure during relaxation in habitat structure. *Oikos* **67**: 539–547. https://doi.org/10.2307/3545366

Moran, V. C. and Southwood, T. R. E., 1982. The guild composition of arthropod communities in trees. *Journal of Animal Ecology* **51**: 289–306. https://doi.org/10.2307/4325

Moreno, J. A., 1991. Status survey of the golden coqui *Eleutherodactylus jasperi*. Publicación Cientifica Miscelánea Departamento de Recursos Naturales de Puerto Rico **1**: 37–41.

Morin, P. J., 2011. *Community Ecology*, Second Edition. Wiley-Blackwell, John Wiley & Sons, Inc., Hoboken, New Jersey.

Morrison, L. W., 1997. The insular biogeography of small Bahamian cays. *Journal of Ecology* **85**: 441–454. https://doi.org/10.2307/2960568

Moses, Not dated a. Genesis, Chapter 1, verses 20–21 in *The Torah. A New* [English] *Translation of The Holy Scriptures According to the Masoretic Text*. The Jewish Publication Society of America, Philadelphia, Pennsylvania, USA.

Moses, Not dated b. Genesis, Chapter 1, verse 24 in *The Torah. A New* [English] *Translation of The Holy Scriptures According to the Masoretic Text*. The Jewish Publication Society of America, Philadelphia, Pennsylvania, USA.

Nordenskiöld, E., 1928. *The History of Biology, a Survey*. Tudor Publishing Company, New York, USA.

Ohyama, L., Holt, R. D., Matthews, T. J., and Lucky, A., 2021. The species–area relationship in ant ecology. *Journal of Biogeography* **48**: 1824–1841. https://doi.org/10.1111/jbi.14149

Patiño, J., Whittaker, R. J., Borges, P. A. V., Fernandez-Palacios, J. M., Ah-Peng, C., Araujo, M. B., Avila, S. P., Cardoso. P., Cornuault, J., de Boer, E. J., de Nascimento, L., Gil, A., Gonzalez-Castro, A., Gruner, D. S., Heleno, R., Hortal, J., Illera, J. C., Kaiser-Bunbury, N., Matthews, T. J., Papadopoulou, A., Pettorelli, N., Price, J. P., Santos, A. M. C., Steinbauer, M. J., Triantis, K. A., Valente, L., Vargas, P., Weigelt, P., and Emerson, B. C., 2017. A roadmap for island biology: 50 fundamental questions after 50 years of The Theory of Island Biogeography.

Journal of Biogeography **44**: 963–983. https://doi.org/10.1111/jbi.12986

Peck, A. L. 1963. English translation of Aristotle's Περι Ζωιων Γενεσεωε. *Generation of Animals*. William Heineman Ltd., London, England, UK.

Piechnik, D. A., Lawler, S. P., and Martínez, N. D., 2008. Food-web assembly during a classic biogeographic study: species' "trophic breadth" corresponds to colonization order. *Oikos* **117**: 665–674. https://doi.org/10.1111/j.0030-1299.2008.15915.x

Piovia-Scott, J., Yang, L. H., Wright, A. N., Spiller, D. A., and Schoener T. W. 2017. The effect of lizards on spiders and wasps: variation with island size and marine subsidy. *Ecosphere* **8**: e01909. https://doi.org/10.1002/ecs2.1909

Pokarzhevskii, A. D., Van Straalen, N. M., Filimonova, Z. V., Zaitsev, A. S., and Butovskii, R. O., 2000. Trophic structure of ecosystems and ecotoxicology of soil organisms. *Russian Journal of Ecology* **31**: 190–197. https://doi.org/10.1007/BF02762820

Polis, G. A. and Hurd, S. D., 1995a. Allochthonous input across habitats, subsidized consumers, and apparent trophic cascades: Examples from the ocean-land interface. Ch. 26, pp. 275–285 in *Food Webs, Integration of Patterns and Dynamics*, edited by G. A. Polis and K. O. Winemiller. Chapman & Hall, New York. https://doi.org/10.1007/978-1-4615-7007-3_27

Polis, G. A. and Hurd, S. D., 1995b. Extraordinarily high spider densities on islands: Flow of energy from the marine to terrestrial food webs and the absence of predation. *Proceedings of the Academy of Sciences of the United State of America* **92**: 4382–4386. https://doi.org/10.1073/pnas.92.10.4382

Polis, G. A. and Hurd, S. D., 1996. Linking marine and terrestrial food webs: Allochthonous input from the ocean supports high secondary productivity on small islands and coastal land communities. *The American Naturalist* **147**: 396–423. https://doi.org/10.1086/285858

Pregill, G. K. and Olson, S. L., 1981. Zoogeography of West Indian vertebrates in relation to Pleistocene climatic cycles. *Annual Review of Ecology and Systematics* **12**: 75–98. https://doi.org/10.1146/annurev.es.12.110181.000451

Putman, R. J., 1994. *Community Ecology*. Chapman & Hall, London, England, UK.

Ricklefs, R. E., 2010. Dynamics of colonization and extinction on islands. Pp. 388–414 in *The Theory of Island Biogeography Revisited*, edited by J. B. Losos and R. E. Ricklefs. Princeton University Press, Princeton, New Jersey. https://doi.org/10.1515/9781400831920.388

Ríos-López, N., 1999. Variation in reproductive biology, physiology, and morphology of *Eleutherodactylus coqui* – ANURA: LEPTODACTYLIDAE – along an altitudinal gradient. MS Thesis. University of Puerto Rico, Río Piedras Campus, San Juan, Puerto Rico.

Ríos-López, N., 2007. The structuring of herpetofaunal assemblages in human-altered coastal ecosystems. PhD Dissertation. University of Puerto Rico, Río Piedras Campus, San Juan, Puerto Rico.

Ríos-López, N. and Aide, T. M., 2007. Herpetofaunal dynamics during secondary succession. *Herpetologica* **63**: 35–50. https://doi.org/10.1655/0018-0831(2007)63[35:HDDSS]2.0.CO;2

Rosindell, J., Hubbell, S. P., and Etienne, R. S., 2011. The unified neutral theory of biodiversity and biogeography at age ten. *Trends in Ecology and Evolution* **26**: 340–348. https://doi.org/10.1016/j.tree.2011.03.024

Roslin, R., Várkonyi, G., Koponen, M., Vikberg, V., and Nieminen, M., 2014. Species–area relationships across four trophic levels – decreasing island size truncates food chains. *Ecography* **37**: 443–453. https://doi.org/10.1111/j.1600-0587.2013.00218.x

Ryberg, W. A., Smith, K. G., and Chase, J. M., 2012. Predators alter the scaling of diversity in prey metacomunities. *Oikos* **121**: 1995–2000. https://doi.org/10.1111/j.1600-0706.2012.19620.x

Sax, D. F. and Gaines, S. D., 2011. The equilibrium theory of island biogeography. Pp. 219–240 in *The Theory of Ecology*, edited by S. M. Scheiner and M. R. Willig. The University of Chicago Press, Chicago, Illinois.

Schoener, T. W., 2010. The MacArthur-Wilson Equilibrium Model; A chronicle of what it said and how it was tested. Pp. 52–87 in *The Theory of Island Biogeography Revisited*, edited by J. B. Losos and R. E. Ricklefs. Princeton University Press, Princeton, New Jersey. https://doi.org/10.1515/9781400831920.52

Simberloff, D., 1971. Population sizes of congeneric bird species on islands. *The American Naturalist* **105**: 190–193. https://doi.org/10.1086/282715

Simberloff, D., 1976a. Experimental zoogeography of islands: effects of island size. *Ecology* **57**: 629–649. https://doi.org/10.2307/1936179

Simberloff, D., 1976b. Trophic structure determination and equilibrium in an arthropod community. *Ecology* **57**: 395–398. https://doi.org/10.2307/1934831

Simberloff, D., 1978. Using island biogeographic distributions to determine if colonization is stochastic. *The American Naturalist* **112**: 713–726. https://doi.org/10.1086/283313

Simberloff, D. and Collins, M. D., 2010. Birds of the Solomon Islands, the Domain of the dynamic equilibrium theory and assembly rules, with comments on the taxon cycle. Pp. 237–263 in *The*

Theory of Island Biogeography Revisited, edited by J. B. Losos and R. E. Ricklefs, Princeton University Press, Princeton, New Jersey. https://doi.org/10.1515/9781400831920.237

Simberloff, D. S., 1974. Equilibrium theory of island biology and ecology. *Annual Review of Ecology and Systematics* **5**: 161–182. https://doi.org/10.1146/annurev.es.05.110174.001113

Simberloff, D. S. and Wilson, E. O., 1969. Experimental zoogeography of islands: The colonization of empty Islands. *Ecology* **50**: 278–296. https://doi.org/10.2307/1934856

Simberloff, D. S. and Wilson, E. O., 1970. Experimental zoogeography of islands: A two-year record of colonization. *Ecology* **51**: 934–937. https://doi.org/10.2307/1933995

Strong, D. R. Jr., Sziska, L. A., and Simberloff, D. S., 1979. Tests of community-wide character displacement against null hypotheses. *Evolution* **33**: 897–913. https://doi.org/10.1111/j.1558-5646.1979.tb04743.x

Strong, D. R., Simberloff, D., Abele, L. G., and Thistle, A. B. (eds), 1984. *Ecological Communities: Conceptual Issues and the Evidence.* Princeton University Press, Princeton, New Jersey. https://doi.org/10.1515/9781400857081

Temperton, V. M., Hobbs, R. J., Nuttle, T. and Halle, S., 2013. *Assembly Rules and Restoration Ecology, Bridging the Gap between Theory and Practice.* Island Press, Washington D. C.. E-book format.

Terborgh, J., 2010. The trophic cascade on islands. Pp. 116–142 in *The Theory of Island Biogeography Revisited,* edited by J. B. Losos and R. E. Ricklefs, Princeton University Press, Princeton, New Jersey. https://doi.org/10.1515/9781400831920.116

Tjørve, E., Matthews, T. J., and Whittaker, R. J., 2020 The history of the species–area relationship. Ch. 2, pp. 20–48 in *The Species–Area Relationship, Theory and Application,* edited by T. J. Matthews, K. A. Triantis, and R. J. Whittaker. Cambridge University Press, Cambridge, England, UK. https://doi.org/10.1017/9781108569422.005

Triantis, K. A. and Matthews, T. J., 2020. Biodiversity theory backed by island bird data. *Nature* **579**: 36–37. https://doi.org/10.1038/d41586-020-00426-5

Triantis, K. A., Vardinoyannis, K., Tsolaki, E. P., Botsaris, I., Lika, K., and Mylons, M., 2006. Re-approaching the small island effect. *Journal of Biogeography* **33**: 914–923. https://doi.org/10.1111/j.1365-2699.2006.01464.x

Valente, L., Phillimore, A. B., Melo, M., Warren, B. H., Clegg, S. M., Havenstein, K., Tiedmann, R., Illera, J. C., Thébaud, C., Aschenbach, T., and Etienne, R. S., 2020. A simple dynamic model explains the diversity of island birds worldwide. *Nature* **579**: 92–96. https://doi.org/10.1038/s41586-020-2022-5

Vandermeer, J., Lazarus, J., Ludwig, C., Lyon, J., Schultz, B., and Yih, K., 1980. Migration as a factor in the community structure of a macroarthropod litter fauna. *The American Naturalist* **115**: 606–612. https://doi.org/10.1086/283585

van Noordwijk, C. G. E., Verberk, W. C. E. P., Turin, H., Heijerman, T., Alders, K., Dekoninck, W., Hannig, K., Regan, E., McCormack, S., Brown, M. J. F., Remke, E., Siepel, H., Berg, M. P., and Bonte, D., 2015. Species–area relationships are modulated by trophic rank, habitat affinity, and dispersal ability. *Ecology* **96**: 518–531. https://doi.org/10.1890/14-0082.1

Vellend, M., 2010. Conceptual synthesis in community ecology. *The Quarterly Review of Biology* **85**: 183–206. https://doi.org/10.1086/652373

Vellend, M. and Orrock, J., 2010. Ecological and genetic models of diversity, lessons across disciplines. Pp. 439–461 in *The Theory of Island Biogeography Revisited,* edited by J. B. Losos and R. E. Ricklefs. Princeton University Press, Princeton, New Jersey. https://doi.org/10.1515/9781400831920.439

Verhoef, H. A. and Olff, H., 2010. Trophic dynamics of communities. Pp. 25–36 in *Community Ecology: Processes, Models, and Applications,* edited by H. A. Verhoef and P. J. Morin. Oxford University Press Inc., New York. https://doi.org/10.1093/acprof:oso/9780199228973.003.0003

Walker, L. R. and Bellingham, P., 2011. *Island Environments in a Changing World.* Cambridge University Press. Cambridge, England, UK.

Warren, B. H., Ricklefs, R. W., Thébaud, C., Gravel, D., and Mouquet, N., 2019. How consideration of islands has inspired mainstream ecology: Links between the theory of islands biogeography and some other key theories. *Encyclopedia of the World's Biomes*: https://doi.org/10.1016/B978-0-12-409548-9.11788-9

Warren, B. H., Simberloff, D., Ricklefs, R. E., Aguilée, R., Condamine, F. L., Gravel, D., Morton, H., Mouquet, N., Rosindell, J., Casquet, J., Conti, E., Cornuault, J., Fernández-Palacios, J. M., Hengl, T., Norder., S. J., Rijsdijk, K. F., Sanmartín, I., Strasberg, D., Triantis, K. A., Valente, L. M., Whittaker, R. J., Gillespie, R. G., Emerson, B. C., and Thébaud, C., 2015. Islands as model systems in ecology and evolution: prospects fifty years after MacArthur-Wilson. *Ecology Letters* **18**: 200–217. https://doi.org/10.1111/ele.12398

Weiher, E. and Keddy, P., 2004. *Ecological Assembly Rules: Perspectives, Advances, Retreats.* Cambridge University Press, Cambridge, England, UK.

Weiher, E., Freund, D., Bunton, T., Stefanski, A., Lee, T., and Bentivenga, S., 2011. Advances, challenges and a developing synthesis of ecological community assembly theory. *Philosophical Transactions of the*

Royal Society B **366**: 2403–2413. https://doi.org/10.1098/rstb.2011.0056

Werner, E. E. and Gilliam, J. F., 1984. The ontogenetic niche and species interaction in size-structured populations. *Annual Review of Ecology and Systematics* **15**: 393–425. https://doi.org/10.1146/annurev.es.15.110184.002141

White, A. D., 1895. *A History of the Warfare of Science with Theology in Christendom*. 1955 reprint by George Braziller, New York, USA.

Whitehead, D. R. and Jones, C. E., 1969. Small islands and the equilibrium theory of insular biogeography. *Evolution* **23**: 171–179. https://doi.org/10.1111/j.1558-5646.1969.tb03503.x

Whittaker, R. J. and Fernández-Palacios, J. M., 2007. *Island Biogeography: Ecology, Evolution, and Conservation*. Second Edition. Oxford University Press, Oxford, England, UK.

Whittaker, R. J., Triantis, K. A., and Ladle, R. J., 2010. A general dynamic theory of oceanic island biogeography: extending the MacArthur-Wilson theory to accommodate the rise and fall of volcanic islands. Pp. 88–115 in *The Theory of Island Biogeography Revisited*, edited by J. B. Losos and R. E. Ricklefs. Princeton University Press, Princeton, New Jersey. https://doi.org/10.1515/9781400831920.88

Whittaker, R. J., Fernández-Palacios, J. M., Matthews, T. J., Borregaard, M. K, and Triantis, K. A., 2017. Island biogeography: Taking the long view of nature's laboratories. *Science* **357**: eaam8326. https://doi.org/10.1126/science.aam8326

Wilson, E. O., 1959. Adaptive shift and dispersal in a tropical ant fauna. *Evolution* **13**: 122–144. https://doi.org/10.1111/j.1558-5646.1959.tb02996.x

Wilson, E. O., 1961. The nature of the taxon cycle in the Melanesian ant fauna. *The American Naturalist* **95**: 169–193. https://doi.org/10.1086/282174

Wilson, J. B., 2007. Trait-divergence assembly rules have been demonstrated: Limiting similarity lives! A reply to Grime. *Journal of Vegetation Science* **18**: 451–452. https://doi.org/10.1111/j.1654-1103.2007.tb02557.x

APPENDIX 1.I. Fifty key questions in the field of insular biogeography that need addressing 50 years after publication of the *Equilibrium Theory of Insular Biogeography* by Robert MacArthur and Edward O. Wilson. Modified from Patiño *et al.* (2017) and used by permission from John Wiley and Sons under License 5466710389261 (12 January 2023).

Macroecology and Biogeography on Islands

Global diversity patterns: (1) What are the relative roles of spatial, historical and ecological processes in driving taxonomic, phylogenetic, and functional diversity patterns of insular systems? (2) How do fundamental biogeographic processes interact through time and space to establish the species–area relationship on islands? (3) How do taxonomic, phylogenetic, and functional diversity compare between islands and ecologically similar continental areas? (4) How important are islands as refuges for now-extinct mainland lineages and/or ecosystems? (5) How important are oceanic islands as generators of biodiversity and for the assembly of continental biota through reverse-colonization and/or colonization *de novo*?

The ontogeny of islands and past climatic change: (6) How do rates of colonization, speciation and extinction change during the ontogeny of islands? (7) How do diversification rates of insular lineages change with islands' age? (8) How important were past geological events and climatic change in promoting colonization of islands and altering dispersal pathways? (9) How has climatic change influenced speciation and extinction within islands?

Island-rules and syndrome: (10) Is the evolution of traits fundamentally different on islands than on continents? (11) How robust are the various rules and syndromes relating to body size, loss of dispersal, colouration, breeding system, woodiness, and clutch size, among others? (12) To what extent are insular populations genetically impoverished, compared to comparable populations on mainlands?

Island-biogeographic theory: (13) How do the dynamics of insular communities scale up to generate the biogeographical patterns predicted by island-biogeographical theories? (14) How can we reconcile island-biogeographic theories with other ecological and evolutionary theories to contribute to a general theory of biodiversity? (15) How applicable are island-biogeographical theories derived from real islands to other forms of insular systems, such as sky-islands and seamounts? (16) How can we best incorporate population genetic and/or phylogenetic data to advance models of the biogeography of islands?

Macroevolution on islands

Immigration-speciation-extinction dynamics: (17) How does the spatial configuration of an archipelago (e.g. intra-archipelagic connectivity) influence colonization, speciation and extinction over time? (18) What is the nature of the relationship between rates of extinction and the isolation, if any? (19) How do the extinction probabilities of species endemic to islands compare to those of non-endemic species? (20) How important are diversity-dependent processes for colonization, speciation, and extinction on islands? (21) How do anthropogenic extinctions affect estimates of speciation and natural extirpation on insular systems?

Speciation and diversification: (22) What functional traits (e.g. relating to dispersal capacity, reproduction, trophic ecology) are associated with high diversification rates within and across insular systems? (23) What traits best predict which groups will undergo adaptive radiation on islands? (24) What is the relative importance of ecological *versus* geographical speciation on islands? (25) What is the influence of gene flow among islands and/or between islands and mainland areas on speciation rates?

Dispersal and colonization: (26) What is the importance of founder effects for the evolution of insular lineages? (27) How frequent is inter-island dispersal and is it enough to form an archipelago-wide metacommunity, or are islands better understood as functionally independent communities? (28) How can palaeoecology contribute to the understanding of species' arrival, establishment, and spread on islands?

Community Ecology on Islands

Community assembly: (29) How do taxonomic, phylogenetic and functional diversities of insular communities change during assembly and disassembly of insular systems? (30) How do area, elevation, and isolation of islands influence the composition and dynamics of communities on islands? (31) What are the relative roles of island-age, phylogenetic group, and functional ecology in determining natural (background) extinction rates among taxa on oceanic islands? (32) How does the order of colonization influence emergent outcomes in the assembly of insular biotas? (33) How important are rare species for the functioning of insular communities? (34) How does *in situ* evolution drive the functioning of insular ecosystems?

Biotic interactions: (35) How do changes in climate and sea-level influence biotic interactions on islands? (36) How do biotic interactions (within and between trophic levels) influence immigration, extinction, and speciation rates on islands?

Conservation and Management on Islands

Global change: (37) How, if at all, do insular biotas differ from continental biotas in their response to global change? (38) Are island species more prone to extirpation than their closest relatives on the mainland, and if so, why? (39) How can we identify which taxa are most at risk from global change and what are their risk-associated traits? (40) What determines anthropogenic extirpation rates among insular taxa? (41) How do anthropogenic changes within islands impact on the capacity of species on islands to respond successfully to climatic change?

Conservation and management policies: (42) How can we identify islands that are more susceptible to loss of biodiversity in the coming decade, and what are the most efficient and cost-effective methods (i.e. policy; education; research; management) for safeguarding their biodiversity? (43) What are the best strategies for *in situ* conservation of insular species that are impacted by non-native species? (44) What are the most effective methods for responding to the crisis of anthropogenic extirpation on islands? (45) How can we best implement long-term monitoring schemes on islands to provide quantitative evidence of changes within insular ecological systems? (46) How can conservation-interests best be integrated with other stakeholders' interests (particularly for tourism) on populated islands?

Invasive alien species (47) What are the impacts of novel biotic interactions between and among alien and native species on their biodiversity of islands and on ecosystem functioning? (48) How does the invasion stage (i.e. colonization, establishment, and long-term adaptation) of alien taxa affect distributional ranges and biotic interactions of native insular biotas? (49) To what extent can alien species act as functional substitutes for extinct native species on islands? (50) How do the ecological effects of introduced species differ from those of naturally arriving colonist species on islands?

APPENDIX 1.II: Definitions of assembly rules and assemblies by different authors in the book *Ecological Assembly Rules: Perspectives, Advances, Retreats* (Weiher and Keddy 2004).

AUTHOR(S)	CHAPTER	DEFINITION
Keddy and Weiher	Introduction	'The term "assembly rules" has been adopted to describe the problem of assembling communities from pools.'
Fox	1	No formal definition given, but Simberloff *et al.* (ch. 2) cites Fox's assembly rule to be 'There is a much higher probability that each species entering a [local] community will be drawn from a different functional group until each group is represented, before the cycle repeats.'
Simberloff, Stone, and Dayan	2	No formal definition given.
Kelt and Brown	3	'For the purposes of this chapter, assembly rules will be defined simply as empirical patterns of community organization, synonymous with what will also be called community structure or community organization.'
Lockwood, Moulton, and Balent	4	No formal definition given.
Wilson	5	'Assembly rules, i.e., patterns due to interactions between species, such as competition.'
Cody	6	'Conceptually, assembly rules are of two basic categories. One type, which will be called "type A", governs community membership or species composition as community [island] size increases or decreases....The extent to which there are identifiable differences between random samples of species and individuals drawn from a species-individuals pool and observed community replicates is the extent to which the second "type B" category of assembly rules operate. Here, the reference is not to communities that vary in size in response to some extrinsic factor, but rather to those which are more or less constant in size yet differ in species composition. Assembly rules of either type, A or B, may each address different components of diversity.'
Booth and Larson	7	Reviews many definitions.
Drake, Zimmerman, Purucker, and Rojo	8	'Assembly is a basic device of nature where a broad range of entities, spanning many levels of scale, interact across time and space to produce oberserved [*sic.*] pattern.' And 'Community assembly is ultimately driven by the invasion [...] and extinction of species played out against a complex background of environmental constraint. While the environment acts as a filter, eliminating some species and promoting others, it also provides spatiotemporal complexities which serve as a resource upon which ecological strategies can be built. Assembly processes and rules which operate within one environment may exhibit entirely different outcomes as a function of even minor environmental variation. Despite such obvious descriptions of the course of nature, the essence of the assembly trajectory remains little more than an elusive metaphor'. Also, 'Assembly is [...] the process of fitting together the dynamically variable pieces which comprise a system, pieces operating at disparate levels of spatial and temporal scale' and finally 'we define an assembly rule as: an operator which exists as a function or consequence of some force, dynamical necessity, or context which provides directionality to a trajectory. The nature of this direction includes movement toward a specific state, some subset of all possible states, or a dynamical realm of definable character. Operators are the mechanics of community assembly.'
Weiher and Keddy	9	'Assembly rules are not recipes for building communities. Rather, they are a set of limits that constrain how species can come together to form assemblages' and 'ecological communities can be thought of as complex adaptive systems, with properties common to both physical energy dissipating systems and species evolving in an adaptive landscape.'

(*Continued*)

APPENDIX 1.II. (Continued)

AUTHOR(S)	CHAPTER	DEFINITION
Lomolino	10	'Here assembly "rules" are defined as patterns in how species composition varies along biogeographical gradients (e.g., increasing area or isolation) and gradients in species richness.'
Strange and Foin	11	'The contemporary meaning of community assembly....de-emphasizes the term "rule" in favour of less-fixed combinations of interactions and outcomes that are known to occur within different communities. Community assembly now encompasses the ecology and short-term, observable evolution of the community from initial colonization to species saturation, and logically includes a number of topics that are referred to by other names, such as trophic cascades [...], top-down vs. bottom-up regulation of trophic webs [...], and the regulation of biodiversity.'
Díaz, Cabido, and Casanoves	12	'Other authors have defined assembly rules in a broader – or looser – sense.....filters of any kind imposed to the regional species pool can be regarded as assembly rules. The objective of assembly rules should then be the prediction of which subset of the total species pool for a given region will occur in a specified habitat. While Diamond [...] associates the idea of assembly rules to "forbidden combinations", thus emphasizing the importance of biotic interactions, Keddy ....links them to the idea of deletion, reinforcing his broader focus. He explicitly mentions climatic conditions, disturbance regime, and biotic interactions as examples of filters. [...] These "filters" fit into the definition of assembly rules in the broad sense.'
Lockwood and Pimm	13	No definition supplied
Keddy	14	'Assembly rules are nested within the habitat template and biogeographic realms as a further set of constraints upon species composition.'

2 Amphibians of the Bahamas and Turks and Caicos Islands

R. Graham Reynolds and Sean T. Giery

Abbreviations and Acronyms

asl	*above sea level*
DD	*Data Deficient*
GBB	*Great Bahama Bank*
IUCN	*International Union for the Conservation of Nature*
kya	*thousand years ago*
LBB	*Little Bahama Bank*
LC	*Least Concern*
mtDNA	*mitochondrial deoxyribonucleic acid*
mya	*million years ago*
Region	*Bahamas Islands and Turks and Caicos Islands*
TCI	*Turks and Caicos Islands*

I. Introduction and geography of the islands

The Bahamas Islands and Turks and Caicos Islands (TCI) are the political units administered by the Commonwealth of the Bahamas and the Turks and Caicos Island Government (the latter an Overseas Territory of the United Kingdom), respectively, that encompass the geographic region referred to as the Lucayan (or Bahamas) Archipelago. This archipelago stretches more than 1,360 km across the

R. Graham Reynolds and Sean T. Giery, 'Chapter 2: Amphibians of the Bahamas and Turks and Caicos Islands' in: *The Conservation and Biogeography of Amphibians in the Caribbean*. Pelagic Publishing (2023). © R. Graham Reynolds and Sean T. Giery. DOI: 10.53061/WNSU3325 northern portion of the greater Caribbean region from the northwest to the southeast. The region is comprised of a complex of large and small carbonate platforms (banks) upon which hundreds of emergent islands occur. These islands atop the banks are generally separated by water no greater than 30–40 m deep, while depths of water between banks can reach thousands of metres. A total of 21 banks are distributed within a region bounded by coordinates between 19° 84' N and 27° 30' N, and between 68° 70' W and 80° 54' W, with the largest banks occurring in the northwest of the region.

The largest banks, the Little Bahama Bank (LBB; 14,769 km^2) and Great Bahama Bank (GBB; 103,000 km^2), are also the northernmost banks and are an order of magnitude larger than others in the region. To the west of these larger banks lies the isolated and uninhabited Cay Sal Bank. This bank lies between peninsular Florida (USA), Cuba, and the GBB. Cay Sal Bank is bounded by the Straits of Florida and the Santaren and Nicholas channels, and its edges consist of a few low-lying strips of land (< 10 m asl), most of them with little vegetation. On islands atop the Cay Sal Bank, the assemblage of vertebrate animals consists of species from the Bahamas and Cuba (Buckner *et al.* 2012), with no species of amphibians.

Moving southeast in the archipelago, the banks become smaller, with some, such as the Rum Cay Bank, supporting only a single vegetated island, and others, such as Hogsty Reef (21° 41.24' N, 73° 48.89' W), supporting a few hundred square metres of vegetated land above the tidal reach and without species of amphibians (Buckner *et al.* 2012). Further southeast are several larger banks, including Inagua Bank (~2,100 km^2) and the Caicos Bank (6,140 km^2). The Turks Bank (324 km^2) is the last bank in the archipelago supporting emergent islands with vegetation. At the southeastern terminus of the archipelago are four banks that are submerged completely (Mouchoir, Silver [two banks], and Navidad), and share a geological origin with the Lucayan Archipelago.

Global climate generates changes in hydric conditions at different timescales in islands within the region. For example, during long temporal scales, changes in global climate generated oscillations in sea level that resulted in periods of emergence and inundation of land areas (Siddall *et al.* 2003; Rohling *et al.* 2009). To illustrate, during the most recent Wisconsin intraglacial period approximately 8–120 kya (Lighty *et al.* 1982; Fairbanks 1989; Clark and Mix 2002; Siddall *et al.* 2003; Hearty *et al.* 2007), the tops of these carbonate banks were exposed, thereby increasing the total land area of the region by an order of magnitude (Sealey 2006; Knapp *et al.* 2011). Many of the current emergent islands formed via wind-blown oolitic deposits during these periods of lower sea level (Sealey 2006). Rising seas then subsequently inundated lower-lying portions of the banks, fragmenting the previously emergent large islands into a complex of smaller ones. Most of the land across the archipelago is presently close to sea level, with the highest point in the region (Mt Alvernia, 63 m asl) occurring on the GBB at the southern end of Cat Island. During shorter temporal scales, global climate generates patterns in rainfall throughout the region. For instance, a gradient in precipitation exists from northwest to southeast in the archipelago, with the southern islands (Samana to Inagua and the TCI) receiving approximately half the annual rainfall of northern islands.

From a zoogeographical perspective, it is expected that long-term and short-term changes in hydrology will influence assemblages of species in the Lucayan Archipelago, particularly the presence of amphibians in the region. Specifically, the changing size of emergent landmasses within the archipelago has had important implications both for the proximity of islands within the archipelago as well as proximity to nearby Greater Antillean islands such as Cuba and Hispaniola with their unique biodiversity. For example, at full emergence during the intraglacial period, the island constituting the GBB would have stretched to within 20 km of Cuba, providing a narrow channel for plants and animals to cross between the fully exposed banks. The Silver and Navidad banks, when exposed, would have lain within ~60 km of Hispaniola. These dynamic fluctuations in island areas are likely responsible for two clear characteristics of the assemblage of amphibians. First, the number species of amphibians is low. Second, the native species of amphibians are derived from Cuban ancestors. During recent time, the gradient in precipitation in the archipelago might result in the southern islands to the east of this gradient (i.e. the drier limit of the gradient) being less amenable to colonization by amphibians. Indeed, there are no native species of amphibians in the TCI (Reynolds 2011a).

The anthropogenic factor is now a significant driving force that, when combined with natural factors, may generate assemblages of amphibians new to the archipelago. Currently, non-native species of amphibians are thriving in the region. This suggests that biogeography (the vicissitudes of dispersal, colonization, and persistence), and the modification of habitats by humans (wells, cisterns, golf-course ponds, irrigation, introductions of amphibians), have contributed to the low level of endemism and to the maintenance and persistence of introduced populations of amphibians, respectively (Reynolds and Niemiller 2010; Reynolds 2011a).

The amphibians of the region, including native and introduced species, occur on eight banks, spanning a reduced portion of the archipelago (20° 53'–27° 30' N and 71° 46'–79° 21' W). Here we describe the native and introduced amphibian species in the Lucayan Archipelago. Additional references exist for the amphibian fauna, and for conservation legislation for the region (Knapp *et al.* 2011; Reynolds 2011a), as well as a comprehensive list of records of occurrence of amphibians for all major and minor islands (Buckner *et al.* 2012). We focus on summarizing the state of knowledge of amphibians in the region, and conclude with a discussion of the outlook for conservation of the native species in the archipelago.

II. The native amphibians

The native amphibians in the region include three recognized species of anurans, one of which is endemic. Two of these species are direct-developing members of the family Eleutherodactylidae, and are generally restricted to the two largest of the Bahamian banks, although they have probably been introduced elsewhere in the region. The third species is a relatively large member of the family Hylidae (*Osteopilus septentrionalis*), and is also native to Cuba and the Cayman Islands. The latter is one of the more invasive species of amphibian, and has been introduced to the TCI. Below we provide accounts of each native species, including information on distribution, taxonomy, and natural history relevant to conservation.

A. *Eleutherodactylus planirostris* (Cope, 1862)

Eleutherodactylus planirostris (Fig. 2.1A) is native to Cuba, the Bahamas, and the Cayman Islands (Schwartz and Henderson 1991). In the Bahamas, it is found mostly on the LBB, although three populations occur on the northern GBB islands of New Providence, Athol, and South Bimini (Buckner *et al.* 2012; their figure 2). As the species occurs nowhere else on the GBB, it seems that its presence on these islands might represent either human introductions, remnant populations from a historically more widespread population, or an incomplete understanding of the boundaries between *E. planirostris* and the endemic *Eleutherodactylus rogersi*.

Recent phylogenetic study (Heinicke *et al.* 2011) has revealed that Bahamian *E. planirostris* share nearly identical mtDNA haplotypes to populations on Grand Cayman Island and coastal Nicaragua. This similarity, combined with a very recent coalescence, suggest recent colonization of (or human introduction to) these regions from Cuba. The Turks and Caicos population of *E. planirostris* shares a mtDNA haplotype present in populations from Jamaica (the species was introduced to Jamaica in 1937), Cayman Brac, and central/eastern Cuba, which implies an even more recent introduction to the TCI. A potentially mislabelled portion of figure 1(a) in Heinicke *et al.* (2011; see supplementary material) suggests that *E. planirostris* was collected on San Salvador Island, yet this likely refers to the collection site for *E. rogersi* used in that study as

Figure 2.1. (A) An introduced *Eleutherodactylus planirostris* from North Caicos, Turks and Caicos Islands. Photograph by Matthew Niemiller. (B) *Eleutherodactylus rogersi* from Eleuthera Island in the Bahamas. Photograph by Joseph Burgess. (C) *Osteopilus septentrionalis* from Eleuthera Island in the Bahamas. Photograph by R. Graham Reynolds. (D) Introduced *Gastrophryne carolinensis* from Abaco Island, Bahamas. Photograph by Sean T. Giery.

E. planirostris is not known to occur on this island (Buckner *et al.* 2012). Taken together, these data suggest that the populations on TCI, and possibly those on GBB, resulted in part from introductions by humans, although comparison with genetic data from the LBB is necessary to determine the origins of Bahamian populations.

Like most *Eleutherodactylus*, *E. planirostris* has direct-development (Hedges *et al.* 2004), and eggs hatch into froglets after a minimum of 13 days under ideal conditions (Lazell 1989). Reproduction occurs during the spring and summer (Stewart 1979), with males producing advertisement calls close to the ground during this period. Females lay clutches of 3–26 eggs in depressions excavated in the soil, attached to rocks in rock walls, attached to the underside of rocks, or buried in damp leaf litter (Grant 1940; Lazell 1989; Seidel and Franz 1994). *Eleutherodactylus planirostris* is capable of using a variety of habitat types, from natural to heavily disturbed. They tend to occupy more mesic microhabitats (e.g. piles of debris or rocks, leaf litter, and caves) within a wide range of macrohabitats, from xeric to mesic forest (Hedges *et al.* 2004). The species is quite capable of living in areas developed by humans, and practices such as irrigation might benefit them. Being a largely nocturnal species, *E. planirostris* is most active on humid nights or during rain, although this species is in general more tolerant of drier conditions than are other members of the genus (Stewart and Martin 1980).

B. *Eleutherodactylus rogersi* Goin, 1955

The only endemic amphibian species in the region, *E. rogersi* (Fig. 2.1B), is largely restricted to the GBB, although it has managed to colonize the San Salvador Bank east of the GBB (Fig. 2.2) (Henderson and Powell 2009; Buckner *et al.* 2012). There is some possibility that this population resulted from human transport, as colonizing the San Salvador Bank (and neither the Rum nor

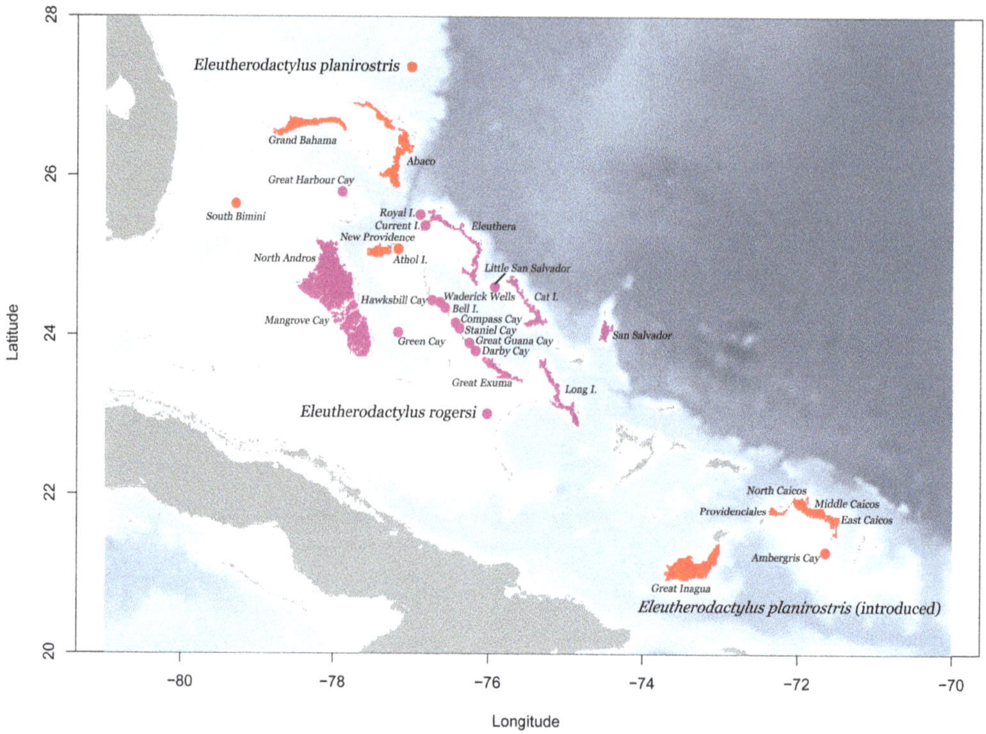

Figure 2.2. Map of the Bahamas and Turks and Caicos Islands, showing the native and introduced ranges of *Eleutherodactylus planirostris* (in red), and the native range of *E. rogersi* (in purple). Note that *E. planirostris* ranges more widely than the study area (not shown). Major islands and other islands mentioned in the text are labelled.

Conception banks) would likely require a long overwater transit perpendicular to prevailing current and to paths of hurricanes. The species is documented from larger island groups on the GBB, except for the Flamingo Cays, Ragged Islands, Bimini islands, South Andros, and New Providence Island (Buckner *et al.* 2012). Absence from the southern extent of the GBB might be explained either by lack of suitable habitat or by limited sampling. On the Bimini islands and New Providence Island, *Eleutherodactylus* populations represent *E. planirostris* and not *E. rogersi* (Heinicke *et al.* 2011; Buckner *et al.* 2012). Absence from South Andros is a sampling artefact, as anecdotal reports indicate the species occurs there.

Following its initial description (Goin 1955), *E. rogersi* was considered a subspecies of the wide-ranging *E. planirostris* (Schwartz 1965), although recent phylogenetic studies based on multilocus sequence data have found that the two species are evolutionarily distinct (Heinicke *et al.* 2007, 2011; Pyron and Wiens 2011). *Eleutherodactylus rogersi* appears to be a member of a clade comprising *Eleutherodactylus goini* from western Cuba, *Eleutherodactylus simulans* from eastern Cuba, and *Eleutherodactylus tonyi* from southern Cuba (Heinicke *et al.* 2007, 2011; Pyron and Wiens 2011; Padial *et al.* 2014). This clade is sister to a clade containing *E. planirostris*, *E. guanahacabibes* from extreme western Cuba, and *E. casparii* from central Cuba (Heinicke *et al.* 2007; Pyron and Wiens 2011; Padial *et al.* 2014; but see Heinicke *et al.* 2011 for alternate topology). Nevertheless, *E. planirostris* and *E. rogersi* are morphologically similar, and the latter was one of the most recent species to be recognized from what was formerly the *E. planirostris* complex (Estrada and Hedges 1997; Lynch and Duellman 1997; Heinicke *et al.* 2007), but now recognized as the *E. planirostris* species group (Hedges *et al.* 2008; Padial *et al.* 2014). *Eleutherodactylus rogersi* likely diverged from other *Eleutherodactylus* species in the Pliocene or late Miocene, and represents an early, independent colonization of the Bahamas from Cuba, followed by the arrival of *E. planirostris* very recently (Heinicke *et al.* 2011).

Little has been published on the natural history *of E. rogersi* (Henderson and Powell 2009), although it probably resembles the natural history of *E. planirostris* in the region. As such, *E. rogersi* would require moist microhabitats for shelter and reproduction, and would likely benefit from irrigation, particularly in the more xeric southern Bahamian islands.

C. *Osteopilus septentrionalis* (Duméril and Bibron, 1841)

Osteopilus septentrionalis (Fig. 2.1C) ranges widely across the Bahamas, Cuba, and Cayman Islands (Schwartz and Henderson 1991; Heinicke *et al.* 2011), and has been introduced throughout the greater Caribbean region (see below). Within the focal region, this species is found on both the LBB and GBB, and has colonized a number of the smaller banks of the Central Bahamas (Rum, Conception, San Salvador, and Crooked-Acklins), as well as Great Inagua (Fig. 2.3). Not all islands on the LBB or GBB have records of this species, such as many of the Exuma Cays, or any of the Flamingo Cays or Ragged Islands (Buckner *et al.* 2012). This might be owing to a lack of suitable habitat on smaller islands, or the lack of comprehensive surveys for this species on these islands. Indeed, recent surveys of the Conception Island bank yielded a new island record for Booby Cay (Reynolds *et al.* 2016). Subfossils are known from sites in the northern Bahamas: Gilpin Point (~0.9 kya) and Ralph's cave (~3.8–1.8 kya) on Abaco Island (Steadman *et al.* 2014, 2015), from several sites on San Salvador Island (Olson *et al.* 1990), and from Banana Hole on New

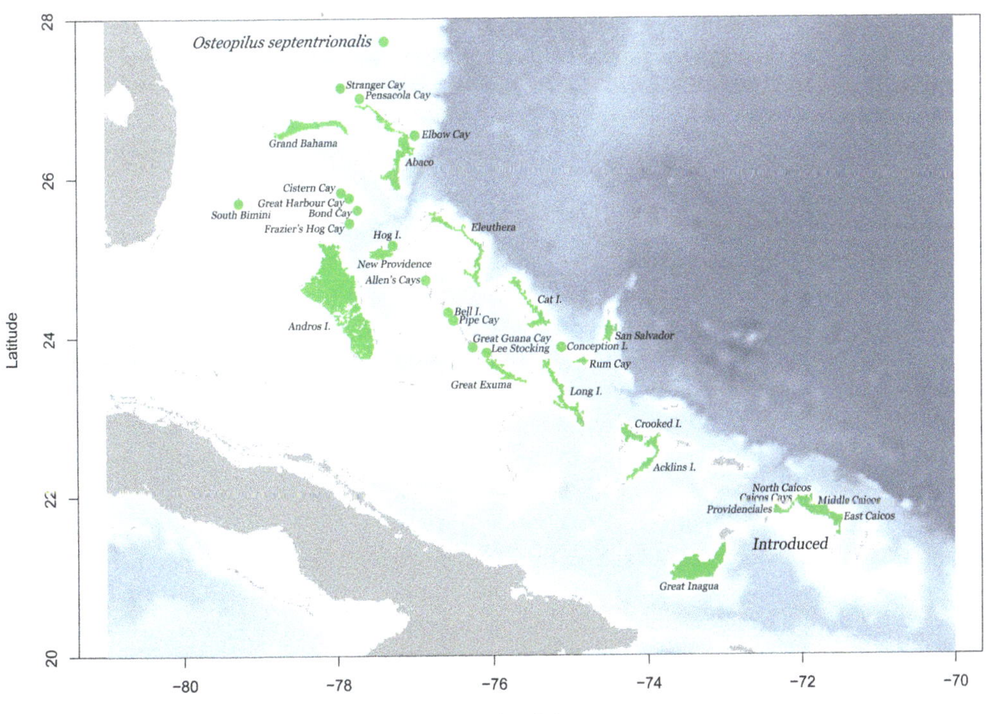

Figure 2.3. Map of the Bahamas and Turks and Caicos Islands, showing the native and introduced ranges of *Osteopilus septentrionalis* in green. Note that *O. septentrionalis* ranges more widely than the study area (not shown). Major islands and other islands mentioned in the text are labelled.

Providence Island (Pregill 1982). Carbon dating from sites on Abaco show that *O. septentrionalis* was present on LBB prior to human settlement, but after the Pleistocene-Holocene transition (Steadman *et al.* 2014, 2015). Demonstrating remarkable continuity across millennia, *O. septentrionalis* continues to breed at these same sites on Abaco.

Fossil data therefore suggest that Bahamian populations of *O. septentrionalis* are likely derived from a colonization dating to the late Pliocene or Pleistocene, and are thought to be one of the earlier diverging lineages of this species (Heinicke *et al.* 2011). These phylogenetic analyses are based only on Bahamian samples from San Salvador Island and the Crooked-Acklins Bank (Heinicke *et al.* 2011); hence, generalization of the phylogenetic relationships of populations on other Bahamian banks, such as the GBB, remains speculative.

Osteopilus septentrionalis has a generalist diet, ranging from macroinvertebrates to vertebrates such as other anurans and *Anolis* lizards (Meshaka 2001; Glorioso *et al.* 2012). The species prefers to shelter in crevices out of direct sunlight, and requires temporary or permanent bodies of water to spawn. *Osteopilus septentrionalis* is easily adaptable to human modification of habitat, and will spawn in water tanks, cisterns, or other artificial environments. Adults often shelter in buildings, refuse piles, or other anthropogenic situations (Henderson and Powell 2009).

III. The introduced amphibians

Presently there are seven non-native amphibian species established (reproducing) in the region, with many of these species occurring on the most populous islands of Grand Bahama (LBB), New Providence (GBB), and Providenciales (TCI). Of these seven, two are native species of amphibians that have been moved beyond their presumed native ranges within the region, establishing populations on the Caicos bank where no native amphibians occurred historically. Three additional non-native species are recorded from the region, although neither apparently has succeeded in establishing a breeding population. Below we discuss both introduction of non-native species to the region, as well as inferred establishment of non-native populations within the archipelago.

A. Native amphibian introductions

1. Eleutherodactylus planirostris

This species, while native to the region, is also widely introduced outside of its native range. It has established populations on other Caribbean islands (i.e. Jamaica; Henderson and Powell 2009). It has also been introduced to the TCI, well south of its native range, likely through the movement of ornamental non-native vegetation (Manco 2008). The species now occurs on the larger islands of the Caicos Bank, where populations largely are found in closed-canopy secondary coppice-forest or irrigated habitats (Schwartz and Henderson 1991; Reynolds and Niemiller 2010; Reynolds 2011a,b).

2. Osteopilus septentrionalis

There is some speculation about whether this species occurs naturally in the southern Bahamas and TCI, or whether these populations were introduced. Heinicke *et al.* (2011) indicated that populations on the Crooked-Acklins and Great Inagua banks are of natural origin, whereas Buckner *et al.* (2012) speculated that the latter population might have been introduced. Genetic data (Heinicke *et al.* 2011) are limited, as only a single mtDNA haplogroup has been characterized from two islands in the Bahamas. Nevertheless, these data support native populations in the central and southern Bahamas owing to the presence of a divergent phylogenetic lineage on San Salvador Island and Crooked Island (Heinicke *et al.* 2011).

The TCI populations of *O. septentrionalis* (Fig. 2.3), are almost certainly the result of introduction, as they do not occur on all islands despite suitable habitat, and are apparently confined to the developed islands of the Caicos Bank (Manco 2008; Reynolds 2011a; Reynolds *et al.* 2011). The pathways of introductions are speculative, although anecdotal evidence suggests that the species was brought to the Turks and Caicos Islands via vegetation and lumber shipped from Florida around the 1970s (Manco 2008). A single individual was found in the cabin of a domestic commercial airline flight between Abaco and San Salvador (Fig. 2.4), demonstrating one potential means of dispersal. While genetic data from these presumably introduced populations were not included in a recent assessment of the origins of invasive populations of this species (Heinicke *et al.* 2011), the presence of two distinct haplotypes in the Floridian population would allow some possibility of ascertaining whether the animals in TCI arrived from Florida. If this is the case, the TCI would represent a stepping-stone pattern of introduction, whereby the introduced animals were themselves derived from an introduced population in Florida.

In its introduced range, *O. septentrionalis* frequently occupies cisterns and old wells where standing water collects (e.g. Perry *et al.* 2006). Singleton males often can be heard calling from these areas. The aquatic life stage is completed in these confined areas, followed by metamorphosis and escape from the deep wells or covered cisterns. In some parts of the TCI, endemic snakes (*Tropidophis greenwayi*) can be found in high densities near these habitats, and are known to prey on metamorphs as they emerge from the water (RGR, per. obs.). Dozens of snakes can

Figure 2.4. (A) Introduced *Osteopilus septentrionalis* from Providenciales, Turks and Caicos Islands. Photograph by Brian Riggs. (B) *Osteopilus septentrionalis* observed on a domestic commercial flight originating on Abaco and terminating in San Salvador, Bahamas. Photograph by Olivia Patterson Maura.

sometimes be seen on the inside walls of old wells, representing a formidable gauntlet for the froglets. It is unknown what effect the species' toxic cutaneous secretions have on the native snakes.

B. Established non-native amphibians

1. Dryophytes squirellus *(Daudin, 1800)*

Dryophytes squirellus (formerly *Hyla squirella*) is native to the southeastern USA, where it is common and widespread. This species was first recorded in the region from Stranger's Cay, ~44 km north of Grand Bahama Island (Barbour 1904), although this was later determined to be an erroneous record as the specimens were misidentified *O. septentrionalis* (Crombie 1972). Specimens collected by Albert Schwartz between 1959 and 1968 from western Grand Bahama (near Freeport and West End), represent the first confirmed records of the species in The Bahamas, and are likely the result of an introduction from Florida (Crombie 1972). The presence of *D. squirellus* on other islands is uncertain. At least one specimen was collected from San Salvador Island in 1978 near the Gerace Research Station and others have reported seeing adults at the same site years later (Hillibrand *et al.* 2011). Additionally, Lee (2004) listed *D. squirellus* as occurring on North Bimini, a reference for which additional information seems lacking. The status of this species on San Salvador and Bimini is unclear (Buckner *et al.* 2012).

Not much is known about the use of habitat by *D. squirellus* in the Bahamas, nor about its current range within Grand Bahama. In its native range, this species uses a wide variety of natural and anthropogenic habitats. Notably, larvae and adults appear somewhat tolerant of saline conditions and often can be encountered in coastal habitats subject to intrusion by salt water, including mangrove swamps in southern Florida (Duellman and Schwartz 1958) and brackish marshes along the Gulf Coast of the USA (Neill 1958).

In Florida, USA, it has been shown that introduced *O. septentrionalis* reduce population densities of *D. squirellus*, possibly owing to direct predation (Meshaka 2001; Kraus 2009). Furthermore, tadpoles of *O. septentrionalis* prey on tadpoles of *D. squirellus* when competition for food is high (Smith 2005a,b). Hence, it is possible that the presence of *O. septentrionalis* in its native range might make colonization by *D. squirellus* challenging. Nevertheless, given that *D. squirellus* occurs within a wide range of environmental conditions, further expansion within the northern Bahamian islands seems likely, especially on heavily disturbed New Providence and Abaco islands.

2. Gastrophryne carolinensis *(Holbrook, 1835)*

Introduced populations of *Gastrophryne carolinensis* were first recorded on Grand Bahama (LBB) in 1972 (Jacobs 1973a), followed by records from New Providence (GBB) in 1978 (Crother 1986). Populations remain extant and established on both of these islands (Buckner *et al.* 2012). Recent observations (2015 and 2016; Fig. 2.1D) of this species from Marsh Harbour (LBB) extend the known introduced range eastward to Abaco Island (Giery *et al.* 2017). Given that these are highly fossorial frogs, a likely mode of introduction and transport is via plants used for landscaping (Kraus 2009). The species also has been introduced elsewhere in the Caribbean, including to the Cayman Islands (Schwartz and Henderson 1991).

Use of habitat by *G. carolinensis* in its native range, the southeastern USA, is highly variable. Populations are commonly found in coastal areas, including the Florida Keys and barrier islands along the Carolinas, suggesting some tolerance of saline conditions (Neill 1958; Dodd 2013). Populations also appear broadly tolerant of anthropogenically altered habitats (Duellman and Schwartz 1958). For example, the population on Abaco breeds in shallow, ephemeral pools along roadways that form during the rainy season (May–August) and thus are not tied to permanent bodies of water. These factors suggest that *G. carolinensis* is likely capable of surviving on a number of islands in the region should they become introduced.

3. Lithobates grylio *(Stejneger, 1901)*

Lithobates grylio was first introduced into New Providence Island (GBB) in the 1950s and was commonly found in the brackish waters of Lake Killarney in central New Providence. Soon thereafter it was found in a few areas of eastern Andros Island (Campbell 1978). Unlike many of the other introduced species in the region, *L. grylio* was introduced intentionally (Neill 1964; Franz *et al.* 1996), presumably as a source of food for humans. This species apparently has spread across New Providence and Andros from these initial populations, although the extent of the invasion seems to have been limited, perhaps by the availability of suitable habitat. Being primarily an aquatic frog, individuals of *L. grylio* are often found floating among emergent vegetation of fresh to slightly brackish ponds (Neill 1958, 1964). This highly aquatic habit coupled with a long duration of the larval stage (3–15 months) probably limits the spread of this species in the region.

While *L. grylio* is a large species (snout–vent length ~10 cm) known to consume a wide variety of invertebrates and vertebrates (Dodd 2013), no known impacts of this species on native fauna are documented. Nonetheless, Lake Killarney is home to several species of small endemic fishes which likely fall prey to adult *L. grylio* (Neill 1964).

4. Lithobates sphenocephalus *(Cope, 1886)*

This medium-sized frog is known only from Grand Bahama Island (LBB) in the region. The first records of this species probably were recorded in Freeport in 1963, but the specimens were identified as *L. pipiens* 'complex'. This mistake was likely rectified by the identification of *L. sphenocephala* in the same area nine years later (Jacobs 1973b). As with other species introduced to Grand Bahama Island, the extent of the geographic range colonized by *L. sphenocephalus* is unknown. Nevertheless, they appear to be firmly established in the area around Freeport.

Native and common throughout the southeastern United States, *L. sphenocephalus* inhabits a broad range of shallow-water habitats including marshes, swamps, and bogs. Less aquatic than *L. grylio*, *L. sphenocephalus* is still largely confined to permanent bodies of fresh water that support their prolonged larval phase (2.5–9 months) (Dodd 2013). Adults and juveniles are capable of dispersing widely during periods of rain. Adults also are generally acknowledged to be among the most salt-tolerant of North American amphibians and frequently can be found in coastal marshes with moderate to low salinities (Duellman and Schwartz 1958; Neill 1958).

The mode of introduction of *L. sphenocephalus* is unknown. Campbell (1978) suggested they may have been stowaways in aeroplanes; however, this seems unlikely given their aquatic habit. Rather, it seems likely that they were either intentionally introduced, or were inadvertently transported from the USA in shipments of ornamental aquatic plants, perhaps as tadpoles (Lee 2004).

5. Rhinella marina *(Linnaeus, 1758)*

Rhinella marina is native to Central America and northern South America east of the Andes (populations west of the Andes may represent a cryptic species, which deserves further study; see Acevedo *et al.* 2016). These large, invasive anurans have been introduced widely and are now established in Australia, Florida, on many Pacific islands, and most of the larger Caribbean islands. Despite the proximity to some long-established introduced populations (e.g. Cuba, Puerto Rico, Florida, and Bermuda) the TCI and the Bahamas only recently have shown evidence of invasion by *R. marina*, and the status of its introduced populations is unclear.

Currently, *R. marina* is restricted to a small number of sites in the region. On the island of Providenciales in the TCI it could potentially reproduce in fresh-water ponds constructed for a golf course at the eastern end of the island and anecdotal evidence suggests they are breeding. On New Providence Island (GBB) in the Bahamas, a large population was discovered in 2013 breeding in golf-course ponds at Lyford Cay (Johnson and Gibson 2018). Efforts to control its population density have been attempted, including the capture of adults in areas surrounding Lyford Cay. The efficacy of this management is unknown at present, but concerted efforts could successfully remove these populations. In addition to this well-known breeding population on New Providence Island, recently verified (2019) reports of adults and eggs from Marsh Harbour, Abaco Island (LBB), indicate that another breeding population is established in the Bahamas. In addition to these established populations, waif individuals have been reported from other islands such as Grand Bahama Island (LBB), Great Guana Cay (LBB), and an additional location on New Providence Island. While some of these reports probably represent lone individuals transported in cargo (e.g. Johnson and Gibson 2018), they nevertheless indicate the potential for further expansion of established populations within the Lucayan Archipelago.

Over much of its introduced geographic range, *R. marina* has become a costly invasive species. Negative impacts from this species arise from poisoning native predators and consuming local fauna. Consumption of adult toads by naïve predators in Australia has caused severe declines in predatory mammals and reptiles. However, deleterious effects of *R. marina* appear less acute for New-World introductions, given that predatory taxa there likely share an evolutionary history with other bufonids (and their toxins), although some negative impact has been suggested on boid snakes in Jamaica (Wilson *et al.* 2011). Individuals of *R. marina* are also likely to consume native terrestrial vertebrates.

Rhinella marina is a highly invasive and damaging species outside of its native range. Efforts to eradicate the few existing populations should be implemented as soon as possible, as should a biosurveillance campaign to prevent further introductions to the Lucayan Archipelago.

C. Unestablished non-native amphibians

1. Dryophytes cinereus *(Schneider, 1799)*

This species was first discovered on New Providence Island in 2015 when a deceased individual was found (Johnson and Yates 2020). Subsequent observations in 2018 and 2019 suggest that the species can be encountered on the island. It is currently unknown whether the species is breeding, or whether these individuals represent propagules transmitted in live plants from Florida.

2. Lithobates catesbeianus *(Shaw, 1802)*

A single collection record for *L. catesbeianus* exists for New Providence Island. A group of specimens in the Philadelphia Academy of Science (ANSP HRP 32805) are tadpoles collected in 1972, but might belong to *L. grylio* instead, although this speculation has not been confirmed. *Lithobates catesbeianus* was also introduced into Cuba, where it is widespread (Schwartz and Henderson 1991; Estrada and Ruibal 1999).

3. Lithobates clamitans *(Latreille, 1801)*

A single male *L. clamitans* was reported in 2001 calling from a pond at a golf course in Freeport, Grand Bahama Island (Lee 2004). No further records of this species are known and the species is not considered to have become established (Kraus 2009).

D. Patterns of invasion

1. Location and dates of introductions

The earliest known introductions of non-native amphibians to the region occurred in the 1950s. Most other successful introductions followed shortly thereafter in the 1960s and 1970s. This pulse of introductions seems to have abated in recent years, with only one additional species (*R. marina*) becoming established between 1972 and 2016.

While the rate of new invasions in the region is quite low, subsequent spread of established species is continuing on two fronts: among-island introductions (secondary introductions), and within-island expansion. Unlike the rather well-documented dates and localities of newly established species in the region, few data are available on the current distribution or expansion rates of introduced species. Given that many of the established species exhibit substantial independence from permanent freshwater habitats (via tolerance of salinity, brief larval period, or direct-development), continuing expansion seems likely for several of the non-native species as long as suitable habitats and hydric conditions persist.

Secondary introductions of amphibians and reptiles appear to be quite common in the Bahamas and TCI (e.g. Giery 2013). For example, both *D. squirellus* and *G. carolinensis* continue to extend their distribution and occupy additional islands within the archipelago, as do two of the archipelago's native species: *O. septentrionalis* and *E. planirostris* (Reynolds and Niemiller 2010). These secondary introductions likely are a consequence of domestic trade among islands, perhaps in ornamental plants (Manco 2008). Nevertheless, it is also possible that some of the additional populations arising on islands actually represent independent introductions (e.g. from the species' native ranges or from invaded areas outside of the region) rather than secondary ones (e.g. from other islands in the region), a possibility that requires genetic studies or direct evidence of transport for clarification (it could also be a combination of introduction modes).

Another aspect of range expansion in the region is the extent of spread within islands. Most (nearly all) records of non-native species come from developed areas, such as Freeport (LBB) and Nassau (GBB), that experienced substantial economic development and expansion during the mid-twentieth century. While these original introductions are extant, the extent to which these species have invaded adjacent undeveloped habitats, which cover the vast majority of the area, is unknown. Clearly, additional research is needed about the post-establishment expansion of non-native amphibians.

2. Are additional invasions by amphibians likely?

All the non-native anurans in the Lucayan Archipelago currently are found in (although not necessarily native to) southern Florida USA, a probable origin of past introductions into the region. This seems to suggest that future invasions are likely to come from the same species pool of southern Floridian amphibians. Using tolerance to salinity as a guide to likely invaders, two species appear candidates for invasion (and establishment; see section III. C. 1.): the Green Treefrog (*Dryophytes cinereus*) and the Southern Toad (*Anaxyrus terrestris* [Bonnaterre, 1789]). Both species show some degree of tolerance to brackish water and are abundant in their native ranges, even in developed areas (Duellman and Schwartz 1958; Neill 1958). As with previous introductions of reptiles and amphibians to the region, a likely mode of new introduction would be via international plant trade for horticultural and landscape businesses (Kraus 2009).

While the USA seems a likely source of future introductions, regions outside southern Florida also include highly invasive species likely to reach the archipelago. One such group is the eleutherodactylid frogs. Species from this large genus, of which two (*E. planirostris* and *E. rogersi*) are native to the region and are proven invaders, both within the Greater Caribbean region and elsewhere (e.g. Hawaii, Bermuda, Brazil, Philippines, Hong Kong, Guam). Two species appear to be likely invaders: Johnstone's Whistling Frog, *Eleutherodactylus johnstonei* Barbour, 1914, and the Puerto Rican Coqui, *Eleutherodactylus coqui* Thomas, 1966. Like their native congeners, these species do not require standing water to complete their lifecycle and are very common within human-altered habitats, both of which are attributes that make them capable invaders.

IV. Threats to the amphibians

A. IUCN status

All three native amphibian species have been characterized according to the protocols of the IUCN Red List. *Eleutherodactylus planirostris* was assessed as Least Concern (LC) by

Hedges *et al.* (2004) and by the IUCN SSC Amphibian Specialist Group (2021a). This designation is owing to a widespread distribution, no evidence of population decline, and occurrence in a number of protected areas throughout its native range. *Eleutherodactylus planirostris* is likely tolerant of human disturbance, and might benefit from human practices such as irrigation. *Eleutherodactylus rogersi* was assessed as being Data Deficient (DD) (Angulo 2008), a category used to designate species of possible conservation concern, but for which there is insufficient data to allow assigning to other categories (IUCN 2016). Re-evaluation in 2021 (IUCN SSC Amphibian Specialist Group 2021b) generated an assessment of Least Concern. Little is known regarding the population status of *E. rogersi* across its fragmented geographic distribution, although it does appear to represent an evolutionarily unique lineage of eleutherodactyline frogs (Heinicke *et al.* 2011). *Osteopilus septentrionalis* is listed as LC by Hedges *et al.* (2010) and by the IUCN SSC Amphibian Specialist Group (2021c), owing to its generally large populations, limited evidence for population declines, and an expanding range.

B. Evolutionary effects: genetic admixture/introgression

While threats to native amphibians in the region seem minor, given their robust, flexible, non-specialized ecology, one threat to local populations seems likely: the admixture/introgression of evolutionarily distinct clades. The consequences of admixture/introgression on fitness at populational or species levels are not always clear. However, a general concern is that introgression can erode local adaptation through outbreeding depression. While certainly not universal, in many cases hybridization and introgression in anurans is expected to be maladaptive (e.g. Lemmon and Lemmon 2010), sometimes to the extent of leading to extinction or of threatening the persistence of populations (Rhymer and Simberloff 1996; Allendorf *et al.* 2001; Holsbeek *et al.* 2008). As of now, such effects on the native amphibians are hypothetical – yet they remain important conservation issues for the region's amphibian fauna. To illustrate, *O. septentrionalis* and *E. planirostris* occupy large ranges within the Greater Caribbean, including Cuba, much of the Bahamas, and the Cayman Islands. Within this region, both species consist of several evolutionarily distinct clades (Heinicke *et al.* 2011). In some cases, these clades are quite divergent. For example, the lineages of Bahamian and Cuban *O. septentrionalis* appear to have diverged in the late Pliocene. Therefore, human-assisted dispersal of individuals within and among regions, including their native geographic range, may result in hybridization between distinct lineages and subsequently in intraspecific genetic introgression. Indeed, introduced populations of *O. septentrionalis* in Florida, USA show evidence of admixture of at least two independent lineages, one from extreme western Cuba and the other from central and eastern Cuba (Heinicke *et al.* 2011). Similarly, Caribbean islands, including Puerto Rico, the Virgin Islands, and the Lesser Antilles, contain introduced *O. septentrionalis* from these same distinct Cuban populations. Although evidence of admixture does not exist in populations from these islands, it seems likely that evolutionary distinct lineages are potentially being introduced throughout the northern Caribbean, which might result in admixture (facilitated by human-mediated secondary contact of divergent lineages).

More proximate, perhaps, is the possibility of interspecific hybridization between *E. planirostris* and *E. rogersi*. These two native species, which are largely allopatric, are recently derived (5 mya) from a common ancestor, and are similar in their ecological requirements and phenotype (Heinicke *et al.* 2011). This recent shared ancestry may not have generated effective reproductive-isolation mechanisms between species. Thus, these species could be at increased risk of introgression by hybridization if they become sympatric by way of future introductions. This similarity also puts these species at risk of deleterious ecological interactions.

C. Ecological effects

Interspecific interactions between native and introduced species can threaten populations of native/endemic amphibians. For instance, nearly all of the anurans introduced into the region have rather broad diets that would likely overlap with those of the natives. If there is a significant dietary overlap between species, then the introduction of non-native species could reduce population sizes of native frogs through competition and/or predation. As mentioned previously, introduction of *E. planirostris* to islands containing *E. rogersi* could potentially increase competition for limited resources between these two species. While such interactions are frequently hypothetical, competitive interaction between two species of eleutherodactyline frogs introduced into Bermuda may be responsible, in part, for the extirpation of *Eleutherodactylus gossei* Dunn, 1926, there (Wingate 1965; Bacon *et al.* 2011). In addition, further expansion of *R. marina* within the region could generate substantial predation pressure on native anurans.

D. Summary

The Lucayan Archipelago has three native species of frogs, of which one is endemic. All have large populations and are widely distributed in the region. Further, the region contains two of the least-populated countries/territories in the Western Hemisphere, with large tracts of undeveloped land, national parks, and protected lands (Knapp *et al.* 2011; Reynolds 2011a). Therefore, the region's native anuran fauna may appear well positioned to withstand the immediate changes of anthropogenic origin that threaten many of the world's amphibians in the foreseeable future, but prospects could be otherwise (see next).

The native anurans in the region do face visible and invisible threats that require particular attention from a conservation perspective. First, increasing coastal development for the construction of international resorts in the northern archipelago (Abaco Island, New Providence Island, Bimini islands) is likely to increase the probability of invasions by non-native species and/or genotypes, as these types of development often import large quantities of ornamental vegetation and inadvertently the amphibians in it. Second, non-endemic species such as *Rhinella marina* are currently expanding their range beyond their previous limits, likely facilitated by human-mediated dispersal. Third, the endemic *E. rogersi* is likely to face increasing competition and hybridization pressures from congeners such as *E. planirostris*. Fourth, genetically distinct populations of *O. septentrionalis* endemic to the Lucayan Archipelago are likely to face threats, such as introgressive hybridization, from the more widespread Cuban lineages. These threats might arise from populations coming directly from Cuba but are perhaps more likely to come from populations that have invaded the Antilles (Greater and Lesser) and/or peninsular Florida. Finally, global climatic change is likely to reduce habitat for all native species. For example, although estimates of the area of land that is likely to be lost because of sea-level rise over the next century vary, the abundance and distribution of terrestrial species on this series of low-lying carbonate platforms undoubtedly will decrease because of a significant loss of habitat (Nichols and Cazenave 2010). A reduced area of habitat – terrestrial and fresh water – will probably result in lower population sizes for all anurans. Unfortunately, corresponding threats of extinction owing to anthropogenic and natural phenomena are not being assessed in the region. Substantial opportunities exist, however, for additional studies on both native and introduced anurans, including detailed characterization of species' ranges, evolutionary relationships, historical biogeography, and climatic–abundance/distribution relationships. Such a research agenda will generate information that can help us understand the influence of these phenomena on the region's anurans for effective conservation efforts in the imminent future.

V. Acknowledgements

We are appreciative of the work done in the region by our numerous colleagues, on which this chapter builds. We thank S. Buckner, S. Johnson, B. Naqqi Manco, and J. Bullard for information and discussions related to this topic. We thank J. Burgess, B. Riggs, O. Patterson Maura, and M. Niemiller for the use of their photographs. We thank the TCI Department of Environment and Maritime Affairs, Turks and Caicos Island Government, Bahamas National Trust, and the Bahamas Department of Agriculture, the Bahamas Environment, Science, and Technology Commission, Ministry of the Environment for research permits. We are grateful to Friends of the Environment, Bahamas National Trust, and the Turks and Caicos National Trust for assistance and support. We thank H. Heatwole, N. Ríos-López, and two anonymous reviewers for helpful comments on earlier drafts of this chapter.

VI. References

Acevedo, A. A., Lampo, M., and Cipriani, R., 2016. The cane toad or marine toad, *Rhinella marina* (Anura, Bufonidae): two genetically and morphologically distinct species. *Zootaxa* 4103: 574–586. https://doi.org/10.11646/zootaxa.4103.6.7

Allendorf, F. W., Leary, R. F., Spruell, P., and Wenburg, J. K., 2001. The problems with hybrids: setting conservation guidelines. *Trends in Ecology and Evolution* 16: 613–622. https://doi.org/10.1016/S0169-5347(01)02290-X

Angulo, A., 2008. *Eleutherodactylus rogersi*. The IUCN Red List of Threatened Species 2008: e.T135989A4223059. Downloaded on 14 October 2016.

Bacon, J. P., Gray, J. A., and Kitson, L., 2011. Status and conservation of the reptiles and amphibians of the Bermuda islands. Pp. 161–182 in *Conservation of Caribbean Island Herpetofaunas, vol. 1: Conservation Biology and the Wider Caribbean*, edited by A. Hailey, B. S. Wilson and J. A. Horrocks. Brill, Netherlands. https://doi.org/10.1163/ej.9789004183957.i-228.58

Barbour, T., 1904. Batrachia and reptilia from the Bahamas. *Bulletin of the Museum of Comparative Zoology* 46: 55–61.

Buckner, S. D., Franz, R., and Reynolds, R. G., 2012. Bahama Islands and Turks & Caicos Islands. Pp. 93–110 in *Island Lists of West Indian Amphibians and Reptiles*, edited by R. Powell and R. W. Henderson. *Bulletin of the Florida Museum of Natural History* 51: 85–166.

Campbell, D. G., 1978. *The Ephemeral Islands: A Natural History of the Bahamas*. MacMillan Publishers, London.

Clark, P. U. and Mix, A. C., 2002. Ice sheets and sea level of the Last Glacial Maximum. *Quaternary Science Reviews* 21: 1–7. https://doi.org/10.1016/S0277-3791(01)00118-4

Crombie, R. I., 1972. The presence of *Hyla squirella* in the Bahamas. *Quarterly Journal of the Florida Academy of Science* 35: 49–52.

Crother, B. I., 1986. A new distribution record for *Gastrophryne carolinensis* in the Bahamas. *Herpetological Review* 16: 114.

Dodd, C. K., 2013. *Frogs of the United States and Canada*, Volume 1. Johns Hopkins University Press, Baltimore.

Duellman, W. E. and Schwartz, A., 1958. Amphibians and reptiles of southern Florida. *Bulletin of the Florida State Museum* 3: 181–324.

Estrada, A. R. and Hedges, S. B., 1997. A new species of frog from the Meseta de Cabo Cruz, eastern Cuba (Leptodactylidae, Eleutherodactylus). *Caribbean Journal of Science* 33: 227–232.

Estrada, A. R. and Ruibal, R., 1999. A review of Cuban herpetology. Pp. 31–62 in *Caribbean Amphibians and Reptiles*, edited by B. I. Crother. Academic Press, San Diego. https://doi.org/10.1016/B978-012197955-3/50014-8

Fairbanks, R. G., 1989. A 17,000-year glacio-eustatic sea level record: influence of glacial melting rates on the Younger Dryas event and deep-ocean circulation. *Nature* 342: 637–642. https://doi.org/10.1038/342637a0

Franz, R., Dodd, C. K., and Buckner, S. D., 1996. A review of herpetology of the Bahamian Archipelago. *Bahamas Journal of Science* 3: 22–30.

Giery, S. T., 2013. First records of the corn snake (*Pantherophis guttatus*) from Abaco Island, The Bahamas, and notes their current distribution in the greater Caribbean region. *IRCF Reptiles & Amphibians, Conservation and Natural History* 20: 36–39. https://doi.org/10.17161/randa.v20i1.13935

Giery, S. T., Richard, D., and Stroud J. T., 2017. Establishment of the Eastern Narrow-mouthed Frog (*Gastrophryne carolinensis*) on Abaco Island in The Bahamas, with notes on the species' current distribution in the West Indies. *IRCF Reptiles & Amphibians, Conservation and Natural History* 24: 139–141. https://doi.org/10.17161/randa.v24i2.14170

Glorioso, B. M., Waddle, J. H., Crockett, M. E., Rice, K. G., and Percival, H. F., 2012. Diet of the invasive Cuban Treefrog (*Osteopilus septentrionalis*) in pine rockland and mangrove habitats in south Florida. *Caribbean Journal of Science* 46: 346–355. https://doi.org/10.18475/cjos.v46i2.a25

Goin, C. J., 1955. Description of a new subspecies of the frog *Eleutherodactylus ricordi* from the Bahama Islands. *American Museum Novitates* 1708: 1–7.

Grant, C., 1940. The herpetology of the Cayman Islands. *Bulletin of the Institute of Jamaica, Science Series* 2: iv–65.

Hearty, P. J., Hollin, J. T., Neumann, A. C., O'Leary, M. J., and McCulloch, M., 2007. Global sea-level fluctuations during the Last Interglaciation (MIS 5e). *Quaternary Science Reviews* 26: 2090–2112. https://doi.org/10.1016/j.quascirev.2007.06.019

Hedges, B., Díaz, L., and Powell, R., 2004. *Eleutherodactylus planirostris*. The IUCN Red List of Threatened Species 2004: e.T56864A11534006. Downloaded on 02 November 2016.

Hedges, S. B., Duellman, W. E., and Heinicke, M. P., 2008. New World direct-developing frogs (Anura: Terrarana): molecular phylogeny, classification, biogeography, and conservation. *Zootaxa* 1737: 1–182. https://doi.org/10.11646/zootaxa.1737.1.1

Hedges, S. B., Díaz, L., Ibéné, B., Joglar, R., Powell, R., Bolaños, F., and Chaves, G. 2010. *Osteopilus septentrionalis*. The IUCN Red List of Threatened Species 2010: e.T55811A11368202. Downloaded on 2 November 2016.

Heinicke, M. P., Diaz, L. M., and Hedges, S. B., 2011. Origin of invasive Florida frogs traced to Cuba. *Biology Letters* 7: 407–410. https://doi.org/10.1098/rsbl.2010.1131

Heinicke, M. P., Duellman, W. E., and Hedges, S. B., 2007. Major Caribbean and Central American frog faunas originated by ancient oceanic dispersal. *Proceedings of the National Academy of Sciences, USA* 104: 10092–10097. https://doi.org/10.1073/pnas.0611051104

Henderson, R. W. and Powell, R., 2009. *Natural History of West Indian Reptiles and Amphibians*. University Press of Florida, Gainesville.

Hillbrand, P. A., Sloan, A. T., and Hayes, W. K., 2011. The terrestrial reptiles of San Salvador Island, Bahamas. *IRCF Reptiles & Amphibians* 18:154–166.

Holsbeek, G., Mergeay, J., Hotz, H., Plötner, J., Volckaert, A. M., and Meester, L., 2008. A cryptic invasion within an invasion and widespread introgression in the European water frog complex: consequences of uncontrolled commercial trade and weak international legislation. *Molecular Ecology* 17: 5023–5035. https://doi.org/10.1111/j.1365-294X.2008.03984.x

IUCN, 2016. The IUCN Red List of Threatened Species. Version 2016-2. www.iucnredlist.org. Downloaded on 02 November 2016.

IUCN SSC Amphibian Specialist Group. 2021a. *Eleutherodactylus planirostris. The IUCN Red List of Threatened Species* 2021: e.T56864A54370037. https://dx.doi.org/10.2305/IUCN.UK.2021-1.RLTS.T56864A54370037.en.

IUCN SSC Amphibian Specialist Group. 2021b. *Eleutherodactylus rogersi. The IUCN Red List of Threatened Species* 2021: e.T135989A172796918. https://dx.doi.org/10.2305/IUCN.UK.2021-1.RLTS.T135989A172796918.en.

IUCN SSC Amphibian Specialist Group. 2021c. *Osteopilus septentrionalis. The IUCN Red List of Threatened Species* 2021: e.T55811A3032751. https://dx.doi.org/10.2305/IUCN.UK.2021-1.RLTS.T55811A3032751.en.

Jacobs, J., 1973a. Geographic distribution: *Gastrophyrne carolinensis* (Eastern narrow-mouthed toad). *HISS News-Journal* 1: 98.

Jacobs, J., 1973b. Geographic distribution: *Rana sphenocephala* (Southern leopard frog). *HISS News-Journal* 1: 98.

Johnson, S. and Gibson, D., 2018. South American Cane Toad (*Rhinella marina*) found on Great Guana Cay, Abaco, The Bahamas. *IRCF Reptiles & Amphibians, Conservation and Natural History* 25: 160–161. https://doi.org/10.17161/randa.v25i2.14286

Johnson, S. and Yates, S. S., 2020. First records of the North American Green Treefrog (*Hyla cinerea*) on New Providence, The Bahamas. *IRCF Reptiles & Amphibians* 26: 257–258. https://doi.org/10.17161/randa.v26i3.14431

Knapp, C. R., Iverson, J. B., Buckner, S. D., and Cant, S. V., 2011. Conservation of amphibians and reptiles in the Bahamas. Pp. 53–88 in *Conservation of Caribbean Island Herpetofaunas, vol. 2: Regional Accounts of the West Indies*, edited by A. Hailey, B. S. Wilson, and J. A. Horrocks. Brill, Leiden, The Netherlands. https://doi.org/10.1163/ej.9789004194083.i-439.21

Kraus, F., 2009. *Alien Reptiles and Amphibians: A Scientific Compendium and Analysis*. Springer, Berlin. https://doi.org/10.1007/978-1-4020-8946-6

Lazell, J. D., Jr., 1989. *Wildlife of the Florida Keys: A Natural History*. Island Press, Washington, D.C.

Lee, D. S., 2004. Additional reptiles and amphibians introduced to the Bahamas: a growing conservation concern. *Bulletin of the Chicago Herpetological Society* 39: 161–164.

Lemmon, E. M. and Lemmon, A. R., 2010. Reinforcement in chorus frogs: lifetime fitness estimates including intrinsic natural selection and sexual selection against hybrids. *Evolution* 64: 1748–1761. https://doi.org/10.1111/j.1558-5646.2010.00955.x

Lighty, R. G., Macintyre, J. G., and Stuckenrath, R., 1982. *Acropora palmata* reef framework: A reliable indicator of sea level in the western Atlantic for the past 10,000 years. *Coral Reefs* 1: 25–30. https://doi.org/10.1007/BF00301694

Lynch, J. D. and Duellman, W. E., 1997. Frogs of the genus *Eleutherodactylus* in western Ecuador. Systematics, ecology, and biogeography. *Special Publication. Natural History Museum, University of Kansas* 23: 1–236. https://doi.org/10.5962/bhl.title.7951

Manco, B. N., 2008. Native fauna of the Turks & Caicos Islands. Pp. 41–58 in *A History of the Turks & Caicos Islands*, edited by C. Mills. Macmillan Education, Oxford.

Meshaka, W. E., Jr., 2001. *The Cuban Treefrog in Florida: Life History of a Successful Colonizing Species*. University Press of Florida, Gainesville.

Neill, W. T., 1958. The occurrence of amphibians and reptiles in saltwater areas, and a bibliography.

Bulletin of Marine Science of the Gulf and Caribbean 8: 1–97.

Neill, W. T., 1964. Frogs introduced on islands. *Quarterly Journal of the Florida Academy of Sciences* 27: 127–130.

Nichols, R. J. and Cazenave, A., 2010. Sea-level rise and its impact on coastal zones. *Science* 328: 1517–1520. https://doi.org/10.1126/science.1185782

Olson, S. L., Pregill, G. K., and Hilgartner, W. B., 1990. Studies on Fossil and Extant Vertebrates from San Salvador (Watling's) Island, Bahamas. *Smithsonian Contributions to Zoology*, No. 508. Smithsonian Institute Press, Washington, D.C. https://doi.org/10.5479/si.00810282.508

Padial, J. M., Grant T., and Frost, D. R., 2014. Molecular systematics of terraranas (Anura: Brachycephaloidea) with an assessment of the effects of alignment and optimality criteria. *Zootaxa* 3825: 1–132. https://doi.org/10.11646/zootaxa.3825.1.1

Perry, G., Powell, R., and Watson, H., 2006. Keeping invasive species off Guana Island, British Virgin Islands. *Iguana* 13: 273–277.

Pregill, G. K., 1982. Fossil amphibians and reptiles from New Providence Island, Bahamas. Pp. 8–21 in *Fossil Vertebrates from the Bahamas*, edited by S. L. Olson. *Smithsonian Contributions to Paleobiology* 48.

Pyron, R. A. and Wiens, J. J., 2011. A large-scale phylogeny of Amphibia including over 2800 species, and a revised classification of extant frogs, salamander, and caecilians. *Molecular Phylogenetics and Evolution* 61: 543–583. https://doi.org/10.1016/j.ympev.2011.06.012

Reynolds, R. G., 2011a. Status, conservation, and introduction of amphibians and reptiles in the Turks and Caicos Islands, British West Indies. Pp. 377–406 in *Conservation of Caribbean Island Herpetofaunas, vol. 2: Regional Accounts of the West Indies*, edited by A. Hailey, B. S. Wilson, and J. A. Horrocks. Brill, Leiden. https://doi.org/10.1163/ej.9789004194083.i-439.151

Reynolds, R. G., 2011b. *Eleutherodactylus planirostris planirostris* (Cuban Flathead Frog), distribution. *Caribbean Herpetology* 27: 1. https://doi.org/10.31611/ch.27

Reynolds, R. G. and Niemiller, M. L., 2010. Island invaders: introduced amphibians and reptiles of the Turks and Caicos Islands. *IRCF Reptiles & Amphibians: Conservation and Natural History* 17: 117–121.

Reynolds, R. G., Niemiller, M. L. Riggs, B. M., and Manco, B. N., 2011. *Osteopilus septentrionalis* (Cuban Treefrog). Distribution. *Herpetological Review* 42: 237–238.

Reynolds, R. G., Herrmann, N. C., Puente-Rolón, A. R., Geneva, A. J., and Aviles-Rodriguez, K. J., 2016. *Osteopilus septentrionalis* (Cuban Treefrog). Distribution. *Herpetological Review* 47: 420.

Rohling, E. J., Grant, K., Bolshaw, M., Roberts, A. P., Siddall, M., Hemleben, C., and Kucera, M., 2009. Antarctic temperature and global sea level closely coupled over the past five glacial cycles. *Nature Geoscience* 2: 500–504. https://doi.org/10.1038/ngeo557

Rhymer, J. M. and Simberloff, D., 1996. Extinction by hybridization and introgression. *Annual Review of Ecology and Systematics* 27: 83–109. https://doi.org/10.1146/annurev.ecolsys.27.1.83

Schwartz, A. and Henderson, R. W., 1991. *Amphibians and Reptiles of the West Indies: Descriptions, Distributions, and Natural History*. University of Florida Press, Gainesville.

Schwartz, A. 1965. Geographic variation in two species of Hispaniolan *Eleutherodactylus* with notes on Cuban members of the ricordi group. *Studies on the Fauna of Curaçao and Other Caribbean Islands* 22: 98–123.

Sealey, N. E., 2006. *Bahamian Landscapes*, 3rd edn. Macmillan-Caribbean, Oxford.

Seidel, M. E. and Franz, R., 1994. Amphibians and reptiles (exclusive of marine turtles) of the Cayman Islands. Pp. 407–433 in *The Cayman Islands: Natural History and Biogeography*, edited by M. A. Brunt and J. E. Davies. Kluwer Academic Publishers, Dordrecht. https://doi.org/10.1007/978-94-011-0904-8_20

Siddall, M., Rohling, E. J., Almogi–Labin, A., Hemleben, C., Meischner, D., Schmelzer, I., and Smeed, D. A., 2003. Sea-level fluctuations during the last glacial cycle. *Nature* 423: 854–858. https://doi.org/10.1038/nature01690

Smith, K. G., 2005a. Effects of nonindigenous tadpoles on native tadpoles in Florida: evidence of competition. *Biological Conservation* 123: 433–441. https://doi.org/10.1016/j.biocon.2005.01.005

Smith, K. G., 2005b. An exploratory assessment of Cuban treefrog (*Osteopilus septentrionalis*) tadpoles as predators of native and nonindigenous tadpoles in Florida. *Amphibia-Reptilia* 26: 571–575. https://doi.org/10.1163/156853805774806313

Steadman, D. W., Albury, N. A., Kakuk, B., Mead, J. I., Soto-Centeno, J. A., Singleton, H. M., and Franklin, J., 2015. Vertebrate community on an ice-age Caribbean island. *Proceedings of the National Academy of Sciences USA* 44: E5963–E5971. https://doi.org/10.1073/pnas.1516490112

Steadman, D. W., Albury, N. A., Maillis, P., Mead, J. I., Slapcinsky, J., Krysko, K. L., Singleton, H. M., and Franklin, J., 2014. Late-Holocene faunal and landscape change in the Bahamas. *The Holocene* 24: 220–230. https://doi.org/10.1177/0959683613516819

Stewart, M. M., 1979. The role of introduced species in a Jamaican frog community. Pp. 113–146 in *Proceedings of the IV Symposium on Tropical Ecology,* vol. I-III edited by H. Wolda. Panama City, Panama.

Stewart, M. M. and Martin, G. E., 1980. Coconut-husk piles – a unique habitat for Jamaican terrestrial frogs. *Biotropica* 12: 107–116. https://doi.org/10.2307/2387726

Wilson, B. S., Koenig, S. E., van Veen, R., Miersma, E., and Rudolph, D. C., 2011. Cane toads a threat to West Indian wildlife: mortality of Jamaican boas attributable to toad ingestion. *Biological Invasions* 13: 55–60. https://doi.org/10.1007/s10530-010-9787-7

Wingate, D. B., 1965. Terrestrial herpetofauna of Bermuda. *Herpetologica* 21: 202–218.

3 The Amphibians of Cuba: integrative insight into the diversity and conservation of a highly vulnerable fauna

Luis M. Díaz, Ansel Fong G., Alberto R. Estrada, Antonio Cádiz, Amnerys González, Rolando Fernández de Arcila, Irelis Bignotte-Giró, and Sixto J. Incháustegui

Luis M. Díaz, Ansel Fong G., Alberto R. Estrada, Antonio Cádiz, Amnerys González, Rolando Fernández de Arcila, Irelis Bignotte-Giró, and Sixto J. Incháustegui, 'Chapter 3: The Amphibians of Cuba: integrative insight into the diversity and conservation of a highly vulnerable fauna' in: *The Conservation and Biogeography of Amphibians in the Caribbean*. Pelagic Publishing (2023).
© Luis M. Díaz *et al.* DOI: 10.53061/PWOP8393

Abbreviations and acronyms

AMNH	*American Museum of Natural History*
ASG	*Amphibian Specialist Group, IUCN*
asl	*above sea level*
Bd	Batrachochytrium dendrobatidis
BIOECO	*Centro Oriental de Ecosistemas y Biodiversidad*
BRSR	*Biosphere Reserve Sierra del Rosario*
CAMP	*Conservation Assessment and Management Plans Workshops for Selected Cuban Species*
CITMA	*Cuban Ministry of Science, Technology, and Environment*
CLC	*complex lifecycles*
CNAP	*Centro Nacional de Áreas Protegidas de Cuba*
D2	*Code for 'very restricted population, capable of becoming Critically Endangered or even Extinct within a very short time'*
EF	*Evergreen Forest*
GAA	*Global Amphibian Assessment*
GPL	*Global Pandemic Lineage*
IBA	*Important Bird Area*
IUCN	*International Union for the Conservation of Nature*
ONEI	*Oficina Nacional de Estadística e Información*
pers. obs.	*personal observation*
PROLAB	*Promoting Research Opportunities for Latin American Biochemists*
SDMF	*Semideciduous microphyllic forest*
snm	*sobre nivel del mar*
SNAP	*Sistema Nacional de Áreas Protegidas*
SVL	*snout-to-vent length*
UNESCO	*United Nations Educational, Scientific and Cultural Organization*

> *'Only if we understand can we care. Only if we care will we help. Only if we help shall all be saved.'*
>
> – Jane Goodall

I. Introduction

The Republic of Cuba (hence forward Cuba) includes the largest island of the Greater Antilles in the Caribbean region and comprises a significant part of one of the most important biodiversity hotspots on Earth (Myers *et al.* 2000; Mittermeier *et al.* 2004, 2011; Smith *et al.* 2005). Furthermore, it has been identified as among the 20 nations worldwide that have the highest number of endemic species and is listed among those with the most endangered amphibian faunas (Stuart *et al.* 2008). Along with its other smaller islands, islets, and cays – the Cuban archipelago – it holds 69 native species of amphibians plus two introduced species which have become established in the wild, all anurans (Frost 2021). Its anurans also represent its group of land vertebrates with the highest endemism (95%), compared to its reptiles (83%), mammals (35%), and birds (7%). Cuba is the second Caribbean Island in terms of amphibian diversity, outnumbered only by Hispaniola with 76 species. However, some newly discovered Cuban species are in the process of being named and described.

Native families of anurans comprise Bufonidae (10 spp.), Eleutherodactylidae (58 spp.), and Hylidae (1 sp.). Two more families are represented by introduced, established species (Borroto-Páez *et al.* 2015; Rodríguez-Cabrera *et al.* 2018): Leptodactylidae (*Leptodactylus fragilis*) and Ranidae (*Lithobates catesbeianus*). One additional species, the Asian Bullfrog *Hoplobatrachus tigerinus* (Dicroglossidae), was introduced by the Ministerio de la Pesca (Ministry of Fisheries) under controlled conditions in the vicinity of Havana but has still not been reported from the wild (Borroto-Páez *et al.* 2015).

We are aware of large information gaps that limit our knowledge of these species, particularly with regard to their ecology, ethology, and taxonomic resolution, which are of paramount importance for their conservation. As stated by Blaustein *et al.* (2011), the conservation of amphibians is dimensionally complex, and requires a deep understanding of the magnitude of changes in diversity and adaptive mechanisms that operate beyond species lists and occasional assessments. For this, we highlight the importance of transdisciplinary approaches that can directly address what underpins population processes, thus providing insight into the causes of biodiversity loss (Walls and Gabor 2019). Our objectives are twofold. First, we provide an updated overview of the Cuban anurans that highlights (1) diversity; (2) distributional patterns; (3) current and potential threats; (4) conservation assessments; (5) the status of some taxonomic, ecological, and life-history traits relevant to conservation; and (6) the representation of anurans in the Cuban System of Protected Areas (*Sistema Nacional de Áreas Protegidas* [SNAP]). Second, by examining information at hand about the diverse, poorly known, and extremely fragile batrachofauna of Cuba, we aim to encourage further studies, the identification of priorities for scientific enquiry, and the implementation of long-term conservation strategies that increasingly supplement the knowledge of multiple aspects of species' biology. In Cuba, as in many other islands in the Caribbean, this knowledge accumulates steadily, albeit very slowly.

II. The Cuban archipelago

The total surface area of the archipelago is 109,884.01 km^2, distributed as follows: the main island of Cuba (104,338.33 km^2), Isla de la Juventud (formerly Isle of Pines; 2,419.27 km^2), and islets and keys (collectively 3,126.41 km^2). It has a coastline of 5,746 km. Its geographic limits are Cayo Cruz del Padre (Matanzas Province: 23° 17' 09" N, 80° 53' 55" W) in the north, and Punta del Inglés (Granma Province: 19° 49' 36" N, 77° 40' 36" W) in the south, Cabo San Antonio (Pinar del Río: 21° 51' 40" N, 84° 57' 54" W) in the west, and Punta de Maisí (Guantánamo Province: 20° 12' 36" N, 74° 07' 52" W) in the east (ONEI 2020).

Two-thirds of the main island consists of plains and low hills surrounding the four highest mountain ranges (Fig. 3.1), located, respectively, in the west (Guaniguanico Range), centre (Guamuhaya Massif), and east (Sierra Maestra and the Nipe-Sagua-Baracoa Massif). The area of river basins covers 73% of the territory (ONEI 2020). The Republic of Cuba comprises 15 provinces and the special municipality Isla de la Juventud (Fig. 3.2).

A. Physiographic regions

The Cuban archipelago was characterized by Acevedo (1989) into three physiographic macroregions: Western, Central, and Eastern, each one with subdivisions. More recently, Mateo Rodríguez *et al.* (2019) defined four major geographic regions: Western, Central, Central-East,

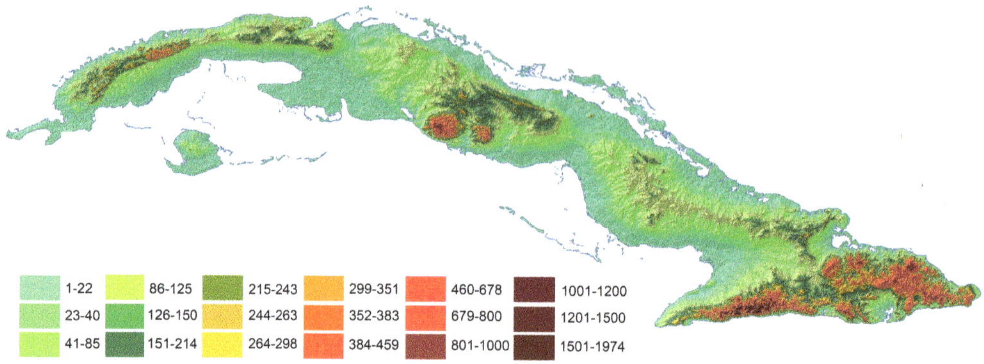

Figure 3.1. Relief map of the Cuban archipelago (m asl). Modified from Castellanos (2005). Courtesy of Enrique A. Castellanos.

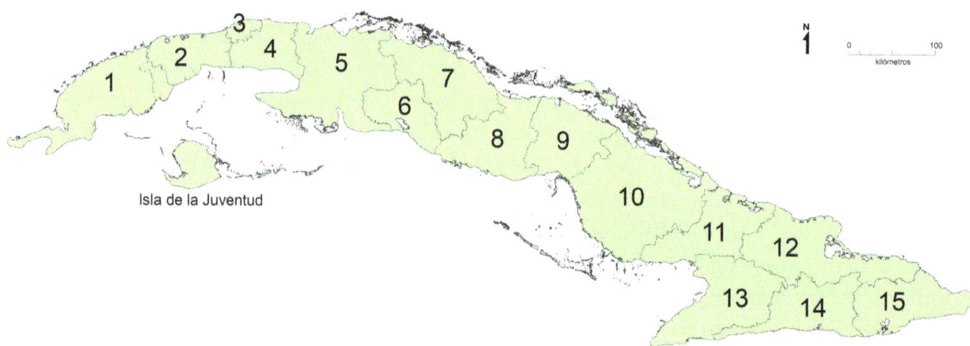

Figure 3.2. Provinces of the Republic of Cuba: (1) Pinar del Río; (2) Artemisa; (3) Havana; (4) Mayabeque; (5) Matanzas; (6) Cienfuegos; (7) Villa Clara; (8) Sancti Spiritus; (9) Ciego de Ávila; (10) Camagüey; (11) Las Tunas; (12) Holguín; (13) Granma; (14) Santiago de Cuba; (15) Guantánamo. Special municipality: Isla de la Juventud (Isle of Youth, formerly Isle of Pines). Map by A. González.

and Eastern. This chapter uses four major regions by considering one of Acevedo's (1989) subdivisions – Camagüey-Maniabón – as a mesoregion, which is consistent with the Central-East region of Mateo *et al.* (2019). A more refined zoogeographic analysis (see section VII. B.) is nonetheless based on the physiographic subdivisions of Acevedo (1989: also see Fig. 3.3). We characterized the major regions briefly as follows:

1. Western

This region covers the provinces Pinar del Río, Artemisa, Havana, Mayabeque, and Matanzas. It contains the Guaniguanico Mountain Range (Pinar del Río and Artemisa Provinces), which is subdivided into Sierra de los Órganos, Sierra del Rosario, and Pizarras Heights (north and south). It is followed in importance by the heights of the north of Havana-Matanzas and the Heights of Mayabeque-Matanzas (or Bejucal-Madruga-Limonar). The highest point in the west (692 m asl) is the Pan de Guajaibón (Sierra del Rosario). The plains extend around the mountain ranges and consist of the karst plain of Guanahacabibes, the alluvial plain of Pinar del Río, the karst plain from Artemisa to Matanzas, and the Plain of Zapata. Offshore to the south lies the Isla de la Juventud, the Los Canarreos Archipelago, and the San Felipe-Los Indios Archipelago.

2. Central

This region contains the provinces Villa Clara, Cienfuegos, Sancti Spiritus, and Ciego de Ávila. The most important mountain range in the region is the Guamuhaya Massif (sometimes referred

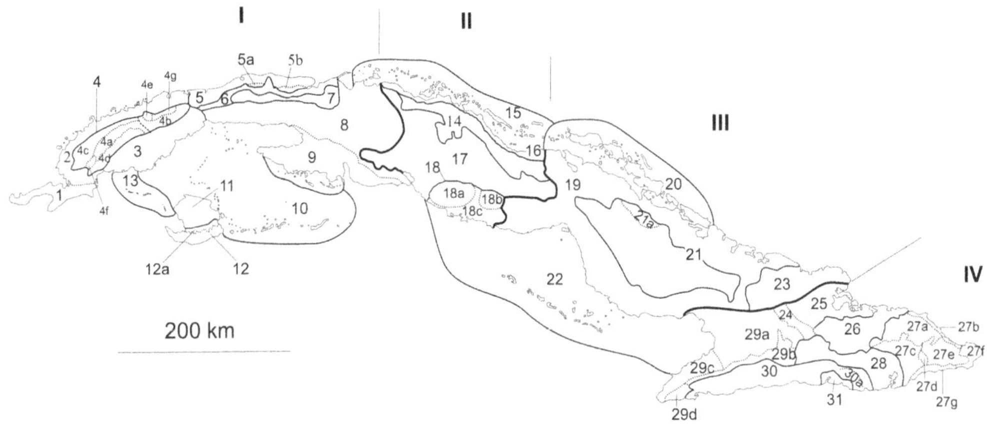

Figure 3.3. Physiographic regions of Cuba and their subdivisions. I: Western region; II: Central region; III: Camagüey-aniabón region; IV: Eastern region. Numbers and lower case are: (1) Península de Guanahacabibes; (2) Llanura norte de Pinar del Río; (3) Llanura sur de Pinar del Río; (4) Cordillera de Guaniguanico; (4a) Sierra de los Órganos; (4b) Sierra del Rosario; (4c) Alturas de Pizarras del Norte; (4d) Alturas de Pizarras del Sur; (4e) Meseta de Cajálbana; (4f) Sierra de Guanes; (4g) Pan de Guajaibón-Sierra Azul; (5) Alturas del norte de La Habana-Matanzas; (5a) Jaruco; (5b) Sierra de Camarones-Pan de Matanzas; (6) Llanuras de Ariguanabo-San Juan; (7) Alturas del centro de la Habana-Matanzas; (8) Llanuras de Artemisa-Colón; (9) Llanuras de Zapata; (10) Archipiélago de los Canarreros; (11) Llanuras y alturas del norte de la Isla de la Juventud; (12) Llanuras del sur de la Isla de la Juventud; (12a) Ciénaga de Lanier; (13) Cayos de San Felipe-Los Indios; (14) Alturas de Cubabacán (alturas del noroeste, del centro, y del este); (15) Archipiélago de Sabana; (16) Llanura Corallilo-Yaguajay; (17) Cubanacán (Manacas, Real Campiña-Cienfuegos, Cubanacán, Santa Clara-Sancti Spiritus); (18) Macizo de Guaymuhaya; (18a) Sierra de Trinidad; (18b) Sierra de Sancti Spiritus; (18c) Llanura de Trinidad-Banao; (19) Llanura periféricas de Camagüey-Maniabón; (20) Archipiélago de Camagüey; (21) Llanuras y alturas centrales de Camagüey; (21a) Sierra de Cubitas; (22) Plataforma insular centro Meridional; (23) Maniabón; (24) Alturas de Báguanos; (25) Llanuras de Nipe; (26) Mayarí (Llanura de Sagua de Tánamo, Montañas de Nipe-Cristal, y Alturas del Segundo Frente); (27) Sagua Baracoa; (27a) Cuchillas de Toya y Moa; (27b) Llanura de Moa-Baracoa; (27c) Meseta del Guaso-Los Montes; (27d)Valle de Puriales de Caujerí; (27e) Cuchillas de Baracoa-Sierra del Purial; (27f) Meseta de Maisi-Zapote; (27g) San Antonio del Sur; (28) Valle Central; (29) Llanuras del Cauto-Guacanayabo; (29a) Llanura del Cauto; (29b) Piedemonte septentrional de la Sierra Maestra; (29c) Llanura de Guacanayabo; (29d) Meseta de Cabo Cruz; (30) Sierra Maestra; (30a) Sierra de la Gran Piedra; (31) Santiago de Cuba (Cuenca y Meseta de Santiago de Cuba). Adapted by Díaz and Cádiz (2008) from Acevedo (1989).

to collectively as Escambray), with its maximum elevation at San Juan Peak (1,156 m asl). Other distinguished mountains are Pico de Potrerillo (931 m asl) and the Loma de Banao (843 m asl). This massif is subdivided into the Sierra de Trinidad and the Sierra de Sancti Spiritus. To the north, the karst belt known as the Alturas de Cubanacán is formed by small mountains and 'mogotes' (mostly karst, limestone hills, although other rocks are also present). The Sierra de Jatibonico stands out at 408 m asl. The rest of the territory consists of undulating plains and low hills.

To the north of the Central region but starting on the eastern limit of Matanzas Province (Western region), and extending to the Camagüey-Maniabón region, is a chain of cays known as the Archipiélago Sabana-Camagüey, or Jardines del Rey. Much of the central-southern island platform has another important group of cays that form the Jardines de la Reina Archipelago.

3. *Camagüey-Maniabón*

This region comprises, totally or partially, the provinces Sancti Spiritus, Ciego de Ávila, Camagüey, Las Tunas, and Holguín. This extensive region is relatively flat, but includes the Sierra de Cubitas, with maximum elevation at the Cerro Tuabaquey (330 m asl), and the Sierra de Najasa, with its highest elevation at the Cerro del Chorrillo (320 m asl). The lower elevations of Maniabón extend between Puerto Padre and Banes to the east. The entire central part of the region is known as the Peniplano Florida-Camagüey-Las Tunas.

4. *Eastern*

The Eastern region contains, totally or partially, the provinces of Holguín, Guantánamo, Granma, and Santiago de Cuba. It is dominated by mountains. The Sierra Maestra extends to the south, from the Cabo Cruz Plateau to the vicinity of Guantánamo Bay (Sierra Larga) and contains the three highest peaks in the country: Turquino (1,974 m asl), Cuba (1,872 m asl), and Suecia (1,734 m asl); its southern slope falls abruptly towards the Bartlett Trench (7,686 m below sea level). To the north, the northern foothills of the Sierra Maestra merge into plains and limestone hills (Carsos de Baire). The plains of Cauto-Guacanayabo extend towards Cabo Cruz farther to the south-west and surround the northern slope of Sierra Maestra.

The Central Valley separates Sierra Maestra from the other eastern mountains: Nipe-Cristal, Alturas del Segundo Frente, and Sagua-Baracoa. These ranges have been referred to as a whole as Sagua-Baracoa or Nipe-Sagua-Baracoa Massif. In general, this complex mountainous system is subdivided into 10–12 orographic regions and, to facilitate the understanding of the rest of this chapter, we will use Nipe-Sagua-Baracoa Massif to cover all of these ranges (Fig. 3.3). The highest elevations in this subregion are the Pico del Cristal (1,231 m asl) and the Pico El Gato (1,181 m asl). The coastal section from the east of Santiago de Cuba to Maisí, the coastline of Cabo Cruz, and some coastal sections to the south of the Sierra Maestra are semi-deserts with marine terraces.

B. Climate

The prevailing climate of Cuba is tropical, warm, and seasonally humid, with a maritime influence and features of semi-continentality. The rainy season is from May to October. Climatic extremes occur in the mountains (humid and colder) and the southern coasts of the Eastern region are dry (Fig. 3.4). The most notable variations in temperature and precipitation are associated with cold fronts, 'souths', hurricanes, and elevational gradients (ONEI 2020).

C. Vegetation

The different vegetation categories in Cuba are shown in Figure 3.5, which is based on the former classification scheme by Capote and Berazaín (1984) and mapped by Estrada *et al.* (2013). The map included herein supports further analyses in this chapter (see section VII. A. *Species richness*).

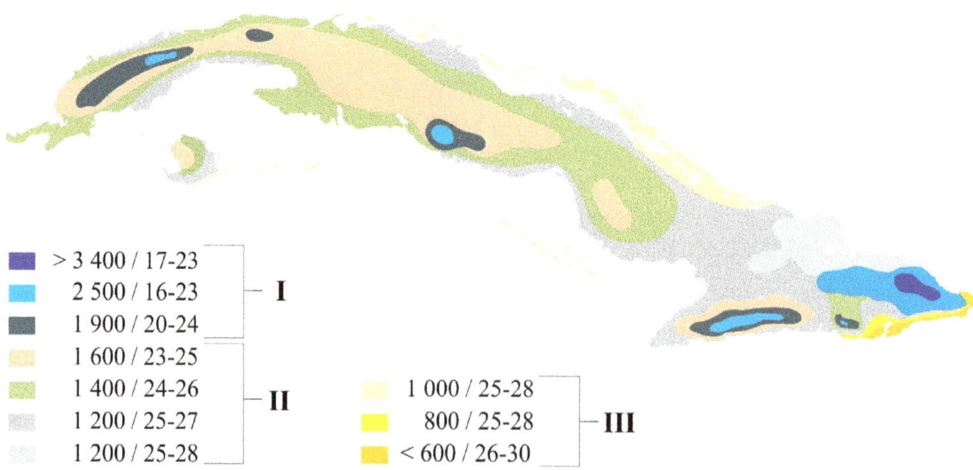

Figure 3.4. Climatic regions of Cuba. Legend data are accumulated rainfall (mm)/temperature (°C). I: Montane climate, with high and stable humidity, low evaporation, and moderate temperatures; II: Plains and heights with relatively stable seasonal humidity, high evaporation, and high temperatures; III: Plains and cays with poor and unstable wetting, high evaporation, and very high temperatures. Map by Díaz and Cádiz (2008) from Díaz Cisneros (1989).

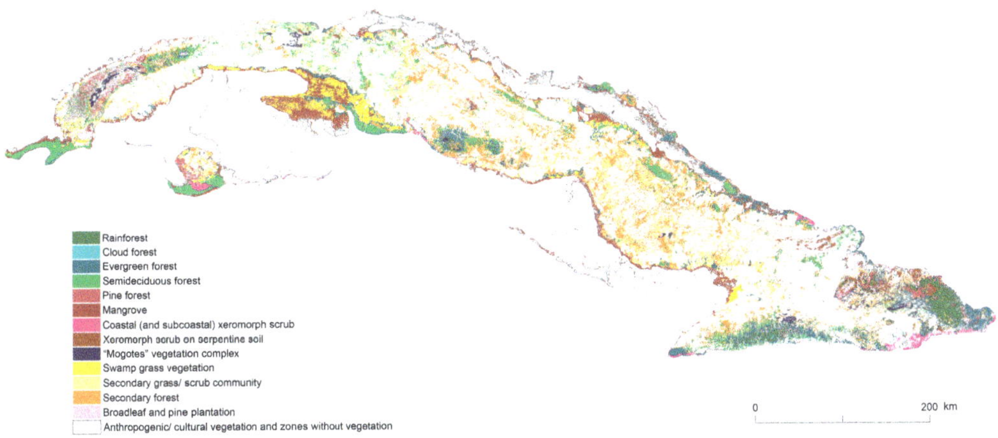

Figure 3.5. Cuban vegetation types, based on Capote and Berazain (1984). Map courtesy of Reinaldo Estrada Estrada.

III. History of discovery and main contributions to the knowledge of Cuban amphibians

A. Chronology of species discovered and described

The description of Cuban amphibians first occurred in the late 1830s (Fig. 3.6) and continued up to the late 1950s; European and North American authors dominated the discovery and description of amphibians. This period of discoveries was initiated by Johann J. von Tschudi, when, in 1838, he named *Peltophryne peltocephala* (formerly *Bufo peltocephalus* [Tschudi, 1838]). In 1841, in their *Erpétologie Générale*, André Marie Constant Duméril and Gabriel Bibron of the Natural History Museum in Paris placed *Hyla septentrionalis* (now *Osteopilus septentrionalis*) as a synonym of *Trachycephalus marmoratus*, a species that Bibron himself included in Ramón Dionisio José de la Sagra y Peris' (1843) *Histoire Physique, Politique et Naturelle de l'Ile de Cuba* (Part Reptilia in a 12-volume series published from 1838 to 1857). Duméril and Bibron also described *Hylodes ricordii* (today *Eleutherodactylus ricordii*) that Alexandre Ricord of the Natural History Museum of Paris collected (Duméril and Bibron 1841). Twenty years later, in 1862, Edward Drinker Cope of the Academy of Natural Sciences of Philadelphia, USA, described *Hylodes dimidiatus, H. auriculatus, H. cuneatus, H. planirostris* (all species currently under the genus *Eleutherodactylus*), and *Phyllobates limbatus* (currently *E. limbatus*); some of the specimens were collected by the botanist Charles Wright (Cope 1862a). Also in 1862, Cope described *Peltophryne empusa* (at the time *Peltaphryne*) from material supplied by Felipe Poey (Cope 1862b). Finally, in 1864, Johann Christoph Gundlach Redberg, a German naturalist based in Cuba, described – in collaboration with Wilhelm C. H. Peters – the last species of amphibian named during the Cuban colonial era: *Hylodes varians* (currently *Eleutherodactylus varians*) (Gundlach and Peters, 1864).

The number of Cuban species of amphibians known by the end of the nineteenth century was only ten (Fig. 3.6). The pace of discovery of new species for science underwent a 30-year hiatus (1868–98) during the wars of liberation from Spain. By the early 1940s, the number of named species had increased to 26. During this period, three species of toads (currently *Peltophryne*) were described: *Bufo longinasus* (Stejneger, 1905), *B. ramsdeni* (Barbour, 1914), and *B. dunni* (Barbour, 1926). The number of known species of the genus *Eleutherodactylus* also increased. Barbour (1914)

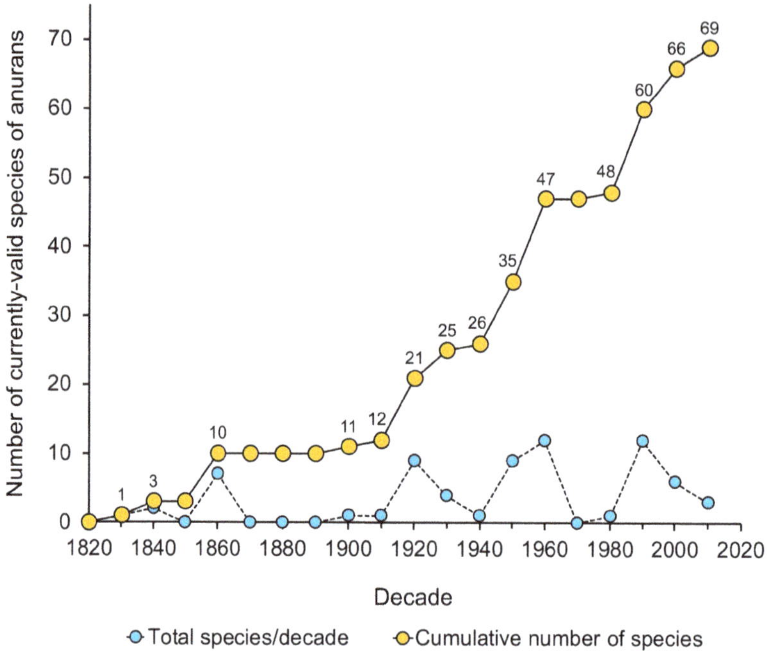

Figure 3.6. Amphibian species discovered and described for Cuba since the 1830s.

described *E. plicatus*, but that name was already given to another species, and Karl Patterson Schmidt (1920) renamed it *E. gundlachi* (Schmidt also described species that were later synonymized, such as *E. brevipalmatus* and *E. sierramaestrae*). Emmett Reid Dunn (1925) described *Eleutherodactylus atkinsi* and *E. varleyi*. The following year, Dunn described five new species, *Eleutherodactylus casparii* (originally *E. planirostris casparii*), *E. eileenae*, *E. emiliae*, *E. greyi*, and *E. pinarensis* (Dunn 1926a). Thomas Barbour and Benjamin Shreve (1937) described five new species, *Eleutherodactylus orientalis* (originally *Sminthillus limbatus orientalis*), *E. turquinensis*, *E. albipes*, *E. intermedius*, and *E. parvus*. Barbour (1942) replaced *E. parvus* with *E. cubanus*, because *parvus* was a name then given to another South American *Eleutherodactylus* (now *Ischnocnema parva* [Girard, 1853]).

From the late 1950s to the mid-1960s the number of known amphibian species in Cuba increased from 26 to 47. The first of these 21 species was described and named by Rodolfo Ruibal, a Cuban-born herpetologist based in the United States. He described a new species of toad originally discovered eight decades earlier by Juan C. Gundlach, *Bufo gundlachi* (Ruibal 1959), today in the genus *Peltophryne*. Then came the numerous contributions of Albert Schwartz, who organized several herpetological expeditions to Cuba. He added 19 new amphibian taxa, including three new species of toads: *Bufo cataulaciceps* (today *Peltophryne* [Schwartz 1959b]), *B. fustiger*, and *B. taladai* (today *Peltophryne* [Schwartz 1960b]). He also added six new *Eleutherodactylus*: *E. symingtoni* Schwartz, 1957; *E. etheridgei* (Schwartz 1958a); *E. zeus* (Schwartz 1958b); *E. varians olibrus* (now *E. olibrus* [Schwartz 1958c]); *E. zugi* Schwartz, 1958(d); *E. thomasi* Schwartz, 1959(c). One of Schwartz's most important contributions was the publication 'Nine new Cuban frogs of the genus *Eleutherodactylus*' (Schwartz 1960a) in which he described *E. acmonis*, *E. bartonsmithi*, *E. bresslerae*, *E. pezopetrus*, *E. ronaldi*, *E. zugi erythroproctus* (now *E. erythroproctus*), *E. planirostris goini* (now

E. goini), *E. varians ionthus* (now *E. ionthus*), *E. varians staurometopon* (now *E. staurometopon*). Schwartz closed this prolific period by describing a new species from the northern slope of the Sierra Maestra, *E. leberi* Schwartz, 1965.

A new hiatus occurred from the mid-1960s to 1980, during which there were no descriptions of new species in Cuba. Main contributions in this period were by Cuban herpetologists from different institutions such as Felipe Poey Museum, the former Institutes of Biology and Zoology, today the Institute of Ecology and Systematics (IES), the Faculty of Biology at the University of Havana, the National Enterprise for the Protection of the Flora and Fauna (Empresa Nacional para la Protección de la Flora y la Fauna), and National Museum of Natural History. In this group of authors are Miguel L. Jaume, Orlando H. Garrido, Mario Buide, and Luis V. Moreno. The latter was a curator of the main Cuban collections of amphibians and reptiles for more than 40 years, an important teacher to several generations of Cuban herpetologists, and one of the most significant collectors of Cuban specimens. In 1980, Alberto Valdés de la Osa of the Zoology division, Institute of Ecology and Systematics (IES), and Fernando N. Ruíz García described a new subspecies, *Bufo longinasus cajalbanensis* (now *P. l. cajalbanensis*, Valdés de la Osa and Ruíz García 1980a). Between 1982 and 1985 Alberto R. Estrada and Julio Novo Rodríguez rediscovered a species originally collected by Alberto Valdés on the Guanahacabibes Peninsula, at the western extreme of Cuba: *Eleutherodactylus guanahacabibes* Estrada and Novo, 1985. In 1987, Estrada also discovered a new species of frog from Cuchillas del Toa, *Eleutherodactylus toa* Estrada and Hedges, 1991. In 1992, three new species of *Eleutherodactylus* were published: *E. guantanamera*, *E. melacara*, and *E. mariposa* (Hedges *et al.* 1992), followed by *E. tetajulia* Estrada and Hedges, 1996(a), *E. iberia* Estrada and Hedges, 1996(b), *E. principalis* Estrada and Hedges, 1997(a), *E. glamyrus* Estrada and Hedges, 1997(b), *E. jaumei* Estrada and Alonso, 1997, and *E. tonyi* Estrada and Hedges, 1997(c). The last discoveries of new Cuban amphibian species in the twentieth century came in 1998 with *Eleutherodactylus blairhedgesi* (Estrada *et al.* 1997) and *E. riparius* Estrada and Hedges, 1998.

The prominence of amphibian research during the twenty-first century passed into the hands of a new generation of Cuban biologists who added to the impetus that their Cuban and foreign predecessors had provided. The total number of valid species discovered and described up to 2010 rose to 66: Luis M. Díaz, of the National Museum of Natural History in Havana, together with Alberto R. Estrada and S. Blair Hedges, described the first new species of this era, *Eleutherodactylus rivularis* (Díaz *et al.* 2001). That same year L. M. Díaz and Ansel Fong Grillo of BIOECO, discovered and named *Eleutherodactylus simulans* Díaz and Fong, 2001. Two years later, in 2003, L. M. Díaz, Antonio Cádiz of the Faculty of Biology of the University of Havana, together with Blair Hedges, published another new species of *Eleutherodactylus* closely related to *E. varleyi*: *E. adelus* (Díaz *et al.* 2003). Another two years would pass until L. M. Díaz, A. Cádiz, and Nils Navarro of the Natural History Museum Carlos de la Torre of Holguín discovered and published a new species from the Sierra Maestra, *Eleutherodactylus maestrensis* (Díaz *et al.* 2005). The same authors discovered another new species of *Eleutherodactylus* in limestone formations on the northern slope of the Sierra Maestra and named it *E. michaelschmidi* (Díaz *et al.* 2007a). The description of a new species of toad closed the first decade of the century: *Peltophryne florentinoi* (Moreno and Rivalta, 2007); both authors were from the IES, Havana.

The second decade of the twenty-first century was also prolific, and the number of valid species described rose to 69. Examples include new species of Cuban anurans, all from the genus *Eleutherodactylus*, including *E. feichtingeri* (Díaz *et al.* 2012) and *E. beguei* Díaz and Hedges, 2015. Ariel Rodríguez and many co-authors are responsible for discovering the most recent Cuban amphibian species, *Eleutherodactylus cattus*, a sister species of *E. glamyrus* from the Sierra Maestra (Rodríguez *et al.* 2017).

Since the first decades of the twentieth century, several species of amphibians not native to Cuba have been introduced. Some have become established as invasive species and are dispersed over a large part of the territory. Fortunately, other introduced species have failed to establish and are not part of the current Cuban fauna. Borroto-Páez *et al.* (2015) summarized the status of the amphibians and reptiles introduced into Cuba. The first species introduced was the American Bullfrog *Lithobates catesbeianus* (formerly *Rana catesbeiana*), which was brought from the United States by Charles Hernández in 1916 and 1917 for an aquacultural project to be developed in the Wajay municipality of Rancho Boyeros, near Havana. This species is present today in almost all freshwater environments of the country: rivers, agricultural irrigation systems, oxidation lagoons, and reservoirs of Cuba and the Isla de la Juventud. It also has been reported from Cayo Coco and Cayo Romano (L. M. Díaz, per. comm., cited by Borroto-Páez *et al.* 2015). The impact of this invasive species is considered significant.

The second introduced species on record is *Rhinella marina* (formerly *Bufo marinus*) (Fig. 3.7). Its first introduction (14 animals) occurred in 1935 from Puerto Rico, to be used in the Agronomic Station of Santiago de Las Vegas, province of Havana, for experiments in the control of insect pests of sugarcane (Bruner 1935). A second introduction (18 animals) took place in 1937 on lands of the Central Lugareño (now Cubitas), in Minas municipality in the province of Camagüey. This introduction was by W. C. Hanson, superintendent of the sugar factory. The same species was introduced for the third time in 1946, this time by Luis Carlos Scaramuzza, who distributed 38 animals among the local sugar cane plantations of Camagüey and Las Mercedes (Borroto-Páez *et al.* 2015). None of these introductions was successful, and there are no records of sightings or specimens deposited in Cuban collections. However, Borroto-Páez *et al.* (2015) drew attention to the existence of specimens collected in Cuba and deposited in the collections of AMNH in New York and Smithsonian National Museum of Natural History in Washington, DC.

Another possible introduction involved *Pseudacris crucifer*, reported for Cuba by Schwartz and Thomas (1975) and Schwartz and Henderson (1985, 1988, 1991). Ruibal and Estrada (1999) dismissed the previous reports, due to lack of sufficient evidence, and did not include it in their list of amphibians for Cuba.

Borroto-Páez *et al.* (2015), citing the National Center for Biological Security in Havana, reported the introduction of *Hoplobatrachus tigerinus* in May 2002 at the Mampostón Aquaculture Center in Havana province. This species was imported from Thailand, but so far there is no evidence that it has become established in the wild; it is used for experimental studies in captivity (Borroto-Páez *et al.* 2015). Powell *et al.* (2011) reported six specimens of hylid frogs of the genus *Osteocephalus*. Borroto-Páez *et al.* (2015) did not find evidence that confirms this introduction, and the current list of Cuban amphibians should not include *Osteocephalus* unless confirmed. The most recently introduced species is *Leptodactylus fragilis* (Rodríguez-Cabrera *et al.* 2018) near Guane, Pinar del Río; its potential impact on native species was analysed by Del Castillo Domínguez *et al.* (2021).

B. Literature review

Although not a comprehensive review of the available literature, this section covers the most notable contributions within a historical context. In all eras, herpetologists have produced highly valued checklists, catalogues, and reviews containing information on taxonomic status,

Figure 3.7. Above: Specimen of *Rhinella marina* at the American Museum of Natural History (AMNH) A45344, collected by Allen Ross near Havana, in May 1937. Below: Subadult individual from the Dominican Republic. Photographs courtesy of AMNH and L. M. Díaz, respectively.

the location of type specimens, series of paratype specimens, distributions, and other data of species in representative groups from a particular region.

Contributions for the period between 1940 and 1980 were made by Pastor Alayo Dalmau, an outstanding Cuban entomologist who published in 1955 the *Lista de anfibios de Cuba*. In this he reported 20 species of anurans then valid: 19 indigenous species and one introduced (Alayo 1955). As of 1966, Miguel L. Jaume, of the newly created Felipe Poey Museum in Havana, published his 'Catálogo de los anfibios de Cuba' and included 55 indigenous species and an introduced one (Jaume 1966). A year later, Mario Buide (1967) published in the Cuban journal *Torreia* the 'Lista de anfibios y reptiles de Cuba' and included 53 species of indigenous anurans and two introduced species.

Albert Schwartz and Richard Thomas (1975) published 'A check-list of West Indian amphibians and reptiles', an exhaustive compilation of the herpetofauna of the West Indies, including very detailed information on holotypes, synonymies, and geographic distribution of all amphibian and reptilian species from Cuba and the rest of the Antillean islands. Later, Schwartz *et al.* (1978) published a supplement to their previous list, which among other data, expanded the distribution of some species.

The Cuban naturalist Orlando H. Garrido with Miguel L. Jaume (1984) published the 'Catálogo descriptivo de los anfibios y reptiles de Cuba'. Albert Schwartz and Robert W. Henderson (1985) of the Milwaukee Public Museum published *A Guide to the Identification of the Amphibians and Reptiles of the West Indies Exclusive of Hispaniola*. The same authors later released 'West Indian amphibians and reptiles: a check-list' (Schwartz and Henderson 1988), a new annotated list of amphibians and reptiles that updated the previous information.

Stephen Blair Hedges and Richard Thomas (1989) updated the distribution records from previous treatment of some Cuban amphibians (e.g. Schwartz and Henderson 1988). *Amphibians and Reptiles of the West Indies: Descriptions, Distributions, and Natural History* by Schwartz and Henderson (1991), perhaps the most complete work of its kind at that point in the twentieth century, is a monumental work including descriptions of species, supplemented by data on natural history and distributional maps of all amphibians and reptiles known from the West Indies at that time.

Robert Powell and Robert W. Henderson edited the book *Contributions to West Indian Herpetology, A Tribute to Albert Schwartz* (Powell and Henderson 1996), a comprehensive work that included more than 30 contributions addressing topics of biogeography, ethnoecology, and conservation of the amphibians and reptiles of the Antilles. That volume was supplemented by 'An annotated checklist of West Indian amphibians and reptiles' (Powell *et al.* 1996).

Rodolfo Ruibal and A. R. Estrada published 'A review of Cuban herpetology', as part of the volume *Caribbean Amphibians and Reptiles*, edited by Brian I. Crother (1999). In this compilation, the Cuban authors outlined a list of 57 species of anurans and included their distribution across the geographic regions of the Cuban archipelago.

Lourdes Rodríguez Schettino, of the IES in Havana, edited *Anfibios y reptiles de* Cuba in 2003. This work presented outstanding pictures of most of the Cuban species (Rodríguez Schettino 2003). Then, Luis M. Díaz and Antonio Cádiz (2008) launched their *Guía Taxonómica de los Anfibios de Cuba*; this treatise is the most comprehensive revision of the anurans of Cuba and includes a complete account of the biology of Cuban species. Vilma Rivalta and others followed and added another list, complemented with distributional maps: 'Amphibians of Cuba: Checklist and Geographic Distributions' (Rivalta González *et al.* 2014). A most recent contribution is a chapter by Alonso Bosch and García-Padrón (2017), a brief compendium of methods for inventories, monitoring, and collection, including a list of the Cuban anuran species.

Since the late 1990s, S. Blair Hedges has kept a website (http://www.caribherp.org) dedicated to reviewing the lists of species of amphibians and reptiles of the Antilles. Recently Hedges *et al.* (2019) published *Definition of the Caribbean Islands Biogeographic Region*, with a checklist and recommendations for standardized common names of amphibians and reptiles. In this contribution, the authors recognized the specific status of *Peltophryne dunni* and *P. ramsdeni*, formerly a subspecies of *P. longinasus*. Unfortunately, the list does not consider any of the local names of Caribbean species.

C. Natural history and evolution

Very little has been published about the ecology and life-history of Cuban amphibians, and the scant literature – primarily generated during the second half of the twentieth century – dealt mainly with reproduction, feeding, and the habitat-use of a few species. The first contributions documenting reproduction are observations by Gundlach (1880) on species of each of the native genera. Others have described the developmental stages of some Cuban species: Barbour (1926, citing field notes provided to him by Emmett R. Dunn) for the larvae of *Peltophryne* (then *Bufo*) *dunni*, Ruíz (1980) and Valdés de la Osa and Ruíz García (1980a) for the tadpoles of *P. longinasus* and *P. l. cajalbanensis*, and Díaz *et al.* (2000) for the larvae of *P. taladai*. Duellman and Schwartz (1958) provided the first data on the quantity and size of eggs and the description of *Osteopilus septentrionalis* tadpoles (but from Florida, USA).

The work of Díaz and Cádiz (2008) documented the larvae of the Cuban amphibians with aquatic reproduction, including a key to identifying tadpoles at the generic level. Notes on the reproductive ecology of *P. peltocephala* were published by Sampedro and Berovides (1982). Some characteristics of breeding habitats of the Cuban species of the family Eleutherodactylidae (eggs with direct development) were documented by Estrada (1987, 1990, 1992) and Novo *et al.* (1987). Other authors who have contributed or compiled data on various aspects of reproduction of Cuban amphibians are Dunn (1926b,c), Goin (1947), Königstedt and Königstedt (1982), Schwartz and Henderson (1991), Estrada and Hedges (1996a,b, 1997b, 1998), Townsend (1996), Díaz *et al.* (2001, 2003), Meshaka (2001), Rodríguez Schettino (2003), Rivalta González *et al.* (2004), Alonso *et al.* (2005), Díaz and Cádiz (2006, 2008), Herrmann (2006), Vargas-Salinas (2006), Sanz-Ochotorena *et al.* (2008, 2009, 2011, 2019), Henderson and Powell (2009), Rodríguez and Alonso (2009), Rodríguez Gómez *et al.* (2009, 2010). Iturriaga *et al.* (2012, 2014), Alonso Bosch *et al.* (2015), Alfonso *et al.* (2017), Iturriaga and Dugo-Cota (2018), and Bignotte-Giró (2019). More recently, Bignotte-Giró *et al.* (2021) studied the reproductive phenology of five endemic *Eleutherodactylus* species that coexist in the montane rainforest of La Gran Piedra, Sierra Maestra.

Studies analysing the feeding habits of Cuban amphibians are scarce in the literature. For *Peltophryne peltocephala*, Sampedro *et al.* (1982) and Sampedro and Berovides (1985) are the primary references on feeding habits. Fong G. and Garcés (2002) listed the items found in three faecal balls of *P. taladai*. The diet of the introduced *Lithobates catesbeianus* was documented by Montañez *et al.* (1996) and more recently by García-Padrón *et al.* (2021b). In *Eleutherodactylus*, Valdés de la Osa and Zayas (1980) were the first authors to analyse the diet of several species; however, their results were not published entirely, and only an abstract is available. More recently, two other articles described the diet of *E. eileenae* (Alonso *et al.* 2001) and *E. atkinsi* (García-Padrón and Borrego Quevedo 2020) in Western Cuba. García-Padrón (2019) summarized the most commonly used methods for dietary analysis in anurans, including some Cuban species. More recently, García-Padrón *et al.* (2021a) and García-Padrón (2022a) reported on the diet of *Eleutherodactylus zeus*. Finally, García-Padrón (2021) provided preliminary observations on the trophic habits of three syntopic species (*Eleutherodactylus atkinsi*, *E. goini*, and *Osteopilus septentrionalis*) occurring in one agroecosystem in Western Cuba. Garcia-Padron *et al.* (2022) described the diet of another introduced anuran, *Leptodactylus debilis*, from suburban ecosystems of Western Cuba. Additionally, scattered observations of dietary components of Cuban anurans have been reported by various authors (e.g. Schwartz 1958d; Armas 1987, 2014; Díaz and Fong 2001; Díaz *et al.* 2001, 2005, 2007a; Alonso and Rodríguez 2003c; Mancina *et al.* 2016; Borroto-Páez and Reyes Pérez 2020; Portal Ríos *et al.* 2020; García Padrón and Rodríguez-Cabrera 2022; García-Padrón and Borrego Quevedo 2022). Cannibalism was reported by Valdés de la Osa and Ruíz García (1980b), Sampedro *et al.* (1985), Crump (1986), Herman *et al.* (2015), Kaiser *et al.* (2016), and Borroto-Páez and Reyes Pérez (2019).

Vertebrate predators have been documented by Torres *et al.* (1985), Abreu *et al.* (1988), Schwartz and Henderson (1991), Henderson and Sajdak (1996), Estrada *et al.* (1997), Hedges *et al.* (1999), Rodríguez Schettino (1999), Fong (2000), Meshaka (2001), Alonso and Rodríguez (2003b,c), Fong G. (2004), Fong G. *et al.* (2009), Henderson and Powell (2009), Alfonso *et al.* (2013), Fong G. *et al.* (2013), Díaz *et al.* (2014), Rodríguez-Cabrera and Rodríguez-Silva (2015), Rodríguez-Machado and Rodríguez-Cabrera (2015), Armas and Iturriaga (2017), Castellón Maure and Rodríguez-Cabrera (2018), Rodríguez-Cabrera (2018), López-Hurtado *et al.* (2020), Rodríguez-Cabrera *et al.* (2020a,b,c) and Rodríguez-Cabrera and Hernández Gómez (2021). Finally, invertebrate predators upon Cuban amphibians have been documented by Schwartz (1958b), Novo *et al.* (1985), Díaz (1999), Alonso and Rodríguez (2003b,c), and Fong G. *et al.* (2012).

Bioacoustics have played an essential role in discovering and describing new species of amphibians from Cuba (Díaz and Cádiz 2008; Alonso Bosch 2013). William Franklin Blair was the first in describing the vocalisations of a Cuban amphibian species, *Osteopilus septentrionalis*, although from an introduced population in Florida (Blair 1958). Later, Duellman and Crombie (1970) depicted the sonogram of some calls of this frog. However, Albert Schwartz was the first author to use a sonogram as a tool to show the differences between *Peltophryne cataulaciceps* and *P. gundlachi* (Schwartz 1959c). Three decades after Schwartz, Hedges *et al.* (1992) published the acoustic emission sonograms of ten Cuban species of the genus *Eleutherodactylus* (including three new species). Acoustic analyses also were decisive in the rediscovery of *E. turquinensis* by Hedges *et al.* (1995). In the following years, most descriptions of new species of Cuban amphibians have included temporal and spectral analyses of acoustic emissions (Estrada and Hedges 1996a,b, 1997a,b,c, 1998; Estrada *et al.* 1997; Díaz and Fong 2001; Díaz *et al.* 2001, 2003, 2005, 2007a, 2012; Díaz and Cádiz 2007; Díaz and Hedges 2015; Rodríguez *et al.* 2017). Other studies describing species' vocalizations and acoustic activity were those of Díaz and Estrada (2000), Alonso Bosch and Rodríguez Gómez (2001, 2003a), Alonso and Rodríguez (2005), Hernández *et al.* (2010), Rodríguez *et al.* (2010a), Alonso Bosch (2012), Alonso Bosch *et al.* (2017), and Bignotte-Giró *et al.* (2019). Alonso Bosch (2013) summarized the use of vocalizations in the studies of Cuban anurans and outlined some possible research topics for the future. Also in the field of bioacoustics, two contributions stand out: *Guía Sonora de los Anfibios de Cuba* by Alonso *et al.* (2007), which included a CD with the voices of 55 species, and the *Guía descriptiva para la identificación de las llamadas de anuncio de las ranas del género Eleutherodactylus* by Díaz and Cádiz (2007), with a synopsis of the main features of spectral and temporary calls of 49 species. In 2008 Díaz and Cádiz provided a CD with the voices of more than 55 species (in 86 tracks) and updated previously published information. These authors provided oscillograms, sonograms, and descriptions of the calls of more than 95% of the valid Cuban amphibian species to date.

Other contributions address various aspects of the biology of Cuban amphibians. Chytridiomycosis was documented for Cuban amphibians by Díaz *et al.* (2007b), Sabino-Pinto *et al.* (2017), and Cádiz *et al.* (2018). Malformations were reported by García-Padrón and Alonso Bosch (2017) in adult *Osteopilus septentrionalis*; anomalies in the mouth of tadpoles (not due to chytrid fungus) were found in *Peltophryne fustiger* (Alonso Bosch *et al.* 2017); eye malformations in *Peltophryne taladai* were reported by Alonso Bosch and Hernández Marrero (2020), and limb abnormalities in *P. florentinoi* by Alonso Bosch *et al.* (2021). Anomalous colouration (piebaldism) was recorded in *Eleutherodactylus zeus* (García-Padrón and Alonso Bosch 2019). Blood parasites were described by Zajicek and Mauri Méndez (1969). The parasitic helminths of amphibians were treated by Pérez Vigueras (1938), Baruš and Moravec (1967), Odening (1968), Baruš (1972, 1973), Coy (1980), Coy and Baruš (1982), Coy and Hernández (1982), Martínez *et al.* (1982), Coy and Ventosa (1984), Coy Otero and Martínez (1987), and Meshaka (1996). Acari were studied by Černý (1969), De la Cruz (1971), Guglielmone *et al.* (2003), and Barros-Battesti *et al.* (2009).

Other authors have documented the use of habitat and the population dynamics of some species (e.g. Dunn 1926b,c; Schwartz and Henderson 1991; Fong G. 1999, 2007, 2009; Alonso *et al.* 2001; Meshaka 2001; Rodríguez Schettino 2003; Fong G. and Hero 2006; Díaz and Cádiz 2008; Henderson and Powell 2009; Fong G. *et al.* 2010). Microhabitat-use, locomotory patterns, or defensive behaviours have been treated by Valdés de la Osa (1978, 1989), Valdés de la Osa and Ruíz García (1979, 1980a), Alonso *et al.* (2009), Iturriaga and Homar (2014), and Alonso Bosh and Hernández Marrero (2021).

Chromosomal studies were first conducted by Bogart (1972, 1981), who also provided valuable information on *Eleutherodactylus* later used by colleagues for naming cryptic species (Hedges *et al.* 1992). However, the most comprehensive review of the chromosomes of *Eleutherodactylus* and other related Terraranan frogs was by Schmid *et al.* (2010), while Ma and Veltsos (2021) documented the sex chromosomes of several Cuban anurans in a phylogenetic context, highlighting *E. caspari, E. glamyrus,* and *E. turquinensis* as Cuban species having transitional sex-determination mechanisms. The karyotypes of frogs in the genus *Osteopilus* were studied by Anderson (1996). Updated comparisons of chromosomes of hylids (Hylidae: Hylinae: Lophyohylini) were provided by Suárez *et al.* (2020). The list of authors who in recent decades have addressed genetic-molecular studies of Antillean amphibians is long; among others, the contributions of S. B. Hedges, Carla Hass, Jennifer B. Pramuk, Matthew P. Heinicke, Ariel Rodríguez, and Roberto Alonso on hypothetical and biogeographic phylogeny are noteworthy (Hedges 1989, 1996; Hass and Hedges 1991; Pramuk *et al.* 2001, 2007; Pramuk 2002, 2006; Heinicke *et al.* 2007, 2011; Rodríguez *et al.* 2011a, 2012, 2017; Alonso *et al.* 2012; Duellman *et al.* 2016; Dugo-Cota *et al.* 2019; Blotto *et al.* 2020). Rivalta *et al.* (2009) analysed the genetic distance of *Peltophryne florentinoi, P. fustiger* and *P. peltocephala* based on enzymatic and protein variability. Some contributions have focused on the studies of biochemistry and skin toxicity of some amphibian species (Cei *et al.* 1972; Delfino *et al.* 2002; Rodríguez *et al.* 2011b, 2013; Perera Córdova *et al.* 2016; Samgina *et al.* 2021). O'Donohoe *et al.* (2019) included two Cuban toads, *Peltophryne empusa* and *P. peltocephala,* in the revision of the morphology and evolution of the paratoid macroglands in Bufonidae. In *Osteopilus septentrionalis,* Bogan and Mitchell (2017) studied the activity of tissue enzymes, and Langowski *et al.* (2019) made a comparative and functional analysis of the digital mucus glands and secretions. Shilton *et al.* (2001) documented the deposition of lipid in the cornea of this species.

Several authors used osteological approaches to address taxonomic revisions of some genera of the families Hylidae, Bufonidae, and Eleutherodactylidae. These contributions changed the generic status of several Cuban species and significantly increased our knowledge about their evolution and phylogenetic relationships. Some of these contributions are by Ivor Griffiths, of Birkbeck College, University of London, who conducted a study of the skeleton, skeletal musculature, and the internal organs of *Eleutherodactylus limbatus* (formerly *Sminthillus limbatus*). He transferred *limbatus* from the family Dendrobatidae to the Leptodactylidae (Griffiths 1959), which is now in Eleutherodactylidae according to subsequent contributions. Linda Trueb and Michael Tyler studied the morphology of the skull of the Antillean components of the Hylidae and established a new combination for the name of the family's only Cuban representative, *Osteopilus septentrionalis* (Trueb 1966; Trueb and Tyler 1974). Gregory K. Pregill (1981) revised the status of the genus *Bufo* and resurrected the genus *Peltophryne* for the indigenous bufonids of the Antilles with the description of osteological traits. Pramuk *et al.* (2001, 2007), Pramuk (2002, 2006), Alonso Bosch (2012), and Alonso *et al.* (2012) provided multidisciplinary supporting evidence for the recognition of *Peltophryne*. The osteology of *Eleutherodactylus* was partially reviewed by Lynch (1971), including some Cuban species, and the state of several osteological characters was listed by Joglar (1986). Cranial osteology and the hyolaryngeal skeleton were used as taxonomic characters in the description of new species of Cuban *Eleutherodactylus* (e.g. Díaz *et al.* 2001, 2003, 2005).

IV. Native anurans and their conservation status

In the following account, and later, in section VII., we recognize four patterns of distribution of Cuban amphibians (Table 3.1). These are (1) widely distributed species, pan-Cuban or quasi-pan-Cuban, either with a continuous or fragmented distribution in suitable habitats throughout most

Table 3.1. List of the native amphibians of Cuba, their population status, and general information of species distribution. IUCN red list categories: DD = Data Deficient; LC = Least Concern; NT = Near Threatened; VU = Vulnerable; EN = Endangered; CR = Critically Endangered. EOO = extent of occurrence estimated following the IUCN criteria by defining one or more polygons connecting known external localities; some data herein offered may differ from the calculation in the official Red List. Zoo. reg. = zoogeographic regions (see Fig. 3.36), which are **W** (Western), **C** (Central), and **E** (Eastern). Distribution (Dist.) patterns according to text: 1 = wide distribution, 2 = mesolocalized distribution, 3 = regional distribution, 4 = microlocalized distribution. * indicates status given based on the taxonomic criteria of four subspecies distributed in different mountain ranges; currently, three of the subspecies have been elevated to the species level (Hedges *et al.* 2019; also supported herein), and the distributional range of each species is much reduced besides their respective local threats. ** = EOO exceeding 20,000 km^2. Species with EOO = 1 km^2 are only known from the type locality but very likely have a wider distribution. — = no data; NA = not assessed. Superscripts (code of a letter and number) under Population Status correspond to citations of published assessments (bottom of the table) in References.

	POPULATION STATUS			GENERAL INFORMATION OF SPECIES DISTRIBUTION			
FAMILY AND *SPECIES*	**GLOBAL IUCN RED LIST (2004–2014)**	**CUBAN RED BOOK (2012)**	**GLOBAL IUCN RED LIST (IN PRESS)**	**ZOO. REG.**	**EOO (KM^2)**	**ELEVATIONAL DISTRIBUTION (M ASL)**	**DIST. PATTERN**
Bufonidae (10 spp.)							
Peltophryne cataulaciceps	EN [a1]	EN [b1]	EN [c1]	W	2,428	0–70	4
Peltophryne dunni	—	—	EN [d1]	C	1,886	400–900	4
Peltophryne empusa	VU [e1]	NA	LC [f1]	W-C-E	**	0–70	1
Peltophryne florentinoi	CR [g1]	VU [h1]	EN [i1]	W-C	82	0–10	4
Peltophryne fustiger	LC [j1]	NA	LC [k1]	W	**	0–650	3
Peltophryne gundlachi	VU [l1]	NA	LC [m1]	W-C-E	**	0–70	1
Peltophryne longinasus	* EN [n1]	EN [o1]	CR [p1]	W	1,564	100–430	4
Peltophryne peltocephala	LC [q1]	NA	LC [r1]	W-C-E	**	0–800	2
Peltophryne ramsdeni	—	—	CR (PE)(D) [s1]	E	34	800	4
Peltophryne taladai	VU [t1]	NA	LC [u1]	C-E	**	0–850	2
Eleutherodactylidae (58 spp.)							
Eleutherodactylus acmonis	EN [v1]	VU [w1]	CR [x1]	E	4	100–560	4
Eleutherodactylus adelus	EN [y1]	VU [z1]	EN [a2]	W	130	130–220	4
Eleutherodactylus albipes	CR [b2]	VU [c2]	CR [d2]	E	75	1,200–1,974	4
Eleutherodactylus atkinsi	LC [e2]	NA	LC [f2]	W-C-E	**	0–1,212	1
Eleutherodactylus auriculatus	LC [g2]	NA	LC [h2]	W-C-E	**	0–1,316	1
Eleutherodactylus bartonsmithi	CR [i2]	VU [j2]	CR [k2]	E	180	30–212	4
Eleutherodactylus beguei	—	—	CR [l2]	E	1	678	4
Eleutherodactylus blairhedgesi	CR [m2]	CR [n2]	CR [o2]	W	4	0–100	4
Eleutherodactylus bresslerae	CR [p2]	VU [q2]	EN [r2]	E	45	30–212	4

(Continued)

Table 3.1 (Continued)

FAMILY AND *SPECIES*	POPULATION STATUS			GENERAL INFORMATION OF SPECIES DISTRIBUTION			
	GLOBAL IUCN RED LIST (2004–2014)	CUBAN RED BOOK (2012)	GLOBAL IUCN RED LIST (IN PRESS)	ZOO. REG.	EOO (KM^2)	ELEVATIONAL DISTRIBUTION (M ASL)	DIST. PATTERN
Eleutherodactylus casparii	EN [s2]	NA	EN [t2]	C	2,365	300–830	4
Eleutherodactylus cattus	—	—	EN [u2]	E	1	800–1,070	4
Eleutherodactylus cubanus	CR [v2]	VU [w2]	EN [x2]	E	237	800–1,830	4
Eleutherodactylus cuneatus	LC [y2]	NA	LC [z2]	E	**	0–1,970	3
Eleutherodactylus dimidiatus	NT [a3]	NA	LC [b3]	W-C-E	**	0–1,375	1
Eleutherodactylus eileenae	NT [c3]	NA	LC [d3]	W-C	**	0–900	2
Eleutherodactylus emiliae	EN [e3]	VU [f3]	EN [g3]	C	2,788	350–1,026	4
Eleutherodactylus erythroproctus	—	—	EN [h3]	W	49	80–200	4
Eleutherodactylus etheridgei	EN [i3]	VU [j3]	VU [k3]	E	1,260	0–151	4
Eleutherodactylus feichtingeri	—	—	LC [l3]	C-E	**	0–900	2
Eleutherodactylus glamyrus	EN [m3]	VU [n3]	EN [o3]	E	279	800–1,972	4
Eleutherodactylus goini	VU [p3]	NA	LC [q3]	W	4,295	70–690	4
Eleutherodactylus greyi	EN [r3]	NA	VU [s3]	C	4,766	30–1,140	3
Eleutherodactylus guanahacabibes	EN [t3]	VU [u3]	LC [v3]	W	909	0–20	4
Eleutherodactylus guantanamera	VU [w3]	NA	VU [x3]	E	8,033	60–1,150	3
Eleutherodactylus gundlachi	EN [y3]	NA	VU [z3]	E	1,964	200–1,970	3
Eleutherodactylus iberia	CR [a4]	VU [b4]	VU [c4]	E	835	0–950	4
Eleutherodactylus intermedius	EN [d4]	NA	EN [e4]	E	709	800–1,970	4
Eleutherodactylus ionthus	EN [f4]	NA	VU [g4]	E	5,590	0–1,250	3
Eleutherodactylus jaumei	CR [h4]	VU [i4]	CR [j4]	E	24	200–950	4
Eleutherodactylus klinikowskii	EN [k4]	NA	EN [l4]	W	3,778	70–200	4
Eleutherodactylus leberi	EN [m4]	VU [n4]	CR [o4]	E	83	138–753	4
Eleutherodactylus limbatus	VU [p4]	NA	LC [q4]	W-C-E	**	50–1,150	1
Eleutherodactylus maestrensis	DD [r4]	VU [s4]	EN [t4]	E	220	900–1,640	4
Eleutherodactylus mariposa	CR [u4]	VU [v4]	CR [w4]	E	88	720–793	4
Eleutherodactylus melacara	EN [x4]	VU [y4]	EN [z4]	E	263	840–1,974	4

(Continued)

Table 3.1 (Continued)

FAMILY AND *SPECIES*	POPULATION STATUS			GENERAL INFORMATION OF SPECIES DISTRIBUTION			
	GLOBAL IUCN RED LIST (2004–2014)	CUBAN RED BOOK (2012)	GLOBAL IUCN RED LIST (IN PRESS)	ZOO. REG.	EOO (KM²)	ELEVATIONAL DISTRIBUTION (M ASL)	DIST. PATTERN
Eleutherodactylus michaelschmidi	EN [a5]	VU [b5]	CR [c5]	E	65	160–600	4
Eleutherodactylus olibrus	—	—	VU [d5]	W	—	0–440	3
Eleutherodactylus orientalis	CR [e5]	VU [f5]	CR [g5]	E	4	100–560	4
Eleutherodactylus pezopetrus	CR [h5]	VU [i5]	EN [j5]	E	75	100–800	4
Eleutherodactylus pinarensis	EN [k5]	NA	VU [l5]	W	10,320	0–380	3
Eleutherodactylus planirostris	LC [m5]	NA	LC [n5]	W-C-E	**	0–720	1
Eleutherodactylus principalis	EN [o5]	NA	EN [p5]	E	1,046	300–1,000	4
Eleutherodactylus ricordii	VU [q5]	NA	VU [r5]	E	11,240	0–1,150	3
Eleutherodactylus riparius	LC [s5]	NA	LC [t5]	W-C-E	**	0–830	1
Eleutherodactylus rivularis	CR [u5]	VU [v5]	CR [w5]	E	52	80–240	4
Eleutherodactylus ronaldi	VU [x5]	NA	NT [y5]	E	10,720	0–1,060	3
Eleutherodactylus simulans	EN [z5]	NA	VU [a6]	E	3,398	0–450	3
Eleutherodactylus staurometopon	—	—	EN [b6]	W	—	20–200	4
Eleutherodactylus symingtoni	CR [c6]	EN [d6]	CR [e6]	W	2,777	70–450	3
Eleutherodactylus tetajulia	CR [f6]	VU [g6]	EN [h6]	E	456	293–900	4
Eleutherodactylus thomasi	EN [i6]	NA	NT [j6]	W-C-E	**	0–390	2
Eleutherodactylus toa	EN [k6]	NA	EN [l6]	E	5,060	0–850	3
Eleutherodactylus tonyi	CR [m6]	VU [n6]	VU [o6]	E	9,205	0–192	3
Eleutherodactylus turquinensis	CR [p6]	VU [q6]	CR [r6]	E	147	450–1,830	4
Eleutherodactylus varians	VU [s6]	NA	LC [t6]	W-C-E	**	0–845	2
Eleutherodactylus varleyi	LC [u6]	NA	LC [v6]	W-C	**	0–800	2
Eleutherodactylus zeus	EN [w6]	NA	EN [x6]	W	3,792	70–500	4
Eleutherodactylus zugi	EN [y6]	NA	EN [z6]	W	4,077	130–600	3
Hylidae (1 sp.)							
Osteopilus septentrionalis	LC [a7]	NA	LC [b7]	W-C-E	**	0–1,974	1

[a1] Hedges and Díaz (2004mm); [b1] Rivalta González (2012a); [c1] IUCN SSC Amphibian Specialist Group (*in press*, a); [d1] IUCN SSC Amphibian Specialist Group (*in press*, b); [e1] Hedges and Díaz (2004nn); [f1] IUCN SSC Amphibian Specialist Group (*in press*, c); [g1] Rivalta (2008); [h1] Rivalta González (2012b); [i1] IUCN SSC Amphibian Specialist Group (*in press*, d);

Table 3.1 (Continued)

[j1] IUCN SSC Amphibian Specialist Group (2014c); [k1] IUCN SSC Amphibian Specialist Group (*in press*, e); [l1] Hedges and Díaz (2004oo); [m1] IUCN SSC Amphibian Specialist Group (*in press*, f); [n1] Hedges and Díaz (2004pp); [o1] Rivalta González (2012c); [p1] IUCN SSC Amphibian Specialist Group (*in press*, g); [q1] Hedges and Díaz (2010i); [r1] IUCN SSC Amphibian Specialist Group (*in press*, h); [s1] IUCN SSC Amphibian Specialist Group (*in press*, i); [t1] Hedges and Díaz (2004qq); [u1] IUCN SSC Amphibian Specialist Group (*in press*, j); [v1] Hedges and Díaz (2004a); [w1] Rodríguez (2012b); [x1] IUCN SSC Amphibian Specialist Group (*in press*, k); [y1] Hedges and Díaz (2004b); [z1] Rodríguez and García (2012a); [a2] IUCN SSC Amphibian Specialist Group (*in press*, l); [b2] Hedges and Díaz (2004c); [c2] Rodríguez (2012c); [d2] IUCN SSC Amphibian Specialist Group (*in press*, m); [e2] Hedges and Díaz (2004d); [f2] IUCN SSC Amphibian Specialist Group (*in press*, n); [g2] Hedges and Díaz (2010a); [h2] IUCN SSC Amphibian Specialist Group (*in press*, o); [i2] Hedges and Díaz (2004e); [j2] Rodríguez (2012d); [k2] IUCN SSC Amphibian Specialist Group (*in press*, p); [l2] IUCN SSC Amphibian Specialist Group (*in press*, q); [m2] Hedges and Díaz (2004f); [n2] Rodríguez (2012e); [o2] IUCN SSC Amphibian Specialist Group (*in press*, r); [p2] Hedges and Díaz (2004g); [q2] Rodríguez (2012f); [r2] IUCN SSC Amphibian Specialist Group (*in press*, s); [s2] Hedges and Díaz (2004h); [t2] IUCN SSC Amphibian Specialist Group (*in press*, t); [u2] IUCN SSC Amphibian Specialist Group (*in press*, u); [v2] Hedges and Díaz (2004i); [w2] Rodríguez (2012g); [x2] IUCN SSC Amphibian Specialist Group (*in press*, v); [y2] IUCN SSC Amphibian Specialist Group (2014a); [z2] IUCN SSC Amphibian Specialist Group (*in press*, w); [a3] Hedges and Díaz (2010b); [b3] IUCN SSC Amphibian Specialist Group (*in press*, x); [c3] Hedges and Díaz (2004j); [d3] IUCN SSC Amphibian Specialist Group (*in press*, y); [e3] Hedges and Díaz (2004k); [f3] Rodríguez (2012h); [g3] IUCN SSC Amphibian Specialist Group (*in press*, z); [h3] IUCN SSC Amphibian Specialist Group (*in press*, aa); [i3] Hedges and Díaz (2010c); [j3] Rodríguez and Urgellés (2012); [k3] IUCN SSC Amphibian Specialist Group (*in press*, bb); [l3] IUCN SSC Amphibian Specialist Group (*in press*, cc); [m3] Hedges and Díaz (2004l); [n3] Rodríguez (2012i); [o3] IUCN SSC Amphibian Specialist Group (*in press*, dd); [p3] Hedges and Díaz (2004m); [q3] IUCN SSC Amphibian Specialist Group (*in press*, ee); [r3] Hedges and Díaz (2004n); [s3] IUCN SSC Amphibian Specialist Group (*in press*, ff); [t3] Hedges and Díaz (2004o); [u3] Rodríguez (2012j); [v3] IUCN SSC Amphibian Specialist Group (*in press*, gg); [w3] Hedges and Díaz (2004p); [x3] IUCN SSC Amphibian Specialist Group (*in press*, hh); [y3] Hedges and Díaz (2010d); [z3] IUCN SSC Amphibian Specialist Group (*in press*, ii); [a4] Hedges and Díaz (2004q); [b4] Rodríguez (2012k); [c4] IUCN SSC Amphibian Specialist Group (*in press*, jj); [d4] Hedges and Díaz (2010e); [e4] IUCN SSC Amphibian Specialist Group (*in press*, kk); [f4] Hedges and Díaz (2010f); [g4] IUCN SSC Amphibian Specialist Group (*in press*, ll); [h4] Hedges and Díaz (2004r); [i4] Rodríguez (2012l); [j4] IUCN SSC Amphibian Specialist Group (*in press*, mm); [k4] Hedges and Díaz (2004s); [l4] IUCN SSC Amphibian Specialist Group (*in press*, nn); [m4] Hedges and Díaz (2004t); [n4] Rodríguez (2012m); [o4] IUCN SSC Amphibian Specialist Group (*in press*, oo); [p4] Hedges and Díaz (2010g); [q4] IUCN SSC Amphibian Specialist Group (*in press*, pp); [r4] Stuart (2006); [s4] Rodríguez (2012n); [t4] IUCN SSC Amphibian Specialist Group (*in press*, qq); [u4] Hedges and Díaz (2004u); [v4] Rodríguez (2012o); [w4] IUCN SSC Amphibian Specialist Group (*in press*, rr); [x4] Hedges and Díaz (2004v); [y4] Rodríguez (2012p); [z4] IUCN SSC Amphibian Specialist Group (*in press*, ss); [a5] Díaz (2008); [b5] Rodríguez (2012q); [c5] IUCN SSC Amphibian Specialist Group (*in press*, tt); [d5] IUCN SSC Amphibian Specialist Group (*in press*, uu); [e5] Hedges and Díaz (2004w); [f5] Rodríguez (2012r); [g5] IUCN SSC Amphibian Specialist Group (*in press*, vv); [h5] Hedges and Díaz (2004x); [i5] Rodríguez (2012s); [j5] IUCN SSC Amphibian Specialist Group (*in press*, ww); [k5] Hedges and Díaz (2004y); [l5] IUCN SSC Amphibian Specialist Group (*in press*, xx); [m5] IUCN SSC Amphibian Specialist Group (2021a); [n5] IUCN SSC Amphibian Specialist Group (*in press*, yy); [o5] Hedges and Díaz (2004z); [p5] IUCN SSC Amphibian Specialist Group (*in press*, zz); [q5] Hedges and Díaz (2010h); [r5] IUCN SSC Amphibian Specialist Group (*in press*, aaa); [s5] IUCN SSC Amphibian Specialist Group (2014b); [t5] IUCN SSC Amphibian Specialist Group (*in press*, bbb); [u5] Hedges and Díaz (2004aa); [v5] Rodríguez (2012t); [w5] IUCN SSC Amphibian Specialist Group (*in press*, ccc); [x5] Hedges and Díaz (2004bb); [y5] IUCN SSC Amphibian Specialist Group (*in press*, ddd); [z5] Díaz (2004); [a6] IUCN SSC Amphibian Specialist Group (*in press*, eee); [b6] IUCN SSC Amphibian Specialist Group (*in press*, fff); [c6] Hedges and Díaz (2004cc); [d6] Rodríguez and García (2012b); [e6] IUCN SSC Amphibian Specialist Group (*in press*, ggg); [f6] Hedges and Díaz (2004dd); [g6] Rodríguez (2012u); [h6] IUCN SSC Amphibian Specialist Group (*in press*, hhh); [i6] Hedges and Díaz (2004ee); [j6] IUCN SSC Amphibian Specialist Group (*in press*, iii); [k6] Hedges and Díaz (2004ff); [l6] IUCN SSC Amphibian Specialist Group (*in press*, jjj); [m6] Hedges and Díaz (2004gg); [n6] Rodríguez (2012v); [o6] IUCN SSC Amphibian Specialist Group (*in press*, kkk); [p6] Hedges and Díaz (2004hh); [q6] Rodríguez (2012w); [r6] IUCN SSC Amphibian Specialist Group (*in press*, lll); [s6] Hedges and Díaz (2004ii); [t6] IUCN SSC Amphibian Specialist Group (*in press*, mmm); [u6] Hedges and Díaz (2004jj); [v6] IUCN SSC Amphibian Specialist Group (*in press*, nnn); [w6] Hedges and Díaz (2004kk); [x6] IUCN SSC Amphibian Specialist Group (*in press*, ooo); [y6] Hedges and Díaz (2004ll); [z6] IUCN SSC Amphibian Specialist Group (*in press*, ppp); [a7] Hedges *et al.* (2008b); [b7] IUCN SSC Amphibian Specialist Group (2021b).

of the island; (2) meso-localized distribution that comprises several localities, extensively in more than one of the defined regions, thus covering an appreciable part of the country, but absent in large areas at the regional scale; (3) regional distribution, with occurrence at several localities covering distant points throughout one defined region; and (4) micro-localized, a relatively restricted distribution within a region (or slightly extended to a contiguous one in a specific kind of habitat), usually by concentrating known localities within a single, restricted ecosystem or a single mountain range. When we refer to zoogeographic regions (see section VII. B., below) these are not equivalent to the four physiographic regions of Figure 3.3 (see also Table 3.1), and we define zoogeographic regions as Western, Central, and Eastern fauna. The conservation status (and related analyses) follow the IUCN's Global Amphibian Assessments (GAA) categories and criteria. A first GAA (GAA1) was completed in 2004, and results from a second GAA were concluded in 2021 (GAA2), but they have not been published by the time of this writing (see section VIII. A.); thus, species assessments according to the most recent IUCN Red List categories and criteria are referred to as "in press" in the main text. Herein, IUCN categories are as follows: CR, Critically Endangered; EN, Endangered; VU, Vulnerable; NT, Near Threatened; LC, Least Concern.

A. Genus *Peltophryne* (Bufonidae)

1. General description

Cuba is the hotspot of this endemic genus of the Caribbean Islands, with 10 of the 14 recognized species (Hedges *et al.* 2019) (Table 3.1). Díaz and Cádiz (2008), Alonso Bosch (2012), and Alonso *et al.* (2012) provided taxonomic revisions. Detailed morphological descriptions were presented by Schwartz (1959b, 1960b), Ruibal (1959), Moreno (1969), Valdés de la Osa and Ruíz García (1980a), and Moreno and Rivalta (2007). Larvae of almost all species, except *Peltophryne ramsdeni*, were briefly reviewed and illustrated by Díaz and Cádiz (2008) from tadpoles obtained in the wild and not captive-reared (as argued by Rodríguez 2009). Their work partly built upon previous work by Dunn (in Barbour 1926), Valdés de la Osa and Ruíz García (1980a), Ruíz (1980, 1987), Schwartz and Henderson (1991), Díaz *et al.* (2000), and Rivalta González *et al.* (2004). Advertisement calls were described by Alonso and Rodríguez (2003a), Díaz and Cádiz (2008), Hernández *et al.* (2010), Alonso Bosch (2012), and Alonso *et al.* (2017). The species' distributions were mapped by Schwartz and Henderson (1991), Díaz and Cádiz (2008), and Rivalta González *et al.* (2014). A comprehensive phylogenetic analysis by Alonso *et al.* (2012) supported the monophyly of *Peltophryne* based on genetic sequences and contributed to the clarification of species' boundaries.

Cuban toads cluster in two major groups according to size (Pramuk *et al.* 2001; Alonso Bosch 2012), one containing the smaller species (< 36 mm SVL; *P. cataulaciceps*, *P. dunni*, *P. gundlachi*, *P. longinasus*, *P. ramsdeni*) (Fig. 3.8), and other with the larger species (60–198 mm SVL; *P. empusa*, *P. fustiger*, *P. florentinoi*, *P. peltocephala*, and *P. taladai*) (Fig. 3.9). *Peltophryne empusa* is medium-sized (up to 90 mm SVL; Díaz and Cádiz 2008), and in this species, the position of the paratoid glands, the temporal structure of calls, tadpoles' morphology, and characteristics of egg clutches are more similar to some of the smaller species than to the larger ones. *Peltophryne cataulaciceps*, *P. gundlachi*, and *P. empusa* (Fig. 3.8) are explosive breeders in open lowlands, most of which are impacted by agriculture. The three species are sympatric in western localities like the quarzitic sandy savannahs covered by *Pinus tropicalis* at Isla de la Juventud and Pinar del Río, and *P. cataulaciceps* is mainly restricted to this habitat. *Peltophryne fustiger*, *P. peltocephala* and *P. taladai* (Fig. 3.9) are more generalist species occurring in different types of forests, and even in deforested and human-impacted areas. Among the largest species, *Peltophryne florentinoi* (Fig. 3.9a) has the most localized distribution. *Peltophryne peltocephala* (Fig. 3.9c) is sympatric with *P. florentinoi* and *P. taladai* along with their respective distributional ranges. Alonso *et al.* (2012) recognized genetic divergence of central populations of *P. taladai* (Fig. 3.9d) and considered them as 'good species

Figure 3.8. Small to medium-sized Cuban species of *Peltophryne* (Bufonidae). (A) *P. cataulaciceps* from the vicinity of Hotel Colony, Isla de la Juventud; (B) *P. dunni* from Topes de Collantes, Sancti Spiritus Province; (C) *P. gundlachi* from plains near Sierra de Cubitas, Camagüey Province; (D) *P. longinasus longinasus* from Cabrillas, Alturas de Pizarras del Sur, Pinar del Río Province; (E) *P. l. cajalbanensis* from Meseta de Cajálbana, Pinar del Río Province; (F) *P. empusa* from the vicinity of Hotel Colony, Isla de la Juventud. Photographs by L. M. Díaz (A, D, E, F), A. Cádiz (B), L. Watson (C).

candidates' (earlier, Ruíz [1987] commented that their morphological and acoustic differences warranted their treatment as valid species).

Peltophryne longinasus is the only polytypic species (Fig. 3.8d-e). This species-complex was reviewed by Valdés de la Osa and Ruíz García (1980a). In their phylogenetic analysis, Alonso *et al.* (2012) demonstrated the paraphyletic, basal position of *P. dunni* (Fig. 3.8b) to all the small-sized toads, and the close relationships of *P. longinasus* and *P. l. cajalbanensis*. *Peltophryne ramsdeni* was last seen in June 1976 (Valdés de la Osa and Ruíz García 1980a) and it is the only Cuban anuran that has not been collected in almost half a century. The external morphology of *P. ramsdeni* is somewhat similar to that of *P. dunni*. Valdés de la Osa and Ruíz García (1980a) also mentioned

Figure 3.9. Large Cuban species of *Peltophryne* (Bufonidae). (A) *Peltophryne florentinoi* from Gironcito, Girón, Zapata, Matanzas Province; (B) *P. fustiger* from the vicinity of Santo Tomás, Moncada, Viñales, Pinar del Río Province; (C) *P. peltocephala* from Sierra de Cubitas, Camagüey Province; (D) *P. taladai* from Banao, Sancti Spiritus Province. Photographs by L. M. Díaz.

the different behaviour of *P. ramsdeni* compared with *P. dunni* and *P. longinasus* and considered that each likely represented a valid species. Although *P. ramsdeni* and *P. dunni* originally were described as full species by Barbour (1914 and 1926, respectively), Barbour (1937) later considered them as subspecies of *P. longinasus*. The subspecific status was maintained until the most recent checklist by Hedges *et al.* (2019), despite the evidence mentioned above. However, this reticence probably originated because *P. ramsdeni* was not available for recent analyses. Differences in the morphology of adults and tadpoles, as well as modes of oviposition, are illustrated herein (see section V. *Reproductive modes*), which also confirms a high level of taxonomic distinctiveness.

2. *Conservation status*

Two species tested positive for chytrid fungus: *Peltophryne dunni* (Díaz *et al.* 2007b; Sabino-Pinto *et al.* 2017; Cádiz *et al.* 2018) and *P. longinasus longinasus* (Cádiz *et al.* 2018). The influence of chytrid fungus on responses of these species at the population level remains unexplored. *Peltophryne ramsdeni* was considered as Critically Endangered Possibly Extinct (CR A2ac) by Cobos and Alonso Bosch (2016), while *P. florentinoi* has a specialized reproductive mode and could be highly sensitive to ongoing habitat disturbance (Cobos and Alonso Bosch 2018). Likewise, *P. cataulaciceps* is restricted to one of Cuba's most endangered lowland habitats: the quarzitic sandy savannahs. Although *P. empusa* and *P. gundlachi* are widespread, they are currently affected by agriculture and very likely by pesticides, slash-and-burn practices, and the construction of roads and human settlements, which may explain extirpation or fragmented populations of these species throughout many areas. Finally, acoustic competition by introduced species (Del Castillo Domínguez *et al.* 2021) and malformations, likely linked to pollution (Alonso Bosch *et al.* 2017,

2021; García-Padrón and Alonso Bosch 2017; Alonso Bosch and Hernández Marrero 2020) represent additional threats to the Cuban toads.

B. Genus *Eleutherodactylus* (Eleutherodactylidae)

1. General description

Most frogs in Cuba (84%) belong to this genus. They are classified into three subgenera: *Eleutherodactylus* (*sensu stricto*; 15 species), *Euhyas* (41 species), and *Syrrophus* (2 species). The size of Cuban *Eleutherodactylus* varies from 9.8/10.0 mm SVL (adult males *E. iberia* and *E. limbatus* of some populations) to 127.4 mm SVL (female *E. zeus*; García 2012). We recognize six ecological tendencies in these frogs (in increasing order of species richness; Fig. 3.10): (1) grass dweller, (2) riparian, (3) arboreal/bromeliad dweller, (4) semi-arboreal, (5) saxicolous/cave dweller, and (6) ground/leaf litter dweller. Dugo-Cota *et al.* (2019) defined microhabitat-use and ecomorphology in a broader regional context, and most of their categories are coincident with the scheme presented herein. Species in the subgenus *Eleutherodactylus* have the highest tendency for arboreality, with nine species with semi-arboreal habits and six being arboreal/bromeliad dwellers. *Euhyas* is the subgenus with the most inclusive radiation among terrestrial habitats (in decreasing order of representation of species): species occurring on the ground and leaf litter (18 spp.); saxicolous/cave-dwelling habits (12 spp.); riparian dwellers (4 spp.); and a few grass dwellers (4 spp.) in forested areas, open areas, or both. *Syrrophus* only has saxicolous species. Species' distributional patterns were described by Díaz and Cádiz (2008) and Rivalta González *et al.* (2014), but these need to be updated. Revisionary bioacoustic descriptions are available in Díaz and Cádiz' (2007, 2008) papers, with detailed contributions also by Alonso Bosch and Rodríguez Gómez (2001), Alonso *et al.* (2001), Rodríguez *et al.* (2010a, 2017), Díaz *et al.* (2012), Díaz and Hedges (2015), Bignotte-Giró (2019), and Cádiz and Iturriaga (2020). A monumental revision of chromosomes is available (Schmid *et al.* 2010); major contributions from phylogenetic studies are treated next.

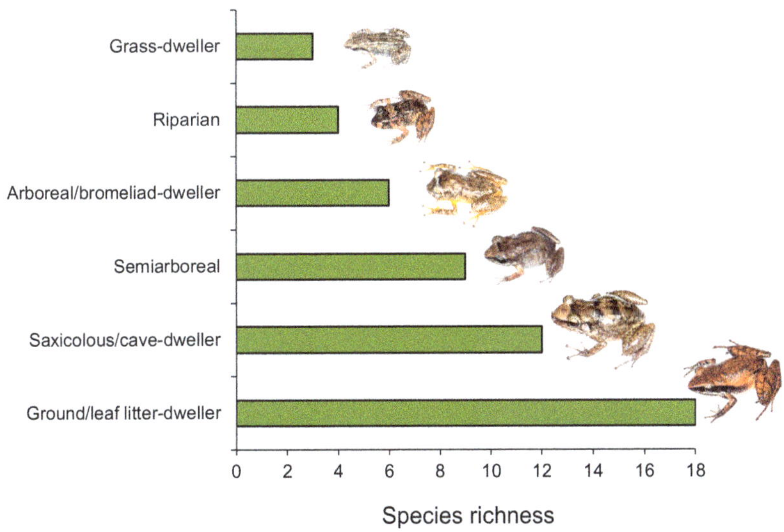

Figure 3.10. Representatives of the six main ecological categories occupied by Cuban species of the genus *Eleutherodactylus*. (A) Grass-dweller (*E. feichtingeri*); (B) riparian (*E. riparius*); (C) arboreal/bromeliad-dweller (*E. ionthus*); (D) semi-arboreal (*E. ronaldi*); (E) saxicolous/cave-dweller (*E. thomasi*); (F) ground/leaf litter-dweller (*E. dimidiatus*). Photographs by L. M. Díaz.

2. Species' account and description of series and groups within series by subgenera

The following account lists eight species-series and 12 species-groups in the three abovementioned subgenera, according to Hedges *et al.* (2008a); some additions, comments, and minor arrangements are suggested, based on subsequent contributions.

(a) *Eleutherodactylus (Eleutherodactylus) auriculatus* species-series

Two species-groups (*auriculatus* and *varians*) and 14 species are included in this series (Figs. 3.11 and 3.12). The most speciose group, *auriculatus,* includes eight species (*E. auriculatus, E. bartonsmithi, E. cattus, E. eileenae, E. glamyrus, E. mariposa, E. principalis,* and *E. ronaldi*) (Fig. 3.11). This group is described next.

General habits: Semi-arboreal species; males call at different heights in the vegetation mostly according to species; nests on the ground, apparently also in rock crevices, and sometimes in low bromeliads; nocturnal, crepuscular, or sometimes call during the daytime depending on weather conditions. **General distribution**: Species-group is represented throughout the main island. *Eleutherodactylus auriculatus* is widely distributed (but see taxonomic notes) and *E. eileenae* is meso-localized (Western and Central Cuba). Eastern Cuba is the centre of diversity with four species in the Sierra Maestra (*E. auriculatus, E. cattus, E. glamyrus,* and *E. ronaldi*), and five species (and a few 'species candidates'; see Taxonomic remarks next) in Nipe-Sagua-Baracoa (*E. auriculatus, E. bartonsmithi, E. mariposa, E. principalis,* and *E. ronaldi*). After GGA 1 distribution of *Eleutherodactylus ronaldi* was extended to Sierra del Cristal (Rodríguez and Alonso 2013) and Gibara (Díaz *et al.* 2014), both in Holguín Province. **Taxonomic remarks**: Rodríguez *et al.* (2010b) obtained phylogenetic evidence of a cryptic radiation in *E. auriculatus.* For the first time, they included *E. principalis* in a phylogenetic analysis, but most of the candidates as 'new species' remain undescribed, which limited a deeper understanding of species' relationships, distributional patterns, and conservation-efforts. Moreover, *E. auriculatus* (*sensu stricto*) has to be redefined, and most likely a neotype should be designated. Finally, Rodríguez *et al.* (2017) described *E. cattus,* a cryptic species related to *E. glamyrus,* and placed it in the *E. auriculatus* species-group based on genetic, morphological, and bioacoustic traits. **Conservation status**: One species (*E. auriculatus*) is widespread throughout the country and is adapted to several forested habitats; lowland populations have been highly affected or extirpated because of human-induced transformations of the land (LC, GAA1 and 2). *Eleutherodactylus ronaldi* is widely distributed but at a regional scale (Eastern region) and occurs in several habitats, including open and human-disturbed situations near forests (VU, GAA1; NT, GAA2; Table 3.1). *Eleutherodactylus eileenae* is meso-localized, always in forested areas. It is mostly extirpated in the lowlands, with localized population declines observed (NT, GAA1; LC, GAA2; see Table 3.1). The rest of the species in the *auriculatus* group have a much-limited distribution and are relegated to montane forest habitats; hence, they are considered threatened (Table 3.1). Loss and fragmentation of habitats are their main threats. *Eleutherodactylus auriculatus* and *E. glamyrus* tested positive for chytridiomycosis (Cádiz *et al.* 2018), but population declines have not been observed.

The *varians* species-group includes seven species, which are classified among two subgroups recognized by Hedges *et al.* (2008a): *leberi* (*E. leberi* and *E. melacara*) and *varians* (*E. guantanamera, E. ionthus, E. olibrus, E. staurometopon,* and *E. varians*) (Fig. 3.12). This second group of species is described next.

General habits: Mostly bromeliad species (except *E. leberi*), also associated with various palms and introduced plants like *Pandanus, Ravenala,* and *Musa,* in which leaf axils are somewhat similar to bromeliads; *E. leberi* inhabits karstic outcrops and has semi-arboreal habits, using shelters in rock crevices; nests in bromeliads and axils of some plants, or inside holes in

Figure 3.11. The *Eleutherodactylus auriculatus* species-group, subgenus *Eleutherodactylus*. (A) *E. auriculatus* from Cupeyal del Norte, A. de Humboldt National Park, Guantánamo Province; (B) *E. bartonsmithi* from the vicinity of La Maquina, Maisí, Guantanámo Province; (C) *E. cattus* from Loma del Gato, Sierra Maestra, Santiago de Cuba Province; (D) *E. eileenae* from Topes de Collantes, Macizo de Guamuhaya, Sancti Spiritus Province; (E) *E. glamyrus* Aguada de Joaquín, Sierra Maestra, Granma Province; (F) *E. mariposa* from Los Hondones, Meseta del Guaso, Guantánamo Province; (G) *E. principalis* from Nuevo Mundo, A. de Humboldt National Park, Guantánamo Province; (H) *E. ronaldi* from Cupeyal del Norte, A. de Humboldt National Park, Guantánamo Province. Photographs by A. Fong (C) and L. M. Díaz (all others).

Figure 3.12. *Eleutherodactylus varians* species-group, subgenus *Eleutherodactylus*. (A) *E. guantanamera* from Boca del Río Yumurí, Guantánamo Province; (B) *E. ionthus* from Alto del Naranjo, Sierra Maestra, Granma Province; (C) *E. leberi* from Tercer Frente, Santiago de Cuba Province; (D) *E. melacara* from El Manguito, Sierra Maestra, Granma Province; (E) *E.* cf. *olibrus* from Guanahacabibes National Park, Pinar del Río Province; (F) *E.* cf. *varians* from Cabo Cruz, Granma Province. Photographs by C. Lukhaup (A) and L. M. Díaz (all others).

rocks (*E. leberi*); mostly nocturnal, sporadically heard during daytime depending on weather conditions; diurnal and nocturnal calls are sometimes different. **General distribution**: Three species (*E. olibrus, E. staurometopon,* and *E. varians*) are in the West; one species (*E. varians*) is in the Central region; the Eastern region is the centre of diversity, with five species particularly in the Sierra Maestra (four species – *E. guantanamera, E. ionthus, E. leberi, E. melacara* – and considering *E.* aff. *varians* from Cabo Cruz according to Díaz and Cádiz 2008). **Taxonomic remarks**: The taxonomy of this group is controversial and needs clarification. For example, Hedges *et al.* (1992, 2008a) elevated subspecies of *E. varians* to the specific level. Hedges *et al.* (1992, 2008a) used the number of chromosomes, external morphological characters, and advertisement calls as diagnostic characters for describing new species (*E. melacara* and *E. guantanamera*) and elevating subspecies to full species (e.g. *E. ionthus* and *E. olibrus*). However, among those formerly considered subspecies of *E. varians*, genetic sequences only from *E. ionthus* and *E. olibrus* were included in the phylogenetic analysis; *E. varians* (*sensu stricto*) has never been subjected to phylogenetic studies. Before these genetic approaches, the only basis for diagnosis of species status were allopatry and Schwartz's (1959a) diagnostic characters. Consequently, the limited information and analyses based on different sources of evidence currently contribute to the limited taxonomic resolution for most species in this group. Moreover, this limitation also influences our understanding of distributional patterns and the extent of occurrence of species of conservation concerns. Finally, bioacoustics provided diagnostic characters for species like *E. guantanamera*: its vocalization consists of multi-note calls sporadically produced at the beginning and the end of the night, sometimes from inside bromeliads. However, we now know that its call is almost indistinguishable from the typical calls of *E. ionthus* (i.e. one to three notes; see Alonso *et al.* 2007; Díaz and Cádiz 2007, 2008), which further limits the accurate identification of *E. guantanamera* in the field only by acoustic surveys. In summary, the taxonomy of the *E. varians* group needs much further study. **Conservation status**: Species are dependent on bromeliads or plants with similar leaf-axillary structure, in which they take refuge and breed; hence, the frogs usually are present in forested situations or arbustive areas with bromeliads. Moreover, bromeliads determine the spatial distribution of frogs and the availability of habitat. García-González *et al.* (2014) found a high preference of frogs for a single species of bromeliad at Guanahacabibes Peninsula; Bignotte-Giró *et al.* (2022) also mentioned preference of different bromeliads depending on the locality inhabited by species throughout its distributional range. *Eleutherodactylus varians* was classified as VU in GAA1 but LC in GAA2. The rest of species in the *varians* group were considered Threatened in GAA2 (Table 3.1). Also, extreme fluctuations in abundance of *E. melacara* after hurricanes in Sierra Maestra have been noted (P. López del Castillo, pers. comm.), and local declines are a concern that deserves further research. Chytrid fungus was reported in *E. ionthus* by Cádiz *et al.* (2018); this species has a wide distribution and the possibility of it being a vector for chytrid fungus to more sensitive species like *E. melacara* should be considered.

(b) *Eleutherodactylus* (*Euhyas*) *dimidiatus* species-series

One species-group (*dimidiatus*) and four species (*E. albipes*, *E. dimidiatus*, *E. emiliae*, and *E. maestrensis*) are included in this series (Fig. 3.13). **General habits**: Terrestrial, ground/leaf litter dwellers, long-legged/jumping frogs (*E. dimidiatus*) to stocky built/short-legged, slowly moving frogs (*E. albipes*, *E. emiliae*, *E. maestrensis*); nests on the ground, under stones and leaf litter. **General distribution**: One species (*E. dimidiatus*) is widely distributed in Cuba and Isla de la Juventud; two species occur in Central Cuba (*E. dimidiatus* and *E. emiliae*) and three in the Eastern region (*E. albipes*, *E. dimidiatus* and *E. maestrensis*), with the Sierra Maestra as the centre of diversity. **Taxonomic remarks**: *E. dimidiatus* is polytypic (*E. d. amelasma* and *E. d. dimidiatus*), although Díaz *et al.* (2005) suggested that *E. d. amelasma* (described from only one individual) is within the morphological variation of other populations. **Conservation status**: Species in this group depend on humid forests; only *E. dimidiatus* is not classified as threatened (Table 3.1). *Eleutherodactylus albipes* is restricted to highland cloud forests, isolated near the highest peaks of Sierra Maestra, and the border with high-elevation montane rainforests. Active searching for this species failed to find it at one locality in Sierra Maestra during the summers of 2015 and 2016. *Eleutherodactylus emiliae* is increasingly rare in Macizo de Guamuhaya and confined to humid montane rainforests; the last record was by Marrero *et al.* (2017a) from Pico San Juan. *Eleutherodactylus maestrensis* is confined to Pico Bayamesa

Figure 3.13. *Eleutherodactylus dimidiatus* species-group, subgenus *Euhyas*. (A) *E. albipes* from Pico Turquino National Park, Sierra Maestra, Granma-Santiago de Cuba Provinces; (B) *E. dimidiatus* from La Melba, A. de Humboldt National Park, Holguín Province; (C) *E. emiliae* from Topes de Collantes, Macizo de Guamuhaya, Sancti Spiritus Province; (D) *E. maestrensis* from Barrio Nuevo, La Bayamesa National Park, Sierra Maestra, Granma Province. Photographs by C. Steinlein (A, C) and L. M. Díaz (B, D).

National Park and is unknown from other parts of Sierra Maestra. The main threats are deforestation and habitat fragmentation. Some of these species are expected to be affected by climatic change in the following few decades (CITMA 2014). Excepting *E. dimidiatus*, the other species have been reported as having a low tolerance to disturbance by humans (Fong G. 2009). The chytrid fungus was reported in *E. dimidiatus* by Sabino-Pinto *et al.* (2017) and Cádiz *et al.* (2018).

(c) *Eleutherodactylus* (*Euhyas*) *greyi* species-series

One species-group (*greyi*) and one species (*Eleutherodactylus greyi*) are included in this series (Fig. 3.14). **General habits**: Saxicolous/cave dweller, on riverbanks, in hollow trunks, and on the forest floor. **General distribution**: Central Cuba, Macizo de Guamuhaya; heights north to Villa Clara Province, to Sierra de Cubitas as the easternmost locality in Camagüey Province, near sea level at Río Jutía, Guajimico, Cienfuegos Province (L.M. Díaz, pers. obs.) to 1,140 m (Torres *et al.* 2017). **Conservation status**: The species is listed threatened (Table 3.1). It is associated with humid, forested habitats. At the lowest elevations, the species commonly is associated with riverbanks. Major threats are loss and fragmentation of habitat.

(d) *Eleutherodactylus* (*Euhyas*) *luteolus* species-series

Three species-groups (*cuneatus*, *riparius*, and *toa*) and five species are included in this series (Fig. 3.15): *cuneatus* (*Eleutherodactylus cuneatus* and *E. turquinensis*) (Fig. 3.15 B, F); *riparius*

Figure 3.14. *Eleutherodactylus greyi* (Subgenus *Euhyas*; *E. greyi* species-series) from Topes de Collantes, Macizo de Guamuhaya, Sancti Spiritus Province. Photograph by L. M. Díaz.

(*E. riparius* and *E. rivularis*) (Fig. 3.15c-d); and *toa* (*E. toa*) (Fig. 3.15e). *Eleutherodactylus gundlachi* is added to this series (see below). **General habits**: Ground/leaf litter dwellers (*E. gundlachi, E. toa*) and riparian species, either in lowland, relatively open areas to well-forested uplands (*E. cuneatus*), including flooded caves and riverbanks; remarkable semi-aquatic habits in forest-covered mountain streams ~800 m asl (*E. turquinensis*); streams at lowland to moderate elevations either in open or forested habitats (*E. riparius* and *E. rivularis*); mostly nocturnal, but *E. cuneatus* and *E. riparius* also vocalize during the daytime. **General distribution**: Eastern region is the centre of diversity with six species; *E. riparius* is widely distributed; *E. cuneatus* and *E. gundlachi* are regionally distributed in the East; *E. rivularis* and *E. turquinensis* are only present in the Sierra Maestra; *E. toa* occurs only in Nipe-Sagua-Baracoa. **Taxonomic remarks**: Díaz and Hedges (2015) mentioned that unpublished genetic data (from S. B. Hedges) after Hedges *et al.* (2008a) indicated that *E. gundlachi* (Fig. 3.15a) actually is a member of the *E. luteolus* species-series instead of being related to *E. varleyi* and the five other species in the former *E. gundlachi* species-group. Hence, they suggested that an *E. varleyi* species-group could be recognized, and we follow that classification here. However, we have no recent information as to which species-group *E. gundlachi* is most related. Examination of type specimens showed sufficient morphological evidence to suggest that the old name *E. brevipalmatus* (Schmidt 1920) is valid and not a synonym of *E. sierramaestrae* Schmidt (= *E. cuneatus sensu* Estrada and Hedges [1998]) as stated by Schwartz (1960a). Instead, *E. brevipalmatus* should be considered the junior synonym of *E. turquinensis*. All the morphological traits considered by Hedges *et al.* (1995) as diagnostic of *E. turquinensis* are present in the types of *E. brevipalmatus* and *E. turquinensis*; these traits differ from both the syntypes of *E. cuneatus* and the type of *E. sierramestrae* (all examined by L. M. Díaz). **Conservation status**: *Eleutherodactylus cuneatus* and *E. riparius* are listed as Least Concern (LC) by the IUCN. However, Cádiz *et al.* (2018) reported presumable declines of *E. cuneatus* along an elevational gradient at the Sierra Maestra, although this species is still common elsewhere. In addition, *Eleutherodactylus rivularis* and *E. turquinensis* were not found at historical localities where they had been commonly found (Cádiz *et al.* 2018), which raises concerns about local declines and

Figure 3.15. *Eleutherodactylus luteolus* species-series, subgenus *Euhyas*. (A) *E. gundlachi* from Pico Turquino National Park, Sierra Maestra, Granma-Santiago de Cuba Provinces; (B) *E. cuneatus* (*E. cuneatus* species-group) from La Gran Piedra, Sierra Maestra, Santiago de Cuba Province; (C) *E. riparius* (*E. riparius* species-group) from Alturas de Pizarras del Sur, Pinar del Río Province; (D) *E. rivularis* (*E. riparius* species-group) from Río Yara, Sierra Maestra, Granma Province; (E) *E. toa* from Cupeyal del Norte, A. de Humboldt National Park, Guantánamo Province; (F) *E. turquinensis* from Cantarrana, Sierra Maestra, Granma Province. Photographs by L. M. Díaz.

extirpations. *Eleutherodactylus riparius* and *E. gundlachi* tested positive for chytrid fungus (Cádiz *et al.* 2018). The geographic range of *Eleutherodactylus rivularis* is not covered by any protected area, being peripherical to the Pico Turquino National Park (see Fong G. *et al.* 2015).

(e) *Eleutherodactylus* (*Euhyas*) *planirostris* species-series

This is the most speciose species-series in Cuba with five species-groups (*atkinsi, limbatus, pinarensis, planirostris,* and *varleyi*) collectively containing 23 species (Figs. 3.16–3.20). The complexity of this series requires a separate synopsis of each species-group, as follows:

Figure 3.16. *Eleutherodactylus atkinsi* (subgenus *Euhyas*; *E. atkinsi* species-group) from Sierra de Cubitas, Camagüey Province. Photograph by L. M. Díaz.

Eleutherodactylus (*Euhyas*) *atkinsi* species-group with one species (*E. atkinsi*) (Fig. 3.16). **General habits**: Mainly ground/leaf litter dwellers; males usually climb to vegetation for calling; nocturnal. **General distribution**: One of the species of the genus with the broadest geographic distribution in Cuba. **Taxonomic remarks**: Polytypic; two subspecies (*E. a. atkinsi* and *E. a. estradai*). **Conservation status**: Least Concern (LC) according to IUCN; widely distributed in different ecosystems, including areas highly altered by humans. The chytrid fungus was reported in this species by Cádiz *et al.* (2018).

Eleutherodactylus (*Euhyas*) *limbatus* species-group consists of six species (*E. cubanus*, *E. etheridgei*, *E. iberia*, *E. jaumei*, *E. limbatus*, and *E. orientalis*) (Fig. 3.17). **General habits**: Ground/leaf litter dwellers (mainly) or saxicolous (*E. etheridgei*); *E. etheridgei* may call from low vegetation at night; most are diurnal species which typically call from the leaf litter. Rodríguez *et al.* (2013) found a mechanism of alkaloid sequestering in frogs of this group and hypothesize on the probable aposematic function of the bright coloration present in most species. **General distribution**: *Eleutherodactylus limbatus* has a fragmented, wide distribution along the island in forested habitats. The Eastern region is the centre of diversity. *Eleutherodactylus cubanus* and *E. jaumei* are endemic to Sierra Maestra; *E. iberia* occurs in the mountains of the orographic system Nipe-Sagua-Baracoa, and *E. orientalis* is endemic to Yunque de Baracoa and its surroundings; *E. etheridgei* inhabits the semidesert subcoastal region throughout Santiago de Cuba and Guantánamo provinces. **Taxonomic remarks**: Rodríguez *et al.* (2012, 2013) published the most comprehensive phylogeny of this group. They suggested further taxonomic revisions of populations of *E. iberia*, *E. jaumei*, and *E. limbatus* and the description of potentially new cryptic species. However, a few comments are pertinent to address future revisions that potentially might represent nomenclatural acts. Rodríguez *et al.* (2013) considered *E. limbatus* (*sensu stricto*) to be the populations from Central and Western Cuba, but the original description was, very likely, based on specimens from Nipe-Sagua-Baracoa Massif (among the many specimens of several species collected by Charles Wright in this mountain range, specifically from the surroundings of Cafetal Monte Verde, which were described later on by E. D. Cope). Hence, according to the most recent phylogenetic analysis and the abovementioned elementary taxonomic fact, one of the candidate species suggested by authors from La Gran Piedra may be more closely related to *E. limbatus* in

Figure 3.17. *Eleutherodactylus limbatus* species-group, subgenus *Euhyas*. (A) *E. cubanus* from Aguada de Joaquín, Pico Turquino National Park, Sierra Maestra, Granma Province; (B) *E. iberia* from El Recreo, A. de Humboldt National Park, Guantánamo Province; (C) *E. jaumei* from the vicinity of Uvero (new locality for the species), Santiago de Cuba Province; (D) *E. limbatus* from La Gran Piedra, Sierra Maestra, Santiago de Cuba Province; (E) *E. orientalis* from Yunque de Baracoa, Guantánamo Province; (F) *E. etheridgei* from Reserva Ecológica Siboney-Jutici, Santiago de Cuba Province. Photographs by C. Tallet (A), N. Navarro (E), L. M. Díaz (all others).

the strictest sense as part of the complex *limbatus* (*sensu stricto*)-*iberia* radiation in this montane region. Although some taxonomic uncertainties were recognized by Rodríguez *et al.* (2012, 2013) for the *jaumei-iberia-limbatus* complex, contact zones and parapatric distribution in different rainforest habitats exist among *E. iberia* and *E. limbatus* in the surroundings of La Melba, Alejandro de Humboldt National Park (Díaz and Cádiz 2008). Certainly, as stated by Rodríguez *et al.* (2013), a new and urgent taxonomic revision of these tiny frogs needs an integrative approach.

Figure 3.18. *Eleutherodactylus pinarensis* species-group (but see text for comments), subgenus *Euhyas*. (A) *E. blairhedgesi* from Boca de Canasí, Mayabeque Province; (B) *E. pezopetrus* from Mella, Santiago de Cuba Province; (C) *E. pinarensis* from Cueva de Saturno, Varadero, Matanzas Province; (D) *E. thomasi* from Sierra de Cubitas, Camagüey Province. Photographs by C. Steinlein (A); E. Alfaro (B); L. M. Díaz (C, D).

Conservation status: *Eleutherodactylus limbatus* was classified VU in GAA1 and LC in GAA2. All other species in its group have been traditionally considered Threatened (Table 3.1) according to the IUCN Red List. Their major threats are loss and fragmentation of habitat (severe, current, and historical). Since previous assessments (see Stuart *et al.* 2008), the distribution of *E. cubanus* and *E. iberia* has become better known (Díaz and Cádiz 2008; Rodríguez *et al.* 2013; Rivalta González *et al.* 2014). For *E. jaumei*, we calculated the EOO (see Table 3.1) to include a new geographic distribution record near Uvero, ~10 km E (air distance), ~140 m asl, from the type locality (Fong G. and Díaz, in preparation). Local extirpations of *E. cubanus* have been reported in Sierra Maestra (L. M. Díaz and A. Fong, pers. obs.). A population fluctuation after a natural phenomenon (hurricane) occurred for *E. limbatus* at Soroa, Artemisa Province (L. M. Díaz, pers. obs.), but extirpations of *E. limbatus* have been occurring for over 80 years, yielding severely fragmented populations throughout the island.

Eleutherodactylus (*Euhyas*) *pinarensis* species-group with four species (*E. blairhedgesi*, *E. pezopetrus*, *E. pinarensis*, and *E. thomasi*) (Fig. 3.18). **General habits**: Mainly saxicolous/cave-dwelling species; nocturnal. **General distribution**: Western Cuba is the centre of diversity, where *E. blairhedgesi*, *E. pinarensis*, and *E. thomasi* (*E. t.* cf. *trinidadensis* and *E. t. zayasi*) occur. *Eleutherodactylus thomasi* is distributed in Central Cuba, including the cays of the Sabana and Camagüey Archipelagos; a population assigned to *E. thomasi*, still under taxonomic study, also occurs in Gibara, Eastern region. *Eleutherodactylus pezopetrus* is distributed along scattered karstic

Figure 3.19. *Eleutherodactylus planirostris* species-group, subgenus *Euhyas*. (A) *E. casparii* from Topes de Collantes, Macizo de Guamuhaya, Sancti Spiritus Province; (B) *E. goini* from Sierra del Rosario Biosphere Reserve, Artemisa Province; (C) *E. guanahacabibes* from Guanahacabibes National Park, Pinar del Río Province; (D) *E. planirostris* from La Habana Province; (E) *E. simulans* from Cupeyal del Norte, A. de Humboldt National Park, Guantánamo Province; (F) *E. tonyi* from Sendero El Guafe, Cabo Cruz, Granma Province. Photographs by L. M. Díaz.

outcrops in the eastern part of Cuba, between Santiago de Cuba and Guantánamo provinces (Díaz and Cádiz, 2008); it was also reported by Marrero Romero *et al.* (2017b) from El Salto del Guayabo, Mayarí, Holguín Province. *Eleutherodactylus blairhedgesi* has the most restricted distribution along the coastline between Mayabeque and Matanzas Provinces. **Taxonomic remarks**: In the phylogenetic analysis by Pyron and Wiens (2011) and Padial *et al.* (2014) *Eleutherodactylus pezopetrus* was recovered as the sister taxon of *E. blairhedgesi, E. pinarensis,* and *E. thomasi*; however, in the phylogeny of Hedges *et al.* (2008a), it was basal to all species in the *E. planirostris* species-series. Divergent results were related to different methods of alignment in generating

Figure 3.20. *Eleutherodactylus varleyi* species-group (but see text for comments), subgenus *Euhyas*. (A) *E. adelus* from Cabrillas, Alturas de Pizarras del Sur, Pinar del Río Province; (B) *E. beguei* from La Munición, A. de Humboldt National Park, Guantánamo Province; (C) *E. feichtingeri* from Río Zaza, Sancti Spiritus Province; (D) *E. intermedius* from Pico Turquino National Park, Granma-Santiago de Cuba Provinces; (E) *E. tetajulia* from Monte Iberia, A. de Humboldt National Park, Guantánamo Province; (F) *E. varleyi*, from Soroa, Sierra del Rosario, Artemisa Province. Photographs by C. Tallet (C), J. Larramendi (E), C. Lukhaup (F), and L. M. Díaz (A, B, D).

phylogenies. The alternative hypothesis of Pyron and Wiens (2011) and Padial *et al.* (2014) is consistent with morphological and bioacoustical similitudes among these species. Hence, we consider *E. pezopetrus* to be a member of the *E. pinarensis* species-group until new updated approaches are able to provide a definitive classification. Rodríguez *et al.* (2011a) found that the small grass frog *Eleutherodactylus adelus* (formerly in the *E. gundlachi* = *E. varleyi* species-group) is the closest relative of *E. pezopetrus*. We recommend replication of this study based on a new comprehensive phylogenetic analysis. Indeed, morphological convergencies and divergencies of *Eleutherodactylus* frogs, apart from their phylogenetic relationships, have occurred multiple times, affecting previous systematic interpretations based only on morphology. However, the particular case of these two species has to be revisited because it is extreme, especially considering the huge hiatus in its distribution. *Eleutherodactylus thomasi* is polytypic, integrated by three subspecies: *E. t. thomasi, E. t. trinidadensis,* and *E. t. zayasi.* The taxonomic status of these populations has to be elucidated (particularly those along the northern coast and cays of Central Cuba and from Gibara in the east). Hedges *et al.* (2008a) proposed that the three subspecies of *E. thomasi* may be full species, which deserves further study. The external morphology and the advertisement vocalizations of the three recognized subspecies are different (see Díaz and Cádiz 2007, 2008), particularly those of the nominal subspecies compared with the other two. We call for a comprehensive transdisciplinary approach to elucidate the variation and distribution of *E. thomasi* and that of its closest relative *E. blairhedgesi* throughout their geographic ranges. **Conservation status**: *Eleutherodactylus thomasi* is NT in GAA2 but it was considered as EN in GAA1. The rest of species in this group are Threatened (Table 3.1). *Eleutherodactylus pinarensis* has a wide, but fragmented, distribution and is strongly associated with caves, particularly those in forests and within patches of vegetation. It has a low tolerance to human disturbances and populations are most likely severely fragmented. *Eleutherodactylus thomasi* is widely distributed, but future taxonomic revisions may modify the distributional patterns of this, and other, closely related taxa that are currently considered subspecies. Their major threats are loss and fragmentation of habitat and disturbance by humans. Many caves are impacted by humans that use them for various purposes. *Eleutherodactylus blairhedgesi* is threatened by activities related to the extraction of petroleum deposits throughout its limited geographic distribution. *Eleutherodactylus pezopetrus* has the second-most restricted distribution in the group and is at higher risk of extinction; at least at one locality (Cantera Mella – formerly Cantera de Miranda – the type locality of this species), open-pit mining is rapidly destroying its habitat.

Eleutherodactylus (Euhyas) planirostris species-group consists of six species (*E. casparii, E. goini, E. guanahacabibes, E. planirostris, E. simulans,* and *E. tonyi*) (Fig. 3.19). **General habits**: Mainly ground/leaf litter dwellers to saxicolous; mainly nocturnal, but they may vocalize in the morning while breeding or inside caves. **General distribution**: three species in Western Cuba (*E. goini, E. guanahacabibes,* and *E. planirostris*); two species in the Central region (*E. casparii* and *E. planirostris*); and three species (*E. planirostris, E. simulans,* and *E. tonyi*) in the East; *E. planirostris* was described originally from Nassau (Bahamas), and there are native populations in southern Florida; it is also an invasive species in several countries (Krysko *et al.* 2011; Contreras-Calvario *et al.* 2018; CABI 2021; Cubillos-Abrahams *et al.* 2021). **Taxonomic remarks**: Heinicke *et al.* (2011) made a phylogenetic analysis of species related to *E. planirostris* and showed that this species is paraphyletic, suggesting that more species may be described shortly. Díaz *et al.* (2014) extended the distribution of *E. tonyi* to Gibara, Holguín Province. **Conservation status**: *Eleutherodactylus goini* and *E. guanahacabibes* are LC in GAA2 but they were classified as Threatened in GAA1. *Eleutherodactylus planirostris* is LC, but the other species are threatened according to the evaluations by the IUCN Red List (Table 3.1). Knowledge about the distribution of species increased after GAA1 (Stuart *et al.* 2008; AmphibiaWeb 2020). The main threats are deforestation and habitat fragmentation. *Eleutherodactylus casparii* tested positive for chytrid fungus (Sabino-Pinto *et al.* 2017; Cádiz *et al.* 2018) as did *E. planirostris* (Cádiz *et al.* 2018).

Eleutherodactylus (Euhyas) varleyi species-group contains six species (*E. adelus, E. beguei, E. feichtingeri, E. intermedius, E. tetajulia,* and *E. varleyi*) (Fig. 3.20). **General habits**: Ground/leaf litter dwellers in montane forested areas (most) or inhabiting the grass stratum in both forested and open habitats (*E. feichtingeri* and *E. varleyi*); diurnal (*E. adelus*), acoustically active late in the afternoon and during part of the night (*E. feichtingeri*), or acoustically active at night and early in the morning, depending on weather conditions (*E. beguei* and *E. varleyi*). **General distribution**: Two species in the West (*E. adelus* and *E. varleyi*); two species in the Central region (*E. varleyi* and *E. feichtingeri*) including some cays in the Sabana and Camagüey Archipelagos; four species in the East (*E. beguei*, *E. intermedius*, *E. feichtingeri*, and *E. tetajulia*); the distribution of *E. intermedius* historically has been extended to the mountains north to Imías because three paratypes (Barbour and Shreve 1937) were collected there by P. J. Darlington in July of 1936 (maps by Díaz and Cádiz 2008; Rivalta González *et al.* 2014), but this geographic record should be viewed with caution. **Taxonomic remarks**: The recognition of an *E. varleyi* species-group was proposed by Díaz and Hedges (2015; also see the taxonomic comments in the *E. luteolus* species-series for *E. gundlachi*), and those in the *E. pinarensis* species-group (for *E. adelus*). Since its original description by Barbour and Shreve (1937), the distribution of *E. intermedius* was considered to be extended to the mountains north to Imías, Guantánamo, by different authors (Schwartz and Henderson, 1991; Díaz and Cádiz, 2008; Rivalta *et al.*, 2014). However, our examination of specimens known from the referred locality led us to conclude that they are morphologically related to *E. tetajulia*, a similar species described by Estrada and Hedges (1996a) distributed at several localities of the Nipe-Sagua-Baracoa Massif. Former taxonomic status has affected conservation assessments that analysed a greater extent of occurrence for *E. intermedius* species than it has. Then, distribution of this species should be considered as only restricted to Sierra Maestra, and the calculation of EOO in Table 3.1 is in accordance with this. **Conservation status**: *Eleutherodactylus feichtingeri* and *E. varleyi* are Least Concern (LC) according to the IUCN Red List's criteria. These species occupy a wide range of habitats distributed from sea level to moderate elevations and shows broad ecological plasticity. The rest of species in the *varleyi* group are threatened (Table 3.1). *Eleutherodactylus beguei* is known only from the type locality, in the pine forests near La Munición, Guantánamo Province, but very likely has a wider distribution to Meseta del Guaso. The major threats are loss and fragmentation of habitat, including large, intense fires in pine forests at localities where *E. adelus* and *E. beguei* occur.

(f) *Eleutherodactylus* (*Euhyas*) *ricordii* species-series

One species-group (*ricordii*) and four species (*E. acmonis*, *E. bresslerae*, *E. michaelschmidi*, and *E. ricordii*) are included in this series (Fig. 3.21). **General habits**: Ground/leaf litter dwellers, partially or entirely saxicolous in the case of *E. bresslerae* and *E. michaelschmidi*; nocturnal. **General distribution**: Eastern Cuba. **Taxonomic remarks**: We recommend a taxonomic revision of populations of *E. ricordii* from the Nipe-Sagua-Baracoa range because they seem to be significantly closer to *E. acmonis*, based on morphology and bioacoustics, than to the 'typical' *E. ricordii* from Sierra Maestra. **Conservation status**: The distribution of *Eleutherodactylus acmonis* originally included Sierra Maestra and several other localities in Guantánamo Province (see Schwartz and Henderson, 1991), where it was listed as Endangered (EN) by Hedges and Díaz (2004a) (Table 3.1). However, its distribution was further restricted to Yunque de Baracoa (e.g. Díaz and Cádiz 2008; Henderson and Powell 2009), which justifies a status of Critically Endangered (CR). In addition, *Eleutherodactylus bresslerae*, *E. michaelschmidi*, and *E. ricordii* are also threatened (Table 3.1); their main threats are loss and fragmentation of habitat.

(g) *Eleutherodactylus* (*Euhyas*) *zugi* species-series

Three species (*E. klinikowskii*, *E. erythroproctus*, and *E. zugi*) are included in this series (Fig. 3.22). **General habits**: Ground/leaf litter dwellers (*E. zugi*), or inhabitants of karstic outcrops covered by forests (*E. klinikowskii* and *E. erythroproctus*); diurnal or nocturnal activity. **General distribution**: Regional, only in the West; *E. klinikowskii* and *E. zugi* occur sympatrically at Sierra de la Güira, San Diego de los Baños, Pinar del Río Province, and parapatrically where Sierra de San Carlos and Alturas de Pizarras del Sur (Cabrillas) are nearly in contact. **Taxonomic remarks:** Hedges

Figure 3.21. *Eleutherodactylus ricordii* species-group, subgenus *Euhyas*. (A) *E. acmonis* from Yunque de Baracoa, Guantánamo Province; (B) *E. bresslerae* from Boca del Yumurí, Guantánamo Province; (C) *E. michaelschmidi* from Tercer Frente, Santiago de Cuba Province; (D) *E. ricordii* from La Gran Piedra, Santiago de Cuba Province. Photographs by N. Navarro (A) and L. M. Díaz (B–D).

et al. (2008a) concluded that *E. klinikowskii* and *E. zugi* were separated by only a short genetic distance; nevertheless, they elevated *E. zugi erythroproctus* to a full species without having genetic evidence. Herein, we report a recently discovered population from Cueva de Saturno, near the airport of Varadero, Matanzas Province; it is morphologically similar to *E. zugi* in shape, length of the vomerine odontophores, and colouration (*sensu* Schwartz 1960a) and its taxonomic status is doubtful. The record not only extends the distribution of *zugi*-like frogs 26 km NE (aerial distance) from Pan de Matanzas but also reports them 2 km south on the coast, near sea level. This is another field of research that deserves further attention in order to assess species' boundaries and phylogeography since the supposed allopatry among some frog populations is perhaps the result of deficient sampling at intermediate localities and/or recent isolation by human-mediated destruction and fragmentation of habitat. **Conservation status**: All species in the *zugi* series are endangered (Table 3.1); local declines in populations of *E. zugi* have been observed after hurricanes (L. M. Díaz, pers. obs.); fires, loss of habitat, and fragmentation of forests are main threats throughout its distribution. Since the first GAA assessment in 2004, knowledge of *E. zugi*'s distribution has increased (Díaz and Cádiz 2008; Rivalta González *et al.* 2014).

(h) *Eleutherodactylus* (*Syrrophus*) *symingtoni* species-series

Two species (*E. symingtoni* and *E. zeus*) are included in this series (Fig. 3.23). The monophyly and basal position of these two species in the subgenus *Syrrophus* was supported by Hernández-Austria *et al.* (2022). **General habits**: Saxicolous/cave dwellers; nocturnal. **General distribution**:

Figure 3.22. *Eleutherodactylus zugi* species-series, subgenus *Euhyas*. (A) *E. klinikowskii* from Moncada, Viñales, Pinar del Río Province; (B) *E. zugi* from Meseta de Cajálbana, Pinar del Río Province; (C) *E. erythroproctus* from Sierra de Camarones, Mayabeque Province; (D) *E.* aff. *zugi* from Cueva de Saturno, Varadero, Matanzas Province. Photographs by L. M. Díaz.

Only in the West. *Eleutherodactylus symingtoni* reaches its easternmost locality at Arana, Sierra de Camarones, Mayabeque Province. Outside Cuba, members of *Syrrophus* occur on mainland Central America. **Conservation status** (Table 3.1): *E. symingtoni* and *E. zeus* are threatened species (Table 3.1). Since GGA1 in 2004, the known distribution of this species was extended to a few other localities like Sierra de la Güira (Rodríguez and Alonso 2006) and Canon del Río Santa Cruz (Rodriguez and Alonso 2006; Rodriguez-Cabrera and López-Silvero 2022). However, the westernmost record from San Ubaldo-Sabanalamar, Pinar Río Province (by Rivalta *et al.* 2014), deserves further study. *Eleutherodactylus symingtoni* have not been found at the type locality, Cueva de Santo Tomás (Moncada, Viñales), since its discovery (Schwartz 1957), nor recently in Soroa or Arana; it is therefore considered a rare species, most likely extirpated from its historical collection sites. With only a few specimens in collections, the main threats of this species are alteration of habitat (deforestation and forest fragmentation), disturbance by human activities, and misuse of caves.

C. Genus *Osteopilus* (Hylidae)

Osteopilus septentrionalis is the only known hylid frog in Cuba (Fig. 3.24). Monophyly of the genus *Osteopilus* and their species' relationships were documented by Duellman *et al.* (2016) and Blotto *et al.* (2020). Tadpoles' variations and types of vocalizations were described briefly and illustrated by Díaz and Cádiz (2008). **General distribution**: Widely distributed in Cuba (Rivalta González *et al.* 2014) in most habitats, including highly anthropized situations and on isolated cays; it is also native to the Bahamas, the Cayman Islands, and the southern cays of Florida (see Henderson and

Figure 3.23. Species of the *Eleutherodactylus symingtoni* species-series, subgenus *Syrrhophus*. (A) *E. symingtoni* from Loma El Salón, Sierra del Rosario, Artemisa Province; (B) *E. zeus* from Moncada, Viñales, Pinar del Río Province. Photographs by L. M. Díaz.

Figure 3.24. *Osteopilus septentrionalis* (Hylidae) from Viñales, Pinar del Río Province. Photograph by L. M. Díaz.

Powell 2009; Heinicke *et al.* 2011; Johnson 2013). This species has been introduced – and is considered invasive – in North America (Krysko *et al.* 2011, 2019), Oahu (Hawaii), Puerto Rico, British Virgin Islands, US Virgin Islands, Anguilla, Turks and Caicos Islands, Antigua, St Martin, and St Barthélemy (Breuil 2004), and Puerto Limón, Costa Rica (Frost 2021). Recently, individuals have been observed in Jamaica, but the species' status there is unknown (S. Pasachnik, pers. comm. and photographs sent to L. M. Díaz). **Conservation status**: Least Concern (LC).

V. Reproductive modes

Nunes-de-Almeida *et al.* (2021) redefined the concept of reproductive modes by considering 11 reproductive traits: (1) reproduction type; (2) oviposition macrohabitat; (3) spawning type; (4) oviposition substrate; (5) egg-surrounding medium; (6) nest construction; (7) oviposition microhabitat; (8) embryonic development; (9) embryonic nutrition; (10) larval and newborn nutrition; and (11) place of larval development. In this section, many of these traits are implicit, but the analysis is more focused on the particularities of the Cuban fauna beyond the global classification of reproductive modes.

Cuban anurans can be grouped into two main reproductive modes by combining the criteria of Wells (2007) and Crump (2015). These modes include (1) biphasic species that lay eggs in the water and have aquatic larvae (16% of native Cuban species; families Bufonidae and Hylidae) either in still or flowing water (Fig. 3.25); and (2) species with direct development that lay eggs in various arboreal and terrestrial habitats (Townsend 1996), 84% of Cuban species, in the family Eleutherodactylidae. In this section, we follow the terminology of Altig and McDiarmid (1999, 2007) for morphological and ecological traits of tadpoles, eggs, and the structure of clutches, and we highlight discrete breeding sites, ovipositional patterns, parental care, and tadpole morphologies that may be useful as an integrative guide for complementary studies, as well as a contribution for designing strategies for conservation. Unfortunately, the information at hand from the vast majority of Cuban anurans is minimal, and, in many cases, it only includes anecdotal and chance observations; the life-history of Cuban anurans thus deserves further detailed studies.

Anurans that (1) lay eggs in water, (2) have free-living aquatic tadpoles, (3) and undergo subsequent metamorphosis, are termed as having complex lifecycles (or CLC; Wilbur 1980). Their biphasic life not only implies an ontogenetic shift in niche (Wells 2007) but also that they are under radically different environmental pressures and stressors during development. Understanding the physiological and ecological requirements and conditions that influence this biphasic life is of paramount importance for successful conservation. The relationships among CLC amphibians and their habitats involve the interaction of biotic and abiotic variables across multiple spatial and temporal scales (Howard *et al.* 2012). The migration to (and from) breeding sites, and subsequent dispersion of juveniles throughout the landscape, challenge the survival of species. The survival at early stages of development depends on hydroperiod (along with other abiotic variables), as well as selection pressures from alteration, fragmentation, and loss of habitat.

A. *Peltophryne* spp.

Reproductive sites, the structure of egg clutches, and larval morphology of Cuban toads are surprisingly diverse (Díaz and Cádiz 2008). Breeding habitats can be grouped into four categories: (1) ephemeral, unpredictable puddles and shallowly flooded plains after heavy rains, with subsequent intense solar radiation and high evaporative loss of water; (2) permanent or semi-permanent montane streams with small pools, flowing through pine or latifoliate forests; (3) edge of slowly moving streams or rivers, or lagoons, and ditches at least seasonally permanent, either

Figure 3.25. Modes of oviposition of Cuban amphibians. (A) Egg masses of *Peltophryne longinasus longinasus* (Bufonidae) at Cabrillas, Alturas de Pizarras del Sur, Pinar del Río Province; (B) rosary of eggs of *P. empusa* at Alta Habana, La Habana Province; (C) uniserial egg strings of *P. peltocephalus* at Río Jauco, near Maisí, Guantánamo Province; (D) uniserial gross strings with scalloped margins of *P. dunni* at Topes de Collantes, Sancti Spiritus Province; (E) single-tiered group of eggs of *Osteopilus septentrionalis* (Hylidae) floating on still water at Bayate, Sierra del Rosario, Artemisa Province; (F) submerged clutch of eggs (with interstices) of *O. septentrionalis* at La Platica, Sierra Maestra, Granma Province; the eggs contain embryos nearly ready to hatch. Photographs by G. Begué (C); L. M. Díaz (all others).

in open or forested areas; and (4) small cavities in karst filled with rainwater and covered by a subcoastal forest. Four oviposition modes in Cuban bufonids are recognizable (Fig. 3.25): (1) Egg mass attached to submerged vegetation and objects; (2) a rosary consisting of lengthy, uniserial groups of eggs in a string with jelly constricted between successive ova; (3) long, narrow uniserial strings of eggs on the bottom of bodies of water; and (4) eggs in uniserial gross strings with scalloped margins, attached to vegetation. All species have aquatic exotrophic, benthic larvae (Altig and McDiarmid 1999: 304; Díaz and Cádiz 2008), but discrete differences in morphology have been observed in different microhabitats.

In the first category of breeding habitats are the burrowing toads *P. cataulaciceps*, *P. empusa*, and *P. gundlachi*. They form dense and strident nocturnal choruses after heavy seasonal rains from May to October. *Peltophryne cataulaciceps* is restricted to a unique and endangered habitat in Cuba: the quarzitic sandy savannahs that are covered by lowland pine forests, palms, grasses, and bushes (Fig. 3.26). The three species are sympatric in this habitat in Pinar del Río and northern Isla de la Juventud, but *P. gundlachi* and *P. empusa* are more generalist and occur in different kinds of natural savannahs, grasslands, and even agricultural areas throughout the archipelago. In the original description of *P. cataulaciceps*, Schwartz (1959b) mentioned that eggs were laid in clumps rather than in strings, but we lack further observations for corroboration.

Peltophryne empusa and *P. gundlachi* lay eggs in 'rosaries' (ovipositional mode 2), interspersed in submerged grass or extended along muddy bottoms. Tadpoles of *P. cataulaciceps*, *P. empusa*, and *P. gundlachi* have exotrophic-lentic-benthic morphology, with high and globose bodies, high tailfins, small oral discs, and less muscular tails (Díaz and Cádiz 2008). Because the habitats occupied by these tadpoles are ephemeral, unpredictable, and with extreme temperatures often exceeding 40°C during the day, metamorphosis is accelerated, with completion in fewer than 15 to 18 days (Díaz and Cádiz 2008; L. M. Díaz, pers. obs.). In 2004, the first author

Figure 3.26. Ephemeral shallow-flooded areas in the sandy savannah at Los Indios, Isla de la Juventud (A), where the tadpoles of the explosive breeders and burrowing toads *Peltophryne cataulaciceps* (B) and *P. empusa* (C) develop. Photographs by L. M. Díaz (A); R. Fernández de Arcila (B, C).

found a number of tadpoles in a small puddle at Los Indios (Isla de la Juventud), supposedly *P. cataulaciceps*, but very small in size and in the earliest stages of development. *In-situ*, those larvae reached stages 35–36 in less than six days (Fig. 3.26b), while the body of water fell in diameter quickly; water temperature reached a high of 47°C. Several tadpoles were collected and preserved, but a few were kept alive for identification in the laboratory after metamorphosis. They completed metamorphosis between 15 and 18 days later at 28–32°C. Most likely, metamorphosis would have taken less time under natural conditions at a higher temperature. Similar results were obtained *in-situ* for *P. empusa* (Fig. 3.26c) and *P. gundlachi* (not shown) in Alta Habana, La Habana Province (L. M. Díaz, pers. obs.). The spatial segregation of interspecific choruses suggests that larval development may occur in isolated bodies of water, presumably reducing competition or other interactions, but this is an open question that awaits future study.

How these species can adjust the speed of metamorphosis depending on the evaporative rate of ephemeral bodies of water and how they adapt to high temperatures deserves further study. Changes in the pattern of precipitation, intensity of rain, retention of soil water, the duration of droughts, and subsequent persistence of the temporary breeding habitats enough to allow the development of tadpoles are factors expected to influence the reproductive success of explosive breeders in unpredictable situations (see Borges Júnior and Duarte Rocha 2013). We also add the influence of disturbances from human activities associated with agriculture, wildfires, and roads and other infrastructure.

Two closely related species, *Peltophryne dunni* and *P. longinasus* reproduce in type-2 breeding habitats (Fig. 3.27). *Peltophryne dunni* inhabit montane latifoliate forests, whereas *P. longinasus*

Figure 3.27. Montane stream habitats of two related Cuban species of *Peltophryne*. (A) Codina, Topes de Collantes, Cienfuegos Province, habitat of *P. dunni*; (B) Cabrillas, Alturas de Pizarras del Sur, Pinar del Río Province, habitat of *P. longinasus longinasus*; (C) male *P. longinasus* floating and calling; (D) tadpole of *P. longinasus longinasus* at Cabrillas; (E) metamorph of *P. l. longinasus* at Cabrillas; (F) tadpoles of *P. dunni* from Codina; (G) metamorph of *P. dunni* at Codina Photographs: A. Cádiz (A, B, F, G); L. M. Díaz (C–E).

longinasus and *P. l. cajalbanensis* are associated with pine forests on sandstone and serpentine rocks, respectively. Pine forests alternate with understorey latifoliate vegetation growing along watercourses. The flow of water in streams is relatively stable year-round, or with seasonally intermittent flow. Water temperatures vary between 18°C and 25°C. In Cabrillas, Alturas de Pizarras del Sur, males of *P. l. longinasus* have been found floating and actively swimming in pools in small streams in the morning, calling either from the shore or emergent objects (e.g. palm fronds, rocks, fallen branches, low vegetation), or from the surface of water while actively wandering and physically interacting with each other (Valdés de la Osa and Ruíz García 1980a; Díaz and Cádiz 2008; Cádiz *et al.* 2018). Pairs in amplexus were seen after midday, with oviposition occurring mostly in the afternoon. One pair deposited several groups of egg masses (ovipositional mode 1) and the real clutch size was the sum of all of them. The same ovipositional mode was observed in *P. l. cajalbanensis* (L. M. Díaz, pers. obs.). Meanwhile, *Peltophryne dunni* lay strings of eggs (ovipositional mode 4) attached to submerged leaves (Valdés de la Osa and Ruíz García 1980a; L. M. Díaz and A. Cádiz, pers. obs.). Tadpoles develop in one to three months in such habitats, where water temperature averages 24–26°C for *P. longinasus longinasus* and 18–23°C for *P. dunni*.

The giant species *P. fustiger*, *P. peltocephala*, and *P. taladai* breed in a wide diversity of bodies of water, but usually in riverine backwaters, lagoons, and ponds (Fig. 3.28). They are generalists and opportunistically use different environmental situations but are 'bad' swimmers compared to the much smaller semi-aquatic stream-breeders of the *P. longinasus* species-complex. Several observations over many years suggest that the Cuban giant toads breed more often when rivers are running slowly or in side-pools in stretches of fast-moving streams. The three species are known to have ovipositional mode 3. Tadpoles are always associated with lentic areas, usually open and totally or partially sunny. Larvae of *P. fustiger* and *P. peltocephala* are very similar (Díaz and Cádiz 2008), but those of *P. taladai* have a conspicuous pale band that disrupts the overall appearance of the tadpole. However, in populations of *P. taladai* from Central Cuba, tadpoles, besides being larger-bodied than those of sympatric *P. peltocephala*, lack this band (L. M. Díaz, personal observations; Febles 2002). Whether or not the observed larval traits match with the phylogenetic divergence of Central Cuban populations (as discussed by Alonso *et al.* 2012) is yet to be demonstrated. Tadpoles of these species develop in variable time from 25 days to two months depending on temperature of the water. Temperature in their breeding habitats ranges from 20–21°C to over 30°C.

Peltophryne florentinoi, one of the largest toads in Cuba, has a unique breeding habitat (type 4) (Fig. 3.29). It breeds in relatively small rain-filled cavities or *casimbas*, throughout the subcoastal sharp limestone of microphyll-evergreen forest (Cobos and Alonso Bosch 2018) in very few isolated localities (1) on the southern coast of Cuba between Girón, Matanzas, and (2) Guajimico, Cienfuegos (Cobos and Alonso Bosch 2016; Díaz *et al.* 2019). At these localities, the species' distributional pattern is associated with the microtopography of the karst substrate and its forest vegetation. Males call at night near selected and defended water-filled limestone depressions. Competition may be intense among calling males (Fig. 3.29d). The structure of egg clutches corresponds to ovipositional mode 3. The tadpoles are unique among Cuban amphibians by having a morphology typical of carnivorous larvae. The larvae were described and illustrated by Díaz and Cádiz (2008) for the first time. At present, the feeding habits of the larvae are a mystery, although it was proposed that they probably feed on worms, insect larvae, and other microinvertebrates that are abundant in their habitat (Díaz and Cádiz 2008). There are no observations of adult females of this species providing feeding-eggs to tadpoles in such breeding sites, as was suggested for *Peltophryne armata* from the Dominican Republic (Landestoy *et al.* 2018), another saxicolous species with a somewhat similar reproductive mode. Cobos and Alonso Bosch (2018)

Figure 3.28. Breeding habitat of (A) *Peltophryne fustiger*, at Bayate, Sierra del Rosario, Artemisa Province; (B) *P. taladai* at Arroyo Bueno, Nipe-Sagua-Baracoa Massif, Guantánamo Province; (C) tadpole of *P. fustiger* from Soroa, Sierra del Rosario, Artemisa Province; (D) tadpole of *P. taladai* from Arroyo Bueno. Males of both species call from shores and pebbles at night. Photographs: A. Cádiz (A); M. Potts (B); R. Fernández de Arcila (C); L. M. Díaz (D).

emphasized the relevance of the specialized breeding habitus of this species for conservation actions, since the reduction in forest cover may increase desiccation, levels of radiation, and temperature. The sympatric and closely related *P. peltocephala* uses larger breeding sites such as *cenotes* (sinkholes; e.g. El Polvorín) and ponds just a few metres apart from the breeding sites of *P. florentinoi*.

Figure 3.29. *Peltophryne florentinoi.* (A) General habitat at Girón, Ciénaga de Zapata, Matanzas Province; (B) calling male (arrow) near a small depression in the limestone where few larvae were found; (C) ventral view of a tadpole showing a very pale, translucent colouration, the development and blood irrigation of the internal gills, and the non-spirally coiled gut; (D) two males in territorial confrontation not far from a potential breeding site (note the human contamination around them as a result of trash-dumping). Photographs by L. M. Díaz.

In conclusion, we suggest the following decreasing order of priority for the conservation of Cuban toads' breeding habitats in terms of levels of specialization, the area of potentially available specific breeding sites along the archipelago, and the human impact on them: *P. florentinoi* > *P. cataulaciceps* > *P. longinasus* species-complex > *P. empusa* + *P. gundlachi* > *P. fustiger* + *P. peltocephala* + *P. taladai*.

B. *Eleutherodactylus* spp.

The direct-developing frogs use a broad spectrum of sites for oviposition in the terrestrial environment. Townsend (1996) considered three major categories: ground-nesting, litter-nesting, and arboreal-nesting. Within arboreal nesters were included sites like bromeliads, tree holes, and palm axils among others. Townsend's classification recognized the main tendencies, but additional subcategories according to species-specific selection of breeding sites may be described in the future.

Eight oviposition sites have been recorded for Cuban *Eleutherodactylus* (Table 3.2): (1) eggs on the ground, under the grass, and associated leaf litter; (2) semi-buried in an excavated depression in the ground, covered by various objects (including artificial ones); (3) inside decaying hollow trunks; (4) inside an apparently accommodated nest chamber under rocks, logs, or other objects, or in the sides of banks with an entrance hole connecting to the chamber; (5) inside bromeliads; (6) in the leaf litter of forests, e.g. under fallen palm fronds; (7) under rocks, logs, and other cover, but without a defined chamber or depression; (8) inside cavities in rocks. Ground and litter nesters (*sensu* Townsend 1996) are dominant over arboreal nesters as expected by the diversity of ground/leaf litter dwellers of Cuban *Eleutherodactylus*. Even semi-arboreal species like *E. auriculatus, E. eileenae,* and *E. glamyrus* are known to lay eggs in the litter. *Eleutherodactylus leberi,* another species with semi-arboreal habits, selects rock cavities as breeding sites (Rodríguez and Alonso 2009). Very likely, *E. mariposa* breeds in similar sites considering its diurnal refuges (Díaz and Cádiz 2008). In some of the ground and litter nesters, eggs are heavily covered by fine soil particles that contribute to their camouflage (Fig. 3.30).

Bromeliad frogs are known to lay a smaller number of eggs for their body size than do other species (Hedges *et al.* 1992; Bignotte-Giró 2019; Bignotte-Giró *et al.* 2022). This fact may be related to the narrow and constricted ovipositional sites available in bromeliads. Adult frogs are morphologically adapted by having depressed bodies (Hedges *et al.* 1992), allowing them to attend relatively large hygroscopic eggs in a small, confined space. Few observations *in-situ* (L. M. Díaz, pers. obs.) suggest that diamond-shaped arrangement of eggs in the clutches of bromeliad frogs result from the position of the male's legs during his protective covering of eggs (Fig. 3.30e–f). In the Guanahacabibes Peninsula, Pinar del Río Province, the bromeliad *Tillandsia utriculata* in semideciduous forest was reported to be the preferred habitat of *Eleutherodactylus* cf. *olibrus* (García-González *et al.* 2014). Also, Bignotte-Giró *et al.* (2022) documented bromeliads of the genera *Catopsis, Guzmania, Tillandsia,* and *Vriesea* that, depending on locality, were most frequently used for oviposition by *Eleutherodactylus guantanamera* and *E. ionthus* (see Table 3.2). Sabagh *et al.* (2017) emphasized the importance of documenting interactions between bromeliad-dwelling frogs and their respective bromeliad hosts to guarantee the conservation of these mutualistic relationships and avoid co-extinctions. However, arboreal/bromeliad-dwelling frogs also inhabit the axils of native palms in open areas like Los Indios, Isla de la Juventud (e.g. *E. staurometopon*); sometimes these frogs are associated with introduced Traveler Trees (*Ravenala madagascariensis*) in Soroa, Artemisa Province (*E. olibrus*); or with banana trees, *Musa* spp., in Alto del Naranjo, Sierra Maestra, Granma Province (*E. ionthus*) and *Pandanus* spp. (*E. guantanamera*) at La Gran Piedra, Sierra Maestra, Santiago de Cuba Province. Very likely, the axils of leaves of those plants also are used as breeding sites. A pair of *E. guantanamera* in Ojito de Agua, Alejandro de Humboldt National Park, Guantánamo Province, was detected inside a tree hole, 1.5 m high, by tracking the male's unusual vocalizations (L. M. Díaz, pers. obs.). The male was deep inside the very narrow cavity, and the female was fore-positioned facing the entrance. Whether or not this constitutes an alternative breeding site for species remains for future study.

Table 3.2. Ecological tendency – of 21 species of Cuban *Eleutherodactylus* – defined as: **1** = grass dweller; **2** = ground/leaf-litter dweller; **3** = semiarboreal; **4** = riparian; **5** = arboreal/bromeliad; and **6** = saxicolous. Also, clutch size (number of eggs), breeding sites, nest attendance (parental care). We use lowercase superscripts (in sequential alphabetical order) to highlight different references when there is more than one observation/datum for a given species.

TENDENCY AND *SPECIES*	CLUTCH SIZE	BREEDING SITE	NEST ATTENDANCE	SOURCE
1 *E. adelus*	3–4	Directly on the ground, under a thick cover of herbaceous vegetation and decayed leaves.	Not observed.	Díaz *et al.* (2003)
1 *E. varleyi*	4	Directly on the ground, under grass and leaf litter.	Female near the eggs.	Estrada (1987)
2 *E. albipes*	42	Inside a nest chamber (70 mm x 40 mm) underneath a rock on the ground, surrounded by leaf litter.	Female covering eggs in protective posture.	Alonso *et al.* (2005)
2 *E. atkinsi*	42–94[a], 17–44[b]	Eggs semi-buried in the ground[a] to fully exposed[c]. Under human objects (metal sheaths, tin cans) or under leaf litter. Apparently uses communal oviposition sites. Clutches with different development stages are found together.	Male or female close to the oviposition site.	[a]Novo *et al.* (1987); [b]Alfonso *et al.* (2017); [c]Y. Alfonso, pers. comm.
2 *E. dimidiatus*	24[d], 52[e], 60[f]	[d]Eggs in a small depression in the ground, under a fallen log. [e]Eggs under fallen fronds of the palm *Prestoaea montana*, directly on the ground. Eggs under leaf litter[f,g] (Fig. 3.30c).	Female partially covering eggs[d]; pair in amplexus covering eggs[e].	[d]Estrada (1987); [e]Bignotte-Giró (2019); [f]Schwartz and Henderson (1991); [g]L. M. Díaz, pers. obs.
2 *E. iberia*	1	Under leaf litter.	Female near egg (Fig. 3.30g).	Estrada and Hedges (1996b); A. R. Estrada, pers. obs.
2 *E. intermedius*	5	Inside a nest chamber (20 mm wide) externally communicated through a hole, in a stream-side bank.	Female in direct contact with eggs; male inside the chamber, calling from near the entrance.	Estrada (1992)
2 *E. limbatus*	1	Under vegetable debris in a rock cavity; under the rotted butt of a fallen bromeliad.	One female near one of the eggs; no mention of care.	Dunn (1926a)
2 *E. planirostris*	21[h], 3–26[i], 4–13[f]	In a fallen bromeliad[h]. Eggs semi-buried in the ground to fully exposed, in tight groups or dispersed[i,j]. In the internal area of lattice bricks lying on the ground[j]. Under stones, vegetal debris, underneath wet piles of cardboard[f]. On the soil of garden pots[g]. Apparently uses communal oviposition sites[j,g].	Males actively defend the territory where several clutches were laid; females have been observed close to or covering eggs with body[i,j,g].	[h]Dunn (1926b); [i]Goin (1947); [f]Schwartz and Henderson (1991); [j]Iturriaga and Dugo-Cota (2018); [g]L. M. Díaz, pers. obs.
2 *E. tetajulia*	6	Inside a hollow tree-fern's decomposing trunk on the ground[k]; inside a hollow, decomposing log[l].	Male and female in the nesting site, in contact with eggs (Fig. 3.30j).	[k]Estrada and Hedges (1996a); [l]Díaz and Cádiz (2008)
2 *E. zugi*	3	Directly on the ground, under leaf litter.	Male near the eggs.	Díaz and Cádiz (2008); L. M. Díaz, pers. obs.

(Continued)

Table 3.2. (Continued)

TENDENCY AND *SPECIES*	CLUTCH SIZE	BREEDING SITE	NEST ATTENDANCE	SOURCE
3 *E. auriculatus*	8[l], 11[m], 20[a]	Eight eggs in the leaf litter[l].	Not observed.	[a]Novo *et al.* (1987); [l]Díaz and Cádiz (2008); [m]Henderson and Powell (2009)
3 *E. eileenae*	26	On the ground, under leaf litter.	Male in body contact with eggs (Fig. 3.30d).	L. M. Díaz, pers. obs.
3 *E. glamyrus*	6–11	Hole in the ground, beneath a stone; pair in amplexus in the leaf litter laid 11 eggs in a collecting bag.	Female in contact with with 6 eggs.	[b]Estrada and Hedges (1997)
3 *E. leberi*	27	Inside a natural cavity in the wall of a limestone outcrop.	Male in body contact with eggs.	Rodríguez and Alonso (2009)
4 *E. cuneatus*	43	Under a rock on the banks of a wet mountain-road.	Not observed.	Estrada and Hedges (1998)
4 *E. riparius*	27, 35, 42	Under a rock, or 'among dirt and gravel', near the water of stream at Soledad, Cienfuegos.	Not observed.	Dunn (1926b)
4 *E. rivularis*	42	Eggs in small depression in the ground, near the river-shore, covered by a piece of cloth and moss sheath (Fig. 3.30i)	Not observed.	Díaz *et al.* (2001)
5 *E. guantanamera*	8–13[n], 12[g]	Inside bromeliads.	Males and females associated with eggs, but most clutches guarded by one male[n]. Male in body contact with 12 eggs[g] (see Fig. 3.30e–f). 'Diamond-shaped' distribution of eggs in the clutch was in correspondence with the position of male's limbs while covering them.	[n]Estrada (1990), as *E. varians*; [g]L. M. Díaz, pers. obs.
5 *E. melacara*	5–13[o], 11[g]	Inside bromeliads (Fig. 3.30h)	Adults in the same axil with eggs.	[o]Hedges *et al.* (1992); [g]L. M. Díaz, pers. obs.
6 *E. zeus*	48–74	On the karstic floor of a cave chamber, semi-buried or exposed (Fig. 3.30k–l)	Female in body contact with eggs; male 1 metre from one clutch.	Alonso Bosch *et al.* (2015)

Parental care in *Eleutherodactylus* seems to be much more widespread than previously thought. Attendance by a parent avoids desiccation, cannibalism, and attacks by insects and fungi (Wells 2007). This behaviour is a complex social trait, and specific aspects of care differ between males and females (Vági *et al.* 2019). Monroe and Alonzo (2014) mentioned that parental care by males is more common in frogs than is care by females, and no *Eleutherodactylus* species displayed biparental care. The available information, however, shows that in Cuban anurans, attendance can be either by males, females, or both. Parental care is definitely more elaborate in some species than in others. Yet in species such as *E. planirostris,* in which parental care was not believed to

Figure 3.30. Oviposition sites, parental care, and comparative size of eggs and adults of several Cuban species of *Eleutherodactylus*. (A) Clutch of *E. auriculatus* in leaf litter; eight eggs in total, only seven are seen in the picture at Nuevo Mundo, north slope of Monte Iberia, Nipe-Sagua-Baracoa Massif, Guantánamo Province; (B) paratypes of *E. beguei* in amplexus surrounded by nine large, recently laid eggs, from La Munición, Nipe-Sagua-Baracoa Massif, Guantánamo Province; (C) part of a clutch of 60 eggs of *E. dimidiatus*, in the leaf litter, from La Gran Piedra, Sierra Maestra, Santiago de Cuba Province; (D) twenty-six eggs of *E. eileenae* guarded by one male *in-situ* at Cabrillas, Luis Lazo, Pinar del Río Province; (E) twelve eggs of *E. guantanamera* inside a bromeliad at Arroyo Bueno, La Melba, Nipe-Sagua-Baracoa Massif, Guantánamo Province; (F) the same clutch as E at an earlier stage of development, protected by one male *in-situ*, (G) paratype *E. iberia* and its egg, at Arroyo Sucio, Monte Iberia, Nipe-Sagua-Baracoa Massif, Guantánamo Province; (H) eggs of *E. melacara* inside a bromeliad at Pico Turquino National Park, Sierra Maestra; (I) forty-two eggs of *E. rivularis* in a depression in the ground, near the shore of Jibacoa River, Las Mercedes, Sierra Maestra, Granma Province; (J) male (back) and female (front) of *E. tetajulia*, both paratype specimens, with a clutch of six eggs under the female inside the hollow trunk of a rotten tree fern, 10 cm above the ground (arrows point toward the eggs), from Monte Iberia, Nipe-Sagua-Baracoa Massif, Guantánamo Province; (K) forty-eight eggs of *E. zeus*, on karstic soil sediment in a depression in the Santo Tomás cave system, Moncada, Viñales, Pinar del Río Province; (L) female *E. zeus* brooding eggs beneath her at the same locality. Photographs by L. M. Díaz (A–D, H, I); M. Potts (E, F); M. Lammertink (G, J); and S. L. del Castillo (K, L).

occur at all (Goin 1947; Schwartz and Henderson 1991; Townsend 1996), there is now evidence of uniparental and biparental attendance at nests (Iturriaga and Dugo-Cota 2018). *Eleutherodactylus planirostris* and *E. atkinsi* (Alfonso *et al.* 2017) have communal aggregations in selected breeding places, where males and females defend territories giving protection to multiple egg clutches and offspring. Male-only attendance has been recorded in arboreal and semi-arboreal species (*E. guantanamera*, *E. melacara*, *E. leberi*, and *E. eileenae*). Female-only attendance was described by Alonso *et al.* (2005) for *E. albipes*. Figure 3.30j shows male and female *E. tetajulia* (same mentioned by Estrada and Hedges 1996a) guarding six eggs inside a hollow, rotten tree-fern trunk about 10 cm above the ground. A similar behaviour was observed by Estrada (1992) in the closely related species *E. intermedius* (see Table 3.2). An alternative explanation for these two species is that a female is only temporarily on the nest compared to the male, who very likely cares for more than one clutch by different females. Another species with apparent biparental care is *E. zeus*, in which the female recognizes the nest and covers the eggs with her body (Fig. 3.30l) while the nearby male probably engages in territorial assertion (Alonso Bosch *et al.* 2015).

Dunn (1926b:143) discovered that *Eleutherodactylus limbatus* (then in the genus *Sminthillus*) had a single-egg clutch in terrestrial situations and indicated that the eggs 'differed from eggs of *Eleutherodactylus* in being alone, and in having a less firm outer coat, so that it gave the impression of being wet and sticky rather than dry and clean'. He observed three eggs (different clutches) in total, two in the wild (Table 3.2) and one in captivity glued to a rock. Estrada and Hedges (1996b) and Hedges *et al.* (1995) concluded that one egg was the typical clutch size in the tiny species of the *E. limbatus* species-group (*E. cubanus*, *E. iberia*, *E. limbatus*, *E. orientalis*). Their observations were supported by the presence of one oviductal egg in most species. In *E. iberia* (Table 3.2; Fig. 3.30g), a female was collected a few centimetres from a single egg in the wild (diameter 3.5 mm versus 10.5 mm of female's SVL; A. Estrada, field notes). However, for *Eleutherodactylus jaumei*, also in the *E. limbatus* species-group, Estrada and Alonso (1997) reported two developed eggs in the right oviduct of two females. Dunn (1926b) mentioned up to four developing eggs inside the body of the female in *E. limbatus*; therefore, it cannot be ruled out that several clutches of one egg each could be produced at short intervals during the same reproductive season.

In other species, there are records of females that have laid eggs in collecting bags, plastic containers, or in terraria under captive conditions, with no information available on breeding sites or parental care from the wild. This is the case for (1) *E. beguei* (nine eggs laid and one oviductal egg retained; Fig. 3.30b; Díaz and Hedges 2015); (2) *E. bresslerae* (42 eggs; Díaz and Cádiz 2008; L. M. Díaz, pers. obs.); (3) *E. guanahacabibes* (9–11 unfertilized eggs in a terrarium; Díaz and Estrada 2000); and (4) *E. planirostris* (19, 21, and 25 eggs in captivity; Dunn 1926c). Hedges *et al.* (1995) mentioned 24 oviductal eggs in *E. turquinensis*. Additional but indirect evidence of clutch size was provided by Bignotte-Giró (2019) through observation of mature ovarian follicles in *E. auriculatus* (n = 13–20), *E. cuneatus* (n = 25–80), *E. dimidiatus* (n = 35–68), *E. gundlachi* (n = 5–9), and *E. intermedius* (n = 7–9 as the regular tendency, but up to 14). Their data are in the range of wild-found clutches in some species but exceed the number of eggs in others. For a number of *Eleutherodactylus*, Henderson and Powell (2009) reported clutch sizes based on unpublished data from S. Blair Hedges, with no comments on breeding sites or parental care: *E. auriculatus* (11 eggs), *E. eileenae* (37 eggs), *E. goini* (27–29 eggs), *E. klinikowskii* (8 eggs), and *E. varleyi* (9 eggs). Ruíz (1987) and Schwartz and Henderson (1991) reported egg clutches for *E. varians* in bromeliads, 30 and 12 eggs respectively, and *E. auriculatus* (20 eggs) in the same breeding site, but due to lack of precise localities the information is confusing because of the taxonomic splitting of these species during the past few decades.

Figure 3.31. Pair of *Osteopilus septentrionalis* (Hylidae) at the moment of oviposition in a side pool at Río Bayate, Sierra del Rosario, Artemisa Province. Photograph by L. M. Díaz.

C. *Osteopilus*

Osteopilus septentrionalis (Fig. 3.31) is the most generalized frog in Cuba, inhabiting many terrestrial habitats, from sea level to the highest elevations. Breeding sites are also very diverse. Larvae develop extreme lentic and lotic morphological adaptations in different habitats (see illustrations by Díaz and Cádiz 2008), from lentic-benthic forms to lotic-benthic-adherent, stream-lined types. Contrastingly, while the tadpoles of the three Hispaniolan species of the genus *Osteopilus* tend to be particularly adapted to different aquatic habitats (Díaz *et al.* 2015), in Cuba, *Osteopilus septentrionalis* is the only apparent species (Heinicke *et al.* 2011), and it occupies all environments. Adults tend to be conservative in morphology throughout Cuba, but the appearance of tadpoles is remarkably variable. Larvae coexist with those of all the other biphasic anurans in Cuba. The ecological and reproductive flexibilities of this frog are, undoubtedly, among the main adaptations that make this species opportunistic, successful, and consequently so well distributed. According to the classification of Altig and McDiarmid (2007), clutches consist of single-tiered groups of eggs floating on still water (Fig. 3.25e). Observed oviposition in Sierra Maestra, however, showed that floating clutches very often sink soon after being laid, forming clumps with interstices in lentic situations even without the intervention of rain (Fig. 3.25f).

VI. Population estimates and patterns of activity

Baseline information on population sizes, dynamics, structure, and natural fluctuations is lacking or poorly documented for most species. The limited data are derived from scattered observations and short-term studies. An understanding of species' demography, reproductive patterns, pathogen-host interactions, and factors affecting them is crucial for establishing long-term monitoring programmes and effective conservation strategies (e.g. Russell *et al.* 2019).

Fong G. (2007) made preliminary estimates of amphibian populations for two years (2002–2004) in a montane rainforest at La Gran Piedra, 1,200 m asl, which is an eastern extension of the Sierra Maestra, Santiago de Cuba Province. He used three methods: (1) visual surveys along 50-m long × 2-m wide × 1.5-m high transects; (2) 8 × 8-m quadrats randomly situated; and (3) acoustic surveys at call-points, every five minutes between 20:00 and 22:00 h. Herein, his data are expressed as the number of individuals/100 m^2 (instead of individuals/quadrat) for allowing comparisons with other contributions. Seven species were surveyed in the montane rainforest: *E. auriculatus, E. dimidiatus, E. guantanamera* (formerly reported as *E. ionthus*), *E. intermedius, E. gundlachi, E. limbatus,* and *E. ricordii.* Highest densities were found in the ground/leaf litter dwellers *E. dimidiatus* and *E. limbatus,* with 2.8–10 and 0.8–8.1 individuals/100 m^2, followed by *E. intermedius* (1.1–2.3 individuals/100 m^2), *E. auriculatus* (1.6–2.0 individuals/100 m^2), and *E. gundlachi* (0.5–1.7 individuals/100 m^2). One individual of *E. ricordii* was detected once in the quadrats, and therefore was the least abundant species.

Along the transects, *E. dimidiatus* was the most commonly detected species, with 9–25 individuals/transect, followed by *E. gundlachi* (2–21 individuals/transect), *E. intermedius* (0–12 individuals/transect), *E. auriculatus* (1–6 individuals/transect), *E. limbatus* (0–6 individuals/transect), and *E. ricordii* (0–4 individuals/transect). The abundance of juveniles and adults varied seasonally, with juveniles doubling the adult numbers during the dry season. During this study, calling activity varied seasonally in every species, with positive peaks in March–April and July–August, and the highest records in July 2003 and March 2004. The acoustic activity was reduced in October to January and ceased by November.

Fong G. *et al.* (2010) surveyed *Eleutherodactylus cuneatus,* a riparian species not included in the report mentioned above, during June 2003 to June 2004, also at La Gran Piedra, along a tributary of the Indio River. They found densities of adult frogs of 17–26 individuals/100 m^2 and 25–61 juveniles /100 m^2. High numbers of visible individuals were found all night, and the peak of activity occurred between 20:00 and 22:00 h. Adults along the stream were observed in all survey periods, but higher numbers were present in the warmer months between March and September. High activity of the adults along the stream was also significantly associated with the calling activity of males away from it. A high proportion of individuals moved along the stream, reaching distances as far as 82 m over long-term periods, but short-term movements were small (up to 4 m in only five days). In this species, the temperature seems to have a greater influence on reproductive activity than does rainfall; however, more extensive surveys are needed over several years to understand the role of these parameters in the life-history of this frog.

Calling behaviour is a good indicator of the composition of assemblages and interspecific isolating mechanisms, as well as of species' presence, daily patterns of activity, breeding apogee, and population size in different habitats at a given time. However, patterns in calling activity, and their relationship with environmental factors, have been documented only for seven species of Cuban *Eleutherodactylus*: *E. eileenae* at Las Terrazas, Artemisa Province (Alonso *et al.* 2001), *E. auriculatus, E. cuneatus, E. gundlachi, E. guantanamera,* and *E. limbatus* at La Gran Piedra, Santiago de Cuba province (Bignotte-Giró *et al.* 2019), and *E. varleyi* at National Botanical Garden, Havana province (Cádiz and Iturriaga 2020) (Fig. 3.32). These patterns were determined for species with sustained, not sporadic, vocal activity.

Alonso *et al.* (2001) described a bimodal pattern of vocal activity in *Eleutherodactylus eileenae* with one peak at dusk (17:00–21:00 h in June, 16:00–19:00 h in October, and 18:00–20:00 h in January) and a second one at about dawn (05:00–08:00 h). In *Eleutherodactylus varleyi,* Cádiz and Iturriaga (2020) observed as many as five peaks of vocal activity overnight in July, with the most intense calling occurring between 02:00 h and 03:30 h. The vocal activity was also high at 20:30

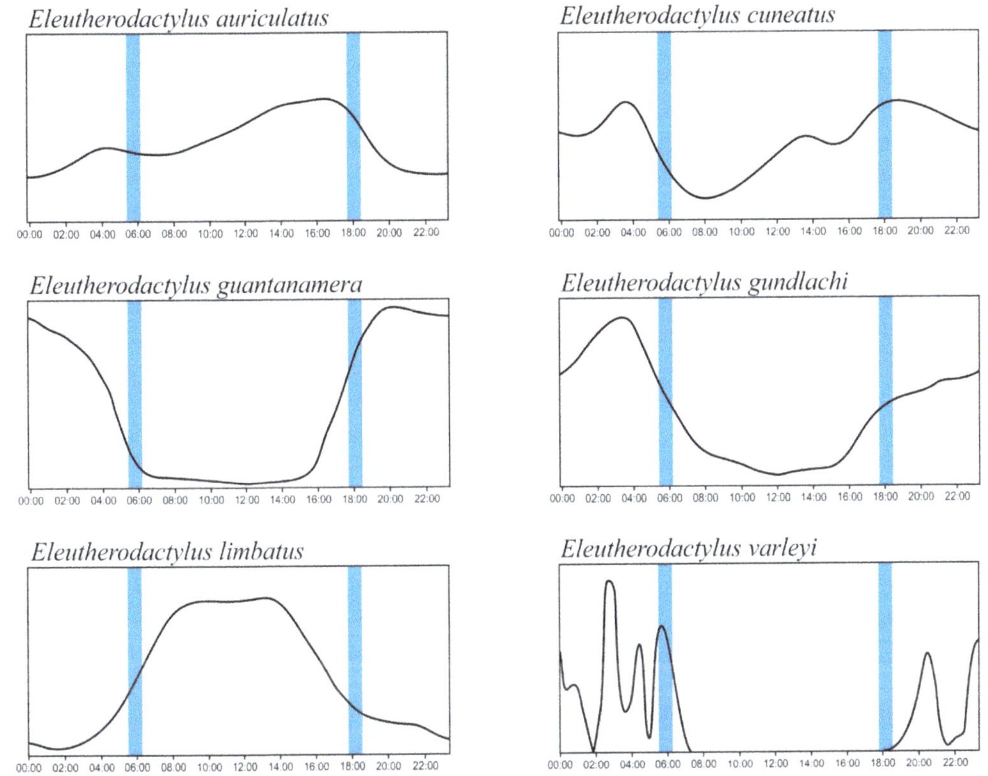

Figure 3.32. Bioacoustical activity patterns in six species of Cuban frogs of the genus *Eleutherodactylus*. Vertical bars represent dawn and dusk. Graphics for the first five species were modified from Bignotte-Giró (2019), based on an average curve obtained from sampling between February and May 2003. The graphic of *E. varleyi* was modified from Cádiz and Iturriaga (2020), based on observations made in July 2007.

h, 23:00 h, and 06:00 h. Although a clear decrease in vocal activity was observed in September, two peaks, one at the beginning and the end of the night, were still evident, demonstrating a tendency toward a bimodal (dusk/dawn) pattern. In a multi-species study, Bignotte-Giró *et al.* (2019) found a tendency for interspecific acoustic overlapping between 03:00–5:00 h (peak just before dawn) and 18:00–22:00 h (peak at dusk). Species that exhibited minimal temporal overlapping with others were *E. guantanamera,* with dominant, sustained nocturnal choruses, and *E. limbatus,* which actively vocalized in the morning between 08:00–12:00 h, and less intensely or sporadically at other times.

Six patterns of calling activity have been identified in Cuban *Eleutherodactylus* (Fig. 3.32) (Alonso *et al.* 2001; Bignotte-Giró *et al.* 2019; Cádiz and Iturriaga 2020): (1) an intense peak before dusk and a lower peak before dawn, with diurnal activity variable and nocturnal activity suppressed sometimes (*E. auriculatus*); (2) two peaks of similar intensity, respectively at dawn and dusk, with some vocal activity variably produced during the daytime (*E. cuneatus,* somewhat similar in *E. eileenae*); (3) sustained vocal activity most of the night, with a discrete peak after dusk (*E. guantanamera*); (4) one intense peak before dawn, with vocal activity starting at dusk and increasing after midnight (*E. gundlachi*); (5) sustained diurnal activity, with sporadic vocalizations at night (*E. limbatus*); (6) multiple peaks of nocturnal activity, with the highest peaks at midnight and before dawn (*E. varleyi*). Depending on favourable weather conditions and the frogs' reproductive frenzy, some of the crepuscular or nocturnal species extend vocal behaviour into daytime.

Anecdotal reference to patterns of calling also has appeared in the original descriptions of species. For example, Estrada and Hedges (1997a) reported data on the acoustic activity of males of *E. principalis* that inhabit the understorey of the pine forests in the mountains of Sagua-Baracoa. *Eleutherodactylus principalis* calls during the early morning, between 00:00 h and 06:00 h, and during the evening twilight. Their choruses are also heard before, during, and after rains. The places used for calling are exposed horizontal surfaces: fern fronds, small leaves, and twigs located 0.8–2.0 m above the ground. Díaz *et al.* (2001) mentioned that males of the riparian frog *E. rivularis* began to call at dusk (after 18:00 h), although one individual was heard in mid-morning. They reported 18 adult individuals of this species, 12 calling males and 6 females, in 100 m^2 of the pebble-covered shore of the Naguas River, Sierra Maestra, in February 2000 (note: there was a previously unnoticed typographic mistake in their original publication, which incorrectly cited 10 m^2 as the area; L. M. Díaz, pers. obs.). In *E. feichtingeri*, peaks of dense choruses, formed by hundreds of individuals in open areas, were registered late in the afternoon (~16:00–19:00 h), remarkably early for lowland situations (Díaz *et al.* 2012). Vocal activity gradually decreases after this period until only a few individuals remained calling, followed by a second, but distinctly lower peak, shortly before dawn. These authors reported 16 calling males in less than 100 m^2 at about 20:00 h in October 2010 at Marea de Limones, Eastern Cuba.

Díaz and Estrada (2000) reported diurnal vocalizations by captive females of *E. guanahacabibes* prior to oviposition. Fong G. and Bignotte-Giró (2021) found a similar behaviour by a female of *E. cuneatus* and interpreted it to advertise receptivity and attract, find, or localize a mate. For *E. guanahacabibes* and perhaps the closely related *E. planirostris*, diurnal vocal behaviour extending to midday may indicate a peak in the breeding activity and the period when most eggs are laid. More incidental information on the calling activity of other Cuban *Eleutherodactylus* was provided by Díaz and Cádiz (2008).

All the evidence previously examined indicates that most of the Cuban *Eleutherodactylus* are crepuscular or nocturnal, but species with diurnal calling activity and species vocally active nearly all day are also present on this island. Slight modifications of these patterns occur, depending on such factors as environmental changes, number of sympatric species, and competition for resources.

For toads in the genus *Peltophryne*, species in the *P. longinasus* complex are among the few Cuban anurans with diurnal habits. The peak of calling by males occurs in the morning in *P. longinasus* (L. M. Díaz and A. Cádiz, pers. obs.), with females arriving from late morning to early afternoon and most amplexus and oviposition occurring in late afternoon before dusk. *Peltophryne dunni* is also diurnal, crawling and calling from leaves of the Butterfly Ginger Lily (*Hedychium coronarium*), and branches and stones associated with mountain streams, between 08:00 h and 18:20 h (Valdés de la Osa and Ruíz García 1980a).

Sampedro *et al.* (1982) and Sampedro and Berovides (1985) reported all-night activity in *Peltophryne peltocephala* at Cienfuegos Botanical Garden, based on the active collection of adult individuals between 21:00 h to 06:00 h in May (rainy season) and November (dry season). Two activity peaks were detected between 22:00–23:00 h and 01:00–03:00 h in the rainy season, which moved about one hour later in the dry period. Although toads, other than those in the *P. longinasus* complex, display a nocturnal pattern of activity, small, recently metamorphosed individuals are active by day, sometimes even exposing themselves to extremely high temperatures in sunny areas. This behaviour has been observed in *P. fustiger*, *P. peltocephala*, and *P. taladai*, of which recently metamorphosed toadlets are concentrated on the beaches of rivers and lagoons from where they begin migrating to surrounding forests. At Los Indios, Isla de la Juventud, (L. M. Díaz, pers. obs.), toadlets of *P. cataulaciceps* of about 0.8 cm to 1 cm long were seen congregated in sections of open, sun-exposed areas at 13:00 h, and their bodies were at a very high temperature (unfortunately, not measured). This is an interesting and pristine field of research in Cuba, since no studies have ever addressed how physiology and activity patterns relate to the environment.

Breeding patterns are known only for a few Cuban anurans, despite the importance of such data for the interpretation of the effect of climatic change on anuran populations (Blaustein *et al.* 2001). Iturriaga *et al.* (2014) documented the reproductive cycle of *Eleutherodactylus planirostris* in Havana, western Cuba. The pattern of reproduction was seasonal and discontinuous in both sexes. Bimodal vitellogenesis was observed, occurring in April–June and September–October. Spermatogenesis began in March, with high spermatogenic activity during June and July, and ending in November. The increase in temperature positively influenced vitellogenic activity. Ovarian mass was positively correlated with photoperiod and ambient temperature, whereas the fat-body mass was negatively correlated with ambient temperature for females and photoperiod for both sexes.

In the Eastern region, Bignotte-Giró *et al.* (2021) documented the reproductive patterns of five species of *Eleutherodactylus* (*E. auriculatus, E. cuneatus, E. dimidiatus, E. gundlachi,* and *E. intermedius*) during 2003–2004 in La Gran Piedra, Sierra Maestra, and found a prolonged reproductive pattern in the five species with a fall in the reproductive intensity in the winter months. The temperature had a stronger influence than rainfall on reproductive cycles in these frogs, as deducted from the decrease in the weight of the ovaries and the number of vitellogenic follicles, absence of mature females or reduction of their abundance, and disappearance of male vocal activity in November to February. More recently, Bignotte-Giró *et al.* (2022) found that *E. guantanamera* bred almost year-round (May to October, but also January, February, and April) in mountainous areas and suggested that the stability of microclimatic conditions and availability of large-sized bromeliads favoured both the continuous breeding activity and the protection of eggs against desiccation.

Fong G. (2007) mentioned that at Siboney, a coastal locality just south of La Gran Piedra and one of the driest areas of Cuba, *E. etheridgei* and *E. ionthus* were opportunistic breeders, increasing calling activities in periods of rainfall. Explosive breeders like *Peltophryne cataulaciceps, P. empusa,* and *P. gundlachi,* also depend on specific patterns and intensity of precipitation. Luis M. Díaz and A. Cádiz suggest that very likely, the two main factors known to affect explosive breeders – the consistency of precipitation during the previous 48 h to 72 h and the amount of rainfall during the previous 24 h (according to Ulloa *et al.* 2019) – are also critical for these species. Valdés de la Osa and Ruíz García (1980a) reported that calls in *P. longinasus longinasus* are produced year-round, mostly between June and August, but activity of males was also observed in stream pools in May (Cádiz *et al.* 2018). In captivity, they always showed lower acoustic activity on rainy days, but were very active afterwards (see section XII). This suggests that they probably avoid the temporary increased stream-flow associated with precipitation.

In conclusion, the patterns of breeding activity seem to vary among species and types of habitats throughout the island. In the mountains of the Eastern region, like La Gran Piedra, the results discussed above suggest that more stable climatic conditions favour continuous or prolonged breeding patterns. Seasonality may be more accentuated in lowland and middle-elevation species that are more affected by seasonal patterns of precipitation. In the Western region of the island, the effect of cold fronts and dry periods also are seasonally more accentuated. We agree on the need for the development of a robust research agenda aimed at understanding how the activity/breeding patterns of Cuban anurans would be affected by the expected changes in patterns of rainfall (seasonality, consistency, and amount of precipitation) and a reduction in the extent of cloud cover in montane rainforests as a result of climatic change.

VII. Distributional patterns

We recognize four patterns of distribution of Cuban amphibians as defined in section IV (Table 3.1). Of the currently known native species, 55% are micro-localized, 22% have a regional distribution, 13% are widely distributed throughout the island, and 10% are meso-localized (Table 3.1). These patterns mean that the Cuban amphibian fauna is characterized by a high level of local endemism, with species restricted to single mountain ranges or particular lowland ecosystems.

The two species with the broadest geographic distribution in Cuba are *Osteopilus septentrionalis* and *Eleutherodactylus atkinsi*. Also among the species with the widest distribution are *Eleutherodactylus auriculatus*, *E. dimidiatus*, and *E. limbatus*, occurring throughout most of the country in suitable forested (even anthropized) habitats, either in lowland or montane areas. They were probably even more cosmopolitan when the island was extensively forested. *Peltophryne empusa* and *P. gundlachi* also are widespread and usually found together in most plains, natural savannahs, and agricultural lands. Finally, *Eleutherodactylus planirostris* is widely distributed, but extensive areas of the country are occupied by closely related species.

The majority of Cuban anurans (81%) is present in mountainous regions where humidity is relatively high and stable year-round, with low evaporative water loss, and lower temperature conditions compared to lowland areas (compare Figs. 3.4 and 3.33). Nearly 51% of the native species are mostly confined to climatic zone I. In Sierra Maestra, the seasonal effect is more evident below 700–800 m asl, but the climate is noticeably more stable, humid, and colder at higher elevations.

About 45% of the species are present in the lowlands and hills where humidity is seasonally stable, with high evaporative rates and warm conditions. In addition, 19% are adapted to conditions in the lowlands and on cays with relatively low, unstable humidity, high evaporative rates, and high temperatures. Of these, only one species (*Eleutherodactylus etheridgei*) is exclusively distributed in the driest area of Cuba (usually with less than 600 mm of rainfall per annum). On the other hand, *Peltophryne peltocephala*, *P. taladai*, *Osteopilus septentrionalis*, *Eleutherodactylus atkinsi*, *E. feichtingeri*, *E. ionthus*, *E. limbatus*, *E. planirostris*, *E. riparius*, *E. ronaldi*, and *E. simulans* are present in either of the three climatic regions.

Figure 3.33. Map of categories of anuran species richness in Cuba. Progressively darker colours represent ascending numbers of species; bold-lined areas indicate the maximum absolute richness (18 species).

A. Species richness

Using a simple summation approach, we represented a species richness map by combining distributional areas of each Cuban native anuran species. These were obtained by drawing a polygon including all the known locations from maps in Díaz and Cádiz (2008), complemented and updated with papers and descriptions of taxa published after that date. As a result, we found that the species richness of native anurans in the Cuban archipelago ranges from zones with no anurans to zones with a maximum of 18 species (Fig. 3.33). Zones without recorded anuran species cover 331 km^2, representing 0.3% of the Cuban territory, and include some small cays (see Estrada 2012). The highest anuran richness in Cuba (14–18 species) is concentrated in the four mountain ranges of the main island, characterizing each massif either partially or in total (Fig. 3.33). As expected, zones with the highest richness are in highlands covered by forests, especially those persistent remnants of mature or well-preserved areas (Díaz and Cádiz 2008).

Species richnesses between 9 and 13 species occur along inland lowlands, the transitional zones to the highlands, and on Isla de la Juventud (Fig. 3.33); lower species richnesses (4–8 species) are located in lowlands, on the northern coastal cays, and the Zapata swamp. Finally, zones with the lowest species richnesses (three or fewer species) are present exclusively in east side coastal ranges and some isolated cays; due to its small area (0.5% of the Cuban land-cover), only a few of the zones with the lowest species richnesses (see end of this section) are visible in Figure 3.33.

The two middle categories of species richness (9–13 species and 4–8 species) are represented in 86% of the Cuban territory (Fig. 3.34). More than one-third of Cuba (37%) includes zones inhabited by 8 and 9 anuran species, each zone with an area of more than 19,000 km^2 (40,090 km^2 total). The two zones with the highest species richness also cover a smaller area (1,617 km^2) but double the area of the four zones that harbour three or fewer species (including zones with no species, 584 km^2) (Fig. 3.34); however, these two high-richness zones each harbour triple the maximum species richness of the former (i.e. 17 and 18 species; Fig. 3.34). The zones with the absolute maximum richness value of 18 species occupy 963 km^2 (1% of the territory; Fig. 3.34) and are limited geographically to the higher elevations of eastern Cuba, specifically to Cuchillas de Toa y Moa in Sagua-Baracoa range and the Sierra del Turquino in Sierra Maestra range (Fig. 3.33).

In summary, rainforests and the vegetation complex of the *mogotes* (limestone or karst hills) are the two types of vegetation with the highest diversity of amphibians in Cuba (25 and

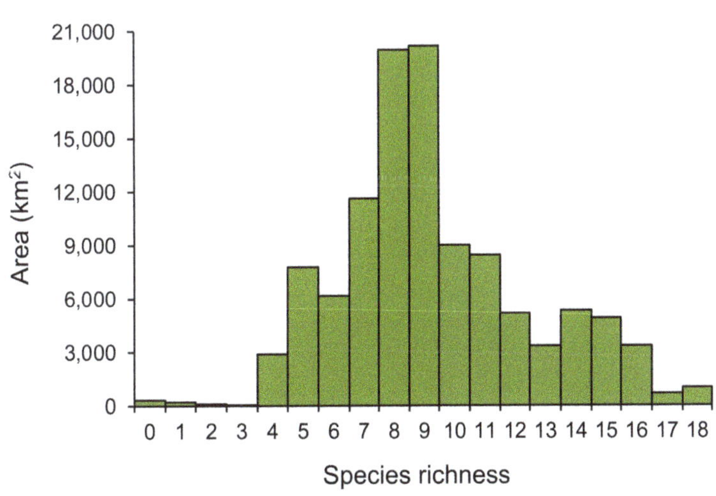

Figure 3.34. Histogram displaying the area (in km^2) occupied by zones of species richness in Cuba.

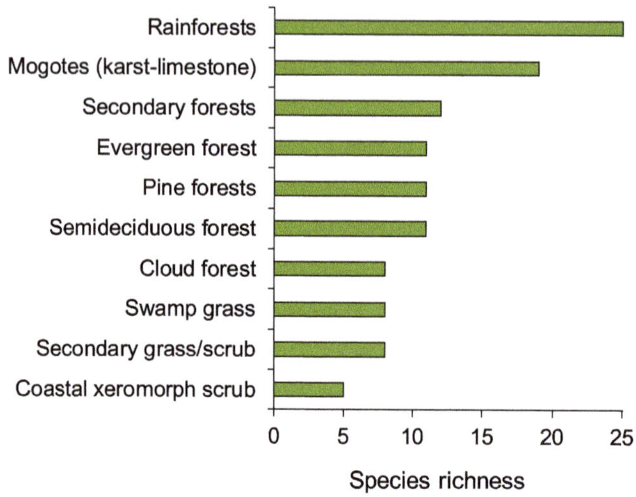

Figure 3.35. Eleven representative vegetation types and their associated amphibian richness.

19 species, respectively), followed by secondary forests (12 species) and semideciduous, evergreen, and pine forests (11 species each). Cloud forests, swamp-grass, and secondary grass-scrub communities hold up to eight species, while a lower diversity of the represented vegetation types is that of coastal and subcoastal scrubs (Fig. 3.35); species richness in mangroves is lowest (or zero).

B. Zoogeography

We analysed similarity, using Ward's distance (Ward 1963; Großwendt *et al.* 2019) based on the absence (0) or presence (1) of species of anurans to assess the relationship between species richness and zoogeography among the 45 physiographic divisions defined by Acevedo (1989). We also used distributional records documented and/or cited by Díaz and Cádiz (2008), Heinicke *et al.* (2011), Rivalta González *et al.* (2014), García-Padrón and Hernández Ledesma (2018), and Díaz *et al.* (2019), as well as unpublished information. We omitted very few of the mapped physiographic areas (Fig 3.3), for example, the Alturas de Pizarras del Norte and some cays south of mainland Cuba, because they have no species or just one (in the case of cays) or have been poorly surveyed and the amphibian fauna is relatively unknown. Nevertheless, the result of this analysis revealed three major zoogeographic regions for anurans that we prefer to call 'faunas' to avoid confusion with physiographic regions: Western, Central, and Eastern (Fig. 3.36).

Hedges (1999) also found three zoogeographic regions but our results differed in various ways. In our study, for example, the Eastern region clusters apart from Western and Central regions due to its very exclusive and highly diverse anuran fauna. Maniabón (Figs. 3.3 and 3.36 [number 23]), which is part of the Camagüey-Maniabón physiographic region of Acevedo (1989), is included in the Eastern zoogeographic cluster. The main mountain ranges are centres of diversity and local endemism; lower hills and plains have different levels of diversity, connectivity, or number of locally endemic species. We noted that the accuracy of these results might be limited because of *a priori* criteria for the delimitation of physiographic regions, gaps in information about species' distributions, and limited taxonomic resolution. Nonetheless, our results revealed a coherent pattern of the structure and richness of the Cuban anurans in a geographic context that may be closer to reality than previously shown. We describe this pattern in detail next.

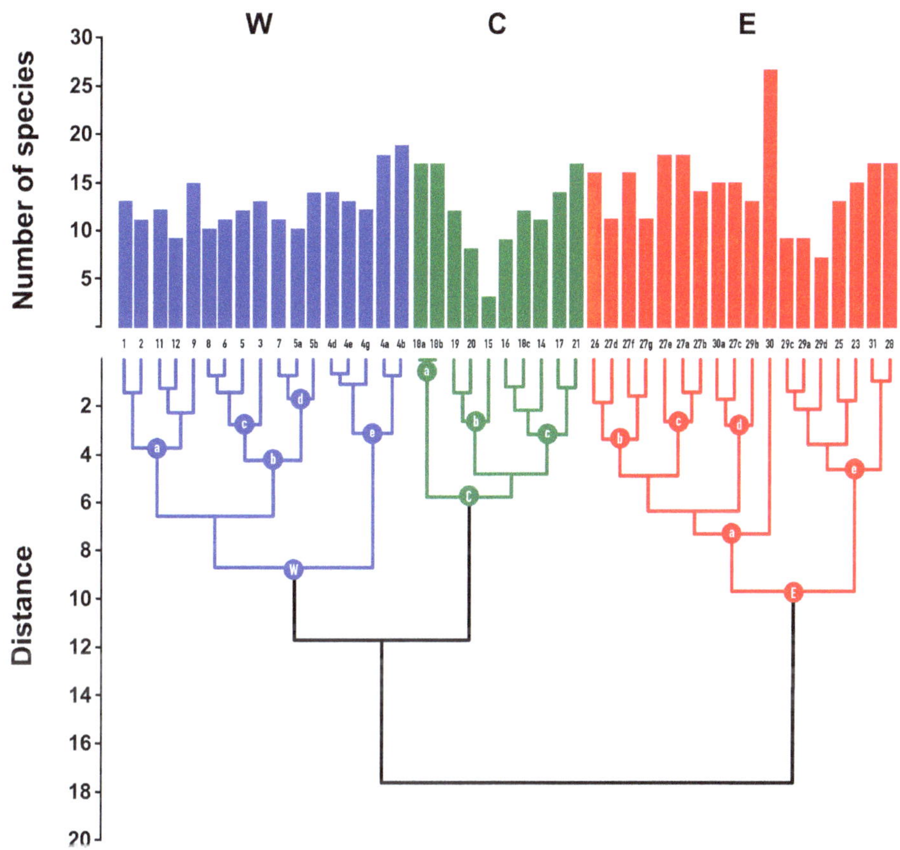

Figure 3.36. Cluster analysis using Ward's method, showing the hierarchical similitude of the batrachofauna from different physiographic regions, subregions, and sections (based on Figure 3.3), and the number of native species per division. (W) Western region, in blue. (W-a) Guanahacabibes (1) + northern savannahs of Pinar del Rio (2), Isla de la Juventud (11 + 12) + Zapata Peninsula (9); (W-b) savannahs and low karstic hills; (W-c) southern savannahs of Pinar del Río (3) + southern and intercalated plains of Artemisa, Havana, Mayabeque, and Matanzas (3, 5, 6, 8); (W-d) centre hills of Mayabeque-Matanzas + Jaruco (5a-7) and Sierra de Camarones-Pan de Matanzas (5b); (W-e) Cordillera de Guaniguanico (4). (C) Central region, in green. (C-a) Massif of Guamuhaya (18a-b). (C-b) Peripheric plains of Camagüey-Maniabón and northern cays (15, 19, 20). (C-c) Northern heights, intermontane plains, and low hills (14, 17, 18c, 17, 21). (E) Eastern region in red. (E-a) Main montane and premontane areas (26, 27, 29b, 30); (E-b), Mayarí (lowlands and highlands) + Puriales de Caujerí, Meseta de Maisí Zapote + San Antonio del Sur (semidesert southern coast); (E-c) Cuchillas de Baracoa-Sierra del Purial + Cuchillas de Toa y Moa + Llanura Moa-Baracoa. (E-e) Peripheric plains, low hills, and basins (23, 25, 28, 29a, 29c, 29d, 31).

1. Western fauna

In the West, there are 30 native species (43.5% of native Cuban amphibians). This region contains a mosaic of habitats, and many amphibians are exclusively linked to some of them. The Cordillera de Guaniguanico (Figs. 3.3 and 3.36: 4a–4e; 20 species) is the centre of diversity. However, the amphibian fauna is not homogeneously distributed there due to the area's intrinsic geological and ecological heterogeneity. Sierra de los Órganos (Figs. 3.3 and 3.36: 4a; 18 species) and Sierra del Rosario (Figs. 3.3 and 3.36: 4b; 19 species) have the highest diversity and they share a large part of their faunas. Meseta de Cajálbana (Figs. 3.3 and 3.36: 4e; 13 species) and Alturas de Pizarras del Sur (Fig. 3.36: 4d; 14 species) are discretely associated within the Cordillera. Plains (Figs. 3.3 and 3.36: 3, 5, 6, 8) and low karst hills (Figs. 3.3 and 3.36: 5a [10 species], 5b [14 species], and 7 [14 species]) extend from Artemisa to Matanzas Provinces forming an inclusive unit with some connectivity with Sierra del Rosario. The lowlands of Zapata Peninsula (Figs. 3.3 and 3.36: 9; 15 species), Isla de la Juventud (Figs. 3.3 and 3.36: 11 [12 species], 12 [9 species]), the Guanahacabibes Peninsula (Figs. 3.3 and 3.36: 1), and the plains in the north of Pinar del Río Province (Figs. 3.3 and 3.36: 2) represent another distinctive group sharing similar composition of their amphibian faunas.

Peltophryne cataulaciceps occurs in a very particular kind of habitat: the quarzitic sand savannahs covered by pine forests and palms, located in the western half of Pinar del Río and the northern part of Isla de la Juventud. The extensive areas of pine forests along the sandstone hills (Alturas de Pizarras) are the only known habitat of *P. longinasus* and *Eleutherodactylus adelus*, only present in a few localities. Similar pine-forest habitats, but on ophiolitic soil, are in Meseta de Cajálbana, and *P. longinasus cajalbanensis* is confined there. Sierra de los Órganos has prominent and dominant *mogotes*. One species, *Eleutherodactylus klinikowskii*, only occurs there. Some *mogotes* are present in the westernmost part of Sierra del Rosario, being the easternmost limit of *E. klinikowskii* so far on record. Sierra de la Güira, in San Diego, is a locality where *E. klinikowskii* and the closely related species *E. zugi* occur sympatrically but, apparently, not syntopically (i.e. incapable of interbreeding). *Eleutherodactylus zeus* and *E. klinikowskii* are exclusively associated with the forested karstic outcrops of Cordillera de Guaniguanico, and the habitat is best represented in Sierra de los Órganos and the westernmost part of Sierra del Rosario. *Eleutherodactylus zeus* seems to be more common in Sierra de los Órganos despite being widely distributed in Guaniguanico.

The Guanahacabibes Peninsula is a relevant lowland in the westernmost tip of Cuba characterized by its relatively preserved forests and biodiversity. There is one locally endemic species in Guanahacabibes and surrounding areas: *Eleutherodactylus guanahacabibes*, a close relative of *E. planirostris* (Hedges *et al.* 2008a; Heinicke *et al.* 2011). Other species in this area are widely distributed either at a regional or national level. The most common habitat in the peninsula, where amphibians mainly occur, comprises the forests associated with limestone. *Eleutherodactylus guanahacabibes* frequents limestone habitats, but also is highly opportunistic and not exclusively restricted to them.

Eleutherodactylus zugi, is present in Sierra del Rosario, Meseta de Cajálbana, and the Alturas de Pizarras. The closely related species *E. erythroproctus*, from Sierra de Camarones and Pan de Matanzas (Mayabeque and Matanzas Provinces, respectively), is extremely similar in morphology to *E. zugi*, and is associated with karst outcrops. Recently, the first author (unpublished data) found a population related to *E. zugi* and *E. erythroproctus* (voucher specimen deposited at the Museo Nacional de Historia Natural de Cuba) at sea level near the coast, in Cueva de Saturno, Península de Hicacos (Matanzas), which suggests some connectivity between the westernmost mountain ranges and the limestone hills and plains in the eastern part of the Western region. This *E. 'zugi'* population, the only one reported near the coast, shares habitat with *E. thomasi* aff. *zayasi* (new locality record), which is also present in Sierra de Camarones and Pan de Matanzas. Another example of a somewhat similar distributional pattern is exhibited by *E. symingtoni*, a relatively rare saxicolous species whose distribution comprises Sierra de los Órganos, Sierra del Rosario, lowlands near the town of Artemisa, and Sierra de Camarones in Mayabeque Province. Another saxicolous species, *E. pinarensis*, exclusive to this region, is the frog with the most bizarre distribution because it is apparently absent in most of the Cordillera de Guaniguanico (see Schwartz 1959a; Schwartz and Henderson 1991; Díaz and Cádiz 2008) but then is found in Península de Guanahacabibes, Isla de la Juventud, Meseta de Anafe, and the limestone lowlands and hills extending near the border with the Central region. During the ancient connection between Isla de la Juventud, Pinar del Río, and the area east to Zapata Peninsula, their habitats (e.g. sandy savannahs, coastal and subcoastal karst) were once continuous, which may explain the allopatric populations of *E. pinarensis* and *P. cataulaciceps*.

The Zapata Peninsula is widely covered by a relatively young swamp over a limestone platform. Not many species of native amphibians are associated with swamp vegetation, but even the semi-arboreal and widely distributed species *Eleutherodactylus auriculatus*, a grass frog (*Eleutherodactylus varleyi*), the common *E. planirostris*, and the riparian *E. riparius* inhabit extensive areas covered by sawgrass *Cladium jamaicensis* and little patches of vegetation that combine trees, palms, and bushes (known as *petenes* and *montañas*), which usually emerge in the middle of the wetland. However, the surroundings of Bahía de Cochinos (Bay of Pigs) and the area from there to the east of the Peninsula, is covered by an extensive forest growing on limestone that is the habitat of the highest diversity of amphibians, the most notable of which is *Peltophryne florentinoi*. This toad is saxicolous and specialized for breeding in small rain-filled depressions or *casimbas*. The species was considered to only occur along the coastal limestone plateau at the surroundings of Girón and Guasasas (Alonso Bosch and Cobos 2016) but, recently, Díaz *et al.* (2019) extended its distribution to Guajimico (Cienfuegos Province) in the Central region.

The subregion of Zapata is an interesting zone of biogeographic confluence and also the eastern and western distributional limit of some species. In fact, the habitat on the eastern part of Zapata is similar to the southern half of Isla de la Juventud and extends to the southern coast of Cienfuegos and Sancti Spiritus. Another apparently specialized species with limited distribution, *Peltophryne florentinoi*, is found throughout a fragmented subcoastal limestone habitat from the eastern extreme of Zapata to Guajimico in Central Cuba (Díaz *et al.* 2019). Therefore, a hypothesis could be that the westernmost distribution of *P. florentinoi* results from its range expansion from the Central region where suitable habitats are best represented; but this has to be confirmed since only one individual was seen in Guajimico. A similar distributional pattern is exhibited by *Eleutherodactylus thomasi trinidadensis*, also present at Girón (Zapata), which extends its range east to the vicinity of Trinidad (Sancti Spiritus Province). *Peltophryne peltocephala* is a very successful large, meso localized toad, widely distributed in the Eastern and Central regions, including cays. Its distribution reaches Zapata and also Isla de la Juventud. On the other hand, the largest amphibian in Cuba (Alonso Bosch *et al.* 2014), *Peltophryne fustiger*, is restricted to the Western region, with its known easternmost limit north to Zapata (Schwartz and Henderson 1991; Díaz and Cádiz 2008; Rivalta González *et al.* 2014).

2. *Central fauna*

This region has the lowest diversity of amphibians of the three analysed zoogeographic regions: 22 species (32% of the total native fauna) and only four regional endemics. Low and relatively arid areas dominate the region. Three species (*Peltophryne florentinoi, Eleutherodactylus eileenae,* and *E. varleyi*) are shared exclusively with the West and two with the East (*Peltophryne taladai* and *Eleutherodactylus feichtingeri*). The Macizo de Guamuhaya (Guamuhaya Massif) is the highest mountain range of the region and the most relevant area for amphibians. It is divided into Sierra de Trinidad and Sierra de Sancti Spiritus (Figs. 3.3 and 3.36: 8a–18b; each with 17 species). These mountains hold three endangered local endemics: *Peltophryne dunni, Eleutherodactylus emiliae,* and *E. casparii*. The second-most important locations in amphibian diversity are in the Sierra de Cubitas and Sierra de Najasa, the main hills at Camagüey Province, part of a huge central plateau (Figs. 3.3 and 3.36: 21; 17 spp.). Most species in the region (55%) are widely distributed throughout the island or are meso-localized (32%). Sierra de Cubitas is the easternmost known limit for *E. greyi* (Schwartz 1959c; Schwartz and Henderson 1991), another species unique to the zoogeographic region. Díaz and Cádiz (2008) were cautious about the status of this species there because there have been no recent sightings; however, voucher specimens (Schwartz 1959c) deposited at AMNH were verified by L. M. Díaz and A. Cádiz. The region is an important zone of sympatry of two meso-localized species: *Eleutherodactylus varleyi* (mostly distributed to the West) and *E. feichtingeri* (mostly distributed to the East), both occurring in the vicinity of Sierra de Cubitas (Díaz *et al.* 2012).

Sierra de Najasa, near the border with Las Tunas Province, has the second-most important regional amphibian assemblage, inclusive of *Eleutherodactylus limbatus*. It is also the easternmost geographic limit of *E. eileenae* (Schwartz and Henderson 1991; Díaz and Cádiz 2008; Rivalta González *et al.* 2014). Nine native species inhabit the northern cays comprising the Archipelago Sabana-Camagüey, and all species are either meso-localized or widely distributed on the mainland. These species are *E. atkinsi, E. feichtingeri, E. planirostris, E. riparius, E. thomasi, Peltophryne empusa, P. gundlachi, P. peltocephala,* and *Osteopilus septentrionalis*. At least on some cays, some of the species could represent recent introductions from the mainland due to the development of tourism in the area (Ángel Arias, pers. comm.). For example, Arias *et al.* (2005) did not find *E. planirostris* in Cayo Santa María at the beginning of hotel construction, but now the species is increasingly common, probably by dispersing with landscaping and building materials.

3. Eastern fauna

This region has 47 native species, 68% of the native Cuban amphibians. The most diverse amphibian faunas of the region (26 species) are located in Sierra Maestra (Figs. 3.3 and 3.36: 30), particularly locations corresponding to Pico Turquino and Pico Bayamesa National Parks, and Nipe-Sagua-Baracoa Massif, with the highest diversity and richness of locally endemic species at Cuchillas de Toa-Moa and Cuchillas de Baracoa-Sierra del Purial (Figs. 3.3 and 3.36: 27a and 27e, each with 18 species). These areas have the best representation of rainforests on the main island. Sierra Maestra and Nipe-Sagua-Baracoa are approximately 45 km apart, and these two mountain ranges share about 34% of the total number known from the whole region. Therefore, the local endemism, mostly restricted to these mountains, is very high, with 13 species found only in Nipe-Sagua-Baracoa and 12 in Sierra Maestra. The region is an important centre of diversity of most lineages of *Eleutherodactylus* (see section IV. B.), of which three species-groups (*cuneatus, toa,* and *ricordii*) are found there exclusively.

The extreme peak of diversity in the Sierra Maestra (highest elevation 1,974 m asl at Pico Turquino), is not only due to its geological heterogeneity, with remarkable variation both in vegetation and climatic conditions, but also because it is the mountain range with the most pronounced elevational transitions of amphibian assemblages. Peaks of diversity are in the transitional zone between 750 m asl and 1,000 m asl (Díaz and Cádiz 2008), but in local assemblages, the maximum number does not exceed 14–18 species. Hedges (1999) also recognized a natural decrease of species richness with elevation and the highest diversity at mid-heights. Four species have been found on the top of Pico Turquino: *Eleutherodactylus albipes, E. glamyrus, E. gundlachi,* and *E. melacara*. *Eleutherodactylus jaumei* is distributed from sea level to 923 m asl (Rodríguez *et al.* 2012) in the transition between the evergreen forest and the montane rainforest. *Eleutherodactylus cubanus* (same species-group) occupies a similar leaf litter ecological niche but at higher elevation. The Sierra Maestra is where more species of bromeliad frogs of the genus *Eleutherodactylus* (*E. ionthus, E. guantanamera,* and *E. melacara*) occur. *Eleutherodactylus ionthus* and *E. melacara* are sympatric between 750 m asl and 1,200 m asl at several localities, but the latter remains as the only bromeliad species at the highest points. The Sierra Maestra has several species distributed above 700–750 m asl. The semi-aquatic and highly endangered *Eleutherodactylus turquinensis* is one of the species distributed above 700–800 m asl in the Sierra Maestra. This mountain range is also unique by having up to three riparian species of *Eleutherodactylus* (*E. cuneatus, E. rivularis,* and *E. turquinensis*).

The limestone premontane area (Figs. 3.3 and 3.36: 29b; 13 species), known as Carsos de Baire, extends along the northern slope of the Sierra Maestra and contains two locally endemic amphibians: *Eleutherodactylus leberi* (semi-arboreal) and *E. michaelschmidi* (saxicolous). The Sierra de La Gran Piedra (Figs. 3.3 and 3.36: 30a; 15 species), considered as the easternmost extension of the Sierra Maestra, clusters with Meseta del Guaso-Los Montes (Figs. 3.3 and 3.36: 27c; 15 species), which is part of the heights of Nipe-Sagua-Baracoa (Figs. 3.3 and 3.36). The similarity is most likely due to some level of peripheral connectivity. Meseta del Guaso-Los Montes also has two species associated with limestone outcrops: *E. mariposa* (a semi-arboreal, locally endemic species) and *E. pezopetrus* (a saxicolous species also present in karstic areas of zone 26, Figs. 3.3 and 3.36). A notable species is *Peltophryne ramsdeni*, which is only known from this area; it has not been collected in almost half a century.

The lowlands (Figs. 3.3 and 3.36: 23, 25, 28, 29a, 29c, 29d, 31) cluster together. The major diversity is in (1) the intermontane valleys and basins, like the Valle Central (Figs. 3.3 and 3.36: 28; 13 species), which divides the Sierra Maestra from the western limit of the complex Nipe-Sagua-Baracoa orographic system; and (2) in the vicinity of Santiago de Cuba, where amphibians occur in the remnants of forest on the lowest levels of the Sierra Maestra. This topographic arrangement reveals some connectivity between montane areas through lowland forests (particularly those along streams) that probably covered the largest areas in the past. Among the few species that are shared with the other amphibian zoogeographic regions, particularly those adapted to lowland forests and/or open areas, are *Peltophryne taladai, P. empusa, P. gundlachi, P. peltocephala, E. auriculatus, E. atkinsi, E. dimidiatus, E. feichtingeri, E. riparius*, and *Osteopilus septentrionalis*. It is very likely that *Peltophryne empusa* and *Eleutherodactylus riparius* from the northern plains extended their range into the Valle Central reaching Guantánamo Province.

The southern coast of Santiago de Cuba and Guantánamo (Figs. 3.3 and 3.36: 27f, 27g) is a semidesert with marine terraces. The relatively high diversity, especially in the vicinities of Sierra del Purial, may be because the narrow semidesert becomes transitional with more humid mountainous slopes or has refugia of forested ravines and creeks. This part of the country, its driest area, is the habitat of the locally distributed endemic *Eleutherodactylus etheridgei*. The plains of Sagua de Tánamo (Figs. 3.3 and 3.36: 26) and the valley of Puriales de Caujerí (Figs. 3.3 and 3.36: 27d) are similar in their amphibian faunas (Figs. 3.3 and 3.36: 27f and 27g). These flatter areas are also surrounded by highlands and have sharp transitions in faunal diversity with the slopes of mountains where the anuran fauna is one of the most diverse in the country.

VIII. Threatened species and assessment of conservation

A. IUCN Red List categories and criteria

The perception of the conservation status of Caribbean amphibians has changed since Hedges' (1993: 298) first approach, who at that time concluded: 'The West Indian frog fauna has not undergone a general decline. No species is known to be extinct, although one Puerto Rican species (*Eleutherodactylus karlschmidti*) has not been seen in recent years despite efforts to locate it. Insufficient information exists on the other seven species that have not been seen recently to draw any conclusions regarding their status.' This situation had changed dramatically based on analyses performed nearly two decades later (Hedges and Díaz 2011). Unfortunately, several species of *Eleutherodactylus* from Puerto Rico now are presumed extinct (Ríos-López *et al.* in this volume, Chapter 8).

The first assessment at the regional scale was performed in Santo Domingo (Dominican Republic) during the first GAA workshop in March 2004 (see Stuart *et al.* 2008). In Cuba, 59 species of anurans were assessed and listed in the IUCN Red List (Table 3.1). Only a few additional species were further evaluated or re-evaluated afterwards (e.g. Díaz 2008; Rivalta 2008). After the first GAA, the IUCN's Red List comprised: 16 species (23% of the total number of species known at that time) Critically Endangered (CR); 24 (35%) Endangered (EN); 9 (13%) Vulnerable (VU); 2 (3%) Near Threatened (NT); and 8 (12%) Least Concern (LC). More recently, the second GAA concluded in 2021 and listed 15 species (22%) as Critically Endangered (CR), 22 (32%) Endangered (EN), 11 (16%) Vulnerable (VU), 2 (3%) Near Threatened (NT), and 19 (28%) Least Concern (LC). Therefore, 70% of the Cuban batrachofauna currently corresponds to species at significant risk of extinction.

The *Libro Rojo de los Vertebrados de Cuba* (Red Book of Cuban Vertebrates), edited by González Alonso *et al.* (2012), listed 28 threatened anurans (41% of the native batrachofauna), and followed the same categories and criteria as the IUCN Red List (Table 3.1). Of the 28 species assessed by González Alonso *et al.* (2012), 23 (82%) were considered Vulnerable (VU), 3 (11%) Endangered (EN), 1 (3.5%) Near Threatened (NT) and 1 (3.5%) Critically Endangered (CR). A generalized criterion of vulnerability was D2: a very restricted population, capable of becoming Critically Endangered or even Extinct within a very short time. Following the Red Book's partial assessment, the Ministerio de Ciencia, Tecnología y Medio Ambiente (Ministry of Science, Technology and Environment of Cuba; CITMA 2014) considered reptiles, instead of amphibians, as the group of Cuban vertebrates at most risk.

In this chapter, we consider a number of threats and stressors that usually act synergistically on amphibians, from loss and fragmentation of habitat as the most important threat, followed by chytridiomycosis, and climatic change. We strongly suggest that most Cuban anurans are actually at greater risk of extinction than other groups of vertebrates over a comparable timeframe, which makes them very important bioindicators and urgent targets for conservation. Declines and extirpations or extinctions may occur in a short period of time. Although not conclusive, first anecdotal evidence of species missing from places in which they were formerly relatively common under similar conditions during the same time of year (Cádiz *et al.* 2018), is an early warning that local declines and extirpations could be happening in Cuba.

Other assessments of conservation are available and were performed much earlier by Perera *et al.* (1994 a, b), Vales *et al.* (1998), and by the Conservation Assessment and Management Plans (CAMP) Workshops for Selected Cuban Species, as compiled by Amaro Valdés (2012). Cobos and Alonso Bosch (2016) suggested categorizing *Peltophryne* (*longinasus*) *ramsdeni* as Critically Endangered, Possibly Extinct (CR: A2ac) until further surveys confirm its status in the wild. Since most taxa in the *P. longinasus* species-complex were elevated to the specific level and the multiple threats they face, including habitat loss and chytridiomycosis, it would be better to extend the category of CR to all of them (Table 3.1). Rodríguez *et al.* (2017) suggested the category of Vulnerable (VU: D2) for *Eleutherodactylus cattus*.

B. Distributional pattern of threatened species

The geographic distribution (estimated as explained in section VII. A.) of threatened species richness – assessed as Critically Endangered, Endangered, or Vulnerable in the IUCN Red List (Table 3.1) – varies from zero to a maximum of 11 species throughout the archipelago (Fig. 3.37). The distribution is not homogeneous across the island, with some areas containing up to 20% of the total number of threatened species for Cuba. A high proportion of the Cuban native species of anurans is threatened (thus the similarity between Figures 3.3 and 3.1). However, zones with medium and high richness have a more restricted distribution than do more widely dispersed ones and the former ones are concentrated in the four mountain ranges and on Isla de la Juventud. Zones with no threatened anurans mainly occupy small geographic ranges (total area 1,357 km^2) located in coastal places dispersed around the main island. The zones with the absolute maximum value are located only at the top of the eastern highlands (828 km^2).

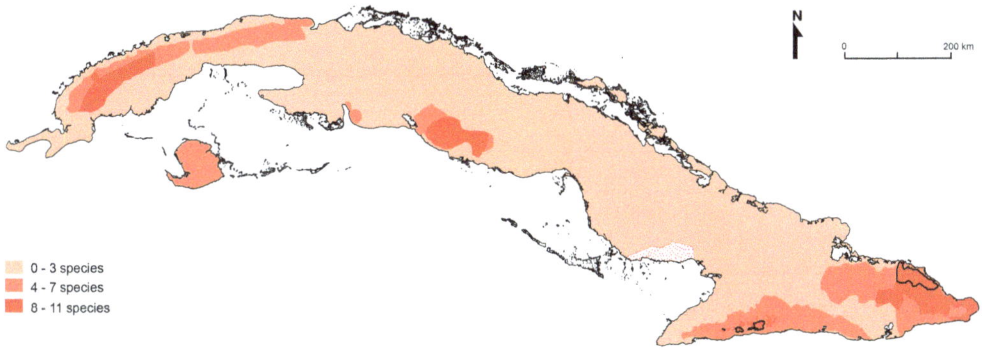

Figure 3.37. Distributional map of the threatened anuran species richness in Cuba. Progressively darker colours represent ascending species numbers; stippled and bold areas indicate the minimum (0 species) and the maximum richness (11 species), respectively.

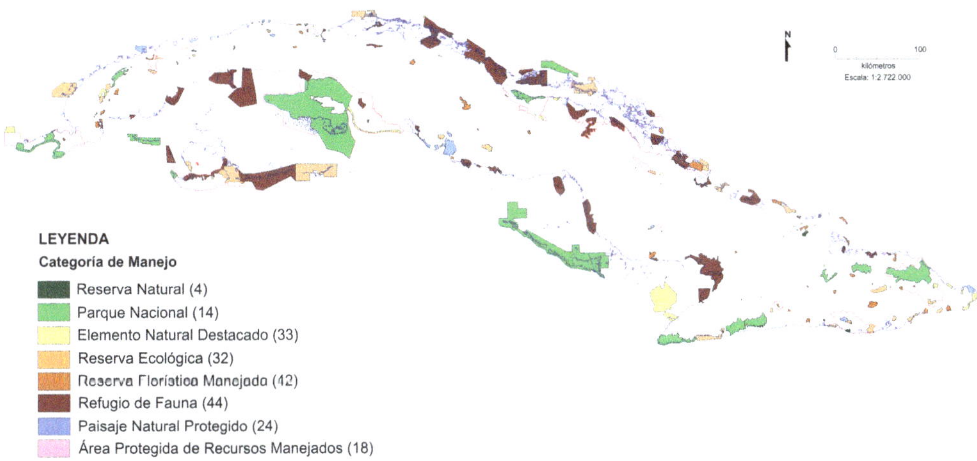

Figure 3.38. The Cuban National System of Protected Areas (SNAP). From Martínez *et al.* (2013).

IX. The National System of Protected Areas (*Sistema Nacional de Áreas Protegidas*)

The National System of Protected Areas or *Sistema Nacional de Áreas Protegidas* (SNAP) has 211 areas covering about 20% of the Cuban territory (Martínez *et al.* 2013; CITMA 2019) (Fig. 3.38; Table 3.3). The number of protected areas will be reviewed after completion of the 2020 new planning consensus. The SNAP Plan is a fundamental tool for the management of protected areas in Cuba. It analyses the existing gaps in biotic and ecological information at the level of landscapes, ecosystems, and species of plants and animals, and defines conservation goals every five years. Of the Cuban protected areas, 77 are considered as Protected Areas of National Significance, according to their well-preserved, unique, and representative natural biota, ecosystems, and landscapes. The other 134 are considered areas of Local Significance, having less geographic extension and lower level of need for conservation, or fewer general values.

Eight categories of management are defined in the Cuban Law, Decree 201 (GORC 1999), following a hierarchical order defined by the level of protection, management, and human impact. The areas forming the SNAP are assigned to any of the categories listed in Figure 3.38, according

Table 3.3. Extension of the National System of Protected Areas (SNAP) of the Republic of Cuba relative to its territory. * = from insular platform.

CUBAN TERRITORY	EXTENT (HA)	SNAP	
		EXTENT (HA)	PER CENT AREA
Terrestrial surface	10,988,410.00	1,885,951.79	17.2%
Marine surface*	6,988,057.65	1,744,394.38	25.0%
Total	17,976,467.65	3,630,346.17	20.2%

to IUCN (1994) criteria. Of these areas, 141 are legally approved by the Council of Ministers, the highest level of the Government having a working staff. Another 5 areas have staff, but no legal support, and the remaining areas have been proposed and delimited but are neither administrated nor legally approved. One of the current efforts to improve the conservation of Cuban biodiversity in protected areas is to provide them with an administrative staff and legal support at the highest level.

The SNAP developed according to several international protocols and agreements. For example, six wetlands of importance for conservation were included in the Ramsar Convention, as they were listed by CNAP (2013). Twenty-eight sites (including 56 protected areas) have been identified as Important Bird Areas or IBAs (Aguilar Mugica 2009). The Guanahacabibes National Park is within the protocols of the Specially Protected Areas and Wildlife, according to Perera (2013). UNESCO has recognized six Biosphere Reserves in Cuba (Sierra del Rosario, Península de Guanahacabibes, Cuchillas del Toa, Baconao, Buenavista, and Península de Zapata) and two Natural World Heritage Sites (Desembarco del Granma National Park and Alejandro de Humboldt National Park).

X. Representation of anurans in the National System of Protected Areas

Fong G. *et al.* (2015) argued that the SNAP does not provide enough protection to every amphibian species and does not satisfy the target of protection for half of the species, but most of the amphibian hotspots were represented in protected areas. The representation of anurans in the SNAP decreased relatively in 2009 when new species were described from unprotected localities. However, the representation increased again in 2014 after gathering new information on species' distributions. In the analysis performed here, we identified 65 species (94% of native species) in the SNAP; 60 of them (58 endemics and 42 endangered) are recorded in protected areas with the highest level of legal support. However, five endemic and endangered species (*Eleutherodactylus blairhedgesi, E. leberi, E. michaelschmidi, E. mariposa, E. pezopetrus*) are in designated areas which are not legally approved at the highest governmental level applicable (Fig. 3.39).

Currently, four species are not covered by any protected areas: *Eleutherodactylus adelus* (Rodríguez and García 2012a), *E. erythroproctus, E. rivularis,* and *Peltophryne ramsdeni.* The last species is reported for the Parnaso-Los Montes Ecological Reserve by Rivalta González (2012c), although as a subspecies of *P. longinasus*. However, the distributional record comes from the only known locality that, although located nearby, does not fall within the limits of the protected area. The habitats of *E. adelus* and *E. erythroproctus* are represented in protected areas close to

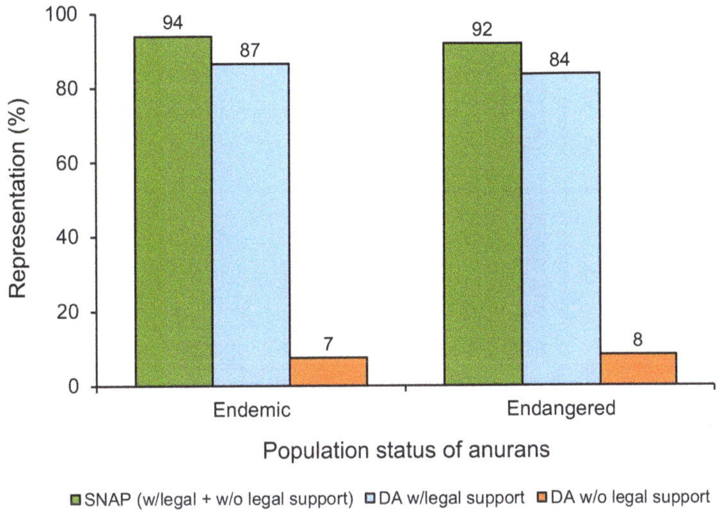

Figure 3.39. Representation (%) of endemic and endangered (CR, EN, VU, according to IUCN Red List) species of anurans in the National System of Protected Areas (SNAP). Columns represent species richness in the SNAP and The Cuban Vertebrates Red Book; in Green: w/ (with) legislative and w/o (without) legislative support. Selected Areas (SA) only with legislative support from the highest level of governmental hierarchy (in Blue: SA w/legal support; most effective for conservation goals), and Selected Areas that have not completed all actions required for legislative support from the highest level of governmental hierarchy (in Orange: SA w/o legal support; at the time, less effective for most conservation goals).

the frogs' distributional range (e.g. Cerro de Cabras and Sierra de Contadores-Cayo Ratones Flora Management Reserves for *E. adelus*; Galindo Flora Management Reserve and the Yumurí Valley Protected Area of Managed Resources for *E. erythroproctus*), but no records of these taxa are available inside the referred areas. Fong G. *et al.* (2015) analysed a peripheral distribution of *E. rivularis* within the northern limits of the Pico Turquino National Park, but most of its distributional range is within highly human-impacted habitats.

Peltophryne longinasus is protected at the species level, but the subspecies *P. l. longinasus* and *P. l. cajalbanensis* deserve different comment. Rivalta González (2012c) stated that *P. longinasus* was protected by the Sierra de San Carlos Ecological Reserve and Viñales National Park, but the known localities and the primary habitat of this taxon are outside the respective limits of those areas. More specifically, the protected areas mostly include karstic landscapes, not the sandstones hills inhabited by the toad. *Peltophryne l. cajalbanensis* is known only from Meseta de Cajálbana, in the periphery of the protected area Sierra Preluda-Cuabales de Cajálbana Flora Management Reserve, a narrow strip of pine forests. Some observations detected the toad at lower elevations in unprotected and severely impacted pine forest. However, this subspecies was also observed at Mil Cumbres Protected Area of Managed Resources (L. Y. García, pers. comm.), which includes part of the Meseta de Cajálbana.

In general, amphibian diversity is highly represented in the SNAP, but knowledge of the biology, ecology, and threats is still lacking for most species (Fong G. *et al.* 2010; Rodríguez 2012a). This deficiency is decidedly a weakness when it comes to implementing appropriate conservation actions and management plans. The design of the SNAP was based largely on vegetation, landscapes, and general values of biodiversity (CNAP 2002). However, a high level

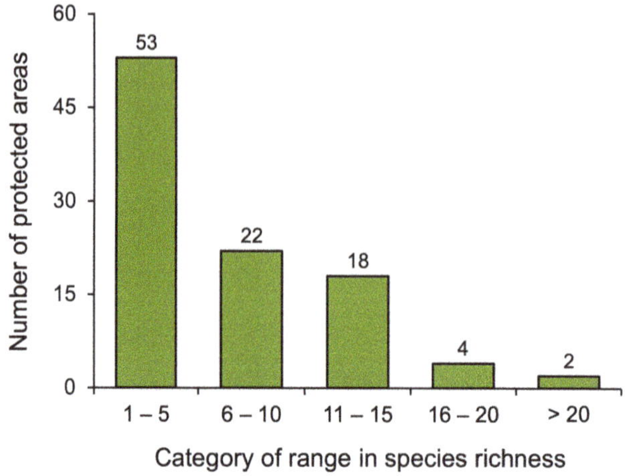

Figure 3.40. The relationship between species richness of anurans and the number of protected areas in the SNAP.

of representation of species of anurans in the SNAP is because landscapes and habitats critical for their conservation are mostly protected. For example, 89% of the types of landscapes identified for Cuba are well represented in protected areas as well as 94% of the vegetation types in the country (CNAP 2013). Moreover, in the main mountain ranges (Guaniguanico, Guamuhaya, Sierra Maestra, and Nipe-Sagua-Baracoa), where the greatest diversity and levels of local endemism occur (Díaz and Cádiz 2008), there are 64 protected areas that together conserve 27% of these montane ecosystems (4,983.7 km^2).

Although the representation of anurans is high, there are records of only one to five species in most areas in the SNAP (Fig. 3.40). Areas protecting more species of anurans are in the mountain ranges of the Eastern region (Cuchillas de Toa y Moa; Sierra Maestra), as reported by Fong G. *et al.* (2015), and within extensive wetlands (Guanahacabibes, Ciénaga de Zapata, north wetlands of Ciego de Ávila) (Table 3.4).

The greatest number of species are present in eight Protected Areas of Managed Resources, eight National Parks, five Ecological Reserves, two Flora Management Reserves, two Protected Natural Landscapes and one Outstanding Natural Feature (Table 3.4). The National Parks, together with Ecological Reserves and Protected Areas of Managed Resources, are areas harbouring the highest species richness, due to their geographic breadth and their diversity of habitats and ecosystems (Fig. 3.41).

Field studies should be carried out in areas for which information on biodiversity is scarce, which would promote a deeper understanding of the distribution and abundance of species and establish new protected areas that increase the extent of coverage for populations of anurans and improve connectivity within the SNAP. Our analysis was based only on the presence of amphibian species inside SNAP, but the percentage of their populations within the limits of these protected areas should be considered in assessing SNAP's coverage of Cuban amphibians (Fong G. *et al.* 2015). It is essential to develop effective management plans that may more effectively conserve ecosystems, habitats, and priority species.

Table 3.4. Protected areas with the highest number of species of anurans in Cuba. Categories of protected areas are: PAMR = Protected Area of Managed Resources; NP = National Park; PNL = Protected Natural Landscape; ER = Ecological Reserve; ONF = Outstanding Natural Feature; FMR = Flora Management Reserve. * indicates coverage of terrestrial area only.

CATEGORY	NAME OF PROTECTED AREA	CUBAN PROVINCES	AREA (KM²)	SPECIES
PAMR	Cuchillas del Toa	Guantánamo – Holguín	2,024.8*	23
NP	Alejandro de Humboldt	Guantánamo – Holguín	684.3*	22
NP	Turquino	Granma – Santiago de Cuba	232.1	19
PNL	Topes de Collantes	Sancti Spíritus – Villa Clara – Cienfuegos	201.4	18
ER	Lomas de Banao	Sancti Spíritus	60.9	16
PAMR	Península de Zapata	Matanzas	5,120.4*	15
NP	Pico Bayamesa	Granma	242.1	16
PAMR	Reserva de Biosfera Sierra del Rosario	Artemisa	255.8	15
PAMR	Reserva de Biosfera Baconao	Santiago de Cuba – Guantánamo	827.7*	15
ER	Loma del Gato – Monte Líbano	Santiago de Cuba	15.4	15
PAMR	Mil Cumbres	Pinar del Río – Artemisa	172.2	15
NP	Viñales	Pinar del Río	111.2	13
PAMR	Península de Guanahacabibes	Pinar del Río	1,019.4*	13
PAMR	Sierra de Cubitas	Camagüey	130.8	13
ER	Sierra de la Güira	Pinar del Río	20.7	13
ONF	Yunque de Baracoa	Guantánamo	21.5	13
PAMR	Humedales del Norte de Ciego de Ávila	Ciego de Ávila	1,275.3*	12
NP	La Mensura - Piloto	Holguín	84.9	12
NP	Pico Cristal	Holguín – Santiago de Cuba	185.4	12
FMR	Pico Galán	Guantánamo	4.4	12
ER	Pico Mogote	Santiago de Cuba	14.9	12
ER	El Salón	Artemisa	5.8	11
PNL	Gran Piedra	Santiago de Cuba	30.7	11
NP	Ciénaga de Zapata	Matanzas	2,818.6*	11
FMR	San Ubaldo-Sabanalamar	Pinar del Río	52.1	10
NP	Guanahacabibes	Pinar del Río	238.8*	10

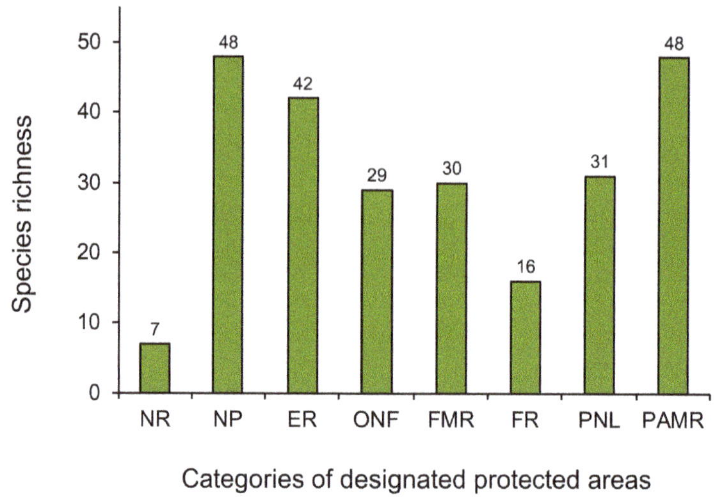

Figure 3.41. Number of amphibian species in each management category of SNAP. See Figure 3.39. (NR) Natural Reserve; (NP) National Park; (ER) Ecological Reserve; (ONF) Outstanding Natural Feature; (FMR) Floral Management Reserve; (FR) Faunal Refuge; (PNL) Protected Natural Landscape; (PAMR) Protected Area of Managed Resources.

XI. Threats

Amphibians are undergoing an unprecedented global decline, arising from an exceedingly complex network of multiple causes that interact in various synergistic and multiplicative ways (Tyler *et al.* 2007; Heatwole 2013; Grant *et al.* 2020). The identified threats to Cuban amphibians are treated *seriatim* below.

A. Loss of habitat

Loss and fragmentation of habitat is the most critical threat to Cuban amphibians. During the colonial period, Spain exploited forests for construction, fuel, obtaining ashes for fertilizers, together with logging and burning to clear for coffee plantations on the hills. Later, in the neocolonial period, forests were devastated by intensive cultivation of sugarcane and livestock.

Thomas Barbour, a prominent herpetologist of the twentieth century, remarked in his book *A Naturalist in Cuba* (Barbour 1945):

> Few countries have changed in appearance more completely during the last two or three decades than Cuba. The demand for sugar and molasses during the last war caused the felling of forests and clearing of land on a tremendous scale. I am glad, therefore, that our first railroad trip through the island took place during the leisurely days of 1908, across a forested country. Our journey was largely through forested country. Now nothing but waving fields of cane may be seen from the railroad.

An estimated 89.2% of forest cover in 1812 fell to 54% by 1900; in 1958, there was only 14% of the original forest cover left (Funes Monzote 2008). After centuries of human impact, Cuba has only about 31.23% of forest cover (CITMA 2019), and deforestation persists at variable rates depending on geographic location. Lowland forests were particularly extirpated in most of the archipelago. These uses of the land have driven amphibian populations into reduced relict areas (Díaz and Cádiz 2008).

Capote *et al.* (2006) reported that the level of fragmentation of forest is medium to high at most localities with high amphibian richness. There is a high degree of fragmentation and isolation of nuclei of natural vegetation; the most natural areas with the greatest representation of terrestrial biota only cover about 10% of the Cuban archipelago (CITMA 2019). Fong G. (2009) pointed out that the density of ruderal paths and roads per square kilometre is a measure of humans' fragmentation of habitat. In the eastern mountain ranges, the amphibian diversity hotspot of Cuba, the average density of ruderal paths is 656 m/km^2 in Sierra Maestra and 677 m/km^2 in Massif Nipe-Sagua-Baracoa, and the respective density of roads is 97 m/km^2 and 49 m/km^2. Even some of the protected areas, such as the Protected Area of Managed Resources Biosphere Reserve Sierra del Rosario (BRSR), in western Cuba, have suffered loss and fragmentation of habitat, largely due to historical impact by humans. Guzmán Menéndez *et al.* (2004) found 904 patches of vegetation in this area, which were mainly isolated trees, living fences, riparian vegetation, and degraded patches of forest with surface areas smaller than 10 km^2; fragments > 10 km^2 and < 100 km^2 were identified, as well as one of > 100 km^2. It is in the latter two that the core areas of the two main sectors of the BRSR are found (e.g. Ecological Reserve El Salón; Table 3.4).

Deforestation and roads create numerous edge zones where patches of forest meet non-forested habitat; edge effects on forests can include altered abiotic conditions, changes in rates of competition and predation, and shifts of community structure (Marsh and Beckman 2004). The limited distribution of most Cuban species, their dominant association with forested and humid environments, together with the cumulative impact of humans on their habitats, makes them particularly susceptible to climatic change and significantly vulnerable to randomly catastrophic natural events. Marroquín-Páramo *et al.* (2021) stated that hurricanes might have major effects on animal populations and assemblages when combined with anthropogenic disturbances and global climatic change. Deforestation affects the connectivity of habitats, which has a key role in enhancing regional viability of amphibian populations (Cushman 2006). In turn, such connectivity affects dispersal of adults and juveniles.

Efforts to reforest Cuba, which include the use of many exotic, fast-growing trees, have been successful at some places like the BRSR. However, many other areas in which reforestation was based on only one or a few species of trees with an absence of connectivity (e.g. corridors) from neighbouring, naturally forested areas do not allow for colonization and development of a rich biota and thus support a considerably lower diversity of amphibians, only one species, or none at all. As a result, the capacity of amphibians to migrate to suitable areas is usually limited or nil.

A study by Fong G. (1999) in the Massif Nipe-Sagua-Baracoa in eastern Cuba reported five species (*Eleutherodactylus atkinsi, E. cuneatus, E. dimidiatus, E. limbatus,* and *E. ricordii*) in a mesophyll-evergreen forest and seven (*Eleutherodactylus atkinsi, E. auriculatus, E. limbatus, E. ionthus, E. ricordii, Osteopilus septentrionalis,* and *Peltophryne taladai*) in a semideciduous microphyllic forest (SDMF). In a coffee plantation located in the SDMF, he found only two (*E. atkinsi* and *E. ionthus*) of the species inhabiting the original vegetation. A forest used for the extraction of timber in an area formerly occupied by mesophyll-evergreen forest was also inhabited by only two species (*E. atkinsi* and *E. dimidiatus*). Pastures located in each of these types of vegetation lacked frogs. These results demonstrated a decrease in species richness in coffee plantations, timber forests, and pastures, compared to original habitats. The reduction or total disappearance of different strata of vegetation in the forest structure together with fragmentation of habitats brings about a decrease of space and food, as well as initiating microclimatic changes such as an increase in temperature and decrease in relative humidity (Blaustein *et al.* 2011).

Fong G. (2009) also compared the species richnesses in areas with different levels of anthropization to that in natural habitats in the mountains of Sierra Maestra and Nipe-Sagua-Baracoa.

He found that the abundance of anurans in coffee plantations and other crops was always lower than in rainforests. However, the abundance in cropland was similar to that of the evergreen and semideciduous forests. The sampled coffee plantations had two to six species, values which can be considered medium to low in comparison with the natural habitat. Sampled plantations had fewer species or an absolute dominance of just one of them. Total abundance in different habitats was correlated with the structure of the vegetation and the leaf litter. The most notable trend is that habitats with poor or absent leaf litter, less cover of trees and bushes, and lower canopy, had the lowest total relative abundance. Deforestation and reduction of understorey vegetation affect microhabitats used by arboreal and semi-arboreal amphibians, which leads to the extirpation of species such as *Eleutherodactylus ionthus* or *E. guantanamera* (depending on locality) and *E. auriculatus*. Of the sampled anurans, two species, *Eleutherodactylus limbatus* and *E. maestrensis*, only used the natural forests and were never found in any of the studied anthropized habitats. In contrast, *E. feichtingeri* and *Osteopilus septentrionalis* were favoured by human-disturbed situations and were significantly less common in natural types of vegetation.

In the Eastern region, forests currently cover 43% of the surface of mountains (having lost 50% of the original cover). Grasslands and crops have displaced the original vegetation in 35.5% of the surface of this territory. Considering the evidence discussed above, 11 species that preferred or had greater abundance in forests have halved their original range. This impact is greater in the strictest forest dwellers, which do not thrive in other types of habitats. By contrast, the survival of species favouring the secondary vegetation types very likely increased and they extended their distributional area by 45% with respect to the original vegetation on these massifs. In particular, *E. feichtingeri* and *O. septentrionalis* may have expanded their ranges by 35%. The great diversity of *Eleutherodactylus* in Cuba involves different combinations of responses to anthropization and makes the proper design of conservation strategies critical since anurans use unique combinations of ecological conditions (Fong G. 2009).

Cobos and Alonso Bosch (2016) analysed the impact of habitat loss in the *P. longinasus* species-complex and concluded that the most significant reduction of forests during 2000–2012 (with a peak in 2008), occurred in the Western region, where *P. longinasus longinasus* and *P. l. cajalbanensis* occur (Fig. 3.42). Significant reduction in forest cover took place in the central (*P. dunni*) and eastern (*P. ramsdeni*) regions in 2005 and 2002 respectively. A chronological view of the rate of deforestation in Central Cuba was presented by Mancina *et al.* (2017). Monitoring the loss of habitat and the management of forests for *P. longinasus* and taxonomically related taxa is also critical for other species that share the same habitats, particularly in the western part of the island where they are most affected.

A critically endangered toad, *Peltophryne florentinoi*, lost ~2% of the forest cover in its restricted habitat at two localities during 2000–2015 (Cobos and Alonso Bosch 2018). Deforestation for agriculture and charcoal production (Fig. 3.42), together with the dumping of rubbish are current threats for this localized, specialized species.

Even for widely distributed lowland toads living on agricultural lands (such as *Peltophryne gundlachi* and *P. empusa*), extensive burning of sugarcane fields and other crops is a highly damaging practice that destroys not only habitats and populations of amphibians but also whole ecosystems and constitutes a source of contamination and carbon emissions (Fig. 3.43).

B. Climatic change

Climatic change nowadays is of serious concern as it directly impacts ecosystems and biodiversity. There are diverse predictive estimates of the effect of climatic change on amphibians based on different models. Nurse *et al.* (2014) considered the Caribbean Islands among the most vulnerable regions in the world. Comparison of current information with the baseline data of 1971–2000,

Figure 3.42. Loss and disturbance of habitat in different areas. (A) Intense deforestation in the surroundings of Río La Mula, Sierra Maestra, Santiago de Cuba Province; (B) forest-use, extraction of precious wood, Rihíto, Yateras, Guantánamo Province, in 2018; (C) residual impacts from past mining surveys, carried out between 1975 and 1997 in the El Toldo Plateau, Guantánamo-Holguín Provinces, photograph taken in 2017; (D) intense deforestation in pine forests at the surroundings of Moncada, Viñales, Pinar del Río Province; (E) charcoal ovens in Península de Zapata, just in front of one of the main localities where *Peltophryne florentinoi* occurs. Photographs by A. Fong (A), Gerardo Begué (B–C), Lázaro Yusnaviel García (D), and L. M. Díaz (E).

Figure 3.43. Intense fires burning sugar cane fields in Florencia, Camagüey, in February 2020. Photograph by L. M. Díaz.

showed the Caribbean climatic domain to be 0.5°C to 1.5°C warmer and 5–10% wetter, except for the north-eastern and south-eastern Caribbean, which are drier, and experience increases in annual warm spells of more than 100 days (Taylor *et al.* 2018). Centella *et al.* (1999) described a general tendency over a few decades towards an increase in temperature, seasonal change in the pattern of rainfall in Cuba, and gradually accentuated droughts, with a potential effect on vegetation and on biodiversity in general. Increasing uncertainty in the intensity, arrival, and duration of seasonal rainfall is one of the critical consequences of climatic change in the tropics (Feng *et al.* 2013). Helmer *et al.* (2019) suggested that in less than 25 to 45 years, declines in cloud immersion will shrink or dry 57–80% of neotropical montane cloud forests, including those of the Caribbean. In eastern Cuba, the projections of Fuentes Marrero *et al.* (2019), based on climate-niche models of climatic scenarios for 2050, indicate a significant reduction in climatically suitable areas and a significant impact on humid forests where amphibians occur. Moreover, these authors predicted that the assemblages with the greatest number of species will be relegated to the higher elevations of the Sierra Maestra and Nipe-Sagua-Baracoa mountains; hence these zones could become climatic shelters for moist forests and their associated amphibians.

Foden *et al.* (2013) assessed the vulnerability of birds, amphibians, and corals to climatic change worldwide. They included 61 Cuban amphibians (52 *Eleutherodactylus*, 8 *Peltophryne*, and *Osteopilus septentrionalis*). As a result, the overall vulnerability of species was low for 87% of the analysed Cuban species and high for 13%. Species listed as having high overall vulnerability were: *Eleutherodactylus bartonsmithi*, *E. emiliae*, *E. jaumei*, *E. leberi*, *E. mariposa*, *E. principalis*, *E. simulans*, and *E. toa*. This study is relevant as a global assessment, but it is decidedly of low resolution as a local approach, partially contradicting other results discussed in this section below.

Cobos and Alonso Bosch (2016) analysed the potential threats of climatic change and additive factors such as habitat loss and chytrid fungus, which could synergistically affect the long-term survival of *Peltophryne longinasus, P. dunni,* and *P. ramsdeni.* They predicted future increases in temperature and a decrease in precipitation throughout the distributional ranges of these species. The Western region was identified as subject to significantly greater impact by climatic change and loss of habitat. Those conclusions mostly coincide with earlier predictions by Centella *et al.* (1999) who suggested that the greatest increases in temperature will occur in the central and eastern parts of Cuba, while the most substantial reduction of precipitation and cloud cover is expected for the western half of the country. They pointed out that in the coming decades the annual variation will include the greatest warming combined with significant reductions of rainfall in the summer. This indicates a warmer and drier climate than at present. The predicted reduction in rainfall is about 10% to 15% in the spring and summer months (except April). The increase in rainfall (by about 15% to 20%) is expected to occur in November–December. Cobos and Alonso Bosch (2018) also estimated that as a result of climatic change, a gradual loss in forest cover would produce significant transformations of the habitats of the toad *Peltophryne florentinoi.* These changes include an increase both in temperature and evaporation rate in the breeding habitats of this toad, together with a higher incidence of ultraviolet radiation.

With the increasing variability in rainfall in tropical regions caused by climatic change, the explosive reproductive patterns and seasonal breeding in amphibians could be significantly affected. Explosive breeders like *Peltophryne cataulaciceps, P. empusa,* and *P. gundlachi,* occupy fragile and ephemeral habitats and their reproduction depends on the amount of precipitation and the timing of rain, which makes them vulnerable to climatic changes despite their wide distributional ranges (Ulloa *et al.* 2019).

There are preliminary studies on how climatic change can affect frogs of the genus *Eleutherodactylus* in Cuba. The report by CITMA (2014) indicates that the geographic ranges of 30 of the species will decline by an average of 85% by 2050 and 97% by 2080. For example, the range of *Eleutherodactylus limbatus* will be progressively reduced until it is restricted to the highest mountain ranges of eastern Cuba. Seven species: *E. glamyrus, E. cubanus, E. albipes, E. jaumei, E. maestrensis, E. melacara,* and *E. turquinensis* could disappear completely by 2050 and another group of 11 species probably will become extinct by 2080. The areas with the greatest core of extinctions by 2080 are predicted to be located in the Sierra del Rosario and the Nipe Sagua-Baracoa mountain ranges, where 10–13 species could disappear. Only two of the analysed species, *E. thomasi* and *E. guanahacabibes,* could increase their distribution as a result of climatic change (according to CITMA 2014).

Other consequences of climatic change are saline intrusions and a gradual rise in sea level (Rodríguez Schettino and Rivalta González 2007). Two Cuban species that have a restricted distribution to coastal areas are highly vulnerable to climatic change. First, *Eleutherodactylus blairhedgesi* is known only from a limited area along the northern coast of Mayabeque Province (Díaz and Cádiz 2008). Although it is a littoral inhabitant, studies have to be carried out to examine its adaptive capacity to increased salinity levels and predict the potential impact of an increase in sea level throughout its distributional range. Second, *Peltophryne florentinoi* only occurs in a few localities along the karstic coastal and subcoastal lines between Girón and Guajimico, Matanzas and Cienfuegos provinces, respectively (Díaz *et al.* 2019). Although Cobos and Alonso Bosch (2018) pointed out that a rise in sea level will not have direct impacts on the range of *P. florentinoi,* we do not know its adaptive capacity to salt water; occasional inundations and saltwater intrusion occur at specific breeding sites used by this toad, which could threaten its survival in the immediate future.

C. Chytrid fungus in Cuba

The widespread fungus *Batrachochytrium dendrobatidis* (*Bd*) is a generalist pathogen that is responsible for some recent global declines in amphibian populations (Lips *et al.* 2005, 2006; Skerratt *et al.* 2007; Vredenburg *et al.* 2010; Lips 2016; Fisher and Garner 2020). Scheele *et al.* (2019) estimated that *Bd* had caused the decline of about 501 species of amphibians. Although chytrid fungus is unquestionably a widespread threat for amphibians around the world, the data and methods implicating the disease chytridiomycosis are deficient for many species of amphibians. Consequently, the complete list of the affected species is probably not fully known (see Lambert *et al.* 2020 *versus* Scheele *et al.* 2019). Furthermore, pathogenicity and virulence vary among fungal strains (Berger *et al.* 2005; Voyles *et al.* 2009; Fisher and Garner 2020). Early in the twentieth century, the origin of chytridiomycosis was traced to East Asia, which coincides with the global expansion of commercial trade of amphibians (O'Hanlon *et al.* 2018). Although there is evidence of relatively old endemic chytrid haplotypes in the Neotropics, the pathogenic incidence of human-mediated invasions of genotypes worldwide resulted in the *Bd* Global Pandemic Lineage (GPL; Jenkinson *et al.* 2016). Additionally, individuals and populations may be more vulnerable to *Bd* when exposed to multiple stressors such as chemical contaminants, modification of their habitats, and climatic alterations (Grant *et al.* 2016; Fisher and Garner 2020).

The global impact of *Bd* is so severe that the IUCN has called it 'the worst infectious disease ever recorded among vertebrates in terms of the number of species impacted, and its propensity to drive them to extinction' (Gascon *et al.* 2007). Even though *Bd* has a pan-global distribution, it is within a set of more restricted geographic regions where it has caused the most extensive loss of amphibian diversity on record. These include regions of eastern Australia, Central America, north-western South America, and western North America (Berger *et al.* 1998; Lips *et al.* 2005; Vredenburg *et al.* 2010; Fisher and Garner 2020). Reports of *Bd* on seven Caribbean islands indicate that several West Indian species are also highly susceptible to infection (Magin 2003; Burrowes *et al.* 2004; Malhotra *et al.* 2007; Díaz *et al.* 2007b; Alemu *et al.* 2013; Hudson *et al.* 2016, 2019; Sabino-Pinto *et al.* 2017; Cádiz *et al.* 2018).

If we consider that nearly 40% of the Cuban endangered species are vulnerable because their ecological requirements closely match optimal conditions for fungal development, it is clear why there is such a high probability of infection. Furthermore, predictive modelling that evaluated the probability of detecting *Bd* in three amphibian families in Cuba suggested a high likelihood of detecting *Bd* in amphibians from the family Hylidae and a low probability of detecting it from Bufonidae and Ranidae (Olson *et al.* 2013). Thus, we do not suggest that these two families are not susceptible to the disease, only that the probability of detecting it in these families may be comparatively lower.

Chytridiomycosis was documented for the first time in Cuba (Fig. 3.44) in one dying individual of the endemic and endangered toad *Peltophryne dunni* (Díaz *et al.* 2007b). Since that discovery, populations of *P. dunni* appear to have declined. For ten years following the initial report of the disease, no additional sampling was conducted to monitor chytridiomycosis in Cuba. Synergism of climatic change, loss of habitat, and the presence of *Bd* in the *P. longinasus* species-complex was analysed by Cobos and Alonso Bosch (2016).

The first assessment of the prevalence and geographic extent of *Bd* in Cuba was provided by Sabino-Pinto *et al.* (2017). The three main mountain ranges of the island were sampled, but most of the sampling was concentrated in localities in the Eastern and Central regions. As a result, *Bd* was detected in six individuals of four species (*E. casparii, E. dimidiatus, Osteopilus septentrionalis,* and *P. l. dunni*), all from the Guamuhaya mountain range, Central Cuba, at three localities very close to each other during the months of March and July, and in very close proximity to

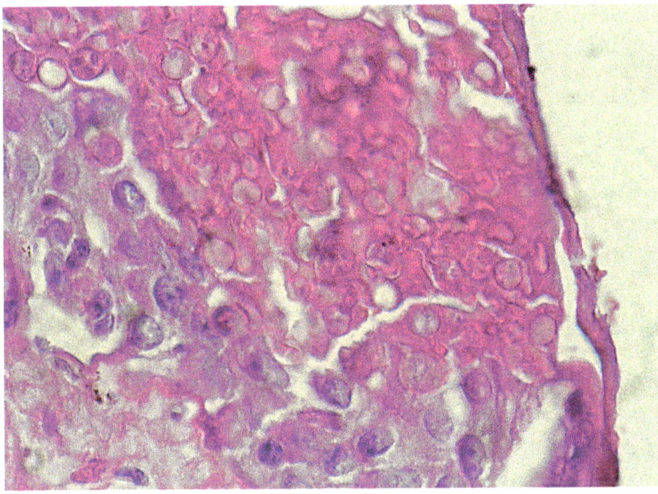

Figure 3.44. A microscopic sample of *Batrachochytrium dendrobatidis* in a dying individual of *Peltophryne dunni* found in Cuba in 2006 (Díaz *et al.* 2007b). Photograph by A. Chong.

the locality where *Bd* was first reported. Sabino-Pinto *et al.* (2017) covered 16 localities, collected 182 swabs from 19 species, and obtained an overall prevalence of 3.2%. They suggested that the central Guamuhaya mountain range might be a core area of *Bd* in Cuba.

A more recent effort to monitor the distribution of *Bd* and to assess the potential impacts of chytridiomycosis in Cuba documented new *Bd* infections in additional species. Cádiz *et al.* (2018) sampled ten localities during the wet season of 2015 and 2016 covering the Western, Central, and Eastern regions of Cuba and collected 220 swab samples from 18 species. They also collected 28 environmental water samples using water filters at seven of those localities. Sampling of water is important for understanding the prevalence of *Bd* because populations are more vulnerable to disease when an environmental reservoir allows a pathogen to persist (Blaustein *et al.* 2012), thereby enhancing indirect transmission to terrestrial species via aquatic ones. In addition, sampling *Bd* in the aquatic environment allows a better understanding of its distribution.

Cádiz *et al.* (2018) found *Bd* infections in ten species of amphibians (16% of the specimens carried a detectable *Bd* load), seven of which (*E. atkinsi, E. auriculatus, E. glamyrus, E. gundlachi, E. ionthus, E. planirostris,* and *E. riparius*) had not been previously reported to be infected. Notably, they reported the presence of *Bd* in a population of *P. l. longinasus*. *Bd* also was detected from environmental water samples at two of the visited localities. Combining the results from swabs and water samples, Cádiz *et al.* (2018) reported *Bd* at five new sites, confirming the distribution of the fungus in the eastern, central, and western mountains, at elevations between 170 m asl and 1,330 m asl. They showed that *Bd* also infected two semi-arboreal, one bromeliad, three terrestrial, and one riparian species, confirming it as a generalized pathogen in Cuba.

Fourteen years after its discovery in Cuba, *Bd* continues to be a significant threat to Cuban amphibian biodiversity, and the above results suggest that there are more reasons to be concerned than before. Since 2007, no explicit conservation strategy has been put into practice for amphibians, and recent increases in touristic activity raise great concern about the future impact of this disease on Cuban anurans. Mountain ranges are important targets for tourism, where much of the nation's natural value is concentrated, including the highest diversity of amphibians. Some tours offer a sequential visit of these mountain ranges. In fact, most localities included in the previous studies experience intense tourism activities, and the movement of people between natural areas could be a mechanism of dispersion for *Bd*.

D. Introduced species

The presence of introduced species is attributable to human activity, with most species having been deliberately or accidentally introduced into a new area from their native range or other sites of introduction; some of them become invasive by affecting ecosystems, habitats, or species when they spread widely (Essl *et al.* 2018). Therefore, biological invasions are a major driver of ecosystem change, like invasive species being credited with causing about one-third of amphibian extinctions (Blackburn *et al.* 2019; Pyšek *et al.* 2020). Moreover, introduced amphibians are considered the group causing a significant decrease in abundance and diversity of native amphibians worldwide (Nunes *et al.* 2019).

In Cuba, Rodríguez-Cabrera *et al.* (2018) emphasized that the recently introduced *Leptodactylus fragilis* (Fig. 3.45) may compete with some locally endemic anurans such as the endangered toad *Peltophryne cataulaciceps*. Considering the kind of habitat occupied by this frog, there is some potential for expanding its distribution. It could represent an increasing threat for native toads by competing with them at breeding sites and by interfering with them acoustically, having potentially adverse effects on mate choice and breeding success, according to Del Castillo Domínguez *et al.* (2021; also see García-Padrón *et al.* 2022). Riparian species of the genus *Eleutherodactylus*

Figure 3.45. Two introduced species of anurans that are now established in the wild in Cuba: *Lithobates catesbeianus* (A) and *Leptodactylus debilis* (B). Photographs by L. M. Díaz (A) and Tomás M. Rodríguez (B).

could also be affected. This frog has likely been introduced accidentally into rice fields in the vicinity of Guane (Pinar del Río Province). The Cuban population is genetically similar to those of Venezuela and Colombia (Rodríguez-Cabrera *et al.* 2018).

The status of the introduced Asian anuran *Hoplobatrachus tigerinus* is unknown. It probably has not yet spread into the wild, but this species is highly invasive and can predate on other species of frog. It is a prolific breeder, and the population could increase rapidly. The tadpoles are carnivorous (Grosjean *et al.* 2004) and feed on larvae of other species of frogs (Surendran and Vasurevan 2013; Mohanty and Measey 2019), which could be a potential problem for many native Cuban species. García-Padrón *et al.* (2021b) considered low trophic competition between the American Bullfrog (*Lithobates catesbeianus*) and the endemic toad *Peltophryne fustiger*, and suggested that very likely, the major impact of this introduced anuran is on the invertebrate fauna.

There are no studies on the influence of the introduced African sharp-tooth catfish *Clarias gariepinus* on native amphibians. Most Cuban frogs are terrestrial breeders, but endemic toads potentially are at major risk of predation by catfish. Assessments are needed to elucidate whether this fish represents a threat to some Cuban anurans. The few isolated anecdotal references to predation by catfish on Cuban anurans are insufficient evidence of deleterious impact but do constitute a reason for concern. García-Padrón (2008) found a male *P. fustiger* in the stomach of a metre-long specimen of this fish. Rodríguez-Machado and Rodríguez-Cabrera (2015) reported an adult *Osteopilus septentrionalis* devoured by the catfish. Díaz and Cádiz (2008) mentioned their concern about catfish's fry in temporary ditches at Los Indios (Isla de la Juventud) in breeding habitats of *Peltophryne cataulaciceps* and *P. empusa*, thereby posing a potential risk of predation on tadpoles of these toads. Some informal but long-term field observations suggest that *Lithobates catesbeianus* (Fig. 3.45) is now less common or has disappeared in places where catfishes were established (Díaz and Cádiz 2008).

Díaz and Cádiz (2008) reported pigs (*Sus scrofa*) disturbing the banks of streams where *Peltophryne longinasus longinasus* reproduce. The pigs constantly rummaged the ground at places where toadlets emerged after metamorphosis. Adult toads were not present in segments of streams with the highest disturbance by pigs (L. M. Díaz, pers. obs.). Finally, we need to assess whether introduced water buffalos *Bubalus bubalis* may locally affect the breeding activity of the widely distributed toads *P. gundlachi* and *P. empusa* in some lowland ephemeral ponds inside farming areas.

E. Pollution

Pollution is one of the main causes of amphibians' malformations and several physiological disorders worldwide. Amphibians are more susceptible to pesticides than are mammals and birds (Islam and Malik 2018). Studies on the impact of pesticides on Cuban amphibians are necessary. According to the recent review by Pérez-Consuegra and Montano-Pérez (2021), in the Official List of Authorized Pesticides of Cuba, 877 formulations are registered in whose composition there are 340 active ingredients, of which 110 are highly hazardous pesticides; these represent 32.6% of the total: 45 insecticides (40.9%); 22 fungicides (20%); 19 herbicides (17.3%); six acaricides (5.4%); six rodenticides (5.4%); a nematicide (0.9%); four fumigants (3.7%); and seven with other modes of action (6.4%). The 12 active ingredients with the highest number of records are: cypermethrin-37 (6.7%); imidacloprid-32 (5.8%); 2,4-D-32 (5.8%); glyphosate-29 (5.3%); mancozeb-27 (4.9%); diuron-22 (4.0%); tetramethrin-19 (3.4%); permethrin-19 (3.4%); deltamethrin-17 (3.1%); chlorpyrifos-15 (2.7%); chlorothalonil-15 (2.7%); and folpet 15 (2.7%).

Tadpoles of many species seem to be more sensitive to pollution than are adults. Species on agricultural lands, and urban and tourist areas are very likely more threatened by different chemicals. Harischandra *et al.* (2011) demonstrated that tadpoles of toads at early stages of development remain as larvae for a more extended period and develop malformations and alteration of bones, webbing of the skin in the hind limbs, ectrodactyly (malformation and absence of digits), and oedemas when exposed to the widely used organophosphate temephos. Junges *et al.* (2017) showed that, of the three products used for the same purpose, permethrin was the most toxic insecticide for toad tadpoles, followed by temephos and *Bacillus thuringiensis*. Permethrin is lipophilic and has low solubility in water, implying a cumulative concentration in biota (Miyamoto *et al.* 2015) and high ecotoxicological risk for anuran larvae (Junges *et al.* 2017). Juveniles are more sensitive than are adults and may die from chronic exposure to pesticides. High risk is also reported in amphibians exposed to glyphosate and other herbicides, and several toxicological effects including genotoxicity, cytotoxicity, neurotoxicity, nuclear aberration, hormonal disruption, chromosomal aberrations, and DNA damage, among others, have been reported (Gill *et al.* 2017). In addition, exposure to pesticides during the metamorphosis of amphibians results in a reduction in growth and in low reproductive activity (Islam and Malik 2018). Lajmanovich *et al.* (2015) showed dose-dependent toxicity of commercial formulations of *Bacillus thuringiensis* var. *israelensis* in tadpoles of *Leptodactylus latrans*. This 'biopesticide' alters enzymes of oxidative stress, induces genotoxicity, and causes intestinal damage.

In Cuba, polymelia was reported in adult *Osteopilus septentrionalis* by García-Padrón and Alonso Bosch (2017), who also mentioned polydactyly, ectrodactyly, and ectromelia in some large-sized Cuban *Peltophryne* toads and the introduced *Lithobates catesbeianus*; these authors suggested that polymelia could be caused in association with eutrophic water and parasitic infections (e.g. Johnson *et al.* 1999, 2007). Alonso Bosch *et al.* (2017) found anomalies in mouthparts of tadpoles of *Peltophryne fustiger*, which were not produced by chytrid fungus; the authors speculated on pollution as a possible cause of developmental abnormalities. Alonso Bosch and Hernández Marrero (2020) reported abnormalities of the eye (opacity or cataracts; excessive yellow colouration or hyperxanthism) and polydactyly in adults of *Peltophryne taladai* in eastern Cuba. These authors suggested that chemical contaminants could be the potential agents of malformations. They highlighted the importance of examining aquatic molluscs for encysted trematodes (*Ribeiroia*), which are generally implicated in malformations in several anurans; however, these potential pathogenic interactions have not been thoroughly studied (or studied at all) on the island. More recently, Alonso Bosch *et al.* (2021) also found malformations in 16–18% of sampled adult individuals of the endangered and endemic toad *Peltophryne florentinoi* at two localities. Abnormalities in hands, toes, and limbs included brachydactyly, amely, ectropody, schizodactyly, and syndactyly (those interested in terminology of malformations in amphibians may consult Henle *et al.* 2017). The causes of these affections remain undetermined, but pollutants and increasing UV radiation were proposed as potential candidates.

F. Tourism

The increasing development of touristic facilities in Cuba is a challenge for the conservation of nature. The growth of tourism is magnified in coastal areas due to the attractiveness of beautiful tropical beaches. However, mountain ranges are also affected. Díaz and Cádiz (2008) referred to the contamination of montane streams with soaps and detergents in a camping village at Sierra Maestra, Eastern Cuba. Tourists' itineraries that include visiting multiple mountain ranges on the island may contribute to the dispersion of chytrid fungus and other pathogens if no biosecurity measures of control (e.g. protocols for the disinfection of footwear and pneumatic tyres) is taken. The use of insecticides to mitigate mosquitos is another threat generated by tourism. Local dumping of trash and inefficient collection and recycling of waste are worrisome problems in many touristic facilities in Cuba, with imminent effects on the biota, of which amphibians are a part. Acoustic contamination is another potential threat in touristic areas where loud music, and other sounds, could affect wildlife, particularly the communication system of some anurans. A further risk of tourism is translocations of native species and the spreading of introduced species, which can affect the dynamics of local ecosystems.

G. Social misconceptions

In general, there is a popular tendency to reject amphibians (García-Padrón *et al.* 2021c). Cuban women, in particular, often show extreme fear of frogs. Such rejection is quickly transmitted to new generations of people and is deeply rooted in culture. Most Cuban frogs of the genus *Eleutherodactylus* are too similar for the general public to identify and are not perceived as charismatic animals. Only a few species have a vernacular name in Spanish. For example, *Eleutherodactylus eileenae* is called 'colín' (onomatopoeic) in Spanish; *Peltophryne empusa* is known as 'guasábalo'; larger toads are known as *'sapos'*, *'cucarros'*, or *'quincarros'*; and greenhouse frogs like *Eleutherodactylus planirostris* are sometimes referred as *'ventorrillas'*. The Monte Iberia Pygmy Frog (*Eleutherodactylus iberia*) or Ranita de Iberia, is iconic because it is one of the smallest anurans in the world. However, people believe that individuals of this species are instead hatchlings of the more common *Eleutherodactylus*, like *E. planirostris*, which occur in cities and towns along the island. Not many schools teach that most Cuban frogs breed in terrestrial habitats and have direct development. Even people who have lived in the countryside since childhood often believe sounds produced by frogs to be those of insects or spiders instead. There is a generalized ignorance about Cuban amphibians, a problem exclusive to the island but disadvantageous for conservation there and elsewhere. Although a few books focusing on the general public have been published (e.g. Ruíz 1987 and Rodríguez Schettino 2003), we believe that more, continued, and better education is mandatory.

XII. Captive breeding of an endangered anuran

Díaz and Cádiz (2006) published a preliminary report on the captive breeding of the Cuban endangered species *Peltophryne longinasus longinasus* (Fig. 3.46). The project was partially supported by an Amphibian Ark seed grant to Luis M. Díaz and his team in 2010 (Díaz 2011). The preliminary study provided some insights on unknown aspects of the biology and husbandry of this species, which could be useful as initial protocols for a long-term *ex-situ* programme. However, no *ex-situ* conservation actions for Cuban amphibians are currently in place. Cuban zoos, for example, basically ignore amphibians and are mostly focused on exotic species, with only sporadic inclusions of native fauna. There is an urgent need for captive-breeding programmes for endangered amphibians, and they have been strongly recommended and contemplated in the Amphibian Conservation Action Plan (Gascon *et al.* 2007; Zippel *et al.* 2011; Johnson *et al.* 2020).

The small-scale captive-breeding experience with *Peltophryne longinasus longinasus* can serve as an example of techniques applicable to captive breeding. The programme started with eight individuals (five females and three males) placed in a 95 × 46 × 50 cm all-glass vivarium. The following protocol proved partially successful (see below), with aquatic and terrestrial sections provided according to the species' breeding habits. The terrestrial section was delimited by stones and slabs and covered by a gravel layer (grain diameter: 4–5 mm) and a sheath of live moss. Some plants (*Scindapsus aureum* and ferns) were planted directly in the gravel or small pots. The aquatic section had a depth of 6 cm, with a gravel layer of about 1 cm. Water movement and filtration was achieved by using a small water-pump that circulated water through the gravel. Submerged thin roots and sticks were anchored to rocks to provide oviposition sites. The vivarium was illuminated with three 20 W halogen bulbs (12 V) and two 18 W (6,500 K) fluorescent tubes, all automatically controlled by a timer that adjusted the photoperiod from 08:00 h. The vivarium was placed in a room with plenty of natural light (not direct sunshine) through

Figure 3.46. Captive rearing and breeding of *Peltophryne longinasus longinasus*. (A) General view of a metre-long vivarium; one breeding male is near the posterior corner, sitting on a rock used as a calling site; movement and filtration of water were produced by an aquarium pump (waves produced by returning water is noticeable in the foreground). Some tadpoles are present on the bottom of the aquatic section; (B) two adult calling males positioned on the stem of *Scindapsus aureum*; (C) amplecting pair attaching their eggs onto a thin anchored branch, with male's hind limbs positioned to repel competing males; (D) small larvae a few hours after hatching, adhered to the vivarium glass; (E) tadpole in Gosner's (1960) stage 36; (F) tadpole in stage 41; (G) recently metamorphosed toadlet on a person's thumb, illustrating the small size of recently metamorphosed individuals. Photographs by L. M. Díaz.

a window and was partially covered by a fine mesh. The temperature inside the terrarium was 27–32°C, descending to 21–23°C in the coldest months. The relative humidity fluctuated between 75% and 90% (average 79%). The water had a pH of 7.6 and a conductivity of 650–700 µS, 12 dGH (tap water of Havana). Toads were fed every other day with small crickets (*Gryllodes sigillatus*) alternating with termites (*Nasutitermes* sp.), isopods, buffalo worms (*Alphitobius diaperinus*), and ants (*Camponotus* sp. and winged *Solenopsis* sp.). Crickets were gut-loaded with a highly nutritious formula and Rep-Cal, 24–48 h before being offered as food. Husbandry protocol for *P. longinasus* is quite similar to that for species of *Atelopus* (Gawor *et al.* 2012).

These toads are diurnal and very active, and they opportunistically reproduced year-round in captivity, either in December–January, among the coldest months, or July–August, the warmest. However, they reduced breeding activity under 22°C. Contrary to what is expected in other toads, they avoided breeding on rainy days. Females that showed abdominal peristaltic movements during amplexus in the water always laid eggs. The maximum number of eggs recorded was 301. Females distributed an entire clutch in a few scattered groups of eggs attached to different sites. Oviposition occurred between 14:00 h and 16:00 h.

Most eggs hatched in about 72 hours. Tadpoles developed in four (as first found by Ruíz [1980]) to six weeks, depending on temperature. They were fed spirulina-based fish flakes, but detritus and algae were available in the gravel. Toadlets metamorphosed between 6 mm and 8 mm SVL and required large amounts of small-sized prey such as springtails. They grew fast and soon accepted 'pin-head' crickets. Males started calling at ages between five and six months.

After four years, there were some issues related to the captive management of the species (L. M. Díaz, personal observations). Hypercalcaemia (see Fig. 3.47) was detected in some adults and it was linked to inadequate water chemistry and excessive mineral/vitamin supplementation as found in other anurans (e.g. Yoshimi *et al.* 1996). In the wild, *P. longinasus* lives in soft and slightly acidic streams (Díaz and Cádiz 2006) with a pH of 6.4 and a conductivity of less than 90 µS (2 to 5 d GH). Hence, very likely, it is not adapted to a high concentration of calcium

Figure 3.47. Signs of hypercalcaemia in a captive male *Peltophryne longinasus longinasus*. Arrows point to endolymphatic sacs and the milky liquid they contain. Photograph by L. M. Díaz.

carbonate, or even phosphates, as these were always provided. Species from similar habitats, like those in the genus *Atelopus*, are susceptible to phosphates (Poole 2006; Odum and Zippel 2008).

In the natural habitat of *Peltophryne l. longinasus*, the air temperature may reach up to 32°C, while the water temperature rarely exceeds 25°C. Adult males and juveniles expose themself to direct sunlight, even when shelters and water are much cooler. In captivity, the temperature was always high in the summer, particularly that of the water. These problems have to be adequately controlled in the future, and gradients of temperature should be provided to replicate the natural microclimatic conditions as much as possible. After oviposition, it is important to separate females from males since males keep trying to mate and interfere with the proper feeding behaviour of females. In nature, females migrate to the forest well apart from breeding sites and later return to them (L. M. Díaz and A. Cádiz, pers. obs.), probably attracted by males' activity. Toadlets required large quantities of tiny prey, like springtails, small fruit flies, and aphids. Pinhead-sized crickets are too big for recently metamorphosed toadlets. The selection of suitable food sources for amphibians is another field of research needed for successful large-scale captive-breeding programmes.

XIII. Future directions

A. Taxonomy and distribution

Taxonomic uncertainty still exists for some Cuban amphibians. It is important that the implementation of an integrative taxonomy combines available modern approaches, like advanced genetics, with traditional morphology and relevant aspects of ethology and ecology. Some of the current taxonomic issues were discussed in section IV. (*Native anurans and their conservation status*). For instance, we need to (1) clarify the status of populations of bromeliad frogs (*E. varians* species-group) throughout the country; (2) review the species' boundaries in the *Eleutherodactylus limbatus* and *E. auriculatus* species complexes as a means of making taxonomic decisions; and (3) review the variation in the *Eleutherodactylus zugi-E. erythroproctus, E. thomasi-E. blairhedgesi* complex, *E. ricordii-E. acmonis, E. planirostris,* and *Peltophryne taladai.* These are critical areas of taxonomic research because the distributional pattern and assessments of the status of conservation of Cuban anurans depend on well-defined boundaries between species. The diversity of Cuban frogs is currently underestimated, with descriptions of a few additional species in progress.

B. Monitoring of species

Baseline information on the size and structure of populations is scant or absent for most species. We also lack information on habitat-use, feeding habits, activity patterns, breeding habits, phenology, behaviour, physiology, and geographic range for most species. Therefore, field data on natural history are critically needed to document species' general biology and ecology and examine fluctuations, resilience, and potential declines of populations in response to environmental stressors and habitat dynamics. They also allow for a broader understanding of the adaptive mechanisms of amphibians and their capacity for overcoming threats to their survival. Finally, they are fundamental for long-term monitoring protocols based on rigorous scientific evidence that allows for unbiased assessments needed for effective conservation. For example, species are becoming endangered at increasing rates due to human activities, which magnify the synergism between natural and anthropogenic phenomena and their influence on responses by anurans in particular, and all species in general. Evolution is an ongoing process that we can neither reverse nor stop. Thus, by acknowledging our influence on natural systems and the species that depend on them for reproduction and survival, we call for the protection of processes that generate and maintain biodiversity, which requires transdisciplinary research and international collaboration and support.

C. Monitoring chytridiomycosis and other emergent diseases

An understanding of the prevalence, resilience, and lethality of chytrid fungus in nature is urgent. Other pathogens, like viruses, Perkinsozoa, and co-infections, have never been detected in Cuban frogs (see Chambouvet *et al.* 2020; Herczeg *et al.* 2021). The synergism between pathogens and environmental stressors is another field that deserves special attention in Cuba. Hudson *et al.* (2019) found significant seasonality in the prevalence and load of *Bd* in amphibians of Dominica and Montserrat, correlated with temperature but not rainfall. They reported that the prevalence of infection decreases in reservoir hosts following epizootic mortality due to chytridiomycosis in disease-susceptible hosts. There is evidence that tadpoles growing in streams with lower temperatures and slower flow of water have the highest prevalence of *Bd* in the Neotropics (das Neves-da-Silva *et al.* 2021). Alonso Bosch and Hernández Marrero (2020) have recommended an adequate research design to examine the influence of potential intermediate hosts of encysted trematodes (*Ribeiroia*) on amphibians; trematodes also are responsible for malformations in several species of amphibians worldwide.

D. Protected areas and conservation

Protected areas in Cuba require more extensive development of facilities and improved infrastructure and human resources to allow monitoring, research, and effective management of natural areas and biodiversity. The National System of Protected Areas (SNAP) needs to continue to fill gaps in knowledge, not merely to attach the name of a species to a protected area but to reach sufficient inclusiveness for most of the populations. Loss of habitat within protected areas, either due to natural or human causes, needs to be assessed and conveniently mitigated or eradicated. In protected areas with public use, the impact of humans should be regulated, along with some aspects such as excess noise, chemical contamination (e.g. detergents, pesticides), rubbish-dumping, black waters, and overload of trails used for nature tourism. In addition, the unnecessary capture and manipulation of living organisms like frogs should be more effectively regulated. Finally, we need to establish long-term protocols for controlling invasive species and dispersion of disease throughout the landscape. Many protected areas of local relevance in Cuba are still not approved nationally by the Council of Ministers. In addition to protected areas, there are other environmental instruments in Cuba, such as the National Environmental Strategy (Estrategia Nacional Ambiental) and the National Programme for Biological Diversity (Programa Nacional de Diversidad Biológica), through which a National Action Plan for Cuban amphibians could be proposed and implemented countrywide.

E. Education

It is imperative to increase the number of educational initiatives to include amphibians and increase awareness in the general public and the education system. Furthermore, fostering capacities in protected areas relevant to amphibians is beneficial for conservation since local people can seldom identify which species are present in their environment nor appreciate their importance. Therefore, education is a fundamental tool for correcting erroneous perceptions about frogs and toads, and for minimizing negative impacts that stem from cultural prejudices; it should begin in schools. Efforts to inform people about the relevance of Cuban amphibians and their diversity, their role in ecosystem processes, importance to humans, as well as details of their behaviours and adaptations should be publicized through popular media such as television, brochures, newspapers, websites, and newsletters from institutions and non-governmental societies.

F. Combined *ex-situ/in-situ* actions

Zoos and other organizations should be more involved in supporting *ex-situ* programmes for amphibians abroad. However, the need to combine and support *ex-situ* and *in-situ* conservation actions at once is crucial, but in many instances, such efforts pursue different trajectories and operate independently. Saving species in captivity is not a permanent solution, nor a goal in itself, because it is also important to rehabilitate the ecological resources for the reproduction and survival of amphibians in the wild. It is our responsibility to allow them to continue natural lives side by side with humans.

XIV. Acknowledgements

We wish to thank Neftalí Ríos and Harold Heatwole for the kind invitation to write this chapter. The World Wildlife Fund (WWF) supported a project to survey chytrid fungus in Cuba in 2004–2006. The Rufford Small Grant Foundation (Project Number 18878-1) and the Mohamed bin Zayed Species Conservation Fund (Project Numbers 14259845, 13056409, and 180517654) funded fieldwork. A research grant from the American Society for Biochemistry and Molecular Biology (PROLAB) and a collaboration with George Amato funded the molecular analysis at AMNH. Amphibian Ark partially supported fieldwork and captive breeding, especially Kevin Zippel and Kevin Johnson. Joseph Burguess and Peter Tolson kindly donated equipment for captive-breeding projects. Ariatna Linares helped coordinate fieldwork and other activities. The Faculty of Biology of Havana University, the National Museum of Natural History of Cuba, and BIOECO provided additional support for expeditions and work-space. Important sampling equipment was provided by Tara Chestnut from Mt. Rainier National Park, Jenny Urbina from Oregon State University, and Masakado Kawata from Tohoku University. The fieldwork was made possible thanks to help provided by Michael Schmid, Wolfgang Feichtinger, Claus Steinlein, Ana Sanz, Yamilka Joubert, Yves Samyn, Eduardo Abreu, Gerardo Begué, Yonder Verdecia, Ángel Céspedes, Barchile Calzada, Luz M. Figueredo, Karina Massieu, and Esther Pérez. Our deepest gratitude goes to the personnel at several protected areas of SNAP. Access to specimens in collections were kindly allowed by Stephen Blair Hedges, David Kizirian, Joseph Martínez, Luis V. Moreno, Laure Pierre, José Rosado, and Christopher Raxworthy. Many photographs were kindly donated by Laura Watson, Chris Lukhaup, Martjan Lammertink, Tomás M. Rodríguez-Cabrera, Gerardo Begué, Lauren Vonnahme, Julio Larramendi, Emilio Alfaro, Nils Navarro, Sergio Luis del Castillo, Lázaro Yusnaviel García, and Mike Potts. Diego Alameda helped with statistical analysis. We are grateful to Kelsey Neam and Ariadne Angulo of the IUCN SSC Amphibian Specialist Group (ASG) for their advice and support during IUCN Red List assessments of Cuban species. Additionally, we are grateful to Roberto Alonso Bosch, Ariel Rodríguez Gómez (both expert contributors) and Marcelo Tognielli (chairman) for their participation in the ASG/IUCN's 2020 workshop for Cuban amphibian assessment. Craig Hilton-Taylor revised and provided proper citations for the yet unpublished IUCN Red List of Cuban amphibians after the second GAA. Useful personal communications were provided by Yasel Alfonso, Ángel Arias, Lázaro Yusnaviel García, Pedro López del Castillo, Carlos Mancina, Stesha Pasachnik, and Christopher Hildreth (Outreach Coordinator of Jane Goodall's Institute). Reynaldo Estrada, Enrique A. Castellanos, and Leodan Roque helped with the maps. Many thanks to Eldis Becquer for his comments and advice on the classification of vegetation types. Finally, our gratitude to Edelbis López Dávila and Marcel Montano for their critical comments and suggestions about pollution (section XI. E).

XV. References

Abreu, R. M., de la Cruz, J., and Rams, A., 1988. Algunos datos sobre la alimentación del almiquí (*Solenodon cubanus*; Insectivora: Solenodontidae) en vida libre. *Garciana* 10: 2–3.

Acevedo, M., 1989. Regionalización geomorfológica, IV.3.2–3, mapa 12. In *Nuevo Atlas Nacional de Cuba, Instituto de Geografía*, Academia de Ciencias de Cuba.

Aguilar Mugica, S., 2009. Cuba. Pp. 157–164 in *Important Bird Areas Americas – Priority Sites for Biodiversity Conservation*, edited by C. Devenish, D. F. Díaz Fernández, R. P. Clay, I. Davidson, and I. Yépez Zabala. BirdLife International (BirdLife Conservation Series No. 16). Quito, Ecuador.

Alayo, P., 1955. *Lista de Anfibios de Cuba*. Museo de Historia Natural Charles Theodore Ramsden, Universidad de Oriente, Santiago de Cuba, Cuba.

Alemu, J. B., Cazabon-Mannette, M. N., Cunningham, A. A., Dempewolf, L., Hailey, A., Mannette, R. P., Naranjit, K. T., Perkins, M. W., and Schmidt-Roach, A. C., 2013. Presence of the chytrid fungus *Batrachochytrium dendrobatidis* in a Vulnerable frog in Trinidad, West Indies. *Endangered Species Research* 20: 131–136. http://doi: 10.3354/esr00485

Alfonso, Y., Rodríguez Gómez, Y., Charruau, P., Sanz-Ochotorena, A., and Krysko, K. L., 2017. Oviposition and gonadal ultrastructure in the Cuban groin-spot frog, *Eleutherodactylus atkinsi* Dunn, 1925 (Anura: Eleutherodactylidae). *Copeia* 105: 781 790. https://doi.org/10.1643/CH-17-633

Alfonso, Y. U., Pellicier López, K., and Armijos-Ojeda, D., 2013. Endemic frog predation by the Cuban Lesser Racer, *Caraiba andreae* (Squamata: Dipsadidae), on La Melba, Alexander Von Humboldt National Park, eastern Cuba. *Herpetology Notes* 6: 91–93.

Alonso Bosch, R., 2012. Origen y diversificación del género *Peltophryne* (Amphibia: Anura: Bufonidae) en Cuba. PhD Dissertation, Editorial Universitaria, Universidad de La Habana, Cuba.

Alonso Bosch, R., 2013. Progresos en el estudio de las vocalizaciones de anfibios cubanos. *Revista Cubana de Ciencias Biológicas* 2: 4–13.

Alonso Bosch, R. and Cobos, M. E., 2016. A rapid survey of the Critically Endangered Cuban toad, *Peltophryne florentinoi* (Anura, Bufonidae). New locality and conservation approaches. *IRCF Reptiles & Amphibians* 23: 10–15. https://doi.org/10.17161/randa.v23i1.14086

Alonso Bosch, R. and García-Padrón, L. Y., 2017. Anfibios. Pp. 348–375 in *Diversidad Biológica de Cuba: métodos de inventario, monitoreo y colecciones biológicas*, edited by C. Mancina and Cruz, D. D. Editorial Agencia de Medio Ambiente (AMA), La Habana, Cuba.

Alonso Bosch, R. and Hernández Marrero, A., 2020. Deformities in the Cuban spotted toad *Peltophryne taladai*: a rare case of abnormal eyes. *The Herpetological Bulletin* 154: 23–25. https://doi.org/10.33256/hb154.2325

Alonso Bosch, R. and Hernández Marrero, A., 2021. *Peltophryne taladai*: (Cuban Spotted Toad), Climbing Behavior. *Herpetological Review* 52(3): 616.

Alonso Bosch, R. and Rodríguez Gómez, A., 2001. Llamadas del repertorio vocal de *Eleutherodactylus eileenae* (Anura: Leptodactylidae) del occidente de Cuba. *Cuadernos de Herpetología* 14: 123–130.

Alonso, R. and Rodríguez, A., 2003a. Advertisement calls of Cuban toads of the genus *Bufo* (Anura: Bufonidae). *Phyllomedusa* 2: 75–82. https://doi.org/10.11606/issn.2316-9079.v2i2p75-82

Alonso, R. and Rodríguez, A., 2003b. Insospechados habitantes de las penumbras. Pp. 20–29 in *Anfibios y Reptiles de Cuba*, edited by L. Rodríguez Schettino. UPC Print, Vaasa, Finland.

Alonso, R. and Rodríguez, A., 2003c. Entre la hojarasca bajo nuestros pies. Pp. 30–37 in *Anfibios y Reptiles de Cuba*, edited by L. Rodríguez Schettino. UPC Print, Vaasa, Finland.

Alonso, R. and Rodríguez, A., 2005. Intra-population variation in the advertisement calls of the Cuban giant toad *Bufo fustiger* (Anura: Bufonidae). *Revista Biología* 10: 89–93.

Alonso, R., Crawford, A. J., and Bermingham, E., 2012. Molecular phylogeny of an endemic radiation of Cuban toads (Bufonidae: *Peltophryne*) based on mitochondrial and nuclear genes. *Journal of Biogeography* 39: 434–451. https://doi.org/10.1111/j.1365-2699.2011.02594.x

Alonso, R., Hernández Quinta, M., and Martínez, Y. C., 2009. *Eleutherodactylus atkinsi* (NCN). Diurnal retreats. *Herpetological Review* 40: 203.

Alonso, R., Rodríguez Gómez, A., and Estrada, A. R., 2001. Patrones de actividad acústica y trófica de machos cantores de *Eleutherodactylus eileenae* (Anura: Leptodactylidae). *Revista Española de Herpetología* 15: 45–52.

Alonso, R., Rodríguez, A., and Hernández, A., 2005. *Eleutherodactylus albipes* (NCN). Reproduction. *Herpetological Review* 36: 433.

Alonso, R., Rodríguez, A., and Márquez, R., 2007. *Guía Sonora de los Anfibios de Cuba*. Sound guide of the amphibians of Cuba. Audio CD + booklet. Alosa sons de la natura, Barcelona, Spain.

Alonso Bosch, R., Rodríguez, A., and Hernández Quinta, M., 2017. Advertisement call variation and individual acoustic distinctiveness in the explosive breeding toad *Peltophryne cataulaciceps* (Anura:

Bufonidae). *Acta ethologica* 20: 197–205. https://doi.org/10.1007/s10211-017-0261-8

Alonso Bosch, R., del Castillo Domínguez, S., Torres Martínez, E. L., and García, L. Y., 2014. El anfibio endémico cubano de mayor talla: conciliando récords, literatura y colecciones. *Revista Cubana de Ciencias Biológicas* 3: 60–64.

Alonso Bosch R., García, L. Y., del Castillo Domínguez S., and Torres Martínez, E. L., 2015. Clutches, nest attendance, and hatching in a rock-cave dwelling frog, *Eleutherodactylus* (*Syrrophus*) *zeus*, from Cuba. *Journal of Cave and Karst Studies* 77: 83–86. https://doi.org/10.4311/2014LSC120

Alonso Bosch, R., García-Padrón, L. Y., Almaguer, M., and Valle, M., 2017. First reports of tadpole mouthpart anomalies in a Cuban toad (Anura: Bufonidae: *Peltophryne*). *Herpetological Review* 48: 58–61.

Alonso Bosch, R., Hernández Marrero, A., Leal Echevarría, J. L., Herrera Estrada, I., Gómez Castillo, L., and Trapero Quintana, A. D., 2021. Limb abnormalities in *Peltophryne florentinoi* (Anura: Bufonidae) from Cuba. *Phyllomedusa: Journal of Herpetology*, 20: 117–123. https://doi.org/10.11606/issn.2316-9079.v20i1p117-123

Altig, R. G. and McDiarmid, R. W., 1999. Body plan. Development and morphology. Pp. 24–51 in *Tadpoles: The Biology of Anuran Larvae*, edited by R. W. McDiarmid and R. Altig. The University of Chicago Press, Chicago, IL, USA.

Altig, R. G. and McDiarmid, R. W., 2007. Morphological diversity and evolution of egg and clutch structure in amphibians. *Herpetological Monographs* 21: 1–32. https://doi.org/10.1655/06-005.1

Amaro Valdés, S., 2012. *Lista Roja de la fauna cubana*. Editorial Agencia de Medio Ambiente (AMA), La Habana, Cuba.

AmphibiaWeb. 2020. <https://amphibiaweb.org> University of California, Berkeley, CA, USA. Accessed 20 September 2020.

Anderson, K., 1996. A karyological perspective on the monophyly of the hylid genus *Osteopilus*. Pp 157–168 in *Contributions to West Indian Herpetology: A Tribute to Albert Schwartz*, edited by R. Powell and R. W. Henderson. Contributions to Herpetology 12, Society for the Study of Amphibians and Reptiles, Ithaca, New York, USA.

Arias, A., Ruíz, E., Rivero, A., and Faife, M., 2005. Especies introducidas de la fauna en Cayo Santa María, Archipiélago Sabana – Camagüey, Cuba. In CDROM V Taller de Biodiversidad, BIOECO, Santiago de Cuba, Cuba.

Armas, L. F. de, 1987. Depredación de arácnidos por dos vertebrados cubanos. *Miscelanea Zoológica* 34: 1–2.

Armas, L. F. de, 2014. Depredación de *Centruroides gracilis* (Latreille, 1804) (Scorpiones: Buthidae) por *Osteopilus septentrionalis* Duméril and Bibron, 1841 (Anura: Hylidae). Revista Ibérica de Aracnología 24: 125–126.

Armas, L. F. de and Iturriaga, M., 2017. Depredación de *Hemidactylus mabouia* (Squamata: Gekkonidae) por *Tropidophis pardalis* (Serpentes: Tropidophiidae). *Novitates Caribaea* 11: 99–102. https://doi.org/10.33800/nc.v0i11.22

Barbour, T., 1914. A contribution to the zoogeography of the West Indies, with especial reference to amphibians and reptiles. *Memoirs of the Museum of Comparative Zoology* 44: 205–359. https://doi.org/10.5962/bhl.title.49187

Barbour, T., 1926. New amphibia. *Occasional Papers of the Boston Society of Natural History* 5: 191–194.

Barbour, T., 1937. Third list of Antillean reptiles and amphibians. *Bulletin of the Museum of Comparative Zoology* 82: 77–166.

Barbour, T., 1942. Two preoccupied names. *Copeia* 1942: 179. https://doi.org/10.2307/1438216

Barbour, T., 1945. *A Naturalist in Cuba*. Little, Brown and Company, Boston. https://doi.org/10.2307/1438367

Barbour, T. and Shreve, B., 1937. Novitates cubanae. *Bulletin of the Museum of Comparative Zoology* 80: 377–387.

Barros-Battesti, D. M., Reyes Hernández, M., Famadas, K. M., Onofrio, V. C., Beati, L., and Guglielmone, A. A., 2009. The ixodid ticks (Acari: Ixodidae) of Cuba. *Systematic & Applied Acarology* 14: 101–128. https://doi.org/10.11158/saa.14.2.3

Baruš, V., 1972. Nematodes parasitizing hosts of the genus *Eleutherodactylus* (Amphibia) in Cuba. *Vestnik Csl. Spolecnosti Zoologicke* 36: 161–168.

Baruš, V., 1973. Nematodes parasitizing hosts of the genus *Bufo* (Amphibia) in Cuba. *Folia Parasitologica* 20: 29–39.

Baruš, V. and Moravec, F., 1967. Systematic studies of parasitic worms found in hosts *Lepisosteus tristoechus* (Gynglimodi: Lepisosteidae) and *Hyla insulsa* (Ecaudata: Hylidae) from Cuba. *Vestnik Csl. Spolecnosti Zoologicke* 31: 1–14.

Berger, L., Hyatt, A. D., Speare, R., and Longcore, J. E., 2005. Life cycle stages of the amphibian chytrid *Batrachochytrium dendrobatidis*. *Diseases of Aquatic Organisms* 68: 51–63. https://doi.org/10.3354/dao068051

Berger L., Speare, R., Daszak, P., Green, D. E., Cunningham, A. A., Goggin, C. L., Slocombe, R., Ragan, M. A., Hyatt, A. D., McDonald, K. R., Hines, H. B., Lips, K. R., Marantelli, G., and Parkes, H., 1998. Chytridiomycosis causes amphibian mortality associated with population declines in the rain forests of Australia and Central America. *Proceedings of the National Academy of Sciences of the United States of America* 95: 9031–9036. https://doi.org/10.1073/pnas.95.15.9031

Bignotte-Giró, I., 2019. Reproducción y vocalizaciones en especies cubanas de ranas del género *Eleutherodactylus* del Oriente de Cuba. PhD Dissertation, Universidad de Alicante, Spain.

Bignotte-Giró, I., Fong G., A., and Begué Quiala, G., 2022. The clutches of two bromeliad-dweller frogs (*Eleutherodactylus guantanamera* and *E. ionthus*) from Cuba. *Reptiles & Amphibians* 29: 39–43. https://doi.org/10.17161/randa.v29i1.15996

Bignotte-Giró, I., Fong G., A., and López-Iborra, G. M., 2019. Acoustic niche partitioning in five Cuban frogs of the genus *Eleutherodactylus*. *Amphibia-Reptilia* 40: 1–11. https://doi.org/10.1163/15685381-17000170

Bignotte-Giró, I., López-Iborra, G. M., Ramírez-Pinilla, M. P., and Fong G., A., 2021. Reproductive phenology of five species of terrestrial frogs (genus *Eleutherodactylus*) from Cuba. *Journal of Herpetology* 55: 127–136. https://doi.org/10.1670/20-027

Blackburn, T. M., Bellard, C., and Ricciardi, A., 2019. Alien versus native species as drivers of recent extinctions. *Frontiers in Ecology and the Environment* 17: 203–207. https://doi.org/10.1002/fee.2020

Blair, W. F., 1958. Call differences as an isolation mechanism in Florida species of hylid frogs. *Quaternary Journal of Florida Academy of Sciences* 21: 32–48.

Blaustein, A. R., Belden, L. K., Olson, D. H., Green, D. M., Root, T. L., Gibbs, J. M., and Breisch, A. R., 2001. Amphibian breeding and climate change. *Conservation Biology* 15: 1804–1809. https://doi.org/10.1046/j.1523-1739.2001.00307.x

Blaustein, A. R., Gervasi, S. S., Johnson, P. T. J., Hoverman, J. T., Belden, L. K., Bradley, P. W., and Xie, G. Y., 2012. Ecophysiology meets conservation: understanding the role of disease in amphibian population declines. *Philosophical Transactions of the Royal Society B* 367: 1688–1707. https://doi.org/10.1098/rstb.2012.0011

Blaustein, A. R., Han, B. A., Relyea, R. A., Johnson, P. T. J., Buck, J. C., Gervasi, S. S., and Kats, L. B., 2011. The complexity of amphibian population declines: understanding the role of cofactors in driving amphibian losses. *Annals of the New York Academy of Sciences* 1223: 108–119. https://doi.org/10.1111/j.1749-6632.2010.05909.x

Blotto, B. L., Lyra, M. L., Cardoso, M. C. S., Trefaut Rodrigues, M., Díaz, L. R., Marciano, E.-Jr., Dal Vechio, F., Orrico, V. G. D., Brandão, R. A., Lopes de Assis, C., Lantyer-Silva, A. S. F., Rutherford, M. G., Gagliardi-Urrutia, G., Solé, M., Baldo, D., Nunes, I., Cajade, R., Torres, A., Grant, T., Jungfer, K-H., da Silva, H. R., Haddad, C. F. B., and Faivovich, J., 2020. The phylogeny of the Casque-headed Treefrogs (Hylidae: Hylinae: Lophyohylini). *Cladistics* 2020: 1–37. https://doi.org/10.1111/cla.12409

Bogan, J. E., Jr. and Mitchell, M. A., 2017. Characterizing tissue enzyme activities of Cuban Tree Frogs (*Osteopilus septentrionalis*). *Journal of Herpetological Medicine and Surgery* 27: 22–28. https://doi.org/10.5818/1529-9651-27.1-2.22

Bogart, J. P., 1972. Karyotypes. Pp 171–195 in *Evolution of the Genus* Bufo, edited by W. F. Blair. University of Texas Press, Austin, TX, USA. and London, UK.

Bogart, J. P., 1981. Chromosome studies in *Sminthillus* from Cuba and *Eleutherodactylus* from Cuba and Puerto Rico (Anura: Leptodactylidae). *Life Science Contributions of the Royal Ontario Museum* 129: 1–22. https://doi.org/10.5962/bhl.title.52080

Borges Júnior, V. N. and Duarte Rocha, C. F., 2013. Tropical tadpole assemblages: which factors affect their structure and distribution? *Oecologia Australis* 17: 217–228. https://doi.org/10.4257/oeco.2013.1702.04

Borroto-Páez, R. and Reyes Pérez, D., 2019. Unusual behavior during cannibalism of Cuban Treefrogs (*Osteopilus septentrionalis*) in a commensal habitat in Cuba. *Reptiles & Amphibians* 26: 117–118. https://doi.org/10.17161/randa.v26i2.14375

Borroto-Páez, R. and Reyes Pérez, D., 2020. Predation by a Cuban Treefrog (*Osteopilus septentrionalis*) and a Domestic Cat (*Felis catus*) on Tropical House Geckos (*Hemidactylus mabouia*) in Central Cuba, with a review of predators and vertebrate prey of Tropical House Geckos. *Reptiles & Amphibians* 27: 120–128. https://doi.org/10.17161/randa.v27i2.14022

Borroto-Páez, R., Alonso Bosch, R., Fabrés, B. A., and Alvarez García, O., 2015. Introduced amphibians and reptiles in the Cuban Archipelago. *Herpetological Conservation and Biology* 10: 985–1012.

Breuil, M., 2004. *Amphibiens et reptiles des Antilles*. PLB Editions, Le Gosier, Guadeloupe.

Bruner, S. C., 1935. La introducción en Cuba del sapo gigante (*Bufo marinus*). *Revista de la Agricultura* 18: 73–75.

Buide, M. S., 1967. Lista de los anfibios y reptiles de Cuba. *Torreia*, Nueva Serie, 1: 1–60.

Burrowes, P. A., Joglar, R. L., and Green, D. E., 2004. Potential causes for amphibian declines in Puerto Rico. *Herpetologica* 60: 141–154. https://doi.org/10.1655/03-50

CABI, 2021. *Eleutherodactylus planirostris* (greenhouse frog). Invasive Species Compendium. Available: https://www.cabi.org/isc/datasheet/84737#tointroductions. Accessed: 13 July 2021.

Cádiz, A. and Iturriaga, M., 2020. Male movement patterns and bioacoustic activity of the Western Cuban grassfrog, *Eleutherodactylus varleyi* (Anura: Eleutherodactylidae). *IRCF Reptiles & Amphibians* 27: 1–9. https://doi.org/10.17161/randa.v27i1.14436

Cádiz, A., Reytor, M. L., Díaz, L. M., Chestnut, T., Burns, J. A., and Amato, G., 2018. The chytrid fungus, *Batrachochytrium dendrobatidis,* is widespread among Cuban amphibians. *EcoHealth* 16: 128–140. doi:10.1007/s10393-018-1383-9

Capote, R. P. and Berazaín, R., 1984. Clasificación de las formaciones vegetales de Cuba. *Revista del Jardín Botánico Nacional* 5: 27–75.

Capote, R. P., Cruz Díaz, R. O., and Vantour, A., 2006. Fragmentación de la vegetación en el Archipiélago Cubano: Conservación de la diversidad biológica y mitigación de la desertificación. Pp. 33–36 in *Memoria 1er Taller Binacional y regional sobre Desertificación,* edited by M. Riveros Caballeros, L. E. Sánchez, and J. Paolini. Editorial IVIC, Caracas, Venezuela.

Castellanos, E., 2005. Processing SRTM DEM data for national landslide hazard assessment. VI Congreso de Geología (GEOLOGÍA 2005), Simposio de Sismicidad y Riesgos Geológicos. Primera Convención Cubana de Ciencias de la Tierra, GEOCIENCIAS 2005. La Habana, 5–8 abril de 2005. https://www.researchgate.net/publication/209803322.

Castellón Maure, A. and Rodríguez-Cabrera, T. M., 2018. First record of vertebrate predation by the Cuban Whiptail, *Pholidoscelis auberi zugi* (Squamata: Teiidae), with a summary of reptilian and amphibian prey of West Indian Whiptails. *Reptiles & Amphibians* 25: 188–191. https://doi.org/10.17161/randa.v25i3.14298

Cei, J. M., Erspamer V., and Roseghini, M., 1972. Biogenic amines. Pp 233–243 in *Evolution of the Genus* Bufo, edited by W. F. Blair. University of Texas Press, Austin and London.

Centella, A., Gutiérrez, T., Limia, M., and Rivero Jaspe, R., 1999. Climate change scenarios for impact assessment in Cuba. *Climate Research* 12: 223–230. https://doi.org/10.3354/cr012223

Černý, V., 1969. Nuevos conocimientos sobre la ixodofauna cubana. *Torreia* 21: 1–12.

Chambouvet, A., Smilansky, V., Jirků, M., Isidoro-Ayza, M., Itoïz, S., Derelle, E., Monier, A., Gower, D. J., Wilkinson, M., Yabsley, M. J., Lukes, J., and Richards, T. A., 2020. Diverse alveolate infections of tadpoles, a new threat to frogs? *PLOS Pathogens* 16: e1008107. https://doi.org/10.1371/journal.ppat.1008107

CITMA (Ministerio de Ciencia, Tecnología y Medio Ambiente de la República de Cuba), 2014. V Informe Nacional al Convenio sobre la Diversidad Biológica. La Habana, Cuba.

CITMA (Ministerio de Ciencia, Tecnología y Medio Ambiente de la República de Cuba), 2019. VI Informe Nacional al Convenio sobre la Diversidad Biológica. La Habana, Cuba.

CNAP (Centro Nacional de Áreas Protegidas), 2002. Sistema Nacional de Áreas Protegidas, Cuba. Plan 2003–2008. Centro Nacional de Áreas Protegidas. Escandón Imp., Sevilla, España.

CNAP (Centro Nacional de Áreas Protegidas), 2013. Plan del Sistema Nacional de Áreas Protegidas 2014–2020. Ministerio de Ciencias Tecnología y Medio Ambiente, La Habana, Cuba.

Cobos, M. E. and Alonso Bosch, R., 2016. Recent and future threats to an endangered Cuban toad: potential additive impacts from climate change and habitat loss. *Oryx* 52: 116–125. https://doi.org/10.1017/S0030605316000612

Cobos, M. E. and Alonso Bosch, R., 2018. Breeding sites of a narrowly distributed amphibian, a key element in its conservation in the face of global change. *Aquatic Conservation: Marine and Freshwater Ecosystems* 28: 1089–1098. https://doi.org/10.1002/aqc.2967

Contreras-Calvario, A. I., Mora-Reyes, A., Parra-Olea, G., and Mendoza, A. M., 2018. New record of the introduced species *Eleutherodactylus planirostris* (Anura: Eleutherodactylidae) in the state of Veracruz, Mexico. *Herpetological Journal* 28: 96–99.

Cope, E. D., 1862a. On some new and little known American anurans. *Proceedings of the Academy of Natural Sciences of Philadelphia* 14: 151–159 + 594.

Cope, E. D., 1862b. Notes upon some Reptiles of the Old World. *Proceedings of the Academy of Natural Sciences of Philadelphia* 14: 337–344 + 594.

Coy, A., 1980. Primer hallazgo del género *Physalopteroides* Wu et Liu, 1940 (Nematoda: Physalopteridae) en Cuba. *Poeyana* 200: 1–6.

Coy, A. and Baruš, V., 1982. Primer hallazgo de *Foleyela* (Nematoda: Oswaldofilariidae) en Cuba. *Poeyana* 244: 1–4.

Coy, A. and Hernández, N., 1982. Lista de los helmintos parásitos de los vertebrados silvestres cubanos. *Poeyana* 235: 1–57.

Coy, A. and Ventosa, L., 1984. Nemátodos parásitos de anfibios cubanos. *Poeyana* 269: 1–20.

Coy Otero, A. and Martínez, J. J., 1987. Nuevo hallazgo en Cuba de larvas del género *Eustrongylides* Jaegerskiold, 1909 (Nematoda: Dioctophymidae) en la rana toro (*Rana catesbeiana*). *Miscelánea Zoológica* 32: 3.

Crother, B. I. (ed.), 1999. *Caribbean Reptiles and Amphibians*. Academic Press, San Diego, CA, USA.

Crump, M. L., 1986. Cannibalism by younger tadpoles: another hazard of metamorphosis. *Copeia* 1986: 1007–1009. https://doi.org/10.2307/1445301

Crump, M. L., 2015. Anuran reproductive modes: evolving perspectives. *Journal of Herpetology* 49: 1–16. https://doi.org/10.1670/14-097

Cubillos-Abrahams, M. V., Montes-Correa, A. C., and Vera-Pérez, L. E., 2021. First record of the greenhouse frog *Eleutherodactylus planirostris* (Anura, Eleutherodactylidae) from San Andrés, Colombian Caribbean islands. *Caribbean Journal of Science* 51: 136–145. https://doi.org/10.18475/cjos.v51i1.a16

Cushman, S. A., 2006. Effects of habitat loss and fragmentation on amphibians: A review and prospectus. *Biological Conservation* 128: 231–240. https://doi.org/10.1016/j.biocon.2005.09.031

das Neves-da-Silva, D., Borges-Júnior, V. N. T., Branco, C. W. C., and Telles de Carvalho-e-Silva, A. M. P., 2021. Effects of intrinsic and extrinsic factors on the prevalence of the fungus *Batrachochytrium dendrobatidis* (Chytridiomycota) in stream tadpoles in the Atlantic Forest domain. *Aquatic Ecology* 55: 891–902. https://doi.org/10.1007/s10452-021-09869-y

de la Cruz, J., 1971. Los ácaros de la subfamilia Lawrencarinae (Acari: Ereynetidae) parásitos de los batracios de Cuba. *Poeyana* 86: 1–9.

del Castillo Domínguez, S. L., Mancina González, C. A., Bandera Fernández, E., Pérez Pelea, L., Cézilly, F., and Alonso Bosch, R., 2021. Predicting the invasion of the acoustic niche: potential distribution and call transmission efficiency of a newly introduced frog in Cuba. *Perspectives in Ecology and Conservation* 19: 90–97. https://doi.org/10.1016/j.pecon.2020.12.002

Delfino, G., Brizzi, R., Nosi, D., and Terreni, A., 2002. Serous cutaneous glands in New World hylid frogs: an ultrastructural study on skin poisons confirms phylogenetic relationships between *Osteopilus septentrionalis* and *Phrynohyas venulosa*. *Journal of Morphology* 253: 176–186. https://doi.org/10.1002/jmor.1119

Díaz, L., 2004. *Eleutherodactylus simulans*. *The IUCN Red List of Threatened Species* 2004: e.T56968A11561298. https://dx.doi.org/10.2305/IUCN.UK.2004.RLTS.T56968A11561298.en. Downloaded on 30 July 2021.

Díaz, L. M., 1999. Planarias y chinches acuáticas (Tricladida, Hemiptera), depredadores de la rana platanera *Osteopilus septentrionalis* (Anura: Hylidae). *Cocuyo* 9: 24–25.

Díaz, L. M., 2008. *Eleutherodactylus michaelschmidi*. *The IUCN Red List of Threatened Species* 2008: e.T135753A4197152. https://dx.doi.org/10.2305/IUCN.UK.2008.RLTS.T135753A4197152.en. Downloaded on 30 July 2021.

Díaz, L. M., 2011. Protecting the Cuban Long-Nosed Toad. *Froglog* 96: 19.

Díaz, L. M. and Cádiz, A., 2006. Pflege und Vermehrung von *Bufo longinasus* Stejneger, 1905: Ein Beitrag zur Erhaltung dieser Art. *Aquaristik Fachmagazin & Aquarium heute* 38: 18–21.

Díaz, L. M. and Cádiz, A., 2007. Guía descriptiva para la identificación de las llamadas de anuncio de las ranas cubanas del género *Eleutherodactylus* (Anura: Leptodactylidae). *Herpetotropicos* 3: 100–122.

Díaz, L. M. and Cádiz, A., 2008. Guía Taxonómica de los Anfibios de Cuba + 1 CD. *AbcTaxa* 4, Bélgica.

Díaz, L. M. and Estrada, A. R., 2000. The male and female vocalizations of the Cuban frog *Eleutherodactylus guanahacabibes* (Anura: Leptodactylidae). *Caribbean Journal of Science* 36: 328–331.

Díaz, L. M. and Fong, A., 2001. A new mottled frog of the genus *Eleutherodactylus* (Anura: Leptodactylidae) from Eastern Cuba. *Solenodon* 1: 76–84.

Díaz, L. M. and Hedges, S. B., 2015. Another new cryptic frog related to *Eleutherodactylus varleyi* Dunn (Amphibia: Anura: Eleutherodactylidae), from eastern Cuba. *Solenodon* 12: 124–135.

Díaz, L. M., Cádiz, A., and Hedges, S. B., 2003. A new grass frog from pine forests of Western Cuba, and description of acoustic and pattern variation in *Eleutherodactylus varleyi* (Amphibia: Leptodactylidae). *Caribbean Journal of Science* 39: 176–188.

Díaz, L. M., Cádiz, A., and Navarro, N., 2005. A new ground dwelling frog of the genus *Eleutherodactylus* from eastern Cuba, with a redefinition of the *E. dimidiatus* group. *Caribbean Journal of Science* 41: 307–318.

Díaz, L. M., Cádiz, A., and Navarro, N., 2007a. A new rock dwelling frog of the genus *Eleutherodactylus* (Amphibia: Leptodactylidae) from eastern Cuba, with comments on other species with similar habits. *Zootaxa* 1435: 51–68. https://doi.org/10.11646/zootaxa.1435.1.6

Díaz, L. M., Estrada, A. R., and Hedges, S. B., 2001. A new riparial frog of the genus *Eleutherodactylus* (Anura: Leptodactylidae) from Eastern Cuba. *Caribbean Journal of Science* 37: 63–71.

Díaz, L. M., Hedges, S. B., and Schmid, M., 2012. A new cryptic species of the genus *Eleutherodactylus* (Amphibia: Anura: Eleutherodactylidae) from Cuba. *Zootaxa* 3220: 44–60. https://doi.org/10.11646/zootaxa.3220.1.3

Díaz, L. M, Moreno, L. V., and Begué, G., 2000. Descripción de las larvas de *Bufo taladai* (Anura: Bufonidae). *Caribbean Journal of Science* 36: 156–160.

Díaz, L. M., Cádiz, A., Chong, A., and Silva, A., 2007b. First report of chytridiomycosis in a dying toad (Anura: Bufonidae) from Cuba: a new conservation challenge for the island. *Eco Health* 4: 172–175. https://doi.org/10.1007/s10393-007-0094-4

Díaz, L. M., Cádiz, A., Villar, S., and Bermúdez, F., 2014. Notes on the ecology and morphology of the Cuban khaki trope, *Tropidophis hendersoni* Hedges and Garrido (Squamata: Tropidophiidae), with a new locality record. *IRCF Reptiles & Amphibians* 21: 116–119. https://doi.org/10.17161/randa.v21i4.14009

Díaz, L. M., Incháustegui, S., Marte, C., and Chong, A., 2015. The tadpoles of the hylid frogs (Anura: Hylidae: *Hypsiboas* and *Osteopilus*) of Hispaniola. *Novitates Caribaea* 8: 1–29. https://doi.org/10.33800/nc.v0i8.39

Díaz, L. M., Lemm, J. M., Briggs, E. M., Clark, Sh., Dannenberg, A. L., Gamber, R. G., Kelly, J., Laughlin, P., Markowski, G., Shelley, E. L., and Van Bakel, A. B., 2019. Citizen science enables a new locality record for a critically endangered toad, *Peltophryne florentinoi* Moreno & Rivalta, 2007. *Novitates Caribaea* 13: 122–124. https://doi.org/10.33800/nc.v0i13.196

Díaz Cisneros, L. R., 1989. Regionalización climática general, VI.4.4, mapa 55. In *Nuevo Atlas Nacional de Cuba*. Instituto de Geografía, Academia de Ciencias de Cuba.

Duellman, W. W. and Crombie, R. I., 1970. *Hyla septentrionalis*. *Catalogue of American Amphibians and Reptiles* 92: 1–4.

Duellman, W. E. and Schwartz, A., 1958. Amphibians and reptiles of southern Florida. *Bulletin of the Florida State A. Museum, Biological Sciences* 3: 181–324.

Duellman, W. E., Marion, A. B., and Hedges, S. B., 2016. Phylogenetics, classification, and biogeography of the treefrogs (Amphibia: Anura: Arboranae). *Zootaxa* 4104: 1–109. https://doi.org/10.11646/zootaxa.4104.1.1

Dugo-Cota, A., Rodríguez, A., and González-Voyer, A., 2019. Ecomorphological convergence in *Eleutherodactylus* frogs: a case of replicate radiations in the Caribbean. *Ecology Letters* 22: 884–893. https://doi.org/10.1111/ele.13246

Duméril, A. M. C. and Bibron, G., 1841. *Érpétologie Générale ou Histoire complète des Reptiles*. Vol. 8. *Librairie Encyclopédique de Roret*, Paris, France.

Dunn, E. R., 1925. New frogs from Cuba. *Occasional Papers of the Boston Society of Natural History* 5: 163–166.

Dunn, E. R., 1926a. Additional frogs from Cuba. *Occasional Papers of the Boston Society of Natural History* 5: 209–215.

Dunn, E. R., 1926b. The habits of *Sminthillus limbatus*. *Copeia* 155: 141–143. https://doi.org/10.2307/1436942

Dunn, E. R., 1926c. Notes on two Cuban frogs. *Copeia* 157: 154–156. https://doi.org/10.2307/1435983

Essl, F., Bacher, S., Genovesi, P., Hulme, P. E., Jeschke, J. M., Katsanevakis, S., Kowarik, I., Kühn, I., Pyšek, P., Rabitsch, W., Schindler, S., van Kleunen, M., Vilà, M., Wilson, J. R. U., and Richardson, D. M., 2018. Which taxa are alien? Criteria, applications, and uncertainties. *BioScience* 68: 496–509.

Estrada, A. R., 1987. Los nidos terrestres de dos anfibios cubanos del género *Eleutherodactylus* (Anura: Leptodactylidae). *Poeyana* 352: 1–9.

Estrada, A. R., 1990. Las puestas de *Eleutherodactylus varians* (Gundlach and Peters). *Revista Biología* 4: 163–167.

Estrada, A. R., 1992. Comentarios sobre *Eleutherodactylus intermedius* (Barbour and Shreve, 1937) de la región oriental de Cuba. *Comunicaciones Breves de Zoología*, Editorial Academia. p. 14.

Estrada, A. R., 2012. The Cuban archipelago. Pp. 113–125 in *Island Lists of West Indian Amphibians and Reptiles*, edited by R. Powell and R. W. Henderson. *Bulletin of the Florida Museum of Natural History* 51: 85–166.

Estrada, A. R. and Alonso, R., 1997. Nueva Especie del grupo *limbatus* (Leptodactylidae: *Eleutherodactylus*) de la Región Oriental de Cuba. *Caribbean Journal of Science* 33: 41–44.

Estrada, A. R. and Hedges, S. B., 1991. Nueva especie de *Eleutherodactylus* (Anura: Leptodactylidae) de la Región Oriental de Cuba. *Caribbean Journal of Science* 27: 139–145.

Estrada, A. R. and Hedges, S. B., 1996a. A new frog of the genus *Eleutherodactylus* from eastern Cuba (Anura, Leptodactylidae). *Herpetologica* 52: 435–439. https://doi.org/10.2307/1447647

Estrada, A. R. and Hedges, S. B., 1996b. At the lower size limit in the tetrapods: a new diminutive frog from Cuba (Leptodactylidae: *Eleutherodactylus*). *Copeia* 1996: 852–859. https://doi.org/10.2307/1447647

Estrada, A. R. and Hedges, S. B., 1997a. Nueva especie de *Eleutherodactylus* (Anura: Leptodactylidae) del Macizo Sagua-Baracoa, Cuba. *Caribbean Journal of Science* 33: 222–226.

Estrada, A. R. and Hedges, S. B., 1997b. A new species of frog from the Sierra Maestra, Cuba (Leptodactylidae, *Eleutherodactylus*). *Caribbean Journal of Science* 31: 364–368. https://doi.org/10.2307/1565664

Estrada, A. R. and Hedges, S. B., 1997c. A new species of frog from the Meseta de Cabo Cruz, eastern Cuba (Leptodactylidae: *Eleutherodactylus*). *Caribbean Journal of Science* 33: 227–231.

Estrada, A. R. and Hedges, S. B., 1998. Sistemática de las ranas ribereñas de Cuba (Leptodactylidae: *Eleutherodactylus*) con la descripción de una nueva especie. *Caribbean Journal of Science* 34: 218–230.

Estrada, A. R. and Novo, J., 1985. Nueva especie de *Eleutherodactylus* del grupo *ricordi* (Anura: Leptodactylidae). *Poeyana* 303: 1–10.

Estrada, A. R., Díaz, L. M., and Rodríguez, A., 1997. Nueva especie de *Eleutherodactylus* (Anura: Leptodactylidae) del litoral norte de La Habana. *Revista Española de Herpetología* 11: 19–24.

Estrada, R., Martín Morales, G., Martínez Fernández, P., Rodríguez Hernández, S. V., Capote, R. P., Reyes Alonso, I., Galano Duverger, S., Cabrera Gutiérrez, C., Martínez Bayón, C., Mateo Rego, L., Guerra Sosa, Y., Batte Hernández, A., and Coya de la Fuente, L., 2013. *Mapa (BD-SIG) de vegetación natural y seminatural de Cuba* V.1 *Landsat ETM 7 SLC-Off Gap Filled,* CIRCA 2011. IV Congreso sobre Manejo de Ecosistemas y Biodiversidad de la VIII Convención Internacional sobre Medio Ambiente y Desarrollo, La Habana, Cuba. http://repositorio.geotech.cu/jspui/bitstream/1234/597/1/Mapa%20de%20Vegetacion%20de%20Cuba_Repositorio.pdf

Febles, I., 2002. Morfología externa de las larvas de siete especies de anuros cubanos (Amphibia: Anura). Tesis de Diploma. Facultad de Biología, Universidad de La Habana, Cuba.

Feng, X., Porporato, A., and Rodríguez-Iturbe, I., 2013. Changes in rainfall seasonality in the tropics. *Nature Climate Change* 3: 811–815. https://doi.org/10.1038/nclimate1907

Fisher, M. C. and Garner, T. W. J., 2020. Chytrid fungi and global amphibian declines. *Nature reviews. Microbiology* 18: 332–343. https://doi.org/10.1038/s41579-020-0335-x

Foden, W. B., Butchart, S. H. M., Stuart, S. N., Vié J-C, Akçakaya, H. R., Angulo, A., DeVantier, L. M., Gutsche, A., Turak, E., Cao, L., Donner, S. D., Katariya, V., Bernard, R., Holland, R. A., Hughes, A. F., O'Hanlon, S. E., Garnett, S. T., Şekercioğlu, C. H., and Mace, G. M., 2013. Identifying the world's most climate change vulnerable species: A systematic trait-based assessment of all birds, amphibians and corals. *PLOS ONE* 8: e65427. https://doi.org/10.1371/journal.pone.0065427

Fong G., A., 1999. Changes in amphibian composition in altered habitats of Eastern Cuba. *Froglog* 36: 2–3.

Fong G., A., 2004. *Antillophis andreae* (Black and White Racer, Jubito): Diet. *Herpetological Review* 35: 57.

Fong G., A., 2007. Monitoring amphibian populations in two sensitive habitats in Cuba. *Froglog* 80: 2–4.

Fong G., A., 2009. Distribución y conservación de los anfibios de los macizos montañosos de la Región Oriental de Cuba. PhD Dissertation, Universidad de Alicante, Spain.

Fong G., A. and Bignotte-Giró, I., 2021. Female vocalization in the Oriente Streamside Frog, *Eleutherodactylus cuneatus* (Eleutherodactylidae), from Cuba. *Reptiles & Amphibians* 28: 40–42. https://doi.org/10.17161/randa.v28i1.15302

Fong G., A. and Garcés, G., 2002. *Bufo taladai* (NCN). Diet. *Herpetological Review* 33: 302.

Fong G., A. and Hero, J.-M., 2006. Population dynamics of the stream-dwelling frog *Eleutherodactylus cuneatus* on La Gran Piedra, eastern Cuba. *Froglog* 77: 2–4.

Fong G., A., Alonso, R., and Rodríguez, A., 2009. *Eleutherodactylus dimidiatus* (NCN) and *Eleutherodactylus intermedius* (NCN). Predator/prey. *Herpetological Review* 40: 203–204.

Fong G., A., I. Bignotte Giró, I., and García, K. M., 2013. Unsuccessful predation on the toad *Peltophryne peltocephala* (Bufonidae) by the Cuban snake *Tropidophis melanurus* (Tropidophiidae). *Herpetology Notes* 6: 73–75.

Fong G., A., Viña, N., and López-Iborra, G. M., 2015. Amphibian hotspots and conservation priorities in Eastern Cuba identified by species distribution modeling. *Biotropica* 47: 119–127. https://doi.org/10.1111/btp.12186

Fong G., A., Hero, J. M., Viña, R., and Bignotte-Giró, I., 2010. Population ecology of the riparian frog *Eleutherodactylus cuneatus* in Cuba. *Biotropica* 42: 348–354. https://doi.org/10.1111/j.1744-7429.2009.00592.x

Fong G., A., Hero, J. M., Bignotte-Giró, I., Rueda-Solano, L. A., and Gutiérrez, Y., 2012. *Eleutherodactylus cuneatus*. Predation. *Herpetological Review* 43: 319–320.

Frost, D. R., 2021. Amphibian Species of the World: An Online Reference. Version 6.1 (8 August 2021). Electronic Database accessible at https://amphibiansoftheworld.amnh.org/index.php. American Museum of Natural History, New York, USA. https://doi.org/10.5531/db.vz.0001

Fuentes Marrero, I. M., González-Oliva, L., Baró Oviedo, I., González Echevarría, M. T., and Mancina, C. A., 2019. Potential effect of climate change on the distribution of plants associated to moist forest of eastern Cuba. *Acta Botanica Cubana* 218: 160–170.

Funes Monzote, R., 2008. *From Rainforest to Cane Field in Cuba: an Environmental History since 1492*. University of North Carolina Press, Chapel Hill, North Carolina, USA. https://doi.org/10.5149/9780807888865_funes_monzote

García-González, A., García-Padrón, L. Y., Delgado Fernández, F., and Riverón-Giró, F. B., 2014. Anfibios y reptiles asociados a tres especies de bromelias de tanque en el Parque Nacional Guanahacabibes, Cuba. *Cuadernos de Investigación UNED* 6: 87–97. https://doi.org/10.22458/urj.v6i1.319

García Padrón, L. Y., 2008. Primer reporte de la presencia de un anfibio endémico en el contenido estomacal del pez gato africano (*Clarias gariepinus*) en Cuba. *ECOVIDA* 1: 121–123.

García, L. Y., 2012. *Eleutherodactylus zeus* (Cuban Giant Frog). Record size. *Herpetological Review* 43: 631.

García-Padrón, L. Y., 2019. Métodos para el estudio de la dieta en anfibios ¿Cuál es el más adecuado para las especies cubanas? *Poeyana* 508: 28–33.

García-Padrón, L. Y., 2021. Diet of a community of frogs in an agroecosystem in western Cuba. *Caribbean Herpetology* 76: 1–8.

García-Padrón, L. Y., 2022a. Diet of juveniles and adults of the Cuban Giant Frog, *Eleutherodactylus zeus* (Anura: Eleutherodactylidae). *Caribbean Herpetology* 85: 1–8. https://doi.org/10.31611/ch.8.

García-Padrón, L. Y. and Alonso Bosch, R., 2017. *Osteopilus septentrionalis* (Cuban Tree Frog). Malformation. *Herpetological Review* 48: 412.

García-Padrón, L. Y. and Alonso Bosch, R., 2019. Anomalous colour in a Cuban cave-dwelling frog: first record of piebaldism in *Eleutherodactylus zeus* (Anura: Eleutherodactylidae). *The Herpetological Bulletin* 147: 1–3. https://doi.org/10.33256/hb147.13

García-Padrón, L. Y. and Borrego Quevedo, C. A., 2020. Dieta de *Eleutherodactylus atkinsi* (Anura: Eleutherodactylidae) en el occidente de Cuba. *Poeyana* 511: 53–58.

García-Padrón, L. Y. and Borrego Quevedo, C. A., 2022. Notes on the Diet of the Cuban Flat-headed Frog, *Eleutherodactylus planirostris* (Eleutherodactylidae) from Two Understudied Habitats in Western Cuba. *Reptiles & Amphibians* 29, 311–313. https://doi.org/10.17161/randa.v29i1.17956

García-Padrón, L. Y. and Hernández Ledesma, N., 2018. Anfibios y reptiles de la Reserva Ecológica "Los Pretiles", Pinar del Río, Cuba. *ECOVIDA* 8: 15–27.

García-Padrón, L. Y. and Rodríguez-Cabrera, T. M. 2022. Diurnal Activity in the Cuban Treefrog, *Osteopilus septentrionalis* (Anura: Hylidae), Associated with an unusually high prey density in Western Cuba. *Reptiles & Amphibians* 29, 308–310. https://doi.org/10.17161/randa.v29i1.17953

García-Padrón, L. Y. and Alonso Bosch, R., and Falero Álvarez, O., 2022. Diet of the recently introduced white-lipped frog *Leptodactylus fragilis* from a suburban ecosystem in western Cuba. *The Herpetological Bulletin* 161, 24–27 https://doi.org/10.33256/hb161.2427

García-Padrón, L. Y., Boligán, M., and Barrero Medel, H., 2021a. Diet of the Cuban giant frog, *Eleutherodactylus zeus* (Anura: Eleutherodactylidae) in Viñales National Park, Cuba. *Cuadernos de Herpetología* 35: 43–51.

García-Padrón, L. Y., Cordero Acosta, J. C., and Baños Utria, R., 2021b. Is the introduced American bullfrog competing for food resources with the Cuban Western giant toad? *Acta Científica: una revista transdisciplinaria de Puerto Rico y el Caribe* 32(2): 26–32.

García-Padrón, L. Y., Lugo Blanco, C. and Álvarez Young, C., 2021c. *Actitud de los estudiantes de secundaria básica hacia los anfibios.* Ciencias Sociales y Educación, 10(20): 197–212. https://doi. org/10.22395/csye.v10n20a9.

Garrido, O. H. and Jaume, M. L., 1984. Catálogo descriptivo de los anfibios y reptiles de Cuba. *Doñana, Acta Vertebrata* 11: 5–128.

Gascon, C., Collins, J. P., Moore, R. D., Church, D. R., McKay, J. E., and Mendelson, J. R. III, (editors), 2007. *Amphibian Conservation Action Plan.* IUCN/SSC Amphibian Specialist Group. Gland, Switzerland and Cambridge, UK.

Gawor, A., Rauhaus, A., Karbe, D., Van Der Straeten, K., Lötters, S., and Ziegler, T., 2012. Is there a chance for conservation breeding? *Ex situ* management, reproduction, and early life stages of the Harlequin toad *Atelopus flavescens* Duméril & Bibron, 1841 (Amphibia: Anura: Bufonidae). *Amphibian and Reptile Conservation* 5: 29–44.

Gill, J. P. K., Nidhi, S., Mohan, A., Datta, S., and Girdhar, M., 2017. Glyphosate toxicity for animals. *Environmental Chemistry Letters* 16: 401–426. https://doi.org/10.1007/s10311-017-0689-0

Goin, C. J., 1947. Studies on the life history of *Eleutherodactylus ricordii planirostris* (Cope) in Florida. *University of Florida Studies, Biological Science Series* 4: i–xi, 1–66.

Gosner, K. L., 1960. A simplified table for staging anuran embryos and larvae with notes on identification. *Herpetologica* 16: 183–190.

González Alonso, H., Rodríguez Schettino, L., Rodríguez, A., Mancina, C. A., and Ramos García, I. (editors), 2012. *Libro Rojo de los Vertebrados de Cuba.* Editorial Academia, La Habana, Cuba.

GORC (Gaceta Oficial de la República de Cuba), 1999. Decreto-Ley 201 del Sistema Nacional de Áreas Protegidas. *Edición Ordinaria* 84 (Año XCVII): 1355–1370.

Grant, E. H. C., Miller, D. A. W., and Muths, E., 2020. A synthesis of evidence of drivers of Amphibian declines. *Herpetologica* 76: 101–107. https://doi.org/10.1655/0018-0831-76.2.101

Grant, E. H. C., Miller, D. A. W., Schmidt, B., Adams, M. J., Amburgey, S. M., Chambert, T., Cruickshank, S. S., Fisher, R. N., Ryan, M. E., Waddle, M. E., Green, D. M., Hossack, B. R., Johnson, P. T. J., Joseph, M. B., Rittenhouse, T. A. G., Ryan, M. E., Waddle, H., Walls, S. C., Bailey, L. L., Fellers, G. M., Gorman, T. A., Ray, A. M., Pilliod, D. S., Price, S. J., Saenz, D., Sadinski, W., and Muths, E., 2016. Quantitative evidence for the effects of multiple drivers on continental-scale amphibian declines. *Science Report* 6: 25625. https://doi.org/10.1038/srep25625

Griffiths, I., 1959. The phylogeny of *Sminthillus limbatus* and the status of the Brachycephalidae (Amphibia: Salientia). *Proceedings of the Zoological Society of London* 132: 457–487. https://doi.org/10.1111/j.1469-7998.1959.tb05531.x

Grosjean, S., Vences, M., and Dubois, A., 2004. Evolutionary significance of oral morphology in the carnivorous tadpoles of tiger frogs, genus *Hoplobatrachus* (Ranidae). *Biological Journal of the Linnean Society of London* 81: 171–181. https://doi.org/10.1111/j.1095-8312.2003.00272.x

Großwendt, A., Röglin, H., and Schmidt, M., 2019. Analysis of the Ward's method. Proceedings of the Thirtieth Annual ACM-SIAM Symposium on Discrete Algorithms. arXiv:1907.05094v1. https://doi.org/10.1137/1.9781611975482.182

Guglielmone, A. A., Estrada-Peña, A., Keirans, J. E., and Robbins, R. G., 2003. Ticks (Acari: Ixodida) of the Neotropical region. *International Consortium on Ticks and Tick-Borne Deseases* (ICTTD-2): 1–173.

Gundlach, J. C., 1880. *Contribución a la Erpetología Cubana*. G. Montiel, La Habana, Cuba. https://doi.org/10.5962/bhl.title.11859

Gundlach, J. C. and Peters, W., 1864. *Hylodes varians* Gundlach et Peter n. sp., 1864. Pp. 390–391 in *Berichtete über einige neue Säugetiere* (*Mormops, Macrotus, Vesperus, Molossus, Capromys), Amphibien (Platydactylus, Otocryptis, Euprepes, Ungalia, Dromicus, Tropidonotus, Xenodon. Hylodes) und Fische (Sillago, Sebastes, Channa, Myctophum, Carassius, Barbus, Capoëta, Poecilia, Saurenchelys, Leptocephalus*), by W. C. H. Peters. Monatsberichte der Königlichen Preuf. Akademie der Wissenschaften zu Berlin, 11. vom 20. Juni 1864. Berlin, Germany.

Guzmán Menéndez, J. M., López, R. C., and Menéndez Carrera, L., 2004. La fragmentación del paisaje en la Reserva de la Biosfera Sierra del Rosario, implicaciones en la gestión de la diversidad biológica. Convención Trópico 2004. I Congreso de Biodiversidad y Ecología. http://repositorio.geotech.cu/jspui/handle/1234/1532.

Harischandra, H. K. S. P., Karunaratne, S. H. P. P., and Rajakaruna, R. S., 2011. Effect of mosquito larvicide Abate® on the developmental stages of the Asian common toad, *Bufo melanostictus*. *Ceylon Journal of Science* 40: 133–140. https://doi.org/10.4038/cjsbs.v40i2.3929

Hass, C. A. and Hedges, S. B., 1991. Albumin evolution in West Indian frogs of the genus *Eleutherodactylus* (Leptodactylidae): Caribbean biogeography and calibration of albumin clock. *Journal of Zoology* (London) 225: 413–426. https://doi.org/10.1111/j.1469-7998.1991.tb03825.x

Heatwole, H., 2013. Worldwide decline and extinction of amphibians. Pp. 259–278 in *The Balance of Nature and Human Impact*, edited by K. Rohde. Cambridge University Press, Cambridge, UK. https://doi.org/10.1017/CBO9781139095075.025

Hedges, S. B., 1989. Evolution and biogeography of West Indian frogs of the genus *Eleutherodactylus*: slow evolving loci and the major groups. Pp. 305–369 in *Biogeography of the West Indies Past, Present, and Future*, edited by C. A. Woods. Sandhill Crane Press, Gainesville, Florida, USA.

Hedges, S. B., 1993. Global amphibian declines: a perspective from the Caribbean. *Biodiversity and Conservation* 2: 290–303. https://doi.org/10.1007/BF00056674

Hedges, S. B., 1996. The origin of West Indian amphibians and reptiles. Pp. 95–128 in *Contributions to West Indian Herpetology: A Tribute to Albert Schwartz*, edited by R. Powell and R. W. Henderson. Contributions to Herpetology 12, Society for the Study of Amphibians and Reptiles, Ithaca, New York, USA.

Hedges, S. B., 1999. Distribution Patterns of Amphibians in the West Indies. Pp. 211–254 in *Patterns of Distribution of Amphibians: A Global Perspective*, edited by W. E. Duellman. The Johns Hopkins University Press, Baltimore, Maryland, USA.

Hedges, B. and Díaz, L., 2004a. *Eleutherodactylus acmonis*. *The IUCN Red List of Threatened Species* 2004: e.T56391A11460741. https://dx.doi.org/10.2305/IUCN.UK.2004.RLTS.T56391A11460741.en. Downloaded on 30 July 2021.

Hedges, B. and Díaz, L., 2004b. *Eleutherodactylus adelus*. *The IUCN Red List of Threatened Species* 2004: e.T56397A11462471. https://dx.doi.org/10.2305/IUCN.UK.2004.RLTS.T56397A11462471.en. Downloaded on 30 July 2021.

Hedges, B. and Díaz, L., 2004c. *Eleutherodactylus albipes*. *The IUCN Red List of Threatened Species* 2004: e.T56403A11469778. https://dx.doi.org/10.2305/IUCN.UK.2004.RLTS.T56403A11469778.en. Downloaded on 30 July 2021.

Hedges, B. and Díaz, L., 2004d. *Eleutherodactylus atkinsi*. *The IUCN Red List of Threatened Species* 2004: e.T56434A11477655. https://dx.doi.org/10.2305/IUCN.UK.2004.RLTS.T56434A11477655.en. Downloaded on 30 July 2021.

Hedges, B. and Díaz, L., 2004e. *Eleutherodactylus bartonsmithi*. *The IUCN Red List of Threatened Species* 2004: e.T56452A11480604. https://dx.doi.org/10.2305/IUCN.UK.2004.RLTS.T56452A11480604.en. Downloaded on 30 July 2021.

Hedges, B. and Díaz, L., 2004f. *Eleutherodactylus blairhedgesi. The IUCN Red List of Threatened Species* 2004: e.T56466A11469913. https://dx.doi.org/10.2305/IUCN.UK.2004.RLTS.T56466A11469913.en. Downloaded on 30 July 2021.

Hedges, B. and Díaz, L., 2004g. *Eleutherodactylus bresslerae. The IUCN Red List of Threatened Species* 2004: e.T56473A11482743. https://dx.doi.org/10.2305/IUCN.UK.2004.RLTS.T56473A11482743.en. Downloaded on 30 July 2021.

Hedges, B. and Díaz, L., 2004h. *Eleutherodactylus casparii. The IUCN Red List of Threatened Species* 2004: e.T29471A9500308. https://dx.doi.org/10.2305/IUCN.UK.2004.RLTS.T29471A9500308.en. Downloaded on 30 July 2021.

Hedges, B. and Díaz, L., 2004i. *Eleutherodactylus cubanus. The IUCN Red List of Threatened Species* 2004: e.T56541A11494742. https://dx.doi.org/10.2305/IUCN.UK.2004.RLTS.T56541A11494742.en. Downloaded on 30 July 2021.

Hedges, B. and Díaz, L., 2004j. *Eleutherodactylus eileenae. The IUCN Red List of Threatened Species* 2004: e.T56574A11498618. https://dx.doi.org/10.2305/IUCN.UK.2004.RLTS.T56574A11498618.en. Downloaded on 30 July 2021.

Hedges, B. and Díaz, L., 2004k. *Eleutherodactylus emiliae. The IUCN Red List of Threatened Species* 2004: e.T56578A11485506. https://dx.doi.org/10.2305/IUCN.UK.2004.RLTS.T56578A11485506.en. Downloaded on 30 July 2021.

Hedges, B. and Díaz, L., 2004l. *Eleutherodactylus glamyrus. The IUCN Red List of Threatened Species* 2004: e.T56616A11505831. https://dx.doi.org/10.2305/IUCN.UK.2004.RLTS.T56616A11505831.en. Downloaded on 30 July 2021.

Hedges, B. and Díaz, L., 2004m. *Eleutherodactylus goini. The IUCN Red List of Threatened Species* 2004: e.T29472A9500450. https://dx.doi.org/10.2305/IUCN.UK.2004.RLTS.T29472A9500450.en. Downloaded on 30 July 2021.

Hedges, B. and Díaz, L., 2004n. *Eleutherodactylus greyi. The IUCN Red List of Threatened Species* 2004: e.T56632A11510693. https://dx.doi.org/10.2305/IUCN.UK.2004.RLTS.T56632A11510693.en. Downloaded on 30 July 2021.

Hedges, B. and Díaz, L., 2004o. *Eleutherodactylus guanahacabibes. The IUCN Red List of Threatened Species* 2004: e.T56636A11511323. https://dx.doi.org/10.2305/IUCN.UK.2004.RLTS.T56636A11511323.en. Downloaded on 30 July 2021.

Hedges, B. and Díaz, L., 2004p. *Eleutherodactylus guantanamera. The IUCN Red List of Threatened Species* 2004: e.T56637A11511445. https://dx.doi.org/10.2305/IUCN.UK.2004.RLTS.T56637A11511445.en. Downloaded on 30 July 2021.

Hedges, B. and Díaz, L., 2004q. *Eleutherodactylus iberia. The IUCN Red List of Threatened Species* 2004: e.T56660A11500015. https://dx.doi.org/10.2305/IUCN.UK.2004.RLTS.T56660A11500015.en. Downloaded on 30 July 2021.

Hedges, B. and Díaz, L., 2004r. *Eleutherodactylus jaumei. The IUCN Red List of Threatened Species* 2004: e.T56682A11504390. https://dx.doi.org/10.2305/IUCN.UK.2004.RLTS.T56682A11504390.en. Downloaded on 30 July 2021.

Hedges, B. and Díaz, L., 2004s. *Eleutherodactylus klinikowskii. The IUCN Red List of Threatened Species* 2004: e.T56693A11508050. https://dx.doi.org/10.2305/IUCN.UK.2004.RLTS.T56693A11508050.en. Downloaded on 30 July 2021.

Hedges, B. and Díaz, L., 2004t. *Eleutherodactylus leberi. The IUCN Red List of Threatened Species* 2004: e.T56708A11519298. https://dx.doi.org/10.2305/IUCN.UK.2004.RLTS.T56708A11519298.en. Downloaded on 30 July 2021.

Hedges, B. and Díaz, L., 2004u. *Eleutherodactylus mariposa. The IUCN Red List of Threatened Species* 2004: e.T56742A11529294. https://dx.doi.org/10.2305/IUCN.UK.2004.RLTS.T56742A11529294.en. Downloaded on 30 July 2021.

Hedges, B. and Díaz, L., 2004v. *Eleutherodactylus melacara. The IUCN Red List of Threatened Species* 2004: e.T56754A11531680. https://dx.doi.org/10.2305/IUCN.UK.2004.RLTS.T56754A11531680.en. Downloaded on 30 July 2021.

Hedges, B. and Díaz, L., 2004w. *Eleutherodactylus orientalis. The IUCN Red List of Threatened Species* 2004: e.T56815A11537567. https://dx.doi.org/10.2305/IUCN.UK.2004.RLTS.T56815A11537567.en. Downloaded on 30 July 2021.

Hedges, B. and Díaz, L., 2004x. *Eleutherodactylus pezopetrus. The IUCN Red List of Threatened Species* 2004: e.T56851A11545260. https://dx.doi.org/10.2305/IUCN.UK.2004.RLTS.T56851A11545260.en. Downloaded on 30 July 2021.

Hedges, B. and Díaz, L., 2004y. *Eleutherodactylus pinarensis. The IUCN Red List of Threatened Species* 2004: e.T56859A11546138. https://dx.doi.org/10.2305/IUCN.UK.2004.RLTS.T56859A11546138.en. Downloaded on 30 July 2021.

Hedges, B. and Díaz, L., 2004z. *Eleutherodactylus principalis. The IUCN Red List of Threatened Species* 2004: e.T7455A12832214. https://dx.doi.org/10.2305/IUCN.UK.2004.RLTS.T7455A12832214.en. Downloaded on 30 July 2021.

Hedges, B. and Díaz, L., 2004aa. *Eleutherodactylus rivularis. The IUCN Red List of Threatened Species* 2004: e.T56920A11553683. https://dx.doi.org/10.2305/IUCN.UK.2004.RLTS.T56920A11553683.en. Downloaded on 30 July 2021.

Hedges, B. and Díaz, L., 2004bb. *Eleutherodactylus ronaldi. The IUCN Red List of Threatened Species* 2004: e.T56922A11554203. https://dx.doi.org/10.2305/IUCN.UK.2004.RLTS.T56922A11554203.en. Downloaded on 30 July 2021.

Hedges, B. and Díaz, L., 2004cc. *Eleutherodactylus symingtoni. The IUCN Red List of Threatened Species* 2004: e.T56988A11552316. https://dx.doi.org/10.2305/IUCN.UK.2004.RLTS.T56988A11552316.en. Downloaded on 30 July 2021.

Hedges, B. and Díaz, L., 2004dd. *Eleutherodactylus tetajulia. The IUCN Red List of Threatened Species* 2004: e.T57001A11562958. https://dx.doi.org/10.2305/IUCN.UK.2004.RLTS.T57001A11562958.en. Downloaded on 30 July 2021.

Hedges, B. and Díaz, L., 2004ee. *Eleutherodactylus thomasi. The IUCN Red List of Threatened Species* 2004: e.T57003A11563368. https://dx.doi.org/10.2305/IUCN.UK.2004.RLTS.T57003A11563368.en. Downloaded on 30 July 2021.

Hedges, B. and Díaz, L., 2004ff. *Eleutherodactylus toa. The IUCN Red List of Threatened Species* 2004: e.T57009A11565575. https://dx.doi.org/10.2305/IUCN.UK.2004.RLTS.T57009A11565575.en. Downloaded on 30 July 2021.

Hedges, B. and Díaz, L., 2004gg. *Eleutherodactylus tonyi. The IUCN Red List of Threatened Species* 2004: e.T29470A9500192. https://dx.doi.org/10.2305/IUCN.UK.2004.RLTS.T29470A9500192.en. Downloaded on 30 July 2021.

Hedges, B. and Díaz, L., 2004hh. *Eleutherodactylus turquinensis. The IUCN Red List of Threatened Species* 2004: e.T57019A11568210. https://dx.doi.org/10.2305/IUCN.UK.2004.RLTS.T57019A11568210.en. Downloaded on 30 July 2021.

Hedges, B. and Díaz, L., 2004ii. *Eleutherodactylus varians. The IUCN Red List of Threatened Species* 2004: e.T57030A11571364. https://dx.doi.org/10.2305/IUCN.UK.2004.RLTS.T57030A11571364.en. Downloaded on 30 July 2021.

Hedges, B. and Díaz, L., 2004jj. *Eleutherodactylus varleyi. The IUCN Red List of Threatened Species* 2004: e.T57031A11571702. https://dx.doi.org/10.2305/IUCN.UK.2004.RLTS.T57031A11571702.en. Downloaded on 30 July 2021.

Hedges, B. and Díaz, L., 2004kk. *Eleutherodactylus zeus. The IUCN Red List of Threatened Species* 2004: e.T57065A11576165. https://dx.doi.org/10.2305/IUCN.UK.2004.RLTS.T57065A11576165.en. Downloaded on 30 July 2021.

Hedges, B. and Díaz, L., 2004ll. *Eleutherodactylus zugi. The IUCN Red List of Threatened Species* 2004: e.T57069A11576647. https://dx.doi.org/10.2305/IUCN.UK.2004.RLTS.T57069A11576647.en. Downloaded on 30 July 2021.

Hedges, B. and Díaz, L., 2004mm. *Peltophryne cataulaciceps. The IUCN Red List of Threatened Species* 2004: e.T54604A11171206. https://dx.doi.org/10.2305/IUCN.UK.2004.RLTS.T54604A11171206.en. Downloaded on 30 July 2021.

Hedges, B. and Díaz, L., 2004nn. *Peltophryne empusa. The IUCN Red List of Threatened Species* 2004: e.T54634A11179047. https://dx.doi.org/10.2305/IUCN.UK.2004.RLTS.T54634A11179047.en. Downloaded on 30 July 2021.

Hedges, B. and Díaz, L., 2004oo. *Peltophryne gundlachi. The IUCN Red List of Threatened Species* 2004: e.T54657A11182838. https://dx.doi.org/10.2305/IUCN.UK.2004.RLTS.T54657A11182838.en. Downloaded on 30 July 2021.

Hedges, B. and Díaz, L., 2004pp. *Peltophryne longinasus. The IUCN Red List of Threatened Species* 2004: e.T54693A11176718. https://dx.doi.org/10.2305/IUCN.UK.2004.RLTS.T54693A11176718.en. Downloaded on 30 July 2021.

Hedges, B. and Díaz, L., 2004qq. *Peltophryne taladai. The IUCN Red List of Threatened Species* 2004: e.T54776A11188121. https://dx.doi.org/10.2305/IUCN.UK.2004.RLTS.T54776A11188121.en. Downloaded on 30 July 2021.

Hedges, B. and Díaz, L., 2010a. *Eleutherodactylus auriculatus. The IUCN Red List of Threatened Species* 2010: e.T56441A11478680. https://dx.doi.org/10.2305/IUCN.UK.2010-2.RLTS.T56441A11478680.en. Downloaded on 30 July 2021.

Hedges, B. and Díaz, L., 2010b. *Eleutherodactylus dimidiatus. The IUCN Red List of Threatened Species* 2010: e.T56562A11483663. https://dx.doi.org/10.2305/IUCN.UK.2010-2.RLTS.T56562A11483663.en. Downloaded on 30 July 2021.

Hedges, B. and Díaz, L., 2010c. *Eleutherodactylus etheridgei. The IUCN Red List of Threatened Species* 2010: e.T56590A11490043. https://dx.doi.org/10.2305/IUCN.UK.2010-2.RLTS.T56590A11490043.en. Downloaded on 30 July 2021.

Hedges, B. and Díaz, L., 2010d. *Eleutherodactylus gundlachi. The IUCN Red List of Threatened Species* 2010: e.T56642A11512696. https://dx.doi.org/10.2305/IUCN.UK.2010-2.RLTS.T56642A11512696.en. Downloaded on 30 July 2021.

Hedges, B. and Díaz, L., 2010e. *Eleutherodactylus intermedius. The IUCN Red List of Threatened Species* 2010: e.T56674A11501507. https://dx.doi.org/10.2305/IUCN.UK.2010-2.RLTS.T56674A11501507.en. Downloaded on 30 July 2021.

Hedges, B. and Díaz, L., 2010f. *Eleutherodactylus ionthus. The IUCN Red List of Threatened Species* 2010: e.T56677A11502570. https://dx.doi.org/10.2305/IUCN.UK.2010-2.RLTS.T56677A11502570.en. Downloaded on 30 July 2021.

Hedges, B. and Díaz, L., 2010g. *Eleutherodactylus limbatus. The IUCN Red List of Threatened Species* 2010: e.T56719A11523010. https://dx.doi.org/10.2305/IUCN.UK.2010-2.RLTS.T56719A11523010.en. Downloaded on 30 July 2021.

Hedges, B. and Díaz, L., 2010h. *Eleutherodactylus ricordii. The IUCN Red List of Threatened Species* 2010: e.T56915A11552149. https://dx.doi.org/10.2305/IUCN.UK.2010-2.RLTS.T56915A11552149.en. Downloaded on 30 July 2021.

Hedges, B. and Díaz, L., 2010i. *Peltophryne peltocephala. The IUCN Red List of Threatened Species* 2010: e.T54728A11195872. https://dx.doi.org/10.2305/IUCN.UK.2010-2.RLTS.T54728A11195872.en. Downloaded on 30 July 2021.

Hedges, S. B. and Díaz, L. M., 2011. The conservation status of amphibians in the West Indies. Pp. 31–47 in *Conservation of Caribbean Island Herpetofaunas*, Vol. 1: *Conservation Biology and the Wider Caribbean*, edited by A. Hailey, B. S. Wilson, and J. A. Horrocks. Koninklijke Brill NV, Leiden, The Netherlands. https://doi.org/10.1163/ej.9789004183957.i-228.23

Hedges, S. B. and Thomas, R., 1989. Supplement to West Indian amphibians and reptiles: a check-list. Milwaukee Public Museum, *Contribution in Biology and Geology* 77: 1–11.

Hedges, S. B., Duellman, W. E., and Heinicke, M. P., 2008a. New World direct-developing frogs (Anura: Terrarana): Molecular phylogeny, classification, biogeography, and conservation. *Zootaxa* 1737: 1–182. https://doi.org/10.11646/zootaxa.1737.1.1

Hedges, S. B., Estrada, A. R., and Díaz, L. M., 1999. A new snake (*Tropidophis*) from western Cuba. *Copeia* 1999: 376–381. https://doi.org/10.2307/1447483

Hedges, S. B., Estrada, A. R., and Thomas, R., 1992. Three new species of *Eleutherodactylus* from eastern Cuba, with notes on vocalizations of other species (Anura: Leptodactylidae). *Herpetological Monographs* 6: 68–83. https://doi.org/10.2307/1466962

Hedges, S. B., González, L., and Estrada, A. R., 1995. Rediscovery of the Cuban frogs *Eleutherodactylus cubanus* and *E. turquinensis* (Anura: Leptodactylidae). *Caribbean Journal of Science* 31: 327–332.

Hedges S. B., Powell, R., Henderson, R. W., Hanson, S., and Murphy, J. C., 2019. Definition of the Caribbean Islands biogeographic region, with checklist and recommendations for standardized common names of amphibians and reptiles. *Caribbean Herpetology* 67: 1–53. https://doi.org/10.31611/ch.67

Hedges, B., Díaz, L., Ibéné, B., Joglar, R., Powell, R., Bolaños, F., and Chaves, G., 2008b. *Osteopilus septentrionalis. 2008 IUCN Red List of Threatened Species.* Available: http://www.iucngisd.org/gisd/speciesname/Osteopilus+septentrionalis

Heinicke, M., Díaz, L. M., and Hedges, S. B., 2011. Origin of invasive Florida frogs traced to Cuba. *Biology Letters* 7: 407–410. https://doi.org/10.1098/rsbl.2010.1131

Heinicke, M. P., Duellman, W. E., and Hedges, S. B., 2007. Major Caribbean and Central American frog faunas originated by oceanic dispersal. *Proceedings of the Nacional Academy of Sciences USA* 104: 10092–10097. https://doi.org/10.1073/pnas.0611051104

Helmer, E. H., Gerson, E. A., Baggett, L. S., Bird, B. J., Ruzycki, T. S., and Voggesser, S. M., 2019. Neotropical cloud forests and páramo to contract and dry from declines in cloud immersion and frost. *PLOS ONE* 14: e0213155. https://doi.org/10.1371/journal.pone.0213155

Henderson, R. W. and Powell, R., 2009. *Natural History of West Indian Reptiles and Amphibians.* University Press of Florida, Gainesville, Florida, USA.

Henderson, R. and Sajdak, R. A., 1996. Diets of West Indian racers (Colubridae: *Alsophis*): composition and biogeographic implications. Pp. 327–338 in *Contributions to West Indian Herpetology: A Tribute to Albert Schwartz*, edited by R. Powell and R. W. Henderson. Contributions to Herpetology 12, Society for the Study of Amphibians and Reptiles, Ithaca, New York, USA.

Henle, K., Dubois, A., and Vershinin, V., 2017. Commented glossary, terminology and synonymies of anomalies in natural populations of amphibians. *Studies on Anomalies in Natural Populations of Amphibians. Mertensiella* 25: 9–48.

Herczeg, D., Kásler, A., Holly, D., and Hettyey, A., 2021. Host–multiparasite interactions in amphibians: a review. *Parasites Vectors* 14: article number 296. https://doi.org/10.1186/s13071-021-04796-1

Herman, J. E., Benton, A., and Irizarry, I. N., 2015. *Osteopilus septentrionalis* (Cuban Treefrog). Cannibalism. *Herpetological Review* 46: 75–76.

Hernández M., Alonso, R., and Rodríguez, A., 2010. Advertisement call of *Peltophryne florentinoi* (Anura: Bufonidae), an endemic toad from Zapata Swamp, Cuba. *Amphibia-Reptilia* 31: 265–272. https://doi.org/10.1163/156853810791069083

Hernández-Austria, R., García-Vázquez, U. O., Grünwald, C. I., and Parra-Olea, G., 2022. Molecular phylogeny of the subgenus *Syrruphus* (Amphibia: Anura: Eleutherodactylidae), with the description of a new species from Eastern Mexico. *Systematics and Biodiversity* 20: 1–20. https://doi.org/10.1080/14772000.2021.2014597

Herrmann, H-J., 2006. Verfluchte Kobolde Kuba-Laubfrösche *Osteopilus septentrionalis. Aquaristik Fachmagazin & Aquarium Heute* 38: 22–23.

Howard, J. H., Baldwin, R. F., and Brown, B. L., 2012. Exploratory analysis for complex-life-cycle amphibians: revealing complex forest-reproductive effort relationships using redundancy analysis. *Forest Ecology and Management* 270: 175–182. https://doi.org/10.1016/j.foreco.2012.01.021

Hudson, M. A., Young, R. P., D'Urban Jackson, J., Orozco-terWengel, P., Martin, L., James, A., Sulton, M., García, G., Griffiths, R. A., Thomas, R., Magin, C., Bruford, M.W. and Cunningham, A. A., 2016. Dynamics and genetics of a disease driven species decline to near extinction: lessons for conservation. *Scientific Reports* 6: 30772. https://doi.org/10.1038/srep30772

Hudson, M. A., Griffiths, R. A., Martin, L., Fenton, C., Adams, S. L., Blackman, A., Sulton, M., Perkins, M. W., Lopez, J., Garcia, G., Tapley, B., Young, R. P., and Cunningham, A. A., 2019. Reservoir frogs: seasonality of *Batrachochytrium dendrobatidis* infection in robber frogs in Dominica and Montserrat. *PeerJ* 7: e7021. https://doi.org/10.7717/peerj.7021

Islam, A. and Malik, M. F., 2018. Impact of pesticides on amphibians: a review. *Journal of Toxicological Analysis* 1: 1–4.

Iturriaga, M. and Dugo-Cota, Á., 2018. Parental care in the Greenhouse Frog *Eleutherodactylus planirostris* (Anura: Eleutherodactylidae) from Cuba. *Herpetology Notes* 11: 857–861.

Iturriaga, M. and Homar, T., 2014. First report of floral diurnal retreat in the Cuban treefrog *Osteopilus septentrionalis* (Amphibia: Anura: Hylidae). *Revista Cubana de Ciencias Biológicas* 3: 74–76.

Iturriaga, M., Sanz, A., and Oliva, R., 2014. Seasonal reproduction of the Greenhouse Frog *Eleutherodactylus planirostris* (Anura: Eleutherodactylidae) in Havana, Cuba. *South American Journal of Herpetology* 9: 142–150.

Iturriaga, M., Rodríguez-Gómez, Y., and Sanz-Ochotorena, A., 2012. Structural and ultrastructural description of the gonads of *Eleutherodactylus planirostris* (Anura: Eleutherodactylidae). *Herpetology Notes* 5: 281–290.

IUCN, 1994. Directrices para las categorías de manejo de áreas protegidas. CPNAP-WCMC. UICN, Gland, Suiza y Cambridge, Reino Unido.

IUCN SSC Amphibian Specialist Group, 2014a. *Eleutherodactylus cuneatus. The IUCN Red List of Threatened Species* 2014: e.T56544A56053938. https://dx.doi.org/10.2305/IUCN.UK.2014-1.RLTS.T56544A56053938.en. Downloaded on 30 July 2021.

IUCN SSC Amphibian Specialist Group, 2014b. *Eleutherodactylus riparius. The IUCN Red List of Threatened Species* 2014: e.T56917A56055054. https://dx.doi.org/10.2305/IUCN.UK.2014-1.RLTS.T56917A56055054.en. Downloaded on 30 July 2021.

IUCN SSC Amphibian Specialist Group. 2014c. *Peltophryne fustiger. The IUCN Red List of Threatened Species* 2014: e.T54644A56053389. https://dx.doi.org/10.2305/IUCN.UK.2014-1.RLTS.T54644A56053389.en. Downloaded on 30 July 2021.

IUCN SSC Amphibian Specialist Group, 2021a. *Eleutherodactylus planirostris. The IUCN Red List of Threatened Species* 2021: e.T56864A54370037. https://dx.doi.org/10.2305/IUCN.UK.2021-1.RLTS.T56864A54370037.en. Downloaded on 30 July 2021.

IUCN SSC Amphibian Specialist Group, 2021b. *Osteopilus septentrionalis. The IUCN Red List of Threatened Species* 2021: e.T55811A3032751. https://dx.doi.org/10.2305/IUCN.UK.2021-1.RLTS.T55811A3032751.en. Downloaded on 30 July 2021.

IUCN SSC Amphibian Specialist Group. *Peltophryne cataulaciceps*. The IUCN Red List of Threatened Species 2022(1): e.T54604A48197610. *In press* (a).

IUCN SSC Amphibian Specialist Group. *Peltophryne dunni*. The IUCN Red List of Threatened Species 2022(1): e.T172817092A172817099. *In press* (b).

IUCN SSC Amphibian Specialist Group. *Peltophryne empusa*. The IUCN Red List of Threatened Species 2022(1): e.T54634A62804197. *In press* (c).

IUCN SSC Amphibian Specialist Group. *Peltophryne florentinoi*. The IUCN Red List of Threatened Species 2022(1): e.T135861A49030821. *In press* (d).

IUCN SSC Amphibian Specialist Group. *Peltophryne fustiger*. The IUCN Red List of Threatened Species 2022(1): e.T54644A62804336. *In press* (e).

IUCN SSC Amphibian Specialist Group. *Peltophryne gundlachi*. The IUCN Red List of Threatened Species 2022(1): e.T54657A62804520. *In press* (f).

IUCN SSC Amphibian Specialist Group. *Peltophryne longinasus*. The IUCN Red List of Threatened Species 2022(1): e.T172817049A49525055. *In press* (g).

IUCN SSC Amphibian Specialist Group. *Peltophryne peltocephala*. The IUCN Red List of Threatened Species 2022(1): e.TT54728A62804659. *In press* (h).

IUCN SSC Amphibian Specialist Group. *Peltophryne ramsdeni*. The IUCN Red List of Threatened Species 2022(1): e.T172817113A172817125. *In press* (i).

IUCN SSC Amphibian Specialist Group. *Peltophryne taladai*. The IUCN Red List of Threatened Species 2022(1): e.T54776A62804821. *In press* (j).

IUCN SSC Amphibian Specialist Group. *Eleutherodactylus acmonis*. The IUCN Red List of Threatened Species 2022(1): e.T56391A49489533. *In press* (k).

IUCN SSC Amphibian Specialist Group. *Eleutherodactylus adelus*. The IUCN Red List of Threatened Species 2022(1): e.T56397A49499059. *In press* (l).

IUCN SSC Amphibian Specialist Group. *Eleutherodactylus albipes*. The IUCN Red List of Threatened Species 2022(1): e.T56403A49580401. *In press* (m).

IUCN SSC Amphibian Specialist Group. *Eleutherodactylus atkinsi*. The IUCN Red List of Threatened Species 2022(1): e.T56434A62804960. *In press* (n).

IUCN SSC Amphibian Specialist Group. *Eleutherodactylus auriculatus*. The IUCN Red List of Threatened Species 2022(1): e.T56441A62805075. *In press* (o).

IUCN SSC Amphibian Specialist Group. *Eleutherodactylus bartonsmithi*. The IUCN Red List of Threatened Species 2022(1): e.T56452A49732571. *In press* (p).

IUCN SSC Amphibian Specialist Group. *Eleutherodactylus beguei*. The IUCN Red List of Threatened Species 2022(1): e.T82410169A165386271. *In press* (q).

IUCN SSC Amphibian Specialist Group. *Eleutherodactylus blairhedgesi*. The IUCN Red List of Threatened Species 2022(1): e.T56466A49737832. *In press* (r).

IUCN SSC Amphibian Specialist Group. *Eleutherodactylus bresslerae*. The IUCN Red List of Threatened Species 2022(1): e.T56473A49752857. *In press* (s).

IUCN SSC Amphibian Specialist Group. *Eleutherodactylus casparii*. The IUCN Red List of Threatened Species 2022(1): e.T29471A62803927. *In press* (t).

IUCN SSC Amphibian Specialist Group. *Eleutherodactylus cattus*. The IUCN Red List of Threatened Species 2022(1): e.T113790215A113790276. *In press* (u).

IUCN SSC Amphibian Specialist Group. *Eleutherodactylus cubanus*. The IUCN Red List of Threatened Species 2022(1): e.T56541A49754465. *In press* (v).

IUCN SSC Amphibian Specialist Group. *Eleutherodactylus cuneatus*. The IUCN Red List of Threatened Species 2022(1): e.T56544A62805199. *In press* (w).

IUCN SSC Amphibian Specialist Group. *Eleutherodactylus dimidiatus*. The IUCN Red List of Threatened Species 2022(1): e.T56562A628053510. *In press* (x).

IUCN SSC Amphibian Specialist Group. *Eleutherodactylus eileenae*. The IUCN Red List of Threatened Species 2022(1): e.T56574A62805491. *In press* (y).

IUCN SSC Amphibian Specialist Group. *Eleutherodactylus emiliae*. The IUCN Red List of Threatened Species 2022(1): e.T56578A49755875. *In press* (z).

IUCN SSC Amphibian Specialist Group. *Eleutherodactylus erythroproctus*. The IUCN Red List of Threatened Species 2022(1): e.T201301A2698791. *In press* (aa).

IUCN SSC Amphibian Specialist Group. *Eleutherodactylus etheridgei*. The IUCN Red List of Threatened Species 2022(1): e.T56590A49763479. *In press* (bb).

IUCN SSC Amphibian Specialist Group. *Eleutherodactylus feichtingeri*. The IUCN Red List of Threatened Species 2022(1): e.T78916944A81107390. *In press* (cc).

IUCN SSC Amphibian Specialist Group. *Eleutherodactylus glamyrus*. The IUCN Red List of Threatened Species 2022(1): e.T113790293A49765501. *In press* (dd).

IUCN SSC Amphibian Specialist Group. *Eleutherodactylus goini*. The IUCN Red List of Threatened Species 2022(1): e.T29472A62804055. *In press* (ee).

IUCN SSC Amphibian Specialist Group. *Eleutherodactylus greyi*. The IUCN Red List of Threatened Species 2022(1): e.T56632A62805602. *In press* (ff).

IUCN SSC Amphibian Specialist Group. *Eleutherodactylus guanahacabibes*. The IUCN Red List of Threatened Species 2022(1): e.T56636A49832425. *In press* (gg).

IUCN SSC Amphibian Specialist Group. *Eleutherodactylus guantanamera*. The IUCN Red List of Threatened Species 2022(1): e.T56637A62805716. *In press* (hh).

IUCN SSC Amphibian Specialist Group. *Eleutherodactylus gundlachi*. The IUCN Red List of Threatened Species 2022(1): e.T56642A62805842. *In press* (ii).

IUCN SSC Amphibian Specialist Group. *Eleutherodactylus iberia*. The IUCN Red List of Threatened Species 2022(1): e.T56660A498336950. *In press* (jj).

IUCN SSC Amphibian Specialist Group. *Eleutherodactylus intermedius*. The IUCN Red List of Threatened Species 2022(1): e.T56674A62805966. *In press* (kk).

IUCN SSC Amphibian Specialist Group. *Eleutherodactylus ionthus*. The IUCN Red List of Threatened Species 2022(1): e.T56677A62806086. *In press* (ll).

IUCN SSC Amphibian Specialist Group. *Eleutherodactylus jaumei*. The IUCN Red List of Threatened Species 2022(1): e.T56682A498547430. *In press* (mm).

IUCN SSC Amphibian Specialist Group. *Eleutherodactylus klinikowskii*. The IUCN Red List of Threatened Species 2022(1): e.T56693A62806204. *In press* (nn).

IUCN SSC Amphibian Specialist Group. *Eleutherodactylus leberi*. The IUCN Red List of Threatened Species 2022(1): e.T56708A50011166. *In press* (oo).

IUCN SSC Amphibian Specialist Group. *Eleutherodactylus limbatus*. The IUCN Red List of Threatened Species 2022(1): e.T56719A62806297. *In press* (pp).

IUCN SSC Amphibian Specialist Group. *Eleutherodactylus maestrensis*. The IUCN Red List of Threatened Species 2022(1): e.T61812A50015225. *In press* (qq).

IUCN SSC Amphibian Specialist Group. *Eleutherodactylus mariposa*. The IUCN Red List of Threatened Species 2022(1): e.56742A59420521. *In press* (rr).

IUCN SSC Amphibian Specialist Group. *Eleutherodactylus melacara*. The IUCN Red List of Threatened Species 2022(1): e.T56754A50132027. *In press* (ss).

IUCN SSC Amphibian Specialist Group. *Eleutherodactylus michaelschmidi*. The IUCN Red List of Threatened Species 2022(1): e.T135753A50132989. *In press* (tt).

IUCN SSC Amphibian Specialist Group. *Eleutherodactylus olibrus*. The IUCN Red List of Threatened Species 2022(1): e.T201302A2698798. *In press* (uu).

IUCN SSC Amphibian Specialist Group. *Eleutherodactylus orientalis*. The IUCN Red List of Threatened Species 2022(1): e.T56815A50136348. *In press* (vv).

IUCN SSC Amphibian Specialist Group. *Eleutherodactylus pezopetrus*. The IUCN Red List of Threatened Species 2022(1): e.T56851A50160923. *In press* (ww).

IUCN SSC Amphibian Specialist Group. *Eleutherodactylus pinarensis*. The IUCN Red List of Threatened Species 2022(1): e.T56859A62806407. *In press* (xx).

IUCN SSC Amphibian Specialist Group. *Eleutherodactylus planirostris*. The IUCN Red List of Threatened Species 2022(1): e.T56864A54370037. *In press* (yy).

IUCN SSC Amphibian Specialist Group. *Eleutherodactylus principalis*. The IUCN Red List of Threatened Species 2022(1): e.T7455A62803818. *In press* (zz).

IUCN SSC Amphibian Specialist Group. *Eleutherodactylus ricordii*. The IUCN Red List of Threatened Species 2022(1): e.T56915A62806514. *In press* (aaa).

IUCN SSC Amphibian Specialist Group. *Eleutherodactylus riparius*. The IUCN Red List of Threatened Species 2022(1): e.T56917A165386756. *In press* (bbb).

IUCN SSC Amphibian Specialist Group. *Eleutherodactylus rivularis*. The IUCN Red List of Threatened Species 2022(1): e.T56920A50165450. *In press* (ccc).

IUCN SSC Amphibian Specialist Group. *Eleutherodactylus ronaldi*. The IUCN Red List of Threatened Species 2022(1): e.T56922A62806671. *In press* (ddd).

IUCN SSC Amphibian Specialist Group. *Eleutherodactylus simulans*. The IUCN Red List of Threatened Species 2022(1): e.T56968A62806790. *In press* (eee).

IUCN SSC Amphibian Specialist Group. *Eleutherodactylus staurometopon*. The IUCN Red List of Threatened Species 2022(1): e.T201303A2698805. *In press* (fff).

IUCN SSC Amphibian Specialist Group. *Eleutherodactylus symingtoni*. The IUCN Red List of Threatened Species 2022(1): e.T56988A50167164. *In press* (ggg).

IUCN SSC Amphibian Specialist Group. *Eleutherodactylus tetajulia*. The IUCN Red List of Threatened Species 2022(1): e.T57001A50200036. *In press* (hhh).

IUCN SSC Amphibian Specialist Group. *Eleutherodactylus thomasi*. The IUCN Red List of Threatened Species 2022(1): e.T57003A62806892. *In press* (iii).

IUCN SSC Amphibian Specialist Group. *Eleutherodactylus toa*. The IUCN Red List of Threatened Species 2022(1): e.T57009A62807005. *In press* (jjj).

IUCN SSC Amphibian Specialist Group. *Eleutherodactylus tonyi*. The IUCN Red List of Threatened Species 2022(1): e.T29470A50201546. *In press* (kkk).

IUCN SSC Amphibian Specialist Group. *Eleutherodactylus turquinensis*. The IUCN Red List of Threatened Species 2022(1): e.T57019A50202934. *In press* (lll).

IUCN SSC Amphibian Specialist Group. *Eleutherodactylus varians*. The IUCN Red List of Threatened Species 2022(1): e.T213360955A62807105. *In press* (mmm).

IUCN SSC Amphibian Specialist Group. *Eleutherodactylus varleyi*. The IUCN Red List of Threatened Species 2022(1): e.T78917076A56086444. *In press* (nnn).

IUCN SSC Amphibian Specialist Group. *Eleutherodactylus zeus*. The IUCN Red List of Threatened Species 2022(1): e.T57065A62807204. *In press* (ooo).

IUCN SSC Amphibian Specialist Group. *Eleutherodactylus zugi*. The IUCN Red List of Threatened Species 2022(1): e.T172816874A62807307. *In press* (ppp).

Jaume, M., 1966. Catálogo de la Fauna de Cuba XV Catálogo de los Anfibios de Cuba. *Museo Felipe Poey, ACC, Trabajos de Divulgación* 35: 1–20.

Jenkinson, T. S., Betancourt Román, C. M., Lambertini, C., Valencia-Aguilar, A., Rodríguez, D., L. Nunes-de-Almeida, C. H., Ruggeri, J., Belasen, A. M., Da Silva Leite, D., Zamudio, K. R., Longcore, J. E., Toledo, L. F., and James, T. Y., 2016. Amphibian-killing chytrid in Brazil comprises both locally endemic and globally expanding populations. *Molecular Ecology* 25: 2978–2996. https://doi.org/10.1111/mec.13599

Joglar, R. L., 1986. Phylogenetic relationships of the West Indian frogs of the genus *Eleutherodactylus*. PhD Dissertation. University of Kansas, Lawrence, KA, USA.

Johnson, S. A., 2013. The Cuban Treefrog (*Osteopilus septentrionalis*) in Florida. WEC218, Second Revision. Department of Wildlife Ecology and Conservation, University of Florida/Institute of Food and Agricultural Sciences Extension, Gainesville, Florida, USA. https://edis.ifas.ufl.edu/pdffiles/UW/UW25900.pdf

Johnson, P., Lunde, K. R., Ritchie, E. G., and Launer, A., 1999. The effect of trematode infection on amphibian limb development and survivorship. *Science* 284: 802–804. https://doi.org/10.1126/science.284.5415.802

Johnson, K. J., Baker, A., Buley, K., Carrillo, L., Gibson, R., Gillespie, G. R., Lacy, R. C., and Zippel, K., 2020. A process for assessing and prioritizing species conservation needs: going beyond the Red List. *Oryx* 54: 125–132. https://doi.org/10.1017/S0030605317001715

Johnson, P. T. J., Chase, J. M., Dosch, K. L., Hartson, R. B., Gross, J. A., Larson, D. J., Sutherland, D. R., and Carpenter, S. R., 2007. Aquatic eutrophication promotes pathogenic infection in amphibians. *Proceedings of the Nature Academy of Science USA* 104: 15781–15786. https://doi.org/10.1073/pnas.0707763104

Junges, C. M., Maglianese, M. I., Lajmanovich, R. C., Peltzer, P. M., and Attademo, A. M., 2017. Acute toxicity and etho-toxicity of three insecticides used for mosquito control on amphibian tadpoles. *Water Air & Soil Pollution* 228: 143. https://doi.org/10.1007/s11270-017-3324-6

Kaiser, H., Chamberlain, M. C., Edwards, T., Nuñez, J. R., Rodríguez-Cabrera, T. M., and Torres, J., 2016. Cannibalism in Cuba: First direct observations of Cuban Treefrogs (*Osteopilus septentrionalis*, Hylidae) feeding on conspecifics in their native habitat, with a brief review of anurophagy and cannibalism in treefrogs. *IRCF Reptiles & Amphibians* 23: 21–27. https://doi.org/10.17161/randa.v23i1.14089

Königstedt, B. and Königstedt, D., 1982. Haltung und Zucht des Kuba Laubfroschs. *Aquarien Terrarien* 29: 280–282.

Krysko, K. L., Enge, K. M., and Moler, P. E., 2019. Amphibians and Reptiles of Florida. University of Florida Press, Gainesville, Florida, USA.

Krysko, K. L., Burguess, J. P., Rochford, M. R., Gillette, C. R., Cueva, D., Enge, K. M., Somma, L. A., Stabile, J. L., Smith, D. C., and Wasilewski, J. A., 2011. Verified non-indigenous amphibians and reptiles in Florida from 1863 through 2010: Outlining the invasion process and identifying invasion pathways and stages. *Zootaxa* 3028: 1–64. https://doi.org/10.11646/zootaxa.3028.1.1

Lajmanovich, R. C., Junges C. M., Cabagna-Zenklusen, M. C., Attademo, A. M., Peltzer, P. M., Maglianese, M., Márquez, V. E., and Beccaria, A. J., 2015. Toxicity of *Bacillus thuringiensis* var. *israelensis* in aqueous suspension on the South American common frog *Leptodactylus latrans* (Anura: Leptodactylidae) tadpoles. *Environmental Research* 136: 205–212. https://doi.org/10.1016/j.envres.2014.10.022

Lambert, M. R., Womack, M. C., Byrne A. Q., Hernández-Gómez, O., Noss, C. F., Rothstein, A. P., Blackburn, D. C., Collins, J. P., Crump, M. L., Koo, M. S., Nanjappa, P., Rollins-Smith, L., Vredenburg, V. T., and Rosenblum, E. B., 2020. Comment on "Amphibian fungal panzootic causes catastrophic and ongoing loss of biodiversity". *Science* 367 (6484): eaay1838. https://doi.org/10.1126/science.aay1838

Landestoy, M. A., Turner, D., Marion, A. B., and Hedges, S. B., 2018. A new species of Caribbean toad (Bufonidae, *Peltophryne*) from southern Hispaniola. *Zootaxa* 4403: 523–539. https://doi.org/10.11646/zootaxa.4403.3.6

Langowski, J., Singla, S., Nyarko, A., Schipper, H., van den Berg, F. T., Kaur, S., Astley, H. C., Gussekloo, S. W. S., Dhinojwala, A., and van Leeuwen, J. L., 2019. Comparative and functional analysis of the digital mucus glands and secretions of tree frogs. *Frontiers in Zoology* 16: 19. https://doi.org/10.1186/s12983-019-0315-z

Lips, K. R., 2016. Overview of chytrid emergence and impacts on amphibians. *Philosophical Transactions Royal Society B* 371: 20150465. http://dx.doi.org/10.1098/rstb.2015.0465

Lips, K. R., Burrowes, P. A., Mendelson, J. R., and Parra-Olea, G., 2005. Amphibian declines in Latin America: widespread population declines, extinctions, and impacts. *Biotropica* 37: 163–165. https://doi.org/10.1111/j.1744-7429.2005.00023.x

Lips, K. R., Brem, F., Brenes, R., Reeve, J. D., Alford, R. A., Voyles, J., Carey, C., Livo, L., Pessier, A. P., and Collins, J. P., 2006. From the cover: emerging infectious disease and the loss of biodiversity in a Neotropical amphibian community. *Proceedings of the National Academy of Sciences of the United States of America* 103: 3165–3170. https://doi.org/10.1073/pnas.0506889103

López-Hurtado, Y., García-Padrón, L. Y., González, A., Díaz, L. M., and Rodríguez-Cabrera, T. M., 2020. Notes on the feeding habits of the Caribbean Watersnake, *Tretanorhinus variabilis* (Dipsadidae). *Reptiles & Amphibians* 27: 147–153. https://doi.org/10.17161/randa.v27i2.14105

Lynch, J. D., 1971. Evolutionary relationships, osteology, and zoogeography of Leptodactyloid frogs. *Miscellaneous Publications of the Museum of Natural History of the University of Kansas* 53: 1–238.

Ma, W.-J. and Veltsos, P., 2021. The diversity and evolution of sex chromosomes in frogs. *Genes* 12:483. https://doi.org/10.3390/genes12040483

Magin, C., 2003. Dominica's frogs are croaking. *Oryx* 37: 406.

Malhotra A., Thorpe, R. S., Hypolite, E., and James, A., 2007. A report on the status of the herpetofauna of the Commonwealth of Dominica, West Indies. *Applied Herpetology* 4: 177–194. https://doi.org/10.1163/157075407780681365

Mancina, C. A., Mas Castellanos, L., and Rodríguez-Cabrera, T. M., 2016. Bat predation by a Cuban Treefrog (*Osteopilus septentrionalis*, Hylidae) and a summary of bat predation by West Indian amphibians and reptiles. *IRCF Reptiles & Amphibians* 23: 95–98. https://doi.org/10.17161/randa.v23i2.14113

Mancina, C. A., Fernández de Arcila Fernández, R., Cruz Flores, D. D., Castañeira Colomé, M. A., and González Rossell, A., 2017. Diversidad biológica terrestre de Cuba. Pp. 8–25 in *Diversidad biológica de Cuba: métodos de inventario, monitoreo y colecciones biológicas*, edited by C. A. Mancina and D. D. Cruz. Editorial Agencia de Medio Ambiente (AMA), La Habana, Cuba.

Marrero, R., Torres López, J., and Rodríguez-Cabrera, T. M., 2017a. *Eleutherodactylus emiliae* (Trinidad

Groin-spotted Frog). Maximum elevation. *Herpetological Review* 48: 160–161.

Marrero Romero, R., Torres, J., and Rodríguez-Cabrera, T. M. 2017b. Geographic distribution: *Eleutherodactylus pezopetrus*. *Herpetological Review* 48: 583.

Marroquín-Páramo, J. A., Suazo-Ortuño, I., Urbina-Cardona, N., and Benítez-Malvido, J., 2021. Cumulative effects of high intensity hurricanes on herpetofaunal assemblages along a tropical dry forest chronosequence. *Forest Ecology and Management* 479: 118505. https://doi.org/10.1016/j.foreco.2020.118505

Marsh, D. M. and Beckman, N. G., 2004. Effects of forest roads on the abundance and activity of terrestrial salamanders. *Ecological Applications* 14: 1882–1891. https://doi.org/10.1890/03-5179

Martínez, A. J., Valdés, J. A., Lorenzo, C., Hernández, A., Perera, S., Díaz, C., and Oquendo, O., 2013. Diagnóstico. Estructura espacial. Pp. 60–63 in *Plan del Sistema Nacional de Áreas Protegidas 2014–2020*, edited by Ministerio de Ciencia, Tecnología y Medio Ambiente. La Habana, Cuba.

Martínez, J. R., Coy, A., and Ventosa, L., 1982. Helmintos de *Rana catesbeiana* Shaw (Ranidae) en Cuba. *Poeyana* 243: 1–10.

Mateo Rodríguez, J. M., Celeiro Chaple, M., Acevedo Rodríguez, P., Hernandez Mestre, D., and Veranes Miranda, A., 2019. 2.9 Paisajes. In *Atlas Nacional de Cuba* LX Aniversario, versión 1.0 (CD). Instituto de Geografía Tropical.

Meshaka, W. E., Jr., 1996. Occurrence of the nematode *Skrjabinoptera scelopori* in the Cuban treefrog, *Osteopilus septentrionalis*: mainland and island comparisons. Pp. 271–276 in *Contributions to West Indian Herpetology: A Tribute to Albert Schwartz*, edited by R. Powell and R. W. Henderson. Contributions to Herpetology 12, Society for the Study of Amphibians and Reptiles, Ithaca, New York, USA.

Meshaka, W. E., Jr., 2001. *The Cuban Treefrog in Florida: Life History of a Successful Colonizing Species*. University Press of Florida, Gainesville, Florida, USA.

Mittermeier R. A., Turner, W. R., Larsen, F. W., Brooks, T. M., and Gascon, C., 2011. Global Biodiversity Conservation: The Critical Role of Hotspots. Pp. 3–22 in *Biodiversity Hotspots*, edited by F. Zachos and J. Habel. Springer, Berlin and Heidelberg, Germany. https://doi.org/10.1007/978-3-642-20992-5_1

Mittermeier, R. A., Robles Gil, P., Hoffmann, M., Pilgrim, J., Brooks, T., Mittermeier, C. G., Lamoreux, J., and da Fonseca, G. A. B., 2004. *Hotspots Revisited*. CEMEX Books on Nature, Toppan Printing Co.

Miyamoto, M., Saito, S., Fujisawa, T., and Katagi, T., 2015. Bioconcentration and metabolism of trans-tetramethrin in fish. *Journal of Pesticide Science* 40: 1–8. https://doi.org/10.1584/jpestics.D15-027

Mohanty, N. P. and Measey, J., 2019. No survival of native larval frogs in the presence of invasive Indian bullfrog *Hoplobatrachus tigerinus* tadpoles. *Biological Invasions* 21: 2281–2286.

Monroe, M. J. and Alonzo, S. H., 2014. Sexual size dimorphism is not associated with the evolution of parental care in frogs. *Ecology and Evolution* 4: 3991–3998. https://doi.org/10.1002/ece3.1263

Montañez, L., Sampedro, A., and Álvarez, M., 1996. Utilización diferenciada de los recursos espacio y alimento por juveniles y adultos de *Rana catesbeiana* en la Ciénaga de Zapata. *Revista Biología* 10: 21–26.

Moreno, L. V., 1969. Situación taxonómica de *Bufo taladai jaumei*. *Serie Biología Academia de Ciencias de Cuba* 13: 1–19.

Moreno, L. V. and Rivalta, V., 2007. Especie nueva de sapo del género *Bufo* (Anura: Bufonidae) de la Península de Zapata, Cuba. *Solenodon* 6: 60–69.

Myers, N., Mittermeier, R. A., Mittermeier, C. G., da Fonseca, G. A. B., and Kent, J., 2000. Biodiversity hotspots for conservation priorities. *Nature* 403: 853–858. https://doi.org/10.1038/35002501

Novo, J., Estrada, A. R., and Alayón, G., 1985. *Eleutherodactylus* (Anura: Leptodactylidae) depredado por un araneido. *Miscelánea Zoológica* 28: 1–2.

Novo, J., Estrada, A. R., and Moreno, L. V., 1987. Apuntes sobre la reproducción de *Eleutherodactylus atkinsi* (Anura: Leptodactylidae). *Miscelánea Zoológica* 29: 2–3.

Nunes, A. L., Fill, J. M., Davies, S. J., Louw, M., Rebelo, A. D., Thorp, C. J., Vimercati, G., and Measey, J., 2019. A global meta-analysis of the ecological impacts of alien species on native amphibians. *Proceedings of the Royal Society B* 286: 20182528. http://dx.doi.org/10.1098/rspb.2018.2528

Nunes-de-Almeida, C. H. L., Haddad, C. F. B., and Toledo, L. F., 2021. A revised classification of the amphibian reproductive modes. *Salamandra* 3: 413–427.

Nurse, L. A., McLean, R. F., Agard, J., Briguglio, L. P., Duvat-Magnan, V., Pelesikoti, N., Tompkins, E., and Webb, A., 2014. Small islands. Pp. 1613–1654 in *Climate Change 2014: Impacts, Adaptation, and Vulnerability. Part B: Regional Aspects*, edited by V. R. Barros, C. B. Field, D. J. Dokken, M. D. Mastrandrea, K. J. Mach, T. E. Bilir, M. Chatterjee, K. L. Ebi, Y. O. Estrada, R. C. Genova, B. Girma, E. S. Kissel, A. N. Levy, S. MacCracken, P. R. Mastrandrea, and L. L.

White. Contribution of Working Group II to the Fifth Assessment Report of the Intergovernmental Panel on Climate Change. Cambridge University Press, Cambridge, UK and New York, NY, USA.

Odening, K. V., 1968. Einige Trematoden aus Fröschen und Schildkröten in Vietnam und Kuba. *Zoologischer Anzeiger* 181: 289–302.

O'Donohoe, M. E., Luna, M. C., Regueira, E., Brunetti, A., Basso, N. G., Lynch, J. D., Pereyra, M. O., and Hermida, G. N., 2019. Diversity and evolution of the parotoid macrogland in true toads (Anura: Bufonidae). *Zoological Journal of the Linnean Society* XX: 1–26. https://doi.org/10.1093/zoolinnean/zlz027

Odum, R. A. and Zippel, K. C., 2008. Amphibian water quality: approaches to an essential environmental parameter. *International Zoo Yearbook* 42: 40–52. https://doi.org/10.1111/j.1748-1090.2008.00053.x

O'Hanlon, S. J., Rieux, A., Farrer, R. A., Rosa, G. M., Waldman, B., Bataille, A., Kosch, T. A., Murray, K. A., Brankovics, B., Fumagalli, M., Martin, M. D., Wales, N., Alvarado-Rybak, M., Bates, K. A., Berger, L., Böll, S., Brookes, L., Clare, F., Courtois, E. A., Cunningham, A. A., Doherty-bone, T. M., Ghosh, P., Gower, D., Hintz, W. E., Höglund, J., Jenkinson, T. S., Lin, C., Laurila, A., Loyau, A., Martel, A., Meurling, S., Miaud, C., Minting, P., Pasmans, F., Schmeller, D. S., Schmidt, B. R., Shelton, J. M., Skerratt, L. F., Smith, F., Soto-Azat, C., Spagnoletti, M., Tessa, G., Toledo, L. F., Valenzuela-Sánchez, A., Verster, R., Vörös, J., Webb, R. J., Wierzbicki, C., Wombwell, E., Zamudio, K. R., Aanensen, D. M., James, T. Y., Gilbert, M. T., Weldon, C., Bosch, J., Balloux, F., Garner, T. W., and Fisher, M. C., 2018. Recent Asian origin of chytrid fungi causing global amphibian declines. *Science* 360 (6389): 621–627. https://doi.org/10.1126/science.aar1965

Olson, D. H., Aanensen, D. M., Ronnenberg, K. L., Powell, C. I., Walker, S. F., Bielby, J., Garner, T. W. J., Weaver, G., and Fisher, M., 2013. Mapping the global emergence of *Batrachochytrium dendrobatidis*, the amphibian chytrid fungus. *PLOS ONE* 8: e56802. https://doi.org/10.1371/journal.pone.0056802

ONEI (Oficina Nacional de Estadística e Información), 2020. Anuario Estadístico de Cuba 2019. Capítulo 1: Territorio. Edición 2020, enero-diciembre. http://www.onei.gob.cu/node/14902. Downloaded, 19 September 2020.

Padial, J. M., Grant. T., and Frost, D., 2014. Molecular systematics of terraranas (Anura: Brachycephaloidea) with an assessment of the effects of alignment and optimality criteria. *Zootaxa* 3825: 1–132. https://doi.org/10.11646/zootaxa.3825.1.1

Perera, A., Berovides, V., Garrido, O., Estrada, A., González, A., and Álvarez, M., 1994a. Criterios para la selección de especies amenazadas de vertebrados. La Habana, Cuba.

Perera, A., Berovides, V., Garrido, O., Estrada, A., González, A., and Álvarez, M., 1994b. Criterios para la selección de especies amenazadas de vertebrados. *Resúmenes*, p. 96, III Simposio de Zoología, La Habana, Cuba.

Perera, S., 2013. Convención para la Protección y el Desarrollo del Medio Marino en la región del Gran Caribe. Pp. 131–132 in *Plan del Sistema Nacional de Áreas Protegidas 2014-2020*. Ministerio de Ciencia, Tecnología y Medio Ambiente. La Habana. Cuba.

Pérez-Consuegra, N. and Montano-Pérez, M., 2021. Los plaguicidas altamente peligrosos en Cuba. IPEN/ACTAF/RAPAL. Editora Agroecológica. La Habana, Cuba.

Perera Córdova, W. H., Leitão, S. G., Cunha-Filho, G., Alonso Bosch, R., Alonso, I. P., Pereda-Miranda, R., Gervou, R., Touza, N. A., Quintas, L. E. M., and Noël, F., 2016. Bufadienolides from parotoid gland secretions of Cuban toad *Peltophryne fustiger* (Bufonidae): inhibition of human kidney Na+/K+-ATPase activity. *Toxicon* 10: 27–34. https://doi.org/10.1016/j.toxicon.2015.11.015

Pérez Vigueras, I., 1938. Nota sobre algunos nemátodos parásitos nuevos de Cuba. Pp. 501–508 in *Livro Jubilar Prof. Travassos*. Rio de Janeiro, Brasil III.

Poole, V., 2006. *Panamanian golden frog husbandry manual*, 2nd edn (online). Available: http://www.ranadorada.org/species-info.html

Portal Ríos, Y., Borroto-Páez, R., and Reyes Pérez, D., 2020. A Cuban Treefrog (*Osteopilus septentrionalis*, Hylidae) preying on a caged Yellow-faced Grassquit (*Tiaris olivacea*) in Cuba. *Revista Latinoamericana de Herpetología* 2: 139–142.

Powell, R. and Henderson, R. W., 1996. *Contributions to West Indian Herpetology: A Tribute to Albert Schwartz*, edited by R. Powell and R. W. Henderson. Contributions to Herpetology 12, Society for the Study of Amphibians and Reptiles, Ithaca, New York, USA.

Powell, R., Henderson, R. W., Adler, K., and Dundee, H. A., 1996. An annotated checklist of West Indian amphibians and reptiles. Pp. 51–93 in *Contributions to West Indian Herpetology: A Tribute to Albert Schwartz*, edited by R. Powell and R. W. Henderson. Contributions to Herpetology 12, Society for the Study of Amphibians and Reptiles, Ithaca, New York, USA.

Powell, R., Henderson, R. W., Farmer, M. C., Breuil, M., Echternacht, A. C., van Buurt, G., Romagosa, C. M., and Perry, G., 2011. Introduced amphibians and reptiles in the Greater Caribbean: patterns and conservation implications. Pp. 63–143 in

Conservation of Caribbean Island Herpetofaunas, Vol. 1: Conservation Biology and the Wider Caribbean, edited by A. Hailey, B. S. Wilson, and J. A. Horrocks. Koninklijke Brill NV, Leiden, The Netherlands. https://doi.org/10.1163/ej.9789004183957.i-228.38

Pramuk, J. B., 2002. Combined evidence and cladistic relationships of West Indian toads (Anura: Bufonidae). *Herpetological Monographs* 16: 121–151. https://doi.org/10.1655/0733-1347(2002)016[0121:CEACRO]2.0.CO;2

Pramuk, J. B., 2006. Phylogeny of South American *Bufo* (Anura: Bufonidae) inferred from combined evidence. *Zoological Journal of the Linnean Society* 146: 407–452. https://doi.org/10.1111/j.1096-3642.2006.00212.x

Pramuk, J. B., Hass, C. A., and Hedges, S. B., 2001. Molecular Phylogeny and Biogeography of West Indian Toads (Anura: Bufonidae). *Molecular Phylogenetics and Evolution* 20: 294–301. https://doi.org/10.1006/mpev.2001.0985

Pramuk, J. B., Robertson, T., Sites, J. K., and Noonan, B. P., 2007. Around the world in 10 million years: biogeography of the nearly cosmopolitan true toads (Anura: Bufonidae). *Global Ecology and Biogeography* 17: 72–83. https://doi.org/10.1111/j.1466-8238.2007.00348.x

Pregill, G. K., 1981. Cranial morphology and evolution of the West Indian toads (Salientia: Bufonidae): resurrection of the genus *Peltophryne* Fitzinger. *Copeia* 1981: 273–285. https://doi.org/10.2307/1444216

Pyron, R. A. and Wiens, J. J., 2011. A large-scale phylogeny of Amphibia including over 2800 species, and a revised classification of extant frogs, salamanders, and caecilians. *Molecular Phylogenetics and Evolution* 61: 543–583. https://doi.org/10.1016/j.ympev.2011.06.012

Pyšek, P., Hulme, P. E. , Simberloff, D., Bacher, S., Blackburn, T. M. , Carlton, J. T., Dawson, W., Essl, F., Foxcroft, L. C., Genovesi, P., Jeschke, J. M., Kühn, I., Liebhold, A. M., Mandrak, N. E., Meyerson, L. A., Pauchard, A., Pergl, J. , Roy, H. E., Seebens, H., van Kleunen, M., Vilà, M., Wingfield, M. J., and Richardson, D. M., 2020. Scientists' warning on invasive alien species. *Biological Reviews* 95: 1511–1534. doi: 10.1111/brv.12627

Rivalta, V., 2008. *Peltophryne florentinoi. The IUCN Red List of Threatened Species* 2008: e.T135861A4212829. https://dx.doi.org/10.2305/IUCN.UK.2008.RLTS.T135861A4212829.en. Downloaded on 30 July 2021.

Rivalta González, V., 2012a. *Peltophryne cataulaciceps* Schwartz, 1959. Pp. 88–89 in *Libro Rojo de los Vertebrados de Cuba*, edited by H. González Alonso, L. Rodríguez Schettino, A. Rodríguez, C. A. Mancina, and I. Ramos García. Editorial Academia, La Habana, Cuba.

Rivalta González, V., 2012b. *Peltophryne florentinoi* Moreno y Rivalta, 2007. Pp. 89–90 in *Libro Rojo de los Vertebrados de Cuba*, edited by H. González Alonso, L. Rodríguez Schettino, A. Rodríguez, C. A. Mancina, and I. Ramos García. Editorial Academia, La Habana, Cuba.

Rivalta González, V., 2012c. *Peltophryne longinasa* Stejneger, 1905. Pp. 90–91 in *Libro Rojo de los Vertebrados de Cuba*, edited by H. González Alonso, L. Rodríguez Schettino, A. Rodríguez, C. A. Mancina, and I. Ramos García. Editorial Academia, La Habana, Cuba.

Rivalta González, V., Rodríguez, L., Mancina, C. A., and Iturriaga, M., 2014. Amphibians of Cuba: checklist and geographic distributions. *Smithsonian Herpetological Information Service* 145: 1–48. https://doi.org/10.5479/si.23317515.145.1

Rivalta González, V., Rodríguez Schettino, L., Torres Barboza, A., Martínez Reyes, M., and Larramendi Joa, J., 2004. *Bufo peltocephalus* (Cuban toad). Diurnal retreat. *Herpetological Review* 35: 157.

Rodríguez, A., 2009. Guía taxonómica de los anfibios de Cuba. *Phyllomedusa* 8: 63–65. https://doi.org/10.11606/issn.2316-9079.v8i1p63-65

Rodríguez, A., 2012a. Anfibios. Pp. 55–59 in *Libro Rojo de los Vertebrados de Cuba*, edited by H. González Alonso, L. Rodríguez Schettino, A. Rodríguez, C. A. Mancina, and I. Ramos García. Editorial Academia, La Habana, Cuba.

Rodríguez, A., 2012b. *Eleutherodactylus acmonis* Schwartz, 1960. Pp. 60–61 in *Libro Rojo de los Vertebrados de Cuba*, edited by H. González Alonso, L. Rodríguez Schettino, A. Rodríguez, C. A. Mancina, and I. Ramos García. Editorial Academia, La Habana, Cuba.

Rodríguez, A., 2012c. *Eleutherodactylus albipes* Barbour y Shreve, 1937. Pp. 62–63 in *Libro Rojo de los Vertebrados de Cuba*, edited by H. González Alonso, L. Rodríguez Schettino, A. Rodríguez, C. A. Mancina, and I. Ramos García. Editorial Academia, La Habana, Cuba.

Rodríguez, A., 2012d. *Eleutherodactylus bartonsmithi* Schwartz, 1960. Pp. 63–64 in *Libro Rojo de los Vertebrados de Cuba*, edited by H. González Alonso, L. Rodríguez Schettino, A. Rodríguez, C. A. Mancina, and I. Ramos García. Editorial Academia, La Habana, Cuba.

Rodríguez, A., 2012e. *Eleutherodactylus blairhedgesi* Estrada, Díaz y Rodríguez, 1997. Pp. 64–65 in *Libro Rojo de los Vertebrados de Cuba*, edited by H. González Alonso, L. Rodríguez Schettino, A. Rodríguez, C. A. Mancina, and I. Ramos García. Editorial Academia, La Habana, Cuba.

Rodríguez, A., 2012f. *Eleutherodactylus bresslerae* Schwartz, 1960. Pp. 65–66 in *Libro Rojo de los Vertebrados de Cuba,* edited by H. González Alonso, L. Rodríguez Schettino, A. Rodríguez, C. A. Mancina, and I. Ramos García. Editorial Academia, La Habana, Cuba.

Rodríguez, A., 2012g. *Eleutherodactylus cubanus* Barbour, 1942. Pp. 66–67 in *Libro Rojo de los Vertebrados de Cuba,* edited by H. González Alonso, L. Rodríguez Schettino, A. Rodríguez, C. A. Mancina, and I. Ramos García. Editorial Academia, La Habana, Cuba.

Rodríguez, A., 2012h. *Eleutherodactylus emiliae* Dunn, 1926. Pp. 67–68 in *Libro Rojo de los Vertebrados de Cuba,* edited by H. González Alonso, L. Rodríguez Schettino, A. Rodríguez, C. A. Mancina, and I. Ramos García. Editorial Academia, La Habana, Cuba.

Rodríguez, A., 2012i. *Eleutherodactylus glamyrus* Estrada y Hedges, 1997. Pp. 69–71 in *Libro Rojo de los Vertebrados de Cuba,* edited by H. González Alonso, L. Rodríguez Schettino, A. Rodríguez, C. A. Mancina, and I. Ramos García. Editorial Academia, La Habana, Cuba.

Rodríguez, A., 2012j. *Eleutherodactylus guanahacabibes* Estrada y Novo Rodríguez, 1985. Pp. 71–72 in *Libro Rojo de los Vertebrados de Cuba,* edited by H. González Alonso, L. Rodríguez Schettino, A. Rodríguez, C. A. Mancina, and I. Ramos García. Editorial Academia, La Habana, Cuba.

Rodríguez, A., 2012k. *Eleutherodactylus iberia* Estrada y Hedges, 1997. Pp. 72–73 in *Libro Rojo de los Vertebrados de Cuba,* edited by H. González Alonso, L. Rodríguez Schettino, A. Rodríguez, C. A. Mancina, and I. Ramos García. Editorial Academia, La Habana, Cuba.

Rodríguez, A., 2012l. *Eleutherodactylus jaumei* Estrada y Alonso, 1997. Pp. 73–74 in *Libro Rojo de los Vertebrados de Cuba,* edited by H. González Alonso, L. Rodríguez Schettino, A. Rodríguez, C. A. Mancina, and I. Ramos García. Editorial Academia, La Habana, Cuba.

Rodríguez, A., 2012m. *Eleutherodactylus leberi* Schwartz, 1965. Pp. 74–75 in *Libro Rojo de los Vertebrados de Cuba,* edited by H. González Alonso, L. Rodríguez Schettino, A. Rodríguez, C. A. Mancina, and I. Ramos García. Editorial Academia, La Habana, Cuba.

Rodríguez, A., 2012n. *Eleutherodactylus maestrensis* Díaz Cádiz y Navarro, 2005. Pp. 75–76 in *Libro Rojo de los Vertebrados de Cuba,* edited by H. González Alonso, L. Rodríguez Schettino, A. Rodríguez, C. A. Mancina, and I. Ramos García. Editorial Academia, La Habana, Cuba.

Rodríguez, A., 2012o. *Eleutherodactylus mariposa* Hedges, Estrada y Thomas, 1992. Pp. 76–77 in *Libro Rojo de los Vertebrados de Cuba,* edited by H. González Alonso, L. Rodríguez Schettino, A. Rodríguez, C. A. Mancina, and I. Ramos García. Editorial Academia, La Habana, Cuba.

Rodríguez, A., 2012p. *Eleutherodactylus melacara* Hedges Estrada y Thomas, 1992. Pp. 77–79 in *Libro Rojo de los Vertebrados de Cuba,* edited by H. González Alonso, L. Rodríguez Schettino, A. Rodríguez, C. A. Mancina, and I. Ramos García. Editorial Academia, La Habana, Cuba.

Rodríguez, A., 2012q. *Eleutherodactylus michaelschmidi* Díaz, Cádiz y Navarro, 2007. Pp. 79–80 in *Libro Rojo de los Vertebrados de Cuba,* edited by H. González Alonso, L. Rodríguez Schettino, A. Rodríguez, C. A. Mancina, and I. Ramos García. Editorial Academia, La Habana, Cuba.

Rodríguez, A., 2012r. *Eleutherodactylus orientalis* Barbour y Shreve, 1937. Pp. 80–81 in *Libro Rojo de los Vertebrados de Cuba,* edited by H. González Alonso, L. Rodríguez Schettino, A. Rodríguez, C. A. Mancina, and I. Ramos García. Editorial Academia, La Habana, Cuba.

Rodríguez, A., 2012s. *Eleutherodactylus pezopetrus* Schwartz, 1960. Pp. 81–82 in *Libro Rojo de los Vertebrados de Cuba,* edited by H. González Alonso, L. Rodríguez Schettino, A. Rodríguez, C. A. Mancina, and I. Ramos García. Editorial Academia, La Habana, Cuba.

Rodríguez, A., 2012t. *Eleutherodactylus rivularis* Díaz, Estrada y Hedges, 2001. Pp. 82–83 in *Libro Rojo de los Vertebrados de Cuba,* edited by H. González Alonso, L. Rodríguez Schettino, A. Rodríguez, C. A. Mancina, and I. Ramos García. Editorial Academia, La Habana, Cuba.

Rodríguez, A., 2012u. *Eleutherodactylus tetajulia* Estrada y Hedges, 1996. Pp. 84–85 in *Libro Rojo de los Vertebrados de Cuba,* edited by H. González Alonso, L. Rodríguez Schettino, A. Rodríguez, C. A. Mancina, and I. Ramos García. Editorial Academia, La Habana, Cuba.

Rodríguez, A., 2012v. *Eleutherodactylus tonyi* Estrada y Hedges, 1997. Pp. 86–87 in *Libro Rojo de los Vertebrados de Cuba,* edited by H. González Alonso, L. Rodríguez Schettino, A. Rodríguez, C. A. Mancina, and I. Ramos García. Editorial Academia, La Habana, Cuba.

Rodríguez, A., 2012w. *Eleutherodactylus turquinensis* Barbour y Shreve, 1937. Pp. 87–88 in *Libro Rojo de los Vertebrados de Cuba,* edited by H. González Alonso, L. Rodríguez Schettino, A. Rodríguez, C. A. Mancina, and I. Ramos García. Editorial Academia, La Habana, Cuba.

Rodríguez, A. and Alonso, R., 2006. *Eleutherodactylus symingtoni*. Geographic distribution. *Herpetological Review* 37: 237.

Rodríguez, A. and Alonso, R., 2009. Natural History: *Eleutherodactylus leberi* (NCN), clutch size and parental care. *Herpetological Review* 40: 204.

Rodríguez, A. and Alonso, R., 2013. *Eleutherodactylus ronaldi*, Cuba: Holguín. *Herpetological Review* 3: 258.

Rodríguez, A. and García, Y., 2012a. *Eleutherodactylus adelus* Díaz, Cádiz y Hedges, 2003. Pp. 61–62 in *Libro Rojo de los Vertebrados de Cuba*, edited by H. González Alonso, L. Rodríguez Schettino, A. Rodríguez, C. A. Mancina, and I. Ramos García. Editorial Academia, La Habana, Cuba.

Rodríguez, A. and García, Y., 2012b. *Eleutherodactylus symingtoni* Schwartz, 1957. Pp. 83–84 in *Libro Rojo de los Vertebrados de Cuba*, edited by H. González Alonso, L. Rodríguez Schettino, A. Rodríguez, C. A. Mancina, and I. Ramos García. Editorial Academia, La Habana, Cuba.

Rodríguez, A. and Urgellés, Y. 2012. *Eleutherodactylus etheridgei* Schwartz, 1958. Pp. 68–69 in *Libro Rojo de los Vertebrados de Cuba*, edited by H. González Alonso, L. Rodríguez Schettino, A. Rodríguez, C. A. Mancina, and I. Ramos García. Editorial Academia, La Habana, Cuba.

Rodríguez, A., De la Nuez, D., and Alonso, R., 2010a. Intraspecific variation in the advertisement call of the cloud-forest frog *Eleutherodactylus glamyrus* (Anura: Eleutherodactylidae). *Journal of Herpetology* 44: 457–466. https://doi.org/10.1670/09-038.1

Rodríguez, A., Vences, M., and García, L. Y., 2011a. Expansion of the known distribution and molecular systematics of *Eleutherodactylus adelus* (Anura: Eleutherodactylidae). *Salamandra* 47: 222–226.

Rodríguez, A., Alonso, R., Rodríguez, J. A., and Vences, M., 2012. Geographic distribution, colour variation and molecular diversity of miniature frogs of the *Eleutherodactylus limbatus* group from Cuba. *Salamandra* 48: 71–91.

Rodríguez, A., Poth, D., Schulz, S., and Vences, M., 2011b. Discovery of skin alkaloids in a miniaturized eleutherodactylid frog from Cuba. *Biology Letters* 7: 414–418. https://doi.org/10.1098/rsbl.2010.0844

Rodríguez, A., Poth, D., Schulz, S., Gehara, M., and Vences, M., 2013. Genetic diversity, phylogeny and evolution of alkaloid sequestering in Cuban miniaturized frogs of the *Eleutherodactylus limbatus* group. *Molecular Phylogenetics and Evolution* 68: 541–554. https://doi.org/10.1016/j.ympev.2013.04.031

Rodríguez, A., Vences, M., Nevado, B., Machordom, A., and Verheyen, E., 2010b. Biogeographic origin and radiation of Cuban *Eleutherodactylus* of the *auriculatus* species group, inferred from mitochondrial and nuclear gene sequences. *Molecular Phylogenetics and Evolution* 54: 179–186. https://doi.org/10.1016/j.ympev.2009.08.023

Rodríguez, A., Dugo-Cota, Á., Montero-Mendieta, S., González-Voyer, A., Alonso, R., Vences, M., and Vilà, C., 2017. Cryptic within cryptic: genetics, morphometrics, and bioacoustics delimitate a new species of *Eleutherodactylus* (Anura: Eleutherodactylidae). *Zootaxa* 4221: 501–522. https://doi.org/10.11646/zootaxa.4221.5.1

Rodríguez-Cabrera, T. M., 2018. The American Bullfrog, *Lithobates catesbeianus* (Ranidae): An exotic prey item of the Cuban endemic Giant Trope, *Tropidophis melanurus* (Tropidophiidae). *Reptiles & Amphibians* 25: 148–150. https://doi.org/10.17161/randa.v25i2.14276

Rodríguez-Cabrera, T. M. and Hernández Gómez, A. 2021. New Prey Records for Two Snakes of the Genus *Tropidophis* (Tropidophiidae) from Urban Habitats in La Habana, Cuba. Reptiles & Amphibians 28: 512–515. https://doi.org/10.17161/randa.v28i3.15858

Rodríguez-Cabrera, T. M. and Lopez-Silvero, R., 2022. New record of the critically endangered Western Spiny Frog, *Eleutherodactylus symingtoni* (Eleutherodactylidae). *Reptiles & Amphibians* 29: 314–315. https://doi.org/10.17161/randa.v29i1.17957

Rodríguez-Cabrera, T. M., and Rodríguez-Silva, R., 2015. Predation on the Rosario Red-legged Frog, *Eleutherodactylus zugi* (Eleutherodactylidae), by the Cuban Mosquitofish, *Gambusia punctata* (Poeciliidae) in western Cuba. *Reptiles & Amphibians* 22: 95–97. https://doi.org/10.17161/randa.v22i3.14057

Rodríguez-Cabrera, T. M., Fong G., A., and Torres, J., 2020a. New dietary records for three Cuban snakes in the genus *Tropidophis* (Tropidophiidae), with comments on possible niche partitioning by Cuban Tropes. *Reptiles & Amphibians* 27: 201–208. https://doi.org/10.17161/randa.v27i2.14177

Rodríguez-Cabrera, T. M., García-Padrón, L. Y., Morell Savall, E., and Torres, J., 2020b. Predation on direct-developing frogs (Eleutherodactylidae: *Eleutherodactylus*) in Cuba: new cases and a review. *Reptiles & Amphibians* 27: 161–168. https://doi.org/10.17161/randa.v27i2.14175

Rodríguez-Cabrera, T. M., Morell Savall, E., Rodríguez-Machado, S., and Torres, J., 2020c. Trophic ecology of the Cuban Boa, *Chilabothrus angulifer* (Boidae). *IRCF Reptiles & Amphibians* 27: 169–200. https://doi.org/10.17161/randa.v27i2.14176

Rodríguez-Cabrera, T. M., García-Padrón, L. Y., Acosta-Galvis, A. R., de Sá, R. O., and Alonso, R., 2018. First record of the genus *Leptodactylus* (Anura: Leptodactylidae) in Cuba: *Leptodactylus fragilis*, a biological invasion? *Journal of Natural History* 52: 1883–1892. https://doi.org/10.1080/00222933.2018.1498549

Rodríguez-Gómez, Y., Sanz-Ochotorena, A., Segura-Valdez, M. L., and Jiménez-García, L. F., 2009. Caracterización ultraestructural del espermatozoide del género *Eleutherodactylus* (Anura: Eleutherodactylidae). *Acta Microscopica* 18: 631–632.

Rodríguez-Gómez, Y., Sanz-Ochotorena, A., Segura-Valdez, M. L., Lara-Martínez, R., and Jiménez-García, L. F., 2010. Morfología del ovario en ranas del género *Eleutherodactylus* (Anura: Leptodactylidae). *Acta Microscopica* 19: 271–278.

Rodríguez-Machado, S., and Rodríguez-Cabrera, T. M., 2015. First record of native amphibian predation by the invasive alien African catfish *Clarias gariepinus* (Siluriformes, Clariidae) in Cuba. *Pan-American Journal of Aquatic Sciences* 10: 254–258.

Rodríguez Schettino, L., 1999. Systematic accounts of the species. Pp. 104–380 in *The Iguanid Lizards of Cuba*, edited by L. Rodríguez. University Press of Florida, Gainesville, Florida, USA.

Rodríguez Schettino, L. (editor), 2003. *Anfibios y Reptiles de Cuba*, UPC Print, Vaasa, Finlandia.

Rodríguez Schettino, L. and Rivalta González, V., 2007. Efectos probables del aumento del nivel del mar sobre la herpetofauna de la Reserva de la Biosfera Ciénaga de Zapata, Matanzas, Cuba. *Poeyana* 495: 8–13.

Ruibal, R., 1959. *Bufo gundlachi*, a new species of Cuban toad. *Breviora* 105: 1–14.

Ruibal, R. and Estrada, A. R., 1999. A review of Cuban herpetology. Pp. 31–62 in *Caribbean Reptiles and Amphibians*, edited by B. I. Crother. Academic Press, San Diego, California, USA. https://doi.org/10.1016/B978-012197955-3/50014-8

Ruíz, F. N., 1980. La larva de *Bufo longinasus* (Stejneger) (Amphibia: Bufonidae) en el laboratorio. *Poeyana* 207: 1–8.

Ruíz, F. N., 1987. *Anfibios de Cuba*, Editorial Gente Nueva, La Habana, Cuba.

Russell, R. E., Halstead, B. J., Mosher, B. A., Muths, E., Adams, M. J., Grant, E. H. C., Fisher, R. N., Kleeman, P. M., Backlin, A. R., Pearl, C. A., Honeycutt, R. K., and Hossack, B. R., 2019. Effect of amphibian chytrid fungus (*Batrachochytrium dendrobatidis*) on apparent survival of frogs and toads in the western USA. *Biological Conservation* 236: 296–304. https://doi.org/10.1016/j.biocon.2019.05.017

Sabagh, L. T., Barbosa Ferreira, R., and Duarte Rocha, C. F., 2017. Host bromeliads and their associated frog species: further considerations on the importance of species interactions for conservation. *Symbiosis* 73: 201–211. https://doi.org/10.1007/s13199-017-0500-9

Sabino-Pinto, J., Bletz, M. C., Iturriaga, M., Vences, M., and Rodríguez, A., 2017. Low infection prevalence of the amphibian chytrid fungus *Batrachochytrium dendrobatidis* (Chytridiomycetes: Rhizophydiales) in Cuba. *Amphibia-Reptilia* 38: 243–249. https://doi.org/10.1163/15685381-00003100

Samgina, T. Y., Tolpina, M. D., Surin, A. K., Kovalev, S. V., Alonso Bosch, R., Pascual Alonso, I., Almeida García, F., González López, L. J., and Lebedev, A. T., 2021. Manual mass spectrometry de novo sequencing of the anionic host defense peptides of the Cuban Treefrog *Osteopilus septentrionalis*. *Rapid Communication in Mass Spectrometry* 35: e9061. https://doi.org/10.1002/rcm.9061

Sampedro, A. and Berovides, V., 1982. Ecología reproductiva del sapo común (*Bufo peltacephalus*). *Reporte de Investigación del Instituto de Zoología* 3: 1–10.

Sampedro, A. and Berovides, V., 1985. Ecología trófica y actividad de *Bufo peltocephalus* (Amphibia: Anura) durante los períodos de seca y lluvia, en el Jardín Botánico de Cienfuegos. *Poeyana* 297: 1–8.

Sampedro, A., Berovides, V., and Torres, O., 1982. Hábitos alimentarios y actividad de *Bufo peltocephalus* Tschudi (Amphibia: Bufonidae) en el Jardín Botánico de Cienfuegos. *Poeyana* 233: 1–14.

Sampedro Marin, A., Montañez Huguez, L., and Suárez Boado, C., 1985. Alimentación de *Rana catesbeiana* en dos zonas de captura de Cuba. *Ciencias Biológicas* 13: 59–66.

Sanz-Ochotorena, A., Segura-Valdés, M. de L., Rodríguez Gómez, Y., Lara Martínez, R., and Jiménez-García, L. F., 2008. Estructura y ultraestructura de la gónada de los machos de *Bufo fustiger* (Anura: Bufonidae). *TIP Revista Especializada en Ciencias Químico-Biológicas* 11: 81–86.

Sanz-Ochotorena, A., Segura-Valdés, M. de L., Rodríguez-Gómez, Y., Lara-Martínez, R., and Jiménez-García, L. F., 2009. Ultraestructura de la gónada del macho de dos subespecies de *Bufo longinasus*. *TIP Revista Especializada en Ciencias Químico-Biológicas* 13: 36–39.

Sanz-Ochotorena, A., Segura-Valdés, M. de L., Rodríguez-Gómez, Y., Lara-Martínez, R., and Jiménez-García, L. F., 2011. Pigmentos en los testículos de cinco anfibios endémicos de Cuba (*Eleutherodactylus turquinensis, E. cuneatus, E. glamyrus, Bufo longinasus y B. longinasus cajalbanensis*). *TIP Revista Especializada en Ciencias Químico-Biológicas* 14: 85–92.

Sanz Ochotorena, A., Rodríguez Gómez, Y., Segura Valdéz, M. de L., Lara Martínez, R., and Jiménez García, L. F., 2019. Atlas de la morfología microscópica de las gónadas de anfibios y reptiles de Cuba. Editorial de la Universidad de La Habana, La Habana, Cuba.

Scheele, B. C., Pasmans, F., Skerratt, L. F., Berger, L., Martel, A., Beukema, W., Acevedo, A. A. , Burrowes, P. A., Carvalho, T., Catenazzi, A., De la Riva, I., Fisher, M. C., Flechas, S. V., Foster, C. N., Frías-Álvarez, P., Garner, T. W. J., Gratwicke, B., Guayasamin, J. M., Hirschfeld, M., Kolby, J. E., Kosch, T. A., La Marca, E., Lindenmayer, D. B., Lips, K. R., Longo, A. V., Maneyro, R., McDonald, C. A., Mendelson 3rd, J., Palacios-Rodríguez, P., Parra-Olea, G., Richards-Zawacki, C. L., Rödel, M.-O., Rovito, S. M., Soto-Azat, C., Toledo, L. F., Voyles, J., Weldon, C., Whitfield, S. M., Wilkinson, M., Zamudio, K. R., and Canessa, S., 2019. Amphibian fungal panzootic causes catastrophic and ongoing loss of biodiversity. *Science* 363: 1459–1463. https://doi.org/10.1126/science.aav0379

Schmid, M., Steinlein, C., Bogart, J. P., Feichtinger, W., León, P., La Marca, E., Díaz, L. M., Sanz, A., Chen, S-H., and Hedges, S. B., 2010. The chromosomes of terraranan frogs: Insights into vertebrate cytogenetics. *Cytogenetic Genome Research* 130–131: 1–568. https://doi.org/10.1159/000301339

Schmidt, K. P., 1920. Some new and rare amphibians and reptiles from Cuba. *Proceedings of the Linnaean Society of New York* 33: 3–8.

Schwartz, A., 1957. A new species of *Eleutherodactylus* (Amphibia: Leptodactylidae) from Cuba. *Proceedings of the Biological Society of Washington* 70: 209–212.

Schwartz, A., 1958a. Four new frogs of the genus *Eleutherodactylus* (Leptodactylidae) from Cuba. *American Museum Novitates* 1873: 1–20.

Schwartz, A., 1958b. A new frog of the *auriculatus* group of the genus *Eleutherodactylus* from western Cuba. *Herpetologica* 14: 69–77.

Schwartz, A., 1958c. Another new species of *Eleutherodactylus* (Amphibia: Leptodactylidae) from western Cuba. *Journal of the Washington Academy of Sciences* 48: 127–131.

Schwartz, A., 1958d. Another new large *Eleutherodactylus* (Amphibia: Leptodactylidae) from Western Cuba. *Proceedings of the Biological Society of Washington* 71: 37–42.

Schwartz, A., 1959a. The status of *Eleutherodactylus pinarensis* and new species of Genus from Western Cuba. *Herpetologica* 15: 61–69.

Schwartz, A., 1959b. A new species of toad, *Bufo cataulaciceps*, from the Isla de Pinos and western Cuba. *Proceedings of the Biological Society of Washington* 73: 45–56.

Schwartz, A, 1959c. A new species of frog of the *Eleutherodactylus ricordi* group from central Cuba. *American Museum novitates* 1926: 1–16.

Schwartz, A., 1960a. Nine new Cuban frogs of the genus *Eleutherodactylus. Reading Public Museum and Art Gallery, Special Publication*, Pennsylvania, 11: 1–60.

Schwartz, A., 1960b. The large toads of Cuba. *Proceedings of the Biological Society of Washington* 73: 45–56.

Schwartz, A., 1965. A new Cuban *Eleutherodactylus* of *auriculatus* group. *Herpetologica* 21: 27–31.

Schwartz, A. and Henderson, R. W., 1985. *A Guide to the Identification of the Amphibians and Reptiles of the West Indies Exclusive of Hispaniola*. Milwaukee Public Museum, Milwaukee, Wisconsin, USA.

Schwartz, A. and Henderson, R. W., 1988. West Indian amphibians and reptiles: a checklist. Milwaukee Public Museum, *Contribution in Biology and Geology* 74: 1–274.

Schwartz, A. and Henderson, R. W., 1991. *Amphibians and Reptiles of the West Indies: Descriptions, Distributions, and Natural History*. University of Florida Press, Gainesville, Florida, USA.

Schwartz, A. and Thomas, R., 1975. A check-list of West Indian amphibians and reptiles. *Carnegie Museum of Natural History, Special Publication* 1: 1–126. https://doi.org/10.5962/bhl.title.123681

Schwartz, A., Thomas, R., and Ober, L. D., 1978. First supplement to a check-list of West Indian amphibians and reptiles. *Carnegie Museum of Natural History, Special Publication* 5: 1–35.

Shilton, C. M., Smith, D. A., Crawshaw, G. J. B., Valdés, E., Keller, C. B., Maguire, G. F., Connelly, P. W., and Atkinson, J., 2001. Corneal lipid deposition in Cuban Tree Frogs (*Osteopilus septentrionalis*) and its relationship to serum lipids: an experimental study. *Journal of Zoo and Wildlife Medicine* 32: 305–319. https://doi.org/10.1638/1042-7260(2001)032[0305:CLDICT]2.0.CO;2

Skerratt, L. F., Berger, L., Speare, R., Scott Cashins, S., Raymond McDonald, K., Dawn Phillott, A., Bryan Hines, H., and Kenyon, N., 2007. Spread of chytridiomycosis has caused the rapid global decline and extinction of frogs. *EcoHealth* 4: 125–134. https://doi.org/10.1007/s10393-007-0093-5

Smith, M. L., Hedges, S. B., Buck, W., Hemphill, A., Incháustegui, S., Ivie, M., Martina, D., Maunder, M., and Ortega, J. F., 2005. Caribbean Islands. Pp. 112–118 in *Hotspots revisited: Earth's biologically richest and most endangered terrestrial ecoregions*, edited by R. A. Mittermeier, P. R. Gil, M. Hoffmann, J. Pilgrim, T. Brooks, C. G. Mittermeier, J. Lamoreux, and G. A. B. da Fonseca. Mexico City, CEMEX, Mexico.

Stejneger, L., 1905. Description of a new toad from Cuba. *Proceedings of the United States National Museum* 28: 765–767. https://doi.org/10.5479/si.00963801.28-1406.765

Stuart, S., 2006. *Eleutherodactylus maestrensis*. The IUCN Red List of Threatened Species 2006: e.T61812A12561025. https://dx.doi.org/10.2305/IUCN.UK.2006.RLTS.T61812A12561025.en. Accessed on 06 February 2022.

Stuart, S. N., Hoffmann, M., Chanson, J. S., Cox, N. A., Berridge, R., Ramani, P., and Young, B. E., (editors), 2008. Threatened Amphibians of the World. Lynx Ediciones with IUCN-The World Conservation Union, Conservation International and Nature Server, Barcelona, Spain.

Suárez, P., Ferro, J. M., Nagamachi, C. Y., Cardozo, D. E., Blasco-Zúñiga, A., Silva, J. B., Marciano, E., Costa, M. A., Orrico, V. G. D., Solé, M., Roberto, I. J., Rivera, M., Wiley, J. E., Faivovich, J., Baldo, D., and Pieczarka, J. C., 2020. Chromosome evolution in Lophyohylini (Amphibia, Anura, Hylinae). *PLOS ONE* 15: e0234331. https://doi.org/10.1371/journal.pone.0234331

Surendran, H. and Vasudevan, K., 2013. Recent introduction and spread of Indian bullfrog *Hoplobatrachus tigerinus* (Daudin, 1802) into the Andaman Islands. *Aliens* 33: 42–43.

Taylor, M. A., Clarke, L. A., Centella, A., Bezanilla, A., Stephenson, T., Jones, J. J., Campbell, J. D., Vichot, A., and Charlery, J., 2018. Future Caribbean climates in a world of rising temperatures: The 1.5 vs 2.0 dilemma. *Journal of Climate* 31: 2907–2926. https://doi.org/10.1175/JCLI-D-17-0074.1

Torres, O., Mujica, L., and Llanes, A., 1985. Alimentación de la Garza Ganadera (*Bubulcus ibis*) en algunas regiones de Cuba. *Ciencias Biológicas* 13: 67–77.

Torres, J., Rodríguez-Cabrera, T. M., and Marrero, R., 2017. *Eleutherodactylus greyi* (Grey's Frog). *Herpetological Review* 48: 161.

Townsend, D. S., 1996. Patterns of parental care in frogs of the genus *Eleutherodactylus*. Pp. 229–239 in *Contributions to West Indian Herpetology: A Tribute to Albert Schwartz*, edited by R. Powell and R. W. Henderson. Contributions to Herpetology 12, Society for the Study of Amphibians and Reptiles, Ithaca, New York, USA.

Trueb, L., 1966. Morphology and development of the skull in the frog *Hyla septentrionalis*. *Copeia* 1966: 562–563. https://doi.org/10.2307/1441083

Trueb, L. and Tyler, M. J., 1974. Systematics and evolution of the Greater Antillean hylid frogs. *Occasional Papers of the Museum of Natural History, University of Kansas* 24: 1–60. https://doi.org/10.5962/bhl.part.11750

Tschudi, J. J., 1838. Classification der Batrachier mit Berücksichtigung der fossilen Thiere dieser Abtheilung der Reptilien. In der buchdruckerei von Petitpierre, Neuchâtel, Switzerland. https://doi.org/10.5962/bhl.title.4883

Tyler, M. J., Wassersug, R., and Smith, B., 2007. How frogs and humans interact: influences beyond habitat destruction, epidemics and global warming. *Applied Herpetology* 4: 1–18. https://doi.org/10.1163/157075407779766741

Ulloa, J. S., Aubin, T., Llusia, D., Courtois, E. A., Fouquet, A., Gaucher, Ph., Pavoine, S., and Sueur, J., 2019. Explosive breeding in tropical anurans: environmental triggers, community composition and acoustic structure. *BMC Ecology* 19: 28. https://doi.org/10.1186/s12898-019-0243-y

Vági, B., Végvári, Z., Liker, A., Freckleton, R. P., and Székely, T., 2019. Parental care and the evolution of terrestriality in frogs. *Proceedings of the Royal Society B* 286: 20182737. https://doi.org/10.1098/rspb.2018.2737

Valdés de la Osa, A., 1978. Utilización del nido de *Speotyto cunicularia* (Aves: Strigiformes) como refugio por *Bufo p. peltocephalus* (Amphibia: Bufonidae). *Miscelánea Zoológica* 7: 4.

Valdés de la Osa, A., 1989. Notas sobre *Sminthillus limbatus* (Cope). *Caribbean Journal of Science* 25: 96–98.

Valdés de la Osa, A. and Ruíz García, F. N., 1979. Manifestación del patrón conductual de aumento de volumen en bufónidos cubanos. *Miscelánea Zoológica* 8: 3–4.

Valdés de la Osa, A. and Ruíz García, F. N., 1980a. Consideraciones sistemáticas sobre *Bufo longinasus* (Anura: Bufonidae) y descripción de una nueva subespecie. *Poeyana* 206: 1–33.

Valdés de la Osa, A. and Ruíz García, F. N., 1980b. Caso de canibalismo en *Eleutherodactylus cuneatus* (Cope) (Salientia: Leptodactylidae). *Miscelánea Zoológica* 6: 4.

Valdés de la Osa, A. and Zayas, L., 1980. Observaciones alimentarias en anuros leptodactílidos de la Sierra del Rosario. *Ciencias Biológicas* 4: 123–124.

Vales, M., Álvarez, A., Montes, L., and Ávila, A., 1998. *Estudio nacional sobre la diversidad biológica en la República de Cuba*. Editorial CESYTA, Madrid, Spain.

Vargas-Salinas, F., 2006. *Osteopilus septentrionalis* (Cuban Tree Frog). Reproduction. *Herpetological Review* 37: 205.

Voyles, J., Young, S., Berger, L., Campbell, C., Voyles, W. F., Dinudom, A., Cook, D., Webb, R., Alford, R. A., Skerratt, L. F., and Speare, R., 2009. Pathogenesis of chytridiomycosis, a cause of catastrophic amphibian declines. *Science* 326: 582–585. https://doi.org/10.1126/science.1176765

Vredenburg, V. T., Knapp, R. A., Tunstall, T. S., and Briggs, C., 2010. Dynamics of an emerging disease drive large-scale amphibian population extinctions. *Proceedings of the National Academy of Sciences of the United States of America* 107: 9689–9694.

Walls, S. C. and Gabor, C. R., 2019. Integrating behavior and physiology into strategies for amphibian conservation. *Frontiers in Ecology and Evolution* 7:234. https://doi.org/10.3389/fevo.2019.00234

Ward, J. H., Jr., 1963. Hierarchical grouping to optimize an objective function. *Journal of the American Statistical Association* 58: 236–244. https://doi.org/10.1080/01621459.1963.10500845

Wells, K. D., 2007. *The Ecology and Behavior of Amphibians*. The University of Chicago Press. Chicago, Illinois, USA.

Wilbur, H. M., 1980. Complex Life Cycles. *Annual Review of Ecology and Systematics* 11: 67–93. https://doi.org/10.1146/annurev.es.11.110180.000435

Yoshimi, D. H., Payne, D. A., and Slavens, F. L., 1996. Maintenance and captive breeding of the Solomon Islands leaf frog (*Ceratobatrachus guentheri*). *Advances in Herpetoculture* 1: 23–32.

Zajicek, D. and Mauri Méndez, M., 1969. Hemoparásitos de algunos animales de Cuba. *Poeyana* 66: 1–10.

Zippel, K., Johnson, K., Gagliardo, R., Gibson, R., McFadden, R., Browne, R., Martínez, C., and Townsend, E., 2011. The amphibian ark: a global community for *ex situ* conservation of amphibians. *Herpetological Conservation and Biology* 6: 340–352.

4 Amphibians of the Southwestern Caribbean Islands

Harold Heatwole and Javier Sunyer

Abbreviations and acronyms

asl	*above sea level*
CIOH	*Centro Investigaciones Oceanográficas e Hidrográficas*
CITES	*Convention on International Trade in Endangered Species*
CORALINA	*Corporación para el Desarrollo Sostenible del Archipiélago de San Andrés, Providencia y Santa Catalina*
DIMAR	*Dirección General Marítima (Autoridad Marítima Colombiana)*
IUCN	*International Union for Conservation of Nature*
JS	*Javier Sunyer*
LC	*Least Concern*
RAMSAR	*An Iranian city where an international governmental treaty called the Convention on Wetlands (Ramsar Convention) was adopted in 1971 and which provides the framework for the conservation and wise use of wetlands and their resources*
UNESCO	*United Nations Educational, Scientific and Cultural Organization*

I. Introduction

The islands of the southwestern Caribbean (Fig. 4.1) are a group of small islands and cays of varying origin, geology, palaeogeography, ecology, and history. The practices of exploitation and development by humans, and potential threats to their amphibian faunas, differ on each of them too. They are either part of, and under the control of, a Central American or South American country, or Great Britain, each with its own policies and laws regarding conservation

Harold Heatwole and Javier Sunyer, 'Chapter 4: Amphibians of the Southwestern Caribbean Islands' in: *The Conservation and Biogeography of Amphibians in the Caribbean*. Pelagic Publishing (2023). © Harold Heatwole and Javier Sunyer. DOI: 10.53061/QEMH9420

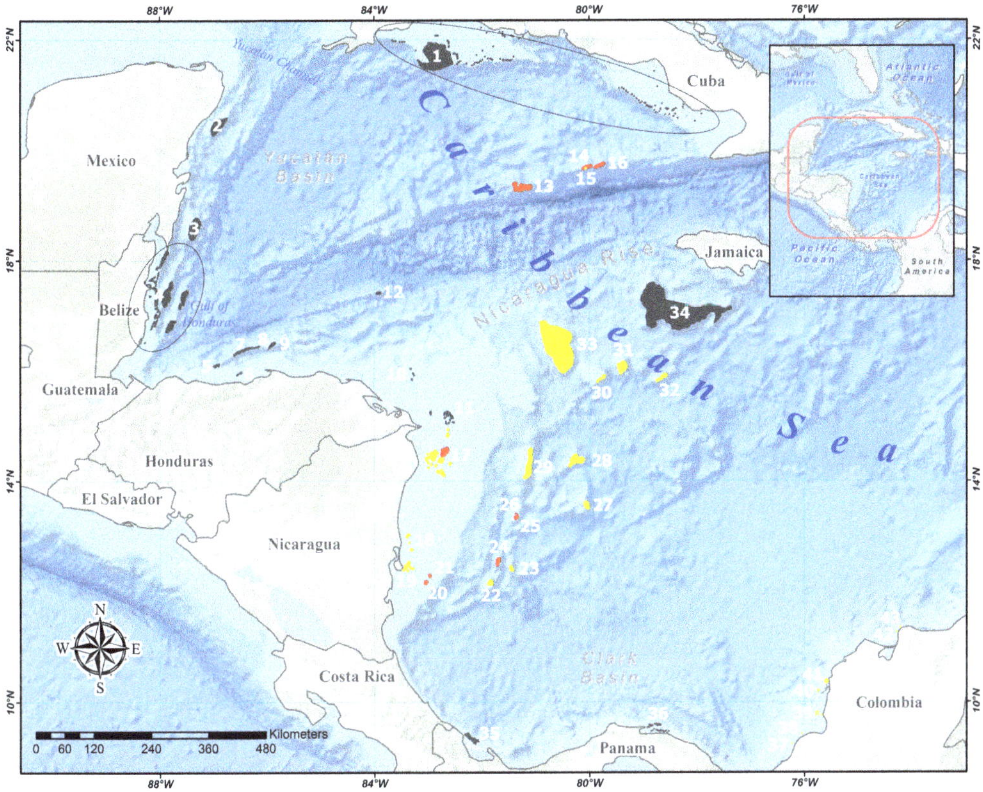

Figure 4.1. Map of southwestern Caribbean Sea showing the location of its major islands, islets, banks, atolls, and cays. Grey colours depict those that are not covered in the present chapter. Yellow colours represent amphibian free islands and red colours are used for those containing a resident amphibian fauna.

NUMBER ON MAP	ENGLISH NAME	SPANISH NAME	COUNTRY	NUMBER OF AMPHIBIANS
1	Isle of Youth[1] (and adjacent Cuban cays)	Isla de la Juventud (y cayos cubanos adyacentes)	Cuba	14, (Estrada 2012)
2	Cozumel Island	Isla de Cozumel	Mexico	To be covered in Vol. 9, Part 6 Central America and Mexico, in preparation.
3	Chinchorro Bank	Banco Chinchorro	Mexico	
4	Islands of Belize: Internal islands in a north-to-south direction: Ambergris Cay, Cay Caulker, Hicks Cays, Cay Chapel, Drowned Cays, Water Cay, Robinson Cays, Rendezvous Cays, Tobacco Cays, Colson Cay, Southern Cay, Columbus Cay, Thatch Cay, Bread & Butter Cays, Wish Bone Cays, and Harvest Cay. External islands in a north-to-south direction: Turneffe Atoll, Lighthouse Atoll, and Glovers Reef.	Islas de Belice	Belize	To be covered in Vol. 9, Part 6 Central America and Mexico, in preparation.

(Continued)

(Continued)

NUMBER ON MAP	ENGLISH NAME	SPANISH NAME	COUNTRY	NUMBER OF AMPHIBIANS
5	Utila Island	Isla de Utila	Honduras	
6	Cochinos Cays	Cayos Cochinos	Honduras	
7	Roatán Island	Isla de Roatán	Honduras	
8	Barbareta Island	Isla Barbareta	Honduras	
9	Guanaja Island	Isla de Guanaja	Honduras	
10	Vivorillo Cays	Cayos Vivorillo	Honduras	
11	(Unnamed)	(Sin nombre)	Honduras	
12	Swan Islands (Great Swan, Little Swan)	Islas del Cisne	Honduras	0, (Powell and Henderson 2012)
13	Grand Cayman	Gran Caimán	Cayman Islands	4
14	Little Cayman	Pequeña Caimán	Cayman Islands	1
15	Owen Island	Isla Owen	Cayman Islands	0
16	Cayman Brac	Caimán Brac	Cayman Islands	4
17	Miskito Cays	Cayos Miskitos	Nicaragua	1
18	Man-o-War Cays	Cayos Guerrero	Nicaragua	0
19	Pearl Cays	Cayos Perlas	Nicaragua	0
20	Little Corn Island	Isla Pequeña del Maíz	Nicaragua	5
21	Big Corn Island	Isla Grande del Maíz	Nicaragua	5
22	Albuquerque Cays (Southsouthwest Cays)	Cayos de Albuquerque (Cayos del Sur-suroeste)	Colombia	0
23	Courtown Atoll (East-southeast Cays)	Atolón Bolívar (Cayos del Estesureste)	Colombia	0
24	San Andrés Island	Isla de San Andrés	Colombia	1
25	Providencia Island	Isla de Providencia	Colombia	1
26	Santa Catalina Island	Isla Santa Catalina	Colombia	1
27	Roncador Cay	Cayos Roncador	Colombia	0
28	Serrana Bank	Banco Serrana	Colombia	0
29	Quitasueño Bank	Banco Quitasueño	Colombia	0
30	Serranilla Cays	Cayos Serranilla	Colombia	0
31	Alice Shoal	Bajo Alicia	Colombia	0, (submerged)
32	New Shoal Bank	Banco Bajo Nuevo	Colombia	0
33	Rosalind Bank	Placer de Rosalinda	In dispute between Colombia, Nicaragua, Honduras, Jamaica, and U.S.A.	0, (submerged)

(Continued)

(Continued)

NUMBER ON MAP	ENGLISH NAME	SPANISH NAME	COUNTRY	NUMBER OF AMPHIBIANS
34	Pedro Bank (NE Pedro Cay, SW Pedro Cay)	Banco de Pedro	Jamaica	0, (Wilson *et al.* 2012)
35	Bocas del Toro Archipelago	Archipiélago de Bocas del Toro	Panama	See Chapter 13, this volume.
36	San Blas Islands	Archipiélago de San Blás	Panama	To be covered in Vol. 9, Part 10 (Central America) of this series, in preparation.
37	Tortuguilla Island	Isla Tortuguilla	Colombia	0
38	Fuerte Island	Isla Fuerte	Colombia	0
39	Archipelago of San Bernardo	Archipiélago de San Bernardo	Colombia	0
40	Coral Islands of Rosario	Islas Corales del Rosario	Colombia	0
41	Tierra Bomba Island	Isla de Tierra Bomba	Colombia	0
42	El Morro Island	Isla El Morro	Colombia	0
43	La Aguja Island	Isla de La Aguja	Colombia	0

[1]Until 1978 this island was called 'Isle of Pines', then changed to 'Isle of Youth' to correspond to the change in the Spanish name from 'Isla de Pinos' to 'Isla de la Juventud'. Not to be confused with the Isle of Pines (Île des Pins) in New Caledonia.

and the protection of the environment and wildlife. In addition to the islands discussed in this chapter, there are various small, near-coastal ones intimately allied with mainland Central America. To serve as an example of these, the Bocas del Toro Archipelago (Panama), is treated by Galeano *et al.* in Chapter 13 of this volume (Islands number 35 in figure 4.1). Other such 'mainland' islands (e.g. San Blas islands: number 36 in figure 4.1), however, will be included under the country of their affiliation in forthcoming books in this series (*Amphibian Biology* Volume 9, Part 6, in preparation). The geographic scope of the present book covers the Caribbean islands of Colombia (Table 4.1), Nicaragua, and the Cayman Islands. The western Caribbean islands belonging to Central American countries farther north (Honduras, Belize, and Mexico) will be treated in Volume 9, Part 6, in preparation). Elsewhere in the Caribbean, small satellite cays associated with major islands are treated with the larger islands with which they are associated (e.g. the near-shore cays off the northern and southern coasts of Cuba, see Chapter 3, this volume).

All amphibians on all Caribbean islands are anurans.

II. The Colombian Caribbean islands

A. The islands

The archipelago of Colombian islands lies in the Los Cayos Basin atop a submarine platform slanting north-northeast in the southwestern Caribbean Sea (Vargas 2004; Baine *et al.* 2007). It consists of two large islands, San Andrés (also called St Andrews Island) and Providencia (also called Old Providence),* flanked by a smaller Santa Catalina Island, as well as by a number of tiny coral islands (Fig. 4.2A to 4.2E and see below) scattered along a stretch of sea of about

* Providencia should not be confused either with the island of New Providence in the Bahamas or Providenciales Island in the Turks and Caicos Islands, both of which are treated in Chapter 2 (this issue).

Table 4.1 Characteristics of the Colombian islands of the southwestern Caribbean Sea. Where there are several values listed, they indicate separate estimates by different authors. Sources of data: Dunn (1945); Pagnacco and Radelli (1962); Corn and Dalby (1973); Defense Mapping Agency Hydrographic/Topographic Center (1976, 1992); National Geospatial-Intelligence Agency (1993); Flórez and Etter (2003); Vargas (2004); Ruiz *et al.* (2005); Miralles (2006); Baine *et al.* (2007); Geister and Diaz (2007); Milliman (2007); Aguilera-Díaz (2010); Calderón-Espinosa and Barragán Forero (2011); Lasso *et al.* (2015); CORALINA (2019), and when all else failed, Wikipedia. Note that several different cays may have the same name, e.g. Low Cay. There are also a number of shallow banks and shoals on which cays potentially could form, e.g. Alice Shoal, Rosalind Bank, and Quitasueño Bank (not listed below).

CATEGORY/NAME	LENGTH × WIDTH	AREA (KM^2)	MAXIMUM ELEVATION (M)	LOCATION	TOPOGRAPHY, GEOLOGY AND USAGE BY HUMANS	TERRESTRIAL VEGETATION AND VERTEBRATES
LARGE ISLANDS						
San Andrés Island	12 × 3 km 13 × 3 km 13 × mean of 2.5 km —	22 24 25 26 27	55 84 87 100 —	12° 32' N; 81° 42' W 12°33' N; 81° 43' W 12°36' N; 81°42' W 12.55° N; 81.71° W 12.5450° N; 81.7076° W 200 km east of Nicaragua	There are three major morphological units: (1) coastal reefs, beaches, and mangroves, (2) emergent reef platform, and (3) central morphology of hills and escarpments. There is a pond and a freshwater lake. Now has a desalination plant. Also see text.	Original vegetation was a lush forest with numerous species that lose their leaves during the dry months. Now vegetation consists of native trees including mangroves, thickets of *Lantana* and other species, as well as coconut palms and breadfruit trees; there are pastures and farmlands. Three species of snakes and 7 of lizards. Brown rats are a problem. 179 spp. of birds, including temporary migratory ones. Also see text.
Providencia	7 × 4 km	15 17 18 —	350 360 363 550	13° 19' to 13° 24' N and 81° 21' to 81° 24' W 250 km E of Nicaraguan coast; 87 km N-NW of San Andrés	Mountainous and very steep except for a narrow coast of relatively level land suitable for agriculture.	1 turtle, 2 snakes, 6 lizards, 10 birds and a 'few' mammals. Also see text.
Tierra Bomba Island (Isla de Tierra Bomba)	7 × 5.8 km	20	20	10° 21'11" N; 75° 34'12" W 2 km SW from the city of Cartagena de Indias	Mostly in hands of private owners with over 9,000 inhabitants, it is also an important touristic attraction.	—

SMALL ISLANDS						
Santa Catalina	—	—	133	13° 22' 48" N; 82° 22' 19" W 13° 23' 17.99" N; 81° 22' 25.00" W 13.3875° N; 81.3757° W 13.39° N; 81.38° W	Rocky island. Numerous gullies. Satellite islet north of Providencia to which it was once connected but now separated from it by 150 m of very shallow water; a wooden bridge across the channel maintains an artificial connection.	See text.
Albuquerque Cays (Cayos de Albuquerque)	—	—	—	12° 10' N; 81° 51' W 12° 11.34' N; 81° 51,84' W 12° 33' N; 81° 43' W	Has two cays, North Cay and South Cay, which is used as a base by fishermen.	Both cays vegetated by palm trees.
Cayo del Norte	250 × 125 m	2	2	12° 10' N 81° 50' W, Lighthouse present.	Has a military base and fishermen's shacks. Has beach rock. Rich, loamy soil.	Heavily wooded with palms, figs, and shrubs. Boobies, terns, migratory birds, and a species of hummingbird present.
Cayo del Sur	180 × 70 m	1	—	A few hundred meters S of Cayo del Norte	Used by fishermen. Has beach rock. Rich, loamy soil.	Heavily wooded with palms, figs, shrubs, and mangroves. Boobies, terns, migratory birds, and a species of hummingbird present.
Bajo Nuevo Bank	—	—	—	15° 33' N; 78° 33' W	Some rocky heads awash and sand ridges barely awash.	—
Low Cay	300 × ? 40 m	0.01	2	15° 50' N; 79° N 15° 32' N; 78° 39' W	Sand, shingle, and driftwood	Barren.
Cayo Crab (Cayo Congrejos)	—	0.02	—	13.3667° N; 81.3500° W East of Providencia	Rocky volcanic islet.	Vegetated with trees. A species of lizard present.
Cayo Rocoso (Rocky Cay)	—	—	—	12° 32' 33" N; 81° 42' 07" W Just E of San Andrés	Rocky islet. Visited by tourists; can be reached by walking.	Palm trees present.
Courtown Atoll (Eastsoutheast Cays)	—	—	—	12° 24' N; 81° 28' W 22 km ESE of San Andrés.	Two small cays and a tiny sandspit occur on the atoll.	—
East Cay + Sandy Cay[1]	800 m × ?	—	*ca.* 2	—	Visited by fishermen and tourists; few wooden huts. Has beach rock.	Palm trees, bushes and grasses; mangroves.
Cayo Bolívar (West Cay)	—	—	*ca.* 2	12.4206° N; 81.4769° W 25 km SE of San Andrés	Has a lighthouse and a post for the Colombian navy. Two concrete buildings.	Palm trees and bushes; mangroves.
Grasey Cay	—	—	—	—	—	—

(Continued)

Table 4.1. (Continued)

CATEGORY/NAME	LENGTH × WIDTH	AREA (KM^2)	MAXIMUM ELEVATION (M)	LOCATION	TOPOGRAPHY, GEOLOGY AND USAGE BY HUMANS	TERRESTRIAL VEGETATION AND VERTEBRATES
Grunt Cay	—	—	—	—	—	—
Haynes (Haines) Cay; also called Cayo Córdoba	—	—	—	12° 33' 4.00" N; 81° 41' 22.99" W	Formed by Sangamon coral limestone covered by sandy Holocene storm deposits. Facilities for day-tourists; cruise ships dock.	Vegetated; palm trees.
Johnny Cay (Cayo Johnny; Cayo Sucre)	—	—	—	12.6000° N; 81.6897° W 1.5 km ENE of San Andrés	Sandy islet on the lagoon terrace formed after Holocene rise in sea level by accumulation of coral shingle and sand during storms. Destination for day-tours for tourists to enjoy the beach and reef.	Covered by palm trees.
Low Cay	—	—	low	13° 32' N; 81° 21' W *ca*. 10 km north of Providencia	—	Barren.
New Shoal (also called Bajo Nuevo and Isla Petrel) Low Cay	340 × 40 m	—	2	15° 53' N; 78° 38' W 15° 51' N; 78° 38' W	Has a lighthouse.	Grass-covered.
Roncador Cay	400 m	—	Low	13° 32' N; 80° 3' W 150 km east-northeast of Providencia Island	Formed of coarse coral debris thrown up by hurricane or tsunami. Previous commercial exploitation of guano. Wall built of coral blocks. There are two barren sandspits, each 15 m in diameter.	Sparsely vegetated; palm trees. A species of lizard present.
Rose Cay (El Acuario)	—	—	Low	12° 33' 00" N; 81° 41' 00" W 12.55° N; 81.683333° W	A small sandspit formed on the lagoon terrace by sediment deposited by converging wave trains refracted round Haine Cay; beachrock is conspicuous along its shore. Contains facilities for touristic day-visits.	Sparse vegetation of shrubs (*Tournefortia*) and creeping vines on part of the island.

Santandar Cay (also called Cotton Cay)	—	—	—	12.58° N; 81.69° W < 1 km S of San Andrés	Islet is a patch of the coral *Acropora cervicornis* that rose above the surface.	—
Serrana Bank	—	—	—	14° 24' N; 80° 16' W	Contains 6 small islets that were commercially exploited for guano.	—
East Cay	80 × 40 m	—	1	—	Covered with coral and shell rubble.	Sparse vegetation.
South Cay	150 × 25 m	—	—	—	Coral Rubble and sand; beachrock.	Sparsely vegetated with *Tournefortia, Ipomoea*, and *Euphorbia*).
Little Cay	< 100 m diam.	—	—	—	Fine sand and rubble.	Covered by bushes and shrubs. Numerous nesting seabirds.
Narrow Cay	—	—	> 10	—	Coral rubble.	No vegetation.
Serrana (Southwest) Cay	500 × 200 m	—	> 10 (dunes)	14° 16' N; 80° 24' W	Sandy Cay. Dilapidated huts. Has a lighthouse and outpost of the Colombian Navy. Formerly exploited for guano.	Grass, brushwood; creeping vines, shrubs, and a few palms. Marine bird resting place. Many nesting seabirds.
North Cay	270 × 135 m (later reduced to 150 × 80 m)	—	Low	14° 28' N; 80° 17' W	Sand and rubble	Sparse *Tournefortia* shrubs and creeping vines. Many seabirds (Fregatidae and Sulidae) nesting.
Little Cay	< 100 m × ?	—	Low	—	—	—
Serranilla Bank	40 × 32 km	—	—	15° 50' N; 79° 50' W 15° 55' N; 78° 33' W 130 km NNE of Providencia There are two breakers awash; one is 0.6 m high.	Mostly submerged but has 2 rocks and 3 small sand and shingle cays, Serranilla Cay, East Cay, and Beacon Cay.	Sparse vegetation of bushes and trees.
Beacon Cay	900 m × ?	—	2.4	—	Sand and shingle cay. Has a lighthouse and is used as a post by the Colombian Navy and Coast Guard.	Densely covered by small weeds (*Portulaca capilosa*) and there are numerous coconut trees. Seabirds are not frequent.
East Cay	—	—	—	15° 52' N; 79° 44' W	—	Covered with bushes.
West Breaker	—	—	—	—	—	—
Middle Cay	—	—	—	—	—	—

(Continued)

Table 4.1. (Continued)

CATEGORY/NAME	LENGTH × WIDTH	AREA (KM^2)	MAXIMUM ELEVATION (M)	LOCATION	TOPOGRAPHY, GEOLOGY AND USAGE BY HUMANS	TERRESTRIAL VEGETATION AND VERTEBRATES
Three Brothers Cay (Cayo Tres Hermanos)	—	—	—	East of Providencia	Emergent volcanic islet.	—
Isla Tortuguilla	*ca.* 0.6 × 0.3 km	0.1	2	09° 01' 50" N; 76° 20' 40" W	—	—
Isla Fuerte	2.45 × 1.45 km	2.9	12	9° 23' N; 76° 11' W	—	Presence of mangroves and trees corresponding to the families Moraceae, Sterculiaceae, and Bombaceae.
Archipelago of San Bernardo	—	—	2	9° 45' 28" N; 75° 46' 24" W	With over 1,300 inhabitants.	Within the Parque Nacional Natural Islas Corales del Rosario y San Bernardo. Secondary vegetation of Lowland Dry Forest dominated by mangroves and cultures of coconuts.
Isla Boquerón	—	—	—	—	—	—
Isla Palma	—	—	—	—	—	—
Isla Panda	—	—	—	—	—	—
Isla Mangle	—	—	—	—	—	—
Isla Ceycén	—	—	—	—	—	—
Isla Cabruna	—	—	—	—	—	—
Isla Tintipán	—	—	—	—	—	—
Isla Maravilla	—	—	—	—	—	—
Isla Múcura	—	—	—	—	—	—
Santa Cruz del Islote	—	—	—	—	—	—
Coral Islands of Rosario	—	0.2	1	10° 10' 30"N; 75° 45' 00" W 10° 10' 55"N; 75° 44' 06" W 10° 10' 48"N; 75° 45' 06" W 10° 11' 00"N; 75° 46' 00" W 10° 14' 05"N; 75° 44' 19" W	—	Within the Parque Nacional Natural Islas Corales del Rosario y San Bernardo. Secondary vegetation of lowland Dry Forest, mangroves, and coconuts.

Isla Grande	—	—	—	—	—	—
Isla Marina	—	—	—	—	—	—
Isla de Roberto	—	—	—	—	—	—
Isla del Tesoro	—	—	—	—	—	—
Isla Caribarú	—	—	—	—	—	—
Isla Arena	—	—	—	—	—	—
Isla Rosario	—	—	—	—	—	—
Isla Pirata	—	—	—	—	—	—
Isla Fiesta	—	—	—	—	—	—
Isla Cagua	—	—	—	—	—	—
Isla Macaví	—	—	—	—	—	—
Isla San Quintín	—	—	—	—	—	—
Isla Pavitos	—	—	—	—	—	—
Isla El Peñón	—	—	—	—	—	—
Isla Majayura	—	—	—	—	—	—
Isla María del Mar	—	—	—	—	—	—
Isla Pajarales	—	—	—	—	—	—
Isla San Martín	—	—	—	—	—	—
Isla Gigi	—	—	—	—	—	—
Isla Coralina	—	—	—	—	—	—
Isla Pescadito	—	—	—	—	—	—
Isla Tijereto	—	—	—	—	—	—
Isla Rosa	—	—	—	—	—	—
Isla El Morro	*ca.* 0.35 × 0.10 km	—	15	11° 15' 00" N; 74° 13' 48" W	—	—
Isla de la Aguja	*ca.* 1.6 × 0.4 km	—	125	11° 19' 13" N; 74° 11' 49" W	—	—

[1] East Cay and Sand Cay were once two separate cays, but they have now coalesced. A former Middle Cay has disappeared, but before that happened it was visited by fishermen who built shacks on it.

Figure 4.2. Colombian Caribbean islands photographed in February 2020: (**A**) Pig Pond Lagoon, San Andrés Island; (**B**) Constructed northern portion of San Andrés Island (photograph taken from Johnny Cay); (**C**) Johnny Cay (photograph taken from the northern portion of San Andrés Island); (**D**) Providencia Island (photograph taken from Santa Catalina Island); and (**E**) Northeastern side of Santa Catalina Island. Photographs **A**, **C** by S. A. Pérez Villegas; **B** by D. M. Galindo-Uribe, and **D-E** by L. L. Navarrete González.

500 km in length (Fig. 4.1) and occupying a collective land area of 57 km^2 (Vargas 2004; Baine *et al.* 2007). Whereas San Andrés is an elevated Tertiary atoll, Providencia was formed by a now extinct Miocene volcano (Geister 1992). Their geotectonic evolution originated from two eroded volcanic cones associated with the Nicaraguan Rise and part of the Caribbean Plate (Geister 1992; Álvarez-Gutiérrez *et al.* 2014). This Rise was formed by eruption of submarine lavas that subsequently tilted and lifted; this elevation led to erosion followed by sedimentation of volcanic conglomerates and deposition of tuffs. Volcanic activity was then resumed, and adventive cones were formed that extruded more lava. Renewed tectonic activity elevated blocks vertically to form the basement of the islands (Vargas 2004).

San Andrés has a volcanic base topped by a thick layer of hard limestone. Moreover, San Andrés and Providencia are thought to have been uplifted independently and to never have had a land bridge either between them, or from them to the mainland (Smith *et al.* 1989; Miralles 2006). The Rise accounts for about 10% of the area of the Caribbean Sea and its maritime borders abut those of Jamaica, the Cayman Islands, Honduras, Nicaragua, Costa Rica, and Panama (Baine *et al.* 2007). The latter three countries form a wide cul-de-sac which semi-encloses the Rise (Marriaga 2009).

Politically, these islands constitute the Department of San Andrés, Providencia, and Santa Catalina, the only one of Colombia's 32 Departments to consist solely of islands. It is also the northernmost and smallest (in terrestrial area) Department of the country, and except for Bogotá within the Department of Cundinamarca, is the area with the highest density of humans, with an estimated population of nearly 80,000 inhabitants (Departamento Administrativo Nacional de Estadística 2007). It is Colombia's only English-speaking (Creole) Department (Baine *et al.* 2007; Sánchez-Jabba 2012).

San Andrés is the most densely populated oceanic island in the Caribbean (Téllez Ramirez and Llurdees Coit 2014) with about 75,000 inhabitants mostly located in the urban center at the northeastern end of the island. In contrast, Providencia and Santa Catalina are among the least environmentally and culturally degraded areas in Colombia, with a population of about 5,000 inhabitants (Baine *et al.* 2007; Departamento Administrativo Nacional de Estadística 2007; Sánchez-Jabba 2012; Ortiz Royero *et al.* 2015). The main economic activities of these islands are tourism, commerce, government employment, traditional agriculture, artisanal fishing, and home-based industries (Baine *et al.* 2007). San Andrés is one of the most important tourist destinations in the Caribbean Sea, welcoming approximately 500,000 tourists per year. Commerce has been favoured since the creation of San Andrés as a free port in 1953 and agriculture there is now minimal.

Coral reefs formed on submerged parts of the volcanic platform and have contributed in several different ways to the geomorphology of the area. Many remained submerged, such as Rosalind Bank (16° 26' N; 80° 30.60' W) and Alice Shoal (16° 5' N; 79° 22' W), but others spawned independent sandy or coralline cays, atolls, or emergent banks. These are exceptionally varied in their cover, ranging from barren to densely clothed in vegetation including trees and shrubs (Table 4.1). A few of them support tourism, or have a military installation, which inevitably subjects them to frequent disturbance compared to the islands that lack permanent residents and only receive occasional fishermen or other itinerant visitors. These small islands have a total land area of about 8 km^2, collectively much less area than either of the two largest islands alone.

Finally, some of the reefs became integrated with the volcanic components of the larger islands. Providence Island, for instance, consists of a ridge of massive lavas of andesitic basalts and basalts that form steep, rugged mountains flanked by gentler slopes of fluidal lavas and conglomerates containing volcanic elements and interbedded tuffs, and there are a few adventive cones with andesitic lavas; around the coast are areas of emergent coralline limestone and there are sandy beaches (Pagnacco and Radelli 1962).

1. Climate

The climate of this tropical archipelago is hot and humid (Marriaga 2009) with a pronounced wet and dry season (Ruiz *et al.* 2005). The climate is influenced by the island's physiographic characteristics as well as by the action of the trade winds that blow from the northeast-east throughout the year. San Andrés can serve as an example: its temperatures range from 17°C to 35°C, with average monthly values of 27.5°C on the coast and 26.5°C in the hilly portions of the island; the mean variation throughout the year is less than 2°C (Vargas 2004; Marriaga 2009). The strongest gusts of wind are observed between August and November, with values ranging between 45 km/h and 70 km/h, but values exceeding 120 km/h have been recorded during hurricanes (Vargas 2004; Ortiz Royero *et al.* 2015).

The archipelago is located within the Caribbean hurricane belt, with the storms approaching from the south-east. In a list of the 245 hurricanes that were reliably reported for the Caribbean Sea from 1494 to 1800, at least five (in 1528, 1531, 1605, and possibly 1681 and 1708) were deemed to have passed over the Colombian Caribbean islands (Millás 1968). More recent registrations occurred in 1818, 1876, 1877, 1906, 1932, 1935, 1940, 1961, 1970, 1971, 1984, 1988, 1996, 1999, and in 2005 (Vargas 2004; Marriaga 2009; Ortiz Royero *et al.* 2015). On 3 September 2007, the eye of Category-5 Hurricane Felix passed over Serrana Bank, and on 15 and 16 November 2020, Hurricane Iota passed over San Andrés, Providencia, and Santa Catalina, causing unprecedented damage to the islands and damaging up to 98% of the infrastructure of Providencia. San Andrés has been affected by at least one hurricane every 10 years for the past 50 years and has been identified as a zone of high vulnerability during hurricanes (Ortiz Royero *et al.* 2015).

Mean annual rainfall on the Island is 1,850 mm and is characterized by a period of low rains (dry or windy season) between December to April and a period of high rains (wet or rainy season) between May and November (Corn and Dalby 1973; Marriaga 2009; Ortiz Royero *et al.* 2015). Between January and April 8% of the annual total rainfall occurs, with 75% falling between June and November. Severe droughts also have been recorded (Forero-Medina *et al.* 2007).

The average annual relative humidity at San Andrés is 81% and the average monthly records vary between 78% and 84% throughout the year, the lowest values being between January and April and the highest between June and October. The average annual potential evapotranspiration at San Andrés is 1,740 mm and the average monthly values vary from a maximum of 177 mm in March and a minimum of 116 mm in October. The average annual solar brightness is 2,650 hours, which represents an average annual sunshine of 61% (Vargas 2004; Ortiz Royero *et al.* 2015).

To summarize, these islands present three different seasons during the year (Marriaga 2009): first, a dry season or summer, from December to April, characterized by strong winds from the northern to northeastern sector and scarce rainfall; second, a period called 'transition', which is between May and July and is characterized by trade winds of low intensity, which increase during July, a time known as the 'Veranillo de San Juan' (midsummer drought); third, a wet or winter season, occurring between August and November, which is characterized by a regime of weak winds of varying orientation and an increased rainfall.

2. Hydrology

There are no permanent surface streams on San Andrés, although the karst layer has several freshwater ponds without definite outlets and there are several small, but well-developed caverns (Corn and Dalby 1973). Occasionally, during rainy periods, small drainages are formed and used as one of the sources of drinking water on the island. Smaller pipes less than one kilometre long that originate in the hills in the central part of the island and drain into the swampy area of the east, recharge deep aquifers in some parts of the island during the rainy season (Vargas 2004). Freshwater ponds on the island vary in size from 0.33 ha to 1.85 ha; only a few of them are permanent, whereas others form only during the rainy season (Forero-Medina *et al.* 2007).

Providencia is the only island in the archipelago that has small streams, locally known as 'gullies', which may become intermittent during the dry season, but there are no freshwater ponds (Corn and Dalby 1973; Lasso *et al.* 2015). Many of the streams do not flow into the sea but infiltrate coastal depressions. Those on the western slope are of greater length, persistence, and flow, and have higher humidity and more surrounding vegetation than do those of the eastern slope (CORALINA-INVEMAR 2012; Cortés-Guzmán and Linares 2016).

3. Vegetation

In addition to diverse marine habitats, the large islands have various terrestrial coastal ecosystems including forests, wetlands, swamps, remnants of mangroves and (on the beaches) pioneer vegetation (Baine *et al.* 2007; Sánchez-Jabba 2012; Garzón 2016; Murcia *et al.* 2016).

San Andrés and Providencia differ markedly in vegetation, topography, and superficial geology (Corn and Dalby 1973). The natural habitats on San Andrés are mainly tropical dry forest, xerophytic habitats, freshwater ponds, and mangrove forests (Forero-Medina *et al.* 2007). During the seventeenth and eighteenth centuries, there apparently were extensive natural stands of 'cedar', possibly *Cedrela odorata* (Meliaceae), which were decimated early in the island's colonization. By the twentieth century, most of the island was devoted to the growing of coconut palms (Corn and Dalby 1973).

In contrast, other than seashore vegetation, coastal mangroves, and some bogs, the native vegetation of Providencia is Dry Tropical Forest (Pagnacco and Radelli 1962; Ruiz *et al.* 2005).

Although most of the human population is concentrated on the flatter, coastal areas of the island, selective logging has taken place generally and may have been responsible for the present rarity of cedar, dominance of Myrtaceae, lower representation of Fabaceae, and differences in the ranking of some other families than in remnant stands of mature forest on mainland Central America (Gillespie *et al.* 2000; Ruiz *et al.* 2005).

In the higher regions on Providencia, bracken and tree ferns are common, and in the steeper areas, there are dense thickets of fig (*Fourcroya* sp.), calabash (*Crescentia* sp.), and other trees. In the lower valleys, introduced guinea grass (*Panicum maximum*) serves for cattle-ranching. The native vegetation can recover relatively rapidly after agriculture and cattle-ranching cease; Ruiz *et al.* (2005) showed that succession leading to restoration of Dry Tropical Forest proceeded rapidly during the first 56 years following abandonment of farmland or pastures. Currently, coconut palms are found only infrequently, and there are few agricultural areas. Extensive mangrove swamps occupy much of the northeastern shore (Corn and Dalby 1973).

4. Historical background

The nature of the influence of humans on the terrestrial ecology of the archipelago changed with successive occupancy by different nationalities and cultures (Aguilera-Díaz 2010; Sánchez-Jabba 2012; Salas Betin 2015; Dagua *et al.* 2018; James Cruz and Soler Caicedo 2018). Known history began in Pre-Columbian times when the Miskitos, indigenous people from coastal Central America, came to harvest the islands' fine woods and marine resources. In the sixteenth century, the Spanish paid little attention to the islands due to the dangers of stranding on multiple reefs. In 1631, however, a company of Puritan adventurers arrived, and the presence of the English quickly developed into an open confrontation with the Spanish because it threatened their transit of vessels between the Greater Antilles and the coasts of Central America. Subsequent control of the islands shifted between the English, Spanish, and Jamaican pirates for more than a century (Milliman 1969).

Extensive settlement of the islands occurred in the seventeenth century and was accompanied by the growing of cotton, sugarcane, tobacco, and other utilitarian crops, supported by slave labour. Providencia, with its freshwater resources and fertile soils, quickly became the agricultural centre, while San Andrés was abandoned (except for occasional visits to collect wood) because of shortages of water and the difficulty of military defence. The official transfer of territorial sovereignty of the islands to Spain via the Treaty of Versailles in 1786, did not affect the island's inhabitants, who remained deeply linked to their English tradition, and the economy continued to focus on agriculture and fishing.

In 1821, the archipelago became part of Gran Colombia and, once again, its inhabitants continued to preserve their English culture. At the end of the first half of the nineteenth century the abolition of slavery was signed, and the island's economy shifted to the monoculture of coconuts until the first third of the twentieth century, when constant droughts and the proliferation of rats damaged the crops and led to the abandonment of the islands. In 1912, the archipelago segregated from the Bolívar Department and achieved its own administrative autonomy.

Colombia declared San Andrés a free port in 1953, which brought migration and a rapid commercial and touristic boom. The social transformation of the islands from then on, particularly in San Andrés, was negatively reflected in an excess of garbage, wastewater, disordered urbanization, unequal opportunities and isolation of the descendants of enslaved Africans. It was not until 1993 that a local environmental authority, Corporación para el Desarrollo Sostenible del Archipiélago de San Andrés, Providencia y Santa Catalina (CORALINA), was created. The vision from then on was focused on natural and cultural protection and on sustainable tourism with the creation of all its protected areas in combination with immigration and controls of land-use. Petroleum and natural gas are potential sources of future industrial development (Jiménez Castilla and Huertas Cardozo 2014).

There has been a longstanding dispute between Colombia and Nicaragua over the islands of the southwestern Caribbean. Nicaragua claimed that since the islands were on the Nicaraguan continental shelf they should be under Nicaraguan sovereignty. The matter finally was settled by the International Court of Justice in 2007 by awarding all of the islands covered in the present chapter to Colombia.

B. The amphibians

The archipelago has three amphibian species: two recently detected exotic species on the island of San Andrés, *Eleutherodactylus planirostris* and *Leptodactylus fragilis* (Cubillos-Abrahams and Montes-Correa, 2021; Cubillos-Abrahams *et al.*, 2021), and a single native species, *Leptodactylus insularum* (Fig. 4.3), which, although it is a frog, its local name is 'toad'. It occurs both on San Andrés and Providencia (Dunn 1944, 1945; Dunn and Saxe 1950; Cochran and Goin 1970; García-Hansen 1998; Forero-Medina *et al.* 2007; Quiceno Mesa *et al.* 2009; Heyer and de Sá 2011; Romero and Lynch 2012; Caicedo-Portilla 2014). In addition, Dunn and Saxe (1950) stated that *L. insularum* (as *L. bolivianus*) had been reported to occur on Santa Catalina.

On mainland Central America *Leptodactylus insularum* also occurs on the Pacific versant from Costa Rica to lowland Panama and in the Caribbean drainages of Colombia and Venezuela, as well as on Trinidad and Tobago at elevations below 600 m asl (Heyer and Heyer 2013; de Sá *et al.* 2014; Señaris *et al.* 2018). Although Savage (2002) included Nicaragua in the distribution of *L. insularum* (as *L. bolivianus*), apparently no voucher specimens are available to support this assertion (Sunyer and Köhler 2010).

There may be some morphological differences between the populations of *Leptodactylus insularum* on San Andrés and Providencia compared to the Central American mainland. Cochran and Goin (1970) noted that specimens from Providencia tend to have a slightly wider head and a longer femur than those from the mainland, although there are no differences in colouration (Fig 4.3).

Figure 4.3. *Leptodactylus insularum* from: (**A**) Old Town, Providencia Island, Colombia (photograph taken in February 2020) and (**B**) Barrancabermeja, mainland Colombia (photograph taken in November 2019). Photograph **A** by L. L. Navarrete González and **B** by J. Sunyer.

Adult males reach up to 170 mm and adult females up to 154 mm in length. They are primarily nocturnal, feed on arthropods and small vertebrates, and usually are associated with permanent bodies of water (which, however, are absent on San Andrés). Males call at night, producing a characteristic 'wuuuuup' that can be heard from temporary ponds and marshes from which females may lay up to 2,000 eggs in foam nests in shallow waters, often hidden in vegetation (Savage 2002; Romero and Lynch 2012; Señaris *et al.* 2018).

In San Andrés, this species has been taken under piles of coconut husks (Dunn and Saxe 1950) and has been recorded from the 'mangrove of the Bahía Hooker–Bahía Honda, in the NE portion of the island', where the salinity varies between 1.0 ppm and 43.3 ppm, and temperature varies between 22.5°C and 28.0°C (García-Hansen 1998). It is also known from near Punta Paraíso (Heyer and de Sá 2011) and near Laguna de Big Pond (12°32.91' N; 81°43.17' W; elevation 46 m asl) (Caicedo-Portilla 2014).

In Providencia, this frog has been recorded from 'Agua Mansa' and 'reservoir, 80 m' (Heyer and de Sá 2011), 'Provision Ground Brook, the most southern stream of Providencia' (Dunn and Saxe 1950), and in all five watersheds that surround The Peak, a Natural Regional Park (Quiceno Mesa *et al.* 2009).

Its previous complex taxonomic history makes it difficult to assess this species' population status, but on the basis of the large number of localities where it has been recorded, and its ability to inhabit brackish water, it likely is best included in the category of LC (Least Concern) in the IUCN Red List (Solís *et al.* 2008; Heyer and Heyer 2013). Lack of monitoring makes it difficult to assess any changes in population size.

The apparent absence of this species, or indeed any amphibians, on the smaller islands could be partly an artefact of lack of data rather than of animals. However, herpetofaunal surveys have been conducted on some of them, as evidenced by reports of lizards, even on some of the very small cays (Table 4.1). The regular transport of goods to the few islands with military personnel or tourists would provide opportunities for inadvertent transport of frogs; yet none are recorded from those islands. Some of the islands receive a very high traffic of day tourists or even the docking of cruise ships, and the supply of materials from external sources to cater for such a large number of people may facilitate accidental introductions. Consequently, is the lack of amphibians on these very small islands because they do not reach a minimum insular size for self-sustaining populations, because of the paucity of standing freshwater, or because they constitute small physical targets for successful instances of overwater transport? The answer currently is unclear, and the remainder of the discussion is devoted to the three larger islands that do harbour amphibians.

C. Threats

1. Natural phenomena

The archipelago is located within the hurricane belt of the Caribbean (see section II. B. *Climate*). Hurricanes cause destruction of habitat by flooding, landslides, erosion of sandy shorelines, and erosive undercutting of rocky ones, and may even result in complete wash-over on small islands (Flood and Heatwole 1986; Geister and Díaz 2007); such events have occurred at the Colombian islands and some sandy cays have even disappeared (Table 4.1). On the other hand, hurricanes can have a positive effect. Blocks of coral wrested from the reef are cast as far as 100 m inland on areas as much as two metres above normal sea level (Geister and Díaz 2007). These offer shelter for small animals and would be especially important on small sandy islands where other cover is unavailable.

Tsunamis can have similar effects as hurricanes but are much less common in the Caribbean (Geister and Díaz 2007). Furthermore, severe droughts have been recorded from the archipelago (Forero-Medina *et al.* 2007). Their effects, if any, on *Leptodactylus insularum* are not known.

2. *Anthropogenic impacts*

Most threats to the terrestrial ecosystems result from the exploitation of natural resources by humans. San Andrés contains large urban infrastructure, poorly planned development, intense pressure on already depleted natural resources and ecosystems, political and economic marginalization, and the resulting loss of self-reliance and cultural identity (Baine *et al.* 2007; Coca-Domínguez *et al.* 2019; also see related works at http://www.invemar.org.co/publicaciones and http://www.coralina.gov.co/nuevositio/en/). It also has suffered from pollution, erosion, and the effects of tourism, which collectively have overloaded its ecosystem's capacity. Commerce has been favoured since it became a free port in 1953 and agriculture now is minimal.

In contrast, Providencia and Santa Catalina, collectively with a population of only slightly more than 5,000 inhabitants, are among the least environmentally and culturally degraded areas in Colombia (Baine *et al.* 2007; Sánchez-Jabba 2012; Ortiz Royero *et al.* 2015); the level of urbanization that characterizes San Andrés has not yet occurred on these two islands. Instead there is great potential for sustainable development (Sánchez-Jabba 2012). However, natural ecosystems are threatened by the sedimentation resulting from grazing by cattle on montane slopes and by slash-and-burn for agricultural purposes.

All three islands suffer from deforestation and inadequate disposal of wastewater and solid waste. These improperly conducted human activities are aggravated by social problems, such as poverty, poor land-management, scarce sources of freshwater, and the filling in of wetland areas. Several of these issues, however, are being addressed since the creation of the Biosphere Reserve Seaflower in 2000 (Botero 2001; Vargas 2004; Santos-Martínez *et al.* 2009; Lasso *et al.* 2015; Salas Betin 2015; see http://www.coralina.gov.co/nuevositio/en/).

In San Andrés, humans have recently introduced some predators, such as spectacled caimans (*Caiman crocodilus*), tegus (*Tupinambis teguixin*), and rats (*Rattus norvegicus*), as well as domestic cats (*Felis catus*) (Forero-Medina *et al.* 2007; Caicedo-Portilla and Lynch 2015).

D. Conservation

Six protected areas have been established in the archipelago (Baine *et al.* 2007; Santos-Martínez *et al.* 2009; Aguilera-Díaz 2010; CORALINA-INVEMAR 2012; Jiménez Castilla and Huertas Cardozo 2014; see http://www.coralina.gov.co/nuevositio/en/). Most of these areas were for the protection of marine resources, but in two of them *Leptodactylus insularum* has been recorded. The Peak Regional Park, which was set aside in 2007 in the Lowland Dry Forest of the central highlands of Providencia and includes portions of the forested higher elevations of the island where there are many springs, and therefore favourable breeding habitat for *L. insularum* that would constitute a refuge from further destruction of its habitat elsewhere. Moreover, the Old Point Regional Mangrove Park (mangrove of the Bahía Hooker–Bahía Honda) was declared in 2001 and is situated in the northeastern portion of San Andrés. There are also additional, smaller protected areas, such as the National Natural Park Old Providence McBean Lagoon, established in 1996 and which includes a cay with a small steep hill 150 m in elevation (Iron Wood Hill). The Regional Park Johnny Cay, established in 2001 and Colombia's most touristic cay, coincidentally might benefit this anuran species, despite not having natural springs.

III. The Nicaraguan Caribbean islands

A. The Islands

The Nicaraguan Caribbean islands are a series of small islands and cays arranged in a north-to-south direction: Miskito Cays constitute an archipelago of 76 cays and reefs, with Cayo Mayor being the largest of the islands (~37 km^2), Man-o-War Cays (including Cayo King and Cayo King Pequeño; collectively ~0.01 km^2 in area, also known as 'Cayos Guerrero' in Spanish (not to be confused with Man O' War Cay in the Bahamian Archipelago), Pearl Cays (~0.59 km^2), and Great and Little Corn Islands (collectively 12.9 km^2). These are included in the Nicaraguan departments Atlántico Norte and Atlántico Sur, formerly collectively known as the Department of Zelaya.

The Reserva Biológica Cayos Miskitos was declared in 1991 and UNESCO declared the Cayos Miskitos y su Franja Costera Inmediata as a RAMSAR site in 2001. Currently, the official category within the Nicaraguan protected areas is Reserva Biológica Marina Cayos Miskitos y Franja Costera Inmediata. These contain a few fisherman and rudimentary military wooden houses that are built above the shallow waters that surround Miskito Great Cay and are used only during the six months when fishing activities are not prohibited. The Man-o-War Cays are rarely visited.

The Pearl Cays consist of a group of 36 cays and reefs, where the Refugio de Vida Silvestre Cayos Perlas was declared in 2010. Currently, some of these cays are privately owned and some are used as a scenario for a foreign reality show. In general, although some of these Nicaraguan islands and cays harbour anurans (Table 4.2), neither the Man-o-War Cays nor the Corn Islands have protected areas, despite the high endemism of herpetofauna in the latter (Sunyer and Köhler 2010; Sunyer *et al.* 2013).

The Corn Islands (or Islas del Maíz) (Fig. 4.4) are two small islands approximately 70 km off the central Caribbean coast of Nicaragua in the Department Atlántico Sur. Great Corn Island has an area of about 10 km^2, with Mount Pleasant (113 m asl) being the highest point. Little Corn Island is located 15 km to the northeast and comprises 2.9 km^2, reaching a maximum height of 38 m asl. Both islands are made up of Tertiary (Miocene-Pliocene) basalts that protrude through late Tertiary carbonates, and present a low general topography (Ryan *et al.* 1998); approximately half of Great Corn Island lies beneath the 2-m contour line, and a sea-level rise of only 0.5 m would inundate about one-third of the island.

The Corn Islands (and by extension all other Nicaraguan Caribbean islands) correspond to tropical Lowland Moist Forest formations (Holdridge 1967), with an annual average air temperature between 27–28°C, an average relative humidity of 81%, and an average rainfall of 3.5 m per year. The rainy season is between May to December and there is a relatively dry season between

Table 4.2. Amphibian species known from the Nicaraguan islands in the Caribbean.

FAMILY AND *SPECIES*	GREAT CORN ISLAND	LITTLE CORN ISLAND	MISKITO GREAT CAY
BUFONIDAE			
Incilius valliceps	X	X	—
ELEUTHERODACTYLIDAE			
Eleutherodactylus planirostris	—	—	X
HYLIDAE			
Agalychnis callidryas	X	X	—
Dendropsophus phlebodes	X	X	—
Scinax staufferi	X	—	—
Tlalocohyla loquax	X	X	—
RANIDAE			
Lithobates miadis	—	X	—
TOTAL SPECIES	**5**	**5**	**1**

Figure 4.4. An adult male of the endemic *Lithobates miadis* from Little Corn Island, Nicaragua (photograph taken in June 2010). Photograph by J. Sunyer.

January and April, which is coincident with the strongest trade winds that blow steadily from the east-northeast with a steadiness factor of 90% (Roberts and Suhayda 1983; Ryan *et al.* 1998). The highest global radiation occurs in April and the lowest in December (López de la Fuente 1994), and the highest evaporation rates in the islands occur during the months of April and May and the lowest during November (Peralta-Williams 1991). Although the Nicaraguan Caribbean islands lie practically outside the hurricane belt (Peters 1929), tropical storms, depressions, and hurricanes (infrequently) impact them, mostly during the second half of the rainy season (Ryan *et al.* 1998).

Original inhabitants of the Corn Islands are afrodescendant and speak Creole. In 1927, and during the first biological expedition to the Corn Islands, the human population on Great Corn Island was estimated at 800 inhabitants (Peters 1929). In 2019, the population was about 9,500, mostly living on the coastline. This is by far the most populated Nicaraguan Caribbean island. Fishing, tourism, and home-based industries are the island's main activities and a significant portion of Mount Pleasant is devoted to agriculture.

B. The amphibians

There have been two major contributions to the species of anurans on the Corn Islands: one by Barbour and Loveridge (1929), who studied the material obtained by James Lee Peters and Edward Bangs in December 1927 and January 1928, and the other by Villa (1972a), who made a compilation of the local batrachofauna. There are several other contributions, but these each generally have dealt with only one species (Savage and Heyer 1967; Zweifel 1967; Duellman and Fouquette 1968; León 1969; Duellman 1970; Villa 1971). Books covering the entire Nicaraguan amphibian fauna and therefore mentioning species from the Corn Islands include those by Villa (1972b), Villa *et al.* (1988), Köhler (2001), Ruiz Pérez and Buitrago (2003), HerpetoNica (2015), and Robleto-Hernández *et al.* (2017). Finally, based on observations made in 1992, Villa (2015) recorded an exotic species, *Eleutherodactylus planirostris*, introduced into Cayo Mayor in the Reserva Biológica Marina Cayos Miskitos y Franja Costera Inmediata. There are no known records of amphibians from either the Pearl Cays or the Man-o-War Cays.

1. Incilius valliceps *(Wiegmann, 1833)*

This toad is extremely common on both Corn Islands and reaches very large sizes. Individuals have been observed active after brief rains, and found under logs, abandoned termite nests, and in trees up to 2.5 m above ground level (Barbour and Loveridge 1929; Villa 1971, 1972a). Barbour and Loveridge (1929) referred to this population as *Bufo valliceps valliceps* and Villa (1971, 1972a) as *B. valliceps*.

2. Tlalocohyla loquax *(Gaige and Stuart, 1934)*

This frog was recorded on both of the Corn Islands by Duellman (1970) and Villa (1972a), who remarked that the specimens exhibited reddish-brown dorsal surfaces with little or no melanophores – in contrast to those from the northern mountains of mainland Nicaragua. Duellman (1970) and Villa (1972a) referred to this population as *Hyla loquax*.

3. Dendropsophus phlebodes *(Stejneger, 1906)*

This frog also was recorded on both of the Corn Islands by Barbour and Loveridge (1929) and Villa (1972a) from emergent vegetation. Barbour and Loveridge (1929) referred to this population as *Hyla miotympanum* and Villa (1972a) as *H. phlebodes*. Solano-Flórez (2012), based on genetic evidence, tentatively designated the Corn Islands' populations of this species as *D. microcephalus*. It is unclear whether both of these species, or only one of them, occurs on these islands.

4. Scinax staufferi *(Cope, 1865)*

This species was recorded on Great Corn Island by León (1969) and Villa (1972a). It was observed on leaves about 20 to 50 cm above water, under a decaying termite nest inside an old house, under moist logs, and in crevices and holes in trees and posts (Villa 1972a). Both León (1969) and Villa (1972a) referred to this population as *Hyla staufferi staufferi*.

5. Agalychnis callidryas *(Cope, 1862)*

This distinctively coloured frog has been recorded on both of the Corn Islands by Barbour and Loveridge (1929), Savage and Heyer (1967), and Villa (1972a). Barbour and Loveridge (1929) noted a striking colour variation between specimens from Little Corn Island compared to those on mainland Panama (a single specimen from San Pablo, Panama), and Villa (1972a) noted the numerous bright yellow tubercles of a single specimen found on the forest floor on Great Corn Island in comparison with inland Nicaraguan populations that have a few small, white spots (not tubercles) on the dorsum. McCranie *et al.* (2019) noted the morphological distinctiveness of the Corn Island populations of this species based on observations in 2010 by JS of two individuals from Little Corn Island that were characterized by an exceptional number of bars on the flanks, a lateral general purplish colouration, the presence of whitish tubercles on the dorsum, and an extremely thin appearance, even by this species' standards. However, and despite the distinct morphological characteristics of the Corn Island populations, Solano-Flórez (2012) demonstrated a close genetic relationship of this population with other Nicaraguan populations of this species, which we believe should be addressed as *A. helenae* (type locality 'Nicaragua' Cope [1885]; also see McCranie *et al.* [2019]). This species is listed in appendix II of CITES.

6. Lithobates miadis *(Barbour and Loveridge, 1929)*

Barbour and Loveridge (1929) described *Rana miadis* from Little Corn Island on the basis of a single specimen; a nematode worm was found free in the body cavity of the holotype. The taxonomic identity of this species has been questioned by Zweifel (1967) who considered it to be conspecific with *L. forreri* (as *Rana pipiens*, name also used by Villa [1972a] to refer to this population). Genetic analyses are needed in order to establish the taxonomic identity of this insular population. This species is listed as Vulnerable (VU: D2) under the IUCN Red-List categories because it is known from only a single location. However, its population is considered stable.

Threats to this species include loss and degradation of habitat due to the increasing development of infrastructure related to the growing tourism industry around Little Corn Island.

7. Eleutherodactylus planirostris *(Cope, 1862)*

Eleutherodactylus planirostris occurs on Cuba and the Isla de Juventud; also on the Cayman Islands, Caicos Islands, Jamaica, and in the USA in Florida, Alabama, Georgia, Louisiana, Hawaii, and Guam (Hedges *et al.* 2004; Frost 2020). It has also been introduced into southern Mississippi and southeastern South Carolina, as well as into Hong Kong, the Philippines, Honduras (Isla de Guanaja), Surinam, Mexico (Yucatán, Veracruz), and Isla San Andrés, Colombia. In 1992 Villa (2015) saw large numbers of individuals in mangroves, under rotten logs, and in leaf litter at Miskito Great Cay, in the Reserva Biológica Marina Cayos Miskitos y Franja Costera Inmediata. However, there are no vouchers or photographic evidence for these records.

C. Threats

Threats to the amphibian fauna in the Nicaraguan Caribbean islands are identified as similar to those from the neighbouring islands mentioned in this chapter, and include: (1) natural phenomena such as hurricanes, tsunamis, and severe droughts; and (2) anthropogenic causes, mostly derived from over-exploitation of natural resources associated with poorly planned development, such as destruction and pollution of natural habitat, land-filling of wetland areas, introduction of exotic species and pathogens, and deliberate and accidental kills. The high endemism of herpetofauna in the Corn Islands (Sunyer and Köhler 2010; Sunyer *et al.* 2013) requires terrestrial protected areas in order to guarantee the conservation of these islands' herpetofaunal populations and, in general, all amphibian populations should be monitored by regular surveys to closely monitor the status of all species. Recently, the Nicaraguan government approved a law that declares and defines Great and Little Corn Islands and Blowing Rock as terrestrial and marine protected areas.

IV. The Cayman Islands (British Overseas Territory)

Echternacht *et al.* (2011) reviewed the conservation issues facing the herpetofauna of the Cayman Islands and the present chapter summarizes their findings as they relate specifically to amphibians.

A. The Islands

The Cayman Islands consist of Grand Cayman (197 km^2; maximum elevation 18 m asl), Cayman Brac (38 km^2; 43 m asl), Little Cayman (28 km^2; 12 m asl), and Owen Island (4.25 ha; < 3 m asl; uninhabited). These limestone islands are about equidistant between Jamaica and Cuba and are emergent peaks on the east–west Cayman Ridge, an extension of the Sierra Maestra mountains of Cuba. The climate is tropical maritime with equable temperatures (varying between 22°C and 28°C), but with a 'summer' wet season from May to November and a 'winter' dry season from November to April. The vegetation consists of dry forest, xerophytic shrubland, and seasonally flooded forests and shrublands, and mangroves.

B. The amphibians

The Cayman Islands have only four species of amphibians (Table 4.3), two of which are introduced from the United States (*Gastrophryne carolinensis* and *Rhinella marina*) and two others that are native (*Eleutherodactylus planirostris* and *Osteopilus septentrionalis*), although some authors have suspected that even these also may have been introduced (Edgar 2010). All four occur on Grand Cayman and Cayman Brac, and *Osteopilus septentrionalis* inhabits Little Cayman. Owen

Table 4.3. Conservation status of native amphibians inhabiting the Cayman Islands.

FAMILY AND *SPECIES*	ISLANDS			STATUS
	GC	CB	LCAY	IUCN RED LIST
NATIVE TAXA				
HYLIDAE				
Osteopilus septentrionalis	X	X	X	LC
ELEUTHERODACTYLIDAE				
Eleutherodactylus planirostris	X	X	—	LC
TAXA INTRODUCED				
BUFONIDAE				
Rhinella marina	X	X	—	LC
MICROHYLIDAE				
Gastrophryne carolinensis	X	X	—	LC

GC = Grand Cayman, CB = Cayman Brac, LCay = Little Cayman; '—' = not documented; Owen Island apparently has no record of amphibians. Population status from IUCN Red-List (IUCN 2022): LC = Least Concern. Abstracted and modified from Table 2 in Ecthernacht *et al.* (2011). These species are not listed in CITES.

Island is devoid of amphibians. Echternacht *et al.* (2011) indicated that most introductions of herpetofauna into the Caiman Islands arrived as stowaways on ships in building materials and live plants.

All four of the amphibian species are listed by the IUCN Red-List of Threatened Species as LC (Least Concern) (Table 4.3) so, unlike most places in the world, conservation concerns for amphibians in the Cayman Islands are minimal. One would have thought, however, that the two introduced species, especially *Rhinella marina*, might have posed a threat to the native species. However, at the time of Echternacht's assessment, it was not certain that *R. marina* had become established.

C. Threats

Threats have been catalogued in a general way for the herpetofauna at large, not specifically for particular species of amphibians, or even amphibians in general (Echternacht *et al.* 2011). However, those listed include (1) modification and destruction of habitat associated with uninformed planning, (2) illegal poaching for the pet trade, (3) introduction of non-native species, (3) road kills, (4) anthropogenic climatic change, and (5) deliberate, senseless killing by humans. The government has attempted to control the illegal trade in wildlife and investigations into offenses have involved the Cayman Islands National Trust; Department of Culture, Customs, and Immigration; Department of Environment; and Civil Aviation.

V. Acknowledgements

We thank Leydi Lorena Navarrete González for photographs 4.2D and E and 4.3A, Sidney Adriano Pérez Villegas for photographs 4.2A and C, Diana María Galindo-Uribe for photograph 4.2B, Julio Cesar Godoy Castillo for photograph 4.5A, and Kevin E. Berry Ingram for photograph 4.5B. We also thank John Fredy Aguilar Brand for the elaboration of Figure 4.1.

VI. References

Aguilera-Díaz, M. M., 2010. Geografía económica del archipiélago de San Andrés, Providencia y Santa Catalina. *Documentos de Trabajo Sobre Economía Regional y Urbana* **133**: 1–83. https://doi.org/10.32468/dtseru.133

Álvarez-Gutiérrez, Y., Amaya-López, C., Barbosa-Mejía, L. A., Builes-Carvajal, J. S., Henao-Casas, J. D., Montoya-Cañola, S. M., Pacheco-Sintura, P. A., Ramírez-Hoyos, L. F., Urrego-Osdorio, S., Zapata-Montoya, A. M., Ordónez-Carmona, O. O., and Restrepo-Álvarez, J. J., 2014. Descripción e interpretación geológica de las islas de Providencia y Santa Catalina. *Boletín de Ciencias de la Tierra* **35**. https://doi.org/10.15446/rbct

Baine, M., Howard, M., Kerr, S., Edgar, G., and Toral, V., 2007. Coastal and marine resource management in the Galapagos Islands and the Archipelago of San Andres: Issues, problems and opportunities. *Ocean & Coastal Management* **50**: 148–173. https://doi.org/10.1016/j.ocecoaman.2006.04.001

Barbour, T. and Loveridge, A., 1929. Reptiles and amphibians. *Bulletin of the Museum of Comparative Zoölogy* **69**: 138–146.

Botero, C., 2001. Manual ecológico y guía botánica de los cayos del Archipielago de San Andrés, Providencia y Santa Catalina. Armada Nacional, Dirección General Marítima, División Litorales – Sección Medio Ambiente. Dimar, Santa Fe de Bogotá, D.C., Colombia.

Caicedo-Portilla, J. R., 2014. Rediscovery of *Mabuya berengerae, Mabuya pergravis* (Squamata: Scincidae) and *Coniophanes andresensis* (Squamata: Colubridae), and assessment of its threatened status in the San Andrés and Providencia Islands, Colombia. *Caldasia* **36**: 181–201. https://doi.org/10.15446/caldasia.v36n1.43899

Caicedo-Portilla, J. R. and Lynch, J. D., 2015. Serpiente de San Andrés, *Coniophanes andresensis* Bailey, 1937. Pp. 91–93 in *Libro rojo de reptiles de Colombia. Instituto de Investigación de Recursos Biológicos Alexander von Humboldt*, edited by M. A. Morales-Betancourt, C. A. Lasso, V. P. Páez, and B. C. Bock. Universidad de Antioquia. Bogotá, D.C., Colombia.

Calderón-Espinosa, M. L. and Barragán Forero, A., 2011. Morphological diversification in solitary endemic anoles: *Anolis concolor* and *Anolis pinchoti* from San Andrés and Providence islands, Colombia. *South American Journal of Herpetology* **6**: 205–210. https://doi.org/10.2994/057.006.0306

Cochran, D. M. and Goin, C. J., 1970. Frogs of Colombia. *Bulletin of the United States National Museum* **288**: 1–655. https://doi.org/10.5479/si.03629236.288.1

Cope, E. D., 1885 '1884'. Twelfth contribution to the herpetology of tropical America. *Proceedings of the American Philosophical Society* **22**: 167–194.

CORALINA-INVEMAR, 2012. Gómez-López, D.I., Segura-Quintero, C., Sierra-Correa, P.C., y Garay-Tinoco, J., (eds). Atlas de la Reserva de Biósfera Seaflower. Archipiélago de San Andrés, Providencia y Santa Catalina. Instituto de Investigaciones Marinas y Costeras 'José Benito Vives De Andréis' -INVEMAR- y Corporación para el Desarrollo Sostenible del Archipiélago de San Andrés, Providencia y Santa Catalina -CORALINA-. Serie de Publicaciones Especiales de INVEMAR # **28**. Santa Marta, Colombia.

Corn, M. J. and Dalby, P. L., 1973. Systematics of the anoles of San Andrés and Providencia Islands, Colombia. *Journal of Herpetology* **7**: 63–74. https://doi.org/10.2307/1563202

Cortés-Guzmán, D. and Linares, E. L., 2016. Gasterópodos dulceacuícolas de Providencia, archipiélago de San Andrés, Providencia y Santa Catalina, Colombia. *Acta Zoológica Mexicana* **32**: 101–110.

Cubillos-Abrahams, M. V. and Montes-Correa, A. C., 2021. On a second invasive frog (Anura: Leptodactylidae) from San Andrés, Colombian Caribbean islands. *Anartia* **32**: 61–66.

Cubillos-Abrahams, M. V., Montes-Correa, A. C., and Vera-Pérez, L. E., 2021. First record of the greenhouse frog *Eleutherodactylus planirostris* (Anura, Eleutherodactylidae) from San Andrés, Colombian Caribbean islands. *Caribbean Journal of Science* **51**: 136–145.

Dagua, C., Torres, R., and Monroy, J., 2018. Condiciones oceanográficas de la reserva de biosfera Seaflower 2014–2016. *Boletín Científico del Centro de Investigaciones Oceanográficas e Hidrográficas* **37**: 53–74. https://doi.org/10.26640/22159045.2018.449

Defense Mapping Agency Hydrographic Center, 1976. *Sailing Directions (Enroute) for the Caribbean Sea.* Sector 12. *Brujas Point to Cabo Gracias a Dios.* First Edition *Defense Mapping Agency Hydrographic Center Publication* No. **144**: 145–368.

Defense Mapping Agency Hydrographic/Topographic Center, 1992. *Sailing Directions Enroute for the Caribbean Sea.* Sector 6. *Coasts of Honduras, Guatemala, Belize, and Mexico-Cabo Gracis a Dios to Cabo Catoche.* Volume 11, Fourth Edition. *Defense Mapping Agency Hydrographic/Topographic Center Publication* No. **148**: 127–153.

Departamento Administrativo Nacional de Estadística, 2007. Proyecciones nacionales y departamentales de población. 2006–2020. Available at http://www.dane.gov.co/files/investigaciones/poblacion/proyepobla06_20/7Proyecciones_poblacion.pdf. 281 pp.

de Sá, R.O., Grant, T., Camargo, A., Heyer, W. R., Ponssa, M. L., and Stanle, E., 2014. Systematics of the Neotropical genus *Leptodactylus* Fitzinger, 1826 (Anura: Leptodactylidae): phylogeny, the relevance of non-molecular evidence, and species accounts. *South American Journal of Herpetology* **9**: S1–S128. https://doi.org/10.2994/SAJH-D-13-00022.1

Duellman, W. E., 1970. The hylid frogs of Middle America. *Monograph of the Museum of Natural History, The University of Kansas* **1**: 1–753. https://doi.org/10.5962/bhl.title.2835

Duellman, W. E. and Fouquette, M. J., 1968. Middle American frogs of the *Hyla microcephala* group. *University of Kansas Publications, Museum of Natural History* **17**: 517–557. https://doi.org/10.5962/bhl.part.7137

Dunn, E. R., 1944. Los géneros de anfibios y reptiles de Colombia. Tercera Parte: reptiles, orden de las serpientes. *Caldasia* **3**: 155–254.

Dunn, E. R., 1945. The amphibians and reptiles of the Colombian Caribbean Islands San Andres and Providencia. *Caldasia* **3**: 363–365.

Dunn, E. R. and Saxe, L. H., Jr., 1950. Results of the Catherwood-Chaplin West Indies Expedition, 1948. Part V. Amphibians and reptiles of San Andres and Providencia. *Proceedings of the Academy of Natural Sciences of Philadelphia* **102**: 141–165.

Echternacht, A. C., Burton, F. J., and Blumenthal, J. M., 2011. The amphibians and reptiles of the Cayman Islands: Conservation issues in the face of invasions. Pp. 129–147 in *Conservation of Caribbean Island Herpetofaunas, Vol. 2. Regional Accounts of the West Indies*, edited by A. Hailey, B. S. Wilson, and J. A. Horrocks. Koninklijke Brill NV, Leiden, The Netherlands. https://doi.org/10.1163/ej.9789004194083.i-439.49

Edgar, P., 2010. The Amphibians and Reptiles of the UK Overseas Territories, Crown *Dependencies and Sovereign Base Areas Species Inventory and Overview of Conservation and Research Priorities*. Amphibian and Reptile Conservation, Bournemouth, UK.

Estrada, A. R., 2012. The Cuban Archipelago. In *Island Lists of West Indian Amphibians and Reptiles*, edited by R. Powell and R. W. Henderson. *Bulletin of the Florida Museum of Natural History* **51**: 113–125.

Flood, P. G. and Heatwole, H. H., 1986. Coral cay instability and species-turnover of plants at Swain Reefs, southern Great Barrier Reef, Australia. *Journal of Coastal Research* **2**: 479–496.

Forero-Medina, G., Castaño-Mora, O. V., and Montenegro, O., 2007. Abundance, population structure, and conservation of *Kinosternon scorpioides albogulare* on the Caribbean island of San Andres, Colombia. *Chelonian Conservation and Biology* **6**: 163–169. https://doi.org/10.2744/1071-8443(2007)6[163:APSACO]2.0.CO;2

Frost, D. R., 2020. Amphibian Species of the World: an online reference. Version 6.1 (*March 3, 2020*). Electronic Database accessible at https://amphibiansoftheworld.amnh.org/index.php. American Museum of Natural History, New York.

García-Hansen, I., 1998. Los bosques de manglar en la Isla de San Andrés. *Revista La Tadeo* **47**: 45–54.

Garzón, A. P., 2012. Estudio y conservación de las aves de la isla de San Andrés. *Conservación Colombiana* **16**: 1–56.

Geister, J., 1992. Modern reef development and Cenozoic evolution of an oceanic island/reef complex: Isla de Providencia (Western Caribbean Sea, Colombia). *Facies* **27**: 1–69. https://doi.org/10.1007/BF02536804

Geister, J. and Díaz, J. M., 2007. *Reef Environments and Geology of an Oceanic Archipelago: San Andrés, Old Providence and Sanat Catalina (Caribbean Sea, Colombia) with Field Guide*. Instituto Colombiano de Geología y Minería (INGEOMINAS), Ministerio de Minas y Energía, República de Colombia. Imprenta Nacional de Colombia, Bogotá, D.C., Colombia. 104 pp.

Gillespie, T. W., Grijalva, A., and Farris, C. N., 2000. Diversity, composition, and structure of tropical dry forests in Central America. *Plant Ecology* **147**: 37–47. https://doi.org/10.1023/A:1009848525399

HerpetoNica, 2015. *Guía Ilustrada de Anfibios y Reptiles de Nicaragua*. Managua: Dirección de Biodiversidad, Ministerio del Ambiente y los Recursos Naturales (MARENA), Managua, Nicaragua.

Köhler, G., 2001. *Anfibios y Reptiles de Nicaragua*. Herpeton-Verlag Elke Köhler, Offenbach, Hesse, Germany.

Hedges, B., Díaz, L., and Powell, R., 2004. *Eleutherodactylus planirostris*. The IUCN Red List of Threatened Species 2004: e.T56864A11534006. https://dx.doi.org/10.2305/IUCN.UK.2004.RLTS.T56864A11534006.en.

Heyer, W. R. and Heyer, M. H., 2013. Systematics, distribution, and bibliography of the frog *Leptodactylus insularum* Barbour, 1906 (Amphibia: Leptodactylidae). *Proceedings of the Biological Society of Washington* **126**: 204–233. https://doi.org/10.2988/0006-324X-126.3.204

Heyer W. R. and de Sá, R. O., 2011. Variation, Systematics, and Relationships of the *Leptodactylus bolivianus* Complex (Amphibia: Anura: Leptodactylidae). *Smithsonian Contributions to Zoology*, **635**: 1–58. https://doi.org/10.5479/si.00810282.635.1

Holdridge, L. R., 1967. *Life Zone Ecology*, revised edition. San José, Tropical Science Center.

IUCN 2022. The IUCN Red List of Threatened Species. Version 2022-1. https://www.iucnredlist.org. Accessed on 30 September 2022.

James Cruz, J. L. and Soler Caicedo, C. S. I., 2018. San Andrés: cambios en la tierra y transformación en el paisaje. *Cuadernos de Geografía: Revista Colombiana de Geografía* **27**: 372–388. https://doi.org/10.15446/rcdg.v27n2.65356

Jiménez Castilla, T. and Huertas Cardozo, N., 2014. El recurso energético como amenaza para la sostenibilidad de los destinos turísticos inmersos en Reservas de la Biosfera. Caso: Archipiélago de San Andrés, Providencia y Santa Catalina. Colombia. Pp. 13–33 in *Mesa: Desarrollo Sostenible–Destinos Turísticos. Congreso internacional de sostenibilidad, competitividad e innovación en destinos insulares*. Vol. II. Foro Internacional de Turismo, Maspalomas Costa Canaria. Servicio de Publicaciones y Difusión Científica, Universidad de Las Palmas de Gran Canaria, Las Palmas, España.

Lasso, C. A., Blanco-Libreros, J. F., and Sánchez-Duarte, P., (editors), 2015. Cuencas pericontinentales de Colombia, Ecuador, Perú y Venezuela: tipología, biodiversidad, servicios ecosistémicos y sostenibilidad de los ríos, quebradas y arroyos costeros. *Serie Editorial Recursos Hidrobiológicos y Pesqueros Continentales de Colombia*, **12**. Instituto de Investigación de Recursos Biológicos Alexander von Humboldt, Bogotá, D.C., Colombia.

León, J. R., 1969. The systematic of the frogs of the *Hyla rubra* group in Middle America. *University of Kansas Publications, Museum of Natural History* **18**: 505–545. https://doi.org/10.5962/bhl.part.19991

López de la Fuente, J., 1994. *Atlas of Solar Radiation in Nicaragua*. Stockholm, Sweden: Report to the Swedish Development Assistance Program.

Marriaga, L., 2009. Características generales del Caribe colombiano. Pp. 17–23 in *DIMAR-CIOH. 2009. Caracterización físico-biótica del litoral Caribe colombiano*. Volume 1. *Dirección General Marítima-Centro Investigaciones Oceanográficas e Hidrográficas*, edited by *DIMAR, Serie Publicaciones Especiales CIOH*. Cartagena de Indias, Colombia.

McCranie, J. R., Sunyer, J., and Martínez-Fonseca, J. G., 2019. Comments and updates to "Guía Ilustrada de Anfibios y Reptiles de Nicaragua" along with taxonomic and related suggestions associated with the herpetofauna of Nicaragua. *Revista Nicaragüense de Biodiversidad* **52**: 1–44.

Millás, J. C., 1968. *Hurricanes of the Caribbean and Adjacent Regions, 1492–1800*. Academy of the Arts and Sciences of the Americas, Miami, FL.

Milliman, J. D., 1969. Four southwestern Caribbean atolls: Courtown Cays, Albuquerque Cays, Roncador Bank and Serrana Bank. *Atoll Research Bulletin* No. **129**: 1–26. https://doi.org/10.5479/si.00775630.129.1

Miralles, A., 2006. A new species of *Mabuya* (Reptilia, Squamata, Scincidae) from the Caribbean island of San Andres, with a new interpretation of nuchal scales: a character of taxonomic importance. *The Herpetological Journal* **16**: 1–7. https://doi.org/10.11646/zootaxa.895.1.1

Murcia, G. A., González, F., García, M. I., and Lasso J., 2016. Angiosperm flag species for mangrove conservation in San Andrés Island (Colombia) are highly vulnerable and locally rare. *Caldasia* **38**: 1–16. https://doi.org/10.15446/caldasia.v38n1.57818

National Geospatial-Intelligence Agency, 1993. Saint Andrews Island: Colombia. Bethesda, Maryland, USA. https://geographic.org/geographic_names/name.php?uni=-842048&fid=1385&c=colombia.

Ortiz Royero, J. C., Plazas Moreno, J. M., and Lizano, O., 2015. Evaluation of extreme waves associated with cyclonic activity on San Andrés Island in the Caribbean Sea since 1900. *Journal of Coastal Research* **31**: 557–568. https://doi.org/10.2112/JCOASTRES-D-14-00072.1

Pagnacco, P. F. and Radelli, L., 1962. Note on the geology of the isles of Providencia and Santa Catalina (Caribbean Sea, Colombia). *Geologia Colombiana* **3**: 125–132.

Peralta-Willimas, L., 1991. *Diagnóstico de los Recursos Pesqueros*. Instituto Nicaraguense de las Regiones Autónomas. Managua: Indera.

Peters, J. L., 1929. Vertebrates from the Corn Islands. *Bulletin of the Museum of Comparative Zoology* **69**: 127–146.

Powell, R. and Henderson, R. W. 2012. Swan Islands. In *Island Lists of West Indian Amphibians and Reptiles*, edited by R. Powell and R. W. Henderson. *Bulletin of the Florida Museum of Natural History* **51**: 91–92.

Quiceno Mesa, M. P., Palacios Nogera, S., Ramos Barón, P., Romero Duque, L. P., Ortego-Rincón, M., and Restrepo Calle, S., 2009. *Caracterización de la Biodiversidad y Lineamientos Generales para la Formulación de un Plan de Manejo del Parque Natural Regional The Peak en La Isla de Providencia*. Final Technical Report presented to *Corporación para el Desarrollo Sostenible del Archipiélago de San Andrés, Providencia y Santa Catalina. Programa de Biología de la Conservación and Programa de Uso y Valoración. Instituto de Investigación de Recursos Biológicos "Alexander von Humboldt"*.

Roberts, H. H. and Suhayda, J. N., 1983. Wave-current interactions on a shallow reef (Nicaragua, Central America). *Coral Reefs* **1**: 209–214. https://doi.org/10.1007/BF00304417

Robleto-Hernández, S., Gutiérrez Rodríguez, A., Otero Ortuño, C., González, E. Y., Leets Rodríguez, L., López Guevara, H., and Sunyer, J., 2017. *Libro Rojo Anfibios y Reptiles de Nicaragua. Conservación de la Diversidad Biológica*. International Union for the Conservation of Nature and Ministerio del Ambiente y los Recursos Naturales, Managua, Nicaragua. 226 pp.

Romero, J. H. and Lynch, J. D., 2012. Anfibios de la Región Caribe. Pp. 677–701 in *Colombia Diversidad*

Biótica XII. La región Caribe de Colombia, edited by J. O. Rangel-Ch. Instituto de Ciencias Naturales, Universidad Nacional de Colombia, Bogotá, D.C., Colombia.

Ruiz, J., Fandiño, M. C., and Chazdon, R. L., 2005. Vegetation structure, composition, and species-richness across a 56-year chronosequence of Dry Tropical Forest on Providencia Island, Colombia. *Biotropica* **37**: 520–530. https://doi.org/10.1111/j.1744-7429.2005.00070.x

Ruiz Pérez, G. A. and Buitrago, F., 2003. *Guía Ilustrada de la Herpetofauna de Nicaragua*. Proyecto Araucaria, Ministerio del Ambiente y los Recursos Naturales, Agencia Española de Cooperación Internacional para el Desarrollo (ARAUCARIA-MARENA-AECI), Managua, Nicaragua.

Ryan, J. D., Miller, L. J., Zapata, Y., Downs, O., and Chan, R., 1998. *Great Corn Island, Nicaragua*; Pp. 95–105 In Caribbean coral reef, seagrass and mangrove sites, edited by B. Kjerfve. Coastal region and small island papers 3, UNESCO. Paris, France.

Savage, J. M. and Heyer, R., 1967. Variation and distribution in the tree-frog genus *Phyllomedusa* in Costa Rica. Central America. *Beitrage zur Neotropischen Fauna* **5**: 11–131. https://doi.org/10.1080/01650526709360400

Smith, D. S., Monroe, E., and Ramos, S. J., 1989. Biogeographical affinities of the butterflies of the southwest Caribbean Islands, San Andrés and Providencia. *Journal of the Lepidopterists' Society* **43**: 274–288.

Salas Betin, J. D., 2015. Tourism, national sovereignty and social transformation in the archipelago of San Andres, Providencia and Santa Catalina, Colombia. *Letras verdes* **18**: 91–112. https://doi.org/10.17141/letrasverdes.18.2015.1665

Sánchez-Jabba, A. M., 2012. Manejo ambiental en Seaflower, Reserva de Biosfera en el Archipiélago de San Andrés, Providencia y Santa Catalina. *Documentos de Trabajo Sobre Economía Regional y Urbana* No. **176** (noviembre): 1–48. https://doi.org/10.32468/dtseru.176

Santos-Martínez, A., Hinojosa, S., and Sierra Rozo, O., 2009. Proceso y avance hacia la Sostenibilidad Ambiental: La Reserva de la Biosfera Seaflower, en el Caribe Colombiano. *Cuadernos de del Caribe* **7**: 7–23.

Savage, J. M., 2002. *The Amphibians and Reptiles of Costa Rica: A Herpetofauna between Two Continents, between Two Seas*. The University of Chicago Press, Chicago, IL.

Señaris, J. C., Padrón, M. M. A., Gil, H. R., and Rojas-Runjaic, F. J. M., 2018. *Guía ilustrada de los anfibios y reptiles del valle de Caracas, Venezuela*. Ediciones Instituto Venezolano de Investigaciones Científicas, Caracas, Venezuela.

Solano-Flórez, L., 2012. *Diversity, Phylogeography and Conservation of two groups of anurans of the family Hylidae in Mesoamerica*. PhD thesis. University of Manchester, UK.

Solís, F., Ibáñez, R., Chaves, G., Savage, J., Jaramillo, C., Fuenmayor, Q., Reynolds, R., Caramaschi, U, Mijares, A., Acosta-Galvis, A., Hardy, J., La Marca, E., Manzanilla, J., and Bolaños, F., 2008. *Leptodactylus bolivianus. The IUCN Red List of Threatened Species* 2008: e.T57114A11582479. https://doi.org/10.2305/IUCN.UK.2008.RLTS.T57114A11582479.en

Sunyer, J. and Köhler, G., 2010. Conservation status of the herpetofauna of Nicaragua. Pp. 488–509 in *Conservation of Mesoamerican Amphibians and Reptiles*, ed L. D. Wilson, J. H. Townsend, and J. D. Johnson. Eagle Mountain Publishing, Eagle Mountain, UT.

Sunyer, J., Nicholson, K. E., Phillips, J. G., Gubler, J. A., and Obando, L. A., 2013. Lizards (Reptilia: Squamata) of the Corn Islands, Caribbean Nicaragua. *Check List* **9**: 1383–1390. https://doi.org/10.15560/9.6.1383

Téllez Ramírez, J. C. and Llurdés Coit, J. C., 2014. Turismo y Desarrollo Sostenible en espacios insulares del Caribe: el caso de San Andrés y Providencia (Colombia). Pp. 146–155 in *Mesa: Desarrollo Sostenible–Destinos Turísticos. Congreso internacional de sostenibilidad, competitividad e innovación en destinos insulares*. Vol. II. Foro Internacional de Turismo, Maspalomas Costa Canaria. Servicio de Publicaciones y Difusión Científica, Universidad de Las Palmas de Gran Canaria, Las Palmas, España.

Villa, J. D., 1971. Arboreal habits of *Bufo valliceps* and other bufonids. *Herpetological Review* **3**: 96. https://doi.org/10.2307/1442852

Villa, J. D., 1972a. Amphibians of the Corn Islands, Caribbean Nicaragua. *Caribbean Journal of Sciences* **12**: 195–198.

Villa, J. D., 1972b. Anfibios de Nicaragua: introducción a su sistemática, vida y costumbres. Instituto Geográfico Nacional & Banco Central de Nicaragua, Managua, Nicaragua.

Villa, J. D., 2015. Las ranitas de Cayos Miskitos. *Revista de Temas Nicaragüenses* **89**: 6–22.

Villa, J. D., Wilson L. D., and Johnson J. D., 1988. *Middle American Herpetology, a Bibliographic Checklist*. University of Missouri Press, Columbia, MO.

Wilson, B., Edwards, T., and Powell, R., 2012. Jamaica. In *Island Lists of West Indian Amphibians and Reptiles*, edited by R. Powell and R. W. Henderson. *Bulletin of the Florida Museum of Natural History* **51**: 126–128.

Zweifel, R. G., 1967. The systematic status of the Central American frog *Rana miadis*. *Herpetologica* **23**: 54–56.

5 Amphibians of Jamaica

Kimberly Stephenson and Byron Wilson

Abbreviations and acronyms

asl	*above sea level*
Bd	Batrachochytrium dendrobatidis *(chytrid fungus)*
CR	*Critically Endangered*
CSGM	*Climate Studies Group, Mona*
EN	*Endangered*
IUCN	*International Union for the Conservation of Nature*
LC	*Least Concern*
NT	*Near Threatened*
PBPA	*Portland Bight Protected Area*
SSC	*The Species Survival Commission (of the IUCN)*
VU	*Vulnerable*

Kimberly Stephenson and Byron Wilson, 'Chapter 5: Amphibians of Jamaica' in: *The Conservation and Biogeography of Amphibians in the Caribbean*. Pelagic Publishing (2023). © K. Stephenson and B. Wilson. DOI: 10.53061/IZDZ9752

I. Introduction

Jamaica is the third-largest island in the Greater Antilles, spanning approximately 230 km from east to west. The country covers an area of approximately 11,500 km^2, with its highest point being Blue Mountain Peak at 2,256 m above sea level (asl). The island is located in the biologically diverse Caribbean region and boasts a high rate of endemism. The endemism and number of critically endangered fauna make it a 'hotspot within a hotspot' and a focal point for conservation (Wilson 2011).

In the late 1800s to early 1900s, Jamaica's diversity drew the attention of a number of scientists. The work of local and foreign taxonomists and natural historians resulted in a vastly greater understanding of the island's rich biodiversity. The result has been a number of descriptive and taxonomic publications coming out of work on Jamaica's amphibians (e.g. Gosse 1851; Barbour 1910; Lynn 1940; Lynn and Dent 1942; Schwartz and Fowler 1973; Crombie 1977; Hedges 1989), but far fewer studies on the ecology of Jamaica's herpetofauna, leaving much of this information to be gleaned from museum specimens and the field notes of biologists (Wilson 2011).

Despite their uniqueness, Jamaica's amphibians have largely been understudied, and the last known records of many species date back to the 1980s (Hedges and Díaz 2011). A survey conducted by Holmes, McLaren, and Wilson in 2010–11 is the most recent assessment of the occurrence of Jamaica's anurans and was the first comprehensive study of the chytrid fungus conducted on the island (Holmes *et al.* 2012).

Many of the island's unique amphibian species are found in habitats that are minimally disturbed, as is to be expected. The most biologically important of these are the Portland Ridge and Hellshire Hills dry forests, Dolphin Head Mountain, Cockpit Country, and the Blue and John Crow Mountains. A brief description of each of these is given next.

Portland Ridge is a hilly peninsula on the southern coast of the island that is connected to the mainland by low-lying plains (Lynn 1954). It has been described as a biological island by Grant (1940) and is inhabited by a high number of endemic and native species. The dry limestone forest of the peninsula increases in moisture at higher elevations but remains an unlikely site for the proliferation of amphibian species (Lynn 1954; Schwartz and Fowler 1973). The exceptions are cave-dwelling species, such as *Eleutherodactylus cavernicola*, for which the moisture and cooler temperatures of the caves are favourable (Lynn 1954; Schwartz and Fowler 1973).

The Hellshire Hills consists of 114 km^2 of karst limestone covered by one of the few remaining intact dry forests in the Caribbean (Wilson 2011). The area is known to support Critically Endangered species, such as the Jamaican iguana (*Cyclura collei*) and a number of other species of reptiles, as well as *Eleutherodactylus cavernicola*, and is considered to be an important site for conservation because species there are at high risk of extinction (Ricketts *et al.* 2005; Wilson 2011; Wilson *et al.* 2016).

Dolphin Head Mountain and surrounds is a region of karst white limestone located in Western Jamaica. The site was once covered in moist tropical forest, although much of the surrounding forest has been removed for lumber and illegal cultivation (Wilson 2011). Still, the area is known for its high number of endemic plants and birds, with new species of endemic plants having been discovered there recently (Wilson 2011; Aguirre-Santoro *et al.* 2016). It also serves as habitat for a number of native reptiles and amphibians, including *E. fuscus*, for which it is the type locality (Wilson 2011; Lynn and Dent 1943).

The Cockpit Country is a karst limestone plateau located in west-central Jamaica. The area is hilly, with elevation ranging from 200 to 600 m (Newman *et al.* 2011). The wet broadleaf forest is of high biological importance and has been declared a Forest Reserve despite having degraded somewhat over the years (Newman *et al.* 2011; Wilson 2011). Despite its significance, the site had no clearly defined conservation boundary until 2017, when the government declared newly defined boundaries for the protected area and a ban on mining (Linton 2017). Nonetheless, it is still threatened by human activities such as illegal harvesting and cultivation (Newman *et al.* 2011; Wilson 2011).

The Blue and John Crow Mountains National Park was established in 1989 and currently is the only Jamaican national park that is not marine. It encompasses the Blue Mountains (up to 2,256 m asl), comprising igneous rock, shale, and limestone, and the John Crow Mountains (up to 1,140 m asl) composed of limestone. The site is largely covered in moist-tropical and wet-tropical forest types. Despite its biological importance, however, the site is threatened by illegal harvesting of timber and by agriculture (Wilson 2011).

II. Native amphibians

All of the native species of amphibians in Jamaica are endemic anurans. The two families represented by these 21 species are Hylidae (four species of *Osteopilus* but see below) and Eleutherodactylidae (17 species of *Eleutherodactylus*). Table 5.1 includes the amphibian species found on the island and the locations to which they are endemic, as outlined by Wilson (2011).

Three of the four species of *Osteopilus* present in Jamaica require relatively undisturbed primary forest and are listed as Vulnerable or Endangered on the IUCN Red List, while the fourth (*Osteopilus ocellatus*) is tolerant of some level of disturbance and is listed as Near Threatened. A fifth species of *Osteopilus* (not included in the descriptions but listed in Table 5.1) is currently being described by S. Blair Hedges, and its IUCN listing will be updated accordingly when the process is complete (Wilson 2011).

The *Eleutherodactylus* species endemic to the island belong to a monophyletic group but can be separated into five species-groups based on closest genetic similarity, as described by Hedges (1989): *luteolus* (*grabhami, luteolus,* and *sisyphodemus*), *gossei* (*fuscus, gossei, junori, pantoni,* and *pentasyringos*), *cundalli* (*cavernicola, cundalli,* and *glaucoreius*), *jamaicensis* (*jamaicensis*), and *nubicola* (*alticola, andrewsi, griphus, nubicola,* and *orcutti*). They also face a substantial threat, with 12 of the 17 species being listed as Endangered or Critically Endangered. All *Eleutherodactylus* species described here lay their eggs on the ground, in soil, or on rocks, and development is direct. Hence, breeding will not be discussed at length in the descriptions of these species. Six of these *Eleutherodactylus* species have not been observed in the past 20 or more years, and one (*E. orcutti*) is considered to be extinct (Wilson 2011).

Although loss and degradation of habitat are regarded as important threats, the fungal disease *Batrachochytrium dendrobatidis* (*Bd*) has recently emerged as a serious threat and is considered the likely cause of extinction of *E. orcutti*. Warmer, drier, and more extreme conditions under climatic change are likely to compound these existing threats. Next, we provide a brief description of the species, their IUCN Red List status, and their habitats. The occurrence of the chytrid fungus is also provided for those species that were collected in the 2010–11 survey. Key conservation concerns are addressed in section IV.

A. Hylidae

1. Osteopilus crucialis *(Harlan, 1826)*

Osteopilus crucialis, also known as the Jamaican snoring frog, is found primarily in bromeliads and on tree trunks in broadleaf forests in the centre of the island, particularly in Cockpit Country, which is among Jamaica's protected areas. It is the largest of the Jamaican

Table 5.1. Endemic and introduced species of amphibians of Jamaica, as shown by Wilson (2011). This table includes IUCN Red List categories and important forested areas in which these species are found. Labels are as follows: Blue and John Crow Mountains National Park (B), The Cockpit Country (C), Dolphin Head Mountain and adjacent forested areas (D), and Hellshire Hills (H). NA = not applicable.

FAMILY ***SPECIES***	**ENDEMIC**	**INTRODUCED**	**LOCATION**	**STATUS CATEGORY**
Bufonidae				
Rhinella marina	—	X	NA	LC
Hylidae				
Osteopilus crucialis	X	—	B, C	VU
Osteopilus marianae	X	—	C	EN
Osteopilus ocellatus	X	—	B, C, D	NT
Osteopilus wilderi	X	—	B, C	VU
Osteopilus sp.	X	—	C	NA
Eleutherodactylidae				
Eleutherodactylus alticola	X	—	B	CR
Eleutherodactylus andrewsi	X	—	B	EN
Eleutherodactylus cavernicola	X	—	H	CR
Eleutherodactylus cundalli	X	—	C, D	NT
Eleutherodactylus fuscus	X	—	C, D	EN
Eleutherodactylus glaucoreius	X	—	B	EN
Eleutherodactylus gossei	X	—	B, C	VU
Eleutherodactylus grabhami	X	—	C, D	EN
Eleutherodactylus griphus	X	—	C	EN
Eleutherodactylus jamaicensis	X	—	B, C	CR
Eleutherodactylus johnstonei	—	X	NA	LC
Eleutherodactylus junori	X	—	C	CR
Eleutherodactylus luteolus	X	—	C, D	EN
Eleutherodactylus nubicola	X	—	B	EN
Eleutherodactylus orcutti	X	—	B	CR (extinct?)
Eleutherodactylus pantoni	X	—	B, C, D	VU
Eleutherodactylus pentasyringos	X	—	B	EN
Eleutherodactylus planirostris	—	X	NA	LC
Eleutherodactylus sisyphodemus	X	—	C	CR
Ranidae				
Lithobates catesbeianus	—	X	NA	LC

Osteopilus species, with brown colouration, mottled pattern, and a very distinctive and loud call (Lynn 1940). Various stages of reproduction occur in different sections of a tree. For example, indentations in branches are used by males for broadcasting mating calls, and bromeliads are used for laying eggs. This species is listed as Vulnerable (VU) on the IUCN Red List due to loss of habitat through various human activities such as tourism, housing, exploitation of resources, and agriculture, as it is not tolerant of degraded habitats. The size of the population is currently decreasing, and observations in the wild are infrequent (IUCN SSC Amphibian Specialist Group 2021a).

2. Osteopilus marianae *(Dunn, 1926)*

Osteopilus marianae, also known as the yellow bromeliad frog, has a restricted range in old-growth deciduous forests along the centre of the island. The species was described initially from very few specimens. Despite this, some variation is noted among individuals, in colour and presence of an interocular bar (Lynn 1940). As is the case in *O. crucialis,* bromeliads are used for laying eggs. Although missing a vocal sac, males create mating calls using vibration. This species is intolerant of disturbance and faces population decline due to agriculture, logging, residential and commercial development, and climatic change. Much of the remaining population is found in a protected area, given that Cockpit Country is within its range. Despite this, the species is listed as Endangered (EN) due to the population declines caused by loss and fragmentation of habitat (IUCN SSC Amphibian Specialist Group 2021b).

3. Osteopilus ocellatus *(Gosse, 1851)*

Osteopilus ocellatus (formerly *O. brunneus*) is also known as the Jamaican laughing frog. This species exhibits wide variation in size, colour, and pattern (Lynn 1940). Individual length is in the range of 39–72 mm, dorsal surface can be shades of brown or grey, and a pattern of spots may or may not be present (Lynn 1940). It is found across much of the island except for the extreme south and coastal areas. Due to its stable population and wide range, it is listed as Near Threatened (NT). This species' tolerance of somewhat degraded or converted habitats allows it to occupy lands used for large-scale or small-scale farms, tourism, and development, while also inhabiting all types of forest. It is arboreal and can flourish in most areas that have high humidity and small areas of trapped water. Eggs are laid in bromeliads, as with other *Osteopilus* species. Despite having a slightly higher tolerance for disturbance, however, threats to this species include logging, mining, tourism and residential development, agriculture, removal of forest resources, and climatic change (IUCN SSC Amphibian Specialist Group 2021c). Of the 54 individuals collected across sites, six were found to be infected with chytrid fungus in the extreme east and west of the island (Holmes *et al.* 2012).

4. Osteopilus wilderi *(Dunn, 1925)*

Osteopilus wilderi, also known as the green bromeliad frog, has a very restricted and fragmented range in the centre of the island. There is minimal variation in the green colouration of this species among individuals and length is typically in the range of 24–30 mm. There is minimal difference in size between sexes, but dimorphism is evident in the presence of a distinctive 'corneous' thumb-pad found only in males, and the interocular bar occurs most frequently in females (Lynn 1940). The range extends to such protected areas as the Blue and John Crow Mountains. It is listed as Vulnerable (VU) due to its declining population but is more commonly encountered than *O. marianae.* It is an arboreal species mainly found in closed-canopy forests in bromeliads, where oviposition and larval development occur. Although this species is tolerant of secondary forest, it is threatened by degradation of habitat due to farming, logging, development, and climatic change (IUCN SSC Amphibian Specialist Group 2021d). Its call was described by Lynn (1940) as a faint ticking that is easily recognized.

B. Eleutherodactylidae

1. Eleutherodactylus alticola *Lynn, 1937*

Eleutherodactylus alticola, common names Jamaican peak frog and Lynn's robber frog, is currently listed as Critically Endangered (CR) due to the ongoing decline of its population and habitat. Prior to the individuals recorded in the 2010–11 survey, the species had not been confirmed as extant since 1986 (Hedges and Díaz 2011; Holmes *et al.* 2012). The species is known for its small size and stocky body, smooth back, and poorly developed digital disks (Lynn 1940). There is some variation in colour and pattern, including mottled or uniform dorsal surface, brown or grey colouration, and an occasional interocular bar. Eggs adhere strongly to each other and the surface on which they are laid (Lynn 1940).

The species can be found in the closed-canopy forest in the Blue Mountains at an elevation of 1,550–2,250 m but has only been confirmed at two separate sites within a small area. Although this species may have a wider range, its occurrence has so far not been confirmed in the John Crow Mountains or any similar sites in which occurrence is expected. Adults often live under rocks or in vegetation at the height of up to one metre. The species is considered to be tolerant of mildly disturbed habitats, such as former logging sites, but is not found in heavily disturbed areas, such as those used for agriculture. Major threats include conversion of natural habitat for coffee plantations and logging, which reduce humidity by removing forest cover and pollute the area with pesticides (IUCN SSC Amphibian Specialist Group 2021e).

2. Eleutherodactylus andrewsi *Lynn, 1937*

Eleutherodactylus andrewsi is commonly called the Jamaican rumpspot frog and is found in the old-growth forest in the Blue and John Crow Mountains. It is known for its black dorsal spots, short legs, smooth body, heels that do not meet when legs are flexed, and minimal variation in colour and pattern among individuals (Lynn 1940). The population is considered to be decreasing, which has contributed to it being listed as Endangered (EN). It is a terrestrial species that can be found in leaf litter or on rocks along streams deep within the forest. It is threatened mainly due to logging, farming, and some tourist activities; these are particularly destructive due to its intolerance of long-term disturbance of habitat (IUCN SSC Amphibian Specialist Group 2021f). Of the six individuals collected across sites in 2010–11, one on the eastern end of the island was found to be infected with chytrid fungus (Holmes *et al.* 2012).

3. Eleutherodactylus cavernicola *Lynn, 1954*

Eleutherodactylus cavernicola is a rare, Critically Endangered (CR) species that was historically only found in two caves in Portland Ridge on the southern coast, making its estimated range approximately 82 km^2 (IUCN SSC Amphibian Specialist Group 2021g). It is similar in appearance to *E. cundalli* but has a larger adult size, longer hind limbs, and no distinct sexual dimorphism in the size of the tympanum (Lynn 1954). It is noted as having plantar, subarticular, and metatarsal tubercles, as well as heels that do not meet when the legs are flexed. Colour varies slightly among individuals from tan to grey, with dark brown dorsal mottling and interocular bar, and yellow tint on the groin. Eggs are laid in fissures inside the caves, whereas adults can be found in exposed areas from the mouth to half a mile (0.85 km) inside the caves (Lynn 1954).

As of 2011, the species also had been sighted in the Hellshire Hills, but its abundance there is not yet known (Wilson 2011). It occupies caves in forested areas. Its already restricted range is severely threatened by climatic change and human activity, particularly destruction of forest and damage to the caves (IUCN SSC Amphibian Specialist Group 2021g).

4. Eleutherodactylus cundalli *Dunn, 1926*

Also known as Cundall's robber frog, *Eleutherodactylus cundalli* is a Near Threatened (NT) species. It is known to be widespread on the western half of Jamaica at an elevation of up to

635 m, but also has been recorded as far east as the Blue Mountains (Lynn 1940; IUCN SSC Amphibian Specialist Group 2021h). Its unique features include dorsal protrusions, long legs with heels that overlap when the legs are flexed, and two outer fingers having enlarged digital disks (Lynn 1940). There is strong sexual dimorphism in the size of the tympanum, with males having a much larger one than females (2/3 and 1/2 diameter of the eye, respectively). There is wide variation in colour and pattern in this species, with the most frequent being a brown dorsal surface, interocular bar, and mottling along the dorsal and lateral surfaces (Lynn 1940). Studies of *E. cundalli* in the Windsor Great Cave in the early 1990s showed persistent brood care by females. Eggs are laid inside the cave, which provides stable conditions for incubation, and metamorphosed young are transported out of the cave by the mothers (Diesel *et al.* 1995).

The species has been recorded at eight sites. Due to degradation of habitat caused by bauxite mining, limestone quarrying, farming, and residential and touristic developments, the population size is decreasing. However, it is still one of the most common *Eleutherodactylus* species on the island. It is found in moist closed-canopy forest, particularly in karst limestone, and in associated caves. It can tolerate moderately disturbed habitats that retain some degree of closed canopy but does not occur where there is minimal or no shade from woody vegetation (IUCN SSC Amphibian Specialist Group 2021h). Of the 29 individuals collected across sites in 2010–11, four were found to be infected with chytrid fungus in the north and west of the island (Holmes *et al.* 2012).

5. Eleutherodactylus fuscus *Lynn and Dent, 1942*

The range of *Eleutherodactylus fuscus* is extremely restricted to a small area of mesic closed-canopy forest in the west of the island and the species is listed as Endangered (EN). It has short legs, a stout body, no dorsal patterns, poorly developed dorsolateral line, and heels that do not meet when the legs are flexed. It was first described from a specimen collected at Dolphin Head, Westmoreland, where specimens were traced by their calls (Lynn and Dent 1943). Calling is conducted from rocks or the ground. Surveys in 2010–11 found eight individuals on the western end of the island, two of which were infected with chytrid fungus (Holmes *et al.* 2012). Prior to this, it had not been seen since 1984 (Hedges and Díaz 2011). It is not currently known whether the population size is decreasing or increasing. However, the species has consistently been under threat from degradation of habitat due to logging, development, and agriculture. Invasive plants such as bamboo are also an issue in regenerating forests (IUCN SSC Amphibian Specialist Group 2021i).

6. Eleutherodactylus glaucoreius *Schwartz and Fowler, 1973*

Eleutherodactylus glaucoreius is listed as Endangered (EN) due to its declining population size and habitat quality. Distinctive features include long hind limbs, small size, heels that overlap when the legs are flexed, interocular bar, and dorsal mottling. It is very similar in appearance to *E. cundalli* of which it was once considered to be a subspecies. However, distinct differences exist between the two species, such as the smaller size of *E. glaucoreius,* and differences in the frequencies of various colour patterns (e.g. mottling) between the two (Schwartz and Fowler 1973).

This species is restricted to eastern Jamaica, can be found on rocks, and has a preference for wet tropical forest, although it tolerates secondary forest. Destruction of habitat due to the cultivation of coffee is a considerable threat, as the species occurs within the Blue and John Crow Mountains National Park (IUCN SSC Amphibian Specialist Group 2021j). Of the 33 individuals collected across sites in 2010–11, in the east of the island a dozen were found to be infected with chytrid fungus (Holmes *et al.* 2012).

7. Eleutherodactylus gossei *Dunn, 1926*

Eleutherodactylus gossei is commonly encountered across Jamaica and adapts easily to different environments. The species shares many characteristics with *E. luteolus* (discussed below), and historically the two have been considered to be the same, or similar, species (Lynn 1940; Crombie 1977). It is listed as Vulnerable (VU), largely due to ongoing habitat loss (IUCN SSC Amphibian Specialist Group 2021k). In the 1890s, *E. gossei* was introduced from Jamaica to Bermuda via imported vegetation. The new population remained stable and limited to the areas in which they were first introduced, except for two colonies that may have originated elsewhere when individuals were accidentally transported in farming manure. By the mid-1990s, it is believed that *E. gossei* was extirpated from Bermuda (Bacon *et al.* 2011). This species occurs in mesic habitats that are preferably undisturbed but also can occupy rural farms and other areas that have been heavily degraded, such as backyards. However, it may be susceptible to agricultural pollutants. Of the 124 individuals collected in 2010–11, 20 from various localities were found to be infected with chytrid fungus (Holmes *et al.* 2012).

8. Eleutherodactylus grabhami *Dunn, 1926*

The range of *Eleutherodactylus grabhami* is restricted to western and central Jamaica. It is a rare species, and its population is considered to be in decline, resulting in it being listed as Endangered (EN). Individuals are small, with females often larger than males. Colour and pattern are consistently brown and mottled, with yellow dorsal warts, white belly, and yellow groin (Lynn 1940). It is often found in moist areas or in arboreal bromeliads in wet limestone forests. Loss of habitat is the primary threat, particularly deforestation for bauxite mining, development, agriculture, and logging (IUCN SSC Amphibian Specialist Group 2021l). The species was last recorded in 2007 (Hedges and Díaz 2011).

9. Eleutherodactylus griphus *Crombie, 1986*

Eleutherodactylus griphus is an Endangered (EN) species that is found only in fragmented areas of Cockpit Country. It was first described in the 1970s and had not been seen since 1985 until the 2010–11 survey, which recorded 25 individuals with no sign of infection by chytrid fungus (Wilson 2011; Holmes *et al.* 2012). Its population is in decline, particularly due to its intolerance of disturbed habitats: it is found in undisturbed closed-canopy forests and is threatened by loss of habitat due to development, agriculture, logging, and the harvesting of firewood (IUCN SSC Amphibian Specialist Group 2021m).

10. Eleutherodactylus jamaicensis *Barbour, 1910*

Eleutherodactylus jamaicensis is listed as Critically Endangered (CR) due to its limited, fragmented range along the centre of the island and its declining population size. There is distinct sexual dimorphism in size, with females larger than males. There is some variation in colour and pattern, but most individuals have an interocular bar and dorsal dark spots. Most notably, the species is known to secrete a bright blue slime from glands on the thighs and sides when handled (Lynn 1940). It is primarily found in bromeliads in moist old-growth forests that are montane or lowland at elevations of 120–1,290 m. Reproduction occurs in bromeliads. It is intolerant of disturbed habitats and is strongly threatened by deforestation due to logging, agriculture, and residential development (IUCN SSC Amphibian Specialist Group 2021n). This species has not been recorded since the 1980s and was not found in the 2010–11 survey. Its semi-aquatic nature makes it particularly susceptible to infection by *Bd*, and it is one of the species most likely to have succumbed to such infection (Holmes *et al.* 2012).

11. Eleutherodactylus junori *Dunn, 1926*

Eleutherodactylus junori is a Critically Endangered (CR) species that is known from only four sites along the centre of the island. This species was first described in 1926, at which time it was already considered rare, and has not been seen or heard in the wild since the late 1990s (Lynn 1940; Hedges and Díaz 2011). Too few individuals have been found for a thorough description to be formulated (Lynn 1940), but distinctive features include short legs, slightly rough skin on the body, and heels that do not meet when the rear legs are flexed. It can be found in upland forests on the edges of pastures or banana groves but is difficult to locate because its vocalization is not loud enough to compete with *E. johnstonei*, an anuran that is invasive on the island (see section III). There is concern that *E. johnstonei* may interfere with *E. junori* both reproductively and ecologically, although this has not been shown definitively (Hedges 1993). Destruction of habitat is another threat to the species, particularly that due to logging, development, and agriculture (IUCN SSC Amphibian Specialist Group 2021o).

12. Eleutherodactylus luteolus *Gosse, 1851*

Eleutherodactylus luteolus is found in coastal and elevated mesic forests in the west of the island. Due to its restricted range and declining population, it is listed as Endangered (EN). It is distinctive in appearance, having long legs, a well-developed dorsolateral line, stout body, red groin, small digits, and heels that meet or overlap when the rear legs are flexed (Lynn 1940). There is wide variation in colour and pattern (Lynn 1940). Due to its similarity to *E. gossei*, past classifications considered *E. luteolus* to be either within the *E. gossei* family of species or simply another morph of *E. gossei* (Lynn 1940; Crombie 1977). However, it is currently considered to be a valid species of Jamaican *Eleutherodactylus*.

The species is tolerant of minimal habitat disturbance but prefers partial to closed-canopy forest. Extensive destruction of habitat has been a continuous threat, particularly that due to agriculture, logging, and development (IUCN SSC Amphibian Specialist Group 2021p). Of the 34 individuals collected in 2010–11, all in Quickstep in the vicinity of Cockpit Country, four were found to be infected with chytrid fungus (Holmes *et al.* 2012).

13. Eleutherodactylus nubicola *Dunn, 1926*

Eleutherodactylus nubicola is found in a limited area in the Blue Mountains in closed-canopy deciduous forest, typically at an elevation between 1,060 and 1,880 m (Lynn 1940; IUCN SSC Amphibian Specialist Group 2021q). It is one of the largest-bodied members of this genus in Jamaica, and its distinctive features include a dark brown uniform body, well-developed dorsolateral line, and long legs with heels that meet or overlap when the rear legs are flexed (Lynn 1940). Like *E. gossei*, this species was originally thought to be a subspecies of *E. luteolus* but has been determined to be an independent species. It can be distinguished from *E. luteolus* by its difference in call, colour, pattern, and size (Lynn 1940).

Due to its restricted range and declining population, it is listed as Endangered (EN). It is intolerant of disturbance to its habitat, which most often takes the form of small-scale farming and invasive flora. Adults often are found beneath rocks in hollows, where eggs are also laid, and where mating calls occur (Lynn 1940; IUCN SSC Amphibian Specialist Group 2021q). Of the six individuals collected across sites in 2010–11, two in the east of the island were found to be infected with chytrid fungus (Holmes *et al.* 2012).

14. Eleutherodactylus orcutti *Dunn, 1928*

Eleutherodactylus orcutti saw a sharp decrease in population size (approximately 80%) in the 1980s, possibly due to chytridiomycosis (IUCN SSC Amphibian Specialist Group 2021r). Like *E. jamaicensis*, this species has not been recorded since the 1980s and was not found in the 2010–11 survey. It is a stream specialist, which makes it highly susceptible to *Bd* infection

(Holmes *et al.* 2012). Its distinctive webbed feet make it unique among Jamaican species of *Eleutherodactylus* (Lynn 1940). It has three primary varieties of pattern: solid with no markings, dark interocular bar and dorsal mottling, or lateral stripes. As with *E. cundalli*, males can be distinguished from females by the substantially larger tympanum of the male (Lynn 1940).

Although *E. orcutti* is listed as Critically Endangered (CR), it is very likely extinct. Historically, it was found in wet forests in the east of the island, including the Blue and John Crow Mountains National Park. Despite the degradation of habitat in its former range, the presence of intact forest implies that its disappearance was not precipitated solely by loss of habitat, but may have been due to other factors, such as invasive predators and chytridiomycosis (IUCN SSC Amphibian Specialist Group 2021r).

15. Eleutherodactylus pantoni *Dunn, 1926*

Despite being a commonly seen species with a widespread distribution across the island, *Eleutherodactylus pantoni* is listed as Vulnerable (VU) due to ongoing decline in habitat extent and quality. Distinctive features include a poorly developed dorsolateral line, red groin, yellow or orange belly, mottled back, and heels that meet or overlap when the rear legs are flexed (Lynn 1940). It is one of the larger species within the genus and often exhibits uniform colouration with no pattern. Individuals have been found inhabiting – somewhat permanently – small burrows (Lynn 1940). Despite no distinct morphological differences among individuals from different areas, a difference in the call has been identified among populations in the western, central, and eastern mountains (Lynn and Dent 1943).

The species occupies a variety of habitats, including riparian, open forest, forest edges, and pastures or banana groves that are close to a forest edge. It can tolerate moderate disturbance but cannot withstand complete destruction of its habitat such as caused by agriculture and development (IUCN SSC Amphibian Specialist Group 2021s). Although the species is found in the same range as *E. nubicola*, the elevational ranges of the two show minimal overlap, with *E. pantoni* occupying areas below approximately 1,219 m asl (Lynn 1940). Notably, individuals have been found with the small lizard *Sphaerodactylus goniorhynchus* in their stomachs, along with the arthropods typically consumed by *Eleutherodactylus* (Lynn 1940). Of the 114 individuals collected across sites in 2010–11, 14 in the west of the island were found to be infected with chytrid fungus (Holmes *et al.* 2012).

16. Eleutherodactylus pentasyringos *Schwartz and Fowler, 1973*

Eleutherodactylus pentasyringos was once considered to be a subspecies of *E. pantoni* (Schwartz and Fowler 1973). Distinctive features include small size; stippled, but pale, ventral surface; minimal dorsal warts; interocular bar; and variation in the dorsal pattern (Schwartz and Fowler 1973). This species occurs in rocky, mesic forests in the Blue and John Crow Mountains and north of these mountains. It is found at high elevations up to 1,275 m. It is listed as Endangered (EN), due to its declining population and a habitat that has been used for agriculture, logging, and human settlement (IUCN SSC Amphibian Specialist Group 2021t). Of the 26 individuals collected across sites in 2010–11, four were found to be infected with chytrid fungus (Holmes *et al.* 2012).

17. Eleutherodactylus sisyphodemus *Crombie, 1977*

Eleutherodactylus sisyphodemus (leaf mimic frog) is listed as Critically Endangered (CR) due to the ongoing decline in its population and its occurrence at only one location. The species was first described from a gravid female found approximately four miles from a cave in Quickstep, Cockpit Country in 1975 (Crombie 1977). It differed in appearance from most *Eleutherodactylus* spp. then known for its developed digital pads, black ventral colouration, dorsal protrusions, and distinct tarsal fringe. At the time, six other specimens existed in collections at the Smithsonian

National Museum of Natural History and at Harvard University's Museum of Comparative Zoology. All six individuals were collected between 1974 and 1975 within the same vicinity on neighbouring ridges in Cockpit Country. Crombie compared these six individuals with the gravid female collected in 1975 and determined that there was very little variation among individuals of the species, particularly given their uniquely limited polymorphism in dorsal pattern (Crombie 1977). It is a small, cryptic species that occupies leaf litter in wet limestone forests at low elevation (Crombie 1977; IUCN SSC Amphibian Specialist Group 2021u). This species was last recorded in 1984 (Hedges and Díaz 2011). Due to its low tolerance of disturbance, it is severely threatened by the degradation of habitat that often follows illegal harvesting of trees in the area that it occupies.

III. Introduced amphibians

There are four species of non-native amphibians that have been introduced and established in Jamaica. Historically, they were introduced for biological control and by accidental importation. They are considered to be detrimental to the native Jamaican fauna in many ways, as outlined next. Additionally, the Cuban treefrog, *Osteopilus septentrionalis,* now appears to be established in Jamaica, having rapidly increased in abundance in recent years, thereby potentially negatively affecting native species (e.g. IUCN SSC Amphibian Specialist Group 2021a).

A. *Rhinella marina* (Linnaeus, 1758)

The cane toad *Rhinella marina* (formerly *Bufo marinus*) is native to a large geographic area spanning the southern United States, Mexico, Central America, and into northern South America. It is a large, toxic member of the family Bufonidae and is considered by the IUCN to be one of the 100 most invasive species in the world (Wilson 2011). Individuals are large (snout–vent length 140–160 mm), nocturnal, and voracious (Lynn 1940). The species was introduced into Jamaica in 1855 as an agent of biological control for sugar-cane pests, and subsequently became a substantial threat to the island's endemic fauna. An example of this is a vulnerable species, the Jamaican boa *Chilabothrus subflavus,* an endemic predator for which a direct link has been established between the toxicity of cane toads and the mortality of snakes ingesting them (Wilson *et al.* 2011). This toad can be found across the island in all types of habitat and can proliferate in degraded areas that native species find challenging (Solís *et al.* 2009; Wilson *et al.* 2011). It is opportunistic in both feeding and breeding and can reproduce almost anywhere that contains a small body of water (Solís *et al.* 2009). The species may also be a vector for diseases such as chytridiomycosis and ranavirus. However, the 14 individuals collected in the 2010–11 *Bd* survey were not found to be infected with chytrid fungus (Holmes *et al.* 2012).

B. *Eleutherodactylus johnstonei* Barbour, 1914

Eleutherodactylus johnstonei, also known as the Lesser Antillean whistling frog, was introduced to Jamaica from the Eastern Caribbean, to which it is native (IUCN SSC Amphibian Specialist Group 2021v; Wilson 2011). This species competes with endemic frogs such as *E. junori* by drowning out their mating calls and consuming available resources, but no definitive studies have been conducted that prove this (Hedges 1993; Wilson 2011). Its increasing population is also aided by its high tolerance for degraded habitats, which will likely allow it to outcompete endemic species in these areas (Wilson 2011). The species may also be a vector for the chytrid fungus. Of the 43 individuals collected across sites in 2010–11, ten from various parts of the island were found to be infected with chytrid fungus (Holmes *et al.* 2012).

C. *Eleutherodactylus planirostris* (Cope, 1862)

The greenhouse frog *Eleutherodactylus planirostris,* is native to Cuba, the Bahamas, and the Cayman Islands and has been introduced to parts of the United States (Heinicke *et al.* 2011; IUCN SSC Amphibian Specialist Group 2021w). The species occurs in all habitat types and can occupy both undisturbed and degraded areas (IUCN SSC Amphibian Specialist Group 2021w). It consumes large quantities of invertebrates but has not been found to compete with native *Eleutherodactylus,* partly because of its preference for disturbed areas (Kraus *et al.* 1999; Wilson 2011). Distribution of this species is largely due to the horticultural trade, as eggs are easily transported in soil and nursery materials (Kraus *et al.* 1999). In the 2010–11 *Bd* survey, one individual collected in Windsor (Cockpit Country) was found to be infected by chytrid fungus (Holmes *et al.* 2012).

D. *Lithobates catesbeianus* (Shaw, 1802)

The American bullfrog *Lithobates catesbeianus* is also considered by the IUCN to be one of the 100 most invasive species in the world (Wilson 2011). It was first introduced to Jamaica in 1967 for commercial purposes and has since spread to most parishes across the island (Mahon and Aiken 1977). Accidental transport may also have occurred frequently via aquaculture (Wilson 2011). This species is an efficient carrier of chytridiomycosis and ranavirus. It is differentially resistant to strains of *Bd* and is suspected of having been the source of introduction of *Bd* to Jamaica (Daszak *et al.* 2004; Schloegel *et al.* 2009; Wilson 2011; Gervasi *et al.* 2013). Studies have also shown that, even when asymptomatic for chytridiomycosis, bullfrogs can still transmit the disease to other anurans via shared bodies of water (Daszak *et al.* 2004; Greenspan *et al.* 2012; Gervasi *et al.* 2013). The individual collected in the 2010–11 *Bd* survey was not found to be infected with chytrid fungus (Holmes *et al.* 2012). *Lithobates catesbeianus* has been found to impact populations of native anurans significantly through predation. Studies of gut contents of American bullfrogs in other parts of the world have definitively shown that local amphibians form part of their diet, with larger bullfrogs consuming more local frogs than do smaller ones (e.g. Da Silva *et al.* 2009; Jancowski and Orchard 2013). Due to the threat that this species poses, the IUCN recommends eradication in its introduced range as a conservation priority (IUCN SSC Amphibian Specialist Group 2015).

IV. Threats to, and conservation priorities for, Jamaica's amphibians

As seen in earlier sections, the primary threat to Jamaican amphibians is degradation and loss of habitat. Much of this loss is due to the conversion of forested land for agriculture, development of infrastructure, bauxite mining, limestone quarrying, and touristic activities. The overuse of forest resources is an issue because, even in areas that are not being converted to agriculture or other development, mature trees are continually being harvested, even in protected areas such as the forest reserves in Cockpit Country. In many cases, these activities are being carried out within or adjacent to protected areas. Most of Jamaica's endemic amphibians occur in areas that are within protected or ecologically important areas. These include Cockpit Country, Blue and John Crow Mountains National Park, Portland Bight Protected Area (PBPA), Black River Lower Morass, and Dolphin Head Mountain (Wilson 2011). Existing laws and regulations that protect Jamaica's natural resources include the Natural Resources Conservation Authority Act 1991, Wildlife Protection Act 1945, Watershed Protection Act 1963, Beach Control Act 1956, and 'National Forest Management and Conservation Plan and Strategic Forest Management Plan' of the Forest Act 1996. Jamaica is also a signatory to the Convention on Biological Diversity and the

Ramsar Convention. Unfortunately, assessments reveal that this does not translate to the protection of species threatened by human activities. It is critical, therefore, that these sites be regulated to prevent overexploitation, deforestation, and other effects of human activity. In the case of dry-forest species, plans are currently underway to manage Goat Islands, off the south coast of the PBPA, as a biodiversity reserve for endemic species from the neighbouring dry forests.

Chytridiomycosis, which is caused by the fungal parasite *Batrachochytrium dendrobatidis*, has probably caused the extinction of various amphibians across the Caribbean. The fungus disrupts cutaneous function and associated processes, resulting in the death of an infected individual (Hedges and Díaz 2011). The survey conducted by Holmes, McLaren, and Wilson in 2010–11 reported that infection rates on the island ranged from 0 to 33% (Holmes *et al.* 2012). All but seven of Jamaica's 21 described endemic amphibians were sampled and tested. Despite chytridiomycosis not being detected in *Osteopilus wilderi*, *Eleutherodactylus griphus*, or *E. cavernicola*, it is uncertain whether the disease is truly absent from their populations, as sample sizes were small. In general, however, chytridiomycosis does not seem to be an immediate threat to native species, given the low loads found across species, at least for the current strain of *Bd*. Two *Eleutherodactylus* species, *E. orcutti* and *E. jamaicensis*, have not been documented since the 1980s and were not found during this study despite extensive efforts. This is a troubling development, as it implies that these species may have fallen victim to *Bd* (Holmes *et al.* 2012). However, a more promising result of the study showed that Jamaica is likely in a post-epidemic phase (Holmes *et al.* 2012). Further studies have since shown that infection rates of *Bd* increase with rainfall during dry months and with temperature during warm months up to a threshold (Holmes *et al.* 2014). This result implies that infection rates are likely to increase in the future under climatic change, as even historically wet and cool areas will become warmer as temperatures rise across the region (Holmes *et al.* 2014).

Climatic change is a global phenomenon that is likely to have far-reaching impacts on local amphibians. Historically, temperatures in Jamaica have been trending upwards. Although rainfall has varied, droughts have occurred frequently in recent years. Projections indicate that Jamaica will likely be up to 4°C hotter and 21% drier by the end of the twenty-first century (CSGM 2017). The result will be unprecedented climatic conditions that surpass any extremes previously known, also known as 'climate departure' (Mora *et al.* 2013). This is of particular concern for the tropics, which will experience such extremes earlier than the rest of the world. For tropical species, particularly herpetofauna that often have a narrow range of optimal conditions, this is an especially detrimental prospect (Araújo *et al.* 2005; Mora *et al.* 2013). Potential effects for amphibians include a possible increase in *Bd* infection rates to epidemic levels, stress and developmental instability, reduction of species' ranges, and shifts in breeding periods (Carey and Alexander 2003; Araújo and Pearson 2005; Daszak *et al.* 2005; Pounds *et al.* 2006; Alford *et al.* 2007). There is much to be done in identifying the risks of extinction and the conservation priorities for local amphibians in the face of climatic change.

For all of Jamaica's native amphibian species, the next critical step is a new wave of surveys to confirm distributions and population sizes. This is particularly important for species that have not been recorded for many years and may have already gone extinct, such as *Eleutherodactylus orcutti*. If small populations of these species are found, it may be possible to establish captive propagation programmes for conserving them (Wilson 2011).

V. References

Aguirre-Santoro, J., Campbell, K. C. S. E., and Proctor, G. R., 2016. A new species of *Hohenbergia* (Bromeliaceae) endemic to the Dolphin Head Mountains in western Jamaica. *Phytotaxa* **247**: 133–137. https://doi.org/10.11646/phytotaxa.247.2.5

Alford, R. A., Bradfield, K. S., and Richards, S. J., 2007. Ecology: Global warming and amphibian losses. *Nature* **447**: E3–4; discussion E5–6. https://doi.org/10.1038/nature05940

Araújo, M. B. and Pearson, R. G., 2005. Equilibrium of species' distributions with climate. *Ecography* **28**: 693–695. https://doi.org/10.1111/j.2005.0906-7590.04253.x

Araújo, M. B., Pearson, R. G., Thuiller, W., and Erhard, M., 2005. Validation of species impact models under climate change. *Global Change Biology* **11**: 1504–1513. https://doi.org/10.1111/j.1365-2486.2005.01000.x

Bacon, J. P., Gray, J. A., and Kitson, L., 2011. Status and conservation of the reptiles and amphibians of the Bermuda islands. Pp. 161–182 in *Conservation of Caribbean Island Herpetofaunas Volume 1: Conservation Biology and the Wider Caribbean*, ed by A. Hailey, B. S. Wilson and J. A. Horrocks. Brill, Leiden, The Netherlands. https://doi.org/10.1163/ej.9789004183957.i-228.58

Barbour, T., 1910. *Notes on the herpetology of Jamaica. Bulletin of the Museum of Comparative Zoölogy* **52**: 271–301 + 2 plates.

Carey, C. and Alexander, M. A., 2003. Climate change and amphibian declines: is there a link? *Diversity and distributions* **9**: 111–121. https://doi.org/10.1046/j.1472-4642.2003.00011.x

Climate Studies Group, Mona (CSGM), 2017. State of the Jamaican Climate 2015: Information for Resilience Building (Full Report). Produced for the Planning Institute of Jamaica (PIOJ), Kingston, Jamaica.

Crombie, R. I., 1977. A new species of frog of the genus *Eleutherodactylus* (Amphibia: Leptodactylidae) from the Cockpit Country of Jamaica. *Proceedings of the Biological Society of Washington* **90**: 194–204.

Da Silva, E. T., Dos Reis, E. P., Feio, R. N., and Ribeiro Filho, O. P., 2009. Diet of the invasive frog *Lithobates catesbeianus* (Shaw, 1802) (Anura: Ranidae) in viçosa, Minas Gerais State, Brazil. *South American Journal of Herpetology* **4**: 286–295. https://doi.org/10.2994/057.004.0312

Daszak, P., Strieby, A., Cunningham, A. A., Longcore, J. E., Brown, C. C., and Porter, D., 2004. Experimental evidence that the bullfrog (*Rana catesbeiana*) is a potential carrier of chytridiomycosis, an emerging fungal disease of amphibians. *Herpetological Journal* **14**: 201–208.

Diesel, R., Bäurle, G., and Vogel, P., 1995. Cave breeding and froglet transport: a novel pattern of anuran brood care in the Jamaican frog, *Eleutherodactylus cundalli*. *Copeia* **1995**: 354–360. https://doi.org/10.2307/1446898

Gervasi, S. S., Urbina, J., Hua, J., Chestnut, T., Relyea, R. A., and Blaustein, A. R., 2013. Experimental evidence for American bullfrog (*Lithobates catesbeianus*) susceptibility to chytrid fungus (*Batrachochytrium dendrobatidis*). *EcoHealth* **10**: 166–171. https://doi.org/10.1007/s10393-013-0832-8

Gosse, P. H., 1851. A Naturalist's Sojourn in Jamaica. Longman, Brown, Green, and Longmans, London. Pp. 508 + xxiv.

Greenspan, S. E., Calhoun, A. J., Longcore, J. E., and Levy, M. G., 2012. Transmission of *Batrachochytrium dendrobatidis* to wood frogs (*Lithobates sylvaticus*) via a bullfrog (*L. catesbeianus*) vector. *Journal of Wildlife Diseases* **48**: 575–582.

Grant, C., 1940. The herpetology of Jamaica. Part II. The reptiles. *Bulletin of the Institute of Jamaica* (Science Series) **1**: 61–148.

Hedges, S. B., 1989. An island radiation: allozyme evolution in Jamaican frogs of the genus *Eleutherodactylus* (Leptodactylidae). *Caribbean Journal of Science* **25**: 123–147.

Hedges, S. B., 1993. Global amphibian declines: a perspective from the Caribbean. *Biodiversity and Conservation* **2**: 290–303. https://doi.org/10.1007/BF00056674

Hedges, S. B. and Díaz, L. M., 2011. The conservation status of amphibians in the West Indies. Pp. 31–47 in *Conservation of Caribbean Island Herpetofaunas Volume 1: Conservation Biology and the Wider Caribbean*, ed. A. Hailey, B. S. Wilson, and J. A. Horrocks. Brill, Leiden, The Netherlands. https://doi.org/10.1163/ej.9789004183957.i-228.23

Heinicke, M. P., Diaz, L. M., and Hedges, S. B., 2011. Origin of invasive Florida frogs traced to Cuba. *Biology Letters* **7**: 407–410. https://doi.org/10.1098/rsbl.2010.1131

Holmes, I., McLaren, K. and Wilson, B., 2012. Surveys for frog diversity and *Batrachochytrium dendrobatidis* in Jamaica. *Herpetological Review* **43**: 219–278.

Holmes, I., McLaren, K., and Wilson, B., 2014. Precipitation constrains amphibian chytrid fungus infection rates in a terrestrial frog assemblage in Jamaica, West Indies. *Biotropica* **46**: 219–228. https://doi.org/10.1111/btp.12093

IUCN SSC Amphibian Specialist Group. 2021a. *Osteopilus crucialis*. The IUCN Red List of Threatened Species 2021: e.T55807A3032124. https://dx.doi.org/

10.2305/IUCN.UK.2021-1.RLTS.T55807A3032124.en. Downloaded 14 January 2022.

IUCN SSC Amphibian Specialist Group. 2021b. *Osteopilus marianae*. The IUCN Red List of Threatened Species 2021: e.T55809A3032423. https://dx.doi.org/10.2305/IUCN.UK.2021-1.RLTS.T55809A3032423.en. Downloaded 14 January 2022.

IUCN SSC Amphibian Specialist Group. 2021c. *Osteopilus ocellatus*. The IUCN Red List of Threatened Species 2021: e.T55806A172794433. https://dx.doi.org/10.2305/IUCN.UK.2021-1.RLTS.T55806A172794433.en. Downloaded 14 January 2022.

IUCN SSC Amphibian Specialist Group. 2021d. *Osteopilus wilderi*. The IUCN Red List of Threatened Species 2021: e.T55812A3032991. https://dx.doi.org/10.2305/IUCN.UK.2021- 1.RLTS.T55812A3032991.en. Downloaded 14 January 2022.

IUCN SSC Amphibian Specialist Group. 2021e. *Eleutherodactylus alticola*. The IUCN Red List of Threatened Species 2021: e.T56408A86895367. https://dx.doi.org/10.2305/IUCN.UK.2021- 1.RLTS.T56408A86895367.en. Downloaded 14 January 2022.

IUCN SSC Amphibian Specialist Group. 2021f. *Eleutherodactylus andrewsi*. The IUCN Red List of Threatened Species 2021: e.T56416A3038632. https://dx.doi.org/10.2305/IUCN.UK.2021-1.RLTS.T56416A3038632.en. Downloaded 14 January 2022.

IUCN SSC Amphibian Specialist Group. 2021g. *Eleutherodactylus cavernicola*. The IUCN Red List of Threatened Species 2021: e.T7143A3137506. https://dx.doi.org/10.2305/IUCN.UK.2021-1.RLTS.T7143A3137506.en. Downloaded 14 January 2022.

IUCN SSC Amphibian Specialist Group. 2021h. *Eleutherodactylus cundalli*. The IUCN Red List of Threatened Species 2021: e.T56543A86895513. https://dx.doi.org/10.2305/IUCN.UK.2021-1.RLTS.T56543A86895513.en. Downloaded 14 January 2022.

IUCN SSC Amphibian Specialist Group. 2021i. *Eleutherodactylus fuscus*. The IUCN Red List of Threatened Species 2021: e.T7144A3137640. https://dx.doi.org/10.2305/IUCN.UK.2021-1.RLTS.T7144A3137640.en. Downloaded 14 January 2022.

IUCN SSC Amphibian Specialist Group. 2021j. *Eleutherodactylus glaucoreius*. The IUCN Red List of Threatened Species 2021: e.T56621A3044897. https://dx.doi.org/10.2305/IUCN.UK.2021-1.RLTS.T56621A3044897.en. Downloaded 14 January 2022.

IUCN SSC Amphibian Specialist Group. 2021k. *Eleutherodactylus gossei*. The IUCN Red List of Threatened Species 2021: e.T56624A3045049. https://dx.doi.org/10.2305/IUCN.UK.2021-1.RLTS.T56624A3045049.en. Downloaded 14 January 2022.

IUCN SSC Amphibian Specialist Group. 2021l. *Eleutherodactylus grabhami*. The IUCN Red List of Threatened Species 2021: e.T56625A3045185. https://dx.doi.org/10.2305/IUCN.UK.2021-1.RLTS.T56625A3045185.en. Downloaded 14 January 2022.

IUCN SSC Amphibian Specialist Group. 2021m. *Eleutherodactylus griphus*. The IUCN Red List of Threatened Species 2021: e.T56633A3045434. https://dx.doi.org/10.2305/IUCN.UK.2021-1.RLTS.T56633A3045434.en. Downloaded 14 January 2022.

IUCN SSC Amphibian Specialist Group. 2021n. *Eleutherodactylus jamaicensis*. The IUCN Red List of Threatened Species 2021: e.T56681A3047028. https://dx.doi.org/10.2305/IUCN.UK.2021-1.RLTS.T56681A3047028.en. Downloaded 14 January 2022.

IUCN SSC Amphibian Specialist Group. 2021o. *Eleutherodactylus junori*. The IUCN Red List of Threatened Species 2021: e.T7145A3137776. https://dx.doi.org/10.2305/IUCN.UK.2021-1.RLTS.T7145A3137776.en. Downloaded 14 January 2022.

IUCN SSC Amphibian Specialist Group. 2021p. *Eleutherodactylus luteolus*. The IUCN Red List of Threatened Species 2021: e.T56733A3048460. https://dx.doi.org/10.2305/IUCN.UK.2021-1.RLTS.T56733A3048460.en. Downloaded 14 January 2022.

IUCN SSC Amphibian Specialist Group. 2021q. *Eleutherodactylus nubicola*. The IUCN Red List of Threatened Species 2021: e.T56797A3049979. https://dx.doi.org/10.2305/IUCN.UK.2021-1.RLTS.T56797A3049979.en. Downloaded 14 January 2022.

IUCN SSC Amphibian Specialist Group. 2021r. *Eleutherodactylus orcutti*. The IUCN Red List of Threatened Species 2021: e.T56813A3050360. https://dx.doi.org/10.2305/IUCN.UK.2021-1.RLTS.T56813A3050360.en. Downloaded 14 January 2022.

IUCN SSC Amphibian Specialist Group. 2021s. *Eleutherodactylus pantoni*. The IUCN Red List of Threatened Species 2021: e.T56824A3050749. https://dx.doi.org/10.2305/IUCN.UK.2021-1.RLTS.T56824A3050749.en. Downloaded 14 January 2022.

IUCN SSC Amphibian Specialist Group. 2021t. *Eleutherodactylus pentasyringos*. The IUCN Red List of Threatened Species 2021: e.T56842A3051459. https://dx.doi.org/10.2305/IUCN.UK.2021-1.RLTS.T56842A3051459.en. Downloaded 14 January 2022.

IUCN SSC Amphibian Specialist Group. 2021u. *Eleutherodactylus sisyphodemus*. The IUCN Red List of Threatened Species 2021: e.T7148A3138071. https://dx.doi.org/10.2305/IUCN.UK.2021-1.RLTS.T7148A3138071.en. Downloaded 14 January 2022.

IUCN SSC Amphibian Specialist Group. 2021v. *Eleutherodactylus johnstonei*. The IUCN Red List of Threatened Species 2021: e.T56684A3047195. https://dx.doi.org/10.2305/IUCN.UK.2021-1.RLTS.T56684A3047195.en. Downloaded 14 January 2022.

IUCN SSC Amphibian Specialist Group. 2021w. *Eleutherodactylus planirostris*. The IUCN Red List of Threatened Species 2021: e.T56864A54370037. https://dx.doi.org/10.2305/IUCN.UK.2021-1.RLTS.T56864A54370037.en. Downloaded 14 January 2022.

IUCN SSC Amphibian Specialist Group, 2015. *Lithobates catesbeianus. The IUCN Red List of Threatened Species* 2015: e.T58565A53969770. http://dx.doi.org/10.2305/IUCN.UK.2015-4.RLTS.T58565A53969770.en. Downloaded 4 August 2019.

Jancowski, K. and Orchard, S., 2013. Stomach contents from invasive American bullfrogs *Rana catesbeiana* (= *Lithobates catesbeianus*) on southern Vancouver Island, British Columbia, Canada. *NeoBiota* **16**: 17–37. https://doi.org/10.3897/neobiota.16.3806

Kraus, F., Campbell, E. W., Allison, A., and Pratt, T., 1999. *Eleutherodactylus* frog introduction to Hawaii. *Herpetological Review* **30**: 21–25.

Linton, L., 2017. "Cockpit Country Boundary Named… No Mining in Protected Area." Jamaica Information Service. Accessed at: https://jis.gov.jm/cockpit-country-boundary-named-no-mining-protected-area/ (04 August 2019).

Lynn, W. G., 1940. The herpetology of Jamaica. Part I. Amphibians. *Bulletin of the Institute of Jamaica* (Science Series) **1**: 1–60.

Lynn, W. G., 1954. Description of a new frog of the genus *Eleutherodactylus* from Jamaica, BWI. *Journal of the Washington Academy of Sciences* **44**: 400–402.

Lynn, W. G. and Dent, J. N., 1943. Notes on Jamaican amphibians. *Copeia* **1943**: 234–242. https://doi.org/10.2307/1438136

Mahon, R. and Aiken, K., 1977. The establishment of the North American bullfrog, *Rana catesbeiana* (Amphibia, Anura, Ranidae) in Jamaica. *Journal of Herpetology* **11**: 197–199. https://doi.org/10.2307/1563140

Mora, C., Frazier, A. G., Longman, R. J., Dacks, R. S., Walton, M. M., Tong, E. J., Sanchez, J. J., Kaiser, L. R., Stender, Y. O., Anderson, J. M., and Ambrosino, C. M., 2013. The projected timing of climate departure from recent variability. *Nature* **502**: 183–187. https://doi.org/10.1038/nature12540

Newman, M. E., McLaren, K. P., and Wilson, B. S., 2011. Use of object-oriented classification and fragmentation analysis (1985–2008) to identify important areas for conservation in Cockpit Country, Jamaica. *Environmental Monitoring and Assessment* **172**: 391–406.

Pounds, J. A., Bustamante, M. R., Coloma, L. A., Consuegra, J. A., Fogden, M. P., Foster, P. N., La Marca, E., Masters, K. L., Merino-Viteri, A., Puschendorf, R., and Ron, S. R., 2006. Widespread amphibian extinctions from epidemic disease driven by global warming. *Nature* **439**: 161–167. https://doi.org/10.1007/s10661-010-1342-6

Ricketts, T. H., Dinerstein, E., Boucher, T., Brooks, T. M., Butchart, S. H., Hoffmann, M., Lamoreux, J. F., Morrison, J., Parr, M., Pilgrim, J. D., and Rodrigues, A. S., 2005. Pinpointing and preventing imminent extinctions. *Proceedings of the National Academy of Sciences* **102**: 18497–18501. https://doi.org/10.1073/pnas.0509060102

Schloegel, L. M., Picco, A. M, Kilpatrick, A. M., Davies, A. J., Hyatt, A. D., and Daszak, P., 2009. Magnitude of the US trade in amphibians and presence of *Batrachochytrium dendrobatidis* and ranavirus infection in imported North American bullfrogs (*Rana catesbeiana*). *Biological Conservation* **142**: 1420–1426. https://doi.org/10.1016/j.biocon.2009.02.007

Schwartz, A. and Fowler, D. C., 1973. The anura of Jamaica: a progress report. *Studies on the Fauna of Curaçao and Other Caribbean Islands* **142**: 50–142.

Solís, F., Ibánez, R., Hammerson, G., Hedges, B., Diesmos, A. and Matsui, M., Hero, J-M., Richards. S., Coloma, L., Ron, S., La Marca, E., Hardy, J., Powell, R., Bolanos, F., Chaves, G., and Ponce, P., 2009. *Rhinella marina. The IUCN Red List of Threatened Species* 2009: e.T41065A10382424. http://dx.doi.org/10.2305/IUCN.UK.2009-2.RLTS.T41065A10382424.en. Downloaded 4 August 2019.

Wilson, B. S., 2011. Conservation of Jamaican amphibians and reptiles. Pp. 273–310 in *Conservation of Caribbean Island Herpetofaunas. Volume 2: Regional Accounts of the West Indies*, ed by A. Hailey, B. S. Wilson, and J. A. Horrocks. Brill, Leiden, The Netherlands. https://doi.org/10.1163/ej.9789004194083.i-439.110

Wilson, B., Grant, T., van Veen, R., Hudson, R., Fleuchaus, D., Robinson, O., and Stephenson, K., 2016. The Jamaican Iguana (*Cyclura collei*): a report on 25 years of conservation effort. *Herpetological Conservation and Biology* **11**: 237–254.

Wilson, B. S., Koenig, S. E., van Veen, R., Miersma, E., and Rudolph, D. C., 2011. Cane toads a threat to West Indian wildlife: mortality of Jamaican boas attributable to toad ingestion. *Biological Invasions* **13**: 55–60. https://doi.org/10.1007/s10530-010-9787-7

6 Status of Amphibians in the Republic of Haiti: a review

Neftalí Ríos-López

Abbreviations and acronyms

asl	*above sea level*
Bd	Batrachochytrium dendrobatidis
CC BY-SA 4.0	*Creative Commons Attribution Share Alike 4.0 International*
CM	*Carnegie Museum of Natural History*
ETOPO1	*Earth Topographic 1 Arc-Minute Global Relief Model*
ha	*hectares*
IUCN	*International Union for the Conservation of Nature*
KU KUH	*University of Kansas Biodiversity Institute Herpetology Collection*
Ma	*mega-annum or million years*
MCZ	*Museum of Comparative Zoology*

Neftalí Ríos-López, 'Chapter 6: Status of Amphibians in the Republic of Haiti: a review' in: *The Conservation and Biogeography of Amphibians in the Caribbean*. Pelagic Publishing (2023). © Neftalí Ríos-López. DOI: 10.53061/YTJG7753

MNHN-RA	*Muséum National D'Histoire Naturelle, Reptiles and Amphibians Collection*
NASA	*National Aeronautics and Space Administration*
NGDC	*National Geophysical Data Center*
S.A.	*Société Anonyme (a French term for a Public Limited Company or equivalent to 'Incorporated' in USA system)*
SBH	*Stephen Blair Hedges*
SRTM3	*Shuttle Radar Topography Mission Version 3.0*
SSC	*Species Survival Commission*
SSG	*Species Specialist Group*
SWBD	*SRTM Water Body Data*
UF	*University of Florida*
USNM	*National Museum of Natural History, Smithsonian Institution*
UTM	*Universal Transverse Mercator*
VMAP	*Vector Map*
UN	*United Nations*
WGS	*World Geodetic System*

I. Introduction

Nowhere in the Caribbean is the condition 'frequently recurrent struggle for existence' (Darwin 1872 [1909]: 23) more evident than in the Republic of Haiti. Greene (2001) and Tarter *et al.* (2016) have outlined Haiti's historical setting, but a few paragraphs are necessary here to explore some general background information. Initially, and in the words of Hill (1898: 289), the

> French and English – took advantage of their [Spaniards'] departure and began to prey upon the island. The French particularly assailed the weaker western end, which was then largely a wilderness; they first established stations, then plantations, and finally, in 1640, organized these irregular settlements into a colony under a governor sent from France. ... it became what was at that time the finest colony in the world.

Spain finally ceded the western third of the island of Hispaniola to France in 1697 (Greene 2001). The colony then was a nomadic harvesting/monoculture-based economy, mostly sponsored by France. Its wealth was driven by slavery (~800,000 slaves brought in from African countries) which accounted for 'a third of the entire Atlantic slave trade' (Federal Research Division 2006: 2) up until the end of the eighteenth century. The slave trade in this colony would not last any longer, nor its colonial status. After about 300 years of occupation by Europeans, Haiti's slave trade and its colonial status ceased when on 2 January 1804 it declared its 'Independence' from France and became the first example of a slave society to earn its nationhood through a revolution (Prou 2005; Pinto 2010).

From colonial times, Haiti had provided other nations with its territory's commodities in abundance. At the end of the nineteenth century, coffee (cultivated in the mountains) was the only crop of importance; other products were harvested from the wild but their 'agricultural possibilities are hardly drawn upon' (Hill 1898). Logwood (considered to be of the best quality; hand-harvested from natural areas) averaged 89,000 tons (US, 'short ton') annually; cocoa, cotton, sugar, and other woods (mahogany [most important], lignum-vitae, bois-jaune [West Indian sandalwood], and bayarondes) followed. Other products, seldom marketed for profit, however, included 'pineapples, peaches, strawberries, blackberries, and other fruits' found in the uplands, while 'oranges, citrons, plantains, bananas, lemons, shaddocks, mangos, artichokes, alligator-pears, and sapodillas', also abounded (Hill 1898).

Wetmore and Swales (1931: 11) considered Haiti to be 'a prosperous French colony' (or 'one of the most productive of colonies' according to Holdridge 1942: 18), yet Hill (1898: 267) concluded: 'From 1804 to the present the moral welfare of Haiti has been largely neglected by other nations and people, who have extended to it neither sympathy, recognition, nor aid.' Moreover, McClintock (2003: 4) highlighted that 'Following the revolution of 1804, the government [of Haiti] was forced to export timber throughout the nineteenth century to pay off a 90 million franc indemnity to France [from losses by the slave-traders].' This debt, estimated at ~21 billion in today's US dollars, was finally settled in 1947 (after 122 years of paying off the debt by Haitians) (Sperling 2017). A passage from Wetmore and Swales (1931: 53–54) is worth quoting:

> Finally, recommendation is made that definite areas in both republics [Haiti and Dominican Republic] be set aside as national parks and established as sanctuaries where original conditions may be preserved for coming generations and the continuance of plants, birds, and other animal life assured. In the present stage of development there remain large areas that as yet have not been commercially exploited where the original xerophytic forests, rain-forests, or stands of pine remain in their primitive condition, but every year sees some curtailment of this condition and some new use proposed for these various types of land. Establishment of park areas and forest reserves, for example in the Quita Espuela district [Dominican Republic], in the high mountain area about Constanza [Dominican Republic], on the ridge of La Selle [Haiti], and in the La Hotte region [Haiti] may now be made without particular difficulty. If delayed too long **the timber will be largely destroyed and a condition similar to that in Porto Rico where it has been possible to establish only one reserve of any extent, that on El Yunque, while elsewhere original conditions have been largely destroyed, will result** [emphasis supplied]. Such park areas, set aside now and guarded from fire and needless cutting, will prove valuable assets for the future.

The twentieth century revealed how different sets of policies (and political affairs) led to different, somewhat opposing results in Haiti and Puerto Rico (see Ríos-López *et al.*, Chapter 8 in this volume). For example, areas under the cover-type of 'forest' in Puerto Rico increased from 6% by the late 1940s to 54.8% in 2014. In contrast, forest conditions in Haiti, estimated at ~30% in 1940, dropped to 8–9% (Burns 1954 cited by Ehrlich *et al.* 1985) or 10% by 1970 (McClintock 2003), 6.4% by 1978, and 2.2% by 1993 (Smucker 2001) (Powell and Henderson [1996] mentioned < 1% of Haiti's 'original forest cover'). In the 2000s, estimates of the cover of forests in Haiti were between 1 and 2% (Michel 2001 cited by McClintock 2003; most recently to ~1% of its 'original primary forest' according to Hedges *et al.* 2018).

In Haiti (the mainland and all associated islands, cays, and islets), the present-day distribution of anurans also reflects the legacy of human affairs upon which the country's history was built. The conservation of Haitian anurans (and Haiti's biodiversity) thus rests in the hands of conservation of the country's landscape, which must include 'physical, social, economic, and institutional components. Otherwise, prospects for success are negligible' (Ehrlich *et al.* 1985: 31–32). Fortunately, the recommendations made by Wetmore and Swales almost a century ago on establishing national parks and sanctuaries at Massif de la Selle and Massif de la Hotte went beyond words alone. A new suite of ministries (agriculture, environment), policies, regulations, and natural areas established by governmental and private initiatives provides for a brighter future for the conservation of Haiti's biodiversity and its natural resources. Indeed, a national effort aimed at establishing and maintaining natural reserves for education, research, increased public awareness, and conservation of 'Key Biodiversity Areas' was initiated in collaboration with national and international partners (Gouvernement de la République d'Haïti 2016).

Nevertheless, there is a lot more to do. There have been centuries of exploitation of timber, monoculture, and slash-and-burn practices, while minimal planning and negligible investment in education, health, agriculture, potable water, and infrastructure have continued since colonial times. The first half of the twentieth century saw a transformation of remnant stands of forest for rubber plantations. In addition, crippling debts and cyclic-practices of vicious loans, and a persistent threat of military intervention, have combined with complex socioeconomical and political affairs (both domestic and international). All these factors, in synergy with natural phenomena (droughts, earthquakes, hurricanes), have exerted long-lasting influence on Haiti's natural resources (see Fig. 1 in Hedges 2006: 287) and other organisms that depend on them for reproduction and survival. Haiti's amphibians are no exception in this tug-of-war, but 'it should be remembered always that there is an essential difference between food and those wrought commodities, the raw materials of which are in great plenty ... The demand for an increased quantity of subsistence is, with few exceptions, constant everywhere, yet we see how slowly it is answered in all those countries that have been long occupied' (Malthus 1798: 28–29; also see Ehrlich *et al.* 1985). It is within Haiti's physical landscape, but also within Haiti's historical landscape that the topic of the conservation of Haitian anurans must be addressed.

In the words of Ernest E. Williams (1999: 1), 'with Schwartz and Henderson's massive 1991 compendium on the West Indian herpetofauna [I now add Henderson and Powell's 2009 expanded and updated compendium], an epoch primarily devoted to collection and description has come to an end. The goals will be quite different in the future.' Fourteen years have passed since Henderson and Powell's 2009 landmark update but, apart from taxonomy, systematics, and regional species accounts, we still lack a great deal of valuable information for the conservation and management of virtually all species of anurans in Haiti. For example, there is much to learn about the factors and variables that generate responses of species leading to population trends, as well as ecological data deemed fundamental for effective conservation efforts (e.g. natural history, reproductive phenology, interspecific interactions, population/community ecology). Likewise, projected changes in climate justify studies on the physiological responses of Haitian anurans to changing environmental conditions (e.g. shifts in elevational distribution), and migration potential in response to sea-level rise and intrusion of salt water in coastal areas. Currently, and fortunately, several areas in Haiti serve as natural laboratories for many studies of anurans, while international collaborations provide for *ex-situ* husbandry and reproduction for the reintroduction of critically endangered Haitian anurans in rehabilitated landscapes (e.g. Murphy and Gratwicke 2017). Research of Haitian anurans is still in its early stages; this is a marvellous opportunity for diverse research and educational agendas by current and future generation of scientists – for instance, see efforts in Martínez Rivera and Rodríguez Plaza (2015) and Mayer (2019).

Prou (2005: 192) indicated that 'the future of Haiti arguably hinges on the capacity and ability of Haitians to solve their own political, social, and economic problems.' Consequently, capacity-building, training, and higher education of Haitians are of the utmost importance (IUCN 1980). These fellow-Antillean citizens represent not only Haiti's most precious natural resource, but they will form current and future generations of scientists in charge of continuous research projects on anurans in their country, and for their own country's benefit.

In this chapter, I review the available literature at hand and pay particular attention to the endemic species of anurans – following Frost's (2022) nomenclature – from the Republic of Haiti.

I will also mention, rather briefly, the other species shared with its neighbouring country, the Dominican Republic (see Incháustegui and Díaz, Chapter 7 in this volume).

A. Geography, physiography/topography, generalized climate, and Holdridge's life zones

Haiti consists roughly of the western third of the island of Hispaniola (36.3%; ~27,750 km^2 [land = 27,560 km^2 and water = 190 km^2]; Smucker 2001; Central Intelligence Agency 2021) (the Dominican Republic occupies the other two-thirds), along with several large islands (in decreasing order of size: Île de la Gonâve, Île de la Tortue, Île à Vache, Îles Cayémites), and many smaller islands and cays (presumably devoid of anurans; see Powell 2012); they are collectively centred on approximately 19° 03' N: 73° 03' W (Fig. 6.1). Haiti also has several natural lakes, among them the second-largest lake (saline) in the Caribbean, Lake Azuei (or Étang Saumâtre; 18° 34' N: 71° 59' W), which

Figure 6.1. Topographic and political map of Haiti (in French; includes administrative divisions), within the Caribbean (inset). Scale bar = 50 km. Elevation and bathymetry in metres relative to the surface of the sea. UTM projection, WGS84 datum. From (title) 'Haiti topographic map-fr.svg' (https://commons.wikimedia.org/wiki/File:Haiti_topographic_map-fr.svg). Author: ©Rémi Kaupp (https://commons.wikimedia.org/wiki/User:Korrigan) – Own work. Created 6 March 2009. Licensed under CC BY-SA 4.0 International (https://creativecommons.org/licenses/by-sa/4.0). Sources of data: NASA SRTM3 (ftp://e0srp01u.ecs.nasa.gov/srtm/version2/SRTM3/) for terrain (public domain); NASA SWBD (ftp://e0srp01u.ecs.nasa.gov/srtm/version2/SWBD/) for coastlines and lakes; NGDC ETOPO1 (http://www.ngdc.noaa.gov/mgg/global/global.html) for bathymetry (public domain); NGDC World Data Bank II (http://rimmer.ngdc.noaa.gov/mgg/coast/getcoast.html) for borders and rivers (public domain); VMap-0 (http://www.mapability.com/info/vmap0_index.html) for miscellaneous data (public domain); UN map of Haiti (http://www.un.org/Depts/Cartographic/map/profile/haiti.pdf) for checking, 2008 Locator map: File:Karibik Haiti Position.png (https://commons.wikimedia.org/wiki/File:Karibik_Haiti_Position.png). Created using the tutorial (https://fr.wikipedia.org/wiki/Wikip%C3%A9dia:Atelier_graphique/Didacticiels_cartographiques/Didacticiel_de_cr%C3%A9ation_de_carte_topographique_vectorielle) by User:Sting (https://commons.wikimedia.org/wiki/User:Sting).

lies east in the Plaine du Cul-de-Sac to the south, and Lake Miragoâne (or Étang de Miragoâne), 'one of the largest, natural freshwater lakes in the Caribbean' (Brenner and Binford 1988: 86) located further south, in centre-north of the Tiburon Peninsula (18° 24' N: 73° 03' W).

According to Mann *et al.* (1991: 5), the outcrops of Cretaceous–Eocene basement rocks of the Republic of Haiti (and Hispaniola at large) can be divided into four groups based on origin, composition, and tectonic affinities (Fig. 6.2). In turn, four morphotectonic zones are represented within these groups (Mann *et al.* 1991: 6). These zones, from north to south Haiti (Fig. 6.2), include Zone Massif du Nord (covers: Plain du Nord [2, which extends to Valle del Cibao {i} in the

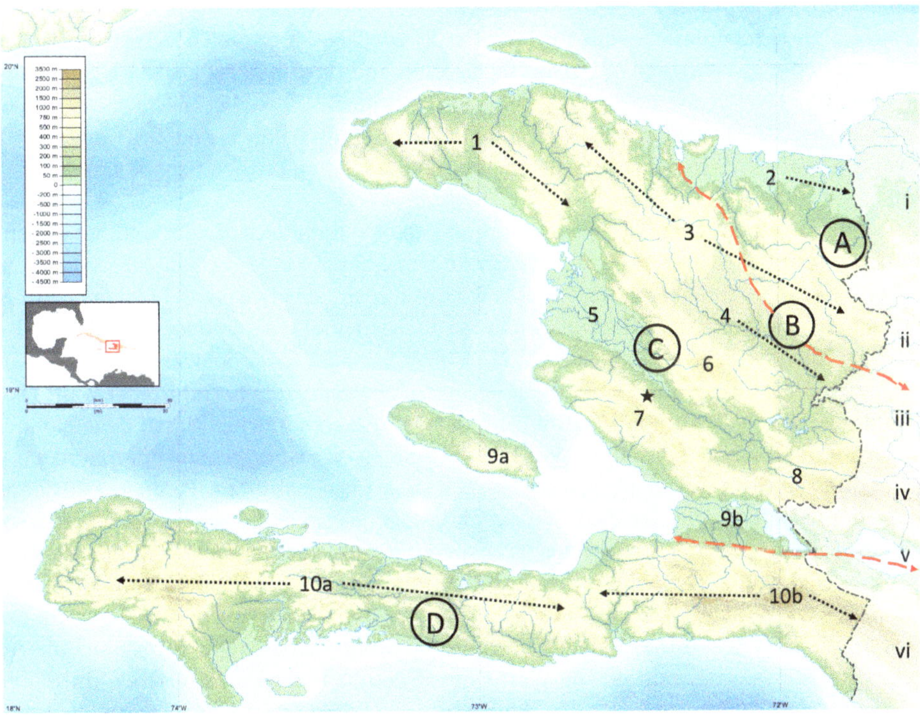

Figure 6.2. Topographic map of Haiti with outcrops of Cretaceous-Eocene basement rocks (encircled upper-case letters and long-dashed arrow lines delimiting extension of outcrops; outcrop B, which delimits outcrops A and C, is narrow in extension and runs diagonally along the Haiti/Dominican Republic northern palaeoisland; outcrop D conforms to the Haiti/Dominican Republic south palaeoisland). In Haitian territory, Arabic numerals with lower-case letters, and dotted arrow lines associated with these numerals, represent approximate extension of the main topographic features, as known in French, within the four morphotectonic zones described in the text; in the Dominican Republic, Roman numerals in lower-case represent the extension of topographic features from Haiti, as known in the Spanish-speaking country). Scale bar = 50 km. Elevation and bathymetry in metres relative to the surface of the sea. UTM projection, WGS84 datum. Modified from (title) 'Haiti blank map with topography.svg' (https://commons.wikimedia.org/wiki/File:Haiti_blank_map_with_topography.svg). Author: ©Rémi Kaupp (https://commons.wikimedia.org/wiki/User:Korrigan) - Own work. Created 11 March 2009. Licensed under CC BY-SA 4.0 International (https://creativecommons.org/licenses/by-sa/4.0). Sources of data: NASA SRTM3 (ftp://e0srp01u.ecs.nasa.gov/srtm/version2/SRTM3/) for terrain (public domain); NASA SWBD (ftp://e0srp01u.ecs.nasa.gov/srtm/version2/SWBD/) for coastlines and lakes NGDC ETOPO1 (http://www.ngdc.noaa.gov/mgg/global/global.html) for bathymetry (public domain); NGDC World Data Bank II (http://rimmer.ngdc.noaa.gov/mgg/coast/getcoast.html) for borders and rivers (public domain); VMap-0 (http://www.mapability.com/info/vmap0_index.html) for miscellaneous data (public domain); UN map of Haiti (http://www.un.org/Depts/Cartographic/map/profile/haiti.pdf) for checking, 2008 Locator map: File:Karibik Haiti Position.png (https://commons.wikimedia.org/wiki/File:Karibik_Haiti_Position.png). Created using this tutorial (https://fr.wikipedia.org/wiki/Wikip%C3%A9dia:Atelier_graphique/Didacticiels_cartographiques/Didacticiel_de_cr%C3%A9ation_de_carte_topographique_vectorielle) by User:Sting (https://commons.wikimedia.org/wiki/User:Sting).

Dominican Republic] and Massif du Nord [3, which extends to the Cordillera Central {ii} in the Dominican Republic]). Next is Zone northwestern-south-central (in Haiti: Presqu'île du Nord-Ouest within the Montagnes du Nord-Ouest [1], Plateau Central [4, which extends to Sierra de San Juan {iii} in the Dominican Republic], Plaine de l'Artibonite [5], and three mountain ranges represented by Montaignes Noires [6], Chaînes des Matheux [7], and Chaîne du Trou d'Eau [8, which extends to the Sierra de Neiba {iv} in the Dominican Republic]). Next is Zone Île de la Gonâve-Plaine de Cul-de-Sac (in Haiti: Île de la Gonâve [9a] and Plaine de Cul-de-Sac [9b, which extends to Valle de Neiba {v} in the Dominican Republic]). Finally, Zone southern peninsula (in Haiti: Massif de la Hotte [10a] and Massif de la Selle [10b, which extend to Sierra de Bahoruco {vi} in the Dominican Republic]).

Haiti's highest mountain peak is Pic la Selle (18° 21.60' N: 71° 58.56' W; located at east Massif de la Selle, southern Haiti) at 2,680 m asl; the second- and third-highest peaks (respectively: Pic de Macaya, 18° 22.98' N: 74° 1.50' W, at 2,347 m asl; Morne du Cibao, 18° 21.42' N: 72° 15.54' W, at 2,280 m asl) are also located at Massif de la Hotte in the south palaeoisland. Schwartz (1973: 249; Schwartz 1980a) refers to these massifs as 'the montane spine of the Hispaniolan south pale-oisland' (see below).

Although understandings of the tectonic and geologic dynamics of an island arc in general, and the Caribbean plate in particular, generate some controversy (Bowin 1975; Montgomery *et al.* 1994; Iturralde-Vinent and MacPhee 1999; Powell *et al.* 1999; Powell and Incháustegui 2011), the biogeographic patterns of the herpetofauna in Hispaniola are frequently associated with two major 'islands' *sensu* Williams (1961: 2) or 'paleoislands' *sensu* Schwartz (1978; 1980a). An elongated, but narrow, south palaeoisland occupies approximately two-thirds of Haiti's southern landmass. It consists of the Tiburon Peninsula, which contains the Massif de la Hotte in its western and central regions, and the Massif de la Selle in its eastern region; approximately one-third of this south palaeoisland lies in the Sierra de Bahoruco, within the Península de Bahoruco in the Dominican Republic. Iturralde-Vinent and MacPhee (1999: 89) suggested that this southern landmass collided with, and affixed to Hispaniola's northern landmass (and hence, Haiti in its present-day relative position) during Early Middle Miocene times (16–14 Ma). This northern landmass (*sensu* Williams 1961) is a larger, much wider, and complex 'palaeoisland' that occupies approximately one-third of Haiti's northern landmass currently) and contains the rest of the mountains, plateaus, valleys, and plains in the country; approximately two-thirds of this northern palaeoisland form the rest of the Dominican Republic. The low-lying Plaine de Cul-de-Sac (Haiti)/Valle de Neiba (Dominican Republic) separates the mountainous south palaeoisland from the northern one (Williams 1961; Schwartz 1978, 1980a; Powell *et al.* 1999; Powell and Incháustegui 2011). The composition of herpetofaunal assemblages in Hispaniola (and Haiti) somewhat fit this generalized geological configuration (Schwartz 1980a; Powell *et al.* 1999). Those interested in the origin of the herpetofauna in these palaeoislands may see Hedges (1996a), Iturralde-Vinent and MacPhee (1999), and Hedges (2006), and references within.

Haiti has six types of climate, based on the Köppen-Geiger classification system (see Kottek *et al.* 2006 for reference) (Fig. 6.3). (1) Tropical Rainforest type (**A**f) occurs mostly on the north (Massif du Nord and Île de la Tortue), and the southwest (Îles Cayémites, west Île à Vache, and the western part of the Massif de la Hotte in the Tiburon Peninsula). (2) Tropical Monsoon type (**A**m) occurs mostly on mid-elevation areas adjacent to the Massif du Nord and Plateau Central, eastern Île à Vache, easternmost part of the Massif de la Hotte (centre of Tiburon Peninsula; interspersed areas at mid-elevation areas in western Tiburon Peninsula), and mid-elevation areas in the Massif de la Selle further east. (3) Subtropical Highland type (**C**wb) occurs mostly at eastern Chaînes deu Matheux and most of the Chaîne du Trou d'Eau (centre of Haiti), and high-elevation areas of the Massif de la Selle. (4) Temperate Oceanic type (**C**fb) occurs interspersed on

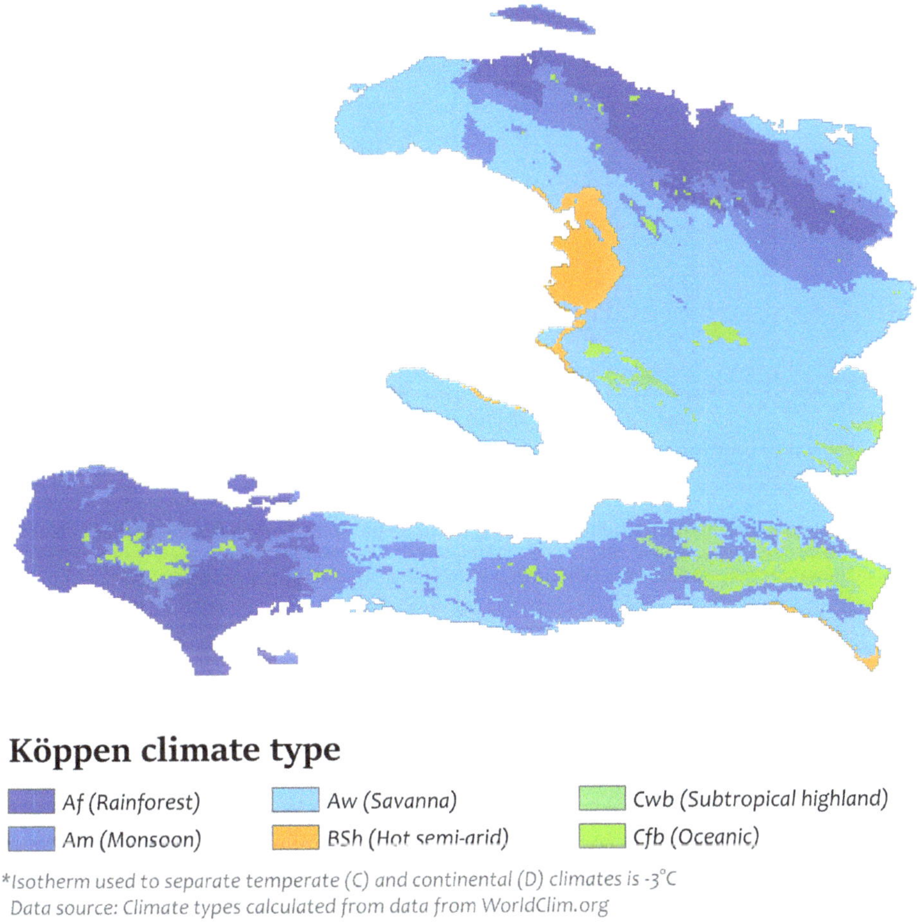

Figure 6.3. Generalized map of Haiti's climates based on the Köppen-Geiger classification. Title: 'Köppen climate types of Haiti' (https://upload.wikimedia.org/wikipedia/commons/a/ac/Haiti_K%C3%B6ppen.svg). Author: Adam Peterson (https://commons.wikimedia.org/wiki/User:Redtitan) – Own work. Created 2 November 2016. Licensed under CC BY-SA 4.0 International (https://creativecommons.org/licenses/by-sa/4.0/deed.en). Sources of data: WorldClim.org (http://worldclim.org/).

the highest peaks of the Massif du Nord and Montaignes Noires, western peaks at Chaînes deu Matheux and on a few scattered peaks at Chaîne du Trou d'Eau, and highest elevation areas at the Massif de la Selle and the Massif de la Hotte. (5) Tropical Savannah type (**A**w) occurs on Île de la Gonâve, Plaine de Cul-de-Sac and the vast majority of the north palaeoisland, central Massif de la Hotte, and lowlands south of Massif de la Selle in the south palaeoisland. (6) Hot Semi-Arid type (**B**Sh) occurs mainly on Plaine de l'Artibonite (central-western north palaeoisland), northernmost coastal Île de la Gonâve, and easternmost coastal areas in the south palaeoisland.

Rainfall and temperature are seasonal in Haiti (Fig. 6.4), and vary with elevation and location relative to the main mountain ranges. Annual rainfall ranges from 300 mm in the northwest to more than 3,000 mm on Pic de Macaya (west Massif de la Hotte) (Smucker 2001). Likewise, depending on the locality, ambient daily temperature ranges between 19°C (winter months: ~December/January–March/April) and 33°C (summer months: ~May–November); temperature fluctuates between 10°C and 8°C (or lower) on the highest peaks of the Massif de la Hotte (Gouvernement

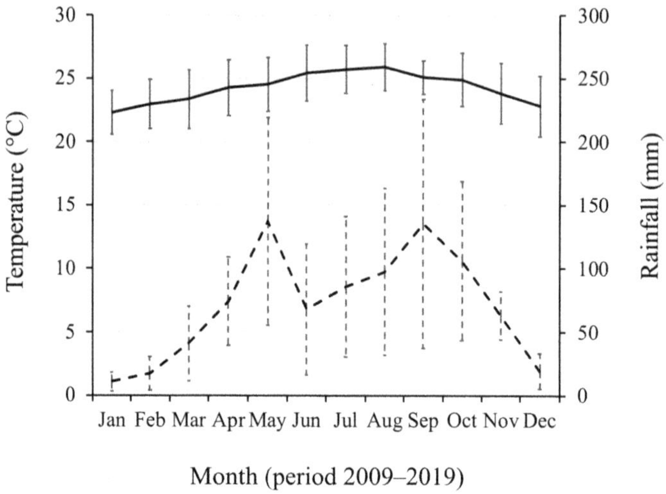

Figure 6.4. Generalized annual pattern in monthly mean ambient temperature (dark line) and its variability (± standard deviation; grey lines with caps) and monthly mean total rainfall (black dashed line) and its variability (± standard deviation; grey dashed lines with caps) at Port-au-Prince, Haiti, for the period 2009–19.

de la République d'Haïti 2016). Data between 2009 and 2019 from Haiti's capital Port-au-Prince (World Weather Online 2021) reveal that annual ambient temperature (grand mean) was 24.3°C ± 1.9°C (max. 33°C; min. 17°C) and mean annual rainfall was 853.9 mm ± 423.6 mm (max. 1,064.1 mm in 2013; min. 316.3 mm in 2018). Being located in the northern Caribbean region, Haiti is also under the influence of larger fluctuations in weather conditions because of natural phenomena such as El Niño Southern Oscillation or simply El Niño (which influence patterns of rainfall and temperature in the region) and seasonal storms (see below).

It is thought that low values of rainfall in Haiti between 2014 and 2018 (and anomalies in temperature) mostly resulted from the influence of El Niño on the Caribbean region. For example, 'very dry conditions were observed in [...] Haiti [...] during May–October [2014]. This resulted in elevated heat stress and water shortages in Haiti' (Stephenson *et al.* 2015: S117); 'Prevailing El Niño conditions were associated with below-normal annual rainfall and above-normal annual mean temperature over much of the region [Caribbean]' during 2015 (Stephenson *et al.* 2016: S181; although Haiti is not mentioned by Stephenson *et al.* 2016, Osborn *et al.* [2016: S33] noted that '[d]rought conditions developed in [...] Haiti, contributing to food insecurity in the region'). Stephenson *et al.* (2017: S183) mentioned that 'warmer-than-average conditions persisted throughout the Caribbean during 2016 ... likely related to the El Niño conditions that lasted through May 2016'. This 2014–16 El Niño was referred to as 'one of the strongest in the 1950–2016 record' (Diamond *et al.* 2017: S93).

Early on 4 October 2016, Hurricane Matthew went over the western part of the Tiburon Peninsula as a category-4 storm on the Saffir–Simpson hurricane wind scale (winds between 209 and 251 km/h). Hurricane Matthew caused the 'largest humanitarian emergency since the 2010 earthquake' (Stephenson *et al.* 2017: S185) in a locality with the highest species richness of anurans documented in Haiti (Massif de la Hotte, with up to 34 species, two of which are introduced, *Lithobates catesbeianus* and *Rhinella marina*) (Powell *et al.* 1999; Hedges 2008; Hedges *et al.* 2008; Gouvernement de la République d'Haïti 2016). Although 'above-normal rainfall dominated much of the Caribbean between July and September [2017] ... likely related to the passage of a number of storms through the region ... Port-au-Prince (Haiti) recorded its driest year (588.3 mm) using records available since 1971' (Stephenson *et al.* 2018: S201–S202). Weak El Niño conditions developed during the last quarter of 2018 with short-term drought conditions reported for some Caribbean islands, including the Bahamas, Cuba, Dominica, Hispaniola, St Lucia, Tobago, and Martinique; Port-au-Prince reported its second driest year on record, with total annual rainfall of 592.0 mm (Stephenson *et al.* 2019: S200).

Leslie Rensselaer Holdridge arrived in Haiti from Puerto Rico in October 1941 as the Manager of the Forestry Division of the Société Haitiano-Américaine de Développement Agricole, an organization mostly created for the development of rubber plantations (Holdridge 1942; Wadsworth 2014). At his arrival, Holdridge administered a '140,000-acre tract [~56,656 ha] of what must have been harvesting Haiti's last pine forest resources' (Wadsworth 2014: 37). While he was studying the relationships between vegetation units (including montane vegetation; Holdridge 1947), his working environment in mature pine forests at Massif de la Selle set the stage for his doctoral dissertation (Wadsworth 2014) and, eventually, his monumental *Life Zone Ecology* (Holdridge 1967). Based on Holdridge's Life Zone Classification System, nine Subtropical Life Zones are identified in Haiti (the following information comes from Ehrlich *et al.* 1985 and Gouvernement de la République d'Haïti 2016): (1) Thorn Woodland (typical of the northwestern peninsula, ~550 mm of annual rainfall); (2) Dry Forest ('the second largest' zone, located mostly below 400 m asl; Plaine du Cul-de-Sac (Département de l'Ouest), in the Départements de Nord-Ouest, Nord-Est, lower areas of l'Artibonite, and along the southern coast; (3) Moist Forest ('the most extensive' life zone; prevails on the Plateau Central, and alluvial plains in north, central, and southern Haiti; almost no original forest of this zone remains because of subsistence agriculture and extensive plantations of avocado and mango trees); (4) Wet Forest (typical of limestone soils, prevails at low-elevation montane ranges and small mountains [= mornes], restricted along coastal areas in northern and southern Haiti and portions of the Plateau Central; areas frequently used for coffee, cacao, and rubber plantations mostly on limestone soils); (5) Rain Forest (primarily at low-elevation areas of the Massif de la Hotte in the western part of the south palaeoisland; its value for biodiversity exceeds that for agriculture yet it is subjected to slash and burn practices); (6) Lower Montane Moist Forest (low-elevation mountains – 800 to 2,000 m asl – north of Massif de la Selle, areas adjacent to Kenscoff town, Département de Ouest); (7) Lower Montane Wet Forest (remaining stands of pine forests); (8) Lower Montane Rain Forest; (9) Montane Wet Forest (the latter two life zones are limited in coverage, mostly occur on slopes at high elevation south of the Massif de la Selle as pine and evergreen broadleaf forests).

II. History of the discovery of species of anurans and their distribution in Haiti, notes on taxonomy, and remarks about naturalized anurans

A. A brief history

Duméril and Bibron (1841) described the endemic Haitian 'crapaud' *Eleutherodactylus oxyrhyncus* (= Cochran's [1935] *E. femur-levis* from Massif de la Hotte, collected by Philip J. Darlington) (Lynch and Schwartz 1971) (Fig. 6.5). Several naturalists, zoologists, and herpetologists afterwards described other species of anurans found in Haiti (e.g. Johann Jakob von Tschudi, Thomas Barbour, Gladwyn Kingsley Noble). It is generally agreed, however, that the history of the herpetology of Haiti – particularly that of the taxonomy of anurans – was formally initiated by Doris Mable Cochran and her landmark work *The Herpetology of Hispaniola* in 1941 (Schwartz and Thomas 1975; Powell *et al.* 1999; Incháustegui and Díaz, Chapter 7, this volume). Cochran (1935: 371) highlighted P. J. Darlington as 'the first zoölogist to ascend Morne La Hotte', who collected several anurans (and reptiles) that were described by Cochran, Benjamin Shreve, and others. Afterwards, the vast majority of Haitian anurans (14 endemics, and depending on the source,

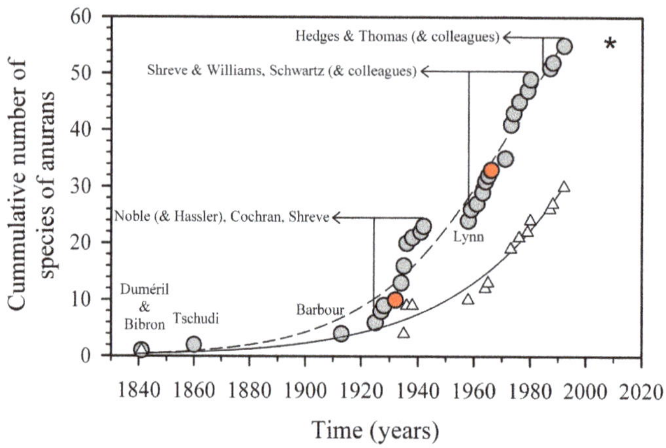

Figure 6.5. Species-accumulation curve of the number of native/endemic species of anurans described and documented from Haiti (grey-filled circles); red-filled circles are naturalized species described in the text. Dashed line shows the best-fitted sigmoidal regression curve. Triangles show the accumulation curve for endemic species only (solid line shows the best-fitted sigmoidal regression curve). Names of authorities are shown; arrows point toward the describers of most of the species for a given period. The asterisk indicates the year of publication of Hedges *et al.* (2008), which resulted in many subspecies being elevated to species status.

up to seven species that Haiti shares with the Dominican Republic). The state of taxonomic knowledge on anurans in Haiti up to the 1990s was summarized by Stephen Blair Hedges, John Paul Richard Thomas, and colleagues. More recently, using molecular data and genetic analyses Hedges *et al.* (2008) elevated several subspecies, most of them from Schwartz's original descriptions, to the species level (see section II. B).

Through herpetological history, a comprehensive checklist that updated Haitian anurans (endemic, native, and introduced) was prepared by Schwartz and Thomas (1975). Based mostly on works of Schwartz from the 1960s to 1980, including this checklist, Schwartz and Henderson (1991), and Henderson and Powell (2009) summarized and expanded significantly most of the information about taxonomy, natural history, and conservation status of all species of anurans known to occur in Haiti (although most of the information frequently referred to Hispaniola instead of to a particular political unit). Currently, the Republic of Haiti is home to 30 endemic species of *Eleutherodactylus* (Table 6.1; IUCN 2022). In addition there are at least 17 species of *Eleutherodactylus*, 4 species of hylids, 1 species of *Peltophryne* (bufonid), and 2 introduced species (*Lithobates catesbeianus* and *Rhinella marina*) that are shared with its neighbouring country, the Dominican Republic (based on published records, distributional maps, and Martínez Rivera and Rodríguez Plaza 2015: 24; Table 6.1). Updated distributional maps and pictures of most species of anurans in Haiti (endemic, native, and introduced) are provided by the IUCN's Red List of Threatened Species (at https://www.iucnredlist.org/; IUCN 2022) and http://www.caribherp.org (© 1999–2022 S. Blair Hedges). I used these checklists, published sources of distributional maps, Hedges *et al.* (2008), species assessments in IUCN (2022), and species lists for 'Key Biodiversity Areas' from the Gouvernement de la République d'Haïti (2016) to summarize data on species richness for the southern and northern palaeoislands, as follows.

The south palaeoisland – and south of Plaine du Cul-de-Sac – is by far the most diverse region in Haiti for anurans, with over 24 of the endemic species (80% endemism) and 42 species overall (80.8%; not including introduced species) (Table 6.1). Although not discussed herein, two species recently described from the Dominican Republic were found relatively close to the

Table 6.1. List of species of anurans in Haiti and category of population-status based on most the recent assessment in IUCN (2022) (in parenthesis[1]; categories as defined by the IUCN: CR = Critically Endangered; EN = Endangered; VU = Vulnerable; NT = Near Threatened; LC = Least Concern; PE = Presumably Extinct), authority, source of picture (specified if in colour or black and white [B&W]) and distributional maps. Most online material is found at http://www.caribherp.org (old and updated distributional maps; © 1999-2022 S. Blair Hedges), and IUCN (2022). * A picture of *Eleutherodactylus glanduliferoides* of live animals is included in Hedges (2008: 104) but its legend reads '*Eleutherodactylus orientalis*' (from Cuba), which also is herein included with the legend '*Eleutherodactylus glanduliferoides*' instead (clearly, the pictures and legends are mismatched, but labelled correctly at www.caribherp.org).

FAMILY AND *SPECIES*	AUTHORITY	SOURCES OF PICTURES (OTHER THAN IUCN 2022)	SOURCES OF DISTRIBUTIONAL MAPS
ENDEMIC			
Eleutherodactylidae			
Eleutherodactylus amadeus (CR)	Hedges, Thomas, and Franz, 1987	Martínez Rivera and Rodríguez Plaza (2015: 111)	IUCN SSC Amphibian Specialist Group 2022d
E. aporostegus (CR[1])	Schwartz, 1965(a)	Fig. 80 in Hedges *et al.* (2008: 86)	Hedges *et al.* 2004f for *E. ruthae* (see section II. B. 1); IUCN SSC Amphibian Specialist Group 2022e
E. apostates (CR)	Schwartz, 1973	Martínez Rivera and Rodríguez Plaza (2015: 115)	IUCN SSC Amphibian Specialist Group 2022f
E. bakeri (CR)	Cochran, 1935	Fig. 61 in Hedges *et al.* (2008: 75)	IUCN SSC Amphibian Specialist Group 2022i
E. brevirostris (CR)	Shreve 1936	Martínez Rivera and Rodríguez Plaza (2015: 120)	IUCN SSC Amphibian Specialist Group 2022j
E. caribe (CR)	Hedges and Thomas, 1992(a)	B&W in Hedges and Thomas (1992a: 192); colour in www.caribherp.org	IUCN SSC Amphibian Specialist Group 2022k
E. chlorophenax (CR)	Schwartz, 1976(a)	Holotype (CM-56833) in colour at https://www.idigbio.org/portal/records/c362c334-ba9a-465f-adda-64ae0ae5521d#media; colour of live animals at www.caribherp.org	IUCN SSC Amphibian Specialist Group 2022l
E. corona (CR)	Hedges and Thomas, 1992(b)	B&W in Hedges and Thomas (1992b: 352; colour in Martínez Rivera and Rodríguez Plaza (2015: 124)	IUCN SSC Amphibian Specialist Group 2022m
E. counouspeus (EN)	Schwartz, 1964	Fig. 81 in Hedges *et al.* (2008: 86); in the wild, Martínez Rivera and Rodríguez Plaza (2015: 125)	IUCN SSC Amphibian Specialist Group 2022n
E. darlingtoni (CR)	Cochran, 1935	Holotype (MCZ-19847) in colour at https://mczbase.mcz.harvard.edu/guid/MCZ:Herp:A-19847; colour of live animals at www.caribherp.org	IUCN SSC Amphibian Specialist Group 2022o
E. diplasius (CR)	Schwartz, 1973	Only at www.caribherp.org	IUCN SSC Amphibian Specialist Group 2022p
E. dolomedes (CR)	Hedges and Thomas, 1992(b)	B&W in Hedges and Thomas (1992b: 352; colour of live animals at www.caribherp.org	IUCN SSC Amphibian Specialist Group 2022q

(*Continued*)

Table 6.1. (Continued)

FAMILY AND *SPECIES*	AUTHORITY	SOURCES OF PICTURES (OTHER THAN IUCN 2022)	SOURCES OF DISTRIBUTIONAL MAPS
E. eunaster (CR)	Schwartz, 1973	Colour of live animals at www.caribherp.org	IUCN SSC Amphibian Specialist Group 2022r
E. glandulifer (CR)	Cochran, 1935	Fig. 62 in Hedges *et al.* (2008: 75)	IUCN SSC Amphibian Specialist Group 2022u
E. glanduliferoides (CR [PE])	Shreve, 1936	* Colour of live animals at www.caribherp.org	IUCN SSC Amphibian Specialist Group 2022v
E. glaphycompus (EN)	Schwartz, 1973	Colour of live animals at www.caribherp.org	IUCN SSC Amphibian Specialist Group 2022w
E. grahami (EN)	Schwartz, 1979	Colour of live animals at www.caribherp.org	IUCN SSC Amphibian Specialist Group 2022x
E. lamprotes (CR)	Schwartz, 1973	Colour of live animals at www.caribherp.org	IUCN SSC Amphibian Specialist Group 2022cc
E. limbensis (CR [PE][1])	Lynn, 1958	No pictures or illustrations found.	www.caribherp.org; IUCN SSC Amphibian Specialist Group 2022ee
E. lucioi (CR [PE])	Schwartz, 1980(b)	Holotype (CM-60537 from Schwartz 1980b) in colour at https://www.idigbio.org/portal/records/6d049d4e-a221-4a4f-9a39-81279ad75470#media	IUCN SSC Amphibian Specialist Group 2022ff
E. oxyrhyncus (CR)	(Duméril and Bibron, 1841)	Holotype (MNHN-RA-0.753) in colour at http://coldb.mnhn.fr/catalognumber/mnhn/ra/0.753; Fig. 64 in Hedges *et al.* (2008: 75); Martínez Rivera and Rodríguez Plaza (2015: 90–91)	IUCN SSC Amphibian Specialist Group 2022ii
E. parapelates (CR)	Hedges and Thomas, 1987	https://www.iucnredlist.org/species/56827/11541207; Martínez Rivera and Rodríguez Plaza (2015: 139–140); www.caribherp.org	IUCN SSC Amphibian Specialist Group 2022kk
E. paulsoni (CR)	Schwartz, 1964	Fig. 65 in Hedges *et al.* (2008: 75)	IUCN SSC Amphibian Specialist Group 2022ll
E. poolei (CR)	Cochran, 1938	Colour of live animals at www.caribherp.org	IUCN SSC Amphibian Specialist Group 2022nn
E. rhodesi (CR)	Schwartz, 1980(b)	Colour of live animals at www.caribherp.org	IUCN SSC Amphibian Specialist Group 2022oo
E. sciagraphus (CR)	Schwartz, 1973	Colour of live animals at www.caribherp.org	IUCN SSC Amphibian Specialist Group 2022rr
E. semipalmatus (CR)	Shreve, 1936	Holotype (MCZ A-21561) in colour at https://mczbase.mcz.harvard.edu/guid/MCZ:Herp:A-21561; Colour of live animals at https://www.facebook.com/haititrust/photos/a.1759048367694643/2316010765331731/?type=3&theater (http://www.haititrust.org/), photo by Eladio Fernández (International League of Conservation Photographers [iLCP]); colour of live animals at www.caribherp.org	IUCN SSC Amphibian Specialist Group 2022ss

E. thorectes (CR)	Hedges, 1988	Colour of live animals at www.caribherp.org; Martínez Rivera and Rodríguez Plaza (2015: 145–146)	IUCN SSC Amphibian Specialist Group 2022uu
E. ventrilineatus (CR)	Shreve, 1936	Colour of live animals at www.caribherp.org; Martínez Rivera and Rodríguez Plaza (2015: 147–148)	IUCN SSC Amphibian Specialist Group 2022vv
E. warreni (CR)	Schwartz, 1976(b)	Holotype (CM54138) in colour at https://www.idigbio.org/portal/records/d48362a7-8c12-40cc-b624-4d4245e79e3e#media	IUCN SSC Amphibian Specialist Group 2022ww
NATIVE Eleutherodactylidae			
E. abbotti (LC)	Cochran, 1923	Fig. 32 in Hedges *et al.* (2008: 55)	IUCN SSC Amphibian Specialist Group 2022b
E. alcoae (LC)	Schwartz, 1971	Colour of live animals at www.caribherp.org; Martínez Rivera and Rodríguez Plaza (2015: 62)	IUCN SSC Amphibian Specialist Group 2022c; (Note: IUCN SSC Amphibian Specialist Group 2013a had it EN)
E. armstrongi (EN)	Noble and Hassler, 1933	Fig. 49 in Hedges *et al.* (2008: 67); Martínez Rivera and Rodríguez Plaza (2015: 64–65)	IUCN SSC Amphibian Specialist Group 2013b; IUCN SSC Amphibian Specialist Group 2022g
E. audanti (VU)	Cochran, 1934	Colour of live animals at www.caribherp.org; Martínez Rivera and Rodríguez Plaza (2015: 66–67); Figs. G–L in Díaz *et al.* (2018: 29)	Hedges *et al.* 2004c; IUCN SSC Amphibian Specialist Group 2022h
E. fowleri (CR)	Schwartz, 1973	Colour of live animals at www.caribherp.org; Martínez Rivera and Rodríguez Plaza (2015: 70–71)	Hedges *et al.* 2004d; IUCN SSC Amphibian Specialist Group 2022s
E. furcyensis (CR)	Shreve and Williams, 1963	Colour of live animals at www.caribherp.org	IUCN SSC Amphibian Specialist Group 2022t
E. heminota (EN)	Shreve and Williams, 1963	Colour of live animals at www.caribherp.org; Martínez Rivera and Rodríguez Plaza (2015: 76–77)	IUCN SSC Amphibian Specialist Group 2022y
E. inoptatus (NT)	(Barbour, 1914)	Fig. 78 in Hedges *et al.* (2008: 86); Martínez Rivera and Rodríguez Plaza (2015: 80–81)	IUCN SSC Amphibian Specialist Group 2022aa; (Note: Hedges and Thomas 2004 had it LC)
E. jugans (CR)	Cochran, 1937	Fig. 63 in Hedges *et al.* (2008: 75); Martínez Rivera and Rodríguez Plaza (2015: 82–83)	IUCN SSC Amphibian Specialist Group 2022bb
E. leoncei (EN)	Shreve and Williams, 1963	Colour of live animals at www.caribherp.org; Martínez Rivera and Rodríguez Plaza (2015: 84–85)	IUCN SSC Amphibian Specialist Group 2022dd; (Note: Hedges *et al.* 2010a had it CR)
E. nortoni (CR)	Schwartz, 1976(a)	Fig. 79 in Hedges *et al.* (2008: 86); Martínez Rivera and Rodríguez Plaza (2015: 88–89)	Hedges *et al.* 2010b; IUCN SSC Amphibian Specialist Group 2022gg

(Continued)

Table 6.1. (Continued)

FAMILY AND *SPECIES*	AUTHORITY	SOURCES OF PICTURES (OTHER THAN IUCN 2022)	SOURCES OF DISTRIBUTIONAL MAPS
E. notidodes (EN)	Schwartz, 1966	Colour of live animals at www.caribherp.org; Fig. N in Díaz *et al.* (2018: 29)	IUCN SSC Amphibian Specialist Group 2013c; IUCN SSC Amphibian Specialist Group 2022hh
E. parabates (EN)	Schwartz, 1964	Colour of live animals at www.caribherp.org	IUCN SSC Amphibian Specialist Group 2022jj; (Note: Hedges *et al.* 2010c had it CR)
E. pictissimus (LC)	Cochran, 1935	Colour of live animals at www.caribherp.org; Martínez Rivera and Rodríguez Plaza (2015: 92–93)	IUCN SSC Amphibian Specialist Group 2022mm; Hedges *et al.* 2004a had it VU; also see text for discussion on the distributional map in www .caribherp.org
E. sommeri (EN)	Schwartz, 1977	Holotype (CM-56837) in colour at https://www.idigbio.org /portal/records/3a02f636-1f79-4ae1-92b4-582c35e791bc #media	IUCN SSC Amphibian Specialist Group 2013d; IUCN SSC Amphibian Specialist Group 2022tt
E. weinlandi (LC)	Barbour, 1914	Colour of live animals at www.caribherp.org	IUCN SSC Amphibian Specialist Group 2022xx; Hedges *et al.* 2004b, but see text for discussion on the distributional map in www.caribherp.org
E. wetmorei (VU)	Cochran, 1932	Fig. 47 in Hedges *et al.* (2008: 63)	IUCN SSC Amphibian Specialist Group 2022yy
Hylidae			
Boana heilprini (VU)	(Noble, 1923)	Colour of live animals, larvae, metamorph, and adult, in Díaz *et al.* (2015: 4, 6)	IUCN SSC Amphibian Specialist Group 2013e; IUCN SSC Amphibian Specialist Group 2022a
Osteopilus dominicensis (LC)	(Tchudi, 1838)	Colour of live animals, larvae, metamorph, and adult, in Díaz *et al.* (2015: 4, 6)	IUCN SSC Amphibian Specialist Group 2014; IUCN SSC Amphibian Specialist Group 2022aaa
O. pulchrilineatus (VU)	(Cope, 1870)	Colour of live animals, larvae, metamorph, and adult, in Díaz *et al.* (2015: 4, 6); colour of live animals, juvenile, in Martínez Rivera and Rodríguez Plaza (2015: 151–152)	IUCN SSC Amphibian Specialist Group 2013f; IUCN SSC Amphibian Specialist Group 2022bbb
O. vastus (VU)	(Cope, 1871)	Colour of live animals, larvae, metamorph, and adult, in Díaz *et al.* (2015: 4, 6)	IUCN SSC Amphibian Specialist Group 2013g (but see text for discussion on the distributional map in www.caribherp.org); IUCN SSC Amphibian Specialist Group 2022ccc

Bufonidae			
Peltophryne guentheri (LC)	(Cochran, 1941)	Colour of live animals at www.caribherp.org	IUCN SSC Amphibian Specialist Group 2022ddd; Hedges *et al.* 2004e had it VU; also see text for discussion on the distributional map in www.caribherp.org
NATURALIZED			
Bufonidae			
Rhinella marina (LC)	(Linnaeus, 1758)	Colour of live animals at www.caribherp.org	Solís *et al.* 2009; IUCN SSC Amphibian Specialist Group (*in press*)
Ranidae			
Lithobates catesbeianus (LC)	(Shaw, 1802)	Colour of live animals at https://www.iucnredlist.org/search?query=Lithobates%20catesbeianus&searchType=species	IUCN SSC Amphibian Specialist Group 2022zz

[1]Subspecies elevated to species status relatively recently (see text).

Haiti/Dominican Republic border on the south palaeoisland; future field surveys may yield new records from within the territory of Haiti (e.g. *Eleutherodactylus neiba* and *Peltophryne armata*). In a similar way to other mountain ranges in Haiti, areas at mid and high elevations on this palaeoisland are most valued for coffee plantations. In contrast, the more extensive north palaeoisland harbours only six endemic species (20% endemism) and 17 species overall (32.7%; not including introduced species). Haiti harbours seven species with palaeoislands-wide distributions and one restricted to the second-largest island, Île de la Tortue (north Haiti). The taxonomic resolution of anurans in Haiti is far from complete and additional species could be described in the near future (e.g. several undescribed species of frog were recently found at Haiti's first private nature reserve of more than 485 hectares in the Grand Bois mountains, west Massif de la Hotte [Mayer 2019; Hedges and Timyan 2014; S. B. Hedges, pers. comm.]), presumably mostly based on newly explored areas (see above; Hedges *et al.* 2008). We require, however, clarification about the trajectory of the taxonomy of several species, the distributional range of others, and collection dates (i.e. of the first documentation of introduction of naturalized species). When needed, I consulted the portals https://www.vertnet.org (primarily) and https://www.idigbio.org (pictures of some specimens not available at the former can be found there) for specific information about collections of Haitian anurans.

B. Notes on taxonomy

The taxonomic trajectory of species is relevant for the following Haitian endemics: (1) *Eleutherodactylus aporostegus* Schwartz, 1965(a); (2) *E. parapelates* Hedges and Thomas, 1987; (3) *E. diplasius* Schwartz, 1973; (4) and *E. limbensis* Lynn, 1958. Likewise, for the following species shared between Haiti and the Dominican Republic: (1) *E. audanti* Cochran, 1934; (2) *E. notidodes* Schwartz, 1966; (3) *E. hypostenor* Schwartz, 1965(a); (4) *E. schmidti* Noble, 1923; (5) *E. sommeri* Schwartz, 1977; (6) *E. weinlandi* Barbour, 1914; (7) *E. wetmorei* Cochran, 1932; (8) and *E. pictissimus* Cochran, 1935. Finally, the following species were resolved within groups of species with relationships with Haiti, but currently only documented for the Dominican Republic: (1) *E. melatrigonum* Schwartz, 1966; (2) *E. rucillensis* Cochran, 1939; (3) *E. paralius* Schwartz, 1976(b); (4) and *E. probolaeus* Schwartz, 1965(b).

1. Group Eleutherodactylus ruthae *Noble, 1923*

Group 1: *Eleutherodactylus ruthae* Noble, 1923; Type from Samaná, Dominican Republic. Schwartz (1965a) described two subspecies, *E. r. aporostegus* from Massif de la Hotte, and *E. r. hypostenor*, a widely distributed form on south palaeoisland. Hedges *et al.* (2008: 87) elevated the former subspecies to *E. aporostegus* Schwartz, 1965(a), which is now endemic to Haiti. Schwartz (1973: 266) elevated the latter subspecies to *E. hypostenor* Schwartz, 1965(a). Hedges and Thomas (1987), however, considered the western population of *E. hypostenor* – from Massif de la Hotte – to be different, and described it as a new species, *E. parapelates* Hedges and Thomas, 1987, which is now endemic to Haiti. Hedges *et al.* (2008) considered *E. hypostenor* Schwartz, 1965(a), from Massif de la Selle (Haiti)/Sierra de Bahoruco (Dominican Republic), to be a valid species (also see Hedges *et al.* 2004g). However, the IUCN (2022z) generated a new geographic-distributional map of *E. hypostenor* that restricts the species' distribution to the Dominican Republic in a rather unnatural design, in my opinion, along the southern Haitian–Dominican political border at the Bahoruco Occidental within the Sierra de Bahoruco. The justification for this distribution was that 'Surveys elsewhere have not recorded this species; thus, it is considered to have a genuinely restricted distribution' (IUCN 2022z). As a result, *E. hypostenor* is now treated as endemic to the Dominican Republic (IUCN 2022z), with no indication about the species being extirpated from Haiti.

Currently, *E. ruthae* Noble, 1923, also is valid (endemic to the Dominican Republic), and Hedges *et al.* (2004f) provided a distributional map for *E. ruthae*. This map, however, may have influenced Hedges' *et al.* (2008: 87) remark: 'Based on call differences in other isolated populations of this complex [their *Eleutherodactylus* (*Pelorius*) *ruthae* Species Series], it is likely that additional species remain to be discovered and described (Hedges and Thomas 1987).' More recently, Hedges' portal http://www.caribherp.org (updated 26 April 2022; accessed on 3 August 2022) includes a distribution locality for *E. ruthae* within Buena Vista, Jarabacoa, La Vega Province (central Dominican Republic) that it is nonetheless omitted in IUCN's (2022pp) geographic distribution for the species.

(Note: Most likely, the generation of distributional maps [and also taxonomic resolution] by different sources is highly dynamic and rapidly changing for Haitian amphibians, a phenomenon that is difficult to follow due to lag-time responses between published updates and the frequency of fieldwork/information gathering, or whenever justifications for changes might not be clearly stated [or both]. I mention and discuss several instances throughout this chapter where discrepancies among dates of a given assessment best illustrate this phenomenon because, I think, it has profound consequences for conservation efforts through time.)

2. Group **Eleutherodactylus schmidti** *Noble, 1923*

Group 2: *Eleutherodactylus schmidti* Noble, 1923; collected in 1922, the Type came from Río Los Bracitos, Distrito Pacificador – changed to Provincia de Duarte in 1925 – in the Dominican Republic. Cochran (1939) described a subspecies, *E. s. rucillensis*, from Loma de Rucilla, northern Cordillera Central, in the Dominican Republic. Hedges *et al.* (2008: 68) elevated it to *E. rucillensis* Cochran, 1939, which is endemic to the Dominican Republic. A population in northwestern Haiti was described in 1958 as a subspecies, *E. s. limbensis*, collected at Limbè River (Rivière du Limbè, Département du Nord) by William Gardner Lynn. Hedges *et al.* (2008: 68) also elevated it to *E. limbensis* Lynn, 1958, mostly based on morphology and its isolated distribution relative to *E. rucillensis* and *E. schmidti*. As a result, *E. limbensis* is endemic to Haiti. Hedges *et al.* (2010d), however, provided a distributional map for *E. schmidti*, which merges the geographic distributions of *E. rucillensis* and *E. limbensis* into one under that of *E. schmidti*, thus generating a Haiti-Dominican Republic distributional map. Similar distributional maps were updated for these three species on http://www.caribherp.org (updated 17 January 2020; accessed 19 January 2020). Also, Hedges *et al.* (2008) concluded: 'One of us (SBH) is aware of an undescribed species belonging to this group [*Eleutherodactylus* (*Euhyas*) *schmidti* Species Group] from the Sierra de Neiba in the Dominican Republic.', which revealed the complexity in taxonomical studies to come. Recently, Incháustegui *et al.* (2015) described *E. neiba* from the Sierra de Neiba in the Dominican Republic – adjacent to the Dominican Republic/Haiti border – and morphologically (and based on the criteria of Hedges 1989), this species shows characters attributable to the subgenus *Euhyas* (Incháustegui *et al.* 2015: 148). Also, as for the geographic distribution of *E. schmidti*, the IUCN (2022qq) settled discrepancies and generated a distributional map of *E. schmidti* restricting it as endemic to the Dominican Republic; the geographic-distributional map in Hedges' updated portal http://www.caribherp.org agrees (updated 26 April 2022 and accessed on 3 August 2022).

3. Group **Eleutherodactylus wetmorei** *Cochran, 1932*

Group 3: *Eleutherodactylus wetmorei* Cochran, 1932; Type collected in 1927 by ornithologist Frank Alexander Wetmore at Fond-des-Nègres (Département de Nippes), mid-central south palaeoisland, Haiti. Schwartz (1968) described two subspecies, *E. w. williamsi*, from northwestern south palaeoisland (northwest Tiburon Peninsula), and *E. w. ceraemerus* from eastern south palaeoisland and into the Dominican Republic (*E. w. wetmorei* Cochran, 1932, would remain and occurs in western south palaeoisland). Schwartz (1973: 250) changed the name of the former to *E. w. diplasius* after a 'careless use of the name *williamsi* twice in the genus *Eleutherodactylus* had been drawn to his attention'. Hedges *et al.* (2008: 64) elevated it to *E. diplasius* Schwartz, 1973, thereby making it an endemic Haitian species. Schwartz (1977) described yet another subspecies, *E. w. sommeri* along Carrefour Marmelade, (Département de l'Artibonite), north palaeoisland, Haiti. Hedges *et al.* (2008: 64) elevated it to *E. sommeri* Schwartz, 1977, and its geographic distribution extends into the western Cordillera Central in the Dominican Republic (IUCN SSC Amphibian Specialist Group 2013d, 2022tt). Data (e.g. size and pattern of colouration) from new collections of specimens led Schwartz (1977: 67, 71) to suggest that *E. w. wetmorei* (western south palaeoisland) and *E. w. ceraemerus* (eastern south palaeoisland) intergrades near La Montagne, (east of Grand Bangnin and west of the coastal city of Jacmel, Département du Sud-Est), Haiti.

Finally, Hedges *et al.* (2008) highlighted two ideas that deserve attention. First, they mentioned that the geographic distribution of *E. diplasius* and *E. wetmorei* (*E. w. wetmorei*) overlap at Massif de la Hotte (western south palaeoisland) but that 'specimens from the area of contact show no signs of intergradation'. Second, they left *E. wetmorei* as a full species, but despite intergradation between *E. w. wetmorei* and *E. w. ceraemerus* (see Schwartz 1977), they also concluded that 'the overall differences between those two subspecies exist, and further study may justify recognition of *E. w. ceraemerus* as a full species', which may lead to *E. wetmorei* endemic to Haiti.

4. *Group* Eleutherodactylus audanti *Cochran, 1934*

Group 4: *Eleutherodactylus audanti* Cochran, 1934; Type (MCZ A-19704) collected at Peak La Selle (at Massif de la Selle, western south palaeoisland, Haiti) by André Audant. Audant presumably collected this specimen in August 1933 (see Cochran [1941: 192] for a presumed date of collection by Audant: information for Type specimen in the Museum of Comparative Zoology, Harvard University, lists 'January 1, 1889' as the date of collection, but Audant was born in 1909; more on André Audant in section II. C). Schwartz (1966) described two subspecies, *E. a. notidodes,* from north-central Sierra de Neiba (Dominican Republic), and *E. a. melatrigonum,* which occurs in the Cordillera Central (Provincia de Independencia) and is endemic to the Dominican Republic. Hedges *et al.* (2004c) provided a distributional map for *E. audanti* and merged the distributions of all three subspecies. On the other hand, Lynch and Duellman (1997: 219, 229) elevated the former to *E. notidodes* Schwartz, 1966, which occurs in the area of Chaîne du Trou d'Eau (central-eastern Haiti)/Sierra de Neiba (central-southwestern Dominican Republic) (IUCN SSC Amphibian Specialist Group 2013c). Likewise, Hedges *et al.* (2008: 54) elevated the latter to *E. melatrigonum* Schwartz, 1966, which is endemic to the Dominican Republic.

Currently, *Eleutherodactylus audanti* remains valid. It occurs in south palaeoisland (extending from Haiti to the Dominican Republic), but Hedges *et al.* (2008: 54–55) mentioned that it occurs only 'in three disjunct regions on the Hispaniolan South Paleoisland – the Sierra de Bahoruco, the Massif de La Selle, and the Massif de la Hotte. These require further study to determine if differentiation has occurred among these isolates'. Hedges *et al.* (2008) also suggested that the population of Massif de la Hotte and (at least) that from Massif de la Selle, could be attributable to different higher-taxonomic units (based also on the distributional map for *E. audanti* in http://www.caribherp.org, updated 17 January 2020; accessed 19 January 2020). A minor discrepancy in the visual representation of distributional maps for *E. audanti* exists between Hedges' updated portal (http://www.caribherp.org; updated 26 April 2022, accessed on 3 August 2022) and IUCN (2022h): Hedges provided two disjunct distribution localities along the Tiburon Peninsula (Haiti) to Sierra de Bahoruco (Dominican Republic), while IUCN (2022h) provides three disjunct distribution localities. (Note: Schwartz [1965c] described *E. neodreptus* from one individual collected in the Dominican Republic; it was synonymized with *E. audanti* by Hedges [1996b]).

5. *Group* Eleutherodactylus weinlandi *Barbour, 1914*

Group 5: *Eleutherodactylus weinlandi* Barbour, 1914; Type collected in 1881 by Monsieur Abbott Frazar at Provincia de Puerto Plata, northern Dominican Republic, Haiti. Schwartz (1965b) described one subspecies, *E. w. chersonesodes*, from Samaná, northeastern Dominican Republic, and despite his recognition of a substantial intergradation between *E. w. chersonesodes* and *E. w. weinlandi* in the northern region of the Dominican Republic (see Fig. 95, p. 120, in Schwartz 1965b; also Hedges *et al.* 2004b for *E. weinlandi*). Schwartz (1965b: 120) also identified an isolated region of *E. w. weinlandi*, within the Chaîne du Trou d'Eau (central-eastern Haiti)/Sierra de Neiba southwest and into Haiti/Dominican Republic, that still deserves study. Schwartz (1976b) described another subspecies, *E. w. paralius*, from the southeastern Dominican Republic; Hedges *et al.* (2008: 82) elevated it to *E. paralius* Schwartz, 1976(b) – which is endemic to the Dominican Republic – because it 'has not been found to intergrade with either of the other taxa despite close proximity of their ranges'. Schwartz and Henderson (1991), and Henderson and Powell (2009), refer to subspecies of *E. weinlandi* in their compendia.

Hedges *et al.* (2008: 82) recognized *E. weinlandi* to be valid but mentioned that '[o]ne of us (SBH) has collected both *E. pictissimus* and *E. weinlandi* throughout Hispaniola, and has encountered many specimens that do not conform to either species, suggesting that species group [*Eleutherodactylus* (*Euhyas*) *lentus* Species Group] is larger than currently recognized.' Hedges' updated portal (http://www.caribherp.org; updated 26 April 2022 and accessed 3 August 2022) and IUCN (2022xx) provide distributional maps that merge former distribution accounts into a continuous distribution. Unfortunately, none of these sources brings light to the taxonomy of *E. weinlandi* in its isolated locality at Chaîne du Trou d'Eau/Sierra de Neiba.

6. Group Eleutherodactylus pictissimus *Cochran, 1935*

Group 6: *Eleutherodactylus pictissimus* Cochran, 1935; Type collected in 1934 by Darlington at Tardieu (Département de Grand'Anse), northern slopes of Massif de la Hotte, Haiti, Provincia de Puerto Plata, northern Dominican Republic. Schwartz (1965b) described three subspecies. First, *E. p. probolaeus* from La Romana (Provincia de La Romana), southeasternmost Dominican Republic; Schwartz and Thomas (1975) elevated it to *E. probolaeus* Schwartz, 1965(b), endemic to the Dominican Republic. Second, *E. p. eremus* from the mountains east of Llano de Azua, Provincia de Azua, south Dominican Republic. Third, *E. p. apantheatus* from the Provincia de Independencia (Dominican Republic), but its distribution extended along the northern lowlands (Plaine de Cul-de-Sac/Valle de Neiba) and farther west of south palaeoisland into Haiti. Finally, *E. p. pictissimus* was relegated to the western two-thirds of south palaeoisland, with an intergradation zone between *E. p. apantheatus* (eastern) and *E. p. pictissimus* (western) on both sides of the Rivière Momance and adjacent areas farther east (Furcy) in northern Département de l'Ouest, Haiti (Schwartz 1965b).

Collection data provided new information about the extent of the intergradation zone between *E. p. apantheatus* and *E. p. pictissimus*: Schwartz and Thomas (1975: 31), and (Schwartz 1976b: 5) indicated that both subspecies intergrade at Los Patos, eastern Península de Barahona (easternmost south palaeoisland into the Dominican Republic). Schwartz (1979b: 5) provided a detailed distributional map for the three subspecies, showing the isolated occurrence of *E. p. eremus* in Azua, whereas Hedges *et al.* (2004a) provided a generalized map of continuous distribution for *E. pictissimus* (but see http://www.caribherp.org for *E. pictissimus*, in which *E. pictissimus* is restrictred to Île-à-Vache, southeastern Haiti—updated 17 January 2020 [accessed 19 January 2020] and updated 26 April 2022 [3 August 2022]; presumably a mistake). IUCN (2022mm), however, provides a wide, geographic distribution along southern Hispaniola, instead. (Note: Schwartz and Henderson [1991] and Henderson and Powell [2009] mentioned three subspecies for *E. pictissimus*).

Finally, Hedges *et al.* (2008: 82) commented that 'two subspecies of *E. pictissimus* also were described (Schwartz 1965b); additional study is needed to determine their taxonomic status [as for *E. weinlandi* above]. One of us (SBH) collected both *E. pictissimus* and *E. weinlandi* throughout Hispaniola and has encountered many specimens that do not conform to either species, suggesting that species group [*Eleutherodactylus* (*Euhyas*) *lentus* Species Group] is larger than currently recognized.' A final note: the specimen used as representative of *E. pictissimus* (USNM-266310) for molecular analyses by Hedges *et al.* (2008) comes from Los Patos, Península de Barahona, an intergradation zone between *E. p. pictissimus* and *E. p. apantheatus* (see above).

7. Group Eleutherodactylus chlorophenax *Schwartz, 1976(a)*

Group 7: Particular attention should be devoted to *Eleutherodactylus chlorophenax* Schwartz, 1976(a); Type collected on 26 June 1971, by a 'native collector' (Schwartz 1976a: 168) in the vicinity of Castillon, Département du Sud, Haiti (precise location of collection, description of microhabitat, and details of its habits are unknown; description of species based on one specimen, a male). I should initiate my analysis citing Thomas's (1996: 18) recollection of one of his many conversations with Al Schwartz: 'Well, remember, all you need is a holotype!' He frequently said that you should go ahead and take a chance. By this he meant that you shouldn't dither around too much if you have only a few specimens but think you have something undescribed. The reader may appreciate why – because of this and the evidence examined next – some researchers contest that *E. chlorophenax* may be a junior synonym of *E. inoptatus* (or *E. nortoni*) instead (e.g. Martínez Rivera and Rodríguez Plaza 2015: 122).

From Schwartz (1976a: 170) it is evident that he based most of his justification of *E. chlorophenax* as a new species by comparing its body colouration with that of *E. inoptatus* (and presumably, because 'you are experienced with these animals and your insight is probably better than anyone else's on the subject', Thomas [1996: 18] while recollecting one of Schwartz's recommendations to him). Schwartz proceeded by comparing measurements (snout–vent length, diameter of the eye, snout length) between this one specimen of *E. chlorophenax* and specimens of *E. inoptatus*, a comparison presumably limited because of small sample size ($n = 1$ for *E. chlorophenax*). In fact, Schwartz admitted that 'although this difference will surely be obliterated with the acquisition of additional *E. chlorophenax*'. Nonetheless, Schwartz emphasized the overall smaller size of the single *E. chlorophenax* compared to *E. inoptatus*. Hedges *et al.* (2008: 14, 85) found significant resolution (see their Fig. 2, '*Eleutherodactylus* (*Pelorius*) *inoptatus* Species Series'), which indeed supported Schwartz's insight about the distinction of *E. inoptatus* from *E. chlorophenax*.

Notably, Schwartz (1976a) also described *E. nortoni* as a new species, and in the same publication. From Schwartz (1976a: 170) it also is evident that he based most of his justification of *E. chlorophenax* as a new species by comparing its bodily colouration with that of *E. nortoni*: *E. chlorophenax* differs from *E. nortoni* 'in [*E. nortoni*] having a dorsal pattern of dark green blotches on a green ground'. In contrast, Schwartz (1976a: 170) also compared measurements from the specimen of *E. chlorophenax* with those from *E. nortoni* (snout–vent length and tibia/snout–vent length ratios) but found that both species are comparable. Hedges and Thomas (1987: 272) described *E. parapelates* and mentioned that 'Additionally, *E. chlorophenax* and *E. nortoni* have larger digital pads and *are often green (some* E. chlorophenax *are tan)*' (emphasis added). In contrast, Martínez Rivera and Rodríguez Plaza (2015: 122) displayed a picture of individuals of *E. nortoni* that are also tan in colouration of the body and vary in eye colouration. Finally, according to Hedges *et al.* (2008: Fig. 2, on p. 14), the closest relative of *E. chlorophenax* (sample from a female, USNM-257730, collected on 16–17 July 1985: shown alive in Fig. 1f of Hedges and Thomas 1987) is not *E. inoptatus*, but *E. nortoni*; in fact, the species-pair *E. chlorophenax*/*E. nortoni* shows the closest phylogenetic relationship among all species-pairs in Haiti (Hedges *et al.* 2008).

Hedges *et al.* (2008: 85–86) remarked:

> *E. chlorophenax* is poorly known. It was described from a single specimen from the north slope of the Massif de La Hotte and no additional specimens have been

> collected in that region. Hedges and Thomas (1987) collected a specimen [presumably that on 16–17 July 1985; a male, USNM-257729, but the name of collector{s} not specified] on the south slope of the La Hotte [Plain Formon, vicinity of, in adjacent Limestone hills] but were unable to record its call.

However, Franz and Cordier (1986: 16–17) also collected *E. chlorophenax* at the same locality and elevation at Plain Formon (UF-56797, UF-56799, UF-61600–61603) a year before, in 1984. At that same locality, south of Massif de la Hotte, and elevation (1,000 m asl) documented by Hedges and Thomas (1987) for *E. chlorophenax*, Franz and Cordier (1986: 16) also collected *E. nortoni* and mentioned that '[*E. chlorophenax*] was found in association with *Eleutherodactylus nortoni*. Both frogs show similar behaviours in these caves, hopping about or hiding under rocks and organic debris on the bottom, or wedging themselves into nooks and crannies on the vertical walls of the sink.' Finally, Martínez Rivera and Rodríguez Plaza (2015: 122) mentioned that in their study of Haitian anurans since 2012 (and in collaboration with several colleagues) 'the first specimens encountered at Massif de la Hotte [Duchity, east Plain Formon] and identified as *E. chlorophenax* were, in fact, *E. nortoni*' (translated freely from Spanish).

To conclude, in addressing the original question – is the available evidence sufficient to support *E. chlorophenax* as a valid species, or is it a junior synonym of *E. nortoni* instead? – the evidence at hand may be equivocal, and future studies may put this controversy to rest in the herpetological literature.

8. Summary of Part B and comments

Among the groups 1–6 of taxonomically complex material discussed above, 14 subspecies, related to records from Haiti, were described and a population within the distribution of one of these subspecies was described as a Haitian species (i.e. *E. parapelates* Hedges and Thomas, 1987). Albert Schwartz Jr was the author with the most subspecies mentioned herein: 12 subspecies (besides the description of original species in each group; Lynn and Cochran each described one subspecies that were elevated by Hedges *et al.* 2008). In addition, eight of Schwartz's subspecies were elevated to species by Hedges *et al.* (2008) ($n = 5$), the rest distributed among Schwartz (1973), Schwartz and Thomas (1975), and Lynch and Duellman (1997) (currently, four remain as subspecies). Overall, Hedges *et al.* (2008) were responsible for most of the subspecies elevated to species status among these groups ($n = 7$). Finally, Hedges *et al.* (2008) specifically mentioned the possibility of additional species from within their *Eleutherodactylus* (*Euhyas*) *schmidti* Species Group and *Eleutherodactylus* (*Euhyas*) *lentus* Species Group.

Several species with geographic distributions shared between Haiti and the Dominican Republic are worth mentioning; these species (several *Eleutherodactylus* spp., *Boana heilprini* [Noble, 1923], all *Osteopilus* spp., and *Peltophryne guentheri* [Cochran, 1941] would be grouped as being documented either in south palaeoisland, north palaeoisland, or widely distributed throughout the island of Hispaniola. Accounts from groups of species with complex distributional patterns in Haiti/Dominican Republic, and a list of species in satellite islands of Haiti – mostly from Powell 2012 – completes this section. It has not been clarified – as far as I know – whether the following accounts of species' distributions reflect artefacts of limited sampling between areas of known occupancy, the result of extirpation between areas of known occupancy, or a distributional pattern attributable to different taxonomic units.

(a) South palaeoisland

Eleutherodactylus armstrongi Noble and Hassler, 1933, has a fragmented geographic distribution: two widely separated regions in south palaeoisland, each one in a different country: Haiti

(Massif de la Selle) and the Dominican Republic (Península de Barahona) (Hedges *et al.* 2008; IUCN SSC Amphibian Specialist Group 2013b). Later, discrepancies in the visual representation of the geographic distribution of this species were generated between Hedge's portal (http://www.caribherp.org for *E. armstrongi*, updated 17 January 2020 [accessed 19 January 2020] and updated 26 April 2022 [accessed 3 August 2022]) where the distribution is continuous, and IUCN (2022g) where four disjunct localities are shown instead.

Eleutherodactylus audanti Cochran, 1934, shows a fragmented geographic distribution: one region in Massif de la Hotte (west Tiburon Peninsula, Haiti) and a broader region in Massif de la Selle/Sierra de Bahoruco (Hedges *et al.* 2004c; IUCN 2022h; see section II. B. 4).

Eleutherodactylus fowleri Schwartz, 1973, shows a fragmented geographic distribution: two small regions, one west Massif de la Selle (Parc National La Visite) and one Sierra de Bahoruco (Hedges *et al.* 2004d; but see http://www.caribherp.org for *E. fowleri*, updated 17 January 2020 [accessed 19 January 2020] and updated 26 April 2022 [accessed 3 August 2022], and IUCN [2022r], where this distribution is shown continuous).

Eleutherodactylus nortoni Schwartz, 1976(a), also shows a fragmented geographic distribution: two widely separated regions in south palaeoisland, one population at Massif de la Hotte, the other population at east Massif de la Selle (Haiti)/Sierra de Bahoruco (Dominican Republic) (Hedges *et al.* 2010b; IUCN 2022gg).

(b) North palaeoisland

Eleutherodactylus parabates shows a fragmented distribution within the region of Chaîne du Trou d'Eau/Sierra de Neiba: a population in eastern Chaîne du Trou d'Eau (Haiti) and a population in eastern Sierra de Neiba (Dominican Republic) (Hedges *et al.* 2010c; but see http://www.caribherp.org for *E. parabates*, updated 17 January 2020 [accessed 19 January 2020] and updated 26 April 2022 [accessed 3 August 2022], and IUCN [2022jj], where this distribution is shown to be continuous). Similarly, *E. weinlandi* shows a fragmented distribution, one large area in north/central-east of the Dominican Republic and a smaller area of occupancy shared between Haiti and the Dominican Republic in the centre of north palaeoisland (Hedges *et al.* 2004b; but see http://www.caribherp.org for *E. weinlandi*, updated 17 January 2020 [accessed 19 January 2020] and updated 26 April 2022 [accessed 3 August 2022], and IUCN [2022xx], where this distribution is shown to be continuous).

(c) Island-wide Hispaniola

Osteopilus dominicensis (Tschudi, 1838), occurs throughout Haiti/Dominican Republic except in the driest region in the northwest, north peninsula, Haiti (htpp://www.caribherp.org, updated 26 April 2022, accessed 3 August 2022). In contrast, IUCN (2022aaa) includes this driest region as part of the species' distribution.

(d) Complex accounts

Boana heilprini (Noble, 1923), (see Dubois 2017: 27, for generic rank) shows a fragmented distribution within a large region in north palaeoisland within Haiti/Dominican Republic, and a large region mainly in the south palaeoisland, from west Tiburon Peninsula (westernmost Massif de la Hotte) to west Massif de la Selle, Haiti (see IUCN 2022a). However, most of its predicted distribution needs in-field corroboration.

Osteopilus pulchrilineatus (Cope, 1870), shows a fragmented distribution with various localities within a large region of potential occurrence in the north palaeoisland (Haiti/Dominican Republic), and two small and isolated regions of potential occurrence on the south palaeoisland (Massif de la Hotte and west Massif de la Selle, Haiti) (IUCN SSC Amphibian Specialist Group 2013f, 2022bbb). Similar to *B. heilprini*, most of its predicted distribution needs in-field corroboration.

Osteopilus vastus (Cope, 1871), also has a complex distribution (IUCN SSC Amphibian Specialist Group 2013g; 2022ccc), with a large region on the north palaeoisland within most of the Dominican Republic and a smaller region at Massif du Nord, north palaeoisland, Haiti. In addition, there is one large region on the south palaeoisland, within Massif de la Selle/Sierra de Bahoruco, and a smaller region west Massif de la Hotte, south palaeoisland, Haiti. In contrast, a few discrepancies in the visual representation of the species' distribution from IUCN (2022ccc) exist in the portal http://www.caribherp.org for *O. vastus* (updated 17 January 2020 [accessed 19 January 2020] and updated 26 April 2022 [accessed 3 August 2022]). Here, the portal has the species' distribution as continuous in the north, while those areas in south palaeoisland are shown as isolates, one at Massif de la Hotte, one west Massif de la Selle, and one mostly in Sierra de Bahoruco (Dominican Republic), into eastern Massif de la Selle (Haiti).

Peltophryne guentheri (Cochran, 1941), has a fragmented distribution. Initially, records had it with a large region along northern north palaeoisland within Haiti/Dominican Republic and a vast region in the lowlands from Plaine du Cul-de-Sac (Haiti)/and Valle de Neiba/Valle de Azua/ west Peninsula de las Calderas (Dominican Republic) and extending north of Azua into Plateau Central (Haiti) (Hedges *et al.* 2004e; also see http://www.caribherp.org, accessed 3 August 2022). A small locality at Higüey, La Altagracia Province, eastern Dominican Republic, which was attributed to the distribution of *P. fracta* (Schwartz, 1972), was added to the distribution of *P. guentheri* in IUCN (2022ddd) when *P. fracta* was synonymized with the latter by Landestoy *et al.* (2018).

(e) Anurans in satellite islands of Haiti

From north to south Haiti: (1) Île de la Tortue: *Eleutherodactylus inoptatus* (Barbour, 1914), *E. warreni* Schwartz, 1976b, *Osteopilus dominicensis*; (2) Île de la Gonâve: *O. dominicensis*; (3) Île Grande Cayémite: *O. dominicensis*; (4) and Île-â-Vache: *O. dominicensis* (Powell 2012). As mentioned in section II. B. 6., the portal http://www.caribherp.org for *E. pictissimus* includes Île-â-Vache as the only location for *E. pictissimus*, which contrasts with the account of its distribution in Hedges *et al.* (2004a) and IUCN (2022mm).

C. Remarks for naturalized anurans

Two anurans are documented as introduced into Haiti (and broadly into Hispaniola) and are now naturalized: *Lithobates catesbeianus* (Shaw, 1802), and *Rhinella marina* (Linnaeus, 1758). Lever (2003) does not mention the date of introduction in Haiti for either species. The recent herpetological literature seems to be somewhat inaccurate about this (as far as I am aware). The date of first documentation for the introduction of both species in Haiti may be clarified herein.

1. Lithobates catesbeianus *(Shaw, 1802)*

The only published reference I found about the date of first documentation of this species as introduced into Haiti comes from Neils and Bugbee (2007: 443) who gave the following reference: '*Rana catesbeiana* were first reported in Haiti in 2002 by the Invasive Species Specialist Group (ISSG).' Regardless of the source of this statement, a series of five specimens collected in northern Haiti – Cap-Haitien, Département du Nord – on 4 September 1966, with collection numbers KU KUH-265908–265912 (all with the note, recorded by 'native'; at the University of Kansas Biodiversity Institute and Natural History Museum) exists and represents the oldest record of *L. catesbeianus* in Haiti (see http://portal.vertnet.org/search?q=Rana+catesbeiana+Haiti).

2. Rhinella marina *(Linnaeus, 1758)*

Shreve and Williams (1963: 307) included a passage that reads '*B. marinus* is said to have been introduced into Haiti by a Dr Freeman (hence called locally "Freeman")'. Shreve and Williams (1963) included a footnote to their passage: 'Another story is given by Anthony Curtiss in a letter of September 8, 1944. He states that *Bufo marinus* was introduced into Port-au-Prince region twelve years before by M. Audant.'

As for Anthony Curtiss, Shreve and Williams (1963: 295, 301) described him as 'an amateur naturalist resident in the region [Port-au-Prince]' that 'had unrivalled opportunities for field study and used them well' with 'very perceptive and accurate field observations on the herpetofauna of the Port-au-Prince region' and 'whose area of investigation was again the general Port-au-Prince region'. Shreve and Williams (1963) also highlighted that he kept frequent written correspondence with Cochran from 1942 to 1950.

On the other hand, M. (Monsieur) Audant (André Audant) mentioned by Anthony Curtiss above, was a Haitian entomologist at the Service Technique at Damien, also at Port-au-Prince (Blackwelder 1943: 5; also see Cochran 1941: 188, which refers to Audant as 'zoologist at the Government Agricultural School at Damien'). Tanner (1948: 1) referred to Audant as 'former entomologist of the Haitian Agricultural College at Damien'. Cochran (1941: 69) mentioned the many specimens of herpetofauna from Haiti sent by Audant for her inspection and study, but the only date of reference for any specimen collected by Audant and examined by Cochran is that of 1933 (30 August; Cochran 1941: 192); certainly, Audant was also very active in herpetofaunal-related activities in Haiti during the early 1930s.

Port-au-Prince harboured several sugarcane plantations, including the most extensive plantations established there from the early 1900s to the late 1980s (Haitian American Sugar Company, S.A. [or HASCO]; see also The New York Times 1987). Therefore, it is plausible that entomologist André Audant did introduce *R. marina* into Haiti (Port-au-Prince region) in about the year mentioned by Anthony Curtiss (1932). It is also most probable for the same reasons and as a generalized strategy followed by entomologists related to the agriculture-based economy in the Caribbean at the time: introduced as biocontrol for grubs (or other insects) in sugarcane plantations elsewhere. Therefore, *R. marina* could have been first noticed as intentionally introduced into Haiti in 1932 as Curtiss had communicated with Cochran in the letter of 8 September 1944 examined by Shreve and Williams (1963). Surprising, however, is the passage from a footnote by Reinhardt and Lütken (1863: 202) that reads (translated freely from Old Dutch):

> Of our *Bufo marinus* and *Phyllomedusa bicolor* there are specimens in our Museums that are said to be from the Danish-West Indies. The latter is, according to Mr. Riise, completely unknown as a West Indian, and since it is not listed as West Indian anywhere in the Literature, its indication does likely depend on a mistake. *B. marinus*, on the other hand, is sometimes brought there [St. Croix, St. Thomas, St. John] with timber from Vieques [PR] and Haiti.

Note: Ríos-López *et al.* (Chapter 8, this volume) found records of specimens of *R. marina* deposited in the Natural History Museum of Denmark, which were collected in St Croix (1835) and St Thomas (1848). If this evidence is supportive of Reinhardt and Lütken (1863) then the records of *R. marina* in 'Haiti' (either Hispaniola as frequently referred to in the old literature or the territory of the Republic of Haiti proper) would be more than 90 years older than acknowledged by Anthony Curtiss and elsewhere; its introduction – presumably inadvertently – into Haiti would then be something of an enigma.

III. Major threats to anurans in Haiti

In Haiti, similar to other islands in the Caribbean (e.g. Puerto Rico), the present-day distribution of anurans throughout its landscape reveals a footprint of changes in types of land-cover (e.g. woodlands, forests) directly related to Haiti's history of uses of its land since colonial times (see section I. *Introduction*). We also know that many natural and anthropogenic phenomena act synergistically. For instance, the influence of droughts, storms, and earthquakes on Haitian anurans are all the more unpredictable in an anthropogenic-dominated landscape – which is currently mostly deprived of mature vegetation and forests (Hedges *et al.* 2018). Responses of these anurans to the influence of these natural phenomena would have been different if they were in a natural landscape – one covered by mature forests and other vegetation types with little (or without) the influence of humans – that had developed undisturbed over thousands of years. A fact in agreement with Hedges *et al.* (2018) is that what is considered to be a 'mature forest' today (regardless of how different authors define it) (see Churches *et al.* 2014) may not fulfil all the ecological requirements of Haitian anurans that many decades ago inhabited the 'primary forest' that existed throughout Haiti's landscape. This fact has brought consequences to our understanding of nature and present-day ecological patterns that go beyond one's lifetime.

The herpetological literature about Haitian anurans, however, mostly focuses on taxonomy and phylogenetic relationships. It offers little information about estimates of population size, population dynamics, resource partitioning, interspecific interactions, ecological physiology, basic ecological requirements, and responses to natural phenomena related to climatic change, among other topics. As a result, I understand that the few herpetologists – mainly taxonomists and systematists – who have made Haiti their working ground have had a difficult time distinguishing between species of Haitian anurans that are naturally rare and those deemed rare because of anthropogenic threats (topics within the area of herpetofaunal ecology instead; see below). Nonetheless, there are several areas in Haiti most suitable for ecological studies that will help us better understand the biology and ecology of Haitian anurans. These will also help us to add to current conservation efforts, including reforestation and soil rehabilitation whenever possible.

In the following sections, I present the available data on the most common threats to the anurans in Haiti (endemic, native, and naturalized), other than deforestation (carried out mostly for charcoaling, slash-and-burn agriculture, and livestock farming; see section IV. A.; also Martínez Rivera *et al.* [2011], Martínez Rivera and Rodríguez Plaza [2015], and references elsewhere).

A. Pathogens

Martínez Rivera and Rodríguez Plaza (2015: 11) mentioned that the chytrid fungus, *Batrachochytrium dendrobatidis* (*Bd*), was found on individuals of *Eleutherodactylus furcyensis* sampled from Furcy (western Massif de la Selle), Haiti. I could not find published works about *Bd* found in other anurans sampled from Haiti. Nonetheless, *Bd* has been found in anurans shared between Haiti and the Dominican Republic, but only from samples from the latter country, which in many instances may show environmental conditions somewhat different than those in Haiti. The first record of a species deemed positive for *Bd* is *Osteopilus vastus* in 2004 (Joglar and Burrowes 2005; Joglar *et al.* 2007 [reprinted 2011]; Hedges and Díaz 2011; Galvis 2013, cited by Ministerio de Medio Ambiente y Recursos Naturales 2014; IUCN SSC Amphibian Specialist Group 2013g; 2022ccc); later, *Eleutherodactylus flavescens* and *E. weinlandi*, along with *Boana heilprini*, *O. pulchrilineatus*, and *O. vastus* were found to be positive for *Bd* (Galvis 2013; for *B. heilprini* see IUCN SSC Amphibian Specialist Group 2013e, and Díaz *et al.* 2015 for *Bd* in larvae; for *O. pulchrilineatus* see IUCN SSC Amphibian Specialist Group 2013f). On the other hand, *E. abbotti*, *E. inoptatus*, *Lithobates catesbeianus*, and *O. dominicensis* sampled negative for *Bd* (Galvis

2013, as above; IUCN SSC Amphibian Specialist Group [2014, 2022aaa] does not mention *Bd* in its assessments for *O. dominicensis*). In conclusion, these species – other than *Eleutherodactylus furcyensis* – were sampled for *Bd* in the Dominican Republic, but not in Haiti; for a discussion of sampling considerations and interpretation of prevalence for conservation, see Incháustegui and Díaz (Chapter 7 in this volume) and Ríos-López *et al.* (Chapter 8 in this volume).

B. Endoparasites

In the herpetological literature, Schwartz (1983: 1) mentioned that 95% of the 104 specimens of the Haitian *Eleutherodactylus glandulifer* collected by Thomas (along with Elie Joseph and a 'native' person at two localities, south Marche Leon and south Castillon, Département du Sud) 'had light to very heavy infestations of encysted nematodes on the thighs and near the vent, and occasionally elsewhere on the body skin'. (Note: in their text, Henderson and Powell [2009: 52] cross-referenced this observation to their Schwartz [1983b], which is about *E. furcyensis* [p. 471], but should be their Schwartz [1983c], which is about *E. glandulifer*.)

Nematodes have also been found in two species of *Eleutherodactylus* that occur in both Haiti and the Dominican Republic, *E. abbotti* and *E. armstrongi*; but the actual records came from frogs collected in the Dominican Republic (Powell *et al.* 1998). Specifically, larvae of the nematode *Terranova caballeroi* (Nematoda; Anisakidae), a juvenile 'possibly an oxyuroid' (unidentified), and two immature nematodes 'apparently spiruroids' were found in *E. abbotti*. The nematodes *T. caballeroi* and an adult *Oswaldocruzia lenteixeirai* (Nematoda; Molineidae) were encountered in *E. armstrongi*.

Duer *et al.* (1992: 228) documented the stomach contents of *Osteopilus dominicensis* collected during summertime, between 1971 and 1979, in several localities in Haiti. Out of 159 stomachs examined, they found 'two female nematodes (probably Cosmocercoidea)' recovered from the stomach of a male *O. dominicensis* (Albert Schwartz Field Series V38277; collected at Carrefour Marmelade, Département l'Artibonite; Catalogue number 264765 at the University of Kansas, Biodiversity Institute and Natural History Museum). Duer *et al.* (1992: 228) added: 'Nematodes are very common in amphibians (Baker, 1987), but cosmocercoids are almost always found in the rectum or large intestine. These specimens may have used arthropod hosts and might have been spurious parasites of the frog', which appears a product of a fortuitous event of ingestion.

Parsons *et al.* (1995: 31) examined 204 individuals of *P. guentheri* collected from Haiti (Artibonite, near Los Poteaux) and the Dominican Republic (Independencia, near Lago Enriquillo; Monte Cristi, near Copey). These authors found 'immature nematodes (*Skrjabinoptera* sp., Physalopteroidea) mixed with ingesta' in three (not identified) of 204 stomachs examined (from 90 juveniles, 65 adult females, and 49 adult males; Parsons and Powell 1994: 3). Parsons *et al.* (1995: 31) mentioned that 'the low prevalence and the fact that all parasites were juveniles may reflect the inadequacy of *P. guentheri* as a final host'.

Other than documenting instances of endoparasites found in these anurans, the influence of these parasites (and presumed parasites) on the population ecology of anurans in Haiti is unknown. I have not found records of ectoparasites from Haitian anurans.

C. Naturalized species

I found no records of the influence of naturalized species of fishes, amphibians, reptiles, birds, or mammals on anurans in Haiti. Instead, there is one work documenting a fortuitous observation of a predatory anuran in Haiti: Neils and Bugbee (2007) documented an adult *Lithobates catesbeianus* ingesting an adult *Osteopilus dominicensis* in Deschapelles, Département l'Artibonite, Haiti. As above, the influence of this species on the population ecology of native/endemic species of anurans is unknown.

The few studies found that document predation upon anurans in Haiti refer to native species, mainly colubrid snakes: Henderson and Schwartz (1986) for *Ialtris haetianus* preying upon the Haitian *Eleutherodactylus eunaster, E. glandulifer, E. glaphycompus, E. oxyrhyncus,* and *E. sciagraphus* (and the natives *E. abbotti, E. armstrongi,* and *E. heminota*); Henderson *et al.* (1987a) for *Uromacer* spp. on *Osteopilus dominicensis* and *O. pulchrilineata*; Henderson *et al.* (1987b) for *Hypsirhynchus parvifrons* on the natives *E. abbotti, E. alcoae, E. hypostenor* (*E. parapelates*?), *E. inoptatus, E. ruthae,* and *O. dominicensis*; Landestoy (2017) for *Haitiophis anomalus* on *Peltophryne guentheri* (in the Dominican Republic). For birds, Wetmore and Swales (1931: 234–235) mentioned *Osteopilus dominicensis* as prey of the Hispaniolan *Tyto glaucops* (Ashy-faced Owl); their records are from Diquini (near Port-au-Prince) and 'in a cave near L'Acul', northern Département l'Artibonite, Haiti.

D. Harvesting amphibians

In a comprehensive study, Powell *et al.* (2011) examined patterns of importation and exportation of amphibians (and reptiles) in the Caribbean. For anurans exported from Haiti to the USA between 1998 and 2008, Powell *et al.* (2011: 138, appendix 4) listed 4,648 anurans termed '*Hyla* sp.', 888 *Osteopilus vastus* (which were so identified for years 2007 and 2008 only), 616 termed 'amphibian', and 27 *Hyla cinerea* [= *Dryophytes cinereus*], which does not occur in Haiti). *Osteopilus vastus* represented at least 14.4% of total decadal exportations, but 87.6% of total exportations of anurans from Haiti between 2007 and 2008. In their appendix 5, Powell *et al.* (2011: 141) also presented the number of individuals listed 'amphibian' (presumably native) exported from Haiti to the USA only to be re-exported to other markets, mostly European, Canadian, Asian, or back to Caribbean markets: 298 termed '*Hyla* sp.' and 86 *Osteopilus vastus* were included. A brief search on the internet revealed that the selling price for *Osteopilus vastus* in 2019 was US $29.99 (marketed as 'Haitian Giant Tree Frog') and US $34.99 for '*Hypsiboas heilprini*' ('Blue Toed Tree Frog'; not listed by Powell *et al.* 2011); not every source showed prices online. Finally, data from Powell *et al.* (2011) revealed that re-exportation of *O. vastus* only represents 9.7% of the initial exportations to the USA; and USA seems to be the major consumer of anurans from Haiti, via the pet trade.

E. Systematic collection as a proxy of population status in assessments: beyond the 'Precautionary Principle'

Albert Schwartz was the most prolific and productive herpetologist working on the taxonomy of Caribbean herpetofauna in the twentieth century. His contribution was (and still is) fundamental for the systematics of the Caribbean herpetofauna, particularly Haiti. S. B. Hedges (along with J. P. Richard Thomas) continued Schwartz's legacy for Haiti's herpetofauna and elevated his contribution through their own contributions. I must clarify that far from seeking to criticize the outstanding work of early and current herpetologists working in Haiti, I simply highlight an area that still needs particular attention from the perspective of herpetofaunal ecological studies, which remain in their very early stages, yet they have profound implications for current and future conservation efforts for anurans in Haiti.

1. The history of the use of current language in assessments for conservation

Initially, Schwartz (1983) commented – about the series of *Eleutherodactylus glandulifer* collected by R. Thomas at Castillon, Haiti – 'His 104 specimens indicate that *E. glandulifer* is not uncommon in this region.' Hedges and Thomas (2010) added that 'at the time of the last survey in 1991 it was known to be moderately common'. Assessments of many other anurans in Haiti addressed their 'abundance' in similar terms. Some terms include: 'very common in suitable habitat'; 'very common within its restricted habitat'; 'common in suitable habitat'; 'once common in good habitat, which probably no longer exists' or 'that no longer exists'; 'formerly common in good habitat, which no longer exists'; 'locally uncommon and patchily distributed'. Also: 'relatively common in suitable habitat'; 'moderately common'; 'moderately common in suitable habitat'; 'moderately common in original habitat, which largely disappeared'; 'moderately common in suitable habitat, although this habitat is declining rapidly' or 'is rapidly disappearing' or 'has largely disappeared', among others. Assessments made since 2004 for anurans in Haiti revealed several species not found for years, decades, or even since first described. Several fundamental questions about assessments for conservation purposes arise: (1) is it that species 'not found' only appear to be absent despite extensive efforts from surveys in different localities, other than historical collection sites, and at different seasons throughout the year and among years?; (2) or is it that these species were sighted the last time a researcher visited a given known historical area of occupancy, but that no additional surveys at that locality were ever conducted again? How have 'suitable/original/good habitat' and 'very common', 'relatively common', 'moderately common', or 'rare' been quantified? It is beyond the scope of this review to provide answers to all of these questions – as the data on which such comments are based rarely are explicitly defined – but I will try to make sense of Schwartz's (1983) comment about the abundance of *E. glandulifer* above.

2. On the abundance of species based on collection records

Ríos-López *et al.* (Chapter 8 in this volume) examined this topic for anurans in the Puerto Rican Archipelago in detail. These authors discussed the collection records of the Puerto Rican stream-dweller *Eleutherodactylus karlschmidti,* and highlighted that records of the number of specimens collected are not indicative of the abundance of its species; instead, such records mostly reflect the experience, efficiency, and preferences for a given species by the herpetologist doing the collecting. As for *E. glandulifer* in Haiti, Schwartz's (1983) comment originated from the collection effort by Thomas, Joseph, and 'native' people between June and July 1971 in the vicinity of Castillon (northern Massif de la Hotte), an area Schwartz (1973: 269; Schwartz 1978) would refer to as of high 'ecological diversity' (Table 6.2). The 1971 collection-period at Castillon especially deserves analysis because the area harboured a high ecological diversity and its anurans were of particular interest to herpetologists for various reasons (see below).

Presumably, three species of anurans in Haiti were major targets for collection at Castillon. *Eleutherodactylus wetmorei,* known from Haiti from a few specimens, was first collected by Thomas (and Leber) in the Dominican Republic (Schwartz 1968) and prompted the description of *E. diplasius* later on (i.e. Schwartz 1973). It is a species 'predominantly arboreal, it is difficult to collect, and often the species has been vocalizing in areas where specimens were impossible to secure' (Schwartz 1968: 12). Therefore, previous experience (i.e. a fixed searching-image and effective identification of its calling) fuelled by a major contribution to the taxonomy of anurans in Haiti, would motivate the collecting of an adequate series of specimens if found at Castillon in 1971. Second, *E. oxyrhyncus* was described from one specimen of unknown origin (Duméril and Bibron 1841; but see Lynch and Schwartz 1971: 107), one additional individual collected in 1934 by Darlington, and > 27 individuals collected in Castillon just in 1969 by T. Preston Webster and A. R. Keister (Lynch and Schwartz 1971), but presumably 'much of their material was poorly preserved' (Schwartz 1973: 249). Therefore, a motive for an adequate series of this species for its redescription was latent and, similar to the case of *E. wetmorei,* such a series would logically be much appreciated at the time. Third, *E. glandulifer* was known from one female collected in 1934 by Darlington higher in the Massif de la Hotte, south of Castillon, and described by Cochran in 1935. Background – yet limited – information about this species also would make it a particular goal for securing an adequate series once encountered in the field. Evidently, in common was the background information about these species and the impact they had on taxonomic knowledge at the time; for an experienced herpetologist, these species would be easier to detect and collect compared to new species for which no information existed. Finally, it was common

Table 6.2. Some *Eleutherodactylus* spp. collected between 21 and 27 June, and between 22 and 24 July 1971 (as in Schwartz 1973) – arranged in chronological order by date of collections – in the vicinity of Castillon, northern Massif de la Hotte, Haiti. Specimens collected during this brief period resulted in the description of seven species of *Eleutherodactylus*. * = *E. apostates, E. chlorophenax, E. eunaster, E. glaphycompus, E. lamprotes, E. parapelates, E. sciagraphus*) and the 'first adequate series' of species known from one or very few specimens since first described – particularly, *E. oxyrhyncus* since 1841 (Schwartz 1973), *E. wetmorei* since 1932 (Schwartz 1968), and *E. glandulifer* in 1935 (Schwartz 1973), among others. ** = unknown collector. Data were compiled from http://vertnet.org/, VertNet © version 2016-09-29 (accessed on 27 January 2020).

LOCATION, DATE, *SPECIES*	COUNTS		% OF COLLECTION
South Marche Leon, northern Castillon **June 21** (*n* = 124)			
Species	**Count: by investigator(s)**	**plus count by assistant(s)**	**% collection**
E. abbotti	1; (R. Thomas)		0.8
* *E. counouspeus*	1; (R. Thomas)	—	0.8
* *E. parapelates*	1; (R. Thomas)		0.8
E. inoptatus	5; (R. Thomas)	5	4.0
E. heminota	—	23	18.5
E. wetmorei	93; (R. Thomas)	93	75.0
June 22 (*n* = 157)			
* *E. eunaster*	5; (E. Joseph, R. Thomas)	—	3.2
E. inoptatus	—	6	3.8
* *E. glaphycompus*	—	** 14	8.9
E. paulsoni	—	19	12.1
E. heminota	—	24	15.3
E. abbotti	—	30 from native people only	19.1
E. oxyrhyncus	51; (E. Joseph, R. Thomas)	51 + 8 from native people only	37.6
June 23 (*n* = 17)			
E. paulsoni	3; (E. Joseph, R. Thomas)	—	17.6
E. abbotti	4; (E. Joseph, R. Thomas)	—	23.5
E. oxyrhyncus	10; (E. Joseph, R. Thomas)	10	58.8
June 24 (*n* = 1)			
* *E. counouspeus*	1; (R. Thomas)	—	100
South Castillon **June 24** (*n* = 99)			
E. audanti	1; (R. Thomas)	—	1.0
E. oxyrhyncus	5; (R. Thomas)		5.1
E. bakeri	6; (E. Joseph, R. Thomas)		6.1
* *E. sciagraphus*	7; (E. Joseph, R. Thomas)		7.1
E. brevirostris	9; (E. Joseph, R. Thomas)		9.1
* *E. apostates*	9; (E. Joseph, R. Thomas)	+ 6 from native people only	15.2
E. glandulifer	56; (E. Joseph, R. Thomas)	—	56.6

(Continued)

Table 6.2. (Continued)

LOCATION, DATE, *SPECIES*	COUNTS		% OF COLLECTION
South Marche Leon, northern Castillon June 25 (*n* = 2)			
E. inoptatus	—	2	100
South Castillon June 26 (*n* = 71)			
E. apostates	—	1	1.4
E. audanti	—	1	1.4
* *E. chlorophenax*	—	1	1.4
* *E. lamprotes*	1; (R. Thomas)	—	1.4
* *E. sciagraphus*	1; (R. Thomas)	—	1.4
E. oxyrhyncus	—	2	2.8
* *E. eunaster*	4; (R. Thomas)	—	5.6
* *E. glaphycompus*	5; (E. Joseph, R. Thomas)	—	7.0
E. bakeri	8; (E. Joseph)	8	11.3
E. glandulifer	47; (E. Joseph)	47	66.2
South Marche Leon, northern Castillon June 25 (*n* = 7)			
E. paulsoni	1; (R. Thomas)	—	14.3
* *E. counouspeus*	—	3	42.9
* *E. glaphycompus*	—	** 3	42.9
South Marche Leon, northern Castillon July 22 (*n* = 4)			
E. pictissimus pictissimus	4; (R. Thomas)	4	100
South Marche Leon, northern Castillon July 24 (*n* = 1)			
E. pictissimus pictissimus	1; (R. Thomas)	—	100

practice to have locals collect animals for herpetologists in Haiti and elsewhere in the Caribbean. For example, Thomas (1996: 17) mentioned that Schwartz 'often paid for specimens if the local people would collect them', although sometimes 'native collectors were hard to encourage to collect frogs' in some situations in the field (Schwartz 1980b: 112).

Certain patterns of collection from the 1971 expedition to Castillon emerged, which give the word 'abundant' a different meaning (Table 6.2). The species mentioned above were the ones with the most specimens collected, and often assistance from locals was instrumental in securing a diverse sample of species in the area; once encouraged and told what to seek, locals would bring specimens of the target species in numbers. In contrast, *E. abbotti*, which is considered a generalist with a wide geographic distribution – despite its small body size – was rarely collected by herpetologists (but mostly by local inhabitants). On the other hand, the number of specimens of new species – those with which professional herpetologists and assistants had no previous experience – was very low, sometimes based on a single individual (Table 6.2). Therefore, the number of collected specimens cannot be taken as a proxy for a species' abundance; it is the result of a highly efficient and specific (motivated) effort based on previous experience and some degree of encouragement. Most likely, similar analyses elsewhere in the Caribbean may yield similar results, and current assessments about the population status of Haitian anurans may need revision and require a somewhat different approach in the immediate future (see next section).

IV. Conservation status of anurans, regulation, and conservation initiatives for anurans in Haiti

A. Conservation status of anurans

According to the most-recent assessments in the IUCN Red List of Threatened Species, 90.0% of endemic species of Haitian anurans (all *Eleutherodactylus* spp; $n = 30$) are considered Critically Endangered (Table 6.1); the rest of the endemics are considered Endangered. In contrast, 23.5% of *Eleutherodactylus* spp. shared between Haiti and the Dominican Republic ($n = 17$, natives) are considered Critically Endangered, 35.3% are Endangered, 11.8% Vulnerable, 5.8% Near Treatened, and 23.5% Least Concern (Table 6.1). Among other native anurans in Haiti (*Boana heilprini, Osteopilus* spp., *Peltophryne guentheri*) 60% are Vulnerable and 40% Least Concern; *Lithobates catesbeianus* and *Rhinella marina* are considered Least Concern (Table 6.1). As mentioned above, major threats for species with a status of Vulnerable and higher are deforestation (mostly for charcoaling, slash-and-burn agriculture, and livestock farming) and lack of effective regulatory mechanisms (see section IV. B.), including within areas locally deemed protected.

Haiti's herpetological research trajectory must be judged in a historical context. While the research agenda and methods in herpetology have changed in most of the Caribbean, at least since the last quarter of the twentieth century – from taxonomically oriented to ecologically oriented – it has somewhat lagged in Haiti. Consequently, it is most logical to assume that for several species of anurans, there have been extirpations throughout the landscape before much was known of their basic ecological requirements and population dynamics; these are topics frequently referenced as 'research needs' for most assessments. I must emphasize, however, that information about species' taxonomy and systematics has been and is fundamental for local and international conservation strategies of 'Haiti's anuran megadiversity' (Hedges 2004). Currently, all assessments claim that information about population ecology and interspecific dynamics of anurans in Haiti is lacking and urgently needed, more so as it is also fundamental for *in-situ* and *ex-situ* conservation of species and for environmental-restoration efforts (Ríos-López *et al.* Chapter 8 in this volume). It has been 19 years since the first IUCN's formal assessment (2004), but the claim still holds in 2023.

Information from the perspective of herpetological ecology will update and expand that for species' taxonomy and systematics (see Hedges and Timyan 2014), which echoes Williams' (1999; see section I. *Introduction*) comment that 'the goals [of herpetological studies other than taxonomy in the West Indies] will be quite different in the future.' Accordingly, outlets for the publication of studies of life-history and population ecology of anurans in Haiti (among other topics that will fill the information gaps highlighted in current assessments) are readily available and affordable for local and international naturalists and researchers interested in Haitian anurans: *Novitates Caribaea* (http://mnhn.gov.do/publicaciones/novitates-caribaea), *Solenodon* (http://www.caribbeanahigroup.org/solenodon.htm), *Caribbean Journal of Science* (http://www.bioone.org/loi/cjos), *Reptiles & Amphibians: Conservation and Natural History* (http://www.ircf.org/journal/), *Caribbean Naturalist* (https://www.eaglehill.us/programs/journals/cana/caribbean-naturalist.shtml), *LIFE: The Excitement of Biology* (https://blaypublishers.com/), *Acta Científica–una revista transdisciplinaria de Puerto Rico y el Caribe* (https://www.fs.usda.gov/treesearch/pubs/54663), and *Caribbean Herpetology* (http://www.caribbeanherpetology.org/), among others.

B. Current regulation framework

Chapter II of the Constitution of Haiti (1987, as amended) sets the guidelines for Haiti's Environment, specifically Articles 253–258, which are presented here (translated freely from French):

Article 254: The State organizes the enhancement of natural sites, ensures their protection and makes them accessible to all.
Article 255: To protect forest reserves and widen plant cover, the State encourages the development of clean forms of energy: solar, wind, and others.
Article 256: As part of the protection of the Environment and Public Education, the State has an obligation to proceed with the creation and maintenance of botanical and zoological gardens in certain points of the Territory.
Article 256.1: The State may, if the need is demonstrated, declare an area of ecological utility.
Article 257: The law determines the conditions for the protection of flora and fauna. It sanctions offenders.
Article 258: No one may bring waste or residues from foreign sources of any kind into the Country.

Victor (1995) summarized environmental laws in place in Haiti, and he noted updated governmental regulations (and local agencies) for the management and protection of Haiti's natural resources; in addition, areas of high biodiversity (i.e. 'Key Biodiversity Area') are now generally set (e.g. Gouvernement de la République d'Haïti 2016). Nonetheless, there is still much work to do as it is frequently acknowledged that a major threat to the anurans (and wider biodiversity) in Haiti originates from an ongoing lack of effective regulatory mechanisms, even within designated national 'protected areas' (Hedges *et al.* 2018). Presumably, this threat – lack of effective regulatory mechanisms – continues unless 'greater protective measures are taken' (Hedges *et al.* 2018: 5). In my opinion, when the basic subsistence needs of Haitians are fulfilled by effective socioeconomic and education policies, this will lead the citizens completely away from logging for energy production and short-term-focused agricultural practices in the countryside (see section IV. C. b., next). Until then, the vast majority of areas harbouring high diversity of Haitian anurans will be targeted as areas of resources for the very basic survival of the people. This threat presents Haiti with a complex scenario for effective regulatory mechanisms in the face of its historical and recurrent socioeconomic challenges.

C. Initiatives for the conservation of anurans in Haiti

1. Captive breeding

Initiatives for the conservation of anurans in Haiti include the successful *ex-situ* captive-care and maintenance programmes at Philadelphia Zoo and Dallas Zoo, which aim to establish assurance colonies for later reintroduction (Martínez Rivera *et al.* 2011; Murphy and Gratwicke 2017). Collectively, these institutions have worked to conserve 11 species of *Eleutherodactylus* that inhabit the south palaeoisland in Haiti (nine endemics and two natives) at least since 2010 and have been able to breed at least four of them (*E. amadeus*, *E. bakeri*, and *E. thorectes* [at Philadelphia Zoo], and *E. inoptatus* [at Dallas Zoo]). Nevertheless, and as encouraging as it is, breeding frogs without adequate environmental conditions for reintroduction might be impractical if the ultimate goal is to establish long-term, self-sustainable populations of target species in nature once more. From the perspective of Haitian amphibians, David Attenborough's words seem appropriate to recall: 'We must *rewild* the world!' (Attenborough 2020: 121) – Haiti's landscape, that is.

2. *Education and reserves*

Efforts to educate people about Haitian anurans are generally outstanding, take diverse formats, and include different stakeholders from local and international origins; these efforts have continued for more than a decade now. The major local environmental organization Société Audubon Haiti has worked with agronomists, social scientists, herpetologists, journalists, naturalists, ecologists, all in collaboration with universities, institutes, societies, foundations, and non-governmental and governmental agencies from Haiti and the Dominican Republic. It has conducted joint efforts to generate workshops and give talks to local villagers and inhabitants within and adjacent to major conservation areas in Haiti (e.g. Parc National Pic Macaya, Massif de la Hotte, among other areas in Haiti and the Dominican Republic), and given educational material (in several languages and formats), presentations, documentaries, and interviews at international level and throughout widespread media. All these efforts are directed toward spreading the word about the importance for the conservation of Haiti's diversity of anurans and its biodiversity in general (see Martínez Rivera *et al.* 2011; Martínez Rivera and Rodríguez Plaza 2015; Gouvernement de la République d'Haïti 2016; Mayer 2019; for examples, see *Haiti: Extinction en cours* at https://www.youtube.com/watch?v=1HSkLbu81g8 and Hedges and Timyan [2014]).

Since 2014, the government of Haiti, with national and international collaborators, has worked on an ambitious conservation plan for the strengthening of more than 68,000 hectares of terrestrial areas protected under their national park programme. Included therein are areas that harbour most of the endemic Haitian anurans at the Parc National La Visite and Parc National Naturel de la Forêt des Pins (Massif de la Selle), and Parc National Naturel de Macaya (Massif de la Hotte), among other areas. Moreover, this plan foresees the establishment of many other new areas for conservation, collectively deemed 'Key Biodiversity Areas', for the conservation of Haiti's biodiversity, anurans included (Gouvernement de la République d'Haïti 2016).

In 2019, and concluding more than seven years of explorations and assessments for the acquisition of lands of high-diversity value, the creation of Haiti's first nature reserve – Park National Naturel Grand Bois, Massif de la Hotte – was announced by the Société Audubon Haiti and its collaborators: area > 485 hectares of forested land within Grand Bois mountain, west of Massif de la Hotte (Hedges and Timyan 2014; Mayer 2019). In this reserve, at least 17 species of anurans were found, including the presumed-extinct stream-dweller *Eleutherodactylus semipalmatus* and several new species currently being described (Hedges and Timyan 2014). Therefore, the combination of efforts mentioned above (captive-breeding programmes, education, strengthening of current reserves and establishment of new ones in Haiti) paves the way to a brighter future for anurans in a rehabilitated Haitian landscape and fosters opportunities for studies in herpetological ecology that will be of great importance to the conservation of anurans in the second-most taxonomically diverse country (Cuba being the first) for this group in the Caribbean.

V. Conclusions and future directions

Changes in land-use and land-cover played a significant role in explaining patterns of distribution and presumed declines of anurans in Haiti: the evidence points nowhere else (e.g. Hedges *et al.* 2018). Understandably, herpetologists generally look for species of interest in places where collection records proved their existence, but several potential areas for anurans in Haiti remained unexplored (see Rivero 1963 and Ríos-López *et al.*, Chapter 8 in this volume, for a discussion on this topic; also see Table 1 [p. 35–36] in Hedges and Díaz 2011 for species that were overlooked in Haiti for a long time, but see below). This fact is attested by the recent discovery of populations of species long-thought to have been extirpated – see, for instance, comments in Henderson and Powell (2009) about sightings of several Haitian species, and also Hedges (2010a,b), Hedges and Díaz (2011), and Mayer (2019). More recently, in 2011, six anuran species that had not been recorded for a long-time were found during exploration in new areas of Haiti after the 2010-earthquake (The Guardian 2011), followed by the discovery in 2019 of the stream-dweller *Eleutherodactylus semipalmatus* in the newly explored Grand Bois mountain (Mayer 2019). Fortunately, in 2017, two Haitian agroforesters and conservationists associated with the Société Audubon Haiti, Mr Anderson Jean and Mr Maxon Fildor, rediscovered a second population of *E. semipalmatus* at Park National Naturel Macaya, but the species was not detected during recent surveys in 2019 and 2020 at Park National La Visite (M. Fildor, pers. comm.). Alongside such highly positive news for Haitian anurans and conservation initiatives, attention needs to be paid to estimates of abundance based on the subjectivity of the collector, collection records of target species, or efforts from population censuses repeated in space and time (see Hedges and Díaz 2011: 39). Undeniably, although more than three decades have now passed since Schwartz and Henderson's (1991) massive compendium, data from empirical ecological research on anurans in Haiti still are lacking. Studies that explicitly link patterns of distribution, estimates of abundance, and population dynamics, with microhabitat-use, behaviour, and ecophysiology also are urgently needed. Such studies may help us understand the mechanisms that influence responses of anurans in Haiti to projected changes in climate, particularly for those species at the highest areas in the mountains and those inhabiting areas closest to the coast; they also help us understand how they will respond to habitat rehabilitation.

The rapid rate of transformation of Haiti's landscape clearly justifies the application of the precautionary principle (Cooney 2004) for the conservation of Haitian anurans (and its biodiversity at large); this gives time for a much-needed reprieve for Haiti's threatened anurans. Nature reserves – governmental and privately owned – already in place in Haiti now provide promising places for fundamental ecological research of anurans in the Anthropocene (Vitousek *et al.* 1997; see Ríos-López *et al.*, Chapter 8 in this volume). Moreover, these areas can facilitate that agenda for anurans in Haiti, for example by the establishment of field research stations that allow for continuous research by Haitians and international scientists alike. Hopefully, national and international efforts aimed at facilitating a broader understanding of Haiti's prospects for conservation will continue (e.g. Gouvernement de la République d'Haïti 2016; Tarter *et al.* 2016), and current and future scientists will generate ecological data of immediate benefit to anurans in Haiti, thereby facilitating their continued existence for future generations to enjoy, appreciate, and value. Afterall, we will protect ourselves by protecting them.

VI. Acknowledgements

I am most grateful to my family (my wife Elsie Rivera and our two daughters, Laura and Sofia) for their unconditional support to me and their loving understanding of the long hours devoted to this contribution.

VII. References

Attenborough, D., 2020. *A Life on Our Planet: My Witness Statement and a Vision for the Future.* Grand Central Publishing, New York.

Barbour, T., 1914. A contribution to the zoögeography of the West Indies, with especial reference to amphibians and reptiles. *Memoirs of the Museum of Comparative Zoölogy* 44: 205–363. https://doi.org/10.5962/bhl.title.49187

Blackwelder, R. E., 1943. Monograph of the West Indian beetles of the family Staphylinidae. Smithsonian Institution. *United States National Museum, Bulletin* 182: 1–658. https://doi.org/10.5479/si.03629236.182.i

Bowin, C., 1975. The Geology of Hispaniola. Pp. 501–552 in *The Oceans Basins and Margins*, Vol. 3., edited by A. E. M. Nairn and F. G. Stehli. Plenum Publishing Corporation, New York. https://doi.org/10.1007/978-1-4684-8535-6_12

Brenner, M. and Binford, M. W., 1988. A sedimentary record of human disturbance from Lake Miragoane, Haiti. *Journal of Paleolimnology* 1: 85–97. https://doi.org/10.1007/BF00196066

Burns, L. V., 1954. *Report to the Government of Haiti on Forest Policy and its implementation.* Report No. 349, Food and Agriculture Organization of the United Nations (FAO), Rome.

Central Intelligence Agency, 2021. Haiti. *The World Factbook.* Available at https://www.cia.gov/the-world-factbook/countries/haiti/. Updated 16 November 2021.

Churches, C. E., Wampler, P. J., Sun, W., and Smith, A. J., 2014. Evaluation of forest cover estimates for Haiti using supervised classification of Landsat data. *International Journal of Applied Earth Observation and Geoinformation* 30: 203–216. https://doi.org/10.1016/j.jag.2014.01.020

Cochran, D. M., 1923. A new species of *Eleutherodactylus* from the Dominican Republic. *Proceedings of the Biological Society of Washington* 36: 93–94.

Cochran, D. M., 1932. A new frog, *Eleutherodactylus wetmorei*, from the Republic of Haiti. *Proceedings of the Biological Society of Washington* 45: 191–194.

Cochran, D. M., 1934. Herpetological collections made in Hispaniola by the Utowana Expedition, 1934. *Occasional Papers of the Boston Society of Natural History* 8: 163–188.

Cochran, D. M., 1935. New reptiles and amphibians collected in Haiti by P. J. Darlington. *Proceedings of the Boston Society of Natural History* 40: 367–376.

Cochran, D. M., 1937. A necessary change in an amphibian name. *Journal of the Washington Academy of Sciences* 27: 312.

Cochran, D. M., 1938. A new species of frog from Haiti. *Proceedings of the Biological Society of Washington* 51: 93–94.

Cochran, D. M., 1939. Diagnoses of three new lizards and a frog from the Dominican Republic. *Proceedings of the New England Zoölogical Club* 18: 1–3.

Cochran, D. M., 1941. Herpetology of Hispaniola. *Bulletin of the United States National Museum* 177. https://doi.org/10.5479/si.03629236.177.i

Cooney, R., 2004. *The Precautionary Principle in Biodiversity Conservation and Natural Resource Management: An issues paper for policy-makers, researchers and practitioners.* IUCN, Gland, Switzerland and Cambridge, UK.

Cope, E. D., 1870 (1869). Seventh contribution to the herpetology of tropical America. *Proceedings of the American Philosophical Society* 11: 147–169.

Cope, E. D., 1871. Ninth contribution to the herpetology of tropical America. *Proceedings of the American Philosophical Society* 23: 200–224.

Darwin, C., 1872 (1909). *On the Origin of Species*, 6th edn. Introductory note by C. W. Eliot (ed.). P. F. Collier & Son, New York.

Diamond, H. J. and Schreck, C. J., III (eds), 2017. The Tropics. Pp. S93–S128 in *State of the Climate in 2016. Bulletin of the American Meteorological Society* 98. https://doi.org/10.1175/2017BAMSStateoftheClimate.1

Díaz, L. M., Incháustegui, S. J., Marte, C., and Chong, A., 2015. The tadpoles of the hylid frogs (Anura: Hylidae: *Hypsiboas* and *Osteopilus*) of Hispaniola. *Novitates Caribaea* 8: 1–29. https://doi.org/10.33800/nc.v0i8.39

Díaz, L. M., Incháustegui, S. J., Marte, C., Köhler, G., Cádiz, A., and Rodríguez, M., 2018. A new frog of the *Eleutherodactylus abbotti* Species Group (Anura: Eleutherodactylide) from Hispaniola, with bioacoustic and taxonomic comments on other species. *Novitates Caribaea* 12: 25–42. https://doi.org/10.33800/nc.v0i12.81

Dubois, A., 2017. The nomenclatural status of *Hysaplesia, Hylaplesia, Dendrobates* and related nomina (Amphibia, Anura), with general comments on zoological nomenclature and its governance, as well as on taxonomic databases and websites. *Bionomina* 11: 1–48. https://doi.org/10.11646/bionomina.11.1.1

Duer, C. K., Cisek, J. M., and Powell, R., 1992. Food habits of *Osteopilus dominicensis* (Anura: Hylidae). *Caribbean Journal of Science* 28: 226–228.

Duméril, A. M. C. and Bibron, G., 1841. *Erpétologie Générale ou Histore Naturelle complète Des Reptiles*, vol. 8. Librairie Encyclopédique de Roret, Paris.

Ehrlich, M., Conway, F., Adrian, N., LeBeau, F., Lewis, L., Lauwerysen, H., Lowenthal, I., Mayda, Y., Paryski, P., Smucker, G., Talbot, J., and Wilcox, E., 1985. *HAITI: Country Environmental Profile, A Field*

Study. Report to US AID and IIED. Contract No. 521- 0122-C-00-4090-00/IIED Contract No. DAN-5517-A-00-2066-00. 120 pp.

Federal Research Division, 2006. Country Profile: Haiti. Library of Congress, US Government, Washington, DC. Available at https://www.loc.gov/rr/frd/cs/profiles/Haiti.pdf

Franz, R. and Cordier, D., 1986. The Herpetofaunas of the Proposed National Parks in Southern Haiti. 1984–1985 Charles A. Woods-Florida State Museum Expeditions: report to US AID/Haiti. Contract No. 521-0169-C-00-3083-00. University of Florida, Gainesville, Florida. 73 pp.

Frost, D. R., 2020. Amphibian Species of the World: an online reference. Version 6.1 (*3 March 2020*). Electronic Database accessible at https://amphibiansoftheworld.amnh.org/index.php. American Museum of Natural History, New York.

Galvis, P. A., 2013. Presencia de *Batrachochytrium dendrobatidis* en la República Dominicana: estudio de caso. Biodiversidad-PVDC. Santo Domingo, República Dominicana.

Gouvernement de la République d'Haïti, 2016. Cinquième Rapport National d'Haïti sur la Mise en Oeuvre de la Convention sur la Diversité Biologique. Ministère de l'Environnement (MDE) / Agence Nationale des Aires Protégées (ANAP) / United Nations Environment Programme (UNEP) / Global Environment Facility (GEF) / Secrétariat de la Convention sur la Diversité Biologique (SCDB). 310 pp. Available at https://www.cbd.int/doc/world/ht/ht-nr-05-fr.pdf.

Greene, A., 2001. Haiti: Historical Setting. Pp. 261–310 in *Dominican Republic and Haiti-country Studies,* 3rd edn, edited by H. C. Metz. Federal Research Division, Library of Congress. US Government Printing Office, Washington, DC.

Hedges, S. B., 1988. A new diminutive frog from Hispaniola (Leptodactylidae: *Eleutherodactylus*). *Copeia* 1988: 636–641. https://doi.org/10.2307/1445381

Hedges, S. B., 1989. Evolution and biogeography of West Indian frogs of the genus *Eleutherodactylus*: slow-evolving loci and the major groups. Pp. 305–370 in *Biogeography of the West Indies: Past, Present and Future,* edited by C. A. Woods. Sandhill Crane Press, Gainesville, Florida.

Hedges, S. B., 1996a. The origin of West Indian amphibians and reptiles. Pp 95–128 in *Contributions to West Indian Herpetology: A Tribute to Albert Schwartz,* edited by R. Powell and R. W. Henderson. Contributions to Herpetology 12, Society for the Study of Amphibians and Reptiles, Ithaca, New York.

Hedges, S. B., 1996b. The Hispaniolan frog *Eleutherodactylus neodreptus* Schwartz (Anura: Leptodactylidae) is a synonym of *E. audanti* Cochran. *Caribbean Journal of Science* 32: 248.

Hedges, S. B., 2004. Haiti: a megadiverse Caribbean country. Pp. 19 in *Disappearing Jewels: The Status of New World Amphibians,* edited by B. E. Young, S. N. Stuart, J. S. Chanson, N. A. Cox and T. M. Boucher. NatureServe, Arlington, Virginia.

Hedges, S. B., 2006. An overview of the evolution and conservation of West Indian amphibians and reptiles. *Applied Herpetology* 3: 281–292. https://doi.org/10.1163/157075406778905072

Hedges, S. B., 2008. Threatened amphibians of the West Indies. Pp. 104–105 in *Threatened Amphibians of the World,* edited by S. Stuart, M. Hoffman, J. Chanson, N. Cox, R. Berridge, P. Ramani and B. Young. Lynx Ediciones, Barcelona, Spain; IUCN, Gland, Switzerland; and Conservation International, Arlington, Virginia, USA.

Hedges, S. B., 2010a. *Eleutherodactylus jugans* (La Selle Dusky Frog). Conservation. *Caribbean Herpetology* 1:1. https://doi.org/10.31611/ch.1

Hedges, S. B., 2010b. *Eleutherodactylus paulsoni* (Hispaniolan Pink-rumped Frog). Conservation. *Caribbean Herpetology* 9:1. https://doi.org/10.31611/ch.9

Hedges, S. B. and Díaz, L. M., 2011. The conservation status of amphibians in the West Indies. Pp. 31–47 in *Conservation of Caribbean Island Herpetofaunas Volume 1: Conservation Biology and the Wider Caribbean,* edited by A. Hailey, B. S. Wilson, and J. A. Horrocks. Brill, Leiden, The Netherlands. https://doi.org/10.1163/ej.9789004183957.i-228.23

Hedges, S. B. and Thomas, R., 1987. A new burrowing frog from Hispaniola with comments on the *inoptatus* group of the genus *Eleutherodactylus* (Anura: Leptodactylidae). *Herpetologica* 43: 269–279.

Hedges, S. B. and Thomas, R., 1992a. A new marsh-dwelling species of *Eleutherodactylus* from Haiti (Anura: Leptodactylidae). *Journal of Herpetology* 26: 191–195. https://doi.org/10.2307/1564861

Hedges, S. B. and Thomas, R., 1992b. Two new species of *Eleutherodactylus* from remnant cloud forest in Haiti (Anura: Leptodactylidae). *Herpetologica* 48: 351–358. https://doi.org/10.2307/1446629

Hedges, B. and Thomas, R., 2004. *Eleutherodactylus inoptatus. The IUCN Red List of Threatened Species 2004*: e.T56672A11515903. http://dx.doi.org/10.2305/IUCN.UK.2004.RLTS.T56672A11515903.en

Hedges, B. and Thomas, R., 2010. *Eleutherodactylus glandulifer. The IUCN Red List of Threatened Species 2010*: e.T56617A11506184. http://dx.doi.org/10.2305/IUCN.UK.2010-2.RLTS.T56617A11506184.en

Hedges, S. B. and Timyan, J. C., 2014. Grand Bois, Haiti. Hotspot of biodiversity. Factsheet. *Caribnature* (http://www.caribnature.org). Available at https://www.cepf.net/sites/default/files/grand_bois_fact_sheet_1_en.pdf).

Hedges, S. B., Duellman, W. E., and Heinicke, M. P., 2008. New World direct-developing frogs (Anura: Terrarana): Molecular phylogeny, classification, biogeography, and conservation. *Zootaxa* 1737: 1–182. https://doi.org/10.11646/zootaxa.1737.1.1

Hedges, B., Inchaustegui, S., and Hernandez, M., 2004a. *Eleutherodactylus pictissimus. The IUCN Red List of Threatened Species 2004*: e.T56858A11546005. http://dx.doi.org/10.2305/IUCN.UK.2004.RLTS.T56858A11546005.en

Hedges, B., Inchaustegui, S., and Thomas, R., 2004b. *Eleutherodactylus weinlandi. The IUCN Red List of Threatened Species 2004*: e.T57053A11575116. http://dx.doi.org/10.2305/IUCN.UK.2004.RLTS.T57053A11575116.en

Hedges, B., Inchaustegui, S., Hernandez, M., and Powell, R., 2004c. *Eleutherodactylus audanti. The IUCN Red List of Threatened Species 2004*: e.T56437A11477985. http://dx.doi.org/10.2305/IUCN.UK.2004.RLTS.T56437A11477985.en

Hedges, B., Inchaustegui, S., Hernandez, M., and Powell, R., 2004d. *Eleutherodactylus fowleri. The IUCN Red List of Threatened Species 2004*: e.T56605A11501638. http://dx.doi.org/10.2305/IUCN.UK.2004.RLTS.T56605A11501638.en

Hedges, B., Inchaustegui, S., Hernandez, M., and Powell, R., 2004e. *Peltophryne guentheri. The IUCN Red List of Threatened Species 2004*: e.T54656A11182697. http://dx.doi.org/10.2305/IUCN.UK.2004.RLTS.T54656A11182697.en

Hedges, B., Inchaustegui, S., Thomas, R., and Powell, R., 2004f. *Eleutherodactylus ruthae. The IUCN Red List of Threatened Species 2004*: e.T56938A11557345. http://dx.doi.org/10.2305/IUCN.UK.2004.RLTS.T56938A11557345.en

Hedges, B., Inchaustegui, S., Thomas, R., and Powell, R., 2004g. *Eleutherodactylus hypostenor. The IUCN Red List of Threatened Species 2004*: e.T56659A11499630. http://dx.doi.org/10.2305/IUCN.UK.2004.RLTS.T56659A11499630.en

Hedges, B., Inchaustegui, S., and Powell, R., 2010a. *Eleutherodactylus leoncei. The IUCN Red List of Threatened Species 2010*: e.T56712A11520830. http://dx.doi.org/10.2305/IUCN.UK.2010-2.RLTS.T56712A11520830.en

Hedges, B., Inchaustegui, S., and Powell, R., 2010b. *Eleutherodactylus nortoni. The IUCN Red List of Threatened Species 2010*: e.T56796A11526362. http://dx.doi.org/10.2305/IUCN.UK.2010-2.RLTS.T56796A11526362.en

Hedges, B., Inchaustegui, S., and Thomas, R., 2010c. *Eleutherodactylus parabates. The IUCN Red List of Threatened Species 2010*: e.T56825A11540414. http://dx.doi.org/10.2305/IUCN.UK.2010-2.RLTS.T56825A11540414.en

Hedges, B., Inchaustegui, S., Hernandez, M., and Powell, R., 2010d. *Eleutherodactylus schmidti. The IUCN Red List of Threatened Species 2010*: e.T56952A11559646. http://dx.doi.org/10.2305/IUCN.UK.2010-2.RLTS.T56952A11559646.en

Hedges, S. B., Thomas, R., and Franz, R., 1987. A new species of *Eleutherodactylus* (Anura, Leptodactylidae) from the Massif de la Hotte, Haiti. Copeia 1987: 943–949. https://doi.org/10.2307/1445557

Hedges, S. B., Cohen, W. B., Timyan, J., and Yang, Z., 2018. Haiti's biodiversity threatened by nearly complete loss of primary forest. *Proceedings of the National Academy of Sciences USA* 115: 11850–11855. https://doi.org/10.1073/pnas.1809753115

Henderson, R. W. and Powell, R., 2009. *Natural History of West Indian Reptiles and Amphibians*. University Press of Florida, Gainesville, Florida.

Henderson, R. W. and Schwartz, A., 1986. The diet of the Hispaniolan colubrid snake, *Darlingtonia haetiana*. *Copeia* 1986: 529–531. https://doi.org/10.2307/1445014

Henderson, R. W., Schwartz, A., and Noeske-Hallin, T. A., 1987a. Food habits of three colubrid tree snakes (genus *Uromacer*) on Hispaniola. *Herpetologica* 43: 241–248.

Henderson, R. W., Crother, B. I., Noeske-Hallin, T. A., Schwartz, A., and Dethloff, C. R. 1987b. The diet of the Hispaniolan snake *Antillophis parvifrons* (Colubridae). *Journal of Herpetology* 21: 330–334. https://doi.org/10.2307/1563975

Hill, R. T., 1898. *Cuba and Porto Rico with the other islands of the West Indies, their topography, climate, flora, products, industries, cities, people, political conditions, etc.* The De Vinne Press, The Century Co., New York. https://doi.org/10.5962/bhl.title.4428

Holdridge, L. R., 1942. The pine forests of Haiti. *Caribbean Forester* 4: 16–22.

Holdridge, L. R., 1947. Determination of world plant formations from simple climatic data. *Science* 105: 367–368. https://doi.org/10.1126/science.105.2727.367

Holdridge, L. R., 1967. *Life Zone Ecology*, with photographic supplement prepared by Tosi, J. A., Jr., revised edn. Tropical Science Center, San José, Costa Rica. 206 pp.

Incháustegui, S. J., Díaz, L. M., and Marte, C., 2015. Dos especies nuevas de ranas del género *Eleutherodactylus* (Amphibia: Anura: Eleutherodactylidae) de la Hispaniola. *Solenodon* 12: 136–149.

International Union for Conservation of Nature (IUCN), 1980. World conservation strategy: living resource conservation for sustainable development. With the advice, cooperation and financial assistance of the United Nations Environment Programme (UNEP) and the World Wildlife Fund (WWF), and in collaboration with the Food and Agriculture Organization of the United Nations (FAO) and the United Nations Educational, Scientific and Cultural Organization (UNESCO). Morges, Switzerland.

IUCN, 2022. *The IUCN Red List of Threatened Species*. Version 2022-1. https://www.iucnredlist.org. Accessed on 3 August 2022.

IUCN SSC Amphibian Specialist Group, 2013a. *Eleutherodactylus alcoae. The IUCN Red List of Threatened Species 2013*: e.T56404A3037948. http://dx.doi.org/10.2305/IUCN.UK.2013-1.RLTS.T56404A3037948.en

IUCN SSC Amphibian Specialist Group, 2013b. *Eleutherodactylus armstrongi. The IUCN Red List of Threatened Species 2013*: e.T56432A3039176. http://dx.doi.org/10.2305/IUCN.UK.2013-1.RLTS.T56432A3039176.en

IUCN SSC Amphibian Specialist Group, 2013c. *Eleutherodactylus notidodes. The IUCN Red List of Threatened Species 2013*: e.T195011A2372949. http://dx.doi.org/10.2305/IUCN.UK.2013-1.RLTS.T195011A2372949.en

IUCN SSC Amphibian Specialist Group, 2013d. *Eleutherodactylus sommeri. The IUCN Red List of Threatened Species 2013*: e.T195014A2373229. http://dx.doi.org/10.2305/IUCN.UK.2013-1.RLTS.T195014A2373229.en

IUCN SSC Amphibian Specialist Group, 2013e. *Hypsiboas heilprini. The IUCN Red List of Threatened Species 2013*: e.T55502A3030511. http://dx.doi.org/10.2305/IUCN.UK.2013-1.RLTS.T55502A3030511.en

IUCN SSC Amphibian Specialist Group, 2013f. *Osteopilus pulchrilineatus. The IUCN Red List of Threatened Species 2013*: e.T55810A3032570. http://dx.doi.org/10.2305/IUCN.UK.2013-1.RLTS.T55810A3032570.en

IUCN SSC Amphibian Specialist Group, 2013g. *Osteopilus vastus. The IUCN Red List of Threatened Species 2013*: e.T54346A3014515. http://dx.doi.org/10.2305/IUCN.UK.2013-1.RLTS.T54346A3014515.en

IUCN SSC Amphibian Specialist Group, 2014. *Osteopilus dominicensis. The IUCN Red List of Threatened Species 2014*: e.T55808A56091184. http://dx.doi.org/10.2305/IUCN.UK.2014-1.RLTS.T55808A56091184.en

IUCN SSC Amphibian Specialist Group, 2022a. *Boana heilprini. The IUCN Red List of Threatened Species* 2022: e.T55502A172794212. https://dx.doi.org/10.2305/IUCN.UK.2022-1.RLTS.T55502A172794212.en. Accessed on 02 August 2022.

IUCN SSC Amphibian Specialist Group, 2022b. *Eleutherodactylus abbotti. The IUCN Red List of Threatened Species* 2022: e.T56387A3037850. https://dx.doi.org/10.2305/IUCN.UK.2022-1.RLTS.T56387A3037850.en. Accessed on 02 August 2022.

IUCN SSC Amphibian Specialist Group, 2022c. *Eleutherodactylus alcoae. The IUCN Red List of Threatened Species* 2022: e.T56404A172795024. https://dx.doi.org/10.2305/IUCN.UK.2022-1.RLTS.T56404A172795024.en. Accessed on 02 August 2022.

IUCN SSC Amphibian Specialist Group, 2022d. *Eleutherodactylus amadeus. The IUCN Red List of Threatened Species* 2022: e.T56409A3038406. https://dx.doi.org/10.2305/IUCN.UK.2022-1.RLTS.T56409A3038406.en. Accessed on 03 August 2022.

IUCN SSC Amphibian Specialist Group, 2022e. *Eleutherodactylus aporostegus. The IUCN Red List of Threatened Species* 2022: e.T195006A2372534. https://dx.doi.org/10.2305/IUCN.UK.2022-1.RLTS.T195006A2372534.en. Accessed on 03 August 2022.

IUCN SSC Amphibian Specialist Group, 2022f. *Eleutherodactylus apostates. The IUCN Red List of Threatened Species* 2022: e.T56428A3039056. https://dx.doi.org/10.2305/IUCN.UK.2022-1.RLTS.T56428A3039056.en. Accessed on 03 August 2022.

IUCN SSC Amphibian Specialist Group, 2022g. *Eleutherodactylus armstrongi. The IUCN Red List of Threatened Species* 2022: e.T56432A172795214. https://dx.doi.org/10.2305/IUCN.UK.2022-1.RLTS.T56432A172795214.en. Accessed on 02 August 2022.

IUCN SSC Amphibian Specialist Group, 2022h. *Eleutherodactylus audanti. The IUCN Red List of Threatened Species* 2022: e.T20976998A3039336. https://dx.doi.org/10.2305/IUCN.UK.2022-1.RLTS.T20976998A3039336.en. Accessed on 02 August 2022.

IUCN SSC Amphibian Specialist Group, 2022i. *Eleutherodactylus bakeri. The IUCN Red List of Threatened Species* 2022: e.T56449A3039778. https://dx.doi.org/10.2305/IUCN.UK.2022-1.RLTS.T56449A3039778.en. Accessed on 03 August 2022.

IUCN SSC Amphibian Specialist Group, 2022j. *Eleutherodactylus brevirostris. The IUCN Red List of Threatened Species* 2022: e.T56475A3040329. https://dx.doi.org/10.2305/IUCN.UK.2022-1.RLTS.T56475A3040329.en. Accessed on 03 August 2022.

IUCN SSC Amphibian Specialist Group, 2022k. *Eleutherodactylus caribe. The IUCN Red List of Threatened Species* 2022: e.T56493A3040728. https://dx.doi.org/10.2305/IUCN.UK.2022-1.RLTS.T56493A3040728.en. Accessed on 03 August 2022.

IUCN SSC Amphibian Specialist Group, 2022l. *Eleutherodactylus chlorophenax. The IUCN Red List of Threatened Species* 2022: e.T56511A3041254. https://dx.doi.org/10.2305/IUCN.UK.2022-1.RLTS.T56511A3041254.en. Accessed on 03 August 2022.

IUCN SSC Amphibian Specialist Group, 2022m. *Eleutherodactylus corona. The IUCN Red List of Threatened Species* 2022: e.T56525A3041860. https://dx.doi.org/10.2305/IUCN.UK.2022-1.RLTS.T56525A3041860.en. Accessed on 03 August 2022.

IUCN SSC Amphibian Specialist Group, 2022n. *Eleutherodactylus counouspeus. The IUCN Red List of Threatened Species* 2022: e.T56527A3042001. https://dx.doi.org/10.2305/IUCN.UK.2022-1.RLTS.T56527A3042001.en. Accessed on 03 August 2022.

IUCN SSC Amphibian Specialist Group, 2022o. *Eleutherodactylus darlingtoni. The IUCN Red List of Threatened Species* 2022: e.T56549A3042547. https://dx.doi.org/10.2305/IUCN.UK.2022-1.RLTS.T56549A3042547.en. Accessed on 03 August 2022.

IUCN SSC Amphibian Specialist Group, 2022p. *Eleutherodactylus diplasius. The IUCN Red List of Threatened Species* 2022: e.T195008A86896206. https://dx.doi.org/10.2305/IUCN.UK.2022-1.RLTS.T195008A86896206.en. Accessed on 03 August 2022.

IUCN SSC Amphibian Specialist Group, 2022q. *Eleutherodactylus dolomedes. The IUCN Red List of Threatened Species* 2022: e.T56567A3042811. https://dx.doi.org/10.2305/IUCN.UK.2022-1.RLTS.T56567A3042811.en. Accessed on 03 August 2022.

IUCN SSC Amphibian Specialist Group, 2022r. *Eleutherodactylus eunaster. The IUCN Red List of Threatened Species* 2022: e.T56592A3042954. https://dx.doi.org/10.2305/IUCN.UK.2022-1.RLTS.T56592A3042954.en. Accessed on 03 August 2022.

IUCN SSC Amphibian Specialist Group, 2022s. *Eleutherodactylus fowleri. The IUCN Red List of Threatened Species* 2022: e.T56605A3043931. https://dx.doi.org/10.2305/IUCN.UK.2022-1.RLTS.T56605A3043931.en. Accessed on 02 August 2022.

IUCN SSC Amphibian Specialist Group, 2022t. *Eleutherodactylus furcyensis The IUCN Red List of Threatened Species* 2022: e.T56608A3044085. https://dx.doi.org/10.2305/IUCN.UK.2022-1.RLTS.T56608A3044085.en. Accessed on 02 August 2022.

IUCN SSC Amphibian Specialist Group, 2022u. *Eleutherodactylus glandulifer. The IUCN Red List of Threatened Species* 2022: e.T56617A3044500. https://dx.doi.org/10.2305/IUCN.UK.2022-1.RLTS.T56617A3044500.en. Accessed on 03 August 2022.

IUCN SSC Amphibian Specialist Group, 2022v. *Eleutherodactylus glanduliferoides. The IUCN Red List of Threatened Species* 2022: e.T56618A3044627. https://dx.doi.org/10.2305/IUCN.UK.2022-1.RLTS.T56618A3044627.en. Accessed on 03 August 2022.

IUCN SSC Amphibian Specialist Group, 2022w. *Eleutherodactylus glaphycompus. The IUCN Red List of Threatened Species* 2022: e.T56620A3044800. https://dx.doi.org/10.2305/IUCN.UK.2022-1.RLTS.T56620A3044800.en. Accessed on 03 August 2022.

IUCN SSC Amphibian Specialist Group, 2022x. *Eleutherodactylus grahami. The IUCN Red List of Threatened Species* 2022: e.T56627A3045337. https://dx.doi.org/10.2305/IUCN.UK.2022-1.RLTS.T56627A3045337.en. Accessed on 03 August 2022.

IUCN SSC Amphibian Specialist Group, 2022y. *Eleutherodactylus heminota. The IUCN Red List of Threatened Species* 2022: e.T56651A3046130. https://dx.doi.org/10.2305/IUCN.UK.2022-1.RLTS.T56651A3046130.en. Accessed on 02 August 2022.

IUCN SSC Amphibian Specialist Group, 2022z. *Eleutherodactylus hypostenor. The IUCN Red List of Threatened Species* 2022: e.T56659A3046418. https://dx.doi.org/10.2305/IUCN.UK.2022-1.RLTS.T56659A3046418.en. Accessed on 02 August 2022.

IUCN SSC Amphibian Specialist Group, 2022aa. *Eleutherodactylus inoptatus. The IUCN Red List of Threatened Species* 2022: e.T56672A3046571. https://dx.doi.org/10.2305/IUCN.UK.2022-1.RLTS.T56672A3046571.en. Accessed on 02 August 2022.

IUCN SSC Amphibian Specialist Group, 2022bb. *Eleutherodactylus jugans. The IUCN Red List of Threatened Species* 2022: e.T56688A3047415. https://dx.doi.org/10.2305/IUCN.UK.2022-1.RLTS.T56688A3047415.en. Accessed on 02 August 2022.

IUCN SSC Amphibian Specialist Group, 2022cc. *Eleutherodactylus lamprotes. The IUCN Red List of Threatened Species* 2022: e.T56698A3047675. https://dx.doi.org/10.2305/IUCN.UK.2022-1.RLTS.T56698A3047675.en. Accessed on 03 August 2022.

IUCN SSC Amphibian Specialist Group, 2022dd. *Eleutherodactylus leoncei. The IUCN Red List of Threatened Species* 2022: e.T56712A3047929. https://dx.doi.org/10.2305/IUCN.UK.2022-1.RLTS.T56712A3047929.en. Accessed on 02 August 2022.

IUCN SSC Amphibian Specialist Group, 2022ee. *Eleutherodactylus limbensis. The IUCN Red List of Threatened Species* 2022: e.T195009A2372811. https://dx.doi.org/10.2305/IUCN.UK.2022-1.RLTS.T195009A2372811.en. Accessed on 03 August 2022.

IUCN SSC Amphibian Specialist Group, 2022ff. *Eleutherodactylus lucioi. The IUCN Red List of Threatened Species* 2022: e.T56730A3048334. https://dx.doi.org/10.2305/IUCN.UK.2022-1.RLTS.T56730A3048334.en. Accessed on 03 August 2022.

IUCN SSC Amphibian Specialist Group, 2022gg. *Eleutherodactylus nortoni. The IUCN Red List of Threatened Species* 2022: e.T56796A3049811. https://dx.doi.org/10.2305/IUCN.UK.2022-1.RLTS.T56796A3049811.en. Accessed on 02 August 2022.

IUCN SSC Amphibian Specialist Group, 2022hh. *Eleutherodactylus notidodes. The IUCN Red List of Threatened Species* 2022: e.T195011A86896731. https://dx.doi.org/10.2305/IUCN.UK.2022-1.RLTS.T195011A86896731.en. Accessed on 02 August 2022.

IUCN SSC Amphibian Specialist Group, 2022ii. *Eleutherodactylus oxyrhyncus. The IUCN Red List of Threatened Species* 2022: e.T56819A3050521. https://dx.doi.org/10.2305/IUCN.UK.2022-1.RLTS.T56819A3050521.en. Accessed on 03 August 2022.

IUCN SSC Amphibian Specialist Group, 2022jj. *Eleutherodactylus parabates*. *The IUCN Red List of Threatened Species* 2022: e.T56825A3050899. https://dx.doi.org/10.2305/IUCN.UK.2022-1.RLTS.T56825A3050899.en. Accessed on 02 August 2022.

IUCN SSC Amphibian Specialist Group, 2022kk. *Eleutherodactylus parapelates*. *The IUCN Red List of Threatened Species* 2022: e.T56827A3051033. https://dx.doi.org/10.2305/IUCN.UK.2022-1.RLTS.T56827A3051033.en. Accessed on 03 August 2022.

IUCN SSC Amphibian Specialist Group, 2022ll. *Eleutherodactylus paulsoni*. *The IUCN Red List of Threatened Species* 2022: e.T56836A3051338. https://dx.doi.org/10.2305/IUCN.UK.2022-1.RLTS.T56836A3051338.en. Accessed on 03 August 2022.

IUCN SSC Amphibian Specialist Group, 2022mm. *Eleutherodactylus pictissimus*. *The IUCN Red List of Threatened Species* 2022: e.T56858A3051629. https://dx.doi.org/10.2305/IUCN.UK.2022-1.RLTS.T56858A3051629.en. Accessed on 02 August 2022.

IUCN SSC Amphibian Specialist Group, 2022nn. *Eleutherodactylus poolei*. *The IUCN Red List of Threatened Species* 2022: e.T56874A3052197. https://dx.doi.org/10.2305/IUCN.UK.2022-1.RLTS.T56874A3052197.en. Accessed on 03 August 2022.

IUCN SSC Amphibian Specialist Group, 2022oo. *Eleutherodactylus rhodesi*. *The IUCN Red List of Threatened Species* 2022: e.T56909A3052745. https://dx.doi.org/10.2305/IUCN.UK.2022-1.RLTS.T56909A3052745.en. Accessed on 03 August 2022.

IUCN SSC Amphibian Specialist Group, 2022pp. *Eleutherodactylus ruthae*. *The IUCN Red List of Threatened Species* 2022: e.T20995600A3053137. https://dx.doi.org/10.2305/IUCN.UK.2022-1.RLTS.T20995600A3053137.en. Accessed on 02 August 2022.

IUCN SSC Amphibian Specialist Group, 2022qq. *Eleutherodactylus schmidti*. *The IUCN Red List of Threatened Species* 2022: e.T213610247A3053309. https://dx.doi.org/10.2305/IUCN.UK.2022-1.RLTS.T213610247A3053309.en. Accessed on 02 August 2022.

IUCN SSC Amphibian Specialist Group, 20224rr. *Eleutherodactylus sciagraphus*. *The IUCN Red List of Threatened Species* 2022: e.T56955A3053579. https://dx.doi.org/10.2305/IUCN.UK.2022-1.RLTS.T56955A3053579.en. Accessed on 03 August 2022.

IUCN SSC Amphibian Specialist Group, 2022ss. *Eleutherodactylus semipalmatus*. *The IUCN Red List of Threatened Species* 2022: e.T7147A3137920. https://dx.doi.org/10.2305/IUCN.UK.2022-1.RLTS.T7147A3137920.en. Accessed on 03 August 2022.

IUCN SSC Amphibian Specialist Group, 2022tt. *Eleutherodactylus sommeri*. *The IUCN Red List of Threatened Species* 2022: e.T195014A86896986. https://dx.doi.org/10.2305/IUCN.UK.2022-1.RLTS.T195014A86896986.en. Accessed on 02 August 2022.

IUCN SSC Amphibian Specialist Group, 2022uu. *Eleutherodactylus thorectes*. *The IUCN Red List of Threatened Species* 2022: e.T57004A3054028. https://dx.doi.org/10.2305/IUCN.UK.2022-1.RLTS.T57004A3054028.en. Accessed on 03 August 2022.

IUCN SSC Amphibian Specialist Group, 2022vv. *Eleutherodactylus ventrilineatus*. *The IUCN Red List of Threatened Species* 2022: e.T57034A3054268. https://dx.doi.org/10.2305/IUCN.UK.2022-1.RLTS.T57034A3054268.en. Accessed on 03 August 2022.

IUCN SSC Amphibian Specialist Group, 2022ww. *Eleutherodactylus warreni*. *The IUCN Red List of Threatened Species* 2022: e.T57052A3054529. https://dx.doi.org/10.2305/IUCN.UK.2022-1.RLTS.T57052A3054529.en. Accessed on 03 August 2022.

IUCN SSC Amphibian Specialist Group, 2022xx. *Eleutherodactylus weinlandi*. *The IUCN Red List of Threatened Species* 2022: e.T20998545A3054645. https://dx.doi.org/10.2305/IUCN.UK.2022-1.RLTS.T20998545A3054645.en. Accessed on 02 August 2022.

IUCN SSC Amphibian Specialist Group, 2022yy. *Eleutherodactylus wetmorei*. *The IUCN Red List of Threatened Species* 2022: e.T20994808A3054754. https://dx.doi.org/10.2305/IUCN.UK.2022-1.RLTS.T20994808A3054754.en. Accessed on 02 August 2022.

IUCN SSC Amphibian Specialist Group, 2022zz. *Lithobates catesbeianus*. *The IUCN Red List of Threatened Species* 2022: e.T58565A193396825. https://dx.doi.org/10.2305/IUCN.UK.2022-1.RLTS.T58565A193396825.en. Accessed on 02 August 2022.

IUCN SSC Amphibian Specialist Group, 2022aaa. *Osteopilus dominicensis*. *The IUCN Red List of Threatened Species* 2022: e.T55808A172794620. https://dx.doi.org/10.2305/IUCN.UK.2022-1.RLTS.T55808A172794620.en. Accessed on 02 August 2022.

IUCN SSC Amphibian Specialist Group, 2022bbb. *Osteopilus pulchrilineatus*. *The IUCN Red List of Threatened Species* 2022: e.T55810A172794792. https://dx.doi.org/10.2305/IUCN.UK.2022-1.RLTS.T55810A172794792.en. Accessed on 02 August 2022.

IUCN SSC Amphibian Specialist Group, 2022ccc. *Osteopilus vastus*. *The IUCN Red List of Threatened Species* 2022: e.T54346A172794001. https://dx.doi.org/10.2305/IUCN.UK.2022-1.RLTS.T54346A172794001.en. Accessed on 02 August 2022.

IUCN SSC Amphibian Specialist Group, 2022ddd. *Peltophryne guentheri*. *The IUCN Red List of Threatened Species* 2022: e.T172918839A3017633. https://dx.doi.org/10.2305/IUCN.UK.2022-1.RLTS

.T172918839A3017633.en. Accessed on 02 August 2022.

IUCN SSC Amphibian Specialist Group. *Rhinella marina. The IUCN Red List of Threatened Species* 2022(1): e.T100099625A2951416. *In press.*

Iturralde-Vinent, M. A. and MacPhee, R. D. E., 1999. Paleogeography of the Caribbean region: implications for Cenozoic biogeography. *Bulletin of the American Museum of Natural History* 238: 1–95.

Joglar, R. L. and Burrowes, P. A., 2005. Greater Antilles Region Progress Report. *Froglog* 72: 1–4.

Joglar, R. L., Álvarez, A. O., Aide, T. M., Barber, D., Burrowes, P. A., García, M. A., León-Cardona, A., Longo, A. V., Pérez-Buitrago, N., Puente, A., Ríos-López, N., and Tolson, P. J. 2007 (reprinted 2011). Conserving the Puerto Rican herpetofauna. *Applied Herpetology* 4: 327–345. https://doi.org/10.1163/157075407782424566

Kottek, M., Grieser, J., Beck, C., Rudolf, B., and Rubel, F., 2006. World map of the Köpper-Geiger climate classification updated. *Meteorologische Zeitschrift* 15: 259–263. https://doi.org/10.1127/0941-2948/2006/0130

Landestoy T., M. A., 2017. New localities and prey records for the Hispaniolan Brown Racer (*Haitiophis anomalus*), with comments on dorsal pattern variation. *IRCF Reptiles & Amphibians* 24: 106–111. https://doi.org/10.17161/randa.v24i2.14159

Landestoy T., M. A., Turner, D. B., Marion, A. B., and Hedges, S. B., 2018. A new species of Caribbean toad (Bufonidae, *Peltophryne*) from southern Hispaniola. *Zootaxa* 4403: 523–539. https://doi.org/10.11646/zootaxa.4403.3.6

Lever, C., 2003. *Naturalized Reptiles and Amphibians of the World.* Oxford University Press Inc., New York.

Linnaeus, C., 1758. *Systema Naturæ, per Regna Tria Naturae, Secundum Classes, Ordines, Genera, Species, cum Characteribus, Differentiis, Synonymis, Locis.* Tomus I. Editio Decima, Reformata. Holmiæ, Impensis Direct Laurentii Salvii (in Latin; Stockholm, Sweden). https://doi.org/10.5962/bhl.title.542

Lynch, J. D. and Duellman, W. E., 1997. Frogs of the genus *Eleutherodactylus* in western Ecuador: Systematics, Ecology, and Biogeography. Special Publication 23, Natural History Museum, Dyche Hall, University of Kansas, Lawrence, Kansas. 236 pp. https://doi.org/10.5962/bhl.title.7951

Lynch, J. D. and Schwartz, A., 1971. Taxonomic disposition of some 19th century Leptodactylid frog names. *Journal of Herpetology* 5: 103–114. https://doi.org/10.2307/1562732

Lynn, W. G., 1958. Some amphibians from Haiti and a new subspecies of *Eleutherodactylus* schmidti. *Herpetologica* 14: 153–157.

Malthus, T., 1798 (reprinted 1998). *An essay on the principles of population, as it affects the future improvement of society with remarks on the speculations of Mr. Godwin, M. Condorcet, and other writers.* J. Johnson, St. Paul's Church-Yard, London. Electronic Scholarly Publishing Project, available at http://www.esp.org.

Mann, P., Draper, G., and Lewis, J. F., 1991. An overview of the geologic and tectonic development of Hispaniola. Pp. 1–28 in *Geological and Tectonic Development of the North America-Caribbean Plate Boundary in Hispaniola*, edited by P. Mann, G. Draper, and J. F. Lewis. *Geological Society of America Special Paper* 262. https://doi.org/10.1130/SPE262-p1

Martínez Rivera, C. C., Bell, J., and Parker, J., 2011. Haiti's disappearing frog. *AArk Newsletter* 16: 15–16.

Martínez Rivera, C. C. and Rodríguez Plaza, C., 2015. Los Anfibios del Sur de la Hispaniola: su historia natural y conservación. Guía de Campo, Diversidad y Estado de Conservación de los Anfibios. Colaboradores: R. Durocher, M. Fildor, S. J. Incháustegui, A. Jean, M. A. Landestoy T., C. Marte Pimentel, and J. Timyan. Zoológico de Filadelfia, ComuEcoMedia, Alianza deFondos para Ecosistemas Críticos, Museo Nacional de Historia Natural de Santo Domingo, Société Audubon Haiti, Grupo Jaragua. 157 pp. Available at http://www.conservaciondeanfibios.org/libro-versioacuten-digital.html

Mayer, L. R., 2019. Haiti's first-ever private nature reserve will safeguard treasure trove of imperiled species. *Froglog* 27: 4–5.

McClintock, N. C., 2003. Agroforesty and sustainable resource conservation in Haiti: a case study. North Carolina State University, Department of Forestry. Agroforesty Working Paper, 11 pp. Available at http://www.ncsu.edu/project/cnrint/Agro/PDFfiles/HaitiCaseStudy041903.pdf

Michel, R. 2001. Informations sur les technologies traditionelles en Haiti. http://www.haiticulture.ch/Env_techno.html

Ministerio de Medio Ambiente y Recursos Naturales, 2014. Quinto Informe nacional de biodiversidad de la República Dominicana. Santo Domingo, República Dominicana. 80 pp. Available at http://www.conservaciondeanfibios.org/uploads/2/1/6/1/21615192/09._presencia_de_batrachochytrium_dendrobatidis_en_la_rep%C3%BAblica_dominicana_estudio_de_caso.pdf

Montgomery, H., Pessagno, E. A., Jr., and Pindell, J. L., 1994. A 195 Ma terrace in a 165 Ma sea: Pacific origin of the Caribbean Plate. *GSA Today* 4: 1–6.

Murphy, J. B. and Gratwicke, B., 2017. History of captive management and conservation amphibian programs mostly in zoos and aquariums. Part I–Anurans. *Herpetological Review* 48: 241–260.

Neils, A. and Bugbee, C., 2007. *Rana catesbeiana* (American Bullfrog). Diet. *Herpetological Review* 38: 443.

Noble, G. K., 1923. Six new batrachians from the Dominican Republic. *American Museum Novitates* 61: 1–6.

Noble, G. K. and Hassler, W. G., 1933. Two new species of frogs, five new species and a new race of lizards from the Dominican Republic. *American Museum Novitates* 652: 1–17.

Osborn, T. J., Barichivich, J., Harris, I., van der Schrier, G., and Jones, P. D., 2016. Monitoring global drought using the self-calibrating Palmer drought severity index. Pp. S32–S36 in *State of the Climate in 2015. Bulletin of the American Meteorological Society* 97.

Parsons, K. J. and Powell, R., 1994. The diet of the endemic Hispaniolan toad, *Peltophryne guentheri*. *Missouri Herpetological Association Newsletter* 7: 3.

Parsons, K. J., Powell, R., and Greve, J. H., 1995. *Peltophryne guentheri* (NCN). Parasitism. *Herpetological Review* 26: 31–32.

Pinto, A. D., 2010. Denaturalizing "natural" disasters: Haiti's earthquake and the humanitarian impulse. *Open Medicine* 4: E193–E196.

Powell, R., 2012. Hispaniola and Navassa. Pp. 129–135 in *Island Lists of West Indian Amphibians and Reptiles*, edited by R. Powell and R. W. Henderson. *Bulletin of the Florida Museum of Natural History* 51: 85–166.

Powell, R. and Henderson, R. W., 1996. Introduction. Pp 11–14 in *Contributions to West Indian Herpetology: A Tribute to Albert Schwartz*, edited by R. Powell and R. W. Henderson. Contributions to Herpetology 12, Society for the Study of Amphibians and Reptiles, Ithaca, New York.

Powell, R. and Incháustegui, S. J., 2011. Conservation of the herpetofauna of the Dominican Republic. Pp. 167–188 in *Conservation of Caribbean Island Herpetofaunas, Volume 2: Regional Accounts of the West Indies*, edited by B. S. Hailey and J. Horrocks. Koninklijke Brill NV, Leiden, The Netherlands. https://doi.org/10.1163/ej.9789004194083.i-439.65

Powell, R., Greve, J. H., and Howard, A. K., 1998. Hispaniolan *Eleutherodactylus* (Anura: Leptodactylidae) as host of immature *Terranova* (Nematoda: Ascaridae), with notes on additional nematodes. *Caribbean Journal of Science* 34: 155–157.

Powell, R., Ottenwalder, J. A., and Incháustegui, S. J., 1999. The Hispaniolan herpetofauna: diversity, endemism, and historical perspectives, with comments on Navassa Island. Pp. 93–168 in *Caribbean Amphibians and Reptiles*, edited by B. I. Crother. Academic Press, San Diego, California. https://doi.org/10.1016/B978-012197955-3/50016-1

Powell, R., Henderson, R. W., Farmer, M. C., Breuil, M., Echternacht, A. C., van Buurt, G., Romagosa, C. M., and Perry, G., 2011. Introduced amphibians and reptiles in the greater Caribbean: patterns and conservation implications. Pp. 63–143 in *Conservation of Caribbean Island Herpetofaunas Volume 1: Conservation Biology and the Wider Caribbean*, edited by A. Hailey, B. S. Wilson, and J. A. Horrocks. Brill, Leiden, The Netherlands. https://doi.org/10.1163/ej.9789004183957.i-228.38

Prou, M. E., 2005. Haiti's condemnation: history and culture at the crossroads. *Latin American Research Review* 40: 191–201. https://doi.org/10.1353/lar.2005.0061

Reinhardt, J. and Lütken, C. F., 1863 (1862). Bidrag til det vestindiske Öriges og navnligen til de dansk-vestindiske Öers Herpetologie. *Videnskabelige Meddelelser fra den Naturhistoriske Forening i Kjöbenhavn* 10–18: 153–291.

Rivero, J. A., 1963. *Eleutherodactylus hedricki*, a new species of frog from Puerto Rico (Salientia, Leptodactylidae). *Breviora* 185: 1–7.

Schwartz, A., 1964. Three new species of frogs (Leptodactylidae, *Eleutherodactylus*) from Hispaniola. *Breviora* 208: 1–15.

Schwartz, A., 1965a. Variation and natural history of *Eleutherodactylus ruthae* on Hispaniola. *Bulletin of the Museum of Comparative Zoology* 132: 479–508.

Schwartz, A., 1965b. Geographic variation in two species of Hispaniolan *Eleutherodactylus*. *Studies on the Fauna of Curaçao and other Caribbean Islands* 86: 98–123.

Schwartz, A., 1965c. A new species of *Eleutherodactylus* (Amphibian: Leptodactylus) from the Sierra de Baoruco, República Dominicana. *Proceedings of the Biological Society of Washington* 78: 165–168.

Schwartz, A., 1966. The relationships of four small Hispaniolan Eleutherodactylus (Leptodactylidae). *Bulletin of the Museum of Comparative Zoology* 133: 371–399.

Schwartz, A., 1968. Geographic variation in the Hispaniolan frog *Eleutherodactylus wetmorei* Cochran. *Breviora, Museum of Comparative Zoology* 290: 1–13.

Schwartz, A., 1971. A new species of *Eleutherodactylus* (Amphibia, Leptodactylidae) from Hispaniola. *Annals of Carnegie Museum* 43: 25–31.

Schwartz, A., 1973. Six new species of *Eleutherodactylus* (Anura, Leptodactylidae) from Hispaniola. *Journal of Herpetology* 7: 249–273. https://doi.org/10.2307/1563010

Schwartz, A., 1976a. Two new species of Hispaniolan *Eleutherodactylus* (Leptodactylidae). *Herpetologica* 32: 163–171.

Schwartz, A., 1976b. Variation and relationships of some Hispaniolan frogs (Leptodactylidae,

Eleutherodactylus) of the *ricordi* group. *Bulletin of the Florida State Museum* 21: 1–46.

Schwartz, A., 1977. A new subspecies of *Eleutherodactylus wetmorei* Cochran (Anura: Leptodactylidae) from northern Haiti. *Herpetologica* 33: 66–72.

Schwartz, A., 1978. Some aspects of the herpetogeography of the West Indies. Pp. 31–51 in *Zoogeography in the Caribbean, The 1975 Leidy Medal Symposium*, edited by F. B. Gill. Academy of Natural Sciences of Philadelphia Special Publications, Philadelphia, Pennsylvania.

Schwartz, A., 1979. A new species of *Eleutherodactylus* (Amphibia, Anura, Leptodactylidae) from northwestern Haiti, Hispaniola. *Journal of Herpetology* 13: 199–202. https://doi.org/10.2307/1563929

Schwartz, A., 1980a. The herpetogeography of Hispaniola, West Indies. *Studies on the Fauna of Curaçao and other Caribbean Islands* 189: 86–127.

Schwartz, A. 1980b. Two new species of *Eleutherodactylus* (Amphibia, Anura, Leptodactylidae) from Hispaniola. *Annals of Carnegie Museum* 49: 103–112.

Schwartz, A., 1983. *Eleutherodactylus glandulifer. Catalogue of American Amphibians and Reptiles* 310: 1.

Schwartz, A. and Henderson, R. W., 1991. *Amphibians and Reptiles of the West Indies: Descriptions, Distributions, and Natural History*. University of Florida Press, Gainesville, Florida.

Schwartz, A. and Thomas, R., 1975. A check-list of West Indian amphibians and reptiles. *Carnegie Museum of Natural* US *History, Special Publication* 1: 1–126. https://doi.org/10.5962/bhl.title.123681

Shaw, G., 1802. *General Zoology or Systematic Natural History*. Volume III, Part 1. Amphibia. Thomas Davison, London.

Shreve, B., 1936. A new Anolis and new Amphibian from Haiti. *Proceedings of the New England Zoological Club* 15: 93–99.

Shreve, B. and Williams, E. E., 1963. The herpetology of the Port-au-Prince region and Gonave Island, Haiti. Part II. The frogs. *Bulletin of the Museum of Comparative Zoology* 129: 302–342.

Smucker, G. R., 2001. Haiti: The Society and Its Environment. Pp. 311–361 in *Dominican Republic and Haiti-country studies*, 3rd edn, edited by H. C. Metz. Federal Research Division, Library of Congress. Government Printing Office Washington, DC.

Solís, F., Ibáñez, R., Hammerson, G., Hedges, B., Diesmos, A., Matsui, M., Hero, J.-M., Richards, S., Coloma, L., Ron, S., La Marca, E., Hardy, J., Powell, R., Bolaños, F., Chaves, G., and Ponce, P., 2009. *Rhinella marina. The IUCN Red List of Threatened Species 2009*: e.T41065A10382424. http://dx.doi.org/10.2305/IUCN.UK.2009-2.RLTS.T41065A10382424.en

Sperling, D., 2017. In 1825, Haiti paid France $21 billion to preserve its Independence -- Time for France to pay it back. Contributor, Capital Flows, Forbes Opinion, Forbes Media LLC. 6 December 2017.

Stephenson, T. S., Taylor, M. A., Trotman, A. R., Marcellin-Honoré, V., Porter, A. O., Hernández, M., Gonzalez, I. T., Boudet, D., Spence, J. M., McLean, N., Campbell, J. D., Shaw, A., Aaron-Morrison, A. P., Kerr, K., Tamar, G., Blenman, R. C., Destin, D., Joyette, S., Jeffers, B., and Stephenson, K., 2015. The Caribbean. Pp. S176–S178 in *State of the Climate in 2014*, edited by J. Blunden and D. S. Arndt. *Bulletin of the American Meteorological Society* 96. https://doi.org/10.1175/2015BAMSStateoftheClimate.1

Stephenson, T. S., Taylor, M. A., Trotman, Etienne-LeBlanc, S., Porter, A. O., Hernández, M., Boudet, D., Fonseca, C., Spence, J. M., Shaw, A., Aaron-Morrison, A. P., Kerr, K., Tamar, G., Destin, D., Van Meerbeeck, C., Marcellin, V., Joseph, A. C., Willie, S., Stennett-Brown, R., and Campbell, J. D., 2016. Caribbean. Pp. S181–S182 in *State of the Climate in 2015*, edited by J. Blunden and D. S. Arndt. *Bulletin of the American Meteorological Society* 97. https://doi.org/10.1175/2016BAMSStateoftheClimate.1

Stephenson, T. S., Taylor, M. A., Trotman, Van Meerbeeck, C. J., Porter, A. O., Etienne-LeBlanc, S., Hernández, M., Gonzalez, I. T., Cutié, V., Boudet, D., Fonseca, C., Willie, S., Spence, J. M., Kerr, K., Aaron-Morrison, A. P., Tamar, G., Stennett-Brown, R., and Campbell, J. D., 2017. Pp. S183–S185 in *State of the Climate in 2016*, edited by G. Hartfield, J. Blunden, and D. S. Arndt. *Bulletin of the American Meteorological Society* 98. https://doi.org/10.1175/2017BAMSStateoftheClimate.1

Stephenson, T. S., Taylor, M. A., Trotman, Van Meerbeeck, C. J., Marcellin, V., Kerr, K., Campbell, J. D., Spence, J. M., Tamar, G., Hernández Sosa, M., and Stephenson, K., 2018. Caribbean. Pp. S200–S203 in *State of the Climate in 2017*, edited by J. Blunden and D. S. Arndt. *Bulletin of the American Meteorological Society* 99. https://doi.org/10.1175/2018BAMSStateoftheClimate.1

Stephenson, T. S., Taylor, M. A., Trotman, Van Meerbeeck, C. J., Campbell, J. D., Brown, A., and Spence, J., 2019. Caribbean. Pp. S198–S200 in *State of the Climate in 2018*, edited by J. Blunden and D. S. Arndt. *Bulletin of the American Meteorological Society* 100. https://doi.org/10.1175/2019BAMSStateoftheClimate.1

Tanner, E. L., 1948. Haiti as a source of crude drugs and essential oils. World Trade in Commodities 6(Part 3, No. 35): 1–2.

Tarter, A., Kennedy Freeman, K. and Sander, K., 2016. A history of landscape-level land management efforts in Haiti: lessons learned from case studies spanning eight decades. The World Bank, Washington, DC. https://doi.org/10.1596/25764

The Guardian, 2011. Haiti earthquake: Rediscovered frog species offer signs of hope – gallery. Available at https://www.theguardian.com/environment/gallery/2011/jan/12/haiti-earthquake-rediscovered-frog. 12 January 2011.

The New York Times, 1987. Another blow for Haiti: a sugar mill closes. April 12, 1987. Available at https://www.nytimes.com/1987/04/12/world/another-blow-for-haiti-a-sugar-mill-closes.html?searchResultPosition=2.

Thomas, R., 1996. Recollections of Albert Schwartz. Pp. 16–18 in *Contributions to West Indian Herpetology: A Tribute to Albert Schwartz*, edited by R. Powell and R. W. Henderson. Contributions to Herpetology 12, Society for the Study of Amphibians and Reptiles, Ithaca, New York.

Tschudi, J. J., 1838. Classification der Batrachier mit Berücksichtigung der fossilen Thiere dieser Abtheilung der Reptilien. In der buchdruckerei von Petitpierre, Neuchâtel, Switzerland. https://doi.org/10.5962/bhl.title.4883

Victor, J. A., 1995. *Code des lois Haïtiennes de l'environnement*. Projet PNUD/ECMU/HAI/92/001, Unité de Coordination et de Suivi de l'Environnement, Programme des Nations Unies pour le Développement (PNUD), Port-au-Prince, Haiti. 325 pp.

Vitousek, P. M., Mooney, H. A., Lubchenco, J., and Melillo, J. M. 1997. Human domination of Earth's ecosystems. *Science* 277: 494–499. https://doi.org/10.1126/science.277.5325.494

Wadsworth, F. H., 2014. A forestry assignment to Puerto Rico. Forestry Memoirs of Frank Wadsworth. Impresos Emmanuelli, Inc., Aguas Buenas, Puerto Rico.

Wetmore, A. and Swales, B. W., 1931. The birds of Haiti and the Dominican Republic. *Bulletin of the United States National Museum* 155, 483 pp. https://doi.org/10.5479/si.03629236.155.i

Williams, E. E., 1961. Notes on Hispaniolan herpetology 3. The evolution and relationships of the *Anolis semilineatus* group. *Breviora, Museum of Comparative Zoology* 136: 1–8.

Williams, E. E., 1999. Over 300 years of collecting in the Caribbean. Pp. 1–30 in *Caribbean Amphibians and Reptiles*, edited by B. I. Crother. Academic Press, San Diego, California. https://doi.org/10.1016/B978-012197955-3/50013-6

World Weather Online, 2021. Port-Au-Prince Climate Weather Averages. Zoomash Ltd, 3rd Floor, 86-90 Paul Street, London, Hackney, EC2A 4NE, UK. https://www.worldweatheronline.com/port-au-prince-weather-averages/ouest/ht.aspx

7 Status of Amphibian Conservation and Decline in the Dominican Republic, Hispaniola

Sixto J. Incháustegui and Luis M. Díaz

Abbreviations and acronyms

AArk/ASG	*Amphibian Ark/Amphibian Specialist Group*
APOD	*Amphibian pod*
CABI	*Centre for Agriculture and Bioscience International*
CATHALAC	*Centro del Agua del Trópico Húmedo para América Latina y el Caribe*
CBD	*Convention on Biological Diversity*
CITES	*Convention on International Trade in Endangered Species*
CONAU	*Consejo Nacional De Asuntos Urbanos*
CR	*Critically Endangered*
CRI	*Climate Risk Index*
CSE	*Comisión de Supervivencia de Especies*
EEUU	*United States of America*
EN	*Endangered*
EP	*En Peligro (In Danger)*
FONDOCyT	*Fondo Nacional de Innovación y Desarrollo Científica y Tecnológico*
GAA	*Global Amphibian Assessment*
GEEI	*Grupo Especialista de Especies Invasoras*
IPBES	*Intergovernmental Science-Policy Platform on Biodiversity and Ecosystem Services*
IPCC	*Intergovernmental Panel on Climate Change*
IUCN	*International Union for the Conservation of Nature*
LC	*Least Concern*
mya	*million years ago*
NA	*Not Assessed*

Sixto J. Incháustegui and Luis M. Díaz, 'Chapter 7: Status of Amphibian Conservation and Decline in the Dominican Republic, Hispaniola' in: *The Conservation and Biogeography of Amphibians in the Caribbean*. Pelagic Publishing (2023). © Sixto J. Incháustegui and Luis M. Díaz. DOI: 10.53061/CXGT6549

NT	*Near Threatened*
OEA	*Organización de los Estados Americanos*
PC	*Peligro Crítico (Critical Danger)*
PE	*Possibly Extinct*
PNUMA	*Programa de Naciones Unidas para el Medio Ambiente*
SAFE	*Saving Amphibians from Extinction*
SSC	*Species Survival Commission*
UASD	*Universidad Autónoma de Santo Domingo*
UICN	*Unión Mundial para la Naturaleza*
UNDP	*United Nations Development Programme*
USAID	*United States Agency for International Development*
VU	*Vulnerable*
ZOODOM	*National Zoo*

I. Introduction

A. The geography of the Dominican Republic: life zones, topography, climate, vegetation, and biogeography

The Dominican Republic, with an area of 48,230.43 km² (Ministerio de Medio Ambiente y Recursos Naturales 2012a), shares with Haiti the island of Hispaniola (77,914 km²), which is located in the intertropical zone, between the parallels 17°36' N and 19°56' N and the meridians 68°17' W and 74°31' W (Martínez 2012). It is a middle-income developing country with a Human Development Index of 0.745, a total human population of 9,445,281, and a population density of 195 persons per km². Neighbouring Haiti is a low-income developing country, with a Human Development Index of 0.503, a total human population of 10,579,230, and a population density of 381 persons per km² (UNDP 2019).

The physiography of the country is rugged, with 44.2% more than 250 m asl (see Table 7.1). Mountain ranges extend west to east, with three main ranges, Cordillera Central, Sierra de Neiba, and Sierra de Bahoruco (Fig. 7.1).

Geotectonically, Hispaniola is a complex island, formed by two paleoislands previously separated by a marine channel that disappeared approximately 10 mya. Marine barriers separated them during the Cenozoic, and they possess biogeographically distinct terrestrial biotas (Brace *et al.* 2012). Hispaniola also includes several isolated satellite islands. These geologic complexities made Hispaniola much richer in habitats and ecosystems than other Caribbean islands.

The Dominican Republic has the highest Caribbean peak (Pico Duarte: 3,175 m asl), the lowest valley, with the largest inland lake, Lago Enriquillo (~280 km²), at 40 m asl (Martínez 2012). The country encompasses nine life zones and seven transitional ones (Subtropical Thorn Woodland, Subtropical Dry Forest, Subtropical Moist Forest, Subtropical Wet Forest, Low Montane Moist Forest, Montane Wet Forest, Low Montane Wet Forest, Low Montane Rain Forest, and Subtropical Rain Forest). Silva (2014) recently reviewed the life zones of the country, but due to the lack of critical climatic data from some ecologically and biogeographically critical areas, we follow previous works (OEA 1967; OEA Secretaría General 1967; Hager and Zanoni 1993; Ministerio de Medio Ambiente y Recursos Naturales 2012a).

Table 7.1. Percentage of the territory of the Dominican Republic occurring at different elevations (After Izzo 2011).

ELEVATION (M)	< 50	50–250	250–500	500–1,000	1,000–2,000	> 2,000
Percent of territory	17.7	38.1	16.8	14.9	11.2	1.3

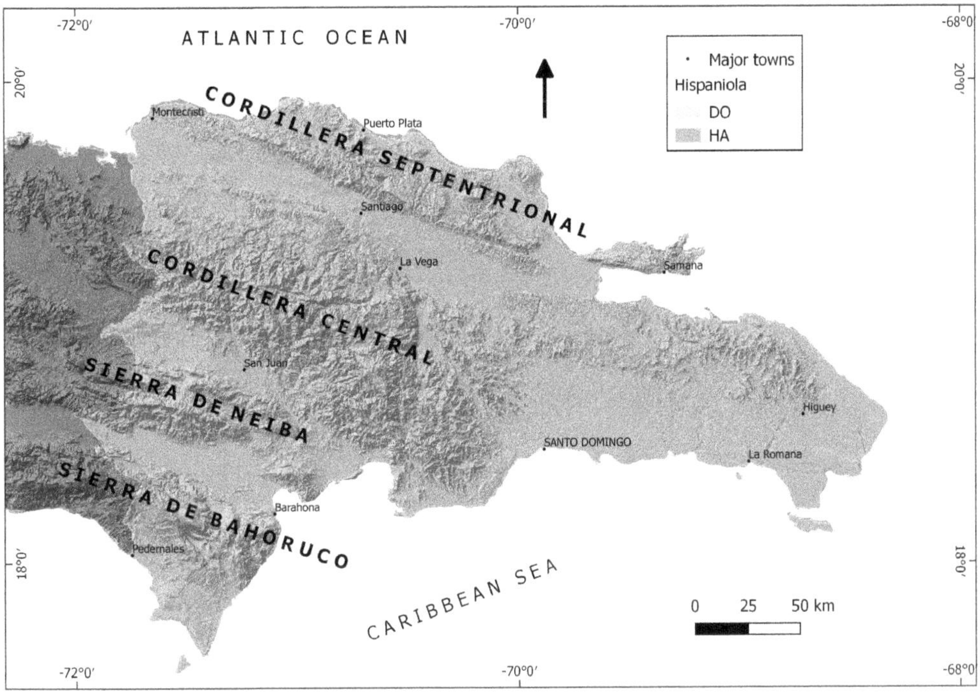

Figure 7.1. Physiography of the Dominican Republic. DO = Dominican Republic; HA = Haiti.

Although in the intertropical zone the climate varies among the different life zones, according to differences in elevation, the temperature usually shows a somewhat regular gradient of −0.6°C/100 m gain in elevation (Izzo 2011) with average annual values from 28.7 to 7.7°C. Average annual precipitation varies from 403 mm to 2,865 mm.

As is presently well known, the climate is changing, mainly due to several factors of anthropogenic origin. Consequently, the United Nations Framework Convention was adopted in 1994, and the Dominican Republic ratified it in 1998 (Secretaría de Estado de Medio Ambiente 2004). The country had a Climate Risk Index for 1999–2018 of 50 and is considered among the ten countries globally expected to suffer most from climatic change (Eckstein *et al.* 2019).

Under the provisions of the United Nations Framework Convention on Climate Change, the Dominican Republic published its first, second, and third national communications. In brief, these studies predicted that temperatures will increase and precipitation will diminish (Secretaría de Estado de Medio Ambiente 2004; Secretaría de Estado de Medio Ambiente and Programa de las Naciones Unidas para el Desarrollo 2009; Izzo 2011; Pérez Fernández 2016; Ministerio de Medio Ambiente y Recursos Naturales 2019). The Intergovernmental Panel on Climate Change (IPCC) analysed the multiple impacts climatic change can have on biodiversity (IPCC 2002). Foden *et al.* (2013) assessed the vulnerability of birds, amphibians, and corals to climatic change, and Anderson *et al.* (2008) analysed the potential impacts on biodiversity specifically for the Dominican Republic.

Hager and Zanoni (1993) proposed a national classification of vegetation, based mostly on its structure and composition. They recognized six major vegetation types, with varying numbers of subtypes. These are characterized by location, geographic characters, climate, physiognomy and composition, and by common and endemic species. The main types are halophytic vegetation, dry forests (including altered forests), semideciduous forests, broadleaf evergreen forests, pine forests, and aquatic and wetland vegetation. Tolentino and Peña (1998) published a national vegetation and land-use map, distinguishing different vegetational units and subunits. At a more macro level, Dinerstein *et al.* (1995) recognized five relevant ecoregions present in Hispaniola: moist broadleaf forest, dry forest, pine forest, flooded grasslands and savannahs, and mangroves, of which four were considered as critically endangered or endangered, and only the flooded grasslands and savannahs, represented mainly by Enriquillo Lake, as vulnerable.

Following the scheme of vegetation and land-use cover by Tolentino and Peña, the Ministry of Environment and Natural Resources, based on information analysed to 2012, presented information on several of those habitats and forest cover for the years 1996 and 2003 (Ministerio de Medio Ambiente y Recursos Naturales 2014a), and more recently, on use of soil and cover (Ministerio de Medio Ambiente y Recursos Naturales 2014a).

Cano Carmona and Cano Ortiz (2012) published a biogeographic profile of the Island of Hispaniola. They recognized 22 different biogeographic areas, based on the distribution of endemic flora and habitats, 13 of which are either entirely or in part within the Dominican Republic. Previously, based on the information then known about the distribution of amphibian species in the insular Caribbean, Hedges (1999) identified the major biogeographic units, recognizing 21 for Hispaniola and 14 for the Dominican Republic. It is interesting to see the similarities between the biogeographic units identified independently through the distribution of plants and of amphibians.

As mentioned above, Hispaniola is composed of two former paleoislands. Merten first recognized that reptilian species in Hispaniola occurred in pairs (Schwartz 1980), whereas Williams (1961) introduced the concept of the two paleoislands, the 'North Paleoisland' and the 'South Paleoisland', in relation to the distributions of anoles. Thereafter, the notion spread, at least among herpetologists; today, the existence of the two paleoislands and its effect on the biogeography of Hispaniola is well recognized, even beyond the herpetological realm (Brace *et al.* 2012).

Schwartz, in his *Herpetogeography of Hispaniola* (1980), considered the herpetofauna of the island within the context of the two paleoislands; whenever feasible, he established the origin of each species. He also recognized that although the South Island comprises about one-eighth of Hispaniola, it has a higher species richness of herpetofauna compared with the North Island.

B. The amphibians of the Dominican Republic

The amphibians of insular Caribbean are very rich in terms of species diversity, and they contribute significantly to the region's status as a global biodiversity hotspot (Myers *et al.* 2000). Although numerically this region has fewer species than continental Mesoamerica or South America, most of the species are endemic. Endemism is, in most cases, island-specific. Its particular geographic circumstance has allowed a limited number of families to be represented, but one in particular, Eleutherodactylidae, has radiated throughout the islands. Besides this, other families present are Bufonidae, Hylidae, Leptodactylidae, and Ranidae, for a total of five (see Appendix 7.I). The last is represented in the Dominican Republic by a single introduced invasive species, *Lithobates catesbeianus* (Powell *et al.* 1999; Powell *et al.* 2011; IUCN SSC Amphibian Specialist Group 2022mm).

A brief review of the history of description and listing of species relevant to the Dominican Republic is justified next. Over the years, several publications have become landmarks in Caribbean herpetology. The first one was Cochran's *Herpetology of Hispaniola* (1941). Cochran presented mostly keys for identification and accounts of species, summarizing the known information and in many cases providing accurate illustrations. Several decades later (1991), Schwartz and Henderson published their *Amphibians and Reptiles of the West Indies: Descriptions,*

Distributions, and Natural History. Schwartz worked actively for more than three decades in the Caribbean, and in this book, he and Henderson summarized all his data and experience. It not only includes descriptions of species but, as the name implies, offers extensive information on species' distributions and natural history. Henderson and Powell (2009) wrote a review that updated Schwartz and Henderson's book. With the advent of digital facilities, Hedges (2020) established the portal caribherp.org, a searchable web-based information site on amphibians and reptiles of the Caribbean Islands. These publications constitute the core sources for information on amphibians of Hispaniola and the Caribbean in general. On a global scale, Frost's (2022) searchable *Amphibian Species of the World* has become the standard guide to amphibian species, and Amphibiaweb (2022), a broader global database. The emphasis has been on taxonomy and systematics, and little else published. However, Powell *et al.* (1999) provided a broader account of the Hispaniolan herpetofauna, and in the same book, Williams (1999) offered a historical overview of herpetological collections in the Caribbean. Finally, Henderson and Powell (1999) summarized the scant information on herpetoecology.

At present, a total of 47 species (or 48; see Frost 2022 and end of this paragraph) in five families are listed for the Dominican Republic (Powell *et al.* 1999; Hedges *et al.* 2008) (Table 7.2). This number has been increasing recently, in part owing to the discovery of entirely new species and partly due to recognition through molecular genetics that some former subspecies should be accorded the status of full species. Mostly previously undescribed, small to diminutive cryptic species are being found. Only two species are known from the Dominican Republic's adjacent islands: *Osteopilus dominicensis* on Saona Island, to the southeast, and *Eleutherodactylus alcoae* on Beata Island to the southwest (Powell 2012). For the purpose of this chapter, we are considering only species that have been formally described. Some of the databases of amphibians differ slightly in the number of species reported from the Dominican Republic. For instance, AmphibiaWeb (2022) listed 50 species and includes *E. aporostegus* and *E. diplasius*, Haitian species not known from the Dominican Republic, and *E. coqui*, from Puerto Rico. The last species has been introduced several times over the past few decades into Santo Domingo city, where its presence is noticed by its very peculiar song, but it always disappears within a few years. Darrel Frost, in his *Amphibian Species of the World* (Frost 2022), also included one Haitian species, *E. dolomedes*, which we believe could be an error because we know of no record of this species on the Dominican territory. Also, Frost (2022) excluded the Haitian *E. darlingtoni* from its previous listing for the Dominican Republic.

In his overview of historical collecting in the region, Williams (1999) expressed the view that after Schwartz and Henderson's (1991) publication, the epoch devoted mostly to collection and description of species had come to an end, and suggested that what he called the 'seventh

Table 7.2. Amphibian families and the number of species listed for the Dominican Republic and considered in this chapter in detail (* from Frost 2022; see main text).

FAMILY	NUMBER OF SPECIES
Bufonidae	4
Eleutherodactylidae	37 (38*)
Hylidae	4
Leptodactylidae	1
Ranidae	1
Total number of species	**47** (48*)

period' in Caribbean Herpetology would be devoted mostly to history, cladogenesis, and systematics. Various taxonomic rearrangements have indeed ensued, including at the species level, mostly based on new molecular techniques and approaches (Hedges *et al.* 2008; Duellman *et al.* 2016). Nonetheless, new species continue to be discovered (Incháustegui *et al.* 2015; Díaz *et al.* 2018; Landestoy *et al.* 2018), albeit not at the same rate.

Three of the five families of amphibians represented in the Dominican Republic include most of the species that exist there. Ranidae is represented only by the introduced and invasive *Lithobates catesbeianus* (Powell *et al.* 2011). Leptodactylids are represented by a single species, *Leptodactylus albilabris*; the Dominican population, with a very narrow geographical range in the northeast of the country, was considered as a separate species endemic to Hispaniola, *Leptodactylus dominicensis*. As such, it was assessed in the IUCN Red List as Endangered (Hedges *et al.* 2004). Later, however, it was synonymized with *L. albilabris* from Puerto Rico and reassessed as of Least Concern (Ríos-López 2010; IUCN SSC Amphibian Specialist Group 2021).

Three families, then, truly include Hispaniolan endemics: Bufonidae (Pramuk *et al.* 2001), Hylidae (Duellman *et al.* 2016), and Eleutherodactylidae (Hedges *et al.* 2008, 2019).

Schwartz (1980), grouped the frogs of the Dominican Republic into 'North Paleoisland' and 'South Paleoisland' species. The North Paleoisland at present represents a bigger portion of the Dominican Republic territory, including several portions of different Hispaniolan mountain ranges. This covers a larger number of species, with more species with lowland geographical distributions than on the South Paleoisland. The South Paleoisland portion of the country corresponds mainly with the Sierra de Bahoruco, to which the majority of species are restricted. It harbours an important group of endemic species, with restricted geographic ranges. A smaller group has a broad, island-wide distribution. Of the 47 species with populations verifiable in the Dominican Republic at the time (excluding the Haitian *Eleutherodactylus dolomedes*; see above), 44 are endemic to Hispaniola, and of those, 22 (50%) are endemic to the Dominican Republic. *Leptodactylus albilabris*, as expressed above, is now considered an introduced species from Puerto Rico (Hedges and Heinicke 2007). *Rhinella marina* and *Lithobates catesbeianus* are widely introduced invasive species across the Caribbean.

In the Dominican Republic, four species of toads are known (Table 7.3) (Fig. 7.2). One, similar in external morphology to the Puerto Rican toad *Peltophryne lemur*, is a recent discovery, *P. armata* (Landestoy *et al.* 2018). For a long time, only one species of toad was known from Hispaniola (Cochran 1941). Schwartz (1972), reviewed the group, describing one new subspecies, and a new species. Later, and as it remains at present, the subspecies (*fractus*) was raised to species level.

Table 7.3. Distribution and level of threat to bufonid species occurring in Hispaniola. H = high; L = low. NA = Not assessed. PE = Possibly extinct. RL = Red List; DR = Dominican Republic; CC Vu = Climatic Change Vulnerability levels.

SPECIES	RANGE	IUCN RL	DR RL[5]	CC VU[6]
Peltophryne armata	Southern slopes of Sierra de Bahoruco	CR[1]	NA	NA
P. fluviatica	Restricted, riverine forest in the NW	CR (PE)[2]	CR	L
P. guentheri	Arid lowlands in the SW and NW	LC[3]	VU	H
Rhinella marina	Broadly distributed in lowlands	LC[4]	NA	NA

[1]IUCN SSC Amphibian Specialist Group (2022qq); [2]IUCN SSC Amphibian Specialist Group (2022rr); [3]IUCN SSC Amphibian Specialist Group (2022ss); [4]IUCN SSC Amphibian Specialist Group (*in press*); [5]Ministerio de Medio Ambiente y Recursos Naturales 2011; [6]Foden *et al.* (2013).

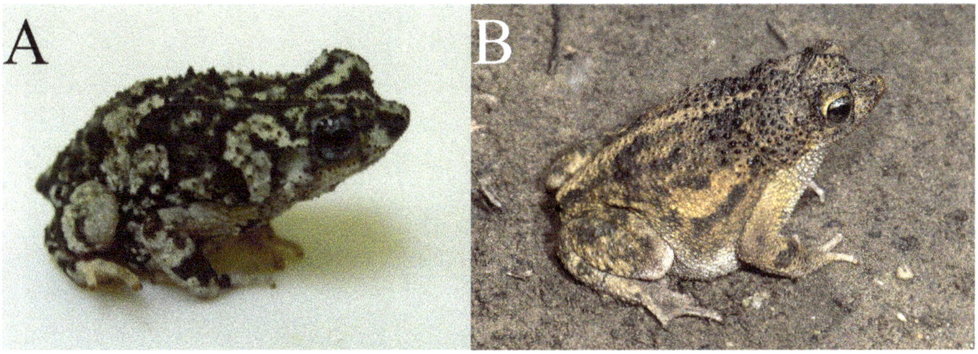

Figure 7.2. Two species of toads from the Dominican Republic: (A) *Peltophryne armata*; (B) *Peltophryne guentheri*.

Landestoy *et al.* (2018), however, do not recognize *fractus* (*fracta*) as distinct from *P. guentheri*. These toads are considered members of the monophyletic West Indian *Peltophryne peltocephala* group (Pramuk *et al.* 2001). *Rhinella marina* is the non-native introduced invasive toad, present since the early 1900s (Cochran 1941; Powell *et al.* 2011). This animal was introduced into many tropical geographic regions, especially onto islands, to act as a biological control of sugar cane beetles, then becoming a pest in itself.

All four hylids (Fig. 7.3) had been described by the early 1900s (Cochran 1941). A significant taxonomic rearrangement consisted of recognizing the Hispaniolan Green Treefrog as part of the genus *Hypsiboas* (*H. heilprini*), with other hylids reassigned to the genus *Osteopilus* (*O. dominicensis, O. pulchrilineata,* and *O. vasta*). More recently, *heilprini* has been reassigned to the genus *Boana* (Dubois 2017). This group is considered to have dispersed over water from South America, with speciation occurring during the Miocene. Hispaniola is the only island that has a representative of the genus *Boana,* also considered of South American origin (Duellman *et al.* 2016; Dubois 2017).

The genus *Eleutherodactylus* (Figs. 7.4–7.9) remained for many years as the largest vertebrate genus (containing over 800 species). In 2008, however, Hedges *et al.* allocated this group of direct-developing frogs into a new taxon, Terrarana, including four different families, one of them Eleutherodactylidae (199 species) containing the genus *Eleutherodactylus* with 185 species (and counting). All of the species from the Dominican Republic fall within the Eleutherodactylidae and the genus *Eleutherodactylus* (Appendix 7.I).

In summary, there are 47 species – 48 if *E. dolomedes* exists on Dominican territory as listed by Frost (2022) – belonging to seven genera, in the Dominican Republic. Four are bufonids (three natives, one introduced); four are hylids, one is a leptodactylid, one an introduced ranid; and the remaining species are in the genus *Eleutherodactylus* (37 species in Hedges 2022; 38 species in Frost 2022). This last group of species (*Eleutherodactylus*) is well diversified and occupies many different habitats, including terrestrial, aquatic, and arboreal ones. All of the bufonids are lowland to mid-elevation species (Table 7.3). Hylids show a broader distribution, usually from sea level to at least 1,000 m asl (Table 7.4). *Osteopilus dominicensis* is the most widespread frog of Hispaniola, reaching elevations up to 2,000 m asl. The eleutherodactylids include broadly distributed species, as well as those with a more restricted geographic distribution (Table 7.5), mostly in particular mountain ranges (Tables 7.6–7.10). The Cordillera Central and the Sierra de Bahoruco in the southwest of the country gather the two biggest assemblages of frog species, including endemics. Only two adjacent islands have amphibians reported from them: *Osteopillus dominicensis* is present on Saona island, in the Southeast, and *Eleutherodactylus alcoae* is on Beata Island

Figure 7.3. The four hylid species of frogs from the Dominican Republic: (A) *Boana heilprini*; (B) *Osteopilus dominicensis*; (C) *Osteopilus pulchrilineatus*; (D) *Osteopilus vastus*.

Figure 7.4. Some widespread species of *Eleutherodactylus* from the Dominican Republic: (A) *Eleutherodactylus abbotti*; (B) *Eleutherodactylus pictissimus*; (C) *Eleutherodactylus inoptatus*.

in the Southwest (Powell and Henderson 2012). The introduced leptodactylid *Leptodactylus albilabris* (Fig. 7.10) is restricted to narrow coastal wetlands in the northeastern part of the country (Table 7.11), and the introduced ranid *Lithobates catesbeianus* also inhabits wetlands but is widely distributed (Table 7.12).

II. Threats

Many insular Caribbean amphibians, and those from Hispaniola and the Dominican Republic, in particular, have restricted or very restricted natural geographic ranges, mostly montane. Challenges to their conservation were discussed by Incháustegui (2014).

Figure 7.5. Some frogs originating from the North Paleoisland of the Dominican Republic: (A) *Eleutherodactylus weinlandi*; (B) *Eleutherodactylus probolaeus*; (C) *Eleutherodactylus ruthae*; (D) *Eleutherodactylus flavescens*.

Figure 7.6. Some frogs from the Cordillera Central of the Dominican Republic: (A) *Eleutherodactylus auriculatoides*; (B–C) *Eleutherodactylus haitianus*; (D) *Eleutherodactylus minutus*; (E) *Eleutherodactylus montanus*; (F) *Eleutherodactylus patriciae*.

Figure 7.7. *Eleutherodactylus neiba*: a frog from the Sierra de Neiba of the Dominican Republic.

Figure 7.8. *Eleutherodactylus alcoae*, a frog restricted to the Barahona Peninsula of the Dominican Republic.

Figure 7.9. Some frogs from the Sierra de Bahoruco: (A) *Eleutherodactylus armstrongi*; (B) *Eleutherodactylus hypostenor*; (C) *Eleutherodactylus jugans*; (D) *Eleutherodactylus leoncei*; (E) *Eleutherodactylus nortoni*; (F) *Eleutherodactylus wetmorei*.

Table 7.4. Elevational distribution and level of threat to hylid frogs of Hispaniola. H = high; L = low. NA = Not assessed. RL = Red List; DR = Dominican Republic; CC Vu = Climatic Change Vulnerability levels.

SPECIES	RANGE	IUCN RL	DR RL[5]	CC VU[6]
Boana heilprini	Island-wide, sea level to 1,856 m asl	VU[1]	VU	L
Osteopilus dominicensis	Island-wide, sea level to 2,000 m asl	LC[2]	NA	H
O. pulchrilineatus	1,091 m asl	VU[3]	EN	H
O. vastus	Patchy island-wide, sea level to 1,697 m asl	VU[4]	EN	H

[1]IUCN SSC Amphibian Specialist Group (2022a); [2]IUCN SSC Amphibian Specialist Group (2022nn); [3]IUCN SSC Amphibian Specialist Group (2022oo); [4]IUCN SSC Amphibian Specialist Group (2022pp); [5]Ministerio de Medio Ambiente y Recursos Naturales 2011; [6]Foden *et al.* (2013).

Table 7.5. Elevational range and level of threat to *Eleutherodactylus* with wide distributions in Hispaniola. L = low. NA = Not assessed. RL = Red List; DR = Dominican Republic; CC Vu = Climatic Change Vulnerability levels.

SPECIES	RANGE	IUCN RL	DR RL[4]	CC VU[5]
Eleutherodactylus abbotti	Wide-spread; sea level to above 1,800 m asl	LC[1]	NA	L
E. inoptatus	Sea level up to 1,697 m asl.	NT[2]	NA	L
E. pictissimus	South Paleoisland; sea level to 1,758 m asl	LC[3]	VU	L

[1]IUCN SSC Amphibian Specialist Group (2022b); [2]IUCN SSC Amphibian Specialist Group (2022o); [3]IUCN SSC Amphibian Specialist Group (2022bb); [4]Ministerio de Medio Ambiente y Recursos Naturales 2011; [5]Foden *et al.* (2013).

Table 7.6. Distribution, elevational range and level of threat to *Eleutherodactylus* from North Island. H = high; L = low. NA = Not assessed. PE = Possibly extinct. RL = Red List; DR = Dominican Republic; CC Vu = Climatic Change Vulnerability levels.

SPECIES	RANGE	IUCN RL	DR RL[7]	CC VU[8]
Eleutherodactylus flavescens	Eastern Dominican Republic to 900 m asl	NT[1]	NA	H
E. paralius	Coastal SE Dominican Republic	NT[2]	EN	NA
E. probolaeus	Lowlands in extreme eastern Dominican Republic	EN[3]	EN	L
E. ruthae	Eastern Dominican Republic	EN[4]	EN	H
E. schmidti	Restricted range in North Paleoisland to above 1,750 m asl	CR (PE)[5]	CR	H
E. weinlandi	Eastern Dominican Republic and central Haiti to above 700 m asl	LC[6]	VU	H

[1]IUCN SSC Amphibian Specialist Group (2022h); [2]IUCN SSC Amphibian Specialist Group (2022z); [3]IUCN SSC Amphibian Specialist Group (2022dd); [4]IUCN SSC Amphibian Specialist Group (2022gg); [5]IUCN SSC Amphibian Specialist Group (2022hh); [6]IUCN SSC Amphibian Specialist Group (2022kk); [7]Ministerio de Medio Ambiente y Recursos Naturales 2011; [8]Foden *et al.* (2013).

In the IUCN's assessment of the amphibians of the Dominican Republic, 80% of the 44 native species were under some category of endangerment: CR, EN, or VU (Young *et al.* 2004; Stuart *et al.* 2008, 2010; see sections III. A. 1–9). Since 2010, seven species have been added to the 40 documented ones, four of them via elevation of former subspecies to the status of full species, and three newly described. The population conservation status of these seven species are currently not assessed; they have restricted natural ranges and are impacted mostly by the same threats affecting the rest of the species in the Dominican Republic. Only two species of broad

Table 7.7. Elevational range and level of threat to *Eleutherodactylus* in the Cordillera Central. H = high; L = low. NA = Not assessed. PE = Possibly extinct. RL = Red List; DR = Dominican Republic; CC Vu = Climatic Change Vulnerability levels.

SPECIES	RANGE	IUCN RL	DR RL[12]	CC VU[13]
Eleutherodactylus auriculatoides	788 to above 1,800 m asl	VU[1]	EN	NA
E. bothroboans	Restricted to localities around Jarabaoa, around 500 m asl	CR[2]	NA	NA
E. geitonos	Southeastern slopes of Cordillera Central, 1,084 m asl	EN[3]	NA	NA
E. haitianus	1,545–2,455 m asl	EN[4]	EN	H
E. melatrigonum	800 to 2,500 m asl	EN[5]	EN	NA
E. minutus	879–2,300 m asl	EN[6]	EN	H
E. montanus	1,270–2,424 m asl	EN[7]	EN	H
E. patriciae	2,000–3,050 m asl	EN[8]	EN	H
E. pituinus	1,212–1,770 m asl	EN[9]	EN	H
E. rucillensis	High elevations Cordillera Central	CR (PE)[10]	NA	NA
E. sommeri	700–1,040 m asl	EN[11]	CR	NA

[1]IUCN SSC Amphibian Specialist Group (2022f); [2]IUCN SSC Amphibian Specialist Group (2022g); [3]IUCN SSC Amphibian Specialist Group (2022k); [4]IUCN SSC Amphibian Specialist Group (2022l); [5]IUCN SSC Amphibian Specialist Group (2022s); [6]IUCN SSC Amphibian Specialist Group (2022t); [7]IUCN SSC Amphibian Specialist Group (2022u); [8]IUCN SSC Amphibian Specialist Group (2022aa); [9]IUCN SSC Amphibian Specialist Group (2022cc); [10]IUCN SSC Amphibian Specialist Group (2022ee); [11]IUCN SSC Amphibian Specialist Group (2022ii); [12]Ministerio de Medio Ambiente y Recursos Naturales 2011; [13]Foden *et al.* (2013).

Table 7.8. Elevational range and level of threat to *Eleutherodactylus* species from the Sierra de Neiba. H = high. NA = Not assessed. RL = Red List; DR = Dominican Republic; CC Vu = Climatic Change Vulnerability levels.

SPECIES	RANGE	IUCN RL	DR RL[5]	CC VU[6]
Eleutherodactylus neiba	Above 1,600 m asl	EN[1]	NA	NA
E. notidodes	1,340–1,830 m asl	EN[2]	CR	NA
E. parabates	1,455–1,870 m asl	EN[3]	CR	H
E. tychathrous	Mid-elevations of the NE Sierra de Neiba	EN[4]	NA	NA

[1]IUCN SSC Amphibian Specialist Group (2022v); [2]IUCN SSC Amphibian Specialist Group (2022x); [3]IUCN SSC Amphibian Specialist Group (2022y); [4]IUCN SSC Amphibian Specialist Group (2022jj); [5]Ministerio de Medio Ambiente y Recursos Naturales 2011; [6]Foden *et al.* (2013).

distribution on the island, *Eleutherodactylus abbotti* and *Osteopilus dominicensis*, are considered not to be impacted by habitat loss.

A. Loss and fragmentation of habitat

Change in land-use is a major driver of the loss of biodiversity in general (IPBES 2020) and destruction of essential habitat by humans is the main threat to amphibian species. In the Dominican Republic, most of this is due to agricultural expansion, charcoal production, and to a lesser degree, livestock production (Table 7.13). Pollution, mainly from agrochemicals, might be having an impact, but very little information is available.

Table 7.9. Elevational distribution and level of threat to *Eleutherodactylus alcoae* from the Barahona Peninsula. L = low. RL = Red List; DR = Dominican Republic; CC Vu = Climatic Change Vulnerability levels.

SPECIES	RANGE	IUCN RL	DR RL[2]	CC VU[3]
Eleutherodactylus alcoae	Sea level to 600 m asl	LC[1]	EN	L

[1]IUCN SSC Amphibian Specialist Group (2022c); [2]Ministerio de Medio Ambiente y Recursos Naturales 2011; [3]Foden *et al.* (2013).

Table 7.10. Elevational range and the level of threat to *Eleutherodactylus* from Sierra de Bahoruco. H = high; L = low. NA = Not assessed. RL = Red List; DR = Dominican Republic; CC Vu = Climatic Change Vulnerability levels.

SPECIES	RANGE	IUCN RL	DR RL[13]	CC VU[14]
Eleuthreodactylus armstrongi	152–1,697 m asl	EN[1]	EN	L
E. audanti	High elevations of Sierra de Bahoruco	VU[2]	EN	L
E. fowleri	1,045–1,303 m asl	CR[3]	EN	L
E. furcyensis	803–2,100 m asl	CR[4]	CR	L
E. heminota	Sea level up to 1,697 m asl	VU[5]	EN	L
E. hypostenor	Sea level up to 1,061 m asl	EN[6]	EN	L
E. jugans	1,242–2,146 m asl	CR[7]	CR	H
E. leoncei	1,182–2,303 m asl	EN[8]	CR	H
E. ligiae	Southern Sierra de Bahoruco, 1,128 m asl	CR[9]	NA	NA
E. nortoni	576–1,515 m asl	CR[10]	CR	L
E. rufifemoralis	Eastern Sierra de Bahoruco, 727–1,370 m asl	EN[11]	CR	H
E. wetmorei	Sea level up to 1,324m asl	VU[12]	VU	L

[1]IUCN SSC Amphibian Specialist Group (2022d); [2]IUCN SSC Amphibian Specialist Group (2022e); [3]IUCN SSC Amphibian Specialist Group (2022i); [4]IUCN SSC Amphibian Specialist Group (2022j); [5]IUCN SSC Amphibian Specialist Group (2022m); [6]IUCN SSC Amphibian Specialist Group (2022n); [7]IUCN SSC Amphibian Specialist Group (2022p); [8]IUCN SSC Amphibian Specialist Group (2022q); [9]IUCN SSC Amphibian Specialist Group (2022r); [10]IUCN SSC Amphibian Specialist Group (2022w); [11]IUCN SSC Amphibian Specialist Group (2022ff); [12]IUCN SSC Amphibian Specialist Group (2022ll); [13]Ministerio de Medio Ambiente y Recursos Naturales 2011; [14]Foden *et al.* (2013).

Most amphibians in the Dominican Republic have restricted natural distributions, confined in most cases to a mountain range (Cordillera Central, Sierra de Bahoruco, Sierra de Neiba) or to broadleaved forests. Although only four species do not have any portion of their distributional range inside any protected area (*Peltophryne fluviatica*, *P. fracta*, *Eleutherodactylus geitonos*, and *Leptodactylus albilabris*), this does not assure proper protection to those that do. Unfortunately, destruction and loss of habitat are taking place over most of the Dominican Republic, including in protected areas. Sangermano (2015) found that the losses of forests and species' habitats within protected areas are of lesser magnitude, but from similar causes, than losses for the country as a whole; the existing network of protected areas is inadequate for the maintenance of the habitats of *Osteopilus pulchrilineatus*, *O. vastus*, *Boana heilprini*, and *Eleutherodactylus flavescens*, the four species they studied.

In *Estudio de Uso y Cobertura del Suelo 2012* (Ministerio de Medio Ambiente y Recursos Naturales 2014a), it was shown that vegetation cover of the country in 2012 was 39%, and has

Figure 7.10. *Leptodactylus albilabris*, a species introduced into the Dominican Republic from Puerto Rico.

Table 7.11. Distribution and level of threat to the leptodactylid, *Leptodactylus albilabris*. L = low. NA = Not assessed. RL = Red List; DR = Dominican Republic; CC Vu = Climatic Change Vulnerability levels.

SPECIES	RANGE	IUCN RL	DR RL[2]	CC VU[3]
Leptodactylus albilabris	NE narrow coastal wetlands	LC[1]	NA	L

[1]IUCN SSC Amphibian Specialist Group (2021); [3]Ministerio de Medio Ambiente y Recursos Naturales 2011; [3]Foden *et al.* (2013).

Table 7.12. Range and level of threat to the ranid, *Lithobates catesbeianus*. L = low. NA = Not assessed. RL = Red List; DR = Dominican Republic; CC Vu = Climatic Change Vulnerability levels.

SPECIES	RANGE	IUCN RL	DR RL[2]	CC VU[3]
Lithobates catesbeianus	Widespread in wet lowlands	LC[1]	NA	L

[1]IUCN SSC Amphibian Specialist Group (2022mm); [2]Ministerio de Medio Ambiente y Recursos Naturales 2011; [3]Foden *et al.* (2013).

Table 7.13. Major land-uses (categories), area covered, and percentage of the total land area of the Dominican Republic that it occupies (Ministerio de Medio Ambiente y Recursos Naturales 2014a).

CATEGORY	AREA (KM^2)	PERCENTAGE
Agriculture	16,959.00	35.0
Livestock	7,241.23	15.0
Urban	1,133.30	2.4
Forest	18,923.45	39.0

been increasing since 1996. It is important to highlight that these are overall percentages, not accounting for which types of forests have suffered the most, and do not necessarily reflect the situation of native forest. The vegetation cover in question includes reforested areas with exotic species and extensive areas of certain invasive species of trees. It can be seen, for instance, that between 2003 and 2012 cloud forest in particular decreased from 3.26% to 1.80%, with national cloud forests losing 55% of their area in only nine years. Martínez (2016, 2017a,b), based on Hansen *et al.* (2013), analysed the major loss of forests in protected areas. Twenty-two areas had lost > 5% of their forest (range 5.06–20.82%). Most of this was due to the expansion of illegal agriculture, or to forest fires. Although forest fires happen mainly in pine forests, the expansion of agriculture occurs primarily in cloud forests and other wet forests, which often are critical habitats for more species classified as endangered, compared to pine forests.

B. Urbanization

The construction and development of cities have a range of impacts on ecosystems and biodiversity. Andrew and McDonnell (2008) reviewed the effects of urbanization on amphibians, as did Lavilla and Heatwole (2010). In general, these types of impacts similarly affect the Dominican Republic, where urban areas occupy only 2.4% of its area of 48,230.43 km^2. This percentage has been increasing rapidly, from 0.82% in 1996, to 1.46% in 2003, and to 2.4% by 2012 (Ministerio de Medio Ambiente y Recursos Naturales 2014a). Santo Domingo City, the capital and largest city, as well as many other cities and touristic enclaves included in this percentage, are near the sea and so affect coastal species. Geo Santo Domingo (PNUMA, CONAU, and UASD 2007) discussed the environmental impacts of Santo Domingo City on the environment, including some considerations relative to biodiversity, referring mostly to habitats, but not in particular to amphibians. The development of tourism traditionally has altered coastal ecosystems significantly, replacing native vegetation by gardens and golf courses, and using high levels of pesticides to control mosquitos, sandflies, and other biting insects. Mining, although only covering 0.02% of Dominican territory, has detrimental impacts locally, sometimes in critical habitats of endangered amphibians.

Related to the process of urbanization is the rapid increase in development projects, including mostly coastal tourist enclaves. The impact of these projects included the issuance of 3,585 environmental permits in 2014, in which 620 permits alone represented an investment of approximately US$ 1,555,834. Moreover, no specific documents exist that report the impact of urban projects on any specific amphibian species in the country (Ministerio de Medio Ambiente y Recursos Naturales 2014b).

C. Pollution

Agrochemicals are known to produce significant impacts on biodiversity, and on amphibians especially (Brühl *et al.* 2013). In the Dominican Republic agrochemicals have been used intensively and extensively, and legal bans exist for many of those that are considered hazardous for human health and the environment (Comité Nacional para la Aplicación de Medidas Sanitarias y Fitosanitarias 2017). In particular, the Constanza Valley, in the Cordillera Central, has been the primary centre of vegetable production for over 60 years, with intensive use of agrochemicals. Marcelino (2015) published the only known study on the potential impacts of agrochemicals on amphibians in the country. She analysed liver tissues of *Osteopilus dominicensis* and water was sampled from the immediate surroundings of the frogs that were collected in the Valley of Constanza. Five pesticides were found to be present in the liver tissues, and ten in the sampled water. None was at high levels, and probably under acceptable limits. Nonetheless, the sample was small (n = 14), and her results are not conclusive.

D. Diseases, especially chytridiomycosis

No known disease of amphibians, such as ranavirus, had been reported from the Dominican Republic prior to infections by the chytrid fungus, *Batrachochytridium dendrobatidis*. Only limited information is available yet on the presence of chytrid on populations of Dominican frogs, or on its potential impact. The first report was from Joglar *et al.* (2007) from three species, *Eleutherodactylus patriciae, E. pituinus*, and *Osteopilus vastus*, from two localities in the Cordillera Central (Ebano Verde, 1,440 m, and Valle Nuevo, 2,500 m). Gail S. Ross (cited as pers. comm., March 2011, in IUCN SSC Amphibian Specialist Group 2013a–c), also reported chytrid fungus in *O. vastus, O. pulchrilineatus*, and *Boana (Hypsiboas) heilprini*. Galvis (2013 as cited by Ministerio de Medio Ambiente y Recursos Naturales 2014c) reported the presence of chytrid in two other species of *Eleutherodactylus* – *E. flavescens* and *E. weinlandi* – from the area of El Llagal, within the Barrick-Pueblo Viejo mining concession, in the central northeast of the Dominican Republic. Sample sizes reported varied from 187 to a few per species for *O. vastus, O. pulchrilineatus*, and *Boana (Hypsiboas) heilprini*. Independent of sample size, the prevalence was low, being higher in *Eleutherodactylus weinlandi* (12.5%, but that sample was small, $n = 8$) and *B. (Hypsiboas) heilprini* with a prevalence of 8.3% ($n = 48$). Díaz *et al.* (2015) reported that most *B. (Hypsiboas) heilprini* tadpoles collected at Ebano Verde had oral deformities due to the presence of the fungus.

E. Climatic Change

The Dominican Republic was ranked tenth among the top countries that suffered the greatest impacts from changes in climate between 1997 and 2016 (i.e. multidecadal period; Eckstein *et al.* 2017), with a CRI (Climate Risk Index) of 11 in 2016 alone. This index is based on four indicators normalized as follows: total deaths (1/6), deaths per 100,000 inhabitants (1/3), total loss in US$ in the parity of purchasing power (1/6), and losses per unit of gross domestic product (1/3) (Eckstein *et al.* 2019). The lower the value of this index, the higher the rank and a country's risk. The Dominican Republic lowered its rank to 12th between 1998 and 2017, with a CRI of 37 in 2017 (Eckstein *et al.* 2018). Likewise, its rank lowered to 50th between 1999 and 2018, with a most recent CRI of 99 in 2018 (Eckstein *et al.* 2019). Other works have been published, mostly by the country's governmental agencies, concerning climatic change. The first was the national Communication to the Convention on Climate Change, published in 2004 (Secretaría de Estado de Medio Ambiente 2004). Others have followed (Anderson *et al.* 2008; Secretaría de Estado de Medio Ambiente y Programa de las Naciones Unidas para el Desarrollo 2009; Izzo 2011; UICN y Ministerio de Medio Ambiente y Recursos Naturales 2013; Pérez Fernández 2016; Ministerio de Medio Ambiente y Recursos Naturales 2019). Several preliminary conclusions were reached with the data from these years. First, the territory of the Dominican Republic is relatively small, but its climate has never been uniform throughout its regions, and there does not seem to be a single pattern of climatic change for the whole country. This affects generalizations made from general patterns of global (or regional) climatic change. Second, although general predictions are that temperatures will increase and precipitation will be diminished, this has not occurred throughout the country. In fact, the region of Enriquillo Lake has seen the opposite (Cleto and Luo 2013). In 2008, CATHALAC (Centro del Agua del Trópico Húmedo para América Latina y el Caribe) predicted that the northeastern part of the country would be the region under the highest climatic and thermal stress by 2020 (Anderson *et al.* 2008). In 2016, CATHALAC's predictions changed to apply to the northwest, although this time for the years 2050 to 2070 (Ministerio de Medio Ambiente y Recursos Naturales 2019). It is necessary, however, to increase support to collect field data obtained by regular monitoring of the climate to increase predictability for effective implementation of mitigation against expected consequences of climatic change.

The Dominican Republic has the highest mountain of the insular Caribbean. Otherwise, only the Blue Mountains of Jamaica reach > 2,000 m asl. Many species of frog in the country are montane, and some inhabit ecosystems either at or above 2,000 m asl, mainly the endemics of the Cordillera Central. It is imperative to examine how projected and real-time climatic change would impact such species.

Foden *et al.* (2013), assessed the vulnerability of birds, amphibians, and corals to climatic change. Twenty-seven species from the Dominican Republic were assessed, with 17 being considered of high vulnerability to climatic change (63%). If we disregard the two invasive species (*Rhinella marina* and *Lithobates catesbeianus*) the percentage increases to 68%, thereby emphasizing that vulnerability to climatic change is not necessarily in concordance with the IUCN Red-List assessment of endangerment. Two of the most widespread species in Hispaniola, *Osteopilus dominicensis* and *Peltophryne guentheri*, are highly vulnerable to climatic change. Among the three native toads, *Peltophryne guentheri* is the only one considered to be so. Of the four hylids, all three *Osteopilus* species (*O. dominicensis, O. pulchrilineatus*, and *O. vastus*), are considered highly vulnerable. Of the 27 eleutherodactylids assessed, 13 (48%) are of high vulnerability, and nine of those (69%) are high-mountain species, mostly from the Cordillera Central, the Sierra de Bahoruco, and the Sierra de Neiba. Incháustegui (2015) discussed vulnerability of critically endangered (CR) Dominican frogs to climatic change, highlighting the need to carefully look both at Red-list Threats and vulnerability to climatic change, since species such as *Osteopilus dominicensis*, the most widely distributed endemic frog of Hispaniola, is deemed a species of least concern (LC) by the Red List, and yet is considered to be highly vulnerable to climatic change.

F. Introduced species

Introduced invasive species are considered one of the main threats to native and endemic species (McNeely 2001; IPBES 2020). Introduction of alien species to Hispaniola began with the arrival of the first Europeans in the fifteenth century. Besides domestic animals, rats and mice were also introduced. At the beginning of the twentieth century, the mongoose (*Herpestes auropunctatus*) was introduced for the control of rats, as also occurred in many other islands. It has had significant negative impacts on the islands' biodiversity, including that of amphibians, and is considered among the 100 most invasive species worldwide (Barun *et al.* 2011; but see contrary assessment in Chapter 8, this volume). The cane toad *Rhinella marina* had a similar story as that of the mongoose in the Caribbean islands. It was introduced early in the twentieth century for the biological control of beetles affecting sugar-cane plantations. Its negative impact on native biota, including other amphibians, has been documented in many places (CABI 2019a). *Lithobates catesbeianus*, the bullfrog, was introduced decades later and it spread to suitable habitats all over the country. It is considered an invasive species with potential impacts on native species of frogs (Lavilla and Heatwole 2010; Snow and Witmer 2010). Powell *et al.* (2011), published a list of amphibians and reptiles introduced onto the Caribbean Islands, including information about the introduction of these two amphibians.

Exotic fishes have been introduced, either for the food market or as a source of tropical aquarium fishes. *Betta splendens* was introduced several decades ago into freshwater ponds and lagoons and still occurs in some areas close to Santo Domingo city. This fish is an efficient predator upon young tadpoles. *Clarias batrachus*, the walking catfish, is a relatively new introduction to the country but, as in many other places where it has been introduced, is spreading rapidly over the whole country (Lowe *et al.* 2004; Ministerio de Medio Ambiente y Recursos Naturales 2012b). It is a voracious fish, potentially impacting native amphibians (CABI 2019b). Despite the introduction of several of these invasive species into the country long ago, there has been no documentation of their impacts on biodiversity in general or on amphibians specifically.

G. Harvesting of amphibians for food or for the pet trade

Frogs from the Dominican Republic do not have a direct economic value; they are not used for food, folk medicine, or anything else. A few species occasionally appear in the pet market, although they might have come from either the Dominican Republic or Haiti since they are mostly hylids of broad distribution in Hispaniola. Exportation occurs of both introduced invasive species: *Rhinella marina* and *Lithobates catesbeianus*. During recent years, approximately one million individuals per year, mostly of *Lithobates*, have been exported (Adames 2003). These two species, which were introduced several decades ago, have broad geographic distributions in the lowlands and must be affecting native species, but no information is available on the extent of their impact.

III. Status of decline of populations

A. IUCN Designation of degree of risk

Formal national assessments of the conservation status of amphibians were late in being carried out for species in the Dominican Republic. This is reflected in the IUCN Red-List database, where not a single species of Dominican amphibian was included prior to the Global Amphibian Assessment (GAA), in 2004 (IUCN 2017). Since 2004, however, two international and one national IUCN assessment have been conducted. The Caribbean workshop for the GAA was carried out in Santo Domingo, 19–24 March 2004, during which 37 amphibian species were assessed, of which 86% were under some level of endangerment: Critically Endangered (CR), Endangered (EN), or Vulnerable (VU). At the same time, 92% of Haitian frogs, the country sharing the island of Hispaniola with the Dominican Republic, also fell within one of these three categories, making the frogs of Hispaniola the most endangered of the world (Young *et al.* 2004). Considering the new species described since then, and the most recent assessment (see points 1–9 below), the percentage of threatened frogs became 80% (Table 7.14).

From 22 to 30 March, the IUCN Red List Assessment workshop for Amphibians of Jamaica and Hispaniola was carried out in Santo Domingo. The main result of this initiative was that the hylid frogs, especially *Boana (Hypsiboas) heilprini* and *Osteopilus vastus*, had their status changed from CR to VU, primarily because of the results of surveys sponsored by the Pueblo

Table 7.14. Summary of the population status of frogs of the Dominican Republic (DR; excluding introduced species). H = high; L = low. Groupings of percentage in 'Combined % threatened' include 1) CR-EN-VU and 2) NT-LC, based on 44 native species assessed in the most-recent assessment (IUCN 2022). * a minor discrepancy from the total due to rounding to integers.

ASSESSMENT	NUMBER AND % OF SPECIES	NUMBER AND PERCENTAGE OF SPECIES PER CATEGORY OF POPULATION STATUS					ASSESSMENT CC VU[1]	
		CR	EN	VU	NT	LC	H	L
Baillie *et al.* (2004)	36 100%	8 22%	14 39%	5 14%	2 6%	7 19%	n = 17	n = 20
DR Red-Listing (2010)[2]	44 100%	13 30%	17 39%	5 11%	2 4%	7 16%		
IUCN RL (2022)*	44 100%	10 23%	18 41%	7 16%	3 7%	6 14%	—	—
Combined % threatened		**80**			**20**		**46**	**54**

[1]Foden *et al.* (2013; Climate Change Vulnerability levels, in which percentages are in relation to the 37 assessed species in their appendix B); [2]Ministerio de Medio Ambiente y Recursos Naturales (2011).

Viejo Biodiversity Program of the Canadian Barrick mining company between 2008 and 2010. Concomitantly, the Conservation Needs Assessment Amphibian Ark workshop took place in 2011 (AArk/ASG 2011), along with national workshops that produced the first national Red List of endangered species, using the methodology of the IUCN for red-listing. It changed CR to PC ('Peligro Crítico') and EN to EP ('En Peligro') (Ministerio de Medio Ambiente y Recursos Naturales 2011). As a result, the number of native species assessed increased to 44 (47 including the introduced species) (Table 7.14). In 2008, Hedges *et al.* (2008) raised several populations that had been described as subspecies to the level of full species (numbers 1–6 below) or described outright as new species (7–9 below). Most of these 'new species' are currently assessed by the IUCN (Table 7.15).

They are:

1. ***Eleutherodactylus audanti* (Telegraph Frog)**

   This species was considered VU by the IUCN before Hedges *et al.* (2008) raised its three subspecies, *audanti, melatrigonum,* and *notidodes,* to full species level. *Eleutherodactylus notidodes* is listed as EN (IUCN SSC Amphibian Specialist Group 2022x) and CR in the national Red list; *E. melatrigonum* is listed EN (Table 7.15).
2. ***Eleutherodactylus schmidti* (Streamside Frog)**

   This species was listed as CR before being elevated from subspecific level to a full species. It is the most aquatic of the Hispaniolan eleutherodactylids and currently is presumably extinct (IUCN SSC Amphibian Specialist Group 2022hh; Table 7.15).
3. ***Eleutherodactylus rucillensis* (Rucilla Streamside Frog)**

   This species was also elevated to full species, having been previously described as a subspecies of *E. schmidti* (Hedges *et al.* 2008). It was considered as CR by the national red list and presumably is now extinct (IUCN SSC Amphibian Specialist Group 2022ee; Table 7.15).
4. ***Eleutherodactylus ruthae* (Burrowing Frog)**

   This species belongs to a group of frogs whose males call from excavated underground chambers. Prior to the splitting of the species, it was assessed as EN. The species now is considered a species complex, including several species both in Haiti and the Dominican Republic. Of these, three are present in Dominican territory, *Eleutherodactylus bothroboans, E. ruthae,* and *E. tychathrous.* The first is listed CR (IUCN SSC Amphibian Specialist Group 2022g), while the latter two are listed EN (IUCN SSC Amphibian Specialist Group [2022gg] and IUCN SSC Amphibian Specialist Group [2022jj], respectively).
5. ***Eleutherodactylus weinlandi* (Red-rumped Frog)**

   This species is LC (IUCN SSC Amphibian Specialist Group 2022kk) but a former subspecies, *Eleutherodactylus paralius,* was elevated to species level. It is restricted to a relatively narrow area along the coast in the southeastern part of the country. It was considered as EN on the national red list and it is listed NT (IUCN SSC Amphibian Specialist Group 2022z).
6. ***Eleutherodactylus wetmorei* (Whistling Frog)**

   This species was listed as VU on the national list (before being elevated to species level) and currently is listed VU (IUCN SSC Amphibian Specialist Group 2022ll).
7. ***Eleutherodactylus ligiae* (Bahoruco Spotted Frog)**

   This species is only known from a very small area within pine forests in the Sierra de Bahoruco. Although inside a national park, it has a tiny known range and the frequency of forest fires in the area justifies listing it CR (IUCN SSC Amphibian Specialist Group 2022r; Table 7.15).

Table 7.15. Summary of IUCN Red-List assessments and vulnerability to climatic change of 47 species of frogs (except *Eleutherodactylus dolomedes*; see main text) in the Dominican Republic. English common names mostly follow Hedges *et al.* (2019). * Species described since 2015. ** Data from tables 7.3–7.12, this Chapter. RL = Red List; DR = Dominican Republic; CC Vu = Climatic Change Vulnerability levels.

FAMILY AND SPECIES	COMMON NAME	IUCN RL**	DR RL**	IUCN CC VU**	DISTRIBUTION (IUCN 2022)
Bufonidae					
*Peltophryne armata**	Armored Toad	CR	NA	NA	Dominican Republic
Peltophryne fluviatica	Crestless Toad	CR (PE)	CR	L	Dominican Republic
Peltophryne guentheri	Southern Crested Toad	LC	VU	H	Hispaniola
Rhinella marina	Marine Toad	LC	NA	L	Hispaniola
Eleutherodactylidae					
Eleutherodactylus abbotti	Common Chirping Frog	LC	NA	L	Hispaniola
Eleutherodactylus alcoae	Barahona Rockfrog	LC	EN	L	Hispaniola
Eleutherodactylus armstrongi	Baoruco Hammer Frog	EN	EN	L	Hispaniola
Eleutherodactylus audanti	South Island Telegraph Frog	VU	EN	L	Hispaniola
Eleutherodactylus auriculatoides	Northern Hammer Frog	VU	EN	H	Dominican Republic
Eleutherodactylus bothroboans	Jarabacoa Burrowing Frog	CR	CR	NA	Dominican Republic
Eleutherodactylus flavescens	Yellow Split-toed Frog	NT	NA	H	Dominican Republic
Eleutherodactylus fowleri	Khaki Bromeliad Frog	CR	EN	L	Hispaniola
Eleutherodactylus furcyensis	La Selle Red-legged Frog	CR	CR	L	Hispaniola
*Eleutherodactylus geinotos**	Cambita Chirping Frog	EN	NA	NA	Dominican Republic
Eleutherodactylus haitianus	Montane Cricket Frog	EN	EN	H	Dominican Republic
Eleutherodactylus heminota	Half-stripe Bromeliad Frog	VU	EN	L	Hispaniola
Eleutherodactylus hypostenor	Baoruco Burrowing Frog	EN	EN	L	Dominican Republic
Eleutherodactylus inoptatus	Hispaniolan Giant Frog	NT	NA	L	Hispaniola
Eleutherodactylus jugans	La Selle Dusky Frog	CR	CR	H	Hispaniola
Eleutherodactylus leoncei	Southern Pastel Frog	EN	CR	H	Hispaniola
*Eleutherodactylus ligiae**	Bahoruco Rockfrog	CR	NA	NA	Dominican Republic
Eleutherodactylus melatrigonum	Cordillera Central Telegraph Frog	EN	EN	NA	Dominican Republic
Eleutherodactylus minutus	Hispaniolan Wheeping Frog	EN	EN	H	Dominican Republic
Eleutherodactylus montanus	Hispaniolan Montane Frog	EN	EN	H	Dominican Republic
*Eleutherodactylus neiba**	Neiba Leaf-litter frog	EN	NA	NA	Dominican Republic
Eleutherodactylus nortoni	Spiny Giant Frog	CR	CR	L	Hispaniola
Eleutherodactylus notidodes	Neiba Telegraphic Frog	EN	CR	NA	Hispaniola
Eleutherodactylus parabates	Neiba Whistling Frog	EN	CR	H	Hispaniola
Eleutherodactylus paralius	Coastal Red-rumped Frog	NT	EN	NA	Dominican Republic
Eleutherodactylus patriciae	Hispaniolan Cordillera Frog	EN	EN	H	Dominican Republic

(Continued)

Table 7.15. (Continued)

FAMILY AND SPECIES	COMMON NAME	IUCN RL**	DR RL**	IUCN CC VU**	DISTRIBUTION (IUCN 2022)
Eleutherodactylus pictissimus	Hispaniolan Yellow-mottled Frog	LC	VU	L	Hispaniola
Eleutherodactylus pituinus	Hispaniolan Melodious Frog	EN	EN	H	Dominican Republic
Eleutherodactylus probolaeus	Boca De Yuma Frog	EN	EN	L	Dominican Republic
Eleutherodactylus rucillensis	Rucilla Streamside Frog	CR (PE)	CR	NA	Dominican Republic
Eleutherodactylus rufifemoralis	Baoruco Red-legged Frog	EN	CR	L	Dominican Republic
Eleutherodactylus ruthae	Eastern Burrowing Frog	EN	EN	H	Dominican Republic
Eleutherodactylus schmidti	Hispaniolan Streamside Frog	CR (PE)	CR	H	Dominican Republic
Eleutherodactylus sommeri	Massif Du Nord Whistling Frog	EN	CR	NA	Hispaniola
Eleutherodactylus tychathrous	Vallejuelo Burrowing Frog	EN	CR	NA	Dominican Republic
Eleutherodactylus weinlandi	Hispaniolan Red-rumped Frog	LC	NA	H	Hispaniola
Eleutherodactylus wetmorei	Tiburon Whistling Frog	VU	VU	L	Hispaniola
Hylidae					
Boana heilprini	Hispaniolan Green Treefrog	VU	VU	L	Hispaniola
Osteopilus dominicensis	Hispaniolan Laughing Treefrog	LC	NA	H	Hispaniola
Osteopilus pulchrilineatus	Hispaniolan Yellow Treefrog	VU	EN	H	Hispaniola
Osteopilus vastus	Hispaniolan Giant Treefrog	VU	EN	H	Hispaniola
Leptodactylidae					
Leptodactylus albilabris	Antillean White-lipped Frog	LC	EN	L	Dominican Republic
Ranidae					
Lithobates catesbeianus	American Bullfrog	LC	NA	L	Hispaniola

8. ***Eleutherodactyulus neiba* (Neiba Frog)**
   Known from a very limited area within the Sierra de Neiba inside a national park, this species inhabits the most endangered montane cloud forests of the country. It is listed EN (IUCN SSC Amphibian Specialist Group 2022v).
9. ***Eleutherodactylus geitonos* (Cambita Chirping Frog)**
   Only known from the type locality, although there is evidence of a broader distribution, at least in the same general area of the Cordillera Central. It is not known from any protected area. It is listed EN (IUCN SSC Amphibian Specialist Group 2022k).

The overall picture of the conservation status of Dominican frogs included 10 species considered Critically Endangered (CR), 18 Endangered (EN), and 7 Vulnerable (VU) (Table 7.15), for a total of 35 species, which represent 80% of the native species present in the country (or 74% including introduced species). IUCN Red-Listing methodology was developed long before considerations were made for the potential vulnerability to climatic change of these species. The IUCN made an assessment of the vulnerability of most of the world's species of frog (Foden *et al.* 2013 and their appendix B), including 37 Dominican frogs. The potential vulnerability was classified into low (L) or high (H), and 46% of the frogs of the country are considered highly vulnerable to climatic change (Table 7.14). It is important to highlight that species with high vulnerability do not coincide necessarily with those considered threatened under IUCN Red Listing. One illustrative case is *Osteopilus dominicensis,* the commonest hylid frog of Hispaniola, occupying almost all habitats on the island and considered to be of Least Concern in the IUCN Red List. Nonetheless, it has a high vulnerability to climatic change. One toad, three of the four treefrogs, and 13 *Eleutherodactylus* are considered highly vulnerable, and 14 of the 17 species evaluated on the basis of Foden's *et al.* (2013) criteria are considered threatened under the Red Listings. Most of these species are montane, from Cordillera Central, Sierra de Bahoruco, and Sierra de Neiba. Four species are of special concern: (1) *Eleutherodactylus jugans* and (2) *Eleutherodactylus leoncei* from Sierra de Bahoruco; (3) *Eleutherodactylus parabates* from Sierra de Neiba; and (4) *Eleutherodactylus schmidti,* a streamside frog. These four species are critically endangered or endangered, and have high vulnerability to climatic change.

Hedges and Diaz (2011), in their paper on the conservation status of the amphibians of the West Indies, presented lists of the last year each species had been seen or collected according to several trusted reporters and their own experience. In the case of the Dominican Republic, there were few field studies conducted by nationals, and many of the observations were by foreign herpetologists, who did not conduct fieldwork continuously. As mentioned above, after Baillie *et al.* (2004), several initiatives increased local knowledge and interest in amphibians. The failure of many species to be recorded during previous years reflected lack of field studies rather than genuine absences. Proyecto RANA team, undertaken by five experienced persons, was able to find most of the species during one night of fieldwork in appropriate habitats. This does not mean in any way that the species are not threatened, but that in some cases they are still locally common. *Peltophryne fluviatica,* one of the native toads, has not been found since its discovery by Schwartz and colleagues in 1971 (Schwartz 1972). Our team and other colleagues have searched for this toad at the type locality and surroundings, to no avail. More intensive searches are needed since this species is semifossorial and can remain inactive for long periods, making them difficult to find (U.S. Fish and Wildlife Service 1992). Similarly, the aquatic eleutherodactylids *Eleutherodactylus schmidti* and *E. rucillensis* have not been seen since 1986 (Hedges and Diaz 2011). Being more aquatic than other species in the genus, they might have been affected more seriously by chytrids, although at the moment there is no evidence to support that supposition. Another frog, *Eleutherodactylus melatrigonum,* from the Cordillera Central has not been reported since 1993 (Hedges and Diaz 2011).

IV. Conservation and monitoring

A. Legislation

At present, no specific legislation exists to protect amphibians in the Dominican Republic. Nonetheless, they are protected legally as part of the biodiversity/fauna of the country. The main legal document is the General Environmental Law, Law 64-00, Article 138, Chapter IV, related to biodiversity; it specifies that the collection or destruction of all wild animals is prohibited, and Article 140 similarly refers to those species under any degree of endangerment. (Gaceta Oficial 2000). Endangered species are listed in the national red list (Ministerio de Medio Ambiente y Recursos Naturales, 2011). Furthermore, the National Constitution protects natural resources and national biodiversity (Gaceta Oficial 2015a). At the international level, the country is a party to the Convention on Biological Diversity (CBD) (ONU 1992), and as such is obliged to protect its biodiversity. Besides the CBD, the country is also party to other multilateral environmental agreements, such as CITES, but this convention regulates only the international traffic of endangered or threatened species with significant trade. None of the Dominican amphibians is in any of the CITES appendices (checklist.cites.org). Access to genetic resources of amphibians, as well as any other component of the country's biodiversity, is regulated under the CBD agreements. The Biodiversity Law, approved in 2015, also offers legal protection to the wildlife of the country (Gaceta Oficial 2015b).

B. Faunal reserves

Mainly since the 1970s, the Dominican Republic has developed an extensive national system of protected areas. At present, it includes 123 conservation units, classified into six categories of management. These cover 12,033.12 km², which represents 25% of the land surface of the country (Ministerio de Medio Ambiente y Recursos Naturales 2012a). All of these areas were selected mainly for the types of particular ecosystems and/or habitats they represent, and in some cases, also because of particular species of flora or fauna that were present. None, however, was established on the basis of specific needs for the protection and conservation of the native amphibians, although they do include populations of 40 of the known species in the Dominican territory. Two endangered species of native toads with limited known geographic distribution (*Peltophryne fluviatica* [CR{PE}] and *P. armata*); *Leptodactyllus albilabris* (LC), the only leptodactylid present in Hispaniola; and three montane eleutherodactylids, *Eleutherodactylus sommeri* (EN), *E. bothroboans* (CR), and *E. tychathrous* (EN), as well as the newly described *E. geitonos* (EN), are not represented within the national system of protected areas.

C. Conservation and captive breeding

The results of the GAA highlighted the need for conservation actions for the frogs of the Dominican Republic. One crucial issue was that the conservation statuses of the species be reviewed to create and strengthen national capacities for their study and conservation. Following this, Grupo Jaragua and the National Museum of Natural History of Santo Domingo, submitted a project proposal to FONDOCyT (Fondo Nacional de Innovación y Desarrollo Científica y Tecnológico), a national fund to support projects in science and technology. The project 'Endangered Amphibians and Climate Change' (proposal number 2008-1-A2-102) was approved to be implemented in collaboration with Luis M. Díaz, from the National Museum of Natural History from Cuba, and Nils Navarro, from the Cuban Zoological Society. This three-year project strengthened national capacities for the study of amphibians. The collection of the National Museum of Natural History increased significantly, and a database on wildlife, a collection of photographs from the laboratory, and recordings of amphibians' advertisement calls were created. General training in amphibian conservation and illustration was offered. Biology students have been trained in field and laboratory techniques, and on the use of some specific software. Publications, including the description of two new species (Incháustegui *et al.* 2015), environmental education, and a national strategy for the conservation of amphibians were produced. Two other smaller projects have been linked to this major RANA RD Project. From 2011 to 2014 the Critical Ecosystems Partnership Fund funded the project 'Building Local Management Capacity and Conservation Plans to Save Endangered Frogs in Four High Priority Key Biodiversity Areas in Hispaniola', implemented with the Philadelphia Zoo and the Société Audubon Haïti. The project activities on the Dominican side of Hispaniola were in the Sierra de Bahoruco. The project increased environmental education activities related to amphibian conservation and strengthened the links among Dominican and Haitian colleagues and students. At present, as part of the SAFE (Saving Amphibians from Extinction) initiative, Grupo Jaragua, the National Museum of Natural History, and the Durrell Wildlife Conservation Trust are studying populations of endangered frogs in critical habitats in Sierra de Bahoruco. From 2012 to 2015 three other training events were offered in the Dominican Republic, complementing the initiatives to promote national studies and conservation measures for amphibians.

The Amphibian Conservation Training Course for the Caribbean was offered 5–10 February 2012 by AArk at the National Zoo (ZOODOM), mostly devoted to *ex-situ* conservation and captive husbandry of frogs. From 18 to 23 April 2012, the Natural Sounds Recording for Biodiversity Monitoring course was offered by Cornell Ornithology Laboratory and the Autonomous University of Santo Domingo. Then from 21 to 27 September 2015, the National Museum of Natural History, Durrell Wildlife Conservation Trust, and Barrick Pueblo Viejo offered the course Amphibians Inventories and Monitoring Techniques in La Hispaniola.

Experience with captive breeding of Dominican species is limited. As far as we know, no international experience exists and national experience is very limited. Barrick's Pueblo Viejo biodiversity programme established an amphibian pod (APOD) on their premises to start gaining experience with the captive breeding of threatened species included in their biodiversity offsetting programme, mainly *Eleutherodactylus flavescens* and three hylids, *Boana* (*Hypsiboas*) *heilprini*, *Osteopilus pulchrilineatus*, and *Osteopilus vastus*. At the same time, as a component of the offset programme, Barick's Pueblo Viejo, with technical support from the Toledo Zoo, built the National Amphibiarium at ZOODOM, which was inaugurated on 10 February 2012. While the Pueblo Viejo APOD was successful at breeding *B.* (*Hypsiboas*) *heilprini* and *O. pulchrilineatus*, the National Amphibiarium has not yet been able to breed any species.

Diaz *et al.* (2014) of the Dominican Frog Project (Proyecto RANA RD) published their experience with the husbandry, captive breeding, and developmental biology of *Osteopilus pulchrilineatus*.

AArk, held its Amphibian Conservation Needs Assessment workshop for the Caribbean, in Santo Domingo, 22–30 March 2011. As a result, it identified the need to rescue four species of Dominican amphibians, *in-situ* conservation for 28 species, *in-situ* research for 11, *ex-situ* research for 20, and conservation education for 22 (AArk/ASG Assessment Workshop 2011; Johnson 2011). These recommendations have not been implemented.

V. Acknowledgements

We are very grateful to Neftalí Ríos and Harold Heatwole for the kind invitation to write this chapter. Most of the information presented, including photographs, was obtained through field and laboratory work supported by the project FONDOCYT 2008-1-A2-102 'Anfibios en Peligro y Cambio Climático en la República Dominicana' (Proyecto RANA RD), implemented by Grupo Jaragua and the Natural History Museum of Santo Domingo, Museo Nacional de Historia Natural 'Profesor Eugenio de Jesús Marcano'. Important contributions to the work of this project were made by the Museo Nacional de Historia Natural de Cuba (Luis M. Díaz) and Sociedad Cubana de Zoología (Nils Navarro). We thank Cristian Marte, Marcos Rodriguez, Eveling Gabot, Kenia NG, and Miguel Angel Landestoy for support in the field and laboratory. We express our gratitude to Yolanda León, Yvonne Arias, and Miguel Abreu, from Grupo Jaragua, and Celeste Mir and Carlos Suriel, from the Museo Nacional de Historia Natural, for support in the implementation of this project.

VI. References

AArk/ASG Assessment Workshop, 2011. Conservation Needs Assessment. Dominican Republic. Available at https://www.conservationneeds.org/master/assessmentsearch

Amphibiaweb, 2022. Dominican Republic. Available at: https://amphibiaweb.org/cgi/amphib_query?rel-isocc=like&orderbyaw=Order&where-isocc=Dominican+Republic. Access 2 August 2022.

Adames, F., 2003. Las ranas criollas saltan hasta el Mercado de EEUU. *Periódico Hoy*. 23 de noviembre 2003.

Anderson, E. R., Cherrington, E. A., Flores, A. I., Pérez, J. B., Carrillo, R., and Sempris, E., 2008. *Potential Impacts of Climate Change on Biodiversity in Central America, Mexico, and the Dominican Republic*. CATHALAC/USAID. Panama City, Panama.

Andrew, J. and McDonnell, M. J., 2008. Amphibian ecology and conservation in the urbanizing world: A review. *Biological Conservation* **141**: 2432–2449. https://doi.org/10.1016/j.biocon.2008.07.020

Baillie, J. E. M., Hilton-Taylor, C., and Stuart, S. N. (eds), 2004. 2004 IUCN Red List of Threatened Species. A Global Species Assessment. IUCN, Gland, Switzerland and Cambridge, UK.

Barun, A., Hanson, C. C., Campbell, K. J., and Simberloff, D., 2011. A review of Small Indian Mongoose management and eradications on islands. Pp. 17–25 in *Island Invasives: Eradication and Management*, edited by C. R. Veitch, M. N. Clout, and D. R. Towns. IUCN, Gland, Switzerland.

Brace, S., Barnes, I., Powell, A., Pearson, R., Woolance, R. W., Thomas, M. G., and Turvey, S. T., 2012. Population history of the Hispaniolan hutia *Plagiodontia aedium* (Rodentia: Capromyidae): testing the model of ancient differentiation on a geotectonically complex Caribbean island. *Molecular Ecology* **21**: 2239–2253. https://doi.org/10.1111/j.1365-294X.2012.05514.x

Brühl, C., Schmidt, T., Pieper, S., and Alscher, A., 2013. Terrestrial pesticide exposure of amphibians: An underestimated cause of global decline. *Scientific Reports* **3**: 1135. https://doi.org/10.1038/srep01135

CABI, 2019a. *Rhinella marina* (cane toad). CABI. Invasive Species Compendium. http://www.cabi.org/isc/datas. Accessed 20 May 2020.

CABI, 2019b. *Clarias batrachus* (walking catfish). CABI. Invasive Species Compendium. http://www.cabi.org/isc/datas. Accessed 20 May 2020.

Cano Carmona, E. and Cano Ortiz, A., 2012. Establishment of Biogeographic Areas by Distributing Endemic Flora and Habitats (Dominican Republic, Haiti R.). Pp. 99–118 in *Global Advances in Biogeography*, edited by L. Stevens. ISBN: 978-953-51-0454-4, InTech, Available at https://cdn.intechopen.com/pdfs/34657/InTech-Stablishmentof_biogeographicareas_by_distributing_endemic_flora_and_habitats.pdf.

Cleto, J. F. and Luo, M., 2013. *Water Balance Analysis: Lake Enriquillo Sensor Network Expansion & Analysis of Lake Bathymetric*. Preliminary Report. The City College of New York. New York, USA.

Cochran, D. M., 1941. The herpetology of Hispaniola. *Bulletin of the United States Natural Museum* **177**: 1–398. https://doi.org/10.5962/bhl.part.14437

Comité Nacional para la Aplicación de Medidas Sanitarias y Fitosanitarias, 2017. Available at http://www.cnmsf.gob.do/Generalidades/CuarentenaVegetal/RegistrodePlaguicidas/PlaguicidasRestringidosyProhibidosenRepDom/tabid/194/Default.aspx. Accessed 20 June 2017.

Díaz, L. M., Incháustegui, S. J., and Marte, C., 2014. Preliminary experiences with the husbandry, captive breeding, and development of the Hispaniolan Yellow Tree Frog, *Osteopillus pulchrilineatus* (Amphibia: Anura: Hylidae), with ecological and ethological notes from the wild. *Herpetological Review* **45**: 52–59.

Díaz, L. M., Incháustegui, S. J., Marte, C., Köhler, G., Cádiz, A., and Rodríguez, M., 2018. A new frog of hte *Eleutherodactylus abbotti* Species Group (Anura: Eleutherodactylide) from Hispaniola, with bioacoustic and taxonomic comments on other species. *Novitates Caribaea* **12**: 25–42. https://doi.org/10.33800/nc.v0i12.81

Dinerstein, E., Olson, D. M., Graham, D. J., Webster, A. L., Primm, S. A, Bookbinder, M. E., and Ledec, G., 1995. *A Conservation Assessment of the Terrestrial Ecoregions of Latin America and the Caribbean*. The World Bank in association with WWF, Washington, DC. https://doi.org/10.1596/0-8213-3295-3

Dubois, A., 2017. The nomenclatural status of *Hysaplesia, Hylaplesia, Dendrobates* and related *nomina* (Amphibia, Anura), with general comments on zoological nomenclature and its governance, as well as on taxonomic databases and websites. *Bionomina* **11**: 1–48. https://doi.org/10.11646/bionomina.11.1.1

Duellman, W. E., Marion, A. B., and Hedges, S. B., 2016. Phyllogenetics, classifications, and biography of the treefrogs (Amphibia: Anura: Arboranae). *Zootaxa* **4104**: 001–109. https://doi.org/10.11646/zootaxa.4104.1.1

Eckstein, E., Künzel, V., and Schäfer, L., 2017. Global Climate Risk Index 2018. Who Suffers Most From Extreme Weather Events? Weather-related Loss Events in 2016 and 1997 to 2016. German Watch, Bonn and Berlin, Germany.

Eckstein, D., Hutfils, M., and Winges, M., 2018. Global Climate Risk Index 2019. Who Suffers Most From Extreme Weather Events? Weather-related Loss Events in 2017 and 1998 to 2017. German Watch, Bonn and Berlin, Germany.

Eckstein, D., Künzel, Schäfer, and Winges, M., 2019. Global Climate Risk Index 2020. Who Suffers Most from Extreme Weather Events? Weather-Related Loss Events in 2018 and 1999 to 2018. German Watch, Bonn and Berlin, Germany.

Foden, W. B., Butchart, S. H. M., Stuart, S. N., Vié, J. C., Akçakaya, H. R., Angulo, A., DeVantier, L. M., Gutsche, A., Turak, E., Cao, L., Donner, S. D., Katariya, V., Bernard, R., Holland, R. A., Hughes, A. F., O'Hanlon, S. E., Garnett, S. T., Şekercioğlu, Ç., and Mace, G. M., 2013. Identifying the World's Most Climate Change Vulnerable Species: A Systematic Trait-Based Assessment of all Birds, Amphibians and Corals. *PLoS ONE* **8**(6):e65427. https://doi.org/10.1371/journal.pone.0065427

Frost, D. R., 2022. *Amphibian Species of the World: an Online Reference.* Version 6.1 (2 August 2022). American Museum of Natural History, New York, USA. Electronic Database accessible at https://amphibiansoftheworld.amnh.org

Gaceta Oficial, 2000. Ley 64-00. Ley General de Medio Ambiente y Recursos Naturales. Gaceta Oficial **No. 10056**. Santo Domingo, República Dominicana.

Gaceta Oficial, 2015a. Constitución de la República Dominicana. Gaceta Oficial **No. 10805**. Santo Domingo. República Dominicana.

Gaceta Oficial, 2015b. Ley Sectorial sobre Biodiversidad, **No. 333-15**. G. O. No. **10822** del 17 de diciembre de 2015. Santo Domingo. República Dominicana.

Galvis, P. A., 2013. Presencia de *Batrachochytrium dendrobatidis* en la República Dominicana: estudio de caso. Biodiversidad-PVDC. Santo Domingo, República Dominicana.

Hager, J. and Zanoni, T. A., 1993. La vegetación natural de la República Dominicana: Una nueva clasificación. *Moscosoa* **7**: 39–81.

Hansen, M. C., Potapov, P. V., Moore, R., Hancher, M., Turubanova, S. A., Tyukavina, A., Thau, D., Stehman, S. V., Goetz, S. J., Loveland, T. R., Kommareddy, A., Egorov, A., Chini, L., Justice, C. O., and Townshend, J. R. G., 2013. High-Resolution Global Maps of 21st-Century Forest Cover Change. *Science* **342**: 850–853. https://doi.org/10.1126/science.1244693

Hedges, S. B., 1999. Distribution patterns in the West Indies. Pp. 211–254 in *Patterns of Distribution of Amphibians: A Global Perspective,* edited by W. E. Duellman. Johns Hopkins University Press, Baltimore, MD.

Hedges, S. B., 2022. Caribherp: West Indian amphibians and reptiles. Available at www.caribherp.org. Access 2 August 2022.

Hedges, S. B. and Díaz, L. M., 2011. The conservation status of amphibians in the West Indies. Pp. 31–47 in *Conservation of Caribbean Island Herpetofaunas Volume 1: Conservation Biology and the Wider Caribbean,* edited by A. Hailey, B. S. Wilson, and J. A. Horrocks. Koninklijke Brill NV, Leiden, The Netherlands. https://doi.org/10.1163/ej.9789004183957.i-228.23

Hedges, S. B. and Heinicke, M. P., 2007. Molecular phylogeny and biogeography of West Indian frogs of the genus *Leptodactylus* (Anura, Leptodactylidae). *Molecular Phylogenetics and Evolution* **44**: 308–314. https://doi.org/10.1016/j.ympev.2006.11.011

Hedges, S. B., Duellman, W. E., and Heinicke, M. P., 2008. New World direct-developing frogs (Anura: Terrarana): molecular phylogeny, classification, biogeography, and conservation. *Zootaxa* **1737**: 1–182. https://doi.org/10.11646/zootaxa.1737.1.1

Hedges, B., Inchaustegui, S., Thomas, R., and Powell, R., 2004. *Leptodactylus dominicensis.* In IUCN 2007. 2007 IUCN Red List of Threatened Species. <www.iucnredlist.org>. Downloaded on 30 September 2016.

Hedges, S. B., Powell, R., Henderson, R. W., Hanson, S., and Murphy, J. C., 2019. Definition of the Caribbean Islands biogeographic region, with checklist and recommendations for standardized common names of amphibians and reptiles. *Caribbean Herpetology* **67**: 1–53. https://doi.org/10.31611/ch.67

Henderson, R. W. and Powell, R., 1999. West Indian herpetoecology. Pp. 222–268 in *Caribbean Amphibians and Reptiles,* edited by B. I. Crother. San Diego. Academic Press, San Diego, CA. https://doi.org/10.1016/B978-012197955-3/50019-7

Henderson, R. W. and Powell, R., 2009. *Natural History of West Indian Reptiles and Amphibians.* University Press of Florida, Gainesville, FL.

Incháustegui, S. J., 2014. Challenges for the conservation of amphibians and reptiles in the Dominican Republic. P. 108 in *Programa y libro de resúmenes del X Congreso Internacional de Investigación Científica.,* edited by P. F. Gómez-Ramírez, C. M. Rodríguez-Pena y S. J. Incháustegui. Santo Domingo.

Incháustegui, S. J., 2015. Ranas en Peligro Crítico y su vulnerabilidad al cambio climático. P. 193 in *Programa y libro de resúmenes del X Congreso Internacional de Investigación Científica,* edited by P. F. Gómez- Ramírez, C. M. Rodríguez-Pena and S. J. Incháustegui. Santo Domingo.

Incháustegui, S. J., Diaz, L. M., and Marte, C. F., 2015. Dos especies nuevas de ranas del genero *Eleutherodactylus* (Amphibia: Anura:

Eleutherodactylinae) de la Hispaniola. *Solenodon* **12**: 136–149.

IPBES, 2020. *Models of Drivers of Biodiversity and Ecosystem Change*. Available at https://ipbes.net/models-drivers-biodiversity-ecosystem-change. Accessed 20 February 2020.

IPCC, 2002. *Cambio Climático y Biodiversidad. Documento Técnico V del IPCC*. Edited by H. Gitay, A. Suárez, R. T. Watson, and D. J. Dokken.

IUCN, 2017. *The IUCN Red List of Threatened Species*. Version 2016-3. <www.iucnredlist.org>. Downloaded on 15 February 2017.

IUCN, 2022. *The IUCN Red List of Threatened Species*. Version 2022-1. https://www.iucnredlist.org. Accessed on 2 August 2022.

IUCN SSC Amphibian Specialist Group, 2013a. *Hypsiboas heilprini. The IUCN Red List of Threatened Species 2013*: e.T55502A3030511. http://dx.doi.org/10.2305/IUCN.UK.2013-1.RLTS.T55502A3030511.en. Downloaded on 20 December 2017

IUCN SSC Amphibian Specialist Group, 2013b. *Osteopilus pulchrilineatus. The IUCN Red List of Threatened Species 2013*: e.T55810A3032570. http://dx.doi.org/10.2305/IUCN.UK.2013-1.RLTS.T55810A3032570.en. Downloaded on 20 December 2017.

IUCN SSC Amphibian Specialist Group, 2013c. *Osteopilus vastus. The IUCN Red List of Threatened Species 2013*: e.T54346A3014515. http://dx.doi.org/10.2305/IUCN.UK.2013-1.RLTS.T54346A3014515.en. Downloaded on 20 December 2017

IUCN SSC Amphibian Specialist Group, 2021. *Leptodactylus albilabris*. The IUCN Red List of Threatened Species 2021: e.T57113A3055043. https://dx.doi.org/10.2305/IUCN.UK.2021-1.RLTS.T57113A3055043.en. Accessed on 2 August 2022.

IUCN SSC Amphibian Specialist Group, 2022a. *Boana heilprini. The IUCN Red List of Threatened Species* 2022: e.T55502A172794212. https://dx.doi.org/10.2305/IUCN.UK.2022-1.RLTS.T55502A172794212.en. Accessed on 02 August 2022.

IUCN SSC Amphibian Specialist Group, 2022b. *Eleutherodactylus abbotti. The IUCN Red List of Threatened Species* 2022: e.T56387A3037850. https://dx.doi.org/10.2305/IUCN.UK.2022-1.RLTS.T56387A3037850.en. Accessed on 02 August 2022.

IUCN SSC Amphibian Specialist Group, 2022c. *Eleutherodactylus alcoae. The IUCN Red List of Threatened Species* 2022: e.T56404A172795024. https://dx.doi.org/10.2305/IUCN.UK.2022-1.RLTS.T56404A172795024.en. Accessed on 02 August 2022.

IUCN SSC Amphibian Specialist Group, 2022d. *Eleutherodactylus armstrongi. The IUCN Red List of Threatened Species* 2022: e.T56432A172795214. https://dx.doi.org/10.2305/IUCN.UK.2022-1.RLTS.T56432A172795214.en. Accessed on 02 August 2022.

IUCN SSC Amphibian Specialist Group, 2022e. *Eleutherodactylus audanti. The IUCN Red List of Threatened Species* 2022: e.T20976998A3039336. https://dx.doi.org/10.2305/IUCN.UK.2022-1.RLTS.T20976998A3039336.en. Accessed on 02 August 2022.

IUCN SSC Amphibian Specialist Group, 2022f. *Eleutherodactylus auriculatoides. The IUCN Red List of Threatened Species* 2022: e.T56440A3039498. https://dx.doi.org/10.2305/IUCN.UK.2022-1.RLTS.T56440A3039498.en. Accessed on 02 August 2022.

IUCN SSC Amphibian Specialist Group, 2022g. *Eleutherodactylus bothroboans. The IUCN Red List of Threatened Species* 2022: e.T195007A2372554. https://dx.doi.org/10.2305/IUCN.UK.2022-1.RLTS.T195007A2372554.en. Accessed on 02 August 2022.

IUCN SSC Amphibian Specialist Group, 2022h. *Eleutherodactylus flavescens. The IUCN Red List of Threatened Species* 2022: e.T56602A3043646. https://dx.doi.org/10.2305/IUCN.UK.2022-1.RLTS.T56602A3043646.en. Accessed on 02 August 2022.

IUCN SSC Amphibian Specialist Group, 2022i. *Eleutherodactylus fowleri. The IUCN Red List of Threatened Species* 2022: e.T56605A3043931. https://dx.doi.org/10.2305/IUCN.UK.2022-1.RLTS.T56605A3043931.en. Accessed on 02 August 2022.

IUCN SSC Amphibian Specialist Group, 2022j. *Eleutherodactylus furcyensis. The IUCN Red List of Threatened Species* 2022: e.T56608A3044085. https://dx.doi.org/10.2305/IUCN.UK.2022-1.RLTS.T56608A3044085.en. Accessed on 02 August 2022.

IUCN SSC Amphibian Specialist Group, 2022k. *Eleutherodactylus geitonos. The IUCN Red List of Threatened Species* 2022: e.T172816631A172816653. https://dx.doi.org/10.2305/IUCN.UK.2022-1.RLTS.T172816631A172816653.en. Accessed on 02 August 2022.

IUCN SSC Amphibian Specialist Group, 2022l. *Eleutherodactylus haitianus. The IUCN Red List of Threatened Species* 2022: e.T56645A3045969. https://dx.doi.org/10.2305/IUCN.UK.2022-1.RLTS.T56645A3045969.en. Accessed on 02 August 2022.

IUCN SSC Amphibian Specialist Group, 2022m. *Eleutherodactylus heminota. The IUCN Red List of Threatened Species* 2022: e.T56651A3046130. https://dx.doi.org/10.2305/IUCN.UK.2022-1.RLTS.T56651A3046130.en. Accessed on 02 August 2022.

IUCN SSC Amphibian Specialist Group, 2022n. *Eleutherodactylus hypostenor. The IUCN Red List of Threatened Species* 2022: e.T56659A3046418. https://dx.doi.org/10.2305/IUCN.UK.2022-1.RLTS.T56659A3046418.en. Accessed on 02 August 2022.

IUCN SSC Amphibian Specialist Group, 2022o. *Eleutherodactylus inoptatus. The IUCN Red List of Threatened Species* 2022: e.T56672A3046571. https://dx.doi.org/10.2305/IUCN.UK.2022-1.RLTS.T56672A3046571.en. Accessed on 02 August 2022.

IUCN SSC Amphibian Specialist Group, 2022p. *Eleutherodactylus jugans. The IUCN Red List of Threatened Species* 2022: e.T56688A3047415. https://dx.doi.org/10.2305/IUCN.UK.2022-1.RLTS.T56688A3047415.en. Accessed on 02 August 2022.

IUCN SSC Amphibian Specialist Group, 2022q. *Eleutherodactylus leoncei. The IUCN Red List of Threatened Species* 2022: e.T56712A3047929. https://dx.doi.org/10.2305/IUCN.UK.2022-1.RLTS.T56712A3047929.en. Accessed on 02 August 2022.

IUCN SSC Amphibian Specialist Group, 2022r. *Eleutherodactylus ligiae. The IUCN Red List of Threatened Species* 2022: e.T82410205A161212051. https://dx.doi.org/10.2305/IUCN.UK.2022-1.RLTS.T82410205A161212051.en. Accessed on 02 August 2022.

IUCN SSC Amphibian Specialist Group, 2022s. *Eleutherodactylus melatrigonum. The IUCN Red List of Threatened Species* 2022: e.T195010A2372818. https://dx.doi.org/10.2305/IUCN.UK.2022-1.RLTS.T195010A2372818.en. Accessed on 02 August 2022.

IUCN SSC Amphibian Specialist Group, 2022t. *Eleutherodactylus minutus. The IUCN Red List of Threatened Species* 2022: e.T56767A3049058. https://dx.doi.org/10.2305/IUCN.UK.2022-1.RLTS.T56767A3049058.en. Accessed on 02 August 2022.

IUCN SSC Amphibian Specialist Groupu 2022u. *Eleutherodactylus montanus. The IUCN Red List of Threatened Species* 2022: e.T56776A3049380. https://dx.doi.org/10.2305/IUCN.UK.2022-1.RLTS.T56776A3049380.en. Accessed on 02 August 2022.

IUCN SSC Amphibian Specialist Group, 2022v. *Eleutherodactylus neiba. The IUCN Red List of Threatened Species* 2022: e.T82410208A161572338. https://dx.doi.org/10.2305/IUCN.UK.2022-1.RLTS.T82410208A161572338.en. Accessed on 02 August 2022.

IUCN SSC Amphibian Specialist Group, 2022w. *Eleutherodactylus nortoni. The IUCN Red List of Threatened Species* 2022: e.T56796A3049811. https://dx.doi.org/10.2305/IUCN.UK.2022-1.RLTS.T56796A3049811.en. Accessed on 02 August 2022.

IUCN SSC Amphibian Specialist Group, 2022x. *Eleutherodactylus notidodes. The IUCN Red List of Threatened Species* 2022: e.T195011A86896731. https://dx.doi.org/10.2305/IUCN.UK.2022-1.RLTS.T195011A86896731.en. Accessed on 02 August 2022.

IUCN SSC Amphibian Specialist Group, 2022y. *Eleutherodactylus parabates. The IUCN Red List of Threatened Species* 2022: e.T56825A3050899. https://dx.doi.org/10.2305/IUCN.UK.2022-1.RLTS.T56825A3050899.en. Accessed on 02 August 2022.

IUCN SSC Amphibian Specialist Group, 2022z. *Eleutherodactylus paralius. The IUCN Red List of Threatened Species* 2022: e.T195012A86896860. https://dx.doi.org/10.2305/IUCN.UK.2022-1.RLTS.T195012A86896860.en. Accessed on 02 August 2022.

IUCN SSC Amphibian Specialist Group, 2022aa. *Eleutherodactylus patriciae. The IUCN Red List of Threatened Species* 2022: e.T56834A3051159. https://dx.doi.org/10.2305/IUCN.UK.2022-1.RLTS.T56834A3051159.en. Accessed on 02 August 2022.

IUCN SSC Amphibian Specialist Group, 2022bb. *Eleutherodactylus pictissimus. The IUCN Red List of Threatened Species* 2022: e.T56858A3051629. https://dx.doi.org/10.2305/IUCN.UK.2022-1.RLTS.T56858A3051629.en. Accessed on 02 August 2022.

IUCN SSC Amphibian Specialist Group, 2022cc. *Eleutherodactylus pituinus. The IUCN Red List of Threatened Species* 2022: e.T56863A3051957. https://dx.doi.org/10.2305/IUCN.UK.2022-1.RLTS.T56863A3051957.en. Accessed on 02 August 2022.

IUCN SSC Amphibian Specialist Group, 2022dd. *Eleutherodactylus probolaeus. The IUCN Red List of Threatened Species* 2022: e.T56877A3052311. https://dx.doi.org/10.2305/IUCN.UK.2022-1.RLTS.T56877A3052311.en. Accessed on 02 August 2022.

IUCN SSC Amphibian Specialist Group, 2022ee. *Eleutherodactylus rucillensis. The IUCN Red List of Threatened Species* 2022: e.T195013A2373222. https://dx.doi.org/10.2305/IUCN.UK.2022-1.RLTS.T195013A2373222.en. Accessed on 02 August 2022.

IUCN SSC Amphibian Specialist Group, 2022ff. *Eleutherodactylus rufifemoralis. The IUCN Red List of Threatened Species* 2022: e.T56931A3052851. https://dx.doi.org/10.2305/IUCN.UK.2022-1.RLTS.T56931A3052851.en. Accessed on 02 August 2022.

IUCN SSC Amphibian Specialist Group, 2022gg. *Eleutherodactylus ruthae. The IUCN Red List of Threatened Species* 2022: e.T20995600A3053137. https://dx.doi.org/10.2305/IUCN.UK.2022-1.RLTS.T20995600A3053137.en. Accessed on 02 August 2022.

IUCN SSC Amphibian Specialist Group, 2022hh. *Eleutherodactylus schmidti. The IUCN Red List of Threatened Species* 2022: e.T213610247A3053309. https://dx.doi.org/10.2305/IUCN.UK.2022-1.RLTS.T213610247A3053309.en. Accessed on 02 August 2022.

IUCN SSC Amphibian Specialist Group, 2022ii. *Eleutherodactylus sommeri. The IUCN Red List of Threatened Species* 2022: e.T195014A86896986. https://dx.doi.org/10.2305/IUCN.UK.2022-1.RLTS.T195014A86896986.en. Accessed on 02 August 2022.

IUCN SSC Amphibian Specialist Group, 2022jj. *Eleutherodactylus tychathrous. The IUCN Red List of Threatened Species* 2022: e.T195015A2373362. https://dx.doi.org/10.2305/IUCN.UK.2022-1.RLTS.T195015A2373362.en. Accessed on 02 August 2022.

IUCN SSC Amphibian Specialist Group, 2022kk. *Eleutherodactylus weinlandi. The IUCN Red List of Threatened Species* 2022: e.T20998545A3054645. https://dx.doi.org/10.2305/IUCN.UK.2022-1.RLTS.T20998545A3054645.en. Accessed on 02 August 2022.

IUCN SSC Amphibian Specialist Group, 2022ll. *Eleutherodactylus wetmorei. The IUCN Red List of Threatened Species* 2022: e.T20994808A3054754. https://dx.doi.org/10.2305/IUCN.UK.2022-1.RLTS.T20994808A3054754.en. Accessed on 02 August 2022.

IUCN SSC Amphibian Specialist Group, 2022mm. *Lithobates catesbeianus. The IUCN Red List of Threatened Species* 2022: e.T58565A193396825. https://dx.doi.org/10.2305/IUCN.UK.2022-1.RLTS.T58565A193396825.en. Accessed on 02 August 2022.

IUCN SSC Amphibian Specialist Group, 2022nn. *Osteopilus dominicensis. The IUCN Red List of Threatened Species* 2022: e.T55808A172794620. https://dx.doi.org/10.2305/IUCN.UK.2022-1.RLTS.T55808A172794620.en. Accessed on 02 August 2022.

IUCN SSC Amphibian Specialist Group, 2022oo. *Osteopilus pulchrilineatus. The IUCN Red List of Threatened Species* 2022: e.T55810A172794792. https://dx.doi.org/10.2305/IUCN.UK.2022-1.RLTS.T55810A172794792.en. Accessed on 02 August 2022.

IUCN SSC Amphibian Specialist Group, 2022pp. *Osteopilus vastus. The IUCN Red List of Threatened Species* 2022: e.T54346A172794001. https://dx.doi.org/10.2305/IUCN.UK.2022-1.RLTS.T54346A172794001.en. Accessed on 02 August 2022.

IUCN SSC Amphibian Specialist Group, 2022qq. *Peltophryne armata. The IUCN Red List of Threatened Species* 2022: e.T172817035A172817040. https://dx.doi.org/10.2305/IUCN.UK.2022-1.RLTS.T172817035A172817040.en. Accessed on 02 August 2022.

IUCN SSC Amphibian Specialist Group, 2022rr. *Peltophryne fluviatica. The IUCN Red List of Threatened Species* 2022: e.T54639A3017021. https://dx.doi.org/10.2305/IUCN.UK.2022-1.RLTS.T54639A3017021.en. Accessed on 02 August 2022.

IUCN SSC Amphibian Specialist Group, 2022ss. *Peltophryne guentheri. The IUCN Red List of Threatened Species* 2022: e.T172918839A3017633. https://dx.doi.org/10.2305/IUCN.UK.2022-1.RLTS.T172918839A3017633.en. Accessed on 02 August 2022.

IUCN SSC Amphibian Specialist Group. *Rhinella marina*. The IUCN Red List of Threatened Species 2022(1): e.T100099625A2951416. *In press.*

Izzo, M., 2011. Analisi del clima e delle dinamiche climatiche nella Repubblica Dominicana e delle relative influenze sul territorio. Altri titoli: Analysis of climate and its dynamics in the Dominican Republic and influences on territory. Tesi di dottorato. Università degli Studi del Molise, Italia.

Joglar, R., Álvarez, A. O., Aide, T. M., Barber, D., Burrowes, P. A., García, M. A., León-Cardona, A., Longo, A. V., Pérez-Buitrago, N., Puente, A., Rios-López, N., and Tolson, P. J., 2007. Conserving the Puerto Rican herpetofauna. *Applied Herpetology* **4**: 327–345. https://doi.org/10.1163/157075407782424566

Johnson, K., 2011. Amphibian Conservation Needs Assessment Workshop for the Caribbean. *Froglog* **96**: 12–13.

Landestoy T., M. A., Turner, D. B., Marion, A. B., and Hedges, S. B., 2018. A new species of Caribbean toad (Bufonidae, *Peltophryne*) from southern Hispaniola. *Zootaxa* **4403**: 523–539. https://doi.org/10.11646/zootaxa.4403.3.6

Lavilla, E. O. and Heatwole, H., 2010. Status of Amphibian Conservation and Decline in Argentina. Chapter 3, pp. 30–78 in Part 1 *Paraguay, Chile and Argentina* of Volume 9 of *Amphibian Biology. Status of Decline of Amphibians: Western Hemisphere.* Surrey Beatty & Sons, Baulkham Hills, New South Wales, Australia.

Lowe S., Browne, M., Boudjelas, S., and De Poorter, M., 2004. 100 de las Especies Exóticas Invasoras más dañinas del mundo. Una selección del Global Invasive Species Database. Publicado por el Grupo Especialista de Especies Invasoras (GEEI), un grupo especialista de la Comisión de Supervivencia de Especies (CSE) de la Unión Mundial para la Naturaleza (UICN), Primera edición, en inglés, sacada junto con el número 12 de la revista Aliens, Diciembre 2000. Versión traducida y actualizada: noviembre 2004.

Marcelino, N., 2015. *Análisis de Plaguicidas en la Rana Reidora de la Hispaniola,* Osteopilus dominicensis *(Tschudi) en el Valle de Constanza, La Vega, República Dominicana.* Tesis para optar por el título de Licenciado en Ecología y Gestión Ambiental. Pontificia Universidad Católica Madre y Maestra. Santo Domingo, República Dominicana.

Martínez, J. R., 2012. *Sierra de Bahoruco Occidental. República Dominicana. Estudio Biogeomorfológico y Estado de Conservación de su Parque Nacional.* Doctorado en Cambios Ambientales y Riesgos Naturales. Tesis Doctoral. Doctorado en Cambios Ambientales y Riesgos Naturales. Departamento de Geografía Física y Análisis Geográfico Regional. Facultad de Geografía e Historia. Universidad de Sevilla. Sevilla, España.

Martínez, J. R., 2016. Las 9 áreas protegidas que mayor cantidad absoluta de cobertura boscosa han perdido entre 2000 y 2024. 29 de marzo del 2016. Geografía Física – República Dominicana. http://www.geografiafisica.org/

Martínez, J. R., 2017a. Áreas protegidas que han perdido bosque entre 2000 y 2015. 7 de febrero del 2017. Geografía Física – República Dominicana. http://www.geografiafisica.org/

Martínez, J. R., 2017b. Actualización: datos anuales de pérdidas de bosques en las 9 áreas protegidas que mayor cantidad absoluta de cobertura boscosa han perdido entre 2000 y 2014. 11 de febrero del 2017. Geografía Física – República Dominicana. http://www.geografiafisica.org/

McNeely, J., 2001. Invasive species: a costly catastrophe for native biodiversity. *Land Use and Water Resources Research* **1**: 1–10.

Ministerio de Medio Ambiente y Recursos Naturales, 2011. *Lista de Especies en Peligro de Extinción, Amenazadas o Protegidas de la República Dominicana (Lista Roja).* Santo Domingo, República Dominicana.

Ministerio de Medio Ambiente y Recursos Naturales, 2012a. *Atlas de Biodiversidad y Recursos Naturales de la República Dominicana.* Santo Domingo, República Dominicana.

Ministerio de Medio Ambiente y Recursos Naturales, 2012b. *Estrategia Nacional de Especies Exóticas Invasoras.* Santo Domingo, República Dominicana.

Ministerio de Medio Ambiente y Recursos Naturales, 2014a. *Estudio de Uso y Cobertura de Suelo 2012. Informe Metodológico y Resultados.* Dirección de Información Ambiental y Recursos Naturales. Santo Domingo, República Dominicana.

Ministerio de Medio Ambiente y Recursos Naturales, 2014b. Memoria Institucional 2014. Santo Domingo, República Dominicana.

Ministerio de Medio Ambiente y Recursos Naturales, 2014c. Quinto Informe nacional de biodiversidad de la República Dominicana. Santo Domingo, República Dominicana. Available at http://www.conservaciondeanfibios.org/uploads/2/1/6/1/21615192/09._presencia_de_batrachochytrium_dendrobatidis_en_la_rep%C3%BAblica_dominicana_estudio_de_caso.pdf

Ministerio de Medio Ambiente y Recursos Naturales, 2019. "Tercera Comunicación Nacional de la República Dominicana ante la Convención Marco de las Naciones Unidas sobre Cambio Climático". Santo Domingo, República Dominicana.

Myers, N., Mittermeier, R. A., Mittermeier, C. G., da Fonseca, G. A. B., and Kent, J., 2000. Biodiversity hotspots for conservation priorities. *Nature* **403**: 853–858. https://doi.org/10.1038/35002501

OEA, Secretaría General, 1967. *Reconocimiento y Evaluación de los Recursos Naturales de la República Dominicana I.* Organización de los Estados Americanos, Washington, DC.

OEA, 1967. Mapa ecológico de la República Dominicana. OEA. Secretaría General Humberto Tasaico, Washington, DC.

ONU, 1992. *Convenio sobre la Diversidad Biológica (CDB).* https://www.cbd.int/undb/media/factsheets/undb-factsheets-es-web.pdf. Accessed 20 February 2017.

Pérez Fernández, J., 2016. *Conclusiones Estudio de Escenarios de Cambio Climático de la República Dominicana.* Cathalac, República Dominicana.

PNUMA, CONAU and UASD, 2007. *GEO Santo Domingo. Perspectiva del Medio Ambiente Urbano.* Santo Domingo, República Dominicana.

Powell, R., 2012. Hispaniola and Navassa. Pp. 129–135 in *Island Lists of West Indian Amphibians and Reptiles,* edited by R. Powell and R. W. Henderson. *Bulletin of the Florida Museum of Natural History* **51**: 85–166. University of Florida, Gainesville, FL.

Powell, R., and Henderson, R. W. (eds), 2012. Island lists of West Indian amphibians and reptiles. *Bulletin of the Florida Museum of Natural History* **51** :85–166.

Powell, R., Ottenwalder, J. A., and Incháustegui, S. J., 1999. The Hispaniolan Herpetofauna: Diversity, Endemism, and Historical Perspectives, with Comments on Navassa Island. Pp. 93–68 in *Caribbean Amphibians and Reptiles,* edited by B. I. Crother. Academic Press, San Diego, CA. https://doi.org/10.1016/B978-012197955-3/50016-1

Powell, R., Henderson, R. H., Farmer, M. C., Breuil, M., Echternacht, A. C., Buurt, G., Romagosa, C. M., and Perry, G., 2011. Introduced amphibians and reptiles in the greater Caribbean: Patterns and conservation implications. Pp. 63–144 in *Conservation of Caribbean Islands Herpetofauna. Vol. 1. Conservation Biology and the wider Caribbean,* edited by A. Hayley, B. S. Wilson, and J. A. Horrocks, Koninklijke Brill NV, Leiden, The Netherlands. https://doi.org/10.1163/ej.9789004183957.i-228.38

Pramuk, J. B., Haas, C. A., and Hedges, S. B., 2001. Molecular phylogeny and biogeography of West Indian toads (Anura: Bufonidae). *Molecular Phylogeny and Evolution* **20**: 294–301. https://doi.org/10.1006/mpev.2001.0985

Ríos-López, N., 2010. *Leptodactylus albilabris.* The IUCN Red List of Threatened Species 2010: e.T57113A11581929.

Sangermano, F., Bol, L., Galvis, P., Gullison, R. E., Hardner, J., and Ross, G. S., 2015. Habitat suitability and protection status of four species of amphibians in the Dominican Republic. *Applied Geography* **63**: 55–65. https://doi.org/10.1016/j.apgeog.2015.06.002

Schwartz, A., 1972. The native toads (Anura, Bufonidae) of Hispaniola. *Journal of Herpetology* **6**: 217–231. https://doi.org/10.2307/1562774

Schwartz, A., 1980. The herpetogeography of Hispaniola, West Indies. *Studies on the Fauna of Curaçao and other Caribbean Islands* **189**: 86–127.

Schwartz, A. and Henderson, R. W., 1991. *Amphibians and Reptiles of the West Indies: Descriptions, Distributions, and Natural History*. University of Florida Press, Gainesville, FL.

Secretaría de Estado de Medio Ambiente, 2004. *Primera Comunicación Nacional. Convención Marco de las Naciones Unidas sobre el Cambio Climático*. Secretaria de Estado de Medio Ambiente y Recursos Naturales. Santo Domingo, República Dominicana.

Secretaría de Estado de Medio Ambiente y Programa de las Naciones Unidas para el Desarrollo, 2009. *Segunda Comunicación Nacional, Proyecto Cambio Climático*. Secretaria de Estado de Medio Ambiente y Recursos Naturales. Santo Domingo, República Dominicana.

Silva, M., 2014. Actualización del Mapa de Zonas de Vida de la República Dominicana. Tesis de Cuarto Nivel para optar por el título de Maestro en 'Ecología y Ambiente'. Departamento de Biología. Facultad de Ciencias. Universidad Autónoma de Santo Domingo, Santo Domingo, República Dominicana.

Snow, N. P. and Witmer, G., 2010. American Bullfrogs as Invasive Species: A Review of the Introduction, Subsequent Problems, Management Options, and Future Directions. Pp. 86–89 in *The Proceedings of the 24th Vertebrate. Pest Conference*, edited by R. M. Timm and K. A. Fagerstone. University of of California, Davis, California, USA. Available at: http://digitalcommons.unl.edu/icwdm_usdanwrc/1288/. https://doi.org/10.5070/V424110490

Stuart, S. N., Hoffmann, M., Chanson, J. S., Cox, N. A., Berridge, R. J. Ramani, P., and Young, B. E. (eds.), 2008. *Threatened Amphibians of the World*. Lynx Editions, Barcelona, Spain.

Stuart, S. N., Chanson, J. S., Cox, N. A., and Young, B. E., 2010. The global amphibian assessment. Pp. 1–16 in *Conservation of Mesoamerican Amphibians and Reptiles*, edited by L. D. Wilson, J. H. Townsend, and J. D. Johnson. Eagle Mountain Publishing, Eagle Mountain, UT.

Tolentino, L. and Peña, M., 1998. Inventario de la vegetación y uso de la tierra en la República Dominicana. *Moscosoa* **10**: 179–203.

UICN y Ministerio de Medio Ambiente y Recursos Naturales, 2013. Áreas protegidas y cambio climático: Perspectivas legales y acciones de gestión en República Dominicana. Available at http://www2.ecolex.org/server2neu.php/libcat/docs/LI/MON-088128.pdf

UNDP, 2019. Human Development Report 2019. Beyond Income, Beyond Averages, Beyond Today: Inequalities in Human Development in the 21st Century. AGS, an RR Donnelley Company, MD.

U.S. Fish and Wildlife Service, 1992. *Recovery Plan for the Puerto Rican crested toad* (Peltophryne lemur). Atlanta, GA.

Williams, E. E., 1961. Notes on Hispaniolan herpetology. 3. The evolution and relationships of the *Anolis semilineatus* group. Museum of Comparative Zoology, *Breviora* **136**: 1–8.

Williams, E. E., 1999. Over 300 Years of Collecting in the Caribbean. Pp. 1–30 in *Caribbean Amphibians and Reptiles*, edited by B. I. Crother. Academic Press, San Diego, CA. https://doi.org/10.1016/B978-012197955-3/50013-6

Young, B. E., Stuart, S. N., Chanson, J. S., Cox, N. A., and Boucher, T. M., 2004. *Disappearing Jewels: The Status of New World Amphibians*. NatureServe, Arlington, VA.

APPENDIX 7.I: Detailed taxonomic and nomenclatural list of 48 species of anurans listed from the Dominican Republic (* from Frost 2022; see main text).

FAMILY ELEUTHERODACTYLIDAE LUTZ, 1954

Subfamily Eleutherodactylinae Lutz, 1954

Genus *Eleutherodactylus* Duméril and Bibron, 1841

Eleutherodactylus abbotti Cochran, 1923
Eleutherodactylus alcoae Schwartz, 1971
Eleutherodactylus armstrongi Noble and Hassler, 1933
Eleutherodactylus audanti Cochran, 1934
Eleutherodactylus auriculatoides Noble, 1923
Eleutherodactylus bothroboans Schwartz, 1965
**Eleutherodactylus dolomedes* Hedges and Thomas, 1992
Eleutherodactylus flavescens Noble, 1923
Eleutherodactylus fowleri Schwartz, 1973
Eleutherodactylus furcyensis Shreve and Williams, 1963
Eleutherodactylus geitonos Díaz, Incháustegui, Marte, Köhler, Cádiz, and Rodríguez, 2018
Eleutherodactylus haitianus Barbour, 1942
Eleutherodactylus heminota Shreve and Williams, 1963
Eleutherodactylus hypostenor Schwartz, 1965
Eleutherodactylus inoptatus (Barbour, 1914)
Eleutherodactylus jugans Cochran, 1937
Eleutherodactylus leoncei Shreve and Williams, 1963
Eleutherodactylus ligiae Incháustegui, Díaz, and Marte, 2015
Eleutherodactylus melatrigonum Schwartz, 1966
Eleutherodactylus minutus Noble, 1923
Eleutherodactylus montanus Schmidt, 1919
Eleutherodactylus neiba Incháustegui, Díaz, and Marte, 2015
Eleutherodactylus nortoni Schwartz, 1976
Eleutherodactylus notidodes Schwartz, 1966
Eleutherodactylus parabates Schwartz, 1964
Eleutherodactylus paralius Schwartz, 1976
Eleutherodactylus patriciae Schwartz, 1965
Eleutherodactylus pictissimus Cochran, 1935
Eleutherodactylus pituinus Schwartz, 1965
Eleutherodactylus probolaeus Schwartz, 1965
Eleutherodactylus rucillensis Cochran, 1939
Eleutherodactylus rufifemoralis Noble and Hassler, 1933
Eleutherodactylus ruthae Noble, 1923
Eleutherodactylus schmidti Noble, 1923
Eleutherodactylus sommeri Schwartz, 1977
Eleutherodactylus tychathrous Schwartz, 1965
Eleutherodactylus weinlandi Barbour, 1914
Eleutherodactylus wetmorei Cochran, 1932

FAMILY BUFONIDAE GRAY, 1825

Genus *Peltophryne* Fitzinger, 1843

Peltophryne armata Landestoy T., Turner, Marion, and Hedges, 2018

Peltophryne fluviatica (Schwartz, 1972)

Peltophryne guentheri (Cochran, 1941)

Genus *Rhinella* Fitzinger, 1826

Rhinella marina (Linnaeus, 1758)

FAMILY HYLIDAE RAFINESQUE, 1815

Subfamily Cophomantinae Hoffman, 1878

Genus *Boana* Gray, 1825

Boana heilprini (Noble, 1923)

Subfamily Lophyohylinae Miranda-Ribeiro, 1926

Genus *Osteopilus* Fitzinger, 1843

Osteopilus dominicensis (Tschudi, 1838)

Osteopilus pulchrilineatus (Cope, 1870)

Osteopilus vastus (Cope, 1871)

FAMILY LEPTODACTYLIDAE WERNER, 1896 (1838)

Subfamily Leptodactylinae Werner, 1896 (1838)

Genus *Leptodactylus* Fitzinger, 1826

Leptodactylus albilabris (Günther, 1859)

FAMILY RANIDAE BATSCH, 1796

Genus *Lithobates* Fitzinger, 1843

Lithobates catesbeianus (Shaw, 1802)

8 Amphibians and their History, Distribution, and Conservation in Puerto Rico and the Virgin Islands

Neftalí Ríos-López, Alberto R. Puente-Rolón, Nicole F. Angeli, Sondra I. Vega-Castillo, and Daniel Dávila-Casanova

Neftalí Ríos-López, Alberto R. Puente-Rolón, Nicole F. Angeli, Sondra I. Vega-Castillo, and Daniel Dávila-Casanova, 'Chapter 8: Amphibians and their History, Distribution, and Conservation in Puerto Rico and the Virgin Islands' in: *The Conservation and Biogeography of Amphibians in the Caribbean*. Pelagic Publishing (2023). © Neftalí Ríos-López *et al.* DOI: 10.53061/ZLYL5129

Abbreviations and acronyms

ARP-R	*Alberto R. Puente-Rolón*
asl	*above sea level*
AZA	*American Zoological Association*
Bd	Batrachochytrium dendrobatidis
BS	*Bachelor of Science (US)*
BVI	*British Virgin Islands*
CBSG	*Conservation Breeding Specialist Group*
CMIP5	*Coupled Model Intercomparison Project Phase 5*
CR	*Critically Endangered*
DD	*Data Deficient*
DD-C	*Daniel Dávila Casanova*
df-S	*subtropical dry forest*
DNER	*Department of Natural and Environmental Resources*
DRNA	*Departamento de Recursos Naturales y Ambientales*
DSA	*Declining in Some Areas*
E or EN	*Endangered*
ENSO	*El Niño Southern Oscillation*
EYNF	*El Yunque National Forest*
Ex	*Extinct*
ha	*hectares*
IPCC	*Intergovernmental Panel on Climate Change*
IUCN	*International Union for the Conservation of Nature*
LC	*Least Concern*
MCZ	*Museum of Comparative Zoology*
mf-S	*Subtropical moist forest*
NOAA	*National Oceanic and Atmospheric Administration*
NF-A	*Nicole F. Angeli*
NR-L	*Neftalí Ríos-López*
NSS	*Need of Status Survey*
PARC	*Partners for Amphibian and Reptile Conservation*
pers. obs.	*personal observation*
PCR	*Polymerase Chain Reaction*

PRRA	*Puerto Rico Reconstruction Administration*
PR	*Puerto Rico*
PRA	*Puerto Rican Archipelago*
PRB	*Puerto Rican Bank*
PRCTWG	*Puerto Rican Crested Toad Working Group*
PRDNER	*Department of Natural and Environmental Resources of the Commonwealth of Puerto Rico (or Department of Natural and Environmental Resources)*
PROMESA	*Puerto Rico Oversight, Management, and Economic Stability Act*
qPCR	*Quantitative Polymerase Chain Reaction*
RCP	*Representative Concentration Pathways*
RF	*radiative forcing*
rf-LM	*subtropical lower montane rainforest*
rf-S	*subtropical rainforest*
SCI	*science citation index*
SEPARC	*Southeast Partners in Amphibian and Reptile Conservation*
SIV-C	*Sondra I. Vega-Castillo*
SSC	*Species Survival Commission*
SVL	*snout-to-vent length*
T	*Threatened*
U	*Unknown*
UNEP-WCMW	*United Nations Environment World Conservation Monitoring Centre*
UPR	*University of Puerto Rico*
UPRM	*University of Puerto Rico at Mayagüez (also U.P.R. museum)*
UPRRP	*University of Puerto Rico at Río Piedras*
USNM	*National Museum of Natural History, Smithsonian Institution*
USFWS	*United States Fish and Wildlife Service*
USVI	*United States Virgin Islands*
UVI	*University of the Virgin Islands*
V	*Vulnerable*
VI	*Virgin Islands*
wf-LM	*subtropical lower montane wet forest*
wf-S	*subtropical wet forest*
WI	*West Indies*

I. Introduction

We are living in the Anthropocene (Vitousek *et al.* 1997; Steffen *et al.* 2007; Lugo 2015; Frishkoff *et al.* 2019). A frequent thread on forums dedicated to conservation and policymaking relates to habitat destruction – mostly deforestation, among other transformations – noting it as the leading cause of losses of amphibians in particular and biodiversity in general (Rohde *et al.* 2013). In the Puerto Rican Archipelago (PRA; herein defined as Puerto Rico [PR], United States Virgin Islands [USVI], British Virgin Islands [BVI], and all associated islands, cays, and islets), the present-day distribution and abundance of amphibians reveal changes in types of land-cover (e.g. forests, shrublands, woodlands, herbaceous vegetation, pastures) that are mostly related to various types of land-use (e.g. agriculture, rural and urban dwellings, industry); this is a fact. The herpetologist Karl Patterson Schmidt noticed this when he described patterns of elevational distribution of the herpetofauna in PR (*Habitat Associations and Faunal Subdivisions* in Schmidt, 1928: 10–11):

> Since nearly two-thirds of the coastal-plain species overlap the coffee belt in distribution, it seems obvious that the distribution in altitude offers a little basis for a faunal division. The changes due to cultivation, it may be assumed, have played an important part in the present distribution. The clearing of lowland forests, for example, has undoubtedly driven species to the coffee belt and to the residual forests, while the clearing of the hills has probably afforded access to the higher altitudes in the case of species originally confined to the more open spaces in the coastal plain.

The native and introduced biota of the PRA have combined to generate the present-day assemblages of amphibians that in any given landscape have no reference conditions in history. Consequently, new ecosystems emerge that are termed 'novel' – mostly with new types of vegetation cover, types of habitat including urban, and abiotic conditions – that might be the natural adaptive response of our biota in the Anthropocene (Davis *et al.* 2011; Lugo 2014; Backstrom *et al.* 2018). As the area of so-called pristine habitats elsewhere has been reduced dramatically by humans, novel ecosystems could represent reference ecosystems for future generations, and as such they may have a greater conservation value than is frequently acknowledged (Rosenzweig 2003; Lugo *et al.* 2012; Lugo 2015; Backstrom *et al.* 2018). Novel ecological assemblages and ecosystems are not trivial, and their denial may have profound consequences for policymaking and the prioritization of conservation efforts for amphibians, particularly when limited financial resources characterize conservation's socioeconomic conditions, especially in areas that need investment the most (e.g. Hoffmann *et al.* 2010).

The study of land-use history and related changes in land-cover are of paramount importance when attempting to understand the distribution and abundance of species today. This is mainly because responses of vegetation (particularly forests) to various activities from humans can be slow relative to the lifespan of an amphibian. Therefore, the effects of human activities on environmental conditions deemed fundamental for the reproduction and survival of amphibians (e.g. ambient temperature and humidity) can last for decades after the cessation of these activities (for the rate of change in land-cover in PR see Pascarella *et al.* 2000; Thompson *et al.* 2002; Grau *et al.* 2003). With this in mind, the historical changes in land-cover that have resulted from land-use by humans influence our perception of persistence and resilience among most of amphibians, and may also help us to understand the extirpations and presumed extinctions in the PRA. However, whether this history and its consequences are a general state of affairs or an exception to the rule remains to be assessed for other taxonomic groups and Caribbean islands (e.g. Henderson 1992; Hedges and Díaz 2011). Notably, the description provided by Schmidt (1928) relating also to the distribution of amphibians and the clearing of forests had no prior reference in the herpetological literature of the PRA and has rarely been mentioned subsequently.

In this chapter, the reader will notice that in the PRA – and probably elsewhere – we still lack information that would be valuable for the conservation and management of the vast majority of amphibians. For instance, we still have much to learn about the natural history, reproductive phenology, population biology, and dynamics of assemblages of species in a variety of environmental situations. We also lack information on the physiological responses of amphibians from the perspective of expected changes in climate, examples being shifts in elevational distribution (Schmidt 1928; Pregill and Olson 1981; Narins and Meenderink 2014) and migration in response to sea-level rise and intrusion of salt water into coastal areas (Ríos-López 2008a). Bridging the gap between herpetology and other disciplines (e.g. successional ecology, social sciences, and geographic information systems) is fundamental if we are to prioritize efforts towards the conservation of amphibians and their ecosystems. Accordingly, we provide a brief description of land-use and changes in land-cover in PR and the VI through history, and summarize the anurans in the PRA. Note that whenever we use 'VI' as a stand-alone name in the text, we mean the USVI and BVI combined as a subset of islands in a geographical context. Finally, it is exciting that the earliest known fossil of an *Eleutherodactylus* (and the earliest fossil of any anuran in the Caribbean) – a humerus approximately 29 million years old – comes from PR (J. Vélez-Juarbe, pers. comm. to NR-L; Blackburn *et al.* 2020). Even more exciting is our hope that this work will inspire current and future generations of herpetologists to conduct fruitful research on amphibians in the PRA and throughout the Caribbean.

A. Geography, topography, generalized climate, and Holdridge's life zones

The PRA consists of the Commonwealth of Puerto Rico under jurisdiction of the USA (islands of PR, Mona, Monito, Desecheo, Vieques, and Culebra, collectively centred on 18° 15' N; 66° 30' W), the USVI (St Thomas, St John, and St Croix; collectively centred on 18° 20' N; 64° 50' W); and the BVI (Jost Van Dyke, Tortola, Virgin Gorda, and Anegada; collectively centred on 18° 30' N; 64° 30' W) (Nytch *et al.* 2015; Government of the British Virgin Islands 2019a), and numerous associated islets and cays scattered throughout. It lies within a tropical geographical belt from Anegada (18° 41.696' N; 64° 16.231' W) at the north-eastern border of the PRA to St Croix (17° 40.437' N; 64° 54.022' W) at the southernmost border, to Monito (18° 9.519' N; 67° 57.095' W) at the westernmost border (Fig. 8.1).

The PRA consists of over 154 islands, islets, and cays for which herpetological data exist (Mayer 2012). This archipelago has also been called the Puerto Rico Island Shelf (Heatwole and MacKenzie 1967) and Greater Puerto Rico (Thomas and Joglar 1996). Currently, the total area of the PRA is approximately 9,497 km^2 – 9,000 km^2, 346 km^2, and 151 km^2 for PR, the USVI, and the BVI, respectively (Nytch *et al.* 2015; Government of the British Virgin Islands 2019a). Except for Monito, Mona, Desecheo, and the St Croix Bank, all the islands in this area were one large island – a paleoisland referred as the Puerto Rico Bank – during the most recent Pleistocene lowering of sea level (Meyerhoff 1927; Robertson 1962, citing Schuchert 1935; see Larue 1994 for a review of the geology of the archipelago). This lowering of sea level occurred between 14,000 and 15,000 years ago (Heatwole and MacKenzie 1967; Thomas 1999), resulting in a long island with an area of about 21,000 km^2 (Thomas 1999). The end of the latest glaciation was accompanied by a subsequent rise in sea level that covered part of that landmass and created the present-day geography of the PRA. This geography of this archipelago has remained relatively unchanged for the past 6,000–7,000 years (Heatwole and MacKenzie 1967; Heatwole *et al.* 1981; Pregill and Olson 1981). As a result, all islands share their herpetofauna – partly or completely – with other islands within the archipelago (Heatwole and MacKenzie 1967; Thomas 1999; Perry and Gerber 2006).

The PRA constantly receives warm masses of air from the trade winds coming from the Atlantic Ocean in the east. Mean annual air temperature reaches a low of 26.5°C on Mona Island (26.7°C on the main island of PR) (López-Marrero and Villanueva-Colón 2006) and a high of 27.6°C in St John (USVI). However, mean annual air temperature reaches 26°C at Paraquita Bay, Tortola, BVI (Island Resources Foundation 2015), and air temperature may fluctuate between 18°C and 35°C on Anegada (LaBastille and Richmond 1973). However, trade winds prevent major monthly fluctuations in annual temperature: approximately a 3°C difference (range: 2.7–3.3°C) between the months of February (coldest) and August (warmest) throughout the area (López-Marrero and Villanueva-Colón 2006; Nytch *et al.* 2015; Government of the British Virgin Islands 2019a; see also Weather Underground [2022] for Virgin Gorda historical weather averages in the BVI).

Topography varies throughout the PRA, from the highest peak in the Central Mountain Range in PR (= Cordillera Central; Cerro Punta at 1,338 m asl in the central west; López-Marrero and Villanueva-Colón 2006) to lower elevations on Anegada, BVI (depending on the author, maximum elevation is 8 m asl [LaBastille and Richmond 1973] or 11 m asl [Island Resources Foundation 2013]). In the USVI, the highest mountain peak (Bordeaux Mountain, St Thomas)

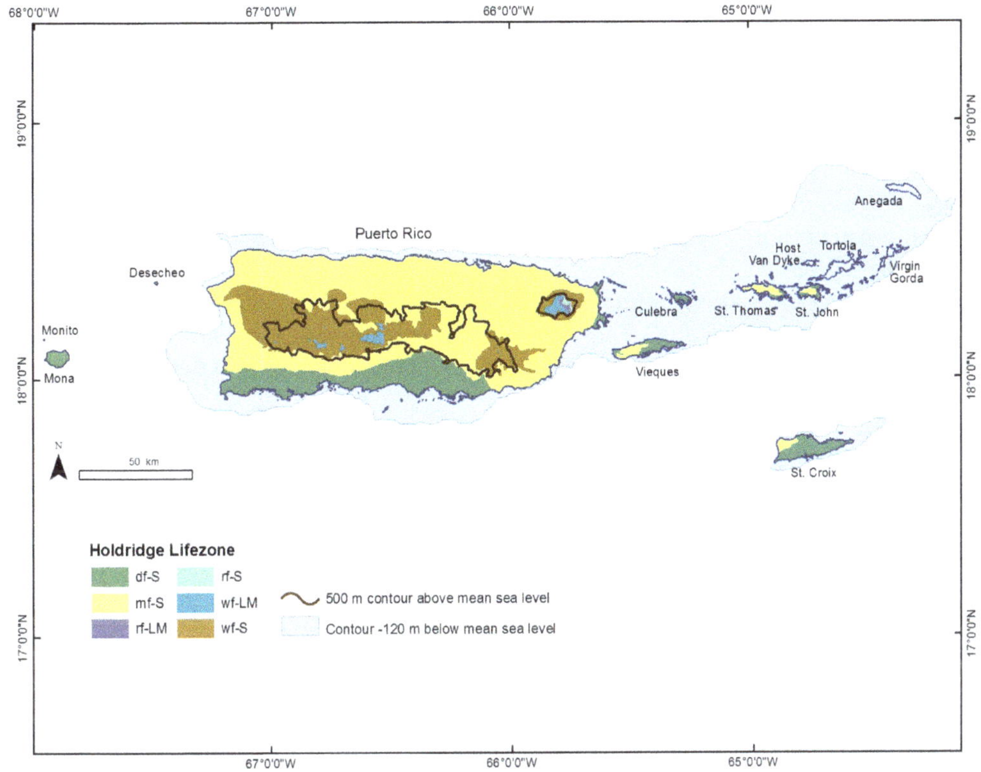

Figure 8.1. Map of the PRA with names of major islands treated in the text and ecological life zones of Holdridge's system of classification: subtropical dry forest (df-S), subtropical moist forest (mf-S), subtropical lower montane rain forest (rf-LM), subtropical rain forest (rf-S), subtropical lower montane wet forest (wf-LM), and Subtropical Wet Forest (wf-S). Kindly prepared and provided by Ms Olga M. Ramos, USDA Forest Service, International Institute of Tropical Forestry, GIS and Remote Sensing Lab, San Juan, Puerto Rico.

attains 474 m asl, while in the BVI, the highest mountain peak reaches 543 m asl (Sage Mountain, Tortola) (Island Resources Foundation 2015).

Topography influences the variation in temperature and rainfall along an elevational gradient. In the PRA, the rate of change in mean air temperature in the rainiest area – the Luquillo Mountains – is ~5°C/1,000 m in elevation (McDowell *et al.* 2012). Also in PR, however, the difference in the mean annual air temperature between the coastal lowlands in the south-western portion of the island and the highest peak in the Central Mountain Range is 8°C to 13°C higher in the lowlands (López-Marrero and Villanueva-Colón 2006). Throughout the archipelago, low-elevation areas and small islands experience warmer conditions and receive less rainfall (see next paragraph) compared with higher elevations and larger islands.

Mean annual rainfall varies with topography, aspect, and size and location of island. In PR, rainfall may reach 1,550 mm in the north, and a peak of 1,780 mm in the Central Mountain Range, but decreases between 910 mm in the south and 986 mm on Mona Island (López-Marrero and Villanueva-Colón 2006; NOAA 2021). However, the maximum annual mean rainfall in the PRA is recorded from the highest mountain peaks of the Luquillo Mountains (1,064–1,074 m asl) in north-eastern PR, with an annual mean rainfall of 4,435 mm per year (range: 3,517–5,457 mm; García-Martinó *et al.* 1996). In addition, moisture deposited from fog occurs in the mountains above 600 m asl and adds 10% more humidity to local water budgets than does rainfall alone in the Luquillo Mountains (Weaver 2012). In the USVI, rainfall may reach 750 mm in the coastal lowlands and 1,400 mm in the highlands (e.g. Nytch *et al.* 2015). On St Croix, rainfall fluctuates between 1,250 mm in the west and less than 750 mm per year in the eastern end of the island (Bowden 1968). In the BVI, rainfall averages approximately 1,273 mm per year on Tortola (Island Resources Foundation 2015), and may fluctuate between 770 mm and 1,020 mm per year on Anegada (LaBastille and Richmond 1973; but see Island Resources Foundation 2013).

The variation in air temperature and rainfall throughout the PRA – because of its location in the Caribbean – is mildly seasonal, with months between January and March having a lower temperature and less rainfall (the 'dry season') compared with months between May and November (the 'wet season'). The combination of topography, island size, and geographical location influences local climate and generates different life zones (see next paragraph), which in turn influence the distribution of amphibians.

Ewel and Whitmore (1973), using Holdridge's life-zones system, classified six life zones in PR. These are (abbreviations from Fig. 8.1): (1) subtropical dry forest (df-S) to the south-south-east (13.8% land-cover); (2) subtropical moist forest (mf-S; 60.5% land-cover); (3) subtropical wet forest (wf-S; 24.1% land-cover, mostly in the Central Mountain Range, Cayey Mountains or 'Sierra de Cayey' in the south-east and the Luquillo Mountains); (4) subtropical lower montane wet forest (wf-LM; 1.2% land-cover; highest peaks in the Central Mountain Range and on the southern aspect of the highest mountains in the Luquillo Mountains); (5) subtropical rainforest (rf-S; 0.2% land-cover); (6) and subtropical lower montane rainforest (rf-LM; 0.2% land-cover). The last two life zones occur on the northern aspect of the highest peaks in the Luquillo Mountains.

West of PR are Mona Island and Desecheo, which contain only subtropical dry forest. Close to PR to the east are the islands of Vieques, with subtropical dry forest (67.5% land-cover) and subtropical moist forest (32.5% land-cover), and Culebra, which consists of subtropical dry forest only. Further east, in the VI, two life zones are mainly represented: (1) subtropical dry forest (33% land-cover on St Thomas and 61% land-cover on St John [USVI]; 100% land-cover on Anegada [BVI]; land-cover is variable among the rest of the major islands) and (2) subtropical moist forest (land-cover variable among islands) (Kennaway *et al.* 2008). On St Croix, subtropical dry forest (72% land-cover) and subtropical moist forest (28% land-cover) are found at various states of post-agricultural regeneration (Morgan *et al.* 2017). In the BVI, the subtropical moist forest is found on Tortola, particularly at higher elevations (approximately 300–523 m asl) on Sage Mountain (Dml'el *et al.* 1997) where 39 ha of former agricultural lands at its peak were designated as protected in 1964 by the National Parks Trust. However, botanist James Beard (1949, *fide* Island Resources Foundation 2015) referred to the vegetation on the summit of Mount Sage as xerophytic rainforest. Today, the Mount Sage National Park in Tortola is considered a unique feature of the VI in general. For a more detailed classification of the major vegetation types in PR and the VI see Kennaway and Helmer (2007) and Kennaway *et al.* (2008) respectively.

II. History of the discovery of species of amphibians in the Puerto Rican Archipelago and definition of their current distributions

The history of the herpetology of PR to the mid-1990s was summarized by Thomas and Joglar (1996). This treatise focused on the herpetofauna at large, and amphibians were not treated in detail. Consequently, we follow Thomas and Joglar's format in describing the history of the Puerto Rican herpetofauna, but focus on the description of new species of amphibians and their distribution in the PRA.

The history was subdivided into three periods, and as we described each of them we paid particular attention to the Early and Mid-Periods, and their contributing authors. The number of authors whose work is relevant has increased over time, and this is particularly true for those documenting new geographical distributional records. For this reason, the Late Period is described decade by decade.

The goal here is to provide thorough historical and current perspectives as a reference for future research by herpetologists, so they can close the gaps in the information about amphibians in the PRA. Other than highlighting specific passages related to historical changes in the taxonomy of native species of amphibians, we follow Frost's (2020) nomenclature. In addition, in Table 8.1 we summarize all major contributions to the taxonomy and geographical distribution of known amphibians. In the following sections, the reader should bear in mind the history of land-use and changes in land-cover that is described in section I., as this places the study of amphibians in the PRA in perspective. Changes in land-use and types of land-cover will be recurrent themes when the conservation of amphibians in the region is discussed.

A. The Early Period

The Early Period began with the description of the first native amphibian, *Leptodactylus albilabris*, by physician and zoologist Albert Charles Lewis Gotthilf Günther in 1859 (referred to as *Cystignathus albilabris* at the time; synonymized as *Leptodactylus* of Fitzinger [1826: 38] by Boulenger [1881], which also used this genus for the first time in the literature of the herpetology of the PRA). He described the species from two specimens (an adult and a smaller individual) collected by the Danish pharmacist and botanist Albert Heinrich Riise in St Thomas. According to Günther (1859: 217), 'Mr Rüse' (*sic*) identified them erroneously as adult '*Hylodes martinicensis*' (Tschudi, 1838) (a genus synonymized as *Eleutherodactylus* by Duméril and Bibron [1841], and used for the first time in the literature of the herpetology of the PRA by Stejneger [1904]) and a smaller *C. ocellatus*; both specimens, however, are '*C. albilabris*' (Günther 1859). Günther (1859: 216–217) reproduced the notes written and sent to him by ornithologist Alfred Newton and his younger brother Edward Newton, an amateur ornithologist, who described the calling and habitus of *L. albilabris* on St Croix, but neither collected nor identified the species. Cope (1868: 311) mentioned that the correspondent of the Smithsonian Institution in PR, George Latimer, sent him 'the following species: [...] *Cystignathus albilabris*' (along with specimens of other species from PR; Stejneger 1904). Peters (1876: 709) mentioned that it occurred throughout PR and that the naturalist Dr Juan Cristóbal Gundlach (Johann Christoph Gundlach was his original German name; Adler 2014a: 53) found eggs, collected them, and followed their development until metamorphosis. The physician and naturalist Dr Agustín Stahl collected five individuals (MCZ A-2186; day and month unknown; Museum of Comparative Zoology [MCZ], Harvard University) of this species in Bayamón, PR, in 1879. Professor Samuel W. Garman also collected 64 individuals (MCZ A-2187) in San Juan, PR, on 19 April 1879. From these accounts, it is reasonable to suggest that the account by Cope (1868) represents a new geographical record of the species from the main island of PR. Later, Gundlach published his description of the foam nest and the development of embryos to metamorphosis, and mentioned that 'this species is very common in the vicinity of Mayagüez in all the ditches, marshes, and shores of the lagoons' (translated freely from Spanish; Gundlach 1881: 313).

The second amphibian described from the PRA, '*Hylodes lentus*', came from one specimen collected from St Thomas by Mr Riise. He sent the specimens to the self-taught palaeontologist and zoologist Edward Drinker Cope for their description (Cope 1862). The third amphibian described from the PRA, '*H. antillensis*', was also based on specimens collected from St Thomas, and sent to zoologists and herpetologists Johannes Theodor Reinhardt and Christian Frederik

Table 8.1. A species checklist that summarizes the geographic distribution of species of amphibians valid in the Puerto Rican Area. Included are the species' names, common names, and locality of the first record with a superscript indicating the reference (located at the bottom of the table. Under Locality: * indicates introduced species for which breeding populations are unknown; ** introduced species which presumably have become locally extirpated; *** native/endemic species which have not been seen/heard in recent decades; *E* indicates species for which extinct status has been explicitly stated in published works; *N* indicates naturalized; *?* indicates species whose origin is unknown; *U* indicates records that cannot be confirmed from the available literature. The reader must consult the main text for *Rhinella marina* because there are specimens collected from the Puerto Rican Area and registered as far back as 1835 in the Natural History Museum of Denmark; also, there are other specimens recorded well before 1920, which is 'the year' frequently accepted in herpetological literature for its introduction into the Puerto Rican Area.

FAMILY, *SPECIES*, AUTHORITY, YEAR	**COMMON NAME**	**LOCALITY**
Bufonidae		
Peltophryne lemur Cope, 1868	Crested Toad	Puerto Rico[a]; Virgin Gorda (*E*) (BVI)[b]
Rhinella marina (Linnaeus, 1758)	Common Toad	Puerto Rico (*N*)[c]; Vieques (PR)[d]; St Croix (USVI)[e]; Culebra Island, Cayo Santiago, Algodones Cay (PR)[f]; St John (USVI)[g]; Tortola[h]; Mona Island and Isla Magueyes (PR), St Thomas (USVI)[i]
Eleutherodactylidae		
Eleutherodactylus antillensis (Reinhardt and Lütken, 1862)	Red-eyed Coqui	St Thomas (USVI)[j]; Vieques (PR)[k]; Puerto Rico, Culebra Island (PR)[l]; St John (USVI)[m]; Tortola (BVI)[kk]; St Croix (?) (USVI)[n]; Piñeros Island (PR)[f]; Guana Island (BVI)[p]; Great Camanoe, Great Thatch Island, Beef Island (BVI)[q]; Jost Van Dyke, Frenchman's Cay, Virgin Gorda, Necker Island (BVI)[i]
E. brittoni Schmidt, 1920	Grass Coqui	Puerto Rico[l]
E. cochranae Grant, 1932	Whistling Coqui	St. John (USVI)[r]; Hassel Island (St Thomas, USVI)[s]; St Thomas (USVI)[e]; Puerto Rico[t]; Palominitos Island (**) (PR)[o]; Tortola (BVI)[f]; Bovony Cay (USVI)[g]; Thatch Cay (St Thomas, USVI)[t]; Culebra Island (PR)[u]; Virgin Gorda (BVI)[h]; Jost Van Dyke, Great Thatch Island (BVI)[p]; Vieques (PR)i
E. cooki Grant, 1932	Guajón	Puerto Rico[w]
E. coqui Thomas, 1966	Common Coqui	Puerto Rico[x]; St Thomas, St Croix (USVI)[g]; Culebra Island (*) (PR)[v]; Vieques (*U*) (PR)[y]; St John[yy].
E. eneidae Rivero, 1959	Eneida's Coqui	Puerto Rico (***)[z]
E. gryllus Schmidt, 1920	Cricket Coqui	Puerto Rico[l]; Note from Tortola: *"The voice of another species, resembling that of* E. gryllus *was heard, but the frogs were not located"* (p. 340 in [kk]).
E. hedricki Rivero, 1963	Hedrick's Coqui	Puerto Rico[aa]
E. jasperi Drewry and Jones, 1976	Golden Coqui	Puerto Rico (***)[bb]
E. johnstonei Barbour, 1914	Lesser Antillean Whistling Frog	Tortola (BVI)[i]
E. juanariveroi Ríos-López and Thomas, 2007	Plains Coqui	Puerto Rico[cc]
E. karlschmidti Grant, 1931	Web-footed Coqui	Puerto Rico (***)[dd]
E. lentus Cope, 1862	Mute Frog	St Thomas (USVI)[ee]; St Croix (USVI)[ff]; St John (USVI)[gg]; Jost Van Dyke (*) (BVI)[hh]; Hassel Island (USVI)[i];
E. locustus Schmidt, 1920	Locust Coqui	Puerto Rico[l]

(Continued)

Table 8.1. (Continued)

FAMILY, *SPECIES*, AUTHORITY, YEAR	COMMON NAME	LOCALITY
E. monensis (Meerwarth, 1901)	Mona Island Coqui	Mona Island[ii]
E. portoricensis Schmidt, 1927	Mountain Coqui	Puerto Rico[jj]
E. richmondi Stejneger, 1904	Richmond's Coqui	Puerto Rico[j]
E. schwartzi Thomas, 1966	Virgin Island Bo-Peep	Tortola, Virgin Gorda (BVI)[x]; St John (*E*) (*, *U*, USVI)[kk]; St Thomas (*?*) and Great Dog (BVI)[f]; St Croix (*, *?*) (USVI)[e, f, but see n]; Beef Island, Frenchman's Cay, Jost Van Dyke (BVI)[q]; Little Thatch Island (BVI)[i]
E. unicolor Stejneger, 1904	Dwarf Coqui	Puerto Rico[j]
E. wightmanae Schmidt, 1920	Melodious Coqui	Puerto Rico[l]
Hylidae		
Dryophytes cinereus (Schneider, 1799)	Green Treefrog	Puerto Rico (**)[g]
Osteopilus septentrionalis (Duméril and Bibron, 1841)	Cuban Treefrog	Puerto Rico (*N*), St Croix (USVI)[g]; St Thomas (USVI)[ll]; St John (USVI)[mm]; Tortola, Necker Island (*), Beef Island, Virgin Gorda (BVI)[nn]; Peter Island (BVI)[oo]; Little St James (*) (USVI)[pp]; Guana Island (*) (BVI)[qq]; Vieques (PR)[zz]; Inner Brass Island and Hassel Island (USVI), Moskito Island (BVI)[i]
Scinax ruber (Laurenti, 1768)	Red-Snouted Treefrog	Puerto Rico (*N*)[rr]
Leptodactylidae		
Leptodactylus albilabris (Günther, 1859)	White-Lipped Frog	St Thomas (USVI)[ss]; Puerto Rico[tt]; Vieques (PR), St Croix (USVI)[j]; Jost Van Dyke (*U*), Tortola, Anegada (BVI)[b]; Culebra Island (PR)[l]; St John (USVI)[m]; Cayo Santiago, Algodones Cay, Piñeros Island, Hicacos (Icacos) (PR)[f]; Isla de Ramos (PR)[uu]; Virgin Gorda (BVI)[h]; Beef Island (BVI)[q]; Luis Peña Cay (PR)[vv]
L. fallax Müller, 1926	Mountain Chicken Frog	Puerto Rico (**)[ww]
Ranidae		
Lithobates catesbeianus (Shaw, 1802)	Bull Frog	Puerto Rico (*N*)[e]; Vieques (PR)[f]; St Croix (*) (USVI)[i]
L. grylio (Stejneger, 1901)	Pig Frog	Puerto Rico (*N*)[xx]
L. pipiens (Schreber, 1782)	Leopard Frog	St Croix (*) (USVI)[e]

[a] Cope (1868); [b] Barbour (1917); [c] Danforth (1925); [d] Grant (1932b); [e] Grant (1937); [f] Heatwole *et al.* (1981); [g] Schwartz and Thomas (1975); [h] MacLean (1982); [i] Mayer (2012); [j] Reinhardt and Lütken (1863); [k] Stejneger (1904); [l] Schmidt (1920); [m] Grant (1932e); [n] Grant and Beatty (1944); [o] Levins and Heatwole (1973); [p] Lazell (1991); [q] Ovaska *et al.* (2000); [r] Grant (1932d); [s] Barbour (1935); [t] Rivero and Mayorga (1963); [u] Philibosian and Yntema (1978); [v] Rivero and Joglar (1979); [w] Grant (1932c); [x] Thomas (1966); [y] Joglar (1998); [z] Rivero (1959); [aa] Rivero (1963); [bb] Drewry and Jones (1976); [cc] Ríos-López and Thomas (2007); [dd] Grant (1931a); [ee] Cope (1862); [ff] Barbour (1915); [gg] Seaman (1961); [hh] Perry (2009b); [ii] Meerwarth (1901); [jj] Schmidt (1927); [kk] Grant (1932f); [ll] Philibosian and Yntema (1977); [mm] Waddle *et al.* (2005); [nn] Owen *et al.* (2005); [oo] Owen *et al.* (2006); [pp] Perry and Platenberg (2007); [qq] Perry (2009a); [rr] Thomas and Joglar (1996); [ss] Günther (1859); [tt] Gundlach (1881); [uu] Schwartz *et al.* (1978); [vv] Ríos-Franceschi *et al.* (2016); [ww] McClelland (1932) and Grant (1932g); [xx] Ríos-López and Joglar (1999); [yy] Waddle *et al.* (2006); [zz] Herrera-Giraldo (2010).

Lütken in 1863 by Mr Riise (Reinhardt and Lütken 1863). Reinhardt and Lütken (1863: 209) wrote (translated freely from Danish):

> Of this kind exist in our Museums some specimens which [are] declared to be from all three Danish Antilles: St Croix, St Thomas and St John; however, is that I notice that Mr Riise, who through a long series of years of utmost skill has endeavoured to spread as much light as possible over the Danish Antilles Fauna, has communicated us that he has received it from St Thomas.

As for the geographical distributions of these species, Reinhardt and Lütken (1863) had doubts about the presence of '*Hylodes antillensis*' on St Croix and St John. Likewise, Reinhardt and Lütken (1863: 205) also had doubts about the presence of '*Cystignathus albilabris*' on St Croix because the passage described by the Newton brothers on the calling of individuals of this species and on their habits after heavy rains in St Croix (as reproduced by Günther 1859 in the *Annals of Natural History*, September 1859) could not be seen or verified. However, Reinhardt and Lütken (1863: 205), included Jost Van Dyke as a new geographical record for '*C. albilabris*' provided to them by Mr Riise. In addition, Reinhardt and Lütken (1863: 209) mentioned that one juvenile of '*H. antillensis*' from Mr Riise's collection was labelled as originating from Vieques; the description of '*H. antillensis*' was based on an adult from St Thomas (Stejneger [1904] pointed out that the report by Reinhardt and Lütken [1863] of the juvenile *Eleutherodactylus antillensis* from Vieques was the first record of the species from the Puerto Rican jurisdiction; see section II. B. 1). Finally, they included a footnote (p. 202; translated freely from Old Danish and corroborated by Bent Erik Kramer Lindow, Collection Manager of Vertebrate Paleontology, Natural History Museum of Denmark, University of Copenhagen, Denmark) that reads:

> Of our *Bufo marinus* and *Phyllomedusa bicolor* there are specimens in our Museums that are said to be from the Danish-West Indies. The latter is, according to Mr Riise, completely unknown as West Indian, and since it is not listed as West Indian anywhere in the Literature, its indication does likely depend on a mistake. *B. marinus*, on the other hand, is sometimes brought there [St Croix, St Thomas, St John] with timber from Vieques and Haiti.
>
> [Note: Grant {1937} would confirm the presence of *Rhinella marina* on Vieques {see below}. Most notably, Daniel Klingberg Johansson, current Collection Manager for Mammalogy and Herpetology (Vertebrate Section, Zoological Museum, Natural History Museum of Denmark) provided NR-L with the following information:
>
> 'We have a few records of *Bufo marinus* from the islands... [although] they mention Riise's find as the first, but we have one older from another collector, but there has been some confusion in older times at the museum since there used to be two Zoological Museums in Copenhagen that merged in 1862... [currently] we have R131059 (old no. CN65) *B. marinus* – St. Croix, registered in the collection 6th of Sep. 1835, the collector is supposedly named Ravn, but when I did a reinspection with a colleague in the old journals, we think the name is wrongly transcribed and might have been called Shoon, or something like that; R13081 (old no. CN85) *B. marinus* – St. Thomas, mounted in the collection 30th of June 1883, but is said to be from before the merger of the two museums (1862), it was collected by Riise; R13119 (old no. CN91) *B. marinus* – St. Thomas, registered in the collection 20th of June 1883, but it is noted that has been sent to the old museum in 1848 by a Mr. Krebs.'

Consequently, from Reinhardt and Lütken {1863} and specimens deposited in the Natural History Museum of Denmark, the record of *R. marina* in the PRA would be at least 85 years older than acknowledged by Wolcott {1924}, Danforth {1925}, Grant {1937}, Turvey {2013}, and elsewhere {see sections II. B. 3. and III. E. 2.a.}; its introduction into the PRA would still be enigmatic.]

The last two species of amphibians described during the Early Period were *Peltaphryne* [*sic*] *lemur* by Cope (1868) (= *Peltophryne lemur*; Pregill [1981] resurrected this genus from Fitzinger [1843: 32] and used it to synonymize *Bufo lemur* for the first time in the literature of the herpetology of the PRA; see section II. C. 3.) and '*Hylodes monensis*' (see Meerwarth 1901), with a new geographical distributional record of '*Hylodes martinicensis*' from PR (Peters 1876). First, one specimen of *P. lemur* was found by the correspondent of the Smithsonian Institution in PR, George Latimer (the locality of the specimen was not specified), who sent it to Cope for its description (Cope 1868). Afterwards, several authors changed the nomenclature and distribution of *P. lemur*; Stejneger (1904) renamed it '*Bufo lemur*', a species 'peculiar to the island' of PR. Specimens of '*H. monensis*' were collected by Mr Berg's engineer Ch. Bock from Mona Island in 1892, but the specimens were described later by zoologist Hermann Meerwarth (1901). Finally, in 1876, the naturalist Dr Wilhelm C. H. Peters, at the time curator of the Berlin Zoological Museum, documented '*H. martinicensis*' from PR as a new locality record for this species. Peters based this new record on specimens sent to him by Gundlach (see Gundlach 1881: 315). Stejneger (1904), Schmidt (1927), and Thomas (1966) further refined the geographical distribution and identity of '*H. martinicensis*' from the PRA (see section II. C. 2).

Naturalists living in PR (or visiting) such as Dr Domingo Bello y Espinosa, Agustín Stahl, and Gundlach (see above) provided significant information on the biology of its amphibians at the time. Specifically, lawyer and resident Bello y Espinosa (1871: 351) first mentioned the reproductive mode of direct development in anurans in the '*coquí*' frog (its common name in Spanish in PR); he described a clutch of about 30 eggs deposited on a lily plant in a garden in Mayagüez, and suggested parental (maternal) care. Bello y Espinosa wrote his description in Spanish but sent it to zoologist Dr E. von Martens who edited it into German and published it on behalf of Bello y Espinosa in 1871. Von Martens, however, had reservations about the account of direct development because Bello y Espinosa saw hatchlings still grouped on site a few days later but did not observe hatching (also see Anonymous 1877: 491).

Peters (1876) remarked that the observation by Bello y Espinosa did not impress the scientific community. As a result, Peters (1876) went on publishing the metamorphosis of the coqui frog, based on observations of specimens and notes sent to him by Bello y Espinosa's law advisee Carl Wilhelm Leopold Krug (a German businessman and later vice consul serving in Mayagüez); Peters' friend Gundlach; and Mr Bavay from Guadeloupe (see note in Peters 1876: 914). Peters (1876) then provided measurements of eggs and embryos and described metamorphosis in detail (he also included a lithograph: p. 714), mostly from specimens sent to him by Gundlach.

Five years had passed since Peters' publication when Gundlach (1881) reproduced the notes he sent to Peters and published them in Spanish, justifying this action on the basis of Bello y Espinosa's description of the direct development in the coqui frog being somewhat shallow, but accrediting the discovery of this reproductive mode to Bello and Espinosa (Gundlach 1881). Gundlach contributed an expanded description of the reproductive biology and natural history of this frog, including amplexus, external morphology of the eggs and embryos during development, colouration of the egg yolk, and the aqueous environment within the gelatinous egg once embryos formed. He also described the changes in morphology of embryos during their development, aspects of the vascularized tail in embryos, and commented on how briefly hatchlings retained the tail. Gundlach (1881: 315) also commented on the species' relative abundance on PR: 'In all parts of the Island I have heard the song *co-quí* or *co-co-quí*; sometimes in the daytime, after showers, but always at night' (translated freely from Spanish). At a similar period, Agustín Stahl (1882: 23) recorded additional observations on the embryology of the coqui frog, but sent his notes also to Peters in Berlin. Finally, the Puerto Rican physician Dr Francisco del Valle Atiles (1887) emphasized the utility of '*Cystignathus typhonius*' (= *Leptodactylus albilabris*), '*Bufo gutturosus*' (= *Peltophryne lemur*), and '*H. martinicensis*' as biological controls of populations of insects considered harmful to agriculture in PR.

In summary, this Early Period lasted 42 years (1859–1901), and six authors contributed to the description of 19.0% of native and endemic species of amphibians (5/21) currently recognized valid for the PRA. Except for Agustín Stahl and Juan Gundlach, the descriptions of species and new geographical distribution records (Table 8.1) were made by authors who had not visited this region.

B. The Mid-Period

The Mid-Period was characterized by an increase in the discovery of amphibians and new distributional records for the PRA. Authors responsible for this increase were visiting herpetologists Leonhard Hess Stejneger, Karl P. Schmidt, and Major Chapman Grant. The herpetologist Thomas Barbour – like several of his predecessors from the early period – did not visit this region; instead, it is known that he 'usually had people bring specimens to him' (Henderson 1992: 6).

We describe this Mid-Period using an author-based format because of the larger number of new species found and the thorough documentation of the geographical distribution of amphibians these authors made.

1. Leonhard Hess Stejneger: 1904 (1902)

Dr Leonhard H. Stejneger visited PR and Vieques from 12 February to 19 April 1900 (the dry season) in the company of ornithologist Dr Charles W. Richmond. They paid special attention to reptiles (Stejneger 1904). Their slow rate of discovery of species of amphibians was probably because of the difficulty of sampling amphibians at night with only the help of an oil lantern (see Gundlach 1881), because an illness kept both men out of the field for almost three weeks (Stejneger 1904: 553), and the period of their visit to PR – during the dry season when the activity of all but one species of amphibian is the lowest. However, Stejneger's work was highly scholarly and detailed, compared with previous works (and a few following). He described two cryptic species, *Eleutherodactylus richmondi* and *E. unicolor*. The former was described as from the Luquillo Mountains (referred to as El Yunque Mountain), at a site ~270 m asl, but Stejneger described its geographical distribution as being along an elevational gradient to the very top of these mountains (~902 m asl); *E. unicolor* was described from the summit of the same mountain range, and it is the species most active during the dry season in PR (see Fig. 6.3 in Joglar 1998: 165).

Stejneger (1904: 591–592) corroborated Reinhardt and Lütken's (1863) geographical distribution of *Eleutherodactylus antillensis* from the town of Isabella Segunda in Vieques and described its abundance there as 'apparently not common'. He echoed Reinhardt and Lütken's (1863) doubts on the occurrence of this species on St Croix and St John, but could not visit those islands for corroboration. However, he documented *Leptodactylus albilabris* from Vieques for the first time and described the species and its tadpole in much detail. Stejneger also commented on the elevational distribution of *L. albilabris* from sea level throughout PR to ~424 m asl at Adjuntas, western Central Mountain Range. He mentioned that *L. albilabris* 'is restricted to the Virgin Islands,

St. Croix, Vieques, and Porto Rico' and that on PR, 'is one of the commonest, most obtrusive, and most easily caught batrachians wherever it occurs' (Stejneger 1904: 561–562).

Stejneger (1904) made a detailed description of an additional specimen of *Eleutherodactylus monensis* that was collected by the ornithologist Mr B. S. Bowdish in 1901 (Bowdish 1902). This specimen was not part of those used by Meerwarth in his 1901 description of the species (Stejneger 1904; Schmidt 1926). Stejneger also provided a detailed description and a more complete summary of the geographical distribution of '*Bufo lemur*' in PR. He based his summary on specimens from Bayamón that were collected by an unidentified friend of Agustín Stahl (Gundlach 1881), Vega Baja (unidentified location), and Arecibo (specimens collected for Stejneger by a young telegraph operator, Enrique L. Brascoechea). These localities occur respectively from east to west in northern coastal PR.

Stejneger made a thorough analysis of the identity and taxonomy of '*Hylodes martinicensis*' in the Caribbean. He considered '*H. martinicensis*' as consisting of more than one species, one from the Lesser Antilles (Martinique, Guadeloupe, St Kitts, and Tobago), and the other from the Greater Antilles (Cuba, Dominican Republic, and PR) (also note that Sampson [1904: 474] included Jamaica in its distribution). Based on his analysis, which included a brief description of the colouration of the iris (a feature that would prove diagnostic for Puerto Rican *Eleutherodactylus* later), Stejneger concluded that Puerto Rican specimens of '*coquíes*' (plural for '*coquí*' in Spanish) belong to the Cuban species *E. auriculatus* instead (Stejneger 1904: 585). The name '*Hylodes*' would not be used for species in the PRA after Stejneger.

2. Thomas Barbour: 1914–1917

Adler (2014a: 85) mentioned that 'Despite his many significant accomplishments and the many honours bestowed upon him—including honorary doctorates from four universities and membership in the U.S. National Academy of Sciences—in some respects Barbour was more a wealthy amateur than a serious professional.' The herpetologists Dr Robert Powell and Dr Robert W. Henderson (1996: 33) add: 'Unfortunately, for observation on natural history, he often relied on second-hand accounts or made offhand judgments on the abundance on many species [...] Nevertheless, Barbour's impact cannot be underestimated.' We agree with criticisms made of Barbour's judgements about species' status, and we also agree that he had a great and long-lasting influence on the herpetology of the PRA and the West Indies (see next and section II. B. 4).

In 1914, Barbour published *A contribution to the Zoögeography of the West Indies, with especial reference to amphibian and reptiles*. This work included corrections to names of species and to previous distributional records, and an annotated list of species of amphibians and reptiles, including introduced species, for the West Indies. It was a reference landmark for all interested in the distribution of these species in the PRA and in the West Indies in general (and expanded that of Garman 1888 [1887], particularly on anurans). He included that the introduced species *Bufo marinis (sic)*, *Hyla rubra*, and *H. squirilla (sic)* occurred on several islands of the West Indies, but at that time none of these anurans was recorded from the PRA (but see section II. A. for *R. marina*). In addition, Barbour (1914: 253) suggested that *Leptodactylus albilabris* was synonymous with *L. pentadactylus* from Martinique and other islands (particularly in the Lesser Antilles). The taxonomy of '*B. lemur*' from PR remained unchanged in this work (later, he would change its taxonomy twice).

Barbour (1914: 247–248) stated of *Eleutherodactylus antillensis* that 'It is definitively known only from St Thomas and Vieques. It may occur upon St. Croix and St. John, but there is no definite record for these islands.' Barbour (1915) added a record for *E. lentus* in St Croix (specimens collected by the herpetologists Dr G. Kingsley Noble and Dr A. G. Ruthven in the summer of 1914). In 1917, Barbour described '*Bufo turpis*' from Virgin Gorda, based on one specimen collected by a young local, who in turn gave it to the ornithologist James Lee Peters. Barbour (1917) also commented on *Rhinella marina* (his *B. marinus*) not having been introduced into Vieques (but see Reinhardt and Lütken 1863; section II. A.) and mentioned having a large series of specimens of *Leptodactylus albilabris* from Tortola and Anegada 'before him' for examination. This would be the first mention of *L. albilabris* from Anegada. He also mentioned *L. albilabris* as being known from Jost Van Dyke, but he did not give the source of this record, nor have we been able to find it.

3. *Karl Patterson Schmidt and Stuart T. Danforth: 1919–28*

Dr Karl Patterson Schmidt visited PR and adjacent islands (Mona, Vieques, and Culebra) from 3 August to 8 October 1919 (the peak of the wet season). He made a contribution as part of the *Scientific Survey of Porto Rico and the Virgin Islands* sponsored by the New York Academy of Sciences, the American Museum of Natural History, and other institutions, which was most significant for years to come. He built upon Stejneger's work and documented the first record of '*B. lemur*' in the southern part of PR (collected at Coamo Springs), and mentioned that this discovery proved this species to be widely distributed (Schmidt 1920). Later, he (1928) commented on the close affinity of '*B. lemur*' with '*B. turpis*' from Virgin Gorda.

Schmidt (1920) also recorded *Eleutherodactylus antillensis* from PR (in both the lowlands and highlands) and on Culebra, and documented the first record of *Leptodactylus albilabris* from Culebra. He further described the foam nests of *L. albilabris* from Coamo Springs, southern PR, in more detail (compared with Gundlach's description in 1881), and considered it to be the most abundant amphibian in PR (the second most abundant being the coqui frog, *E. auriculatus* at the time). In addition, Schmidt (1920) discovered and described four species of *Eleutherodactylus* from the highlands in PR: *E. locustus* (from the Luquillo Mountains) and *E. gryllus*, *E. brittoni*, and *E. wightmanae* (from the Luquillo Mountains and Maricao). He described a fifth species, '*E. cramptoni*' (from the Luquillo Mountains), which was later synonymized with another Puerto Rican *Eleutherodactylus* by Schwartz and Thomas (1975) (see section II. C. 2).

The entomologist and Chief of the Division of Entomology of the Department of Agriculture of Porto Rico at the time, Dr George Norton Wolcott, first mentioned the (deliberate) introduction of *Bufo agua* (= *Rhinella marina*) to PR (Wolcott 1924). Resident ornithologist Stuart Taylor Danforth documented '*B. marinus*' in PR between 1923 and 1924 (at the time, he was a PhD student from Cornell University), and based on two wild-caught individuals (male and female; Danforth 1925). These two specimens were collected by Antonio Ramírez from his home (presumably in Mayagüez, western coastal PR), and he brought them to Danforth; these specimens later were described and measured in detail by Schmidt (1928).

Schmidt (1927) described a new species from PR, *Eleutherodactylus portoricensis,* which he distinguished from the Cuban *E. auriculatus,* recognizing discrepancies between his description of *E. portoricensis* and that of Stejneger's (1904) *E. auriculatus*. Nonetheless, he was convinced that Stejneger's specimens and his own specimens were from the same endemic species, *E. portoricensis* (Schmidt 1927).

Schmidt's (1928) contribution to the amphibians of the PRA was rounded off by his list of species as part of the *Scientific Survey of Porto Rico and the Virgin Islands*. This list was based on his previous work in 1919 and additional material from the mid-1920s available to him for review. By 1928, Schmidt still recognized *Eleutherodactylus auriculatus* from El Yunque (1928: 49), but the naming of this species seemed to be an error as he must have been referring to *E. portoricensis* instead. He also included '*E. cramptoni*' (1928: 55) among the twelve species treated in the list from 1928. In addition, he described Danforth's specimens of '*Bufo marinus*'. Schmidt (1928: 35) said that he remembered hearing reports in 1919 of the introduction of this species, but he did not see any specimens from that time.

4. Major Chapman Grant, Thomas Brown McClelland, and Thomas Barbour: 1930–7

Chapman Grant, Major (US Army), was a self-taught amateur herpetologist (Smith 1986; Adler 2014b) and a productive collector who primarily visited the main island of PR and its adjacent islets and cays, mostly between 1930 and 1932, and chiefly the VI in 1936; his most productive period of herpetological explorations initiated in PR (Smith 1986) when he was appointed to the 65th US Infantry. He was known as having an 'irrepressible iconoclastic drive' towards scientific publications at the time (Smith 1986: 5) but was also increasingly uncomfortable with the academic editorial processes of journals, which he believed (Adler 2014b: 180) 'were frustrating his attempts to publish his work as he saw fit'. Nonetheless, he is considered a scientifically productive person (Smith 1986) and had a significant influence on the herpetology of the PRA.

Thomas and Joglar (1996: 184) highlighted Grant's energetic collecting, and mentioned that 'his discovery of so many species overlooked by others may have been in part the result of increased efficiency and extent of travel that the automobile provided'. Miller (1922) mentioned that PR had 267 km of firm roads (suitable for transportation by car) by 1898, and that a traveller in a car would take 12 hours to cross the island from north to south (i.e. San Juan to Ponce). In contrast, paved roads had increased to 1,300 km by 1922, and a traveller could cover the same route in four hours.

Grant (1931a) discovered the largest of all *Eleutherodactylus* in the PRA (Rivero *et al.* 1963), *E. karlschmidti* (snout–vent length of type = 80 mm), a species that 'lives beneath boulders in the mountain streams of the Luquillo Forest'. This is a specialized river-dwelling species (Rivero *et al.* 1963) that disappeared from all historical collection sites in PR by 1974 (Joglar 1998). For the first time in a publication on herpetology, Grant (1931b) mentioned the introduction, from Dominica, of the 'mountain chicken' *Leptodactylus fallax* (*L. pentadactylus* according to Barbour 1914 and Grant 1931b) into PR in 1929. However, he did not cite the source of this information. In contrast, McClelland (1932) documented all details of this introduction, including the number of specimens first introduced (12), the date of first introduction and the party responsible for it (July 1929; Federal Experiment Station of Mayagüez, western PR), the locality of liberation of these specimens (Cartagena Lagoon, south-western PR), and all details of a second introduction into PR (July 1932; 90 individuals; two localities in Mayagüez). Grant (1932c) cited passages from McClelland's report, but strangely, the manuscripts from McClelland (1932) and Grant (1932c) were published in the same issue of the same journal. Nonetheless, *L. fallax* became the second introduced amphibian documented for the PRA.

In 1932, Grant collected '*Bufo lemur*' at approximately a mile north of Coamo Springs, a historical collection site for the species in southern PR, and provided the first picture of a live specimen (front and side views, Plate IX, Grant 1932a). However, his collection represented the last record of this species from this locality. Grant (1932b) discovered a second species, the specialist-petricolous *Eleutherodactylus cooki*. This species is locally referred to by the Asturian-derived '*guajón*' (in Spanish), referring to the mythological demonic creature '*Guajona*' from the underworld and, hence, its other common name '*Demonio de Puerto Rico*' in Spanish; also see Henry 1958: 186–187). Grant documented *E. cooki* from the granodiorite system of boulders at the Panduras Mountain Range, south-eastern PR. In addition, he discovered a third species, *E. cochranae*, but from St John, and documented its occurrence on Hassel Island, St Thomas (Grant 1932d). This species is frequently found on bromeliads, but occurs on other plants that have bromeliad-like leaf axils that accumulate and retain water (e.g. Joglar 1998; Ríos-López *et al.* 2014). Grant (1932d) also documented *E. antillensis* from St John but did not find it on Hassel Island (St Thomas).

Grant (1932e-g) documented the geographical distribution of several species in PR and the VI and had with him Barbour's (1930b) list of Antillean reptiles and amphibians as a reference.

Grant documented the occurrence of *Leptodactylus albilabris,* also from St John, and suggested that individuals of *L. albilabris* from St John could be of a different species than those in PR (because of differences in colouration of the body; Grant 1932e: 332).

In a list of amphibians and reptiles from Tortola, Anegada, and Virgin Gorda, Grant (1932f) documented the presence of *Eleutherodactylus antillensis* and *'E. portoricensis'* from Tortola. He mentioned having heard *'E. portoricensis'* on nearby St John (Grant 1932f: 340), and did not hesitate to include his claim as a new distributional record for St John (Grant 1932g). Based on his description of Tortolan specimens of *'E. portoricensis'*, which showed little variation among specimens, Grant (1932f: 340) appears to have been the first to support explicitly the notion that two or more species may be included under this name in PR. A note by Grant (1932f: 340) from Tortola is worth reproducing here: 'the voice of another species, resembling that of *E. gryllus* was heard, but the frogs were not located'. Grant (1932f) attributed that sound to a frog, but unfortunately did not specify either the locations or the elevation of sites visited on Tortola. We think that if a remnant population of a small species of frog similar to *E. gryllus* did remain in the 1930s it probably would have occurred in the vicinity of Sage Mountain moist forest (*E. gryllus* from PR occurs in subtropical moist forests to rainforests). However, we can only speculate because the history of land-use on Tortola and of the forest on Mount Sage in particular, between 1900 and the 1940s, indicates that it was 'a mere patch of trees and scrub woodland, a far cry from the original extent of the forest. Surrounding the area were open patches of livestock pasture, crop farming, and fallow and abandoned lands' (Island Resources Foundation 2015: 111).

Thomas Barbour (1935) updated his first list of Antillean reptiles and amphibians published in 1930b. In this update he included *'Eleutherodactylus portoricensis'* from Tortola; *E. cooki* from PR; and *E. cochranae* from St John and Hassel Island (St Thomas). He did not update *E. antillensis* from St John, which was documented by Grant (1932d). Grant (1937) documented a record of *'Bufo marinus'* from Vieques (PR) (presumably introduced sometime after 14 March 1931 [1937: 503], but see Reinhardt and Lütken 1863; section II. A.) and St Croix (presumably introduced in October 1934 [1937: 505], but see Reinhardt and Lütken 1863), and the presence of *'Rana catesbeiana'* (= *Lithobates catesbeianus*) from PR (a roadkill in Río Piedras in 1936). The latter species represents the third introduced species of amphibian documented for the PRA. Grant (1937: 505) also documented a deliberate attempt to introduce two species of anurans into Creque Dam, St Croix. These were *'E. portoricensis'* to St Croix by a Puerto Rican (based on Mr Beatty's 1937 communication to him about an attempt to introduce the coqui frog onto St Croix), and 12 individuals of *Rana* (= *Lithobates*) *pipiens* at 'Criqui [*sic*] dam'. The latter species was the fourth introduced amphibian documented for the PRA; the record for the former species, *'E. portoricensis'*, was invalidated by Grant and Beatty themselves in 1944 when they mentioned that these specimens were *E. antillensis* instead (see section II. C. 1). Grant (1937: 505) said, however, that these species 'cannot yet be included in the fauna'. As far as we know, *L. pipiens* is not currently found on St Croix. Grant added *E. cochranae* from the main island of St Thomas for the first time, but failed to locate it on Hassel Island 'probably on account of the drought' in March 1936 (Grant 1937: 517). He collected a second *'Bufo turpis'* from Virgin Gorda and did not find *Leptodactylus albilabris* on Anegada 'in 1932 or on this trip' (Grant 1937: 520) (this species was first listed for Anegada by Barbour [1917]).

In 1937, Thomas Barbour updated his second list of Antillean reptiles and amphibians published in 1935. The most notable changes for the PRA that were included were the updated geographical distribution of *Bufo marinis* (*sic*) from PR, and of *Eleutherodactylus antillensis* from Culebra and St John. Barbour (1937) provided no justification for several taxonomic changes he made. These were the splitting of *'B. lemur'* into two subspecies, *B. l. lemur* from PR and *B. l. turpis* from Virgin Gorda; the inclusion of *'E. portoricensis'* as a subspecies of *E. auriculatus* from Cuba; and the separation of *E. lentus* from St Thomas and St Croix into four subspecies. One of these subspecies was *E. l. richmondi* from PR. None of these changes remained valid after further taxonomic revisions.

5. *Partial summary*

The Mid-Period covered 33 years (1904–37), and three authors (Stejneger, Schmidt, and Grant) contributed to 46.7% (10/21) of the native species of amphibians currently recognized for the PRA. Adding Barbour to these authors, they delimited the geographical distribution of five native/endemic species: *Eleutherodactylus antillensis, E. cochranae, E. lentus, Leptodactylus albilabris,* and *Peltophryne lemur* (population in southern PR). Danforth, Grant, and McClelland documented the presumably first occurrence of the (intentionally) introduced species *Rhinella marina* in PR, Vieques, and St Croix, *L. fallax* and *Lithobates catesbeianus* in PR, and *L. pipiens* in St Croix. The taxonomic problem of '*E. portoricensis*', '*E. cramptoni*', *E. lentus,* and '*B. lemur*' remained to be solved during the Late Period.

C. The Late Period

The first two periods accounted for the discovery of 15 native species of amphibians and contributed new records of the distribution of several species between 1859 and the late 1930s. The number of herpetological studies documenting new species and distributional records decreased between the mid-1930s and the early 1950s (Fig. 8.2). According to Thomas and Joglar (1996: 184;

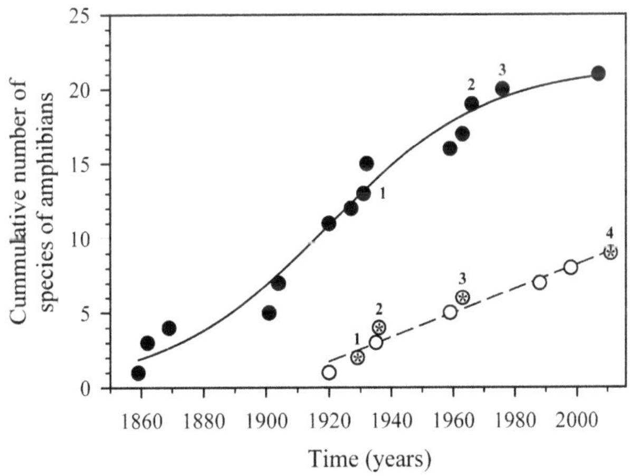

Figure 8.2. Species accumulation curve of the number of native/endemic species of amphibians discovered from the PRA (black-filled circles), between 1859 (*Leptodactylus albilabris*) and 2007 (*Eleutherodactylus juanariveroi*), and the number of introduction attempts of species of amphibians from the first documentation in the PRA (open circles and crossed circles), between 1920 (*Rhinella marina*) and 2011 (*E. johnstonei*). Lines represent the best (sigmoidal) regression fit for native/endemic species,

$$\text{species} = \frac{21.5808}{\left(1+\exp^{\left(\frac{-(\text{Time}-1919.5187)}{25.6632}\right)}\right)}$$

($R_{adj} = 0.9699$, $p < 0.0001$), and the linear regression for introduced species, species = −155.4210 + (0.0819*Time) ($R_{adj} = 0.9550$, $p < 0.0001$). Numbers beside and above filled black circles highlight species extinct (numbers are in order of year of the species' original description): (1) *E. karlschmidti*; (2) *E. eneidae*; and (3) *E. jasperi* (Joglar and Burrowes 1996). Open circles highlight species that became naturalized, and crossed circles with numbers above them highlight species that were introduced but it is believed they did not become established in the Puerto Rican Area: (1) *L. fallax*; (2) *Lithobates pipiens*; (3) *Dryophytes cinereus*; and (4) *E. johnstonei*.

citing the herpetologist Dr Albert Schwartz), the main reason for this pattern was because of the notion that the West Indies in general 'had "limited diversity" of its fauna compared to the mainland [and] they therefore had few species remaining to be discovered'. In fact, more than 80 years earlier, Stejneger (1913: 69) wrote: 'there was consequently some reason for believing that the herpetological fauna might be exhausted, and that at least no conspicuous species would have escaped detection'. The Second World War, however, also must have influenced the hiatus.

Drawing a line that set the boundaries of contributions made between the mid-1930s and the early 1950s may be difficult, but we decided to start our discussion of the third period with contributions made from the 1940s onwards.

The Late Period began in the mid-1940s and was characterized by a combination of significant contributions. First, Grant and Beatty (1944) confirmed the presence of *Eleutherodactylus antillensis* on St Croix (see below) and suggested that this species had been introduced there, a question finally answered in the affirmative by Barker and Rodríguez-Robles (2017). Second, between 1956 and 2005, six endemic species were added to the 'supposedly well-known herpetological fauna' of PR (Schmidt 1927: 2), at a time when 'the finding of a new species of frog in Puerto Rico has seemed very improbable' (Rivero 1963: 1). In addition, this period was characterized by the documentation of distributional records of extant species, the rediscovery of a presumably extirpated species, and the documentation of four introduced species between the 1950s and 2009; three of the last had self-sustained breeding populations in the area. Finally, the number of herpetologists (foreign and local) working in the area increased during this period, and an explosion of studies of amphibians other than on systematic or taxonomic topics was generated from the mid-1960s (Thomas and Joglar 1996).

Unfortunately, this period also was characterized by three species that have not been seen/heard from their historical collection localities in PR (Joglar and Burrowes 1996). These species are *Eleutherodactylus karlschmidti* (Figure 1, plate XV in Rivero 1978), not seen/heard since 1974; *E. eneidae* (Figure 2, plate XVII in Rivero 1978), not seen/heard since 1990; and *E. jasperi* (Fig. 8.3), not seen/heard since 1981 (R. Thomas, pers. comm. to NR-L 2006).

In the VI, populations from several species of amphibians have not been detected from their historical-geographical distribution: *E. schwartzi* (formerly within the '*E. portoricensis*' complex) from St John (Schwartz and Thomas 1975) and *Peltophryne lemur* from Virgin Gorda. The most recent date of collection of *P. lemur* on record is 16 August 1964, from Spanish Town, collected by David C. Leber; Perry and Gerber (2006) declared this species extirpated. Surprisingly, a new population of the endangered *E. lentus* was discovered on Jost Van Dyke, BVI (Perry 2009a), and this record expands its geographical distribution throughout the VI.

Finally, this period includes the introduction of four species of amphibians into the PRA.

1. The 1940s and 1950s

Grant and Beatty (1944) commented on Grant's (1937) documentation of a deliberate attempt to introduce *Eleutherodactylus portoricensis* onto St Croix, and concluded that the introduced specimens were *E. antillensis* instead, clarifying that '*E. portoricensis*' 'was not introduced'. Consequently, Grant and Beatty (1944) documented the record of *E. antillensis* on St Croix and confirmed the presence of *Leptodactylus albilabris* there.

In 1952, the Puerto Rican herpetologist Dr Juan Arturo Rivero-Quintero collected one specimen (a female) of an *Eleutherodactylus* from the lowland forests of Cambalache (northern PR), and named a new species: '*E. ramosi*' (Rivero 1959). He also collected specimens of '*E. cramptoni*' from Cambalache, which later were identified as *E. cochranae* instead (Rivero and Mayorga 1963; see next section). Several authors questioned the taxonomic validity of '*E. ramosi*' (Schwartz and Thomas 1975; Rivero 1978), including Rivero himself (e.g. Rivero and Mayorga 1963). However,

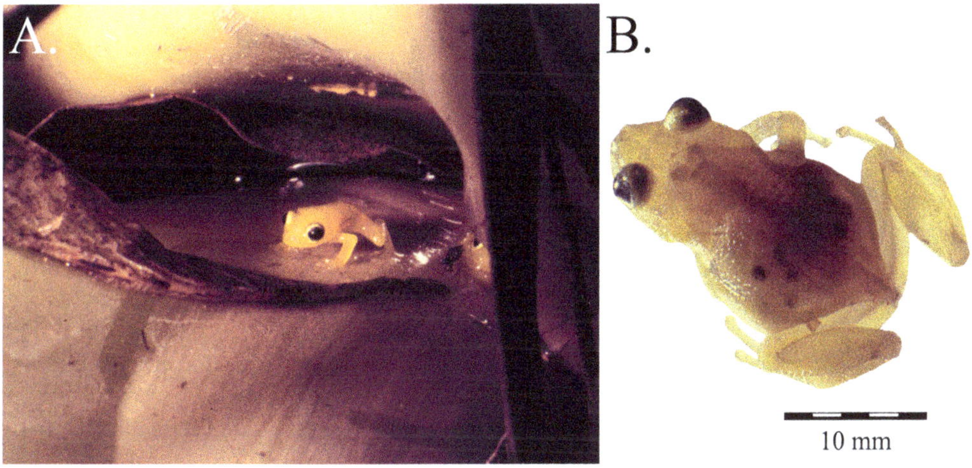

Figure 8.3. (A) Two adults on a leaf axil of a bromeliad and (B) a gravid female of the Puerto Rican *Eleutherodactylus jasperi* (Coquí Dorado, Golden Coquí), the only ovoviviparous frog in the Americas – one of the very few pictures of this species ever taken. The female is illuminated from below to distinguish the eyes and forehead of two embryos visible through the mother's skin, one in each of her flanks. An ecological bromeliad-dwelling specialist, now considered extinct, the Coquí Dorado inhabited a small area of mature forest in the Sierra de Cayey, central-south-eastern Puerto Rico. The female is one of 16 individuals collected for its description (Drewry and Jones 1976), when discovered by George Drewry, Julia R. Clark, and Jasper J. Loftus-Hills during field trips in May and June 1973. Biologist and naturalist José A. Colón, then a research technician at El Verde Research Station in the Luquillo Mountains, northern Puerto Rico, took the original Ektachrome pictures on 2 June 1973. The image in (B) was digitized from the original by José R. Almodovar Rivera, Scientific Instrumentation Specialist at the University of Puerto Rico, Mayagüez Campus, and digitally enhanced by Neftalí Ríos-López.

'*E. ramosi*' remained listed in the accounts of species into the 1970s (e.g. Schwartz and Thomas 1975; MacLean *et al.* 1977), until Joglar and Rivero (1986) settled matters by synonymizing *E. ramosi* with *E. cochranae*.

In 1956, Rivero collected specimens of a new *Eleutherodactylus* from PR, *Eleutherodactylus eneidae*. He considered this species abundant in the highland forests of Villalba, Central Mountain Range, and at El Yunque (Luquillo Mountains; Rivero 1959). According to Joglar and Burrowes (1996) and Joglar (1998), this species was last detected in 1990 from its historical-geographical distribution (from all three major mountain ranges: Central Mountain Range, Cayey Mountains, and the Luquillo Mountains) and presumed it to be extinct (see section III. B. 2).

2. The 1960s and 1970s

From the VI, the naturalist George A. Seaman (1961) added a new geographical record for *Eleutherodactylus lentus* in St John. In addition, he elevated *E. lentus* and *E. portoricensis* to the specific level, rather than the subspecific level recognized by Barbour in 1937. From PR, Rivero (1963) pointed out the tendency to survey localities where amphibians, particularly those found throughout the highlands, were known to occur, and that many potential areas on the island remained unexplored. Consequently, Rivero (1963) discovered the tree-dwelling *E. hedricki*, whose specimens (only males) were collected from El Verde and El Yunque in the Luquillo Mountains in 1962. He also reported this species' egg clutch as being deposited inside a tree trunk and guarded by a male. The species was heard in Toro Negro (state forest), Central Mountain Range.

Rivero and Mayorga (1963) documented the first record of *Eleutherodactylus cochranae* from PR, based on specimens collected by Rivero in 1956. At that time, this species showed a wide distribution in PR and could 'be distinguished from other lowland forms without difficulty' (Rivero and Mayorga 1963: 81). In this work, Rivero highlighted his surprise that this easily distinguished species had not been described from the island by earlier herpetologists.

In addition, Rivero *et al.* (1963) documented the presence of *Eleutherodactylus karlschmidti* at Las Vegas, near Maricao (western PR). The herpetologist Dr Harold Heatwole also collected specimens from Aguas Buenas municipality (east-central PR) (Rivero *et al.* 1963), and the species was collected from the Cayey Mountains in 1963 by Albert Schwartz and in 1965 by the herpetologist Dr John Paul Richard Thomas (at the time an undergraduate student, but became one of the top 25 most productive alpha-taxonomists of reptiles in herpetological history [Uetz 2010]).

In 1966, the Puerto Rican zoologist Dr Julio García-Díaz (1967) documented the rediscovery of the cryptic '*Bufo lemur*' since 1931 (it was thought that the species had been extirpated on PR). William Hast, a student from Isabela, north-western PR, found and collected four specimens of this species from the northern Puerto Rican coast, and gave them to García-Díaz for documentation. With these four specimens, the number of '*Bufo lemur*' collected from PR totalled thirty-two (García-Díaz 1967).

Harold Heatwole and colleagues documented the occurrence of amphibians on islands, islets, and cays associated with PR (eastern PR). Specifically, Heatwole *et al.* (1963a) reported '*B. marinus*' and *Leptodactylus albilabris* from Cayo Santiago, and Heatwole *et al.* (1963b) found '*Bufo marinus*' from Culebra. Heatwole and colleagues' great contribution to the documentation of the geographical distribution of amphibians in the jurisdiction of PR was summarized in 1981 (Heatwole *et al.* 1981), and their data included records until 1972.

In 1966, Richard Thomas described two species included within the '*Eleutherodactylus portoricensis*' complex from the PRA: *E. coqui* from PR, and *E. schwartzi* from Tortola and Virgin Gorda, BVI. Thomas's discovery would end the long taxonomic and geographical problem involving the name '*Hylodes martinicensis*'. Initially, these unknown species were included within '*H. martinicensis*', which was considered as being widely distributed in the Caribbean region (see section II. A). A taxonomic refinement limited these species to *E. auriculatus*, interpreted as being distributed across the Greater Antilles (Stejneger 1904; also see section II. B). Later, those from the PRA were lumped within '*E. portoricensis*' (Schmidt 1927) (Grant 1932f recorded this taxon from Tortola and St John, but see section III. B. 4). Finally, Thomas (1966) separated the complex of '*E. portoricensis*' into three valid species: the already described *E. portoricensis* for PR (the holotype corresponded to the currently valid *E. portoricensis* Schmidt, 1927) and two unnoticed and new species: *E. coqui* from PR and *E. schwartzi* from the VI. Schwartz and Thomas (1975) and Schwartz and Henderson (1988), however, did not find *E. schwartzi* on St John, and listed it tentatively as 'evidently extinct' (= extirpated; see section III. B. 4).

The Cuban treefrog, '*Hyla* (= *Osteopilus*) *septentrionalis*', was introduced into north-western PR (Ramey Air Force Base) (A. Schwartz, pers. comm. to Duellman and Crombie 1970; Schwartz and Thomas 1975). According to Rivero (1998), Frank Juliá was the first to collect specimens from the Aguadilla-Isabela area, north-western PR, in 1959. This species is the fifth introduced amphibian documented for the PR area. Rivero (1998) hypothesized that introduction occurred when pets – brought in by children of the Ramey Base personnel – escaped into the wild.

Heatwole *et al.* (1981) documented new distributional records, but using geographical information documented up to 1972. Their records included *Leptodactylus albilabris* from Algodones Cay, Hicacos, and Piñeros island (all islets off eastern PR); *Eleutherodactylus antillensis* from Piñeros island; *E. cochranae* from Tortola (Philibosian and Yntema [1977: 3] would expand this record as from 'all British Virgin Islands except Anegada'); *E. schwartzi* from Great Dog (BVI), Tortola, and St Croix, but treated the record from St Croix with a tentative 'introduced?' (1981: 52); '*Bufo marinus*' from Algodones Cay; and '*Rana catesbeiana*' from Vieques. Heatwole *et al.* (1981)

did not document *Osteopilus septentrionalis* for PR, and did not include *E. lentus* from St John (but it was included by Seaman [1961]).

Carey (1972) confirmed (by hearing calls) at least one male of *Leptodactylus albilabris* from Anegada, which confirmed Barbour's (1917) record for the species there (Barbour included this geographical record of *L. albilabris* in his 1930b, 1935, and 1937 lists of Antillean reptiles and amphibians). This species was also confirmed from Anegada by Ovaska *et al.* (2000).

Schwartz and Thomas (1975) provided records of *Eleutherodactylus cochranae* from Vieques (PR), Bovoni Cay (St Thomas), and Tortola (BVI), and *E. coqui* from St Thomas and St Croix. Philibosian and Yntema (1976) mentioned that the latter species had been introduced into the VI at least by the mid-1960s as part of the ornamental plant trade from PR. Schwartz and Thomas (1975) documented records of two introduced amphibians: *Osteopilus septentrionalis* into St Croix (verified by Philibosian and Yntema [1976] and the establishment of a breeding population reported by Philibosian and Yntema [1978]); and '*Hyla cinerea*' (= *Dryophytes cinereus*) into north-western PR (Joglar [2005a: 53] timed this introduction as occurring in 1963, but Rivero [1978] noted that Frank Juliá was the first person to collect this species from Guajataca, northwestern PR, in 1954 [MCZ A-28275]). *Dryophytes cinereus* thus became the sixth introduced amphibian documented for the PRA (Thomas and Joglar [1996] and Rivero [1998] suggested that this species may no longer exist in PR). Schwartz and Thomas (1975) and Schwartz and Henderson (1988) did not list *E. schwartzi* in St John and treated it with a tentative 'evidently extinct'. They also removed records of *E. schwartzi* from St Croix and included *Eleutherodactylus lentus* from St John (later, Heatwole *et al.* [1981: 41], would include *E. schwartzi* from St Croix with a tentative '[introduced?]' and exclude *E. lentus* from St John).

Schwartz and Thomas (1975) synonymized several historical taxonomic records and changed the names to those currently valid: '*Bufo lemur turpis*' from Virgin Gorda (Barbour 1937) to '*B. lemur*' of Cope (1868) (Heatwole *et al.* [1981: 40] listed '*B. turpis*' and '*B. lemur*' as distinct species); *E. l. richmondi* from PR (Barbour 1937) to *E. richmondi* of Stejneger (1904); and '*E. cramptoni*' from PR (Schmidt 1920) to *E. locustus* of Schmidt (1920) from PR. Schwartz *et al.* (1978) updated Schwartz and Thomas's (1975) checklist and documented a new geographical distributional record for *Leptodactylus albilabris* from Isla de Ramos, eastern PR (presumably based on material collected by Heatwole).

George E. Drewry (at the time an ecologist and herpetologist for the Puerto Rico Nuclear Center at the University of Puerto Rico) and colleagues collected a most unusual *Eleutherodactylus* (*E. jasperi*) from the Cayey Mountains in 1973 (Drewry and Jones 1976). In their abstract, however, they had its distribution incorrectly labelled as south-western instead of south-eastern PR. This species was a bromeliad-dwelling specialist and the first (so far) ovoviviparous frog known from the Americas (Drewry and Jones 1976; Wake 1978; Fig. 8.3). This discovery increased the number of native/endemic species of amphibians in the PRA to 18, although MacLean *et al.* (1977) inadvertently missed *E. jasperi* on their checklist. *Eleutherodactylus jasperi* was listed as 'Threatened' under the US Endangered Species Act and 'Critical Habitat' was designated in 1977 (US Fish and Wildlife Service 1984). Schwartz *et al.* (1978) included *E. jasperi* in the first supplement of Schwartz and Thomas's (1975) checklist of West Indian amphibians and reptiles. The US Fish and Wildlife Service (1984: 3–4) reported that *E. jasperi* 'occupies a total habitat area of approximately 24 hectares (G. E. Drewry pers. comm.)' and that 'the species is threatened due to past and potential loss of habitat, its obligate bromeliad-dwelling mode of existence, its presumed low reproductive rate, the potential for over collection, and an apparent inability to disperse' (citing Drewry and Jones 1976). Joglar and Burrowes (1996) and Joglar (1998), however, reported that this species was last detected in 1981 from its historical-geographical distribution in PR (Cayey Mountains), and presumed the species to be extinct.

For the VI, Philibosian and Yntema (1977) documented an extension of geographical range for *Osteopilus septentrionalis* to St Thomas. Philibosian and Yntema (1978) added a distributional record for *Eleutherodactylus cochranae* from Thatch Cay (St Thomas), and Rivero and Joglar (1979) newly recorded *E. cochranae* and *E. coqui* from Culebra.

3. *The 1980s and 1990s*

During the 1980s, the rate of the number of studies documenting new species of amphibians in the PRA decreased (Fig. 8.2), but small-bodied species or genetically cryptic species of amphibians may still remain unknown in this region (R. Thomas, pers. comm. to NR-L 2006). As mentioned in section II. A., Pregill (1981) resurrected the name *Peltophryne lemur*, although Schwartz and Henderson [1988: 46; 1991: 157] used '*Peltaphryne*', and considered the name '*Bufo lemur*' as its synonym (see next paragraph). Dr William P. MacLean (1982) published his checklist of reptiles and amphibians of the VI and confirmed the species' geographical distribution documented by Schwartz and Thomas (1975). In addition, MacLean (1982) added *Leptodactylus albilabris* and *Eleutherodactylus cochranae* as occurring on Virgin Gorda; '*Bufo marinus*' on Tortola; and *Osteopilus septentrionalis* on St Thomas, with a detailed description of the pattern of spread into St Croix (between 1974 and 1975, probably escapees from PR, [1982: 20]) and St Thomas (probably in horticultural products late in 1975). MacLean (1982: 20) also predicted the spread of *O. septentrionalis* in St Thomas, and its colonization of St John and the BVI 'probably with man's unwitting assistance'. In 1987, the US Fish and Wildlife Service listed *Peltophryne lemur* as 'Threatened' mainly because of human activities leading to a reduction in the area of its habitat (US Department of the Interior 1987).

Notably, Hedges (1996: 100), based on Hedges *et al.* (1992), questioned the monophyly of Caribbean toads included in *Peltophryne,* which he reckoned 'does not warrant generic recognition', and synonymized *Peltophryne* back to *Bufo*. Several authors followed Hedges' in this (Powell *et al.* 1996; Pramuk 2000; Pramuk *et al.* 2001: Pramuk 2006; Powell and Henderson 2009), other authors used *Peltophryne* (e.g. Thomas and Joglar 1996), and yet others were cautious about its taxonomy and used '*Bufo* (*Peltophryne*) *lemur*' instead (e.g. Perry and Gerber 2006: 241). Alonso *et al.* (2012) used molecular phylogenetic and phylogeographical analyses and found support for the monophyly of *Peltophryne,* which validated Pregill's (1981) resurrected *Peltophryne* currently used for Caribbean toads (e.g. Hedges *et al.* 2019).

The 1980s and 1990s also yielded new distributional records for several amphibians in the PRA. Mayer and Lazell (1988) documented new distributional records for *E. antillensis* from the BVI (Frenchman's Cay and Guana Island) and for *E. lentus* from the USVI (Hassel Island, St Thomas). Reports of geographical distributions of amphibians in the PRA remained unchanged by Schwartz and Henderson (1991). Powell *et al.* (1996: 51) expressed geographical distributions in a widely general format, which prevented identifying localities at a smaller spatial scale, e.g. PRB = Puerto Rican Bank (including Mona Island and the VI) (Powell and Henderson [1999] generated an addendum to the checklist of Powell *et al.* [1996], but did not change the format for geographical names used in their text).

In PR, Joglar and Ríos-López (1995) and Joglar *et al.* (1998) documented new localities for *Osteopilus septentrionalis*. Richard Thomas collected *Scinax rubra* (= *S. ruber*) from Carolina in 1988, but did not publish this record (collected at the Ciudad Deportiva Roberto Clemente, a lowland area of approximately 137 ha in northern PR [Thomas and Joglar 1996; R. Thomas, pers. comm. to NR-L 2006]).

In 1997, the guajón, *E. cooki*, was listed as 'Threatened' based on its limited and small geographical distribution (US Department of the Interior 1997). Joglar (1998: 36) mentioned that *E. coqui* was also introduced to Vieques, but cited Rivero and Joglar (1979; see above) as the reference for this record. The record of *E. coqui* from Vieques was in error because Rivero and Joglar (1979) did not mention Vieques in their publication. Note that Schwartz and Henderson (1991) did not include the record of *E. coqui* from Vieques as part of this species' geographical distribution either. Other authors, however, did include this record: Mayer (2012: 140) included it with a tentative 'Introduced?' but used Powell *et al.* 2011 as one of his references, which in turn cited Joglar 1998 as their reference for Vieques; Henderson and Powell (2009) included Vieques but did not cite the reference for this record. Therefore, published claims of the distribution of *E. coqui* in Vieques included inaccuracies and second-hand information, which were only clarified herein (see section II. C).

Ríos-López (1999) documented breeding populations of '*S. rubra*' in a coastal herbaceous wetland in Toa Baja, northern PR and Richard Thomas (pers. comm. to NR-L 2006) found this species in Dorado, further to the west. Notably, the geographical distribution of *S. ruber* in PR was not included in the checklist of the West Indian amphibians and reptiles published by Hedges and Thomas (1989; evidently not in Schwartz and Henderson [1991], nor in Powell *et al.* [1996]). Rivero (1998) did document this species from PR, with it thus becoming the seventh introduced species of amphibian documented for the PRA (it was restricted to the Bayamón–Carolina municipalities at the time). Finally, Ríos-López and Joglar (1999) documented PR as a distributional record for '*Rana* (= *Lithobates*) *grylio*', making it the eighth introduced species reported for this region (evidently the records for *S. ruber* and *L. grylio* from the PRA were not available for Powell and Henderson's [1999] addenda to Powell's *et al.* [1996] checklist).

4. From 2000 to the present

In 2000, Ovaska *et al.* documented distributional records for four species in the BVI: *Leptodactylus albilabris* from Beef Island; *Eleutherodactylus antillensis* from Great Camanoe, Great Thatch, and Beef islands, and confirmed it from Jost Van Dyke; *E. schwartzi* from Beef Island, Frenchman's Cay, and Jost Van Dyke; and *E. cochranae* from Jost Van Dyke and Great Thatch (these authors did not detect this species on Virgin Gorda whereas MacLean [1982] had listed it from there).

In 2000, Vega-Castillo documented new localities for *Eleutherodactylus cooki* in north-eastern PR and from non-traditional microhabitats – crevices and grottoes in rocky creeks – at several locations in the municipalities of Humacao and Las Piedras (eastern PR). In these creeks, the species was found only in patches of granodiorite boulders in the streambed. Future studies extended the geographical distribution of this species further west-north-west of its historical distribution.

Powell and Henderson (2003) included '*Scinax rubra*' and '*Rana grylio*' in the second set of addenda to the annotated checklist of West Indian amphibians and reptiles by Powell *et al.* (1996). The wide distribution of *Lithobates grylio* on the northern coastal plains of PR led Joglar (2005a: 53) to suggest that this species must have been introduced onto the island in the 1950s or 1960s. For the first time in the herpetological literature, Lever (2003: 182) documented new geographical records for *Osteopilus septentrionalis* from St John (USVI) and Necker Island (BVI), and quoted '(J. D. Lazell, Jr. personal communication to Meshaka 1994)' as the source of these records. However, Waddle *et al.* (2005: 333) provided a distributional record for *Osteopilus septentrionalis* from St John and referred to it as a 'first record for St. John'. Likewise, Owen *et al.* (2005: 76) obtained records for *O. septentrionalis* from several islands of the BVI and referred to them as first records for these islands: Tortola (collected in 1990); Necker Island (collected in 1993); Beef Island (collected in 2002); and Virgin Gorda (collected in 2003). Owen *et al.* (2005: 76) mentioned that 'interviews with BVI residents indicate a decline in native frog populations as the CTF [Cuban Tree Frog] increases in numbers', but estimates of population size were not included. Owen *et al.* (2006) documented *O. septentrionalis* from Peter Island, BVI, for the first time, and provided a detailed account of the first collection of specimens and pattern of spread of this species into Tortola, Necker Island, Beef Island, and Virgin Gorda (BVI).

In 2005, Ríos-López and Thomas collected a new herbaceous wetland dweller, *Eleutherodactylus juanariveroi*, from the coastal plains in PR (Ríos-López and Thomas 2007), making it the twenty-first species of native amphibian documented for the PRA. As with *E. jasperi*, a combination of limited reproductive output, small area of occupancy, and a series of threats to the species' wetland habitat from human activities led the US Fish and Wildlife Service to list *E. juanariveroi* as Endangered in 2012 (US Department of the Interior 2012).

Perry and Gerber (2006) declared *Peltophryne lemur* extirpated on Virgin Gorda. Perry and Platenberg (2007) documented *Osteopilus septentrionalis* and *Eleutherodactylus* spp. (unidentified coqui frog) from Little St James, USVI, but establishment of these species could not be confirmed. Hedges and Heinicke (2007) supported Heyer's (1978) conclusion that *Leptodactylus dominicensis* is a synonym of *L. albilabris*. As a result, the description of *L. dominicensis* by Cochran (1923), collected at Las Cañitas, Hato Mayor province (formerly part of El Seibo province), in the Dominican Republic, represents the first record of *L. albilabris* out of the PRA. Henderson and Powell (2009) updated the work by Schwartz and Henderson (1991); extraordinarily, they provided an account of amphibians and reptiles from the West Indies, including a review of the best scientific information available on the natural history of each species at the time. The geographical distribution of all amphibians in PRA remained unchanged in their checklist of species.

Gad Perry contributed two new geographical records of amphibians from the BVI: *Eleutherodactylus lentus* from Jost Van Dyke (Perry 2009a) and *Osteopilus septentrionalis* from Guana island (Perry 2009b) (the locality of the former was included by Henderson and Powell [2009], who cited Perry as pers. comm.). Whether the presence of *E. lentus* in the BVI is natural has important implications for conservation (see Perry 2009a). Herrera-Giraldo (2010) documented the occurrence of *O. septentrionalis* on Vieques (PR) in the eastern parcel of the Vieques National Wildlife Refuge of the US Fish and Wildlife Service; he presumed this species to have been dispersed in military cargo from other US naval installations in the Caribbean.

Alcides L. Morales documented *O. septentrionalis* (2011a) and *Scinax ruber* (2011b) from the southwestern tip of PR. Perry and Gerber (2011 [2006]; 124, addendum) documented *Eleutherodactylus antillensis* to have been introduced intentionally into the privately owned Necker Island (BVI) by the island's owner. These authors also highlighted the presence of *E. johnstonei* in Tortola (2011: 124, addendum). They mentioned that Alejandro Sanchez (pers. comm. to these authors) 'found several males singing at night on the grounds of a restaurant at sea level' on Tortola in 2008, and thought that this 'seems to point at an introduction in plant produce or the like' (2011: 124, addendum). If proven established, this record would represent the ninth introduced amphibian in the PRA. Sánchez Muñoz (2011: 186) also documented *E. johnstonei* in PR; one male heard calling and 'perched about 75 cm above the ground on the upper-side of a leaf of an exotic garden bush' at a local hotel. Sanchez Muñoz (2011) asserted that 'with this record, the number of introduced amphibian species in Puerto Rico' increased; but there is no evidence of any further introductions nor its establishment (see Dundee 1991 for a discussion about when an 'introduction' may not be an introduction).

Mayer (2012) generated the most recent checklist of the herpetofauna in the PRA, and included species accounts from each island, islet, and cay for which herpetological data exist. He documented new distributional records of *Rhinella marina* from Mona Island and Isla Magueyes (PR); *Lithobates catesbeianus* – with no evidence of its establishment – from St Croix; *Osteopilus septentrionalis* from Inner Brass Island and Hassel Island (USVI), and Moskito Island (BVI); and *Eleutherodactylus schwartzi* from Little Thatch Island (BVI). In his list for the main island of PR, however, he omitted the Puerto Rican endemics *E. coqui*, *E. cooki*, and *E. eneidae*.

In 2013, Greenhawk documented the extension of the geographical range of *Eleutherodactylus cooki* from the municipality of Patillas (PR) westward from the species' historical geographical boundary. Furthermore, Rivera-Lebrón (2014) published a new distributional record for this species in the municipality of Juncos (PR), which was farther to the north-west of the species' previously known distribution. Both studies documented the species from microhabitats similar to those described by Vega-Castillo (2000): patches of granodiorite-boulders in creeks. The studies of Vega-Castillo (2000), Greenhawk (2013), and Rivera-Lebrón (2014) extended the historical distribution of *E. cooki* peripherally within PR.

Ríos-Franceschi *et al.* (2016) extended, by hearing, the known geographical distribution of *Leptodactylus albilabris* to Luis Peña Cay, west of Culebra (PR). Barker and Rodríguez-Robles (2017) corroborated the introduced status of *Eleutherodactylus antillensis* on St Croix. Using molecular data, they concluded that those populations originated either from eastern PR or from the adjacent western VI. As discussed earlier, evidence of the presence of *E. coqui* on Vieques is herein finally presented: the species is restricted to the urban area in the town of Isabel Segunda (ARP-R, pers. obs.). In 2018, Morales-Pérez *et al.* (2022) found a second population of *E. juanariveroi* in north-western PR (NR-L, ARP-R, SIV-C, E. Rivera-Ocasio, and DD-C, pers. obs.; see section VIII).

D. Summary of Part II

The contributions of the authors mentioned above revealed that the amphibians in the PRA are represented only by Anura and consist of 30 species, 21 of which are endemic/native, including three species presumed extinct (*Eleutherodactylus eneidae, E. jasperi,* and *E. karlschmidti*), and three introduced species that seem to have been subsequently extirpated throughout the PRA (*Dryophytes cinereus* [Rivero 1998], *Leptodactylus fallax* [Rivero 1998], and *Lithobates pipiens* [Grant 1937]). The rate of discovery of native/endemic species has declined in recent years, while the rate of non-native species has increased steadily (Fig. 8.2). The batrachofauna in the region consists of five families (Table 8.1; Figs 8.4–8.9): Eleutherodactylidae (20 species), Hylidae (3 species), Bufonidae (2 species), Ranidae (3 species), and Leptodactylidae (2 species).

These islands' geological history, their variety of life zones, and the diversity of habitats have led to the evolution of species adapted to petricolous environments, tropical-montane humid forest, rainforest, lowland limestone humid forest, dry forest, herbaceous dryland, herbaceous wetland, riparian systems, and bromeliads, as well as 'eurytopic' species occupying a range of habitats modified and created by humans. Table 8.1 lists all 30 species, highlights their geographical distributions, and includes the authors who provided documentation. Still pending is the verification of the establishment of breeding populations of: *Eleutherodactylus johnstonei* on PR and Tortola; *E. schwartzi* on St Croix (USVI); *E. lentus* on Jost Van Dyke (BVI); *Lithobates catesbeianus* and *L. pipiens* on St Croix (USVI); *Rhinella marina* on St John (USVI); and *Osteopilus septentrionalis* on Necker Island and Guana Island (BVI).

Works other than taxonomic were provided by Rivero (1978, 1998), Joglar (1998), and Joglar *et al.* (2005), who reviewed most of the literature on the natural history, breeding biology, and ecology of Puerto Rican anurans; provided taxonomic keys, pictures, and drawings (a wonderful addition to these works is a guide to the identification of amphibians and reptiles of the West Indies [exclusive of Hispaniola] by Schwartz and Henderson [1985]); and contributed to the knowledge of the conservation needs of most of the species. Two essential references for the amphibians of the VI are Carey (1972) and MacLean (1982). The former summarized and provided valuable field data on the herpetofauna of Anegada. The latter provided general information on the herpetofauna of the USVI and BVI, also including the islands of Vieques (PR) and

Figure 8.4. Species of amphibians in the PRA (see Table 8.1): (A) *Eleutherodactylus antillensis*; (B) *E. brittoni*; (C) *E. cochranae*; (D) *E. cooki* (male guarding an egg clutch); (E) *E. coqui*; (F) *E. gryllus*. Pictures A–C, F courtesy of Luis J. Villanueva; picture D by Alberto R. Puente; picture E by Neftalí Ríos.

Culebra (PR). MacLean (1982) also included a taxonomic key, a general description of morphology, and geographical distribution, as well as pictures and drawings of each species known to occur on these islands at the time of publication. The works of Carey (1972) and MacLean (1982) were expanded by Rice *et al.* (2005) and Platenberg and Boulon (2006) for the USVI, and by Perry and Gerber (2006) for the BVI from the perspective of the need for conservation of the herpetofauna in the VI. Finally, Henderson and Powell (2009) updated most of the information on the natural history of amphibians in the Caribbean (the PRA included), and treated topics such as species' interactions, population size, use of habitat, and reproductive biology, among others.

Figure 8.5. Species of amphibians in the PRA (see Table 8.1): (A) *Eleutherodactylus hedricki* (male); (B) *E. juanariveroi*; (C) triple egg clutch of *E. juanariveroi*; (D) *E. lentus* (male); (E) *E. locustus*. Pictures A, D, and E courtesy of Alejandro J. Sánchez; picture B by Alberto R. Puente; picture C by Neftalí Ríos.

III. Major threats to anurans in the Puerto Rican Archipelago

Declines of amphibian populations are a worldwide phenomenon (Wake 1991; Heatwole 2013; AmphibiaWeb 2018). Certainly, anthropogenic processes have influenced the distribution and composition of assemblages of amphibians in the PRA in the short term (e.g. from decades to a few hundreds of years compared with geological time), and factors such as the modification or destruction of habitat and the effect of pathogens, introduced species, and climatic change are major threats (Hedges 1993; Blaustein and Wake 1995; Pounds *et al.* 2006; Heatwole and Wilkinson 2009, 2011; Bucciarelli *et al.* 2014). Without estimates of abundance, occupancy, or

Figure 8.6. Species of amphibians in the PRA (see Table 8.1): (A) *Eleutherodactylus monensis*; (B) *E. portoricensis*; (C) *E. richmondi*; (D) *E. schwartzi*; (E) *E. wighmanae*; (F) *E. unicolor* (female). Picture A by Alberto R. Puente; picture B by Neftalí Ríos; pictures C and E courtesy of Luis J. Villanueva; picture D and F courtesy of Alejandro J. Sanchez.

distribution, herpetologists have frequently had a difficult time distinguishing between species of amphibians that are naturally rare and those deemed rare because of anthropogenic threats (Pechman *et al.* 1991; Pounds 2001; Blaustein and Kiesecker 2002; Pimenta *et al.* 2005; Stuart *et al.* 2005). Even more difficult has been identifying and providing general plausible explanations for the apparent reduction in abundance of amphibians, particularly from localities within historical collecting sites (Fournier *et al.* 2019). Certainly, many factors and variables act synergistically, so that identifying the relative contribution of threats on current patterns of distribution and abundance of species of amphibians is difficult (Blaustein *et al.* 2011; Settele *et al.* 2014; Gervasi *et al.* 2017).

Figure 8.7. Species of amphibians in the PRA (see Table 8.1): (A) *Leptodactylus albilabris* (female); (B) foam nest with larvae (arrow) of *L. albilabris*; (C) *Peltophryne lemur* (male); (D) *P. lemur* (female); (E) detail of a mosaic on a retaining wall at the edge of road PR-2 in Quebradillas, north western PR (18.470591° N, 66.933827° W), highlighting the iconic Puerto Rican crested toad, *P. lemur*. Picture A–C and E by Neftalí Ríos; picture D by Alberto R. Puente.

In the following sections, we present and discuss the available data on the most common threats to the amphibians in the PRA. We also provide a detailed description of the history of uses of the land in this region, for two reasons. First, much of the data on the early conditions of the cover of land (e.g. forests and their age and types of soil) in the area is found in non-herpetological literature that, unfortunately, is cited rarely either by local or visiting herpetologists. Second, the PRA in general – and PR in particular – exhibits a pattern of regeneration of woody vegetation and emergent forests that is unparalleled in the Caribbean region, and as such, best illustrates the outcome of secondary succession that renders the understanding of

Figure 8.8. Species of amphibians in the PRA (see Table 8.1): (A) a juvenile of *Peltophryne lemur* showing dorsal coloration pattern; (B) a juvenile of *Rhinella marina* showing dorsal coloration pattern; (C) *R. marina* (female); (D) female *R. marina* with detail of its paratoid gland (arrow); (E) *E. johnstonei* (male); (F) *Scinax ruber* (female). Picture A–D and F by Neftalí Ríos; picture E courtesy of Alejandro J. Sanchez.

the abundance and distribution of amphibians in 'forested areas' difficult. Moreover, the type of land-cover designated as 'forest' is frequently used to refer to primary, secondary, and novel forest conditions indistinctly, which raises an issue for the status of amphibians (and presumably other taxa; e.g. Ríos-López *et al.* 2016) in assessments for conservation purposes. For example, what is considered to be 'mature forest' today (regardless of how different authors define it) may not fulfil all ecological requirements of the species of amphibians that, many decades ago, inhabited the 'primary forest' that presumably existed in that landscape. Efforts for the

Figure 8.9. Species of amphibians in the PRA (see Table 20.1): (A) *Lithobates grylio* (adult); (B) *L. catesbeianus* (female); (C) detail of toes in *L. grylio*, with the fourth toe webbed to its tip; (D) detail of toes in *L. catesbeianus* with the fourth toe not webbed to its tip; *Osteopilus septentrionalis* (females) from PR showing light-cream coloration (E), yellowish-brown coloration (F), and green coloration (G). Pictures A–F by Neftalí Ríos; picture G courtesy of Jesús D. Quiñones.

conservation of amphibians (and other taxa) can benefit by integrating knowledge from fields such as succession ecology (e.g. Davis *et al.* 2001), and the history of land-use and the changes in land-cover that result from socioeconomic dynamics (e.g. Vitousek *et al.* 1997; Steffen *et al.* 2007) to distinguish between historical and biogeographic patterns of present-day abundances and distribution of species.

In this discussion, we also wanted to be as comprehensive as possible for the benefit of current and future herpetologists and ecologists interested in amphibians as models. The information and data on threats to the amphibians in the PRA are contrasted with the available evidence used to assess the conservation status and conservation listings at local, state, and international levels.

A. Environmental modification in relation to the distribution and abundance of amphibians

1. History of land-use

The islands of the PRA are believed to have been virtually forested prior to the arrival of Europeans in the late fifteenth century. Thereafter, a gold-mining economy dominated the first third of the sixteenth century in PR. By the late sixteenth century, a shift in the colonial economy went from mining to agriculture. For instance, a high demand for sugar in Europe promoted a monoculture economy of sugarcane, which was introduced and first planted in 1523 in San Germán, south-western PR (Domínguez-Cristóbal 2000). In the town of San Germán, and later elsewhere, sugarcane plantations required extended areas of deforested land, particularly in the lowlands towards the coast. Intense harvesting of forests for extraction of valuable timbers accompanied the monoculture of sugarcane. Introduced cattle, sheep, and other domesticated animals were soon kept free-roaming in forested areas to be harvested as needed, and some species became feral, most notably pigs and goats, which defoliate shrubs, saplings, and seedlings. Coffee was introduced in 1736 and by the mid-eighteenth century it was considered the most important agricultural product for the Puerto Rican economy (Domínguez-Cristóbal 2006).

In 1797, the island of PR was visited by a French botanist, André-Pierre Ledru, who described the flora and the general landscape (Ledru and Sonnini de Manoncourt 1810), highlighting the rich, varied, and dense vegetation and abundant forests on PR and Vieques. Clearly, Ledru was captivated by the fecundity of the soil and the richness of the forests (Domínguez-Cristóbal 2000). Ledru's fascination with the forests in the Central Mountain Range led him to describe Aibonito (~600 m asl) as follows (Ledru and Sonnini de Manoncourt 1810: 66–68, ch. 22):

> I could not find in Fajardo suitable accommodation for the naturalists of the expedition; and I left this village on the eleventh of November, accompanied by a guide whom Mr. José furnishes me with; but instead of following the ordinary road which leads to San Juan, I took the forest path on the left, in order to approach me to the high mountains of Aibonito, rich in waterfalls, picturesque sites, and the precious trees found there. After five hours of walking, I arrived at their feet. My guide walked forward into the forest, leading our two horses in one hand; I followed him, moving away from there to gather flowers, and I frequently stopped to admire the beauties of these wild places.
>
> What a delicious freshness under these arcades of greenery! The path we followed was lined with trees of extraordinary height and volume [...] I hear the cries of parrots [presumably the Puerto Rican parrot *Amazona vittata*] who abandon the plain and take refuge in the forest.

Ledru and Sonnini de Manoncourt (1810: 174) lamented that 'agriculture, therefore, is far from the state of perfection which it might attain' (also see Flinter [1834: 11], about PR's 'depressed state of agriculture') and added (1810: 175):

> The lack of practical roads, bridges and locks. If the stagnant waters had an outlet, if the island had been crossed by extensive paths that, by facilitating the transportation of goods, would contribute to the circulation of air, [...] Puerto Rico, one of the colonies more fertile in the New World, it would soon become one of the healthiest and richest.

During the nineteenth century, PR experienced a series of socioeconomic and political crises with Spain (Domínguez-Cristóbal 2000). The human population on PR was estimated at 45,000 in 1765 and had more than tripled by 1800 to ~155,000 (Wadsworth 1950). Population growth increased rapidly to nearly 800,000 inhabitants, and a dramatic unregulated deforestation and burning of wood to clear soils for agriculture peaked during the second half of the nineteenth century (Wadsworth 1950; Domínguez-Cristóbal 2000). The maintenance of a nomad-style of land-use because of ineffective agricultural practices worsened the situation (Domínguez-Cristóbal 2000). The cover of forested vegetation diminished dramatically, from almost 100% cover from the beginning of the nineteenth century to 20.4% by the end of that century (Wadsworth 1950; Birdsey and Weaver 1987).

The end of the Spanish-American War in 1898 was followed by a series of soil surveys on PR. Geologist Robert Thomas Hill (1899: 16–17), from the US Geological Survey, described the soils and forests on PR as follows:

> Owing to this soil, the mountains were originally wooded and are now cultivated to their very summits, verticality of slope presenting no obstacle to cultivation in the minds of the natives. The writer has seen the steepest possible slopes cultivated to the highest degree in coffee and tobacco; in fact, the most productive crops of this character are grown upon declivities upon which the American farmer would not risk limb and life.

What Ledru and Sonnini de Manoncourt (1810) described from the Aibonito's forested landscape a century earlier, was now converted to a desolate and deforested landscape (Fig. 8.10).

In 1887, PR's population was 806,708 (Hill 1898) but by 1899 had increased to 953,243 (data from the first American census made by the US War Department in 1899, cited by Miller 1922; see also Wilcox 1902). By 1910, the population had grown to 1.1 million; in the words of Murphy (1916: 22): 'Santo Domingo has a population density of 33 per square mile, Cuba 46, and Porto Rico 325. There is little wonder that Porto Rico is nearly deforested.' Murphy (1916: 23) concluded that a mere 2% of the highland primary forests remained at the time of his visit to PR. Others estimated that between 0.2% (Wiley 1985, 1986) and less than 1% of the original forests were

Figure 8.10. Deforested mountain south of Aibonito (original caption and Plate IV from Hill [1899]).

still unmodified at this time (Snyder *et al.* 1987; Lugo 2004), and fragments of forests in the highlands, e.g. forest stands in Toro Negro in the Central Mountain Range, were used to harvest charcoal and fuel wood (Gilormini 1936; Wadsworth 2012). Between the 1930s and 1940s, during the presidency of Franklin D. Roosevelt in the United States, and with initial support from the Civilian Conservation Corp, the Insular Forest Service, and the Forestry Division of the PR Reconstruction Administration (PRRA), the US Forest Service directed most of its efforts toward the development of tropical forestry internationally and to reforest the landscape of an already over-populated PR with 1.5 million people (Holdridge 1939). Agronomist Mr José Marrero from Utuado was distinguished as 'possibly the most accomplished tree nurseryman and forest plantsman in the American tropics' and was primarily responsible for the reforestation of the largest forest reserve in PR, EYNF (Wadsworth 2014: 19). Initially, these efforts mainly were directed toward the production of timber and for the conservation of soils (Gerhart 1936; Wadsworth 2008, 2012). By the late 1940s, a shift in the Puerto Rican economy, from agriculture to industry, led to rural-to-urban migration and the abandonment of most agricultural lands (Birdsey and Weaver 1987; Dietz 1989), but deforested soils continued their extreme losses in moisture, and critical runoff and erosion limited primary productivity at the time (e.g. Lugo-López 1953). The estimated area under cover-type 'forest' in PR remained low at 6% of the total island by the late 1940s (Birdsey and Weaver 1987; Helmer 2004; Martinuzzi *et al.* 2009) (Fig. 8.11), reaching its lowest value since 1800 (Lugo 2018, 2019).

During the next 60 years, the forests of PR entered a phase of secondary succession due to the combination of abandonment of agriculture, cessation of deforestation, rural-to-urban migration, and assisted and natural reforestation (Figueroa-Colón 1996; Lugo 1996; Kennaway and Helmer 2007) (Fig. 8.12). The transformation of most of the landscape in PR, from agriculture (Fig. 8.12a) to forest (Fig. 8.12b), occurred unexpectedly rapidly, presumably assisted by birds and bats that readily dispersed seeds (e.g. Carlo and Morales 2016), and enhanced by wind-dispersed, pioneer species of trees such as the African Tulip, *Spathodea campanulata* – which was introduced near the beginning of the twentieth century as an ornamental plant (Abelleira and Lugo 2008). Humans planting fruit, shade, and ornamental trees elsewhere also contributed to this transformation (e.g. Marrero 1950). As a result, the land area under cover-type 'forest' had reached 42% by 2000 (Grau *et al.* 2003, Grau *et al.* 2004, Grau and Aide 2007). Many abandoned agricultural areas were converted either to pasture or were replaced by urbanized and industrial lands. Later, many pasturelands were converted to urbanized, touristic, and industrial areas. A new matrix of types of land-cover emerged in the landscape of PR, with remnants of mature forests, young forests, novel forests, urban/industrial areas, agricultural areas, and pasturelands, along with other cover-types of vegetation. Moreover, the human population size in PR more than tripled during the twentieth century and reached 3,725,789 in 2010, but 61% of this population occurred in coastal municipalities (United States Census Bureau 2012).

PR's forests, in general, continue under secondary succession, but at different rates in different areas (Weaver 2000; Lugo 2004; Molina-Colón and Lugo 2006; Marcano-Vega *et al.* 2015; see section III. A. 2). The African Tulip tree is still the most abundant and important species of tree for facilitating succession of novel forests. Since the late 2000s and early 2010s, the percentage cover of forest on PR remained stable above 54% between 2009 (54.7% cover; Marcano-Vega *et al.* 2015) and 2014 (54.8% cover; Marcano-Vega 2017).

The VI experienced a similar process that varied in intensity across islands (Fig. 8.13). In what is currently the USVI, St Thomas was the market centre of the VI but not a centre of agriculture for reasons of topography and for soils prone to runoff and landslides, although fruit trees were popular (Smith 1920). However, Bevan (1940) mentioned that Denmark required extensive extraction of *Guaiacum officinale* ('lignum vitae' in English or '*guayacán*' in Spanish) from St

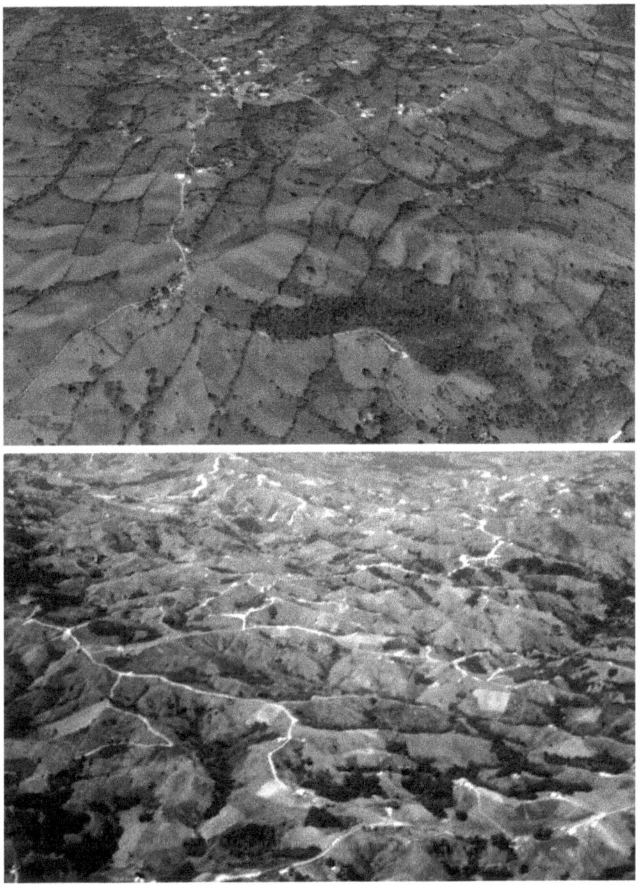

Figure 8.11. Top panel: wide-scale deforestation generates the landscape pattern seen today in Baitoa, Santiago province, Dominican Republic (image from ©Google Earth Pro, approximately at 1 km above the ground). Bottom panel: a similar landscape pattern already seen in the vicinity of the Cayey Mountains, south-eastern PR, in the early decades of the 1900s (Wadsworth 2012: 14). The historical-geographical distribution of the bromeliad-dweller Puerto Rican golden coqui, *Eleutherodactylus jasperi*, was within the Cayey municipality. This species has not been see/heard in its historical collection localities since the early 1980s (Joglar 1998).

Thomas, but that historical records gave no clue of the original composition of the vegetation on this island. Brandeis and Oswalt (2007) estimated the type of forest cover to be 74% on St Thomas. On nearby St John, Bevan (1940: 8) mentioned that 'St John, except for occasional landings, was not inhabited or cultivated to any extent until 1717, when settlers from St Thomas colonized the island. Population increased rapidly and there is every evidence that the island was at one time completely cleared of the original forest.' The emancipation of the slaves by the mid-nineteenth century led to the nearly complete abandonment of the agricultural lands and the beginning of secondary succession leading to forests in St John and St Thomas (Brandeis and Oswalt 2007). By the 1920s, agriculture in St John benefited mostly from the harvest of wild bay leaves (or '*malagueta*' in Spanish), *Pimenta racemosa* (Weaver 2006). Grant (1932e: 331) noted that in St John 'There is no industry and practically no cultivation except where the scattered negro population tends patches of truck.' Bevan (1940: 9) also mentioned that 'forest conditions on St. John today, largely undisturbed, are the direct result of this abandonment and the forest stands should offer an interesting study for the ecologist'. Nearly 75% of the island's land was privately held for

Figure 8.12. Top panel: a view of the landscape at La Plata, Aibonito, eastern Central Mountain Range in PR, in 1938 (from https://www.loc.gov/item/fsa1998023163/PP/). This area was devoted to tobacco plantations, and it is shown here completely deforested to its bare soil. Bottom panel: the same area today, which shows the increase in vegetation cover after abandonment of agriculture since the early 1940s, because of secondary succession (picture by Neftalí Ríos).

recreational purposes, and the abandonment of agriculture led to dense stands of young forests in succession. Since 1956, the mature forests have been holdings of the US National Park Service, after receiving from Laurance Spelman Rockefeller a parcel of approximately 55% of St John (~2,850 ha) of his property for conservation (Weaver 2006). Brandeis and Oswalt (2007) estimated that the type of forest cover is 92% on St John.

The historical nexus with agriculture in the VI was St Croix. Several political maps survive from the Dutch colonization of St Croix in about 1750, and one map includes land-cover and edifices in 1750 (Hopkins 1989). Despite vague uncertainties related to the borders of the island where access was difficult for surveyors, the map is largely similar to the modern digitized topographic forms, extrapolated to the entire VI (Fig. 8.13; Angeli 2017). The old map of land-cover reveals a landscape that was more complex than that of today and covered by nearly 90% subtropical dry and moist forests. Between 1742 and 1754, the acreage of plantations increased (Westergaard 1938). An economic transition from cotton to sugarcane between 1754 and 1800 precipitated the loss of the primary forests (Tyson 1992). In 1917, more than 90% of the native forests and woodlands had been cleared for agriculture or logged for timber (Ward *et al.* 2000). In the 1930s, the forested area reached approximately 16% (Bevan 1940) and the types of forest cover increased with the end of widespread cultivation of sugarcane in the 1940s and 1950s (Seaman 1989). Subtropical

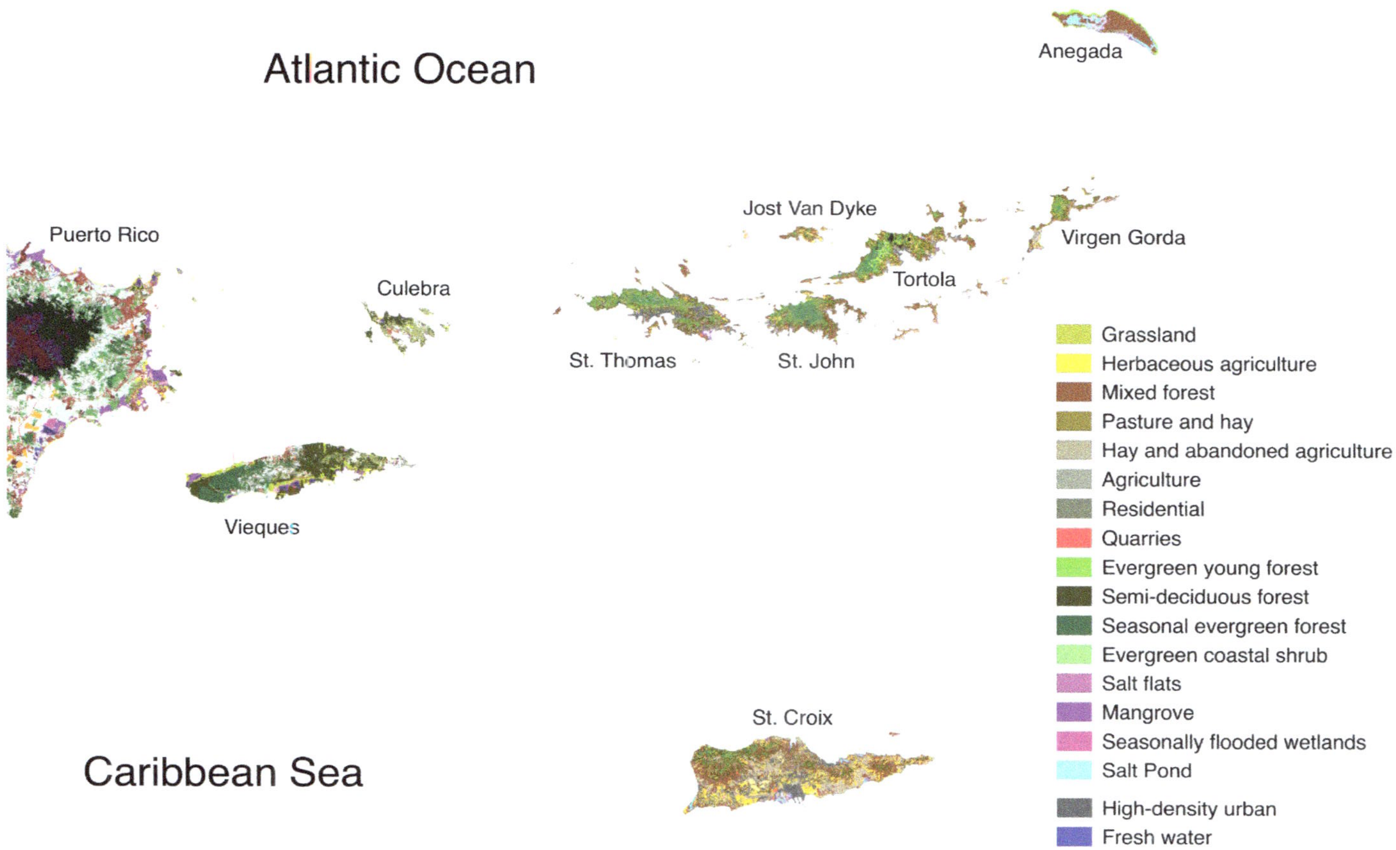

Figure 8.13. Historical landcover of the Virgin Islands including the British and US holdings based on a map from 1750 created by J. Cronenberg and J. von Jaegersber and digitized in Angeli (2017). The green colour scale is darker where denser forests or shrub thicket are found and lighter green in less dense and/or fragmented patches. The brown areas indicate clear-cutting of forests to provide space for houses and buildings.

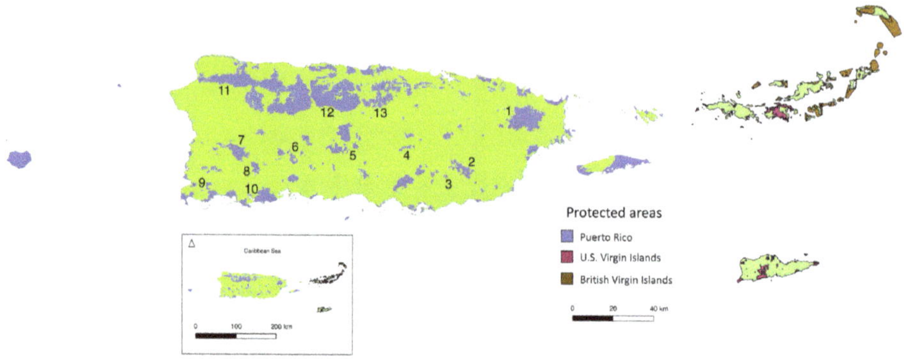

Figure 8.14. Current protected areas of PR and the USVI (colour coded; modified from Protected Areas Conservation Action Team 2018 and Kennaway *et al.* 2008). Also protected are Mona Island and Desecheo (west of PR). In PR, numbers represent the centres of high endemism of plants (see text): (1) Luquillo Mountains, (2) Cayey Mountains, (3) Las Tetas, (4) Torrecilla, (5) Toro Negro, (6) Guilarte-Membrillo, (7) Maricao, (8) Susua, (9) Sierra Bermeja, (10) Guánica, (11) Guajataca, (12) Sabana Hoyos, and (13) Ciales.

secondary forests have returned to St Croix, with naturalized species of trees colonizing formerly agricultural lands and generating novel floral assemblages (Atkinson and Marín-Spiotta 2015). Brandeis and Oswalt (2007) estimated this type of forest cover to be 50%.

The Dutch settled the island of Tortola in 1648. By 1680, the English had control of all of the islands currently recognized as British Overseas Territories, while the Dutch retained what are today the USVI. Until 1834, slaves from Africa imported for agricultural industry comprised the majority of the population at a high of nearly 7,000 occupants over 15 inhabited islands in the British territory in 1844. The population fluctuated around 10,000 people until the early 2000s, when the population increased past 20,000 owing to the diversification of the economy in finance and tourism (Perry and Gerber 2006). The history of land-use on Virgin Gorda is similar to that of St Croix, except that an agriculturally based economy still persists.

The historical deforestation during the agricultural period in PR left only 13 isolated small areas of land that retained the highest species richness of plants (mostly of trees) and endemism (Fig. 8.14) (see also Fig. 2 in Figueroa-Colón 1996). Eight of these areas of vegetation are located in the subtropical lower montane wet forest and subtropical lower montane rainforest life zones, while three areas are located in the subtropical moist forest, and two in the subtropical dry forest life zones. All these small areas became biodiversity refuges and sources of species of plants that assisted in a wide-scale recovery – through secondary succession – of current ecosystems (Figueroa-Colón 1996).

2. Recovery of forests and vegetation during secondary succession

First, we emphasize that the trajectory of the response of an ecosystem to a phenomenon that originates from natural sources (e.g. tropical cyclones, droughts), human sources (e.g. uses of the land), or both, does not necessarily replicate in space and time, nor does it merge with a previous trajectory. Instead, it heads towards a different state during succession, as it is influenced by the rearrangement of the biotic and abiotic components and variables in the landscape (Lugo 2008, 2018, 2019). Furthermore, the rate of the responses of a forest (this as a unit of study of a type of land-cover) depends on the phenomenon and its intensity, frequency, and extent. As a general rule, this rate decreases from the cutting of vegetation, deforestation, and erosion of soils, and to the elimination of soils (A. Lugo, pers. comm. to NR-L 2019). Once in succession, some of the responses of forests include the accumulation of leaf litter, the generation of new soils, the increase in shading from vegetation, and the transformation of local microclimatic conditions. This transformation of microclimatic conditions includes a decrease in air temperature and an increase in humidity in air and soils, as well as a decrease in the coefficient of variation of both climatic variables.

Second, the trajectory of responses of amphibians is influenced by forest succession and includes changes – at the assemblage level – in species richness, species' abundance, and trophic structure in the landscape. Changes in these assemblages occur slowly compared with tangible changes in vegetation, and depend on the specific ecological requirements of the species. Furthermore, the trajectory of responses of amphibians varies with the phenomenon (e.g. deforestation, drought, hurricanes) in different types of land-cover (e.g. mature forest, young forest, grassland, urban areas) and where it has the greatest influences (e.g. uplands, lowlands, hills, valleys). The outcome of these responses during succession is complex (e.g. Ríos-López and Aide 2007; Herrera-Montes and Brokaw 2010; Lugo 2018, 2019; see sections III. B. 3–4). Therefore, the use 'as is' of phrases such as 'increased forest area' and 'increased natural-reserve areas' is not necessarily adequate in assessments for the conservation of species, which justifies herein our discussion of succession to forests and the response of an assemblage of amphibians from PR in the Anthropocene.

In general, most amphibians are philopatric (i.e. 'tend to remain in the location where they were born' [hatched] [Lillywhite 2008: 238]). On PR, 56% of *Eleutherodactylus* (9/16) were frequently found only in highland mature forests (Joglar and Burrowes 1996; Rivero and Seguí-Crespo 1997; Joglar 1998; Ríos-López and Thomas 2007). These species are all endemic and include *E. eneidae*, *E. gryllus*, *E. hedricki*, *E. jasperi*, *E. locustus*, *E. portoricensis*, *E. richmondi*, *E. wightmanae* and *E. unicolor*. Most of these species occur under climatic conditions of low air temperature, high relative humidity, and more precipitation compared with lowland areas; they are ecological specialists that inhabit the interior of forests in the highlands.

In PR, succession to forests occurred with the abandonment of most agricultural lands throughout the island since the late 1940s and 1950s. The emerging forests now have different compositions of tree species compared with the patches of original (e.g. primary) forests (Marcano-Vega *et al.* 2015). However, not all these forests (as a type of land-cover) are similar because the rate and trajectory of responses of forests under succession differ over the landscape, particularly with elevation. For example, it is known that the rate of forest cover is slow to increase on the highest mountain peaks but rapid in the valleys and lowlands. The areas showing the lowest rates of change after deforestation and burning of the vegetation and soils are the elfin forests on the highest mountain peaks in the Luquillo Mountains (Weaver 2000, 2008). We also know that the removal of vegetation rapidly changes local microclimatic variables such as temperature and humidity of the air and in soil; these variables are fundamental to the activity, abundance, distribution, and survival of amphibians (e.g. Heatwole *et al.* 1968; Heatwole and Newby 1972; Heatwole 1982; Heatwole 1983; Hillman *et al.* 2009). Not surprisingly, the influence of different uses of the land on the distribution of species was evident to Schmidt (1928) almost a century ago (see Introduction). However, other phenomena interact and influence the trajectory of responses of amphibians during forest succession and at shorter spatiotemporal scales (see next paragraph).

In 2017, Hurricane María completely removed the canopy cover in most highland forests, which in the short term rapidly transformed the microclimatic conditions in the understorey into conditions typical of open areas (increase in air temperature, decrease in humidity, and increase in variation in both; see section III. G. 4). These new microclimatic conditions in the forests extended over a year and are currently related to changes in the abundance of some endemic and native *Eleutherodactylus* in the landscape of PR. A less obvious phenomenon is drought, which also alters the microclimatic variables that influence the rate and trajectory of responses of amphibian assemblages (see section III. G. 3). The relative contribution of a drought to the trajectory of amphibian responses during succession could be challenging to assess when in combination with other phenomena of human and natural origins (see sections III. B. 3. and III. G. 4. respectively), and among types of land-cover (e.g. mature forest, young forest, shrubland, grassland). In certain instances, however, some combinations of phenomena may lead to a new state that includes the extirpation of a species within an assemblage in the landscape.

In the lowlands, 25% of the native and endemic *Eleutherodactylus* in PR (4/16; *E. antillensis, E. brittoni, E. cochranae, E. coqui*) and *Leptodactylus albilabris* are frequently found (Rivero and Seguí-Crespo 1997; Joglar and Burrowes 1996; Joglar 1998; Ríos-López and Thomas 2007). These species are mostly native, and are ecological generalists associated with environmental conditions of high air temperature, variable relative humidity, and less rainfall than in the highlands. This assemblage of species is also available for examining forest succession in the lowlands, where changes in cover of woody vegetation occur rapidly during succession compared with highland forests – and more so with assisted reforestation (see next paragraph).

Ríos-López and Aide (2007) and Heatwole and Ríos-López (Chapter 1 in this volume) discussed ecological succession and responses of herpetofauna in detail. For example, Ríos-López and Aide (2007) showed that many trees reached 4 m or more in height in about three years after planting in pastures in a lowland limestone valley in PR. This shift in vegetation also influenced changes in temperature and humidity; as vegetation cover increases, values and the coefficient of variation of air temperature and humidity converge to those in adjacent areas used as reference forests. As a rule, these processes increase the connectivity between forest stands that develop in valleys and isolated patches of mature forests at higher elevations. Consequently, species richness of amphibians (and reptiles) increased rapidly in two years after planting, and was facilitated by the proximity of the nearest pool of species for recolonization (i.e. adjacent areas of mature forests in the hills). Changes in the abundance of species was slower and the abundance of species that inhabited pastures decreased (e.g. *Eleutherodactylus antillensis*, highly tolerant of high temperature and low humidity) as the cover of woody vegetation increased. In contrast, species that inhabit mostly forests (e.g. *E. cochranae, E. coqui*, less tolerant of high temperature and low humidity compared with *E. antillensis*), increased as the cover of woody vegetation increased (Ríos-López and Aide 2007). Herrera-Montes and Brokaw (2010) showed that this pattern of association between abundance and distribution of amphibians, and type of land-cover persists in chronosequences that span more than 40 years after the abandonment of agriculture in the lowlands.

To summarize, the response of forests undergoing succession includes the rehabilitation of eroded soils, an increase in the shade, and generation of local climatic conditions such as the lowering of ambient temperature, and increased humidity in soils and air. The expansion of populations of ecological specialists and generalists into these post-agriculture forests in PR depends on the extent of connectivity between the new forests and patches of mature forests, and whether the new biotic characteristics (e.g. of the leaf litter, and of associated assemblages of epiphytic plants) satisfy the ecological requirements of each species of amphibian. This expansion depends strongly on the physiological capacity of a species to cope with the variability in the new microclimatic conditions as well (e.g. Candelas *et al.* 1961; Heatwole *et al.* 1968; Heatwole and Newby 1972; Van Berkum *et al.* 1982; Beuchat *et al.* 1984). In the highlands, for instance, several fragments of forests considered to be biodiversity refuges by Figueroa-Colón (1996) could have facilitated the colonization of some species of amphibians in the new forests within the landscape in these highlands (a metapopulation dynamic mediated by the 'rescue effect'). One of these species is *Eleutherodactylus wightmanae* in Jayuya, Central Mountain Range (NR-L, pers. obs.). At a broader gradient in elevation, however, there are limitations, mediated by physiology, for the colonization of lowland areas by most species adapted to conditions in the highlands. For almost a century of forest succession, Schmidt's (1928) observation of the pattern of elevational distribution of the herpetofauna in PR still holds – species from the lowlands are frequently found in the highlands, but rarely the other way around (see next section).

3. *Expansions of the distributional range of amphibians from the Puerto Rican lowlands to higher elevations*

Figure 8.12 shows the contrast in vegetation after the abandonment of agriculture in a highland area in PR; the upland dispersion of several species of amphibians from the lowlands accompanied the forests in succession throughout this landscape. Following are three examples.

Eleutherodactylus antillensis was documented from the main island of PR in 1920, and despite its generally wide but scattered distribution (Schmidt 1920), this species was not documented from EYNF in the Luquillo Mountains, even though this was a popular forest site for collecting. Today, it is a common and abundant species, not only in the lowland pastures and young forests in succession but up to 750 m asl at EYNF. In this and other highland forests in PR, however, this species is found only along the margins of roads, not in the interior of mature forests in the highest mountain peaks.

Eleutherodactylus cochranae was initially reported from the arid and semi-arid coastal region of south-western PR from specimens collected in 1956 'at the very dry region of Guánica' (Rivero and Mayorga 1963: 81). Currently, it is abundant in herbaceous wetlands at sea level (Ríos-López and Villanueva-Rivera 2013) and in forests within urban areas throughout the landscape. This species also occurs (mostly in bromeliads) at the margins of forests at least up to 650 m asl in the Cayey Mountains and the Central Mountain Range.

Eleutherodactylus brittoni, a species described by Schmidt (1920) and based only on four specimens collected at Maricao (western PR) and EYNF (1920: 180; 'near the Forester cabin', at ~394 m asl). Currently, this species is highly abundant in herbaceous wetlands at sea level (Ríos-López and Villanueva-Rivera 2013), in pastures in most lowland areas (except the arid and semi-arid coastal region), and is frequently found in grasslands and soils densely covered by ferns along the margins of roads in the highland areas across the Central Mountain Range and EYNF. This species also occurs in mature montane forests at 820 m asl in the Cayey Mountains, although at a very low density than its density at other types of land-cover (NR-L, unpublished data).

These examples add to the pattern noted by Schmidt (1928): many species living initially in the lowlands (and living at low densities in mature forests in the highlands) extended their distribution and increased their abundance following changes in the land-cover at higher elevations. Moreover, as connectivity between new forest stands and patches of mature forests in the highlands increased, most species of amphibians from the lowlands overlapped their distribution with that of amphibians living in the mature highland forests in PR. Nonetheless, we see today that the distribution of species from the lowlands is still mostly limited to the new forests, while the distribution of species from the highlands is still mostly limited to the mature forests. Consequently, what Schmidt described from the distribution and abundance of amphibians in PR reflects the legacy of past uses of a land that is still today under succession.

4. *Expansion of a population of an almost extinct frog*

As mentioned earlier, 61% of the 3,725,789 persons in PR (United States Census Bureau 2012) currently occupy coastal municipalities. Not surprisingly, most pressure on Puerto Rican ecosystems are from human activities mainly in those places. This is important because most of the abandoned agricultural lands were transformed into pastures for cattle ranching and for human dwellings and industry. As a result, most of this 'secondary modification of habitat' occurred and is occurring in areas already affected by past agriculture and agriculturally related practices, primarily in coastal areas. One unusual example illustrates the dramatic influence of the abandonment of these lands on a species of amphibian: the Puerto Rican plains coqui (or *'coquí llanero'* in Spanish), *Eleutherodactylus juanariveroi*.

The abandonment of agriculture in a coastal wetland in PR most likely resulted in the persistence and eventual discovery of *Eleutherodactylus juanariveroi* (Ríos-López and Thomas 2007).

This species is the smallest amphibian in PR, and it is a herbaceous-palustrine wetland specialist (Ríos-López and Villanueva-Rivera 2013; Ríos-López *et al.* 2014). It has the lowest breeding output (three eggs per clutch on average) among all *Eleutherodactylus* in the archipelago and uses one species of plant, the bull tongue arrowhead (*Sagittaria lancifolia*), as its only microhabitat for oviposition. This plant is an obligate palustrine species and an indicator of freshwater wetlands throughout PR and continental USA. As far as we knew, *E. juanariveroi* was found nowhere else in PR (but see section VIII.), and it was listed as Endangered by the US Fish and Wildlife Service and the herbaceous-palustrine wetland occupied by this species was designated as a Critical Habitat in 2012 (US Department of the Interior 2012).

Unknown at the time of its discovery, the area of suitable wetland available for this species in the 1930s (approximately 15.2 ha; Fig. 8.15) may have covered approximately 6.1% of the current area (249 ha). Initially, this wetland and adjacent limestone valleys were used extensively for agriculture (Fig. 8.15), the most recent period of high agricultural activity being between 1910 and 1935 (US Naval Security Group Activity Sabana Seca 1999). In 1941 these lands were transferred from the Government of PR to the US Navy, which initially established a naval ammunition depot and a radio receiver facility in 1948 (US Naval Security Group Activity Sabana Seca 1999). The military base then ceased all agricultural activities on its property, particularly in the limestone hills and valleys and the adjacent freshwater wetland farther inland. As a consequence, the herbaceous vegetation that persisted in the two patches of 'natural vegetation' shown in Figure 8.15 (middle panel) spread and recolonized adjacent areas within this wetland and generated the current vegetation cover. Contrary to general belief (e.g. Ríos-López and Thomas 2007), this wetland was transformed by agriculture, but the occupation of the lands by the military halted these activities by the 1940s. In 2005, the naval facility ceased operations (US Department of the Interior 2012). This history of land-use suggests that the isolated populations of *E. juanariveroi* in the two patches seen in Figure 8.15 (middle panel) must have dispersed as the original vegetation recovered throughout the wetland (see Fig. 1 in Ríos-López and Villanueva-Rivera 2013).

The example of the '*coquí llanero*' illustrates the effects that agrarian practices had on amphibians for more than 80 years. For many of us, an important question arises: without the fortuitous recovery of the vegetation, might *E. juanariveroi* have been lost before the time of its discovery? A brief discussion from the flora of PR and Jamaica may help conceptualize this difficult question.

Figueroa-Colón (1996) analysed the percentage of endemic flora in PR and compared it with that from Jamaica. Despite both islands having a similar land area, he noted that PR (2,410 species documented at the time) had one-third the percentage endemism of plants documented from Jamaica (3,003 species documented at the time; 9.9% endemism in PR versus 26.1% endemism in Jamaica). Figueroa-Colón (1996: 100) recognized the lack of long-term documentation of the native flora of PR, but suggested that many species of plants 'were lost (particularly endemics) before proper documentation occurred'. Tentatively, we may think this could have also been the case for extirpation or extinction of amphibians. However, variables other than land area may affect the total number of endemics. For example, Jamaica and PR differ in topography (e.g. mountain plateaus in Jamaica versus steep mountains in several ranges in PR), the extent of elevational gradients (e.g. ~2,256 m asl in Jamaica versus ~1,338 m asl in PR), and types of ecological life zones based on Holdridge's classification. The two places also differ in the distribution and area of ecosystems (e.g. dry and moist forests) and their uses of the land (see Whitbeck 1932, and Asprey and Robbins 1953) during the period of the first large-scale scientific explorations in PR in the twentieth century. Consequently, we may never know the answer to whether or not we could have lost *E. juanariveroi* before the time of its discovery.

Paraphrasing Henderson (1992), and in concordance with Scatena (1993), we believe that the greatest effect that the transformation of habitats by human activities is likely to have on

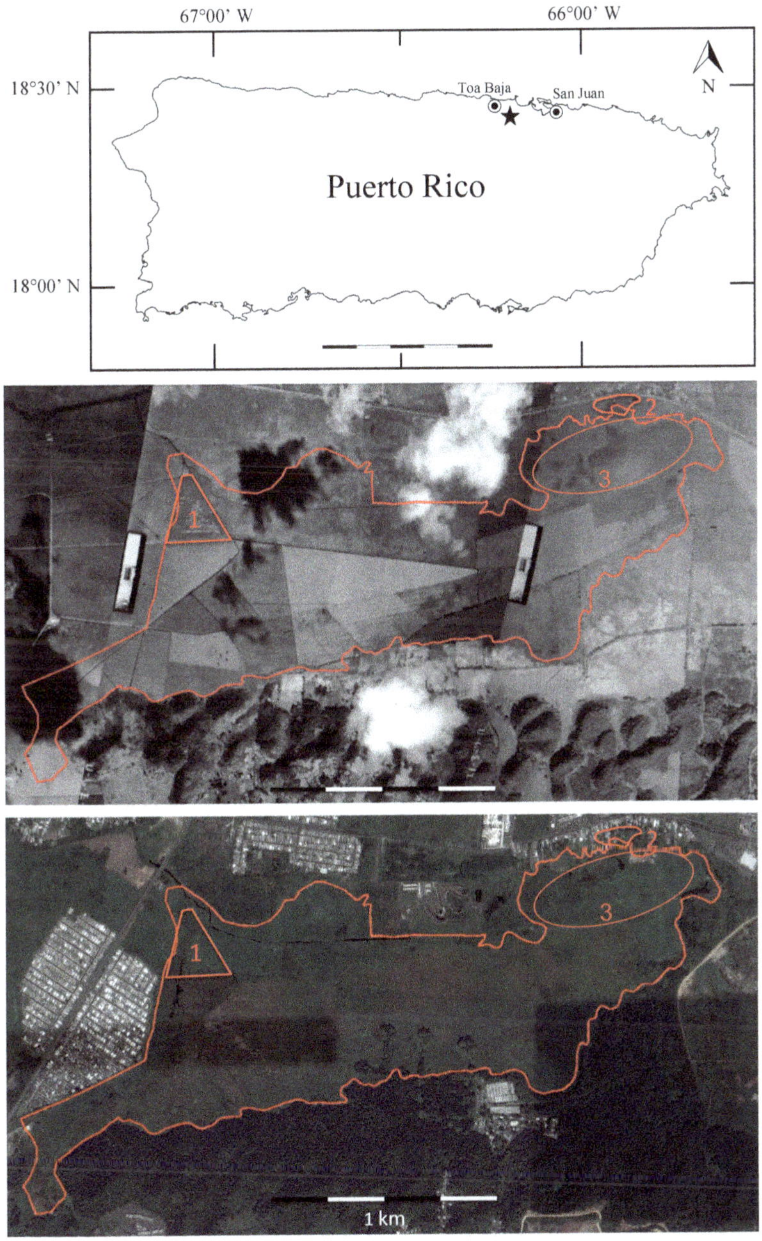

Figure 8.15. Top panel: Map of PR showing the type locality (filled star) of *Eleutherodactylus juanariveroi*, a herbaceous palustrine wetland. Dotted circles highlight the major nearby city, Toa Baja, and the capitol city of San Juan. Scale bar = 50 km; Middle panel: Aerial photograph of the wetland area in 1930 showing the outline of the Critical Habitat for *E. juanariveroi* designated in 2012 by the US Fish and Wildlife Service (US Department of the Interior 2012) and three smaller areas numbered 1 to 3. These were the only areas where herbaceous vegetation (presumably undisturbed) remained within a matrix of agricultural parcels (unidentified crops). Area 1 represented approximately 4 ha, Area 2 represented approximately 1.2 ha, and Area 3 represented approximately 10 ha. Scale bar = 1 km; Bottom panel: the same wetland area by May 2017 (image ©Google Earth Pro; viewed from approximately at 3 km above the ground). The herbaceous palustrine vegetation had spread inland to the base of the limestone mountains in the south and the wetland habitat of *E. juanariveroi* now covers 249 ha (US Department of the Interior 2012); *E. juanariveroi* can be found throughout the wetland area (outlined).

amphibians in PR (and the VI) has already occurred, starting with the island-wide 'massive deforestation' (*sensu* Lugo 2018: 386; Lugo 2019; Wadsworth 1950; see section III. A. 1). The persistence of *E. juanariveroi*, however, may be now procured through regulation of its habitat by various governmental and institutional bodies. Currently, the major threats to this species are the intrusion of salt water and rise in sea level caused by climatic change (e.g. Yu *et al.* 2019; see section III. F. 2). We also believe that humans influence the trajectory of response of amphibians (and other biota and ecological processes) in the Anthropocene in the short term and interact with the long-term influence of natural phenomena. As a result, this interaction generates new species assemblages with no reference in the landscape. Except for *E. juanariveroi*, some combinations of phenomena may lead to a new state that we believe includes the extirpation and extinction of a species within an assemblage in the landscape. The next five examples reflect the complexity of responses of amphibians, which we attribute to a combination of phenomena.

B. Natural and human phenomena in synergy: their influence on the scientific documentation of species and their population status

Historically, many of the best herpetological investigations have been those involving systematic research because collections of specimens were considered 'the most important source of information on the status of tropical populations' (Hedges and Thomas 1991: 56). In general, however, a major shift in paradigm occurred in herpetological work since the 1970s, when monitoring of species commenced and replaced much of the traditional systematic research. Research in the PRA was no exception. Since the 1970s many regulatory mechanisms have been developed and enforced for the scientific collection of specimens. This shift generated heated debates against regarding the collection of 'species that exist in very low densities, have restricted ranges, or for which other aspects of their biology make them especially vulnerable' (Hedges and Thomas 1991: 56).

Such data were lacking for virtually all species because of the ways their abundance and occurrence were estimated (sections III. B. 1–5 for particular cases). To illustrate, some arguments in favour of a less-regulated systematic research included claims such as 'the collection of several hundred specimens of one species from a single locality has no obvious effect on the persistence of the species, as evidenced by repeated collections at the same locality' (Hedges and Thomas 1991: 56). It was also claimed that systematic research would have provided virtually 100% of the information needed for conservation purposes, including records that delineate species' distributions and existing biodiversity elsewhere in the Tropics (Hedges and Thomas 1991). In contrast, arguments in favour of more-regulated systematic research included claims that systematic research was in large part responsible for the population decline and presumable extinction of species of amphibians in PR (Joglar and Burrowes 1991). Specifically, a large number of specimens of species of amphibians, at the time rarely encountered in the wild, were deposited in museums and private collections (Joglar 1998). Therefore, over-collection of specimens of amphibians became an issue concerning the status of several species of amphibians in the PRA. However, other factors for suspected declines of amphibians were proposed by the early 1990s, and Rivero (1991) concluded that evidence supporting one cause (or claim) over another was lacking. In these debates, most attention was focused on several species of amphibians that had not been seen or heard for a relatively long time (e.g. *E. eneidae*, *E. karlschmidti*, *E. schwartzi*, and *Peltophryne lemur*, among others).

Presumably, proponents of the systematic research and conservation sectors in recent decades saw different parts of the same complex problem of disappearances of populations of amphibians. At that time, the history of land-use was not an area of research on amphibians because it deals with longer temporal scales than the generation time of amphibians and describes environmental and climatic processes not obvious in the short term – particularly the timeframe of one person's research career. Pertinent literature was also either difficult to collect or was from somewhat different areas of interest and was consequently not examined in detail. Moreover, with few exceptions, multi-year studies addressing the abundance and distribution of amphibians in the PRA, aimed both at multi-species assemblages and different ecological regions, were mostly lacking. In fact, from the 1970s to the late 1990s, many studies that addressed population censuses were based on visual and acoustic surveys of amphibians and were conducted at small spatiotemporal scales in the EYNF. As a result, the population dynamics and estimates of population size of amphibians in the rest of the island in particular, and the archipelago in general, have barely been documented. Establishing the population status of amphibians for purposes of conservation has been difficult, even today, because there are few consistent, repeated, reproducible records of long-term patterns of population fluctuations and detailed studies of natural history for most species of anurans (Woolbright 1997; Powell and Henderson 2012) (see section VIII).

Bridging the gap between the different proponents of herpetological works in the PRA also required a practical definition of 'abundance' and clarification of the criteria for 'rare' and 'common' (e.g. Holt 1997; Ríos-López *et al.* 2016), or indications whether a species had ever existed at all at a given location. Traditionally, a species has been considered abundant on the basis of a high frequency of collected specimens from its historical distribution (e.g. Barbour [1930b: 113] about *Alsophis* [= *Cubophis*] *fuscicauda*, and Rivero [1959] and Hedges and Thomas [1991] for anurans from PR). Notably, Henderson (1992: 2) mentioned that he was 'aware of the potential shortcomings of this criterion, but unless there is a well-documented written record (e.g., Baskin and Williams 1966; Iverson 1978), it is the only objective approach'. More frequently, the assessment of abundance of several species of amphibians in an area – and therefore of their population status – was made anecdotally (e.g. Joglar and Burrowes 1991). As a result, several species of amphibians that are presumed extinct (see section II. C. 4.) in this area were considered 'common' based on this criterion, which led the scientific community to consider their disappearance 'enigmatic'. Not surprisingly, these anecdotal assessments led to equivocal interpretations about the natural abundance and distribution of species – for instance, contrast the evidence provided by Joglar (1998) and Monroe *et al.* (2017) for *E. wightmanae* in PR. Five examples may illustrate the complexity of different factors acting in synergy on the population status of a particular species, including on how its abundance and occurrence are assessed. These species are *Eleutherodactylus jasperi* (presumed extinct), *E. eneidae* (presumed extinct), *E. karlschmidti* (presumed extinct), *E. schwartzi* (possibly extirpated), and *Peltophryne lemur* (declining).

1. Eleutherodactylus jasperi

Eleutherodactylus jasperi (golden coqui; Fig. 8.3) has not been seen or heard since 1981. The area it inhabited lies for the most part in the deforested landscape depicted in Figures 8.10 and 8.11 (see also Pascarella *et al.* 2000). This species is an obligate bromeliad-dweller, and its geographical distribution is known from a small area (24 ha) in the Cayey Mountains (Drewry and Jones 1976; USFWS 1984). This species is also the only ovoviviparous frog known from the Americas (Drewry and Jones 1976; Wake 1978; Rivero 1998; Wells 2007), and its population size was estimated at between 1,000 and 2,000 individuals (USFWS 1984).

The history of the uses of land and forest in succession after the abandonment of agriculture may provide a plausible explanation for the rarity of *Eleutherodactylus jasperi* at the time of its discovery. Individuals of this species depend on bromeliads for reproduction and survival (Joglar 1998), and a study from the Dominican Republic may help understand the relationship between land-use, secondary succession, and the population status of this species in PR. In particular, Martin *et al.* (2004) showed that dominance of bromeliads in the assemblage of epiphytes was 4.5 times higher in mature highland forests than in secondary forests that developed after ~35 years of abandonment of agriculture (18% versus 4% respectively). The density and species richness of bromeliads were also ~60% higher in mature forests compared with these secondary forests (Martin *et al.* 2004). Consequently, it is suggested that the increase in the dense cover of bromeliads on the surface of trees could take between ~35 to 80 years in tropical areas, or even longer (e.g. in PR, Grau *et al.* 2003; in Costa Rica, Muñíz-Castro *et al.* 2012). The increase in the cover of bromeliads depends on their rate of colonization, which in turn depends on the distance from stands of adjacent mature forests as sources of bromeliad propagules. Consequently, following deforestation of mature forests in the highlands throughout the Cayey Mountains (Figs. 8.10–8.11; Pascarella *et al.* 2000), the development of a dense cover of bromeliads such as *Vriesia* sp., *Hohenbergia* sp., and *Guzmania* sp. (USFWS 1984) suitable for *E. jasperi* may lag considerably behind the growth of trees to large diameter in successional forests.

The fragmentation of, and reduction in, the cover of mature forests and associated bromeliads throughout the landscape also altered local climatic conditions in the highlands, which are vital for the persistence of *E. jasperi* (e.g. accumulation and retention of humidity in bromeliads, low air temperature and its low variation compared with lowland areas). Consequently, the specialized ecological requirements of *E. jasperi*, its low vagility and capacity for dispersal, and low reproductive output (USFWS 1984) could not have been sufficient for a rapid adaptation or for the development of resiliency in the increasingly deforested landscape in which it occurred (Fig. 8.11). Moreover, any natural phenomenon such as droughts or hurricanes for which *E. jasperi* would have been resilient within its primary or mature forest of the past, have now influenced the trajectory of the response of the forest stands in succession differently, and consequently led to a new state for populations of *E. jasperi*. The low estimate of the population size of *E. jasperi* at the time of its discovery throughout the landscape suggests that the species was already following a path toward extinction (see Joglar 1998) and that its population trend is unknown (most likely no longer extant).

2. Eleutherodactylus eneidae

Eleutherodactylus eneidae was last collected in 1984, last heard in 1990 (Hedges 1993) or 1991 (Joglar 1998; an audio recording exists for mid-1990 at EYNF, where presumably a call of this species was heard; A. R. Estrada, a copy of a recording provided to NR-L). This species was described by Rivero from specimens first collected on 26 August 1956, by his wife, Dr Eneida Bordallo Correa (Rivero 1959). Rivero (1959: 6) mentioned: '*Eleutherodactylus eneidae* is quite common in the cloud forests of Doña Juana and El Yunque and it is very strange that the species had not been named before.' Later, Rivero and Mayorga (1963: 81) would refer to *E. eneidae* as 'exceedingly common at the Doña Juana forests' and 'quite common at El Yunque, and perhaps extends throughout the Cordillera Central, where it has been collected as far west as Maricao'. These first references to the species' abundance as 'common' and 'exceedingly common' were presumably based on the subjectivity of the collector (see reviews by Kunin and Gaston [1993] and Holt [1997] on the rarity of a species). However, the initial collection in the cloud forests of Doña Juana, Toro Negro State Forest at Villalba (the Cordillera Central), consisted of 40 individuals of this species. Five years after this discovery, and understandably at the time, various herpetologists and colleagues visited PR and went on collecting *E. eneidae* repeatedly and intensively in this historical locality between 1961 and 1970 (Table 8.2). This area was readily accessible because of construction of recreational facilities and the expansion of access routes through the forest (see Gerhart 1936: 298). These constructions added to the importance of historical changes in the use of lands on the current distribution and abundance of amphibians in PR (see section I).

Table 8.2 summarizes most of the collecting of *E. eneidae* at its type locality since its discovery in 1956, and a pattern emerges from these efforts. First, for a given date of collection, the percentage of the total specimens that were *E. eneidae* is disproportionate compared with the percentage of the total specimens collected from other species of amphibians in the locality.

Table 8.2. Accounts of *Eleutherodactylus eneidae* (and other species) collected at its historical locality in Villalba, Cordillera Central (Doña Juana Recreational Area – Toro Negro State Forest, since its discovery in 1956. For this locality, other entries on record read '9 mi N Villalba, Dona Juana', '10 mi N Villalba', and '14 km N Villalba, *Area Recreacional Dona Juana*' [~9 mi]). Data compiled by DD-C and NR-L from museum records (institutional codes from Sabaj 2016), and http://vertnet.org/, VertNet© version 2016-09-29. The accounts do not include data from other collection sites throughout PR and specimens collected from this locality that may be deposited in private collections (or other institutions). Under 'Number of specimens', we include in parentheses the percentage that *E. eneidae* contributed to the total number of specimens of all species collected on a single date and for which collection data were also available. '*Note' provides the data on specimens collected of other species on the same date, including the total per species (in parenthesis).

COLLECTING DATE AND NOTES	NUMBER OF SPECIMENS	COLLECTOR(S)
1956		
26 August	40 (41.2%)	J. Rivero (mainly) and E. Bordallo
***Note**: Except for the holotype of *E. eneidae* (deposited at the Museum of Comparative Zoology [MCZ]), all paratypes are lodged at the University of Puerto Rico at Mayagüez (UPRM). Other species collected included: *E. brittoni* (*n* = 3), *E. gryllus* (*n* = 16), *E. karlschmidti* (*n* = 1), *E. portoricensis* (*n* = 27), *E. richmondi* (*n* = 4), and *E. wightmanae* (*n* = 6).		
1961		
9 May	104 (43.1%)	D. C. Leber and A. Schwartz
***Note**: '*E. coqui*' (*n* = 30; renamed from *E. portoricensis* in KU KUH (University of Kansas Biodiversity Institute), *E. gryllus* (*n* = 1), '*E. portoricensis*' (*n* = 80), *E. richmondi* (*n* = 10), *E. wightmanae* (*n* = 16).		
1962		
12 June	95 (50%)	R. F. Klinikowski, D. C. Leber, P. E. Leber, A. Schwartz
***Note**: '*E. coqui*' (*n* = 24; renamed from *E. portoricensis* in KU KUH, *E. gryllus* (*n* = 2), *E. locustus* (*n* = 4), '*E. portoricensis*' (*n* = 43), *E. richmondi* (*n* = 13), *E. wightmanae* (*n* = 9).		
1965		
12 March	8 (53.3%)	R. Thomas
***Note**: '*E. coqui*' (*n* = 3; renamed from *E. portoricensis* in KU KUH, '*E. portoricensis*' (*n* = 4).		
1970		
22 and 25 August	27 (37.5%) and 1	Unidentified collectors
***Note**: *E. coqui* (*n* = 20), *E. portoricensis* (*n* = 7), *E. richmondi* (*n* = 7), *E. brittoni* (*n* = 7), *E. wightmanae* (*n* = 4); at USNM (National Museum of Natural History).		

For example, the number of collected *E. eneidae* approximately equalled or doubled that of the most abundant *Eleutherodactylus* in the area. Furthermore, for *E. eneidae*, collected animals represented 45.3% (274/605) of specimens in collections made from three to seven species on any single date (Table 8.2). Second, according to Rivero (1959), *E. eneidae* also can be found on leaves and branches of shrubs 5–6 ft (1.5–1.8 m) above the ground; *E. eneidae* 'seems to prefer mud-banks along trails and roads, but it hides in crevices and holes in the mud and also under rocks, logs, or trash' (Rivero 1998: 321). Thus, it seems that *E. eneidae* was not a species of cryptic habits, but was mostly found on exposed surfaces of vegetation and in mud-banks along trails and road where it was readily accessible to collectors.

Important to highlight is some information on the natural history of other species that were found concurrently with *E. eneidae* at its historical collection site (Table 8.2). These are *E. gryllus*, which is the third small-bodied, arboreal, and bromeliad-dwelling species in the canopy in PR,

E. richmondi, a semi-fossorial species that calls and hides beneath the leaf litter; and *E. wightmanae,* a small-bodied, abundant species found at no more than 40 cm above the ground. This species has very rapid escape responses, and dives into and hides deep beneath the leaf litter. These species were known to science for more than 50 years by 1956 and also collected from the type locality of *E. eneidae* at Toro Negro; collection efforts, however, yielded few specimens there: 1–16 of *E. gryllus*; 6–13 of *E. richmondi*; 4–16 of *E. wightmanae* (Table 8.2). Thus, it seems reasonable to refer to *E. eneidae* as 'common'. However, was *E. eneidae* indeed a common species? Evidence suggests otherwise (see the following paragraphs).

As a reminder, it was frequently noted that the best herpetological work for documenting and protecting the majority of species come largely from systematic collection (e.g. Hedges and Thomas 1991: 57). Yet, at the time, studies based on direct observation of behaviour, systematic non-collection censuses, mark-recapture studies, acoustic monitoring, among other alternate techniques for sampling populations of amphibians and for providing critical information for conservation efforts were rarely conducted in PR other than at EYNF (e.g. Drewry 1970; Turner and Gist 1970; Drewry and Rand 1983; Stewart and Pough 1983; Townsend *et al.* 1984; Lewis and Narins 1985; Woolbright 1985; Narins and Smith 1986). *Eleutherodactylus eneidae* was overlooked for more than 50 years of herpetological work in PR, but at this locality – Doña Juana Recreational Area – it was collected in higher proportions compared with the most common species collected. *E. eneidae* was also collected from readily accessible conditions at this locality. Therefore, the statement made by Rivero (1963) that there seems to be a tendency to survey localities where amphibians are known to occur – particularly throughout the highlands – seemed correct for *E. eneidae* at Toro Negro State Forest. However, the data on abundance presented in Table 8.2 revealed another pattern of collecting activities: a bias towards the collection of an amphibian recently discovered from PR. Presumably, *E. eneidae* attracted much attention as a new species and the type locality in Toro Negro became a popular collecting site for it shortly after the announcement of its discovery in 1959 (Table 8.2).

Collection of specimens contributed to the collapse of the population of *E. eneidae* at its type locality, but collecting was not the sole factor in its demise. For instance, the forest in Doña Juana Recreational Area in Toro Negro visited by Rivero in 1956 was mostly deforested before 1935 (Wadsworth 2008, 2012), but also was one of the areas under the island-wide reforestation programme promoted by the PRRA, the US Forest Service, and the Conservation Civilian Corps (Gerhart 1936). Moreover, in the Toro Negro State Forest alone, more than 5,607,661 small trees from 24 native species and four species of introduced trees were planted by 1943 (this reforestation programme lasted from 1935 to 1943 presumably ceasing because of the advent of the Second World War) (see Wadsworth 2008). Unfortunately, two weeks before Rivero's discovery of *E. eneidae,* Hurricane Santa Clara in Spanish (Betsy in English) crossed the island from south-east to north-west on 12 August 1956. This hurricane struck PR as an estimated category 1 tropical cyclone and its eye passed 10 miles (~16 km) to the north-east of Toro Negro (Dunn *et al.* 1956; Wadsworth and Englerth 1959). Wadsworth and Englerth (1959: 47) described the effects of this hurricane as: 'In even-aged pure forests damage was far less selective than in all-aged mixed stands' (the authors argued that in all-aged stands, dominant, exposed trees generally were affected more than the subordinate ones). They continued: 'Forests of this character, if damaged at all, generally suffered severely, apparently because all trees were so similar as to be about equally susceptible. Extensive damage of this type was suffered by even-aged pure forests at Guajataca, the Scout Camp, Río Abajo, Toro Negro, Aguirre, Mameyes, Cacao, Cambalache, Carite and Luquillo.' Considering that PR had not experienced a hurricane from 1932 to 1956, much of the secondary forests produced by assisted planting had not reached 20 years of age when Hurricane Santa Clara hit the island; these were young forests in succession. As a result, hurricane winds encountered little resistance from these reforested areas below the belt of the few forest-biodiversity refuges that remained in the highlands (e.g. Toro Negro State Forest) (Figueroa-Colón 1996). Consequently, this forest in the Doña Juana Recreational Area was recuperating from previous uses of the land when Hurricane Santa Clara hit it, which may have accelerated the presumable reduction of the local population of *E. eneidae* and other species of amphibians. Although we may never know the relative contribution of natural phenomena influencing the population size and distribution of *E. eneidae*, we suggest that the collection of specimens for scientific purposes also contributed to its population status at the type locality after Hurricane Santa Clara.

Lessons learned from the influence of hurricanes on populations of amphibians were provided by Joglar and Burrowes (1991), Woolbright (1991), Stewart (1995), and Stewart and Woolbright (1996). They emphasized the role of extreme (but infrequent) weather events such as hurricanes and droughts, and suggested that juveniles and egg clutches are life stages most affected from desiccation even though adults can be found, even in higher numbers, after these climatic phenomena. Adults, usually confined to the canopy, are found at ground levels following storms that affect canopies, which makes the frogs more easily found by herpetologists (see Table 8.2) and perhaps by predators. As for juveniles, Joglar and Burrowes (1991), Woolbright (1991), Stewart (1995), and Stewart and Woolbright (1996) emphasized the effects of the short-term droughts that frequently occur after a hurricane. For example, the defoliation of trees after a hurricane allows solar radiation to reach the ground, resulting in increased temperature and lowered humidity in the air, litter, and soil for months (see review by Willig *et al.* 2012). In the leaf litter, juveniles are frequently found, and these changes in microclimate after a hurricane relate to decreases in the abundance of juveniles (e.g. see section III. G. 4.; NR-L, unpublished data). In summary, the removal of individuals (mostly juveniles of *Eleutherodactylus eneidae* from Toro Negro by Hurricane Santa Clara and by the systematic collection of adults in the same locality) resulted in a depression in size of the local population.

The population of *E. eneidae* in Toro Negro presumably could not have recovered rapidly from the effects of this natural phenomena in the type of land-cover that characterized its geographical distribution at the time, and also because of the likely long time for individuals of this species to reach reproductive maturity (see Kunin and Gaston 1993). To illustrate: individuals of another species of *Eleutherodactylus* from the highlands, *E. portoricensis*, kept under laboratory conditions with unlimited food, no risk of predation, and no, or little, risk of disease, and under environmental conditions of low temperature and high humidity that mimicked conditions similar to those at the type locality of *E. eneidae* (e.g. 890 m asl in Doña Juana Recreational Area for *E. eneidae* versus 825 m asl in the Cayey Mountains for *E. portoricensis*), becomes reproductively mature between 1.7 and 1.9 years after hatching (Ríos-López *et al.* 2015a). Therefore, if individuals of *E. eneidae* reached maturity somewhere between 1.5 and two years after hatching, its population at Toro Negro may not have rebounded rapidly from a population depression caused by this hurricane as additional efforts to collect *E. eneidae* at the same locality just a few years afterwards yielded no individuals (Table 8.2).

In the Luquillo Mountains, historical accounts show that *Eleutherodactylus eneidae* was last heard in 1991 at EYNF a few years after category 3 Hurricane Hugo struck the Luquillo Mountains in 1989 (see section III. G. 1), and a record of the call of this species exists for the mid-1990s. However, 1991, 1994, and 1997 were the sixth, third, and second driest years of the twentieth century in PR respectively (Larsen 2000). Moreover, Mote *et al.* (2017) showed that up to 60% less rainfall than expected fell in the Central Mountain Range in the 1994 ENSO event. Intense El Niño years characterized the periods 1990–92 and 1994, a phenomenon which is associated with droughts elsewhere in Central America and the Caribbean (Christensen *et al.* 2013) (see section III. F. 2). Consequently, the years of last hearing of the species in the Luquillo Mountains and throughout PR coincided with the driest years on record, which presumably also contributed to the disappearance of the species during the 1990s.

Finally, the answer to the question whether the systematic collection of *Eleutherodactylus eneidae* could be considered responsible for the extinction of this species in the Cordillera Central (and the Luquillo Mountains) can be yes – as one candidate stressor. This answer is but partly affirmative, however, because factors related to natural phenomena (hurricanes and droughts) acted synergistically with human phenomena and against the status of a population of this species at Toro Negro and in EYNF. Nonetheless, we should refer to the case of *E. eneidae* as a species that is extirpated from its historical type locality because, as mentioned by Rivero (1963: 1), there is a tendency for herpetologists to visit the same locality where a given species is found, and 'many potential areas remained unexplored on the island', particularly in the Central Mountain Range and at EYNF (see end of section VIII).

3. *Eleutherodactylus karlschmidti*

The case of *Eleutherodactylus karlschmidti* mirrors that of *E. eneidae*. It was discovered and collected by Major Chapman Grant in 1931, from beneath boulders in a valley and mountain streams of the Luquillo Mountains, north-eastern PR (Grant 1931a). This is the largest and most aquatic species of *Eleutherodactylus* from PR and one of the largest *Eleutherodactylus* in riparian ecosystems in the Caribbean (Rivero *et al.* 1963; Joglar 1998). The information on the locality of specimens collected by Grant (1931a; Table 8.3) reveals that 79 specimens were collected in only four events at the type-locality before the date of Grant's publication (i.e. 20 July 1931) (Grant 1931a). Table 8.3 shows that during the stay of Grant in PR (from 1931 to 1933), 76.2% (163/214) of all specimens of amphibians he collected during the same date at the same locality were *E. karlschmidti*. Other species are represented by a low number of collected specimens, which by no means are indicative of their population size; the record of specimens in this collection may reveal only the experience, efficiency, and preferences for a given species by the herpetologist doing the collecting. So, Grant was not collecting at random at this locality. In 1931, Grant also collected most of the specimens (adults) during the months of lower rainfall and temperature, which are the months of lower reproductive activity for most species of *Eleutherodactylus* in PR (see sections III. F. 1.b. and G. 4). The next year, most of the collection was made during the peak of the reproductive season for the majority of *Eleutherodactylus* in PR; presumably, Grant collected those adult individuals of *E. karlschmidti* that remained at this locality in 1931 at the peak of the next breeding season in 1932 (e.g. September; Table 8.3). For reasons discussed in section III. A., populations of this species (at least mostly from its historical collection sites in the lowlands) should have been struggling with deforestation of riparian areas for decades (Grant's locality in Canóvanas was no exception). This species, however, remained at a nearby locality where Grant did not collect it: EYNF (see next paragraphs).

The number of specimens of *Eleutherodactylus karlschmidti* collected by Grant during the early 1930s raises an important question: is this information evidence of the commonness of *E. karlschmidti* before its disappearance elsewhere? A new wave of systematic collection was generated elsewhere some 30 years after Grant's collections in PR. For example, Hedges and Thomas (1991: 59) asserted that 'herpetologists later in the 1960s and 1970s observed this species to be quite common in streams on El Yunque. Specimens were easily found near bridges without hiking more than a few decimetres up or down the stream (e.g. R. Thomas, personal observation).' In fact, virtually all collections of this species from its few historical localities occurred in 'the vicinity of bridges over several widely spaced streams within the forest' (Hedges and Thomas 1991: 59). On the one hand, Rivero (1963) warned that there was a tendency to survey localities where amphibians were known to occur, while Joglar (1998: 115;

Table 8.3. Accounts of *Eleutherodactylus karlschmidti* collected by Major Chapman Grant, at the species' type locality at 10 km south of the town of Canóvanas (within the range of the Luquillo Mountains), northern Puerto Rico, from its discovery in 1931 to 1933, which is the first wave of effort to collect this species. Data compiled by NR-L from http://vertnet.org/, VertNet© version 2016-09-29, the University of Michigan Museum of Zoology, and collections from the University of Puerto Rico (UPRM and MZUPRRP). The accounts are incomplete as other specimens from this locality might have been deposited in private collections or other institutions. '*Note' provides the available data on specimens of other species collected in the same locality, when presumably *E. karlschmidti* was not found; the collection's total per species per date occurs in parentheses. Under 'Number of specimens of *E. karlschmidti* collected', we include the percentage contributed by *E. karlschmidti* to the total number of specimens of all species collected on a single date and for which collection data were also available. MCZ = Museum of Comparative Zoology.

COLLECTING DATE AND NOTES	NUMBER OF SPECIMENS OF *E. KARLSCHMIDTI* COLLECTED	OTHER SPECIES COLLECTED ON THE SAME DATE
1931 21 March	19 (95%)	*E. coqui* ('*E. portoricensis*') (*n* = 1)
***Note**: collection of specimens of other species during additional visits to this locality resulted in: '*E. coqui*' (*n* = 3; on 4 March); *E. brittoni* (*n* = 1; on 7 March).		
11 April	14 (100%)	None
19 April	16 (100%)	None
***Note**: *Leptodactylus albilabris* (*n* = 1; 24 April).		
19 July	30 (96.8%)	*E. antillensis* (*n* = 1)
***Note**: *L. albilabris* (*n* = 4; 7 July); *L. albilabris* (*n* = 2; 12 September); '*E. coqui*' (*n* = 2; 13 September); *E. antillensis* (*n* = 1; 8 October); *E. brittoni* (*n* = 1; 13 October); and '*E. coqui*' (*n* = 5; October 18).		
1932		
24 February	11 (64.7%)	*E. antillensis* (*n* = 6)
***Note**: *L. albilabris* (*n* = 18; January 1); *L. albilabris* (*n* = 3; 3 March); *L. albilabris* (*n* = 12), '*E. coqui*' (*n* = 15), and *E. gryllus* (*n* = 1; 1 May); *L. albilabris* (*n* = 3) and *E. antillensis* (*n* = 1; 3 May); '*E. coqui*' (*n* = 1; 5 May); and *L. albilabris* (*n* = 7; 24 July).		
11 July (topotypes at MCZ)	2 (100%)	None
26 July	24 (100%)	None
7 September	19 (30.6%)	'*E. coqui*' (*n* = 43)
19 September	26 (100%)	None
1933		
1 February	At least one (may be two).	None

translated freely from Spanish) hypothesized that 'it is possible that it [*E. karlschmidti*] was not abundant in any locality, but owing to the specialized nature of its habitat and ecology, individuals would become conglomerated in some areas, giving the false impression of being a common species'. Joglar (1998) also posited that *E. karlschmidti* initially might have had low population density judging by the number of specimens from collections in museums (his account: *n* = 473) and those used for dietary analyses (*n* = 51; data from Rivero *et al.* 1963). However, it may also indicate the predominant behaviour of repeated collections at a few preferred localities by experienced herpetologists; if true, this would lead to a rapid population depression of an already declining population. However, Hedges (1993: 297; see also Campos-Cerqueira and Aide 2017) added: 'the disappearance of *E. karlschmidti* has no obvious explanation. Some of the localities are in protected and unaltered forest (Caribbean National Forest) on El Yunque.' Presumably, Hedges' (1993) impression and that of others were confounded by the combination of limited knowledge of the history of land-use and of the influence of climatic factors that generated effects not obvious on the timescale of one person's research career at that time.

Similar to the example of *Eleutherodactylus eneidae* discussed earlier, additional information, presumably overlooked in the earlier herpetological literature, relates closely to the decline of *E. karlschmidti* at EYNF from the 1960s to the last date when the species was found there. First, Peters and Wassenberg (1983) found an inverse relationship between body-size and density of individuals of animals in nature, and thus *E. karlschmidti* – at the high-end portion of the largest *Eleutherodactylus* in the Caribbean – may have occurred at lower overall densities compared with other species of amphibians (mostly mid-sized to small-sized *Eleutherodactylus*) present in its vicinity. Yet rarity (and commonness) is a subjective concept (e.g. Kunin and Gaston 1993; Holt 1997), and other factors may determine the frequency of detection of a species elsewhere (e.g. see Campos-Cerqueira and Aide [2016] for an example of Puerto Rican birds). As mentioned, the collection of *E. karlschmidti* was a species-specific endeavour. Second, *E. karlschmidti* was an ecological specialist, occurring only in creeks and ravines, and with one-third of its diet consisting only of benthic prey (Rivero *et al.* 1963). Presumably, being a species of high philopatry or of low vagility – when PR was mostly deforested – would also make it easy to collect elsewhere (see next paragraphs).

Contrary to general belief (e.g. Hedges 1993; Joglar and Burrowes 1996; Woolbright 1997; Campos-Cerqueira and Aide 2017), the cover of natural forest at El Yunque was restricted to elevations above about 600–700 m asl because of deforestation practices for agriculture and timber extraction from the nineteenth century until at least the mid-1950s (Hill 1899; Murphy 1916; Foster *et al.* 1999; Wadsworth 2012; Weaver 2012; see section I). Deforestation has also accompanied the development of primary access by road and trails on the northern slopes of El Yunque from 1929 to the early 1970s, which has led to access to hard-to-reach areas within EYNF (see Gerhart 1936: 292; Weaver 2012). Of particular importance are roads PR-191 (constructed between 1929 and 1942) and PR-186 (constructed between 1934 and 1957) (see Gerhart 1936; Weaver 2012), and their tributary roads. First, these are routes where most specimens of *Eleutherodactylus karlschmidti* were located and collected since the early 1960s (Table 8.4). Second, Rivero (1978), Burrowes and Joglar (1991), and Joglar (1998) mentioned that this species was relatively abundant between 450 m asl and 700 m asl until 1978, mainly at El Yunque and other highland areas. The date of the most recent specimen collected is 1976, from 'Hwy 191 near El Toro, El Yunque' (Burrowes *et al.* 2004: 154; presumably near Tradewinds trail, which lies in PR-191 and heads to El Toro Wilderness Area). Consequently, the persistence of most populations of *E. karlschmidti* was already limited at the Luquillo Mountains before surveys were conducted in the early 1960s.

The accounts in Table 8.4 revealed a pattern of the frequency of collection and quantity of specimens collected during each visit at EYNF. At the Luquillo Mountains, more than one-third (41.3% or 260/630) of all specimens between two and seven species collected per visit between the 1960s and the 1970s was *E. karlschmidti*. Joglar (1998) mentioned that the majority collected during the early 1960s (from a sample of 188 specimens examined by him) were males, which suggests they could be detected in the field more easily (e.g. by acoustic surveys) than females. In either case, the evidence suggests that this species became very popular to collect by experienced herpetologists in the 1930s (also note in Table 8.4 the shift in preference of collection of a new species [*E. hedricki*] over *E. karlschmidti* during August 1962; other collecting shifts among species also are evident). However, could the systematic collection of *E. karlschmidti*, in a riparian forest in succession, be considered the main phenomenon responsible for the extirpation of this species in the Luquillo Mountains? A third factor – along with deforestation and systematic collection – may now have contributed to the fate of *E. karlschmidti*, this time in a human-dominated landscape: a climatic phenomenon with influence over a wide geographical area in the Luquillo Mountains could facilitate the extirpation of an already vulnerable species with populations in

Table 8.4. Accounts of *Eleutherodactylus karlschmidti* collected by herpetologists and scientists from 1944 to 1976 (second wave of collecting) at the species' historical localities within the boundaries of El Yunque National Forest (EYNF), north-eastern PR (specimens collected from a few other localities in PR are not included). Also included are data on specimens collected from other species in the same locality and date when *E. karlschmidti* was found. Data compiled by NRL from http://vertnet.org/, VertNet© version 2016-09-29, from the Museum of Zoology of the University of Puerto Rico in Río Piedras, the museum at the University of Puerto Rico in Mayagüez, and published accounts. Under 'Number of specimens of *E. karlschmidti* collected' we include the per cent of specimens collected when data on additional species collected on the same date were indicated in the records. '*Note' provides the locality on record and other related information. For simplicity, we omitted the names of the collectors. The accounts do not include specimens from EYNF that may be deposited in private collections (or other institutions) and specimens collected at other localities in PR (e.g. Aguas Buenas, Carite-Guayama, Mayagüez, San Lorenzo-Patillas, Toro Negro).

COLLECTING DATE AND NOTES	NUMBER OF SPECIMENS OF *E. KARLSCHMIDTI* COLLECTED	OTHER SPECIES COLLECTED ON THE SAME DATE
1944 September (without day)	At least one (7.7%).	At least: '*E. auriculatus*' (*n* = 1) and '*E. portoricensis*' (*n* = 3); *E. gryllus* (*n* = 2); *E. locustus* (*n* = 3); *E. richmondi* (*n* = 2); *Leptodactylus albilabris* (*n* = 1).
*Note: the locality is identified as Luquillo/El Yunque Mountains; these records do not specify day and number of specimens collected – it is assumed that one entry in the database represents one specimen collected.		
1961		
15 May	42 (22.7%)	*E. coqui* (*n* = 23), *E. eneidae* (*n* = 15), *E. locustus* (*n* = 54), *E. portoricensis* (*n* = 39), *E. richmondi* (*n* = 8), *E. wightmanae* (*n* = 3), *L. albilabris* (*n* = 1).
*Note: the locality is identified as 'Área de Recreo La Mina, 11.8 km south of the town of Palmer'.		
16 May	3 (8.1%)	*E. richmondi* (*n* = 29), *E. wightmanae* (*n* = 5).
*Note: the locality is identified as '12 km south of the town of Palmer, road to El Yunque', within the 'Área de Recreo La Mina'.		
18 May	10 (100%)	None.
*Note: the locality is identified as '2 mi (3.2 km) south of El Verde, Río Espíritu Santo'.		
1962		
9 June	2 (100%)	None.
10 June	17 (15.9%)	*E. locustus* (*n* = 52), '*E. portoricensis*' (*n* = 32), *E. richmondi* (*n* = 4), *E. wightmanae* (*n* = 2).
4 August **	36 (100%)	*E. hedricki* (at least eight heard).
11 August	***None (0%). ****20 (74.1%)	***E. hedricki* (*n* = 5), '*E. portoricensis*' (*n* = 15); ****'*E. portoricensis*' (*n* = 7).
1 September	3 (74.0%)	*E. eneidae* (at least one).
2 September	2 (9.1%)	*E. locustus* (*n* = 7), '*E. portoricensis*' (*n* = 7), *E. richmondi* (*n* = 4), *E. wightmanae* (*n* = 2).
3 September	*****2 (100%) 25 (100%)	None. None.
*Note: the localities for these dates are identified as '10 km south of the town of Palmer' (10 June), 'El Yunque' (presumably at 10.9 km on road between Palmer and La Mina, elevation 1,765 ft or 535 m asl); **'El Verde' (August 4); ***collected at El Verde but no *E. karlschmidti*, but collectors also went on to ****La Coca Falls and small creeks higher in elevation ('road between Palmer and La Mina'; 11 August). Also, 'La Mina at 1,875 ft in elevation' (568 m asl; 1 September); 'El Yunque, swimming pool area' (2 September); *****'carretera de Río Blanco al Yunque alt. 150' (Naguabo); and '2.2 mi (3.5 km) south-west of Sabana' (3 September), all within EYNF.		
1963		
21 March	53 (98.1%)	*E. eneidae* (at least one).
16 July	2 (100%)	None.
*Note: the localities are identified as 'Big Tree Trail, La Mina' (21 March), and 'El Yunque, La Mina' (16 July).		

(*Continued*)

Table 8.4. (Continued)

COLLECTING DATE AND NOTES	NUMBER OF SPECIMENS OF *E. KARLSCHMIDTI* COLLECTED	OTHER SPECIES COLLECTED ON THE SAME DATE
1964		
9 February	At least one (100%).	None.
30 July	2 (50.0%)	'*E. coqui*' (*n* = 1), *E. wightmanae* (*n* = 1).
*Note: the localities are identified as 'La Mina' (9 February) and at '2 mi (3.2 km) south of El Verde, Río Espíritu Santo' (30 July).		
1965		
11 February	3 (21.4%)	'*E. coqui*' (*n* = 6), '*E. portoricensis*' (*n* = 4), *E. richmondi* (*n* = 1).
25 September	12 (100%)	None.
*Note: the localities are identified as '2 mi (3.2 km) southwest of Sabana' (11 February) and 'El Yunque' (25 September).		
1966		
17 January	At least one (20.0%).	At least one of each: *E. eneidae*, *E. locustus*, *E. portoricensis*, *E. wightmanae*.
25 November	3 (17.6%)	'*E. coqui*' (*n* = 11), *E. eneidae* (*n* = 1), *E. richmondi* (*n* = 1), *L. albilabris* (*n* = 1).
26 November	10 (100%)	None.
*Note: the localities were identified as 'La Mina Recreation Area' (17 January), 'El Yunque Restaurant' (25 November), and 'El Yunque' (26 November).		
1970		
23 May	1 (100%)	None.
*Note: the locality is identified as 'El Verde Field Station, Luquillo Experimental Forest'.		
1971		
10 September	7 (38.9%)	*E. portoricensis* (*n* = 2), *E. richmondi* (*n* = 9).
*Note: the locality is identified as 'Bosque Nacional del Caribe (El Verde), El Verde Field Station, km 22.3 on Rd. 186'.		
1974		
16 May	At least one (100%).	None.
*Note: the locality is identified as 'south of the town of Palmer at 1,400' in elevation' (424 m asl; Zoology Museum of the University of Puerto Rico in Río Piedras).		
1976		
Not Specified	At least three (100%).	None.
*Note: the localities are identified as 'Hwy 191 near El Toro, El Yunque', (Burrowes *et al.* 2004) and 'El Yunque' (estimated locality and elevation; Burrowes *et al.* 2008a).		

decline (see next paragraphs). Figure 8.16 may help us understand the complexity of the disappearance of *E. karlschmidti* from its historical collection sites.

Larsen (2000) analysed a century-old dataset of rainfall and streamflow during the twentieth century to characterize drought and the water resources in PR. His analyses revealed several intense droughts that corresponded to the period of collection of *Eleutherodactylus karlschmidti* at EYNF. The most severe meteorologic drought of the twentieth century was from 1966 to 1968 (three-year drought), and also included the driest year of the century, 1967 (Fig. 8.16). Two

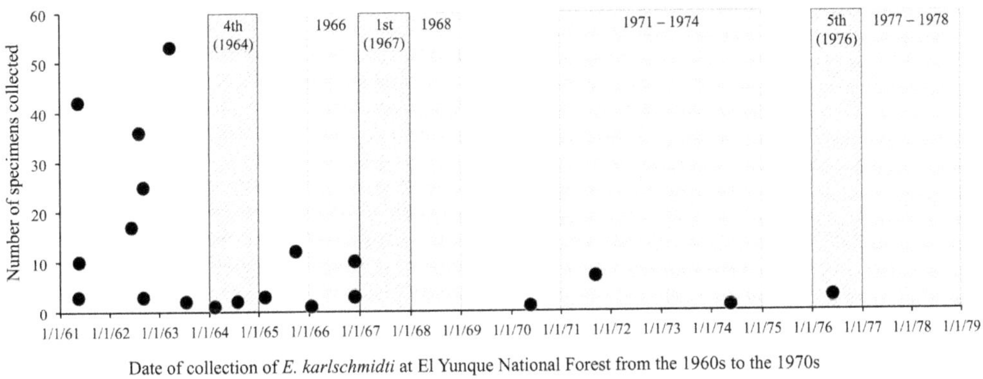

Figure 8.16. Temporal pattern of collection of specimens of *Eleutherodactylus karlschmidti* (black dots) at EYNF in relation to historical severe droughts. Bars filled in grey represent the duration of the drought (when multi-year) and bars filled with lines diagonal highlight the historical ranking of the drought for PR (see text).

multi-year intense droughts occurred from 1971 to 1974 (four-year drought) and from 1976 to 1978 (three-year drought). Since the period of the 1960s to the 1970s, two additional top five driest years were recorded: 1964 (ranked fourth) and 1976 (ranked fifth) (Fig. 8.16; Larsen 2000). Even though other driest years occurred in PR before the 1960s (e.g. 1930 [ranked seventh], 1947 [ranked ninth], 1957 [ranked tenth]), the frequency of these events was low (17 years and 10 years passed between events, respectively). In contrast, the frequency of droughts documented in the 1960s–70s was higher.

Eleutherodactylus karlschmidti was last collected in 1976, and Burrowes *et al.* (2004) suggested that its disappearance may have represented a 'climate-linked epidemic' (*sensu* Pounds and Crump 1994). Accordingly, Burrowes *et al.* (2004: 150) proposed that *E. karlschmidti* (and other *Eleutherodactylus* spp.) 'became stressed for water and tended to move to more humid microhabitats where the pathogenic chytrid fungus [*Batrachochytrium dendrobatidis* or *Bd*] was likely to be found'. More specifically, Burrowes *et al.* (2004) hypothesized that individuals of *E. karlschmidti*, being stream dwellers, clumped 'along shallow pools of water in nearly dried-out streams', which according to Joglar (1998: 115), led to 'the rapid disappearance of the species after being abundant until 1974'. Notably, the chytrid fungus was first detected in PR from anurans collected in 1961 (P. Burrowes, pers. comm. to NR-L), but *E. karlschmidti* already had undergone a decline in abundance since the early 1960s without a specimen having tested positive for *Bd*. Coincidentally, the last specimen of *E. karlschmidti* collected on record in 1976 yielded positive for *Bd*, and this anuran has not been detected ever since. Consequently, these observations provide little support for the hypothesis of disease-driven population declines – as a major explanation for the decline – of this (or other) species in the area (see section III. C. 3). For a stream-dwelling, large-bodied species with terrestrial reproduction such as *E. karlschmidti*, these unusually dry periods must have had consequences for its populations, but droughts might not have acted in isolation. We now discuss a more plausible explanation for the puzzling disappearance of *E. karlschmidti* from its historical collection sites.

The specimens collected al El Yunque revealed an elusive pattern: most were adults from the same few areas (especially the vicinity of Río La Mina); juveniles and egg clutches were rarely collected, if at all. Similar to the pattern of *E. eneidae*, the repeated collection of adults of *E. karlschmidti* from the same localities by experienced herpetologists during the early 1960s may have resulted in a depression of populations because of a lack of recruitment during subsequent years. Not surprisingly, the frequency of collection of specimens at El Yunque along route PR-191 (and also along route PR-186 at El Verde on the western border of EYNF) declined in close concordance with the periods of multi-year severe droughts and the driest years on record for the twentieth century (Fig. 8.16). On the other hand, Heatwole and Newby (1972: 160) mentioned: 'Species from rain forest seldom experience dry conditions [i.e. *E. karlschmidti* and *E. eneidae*], and there would not be strong selection for adaptive behavioral responses to desiccation.' Moreover, individuals of *E. karlschmidti* (and *E. eneidae*) had low activity rates in response to desiccation, which suggests that these specimens have limited searching or escape behaviour with increased desiccation (Heatwole and Newby 1972). Consequently, specimens of *E. karlschmidti* would have dispersed but remained relatively close to their historical collection sites during the extreme drought periods in the 1960s and 1970s (as may also have happened in the 1930s).

Finally, Rivero (1991: 54; translated freely from Spanish) stated:

> I used to visit El Yunque with my students every year. On one occasion, 12 or 15 years ago, I had serious problems in my lower back and sent my assistant along with my students instead. I asked them to collect some *E. karlschmidti* as they were needed for a research project here. When they returned, they informed me that they had not seen or heard any of them. I thought it was inexperience, and I sent them again, this time at the months when *E. karlschmidti* had been collected on other occasions. The result was the same. I still thought it was a lack of experience and I went instead. I did not see it either. Margaret Stewart informed me that she did not see the species. This happened from one year to the next. [Note: Stewart's research agenda was developed during the 1970s to the mid-1990s almost exclusively at El Verde Field Station, western El Yunque.]

According to Rivero's publication in 1991, his recollection of the last time he had seen *E. karlschmidti* was situated precisely during the 1976–78 drought (Fig. 8.16). Hence, these populations were almost certainly at the minimum population size needed for its continued survival (*sensu* Burrowes and Joglar 1991; see Holt 1997). Consequently, the populations of *E. karlschmidti* would have experienced a steady decline from a combination of (unintentional) over-collection in its last, historical collection sites in the Luquillo Mountains and natural phenomena (i.e. record-breaking droughts) that influenced its riparian forests when it would otherwise have the adaptive capacity to deal with natural phenomena alone. The rate of population decline of this species would also have been faster in the riparian forests in succession that were less protected from other human activities and uses of the land throughout PR compared with the Luquillo Mountains (e.g. in Mayagüez to the west and Aguas Buenas to the central east).

We may never know the relative contribution of the systematic collection versus natural phenomena to the disappearance of *E. karlschmidti* from PR. The decline in the number of specimens of Puerto Rican amphibians deposited in museums suggests a shift in the way herpetological research has been conducted since the late 1980s and early 1990s, and probably the influence that increased regulatory mechanisms have had in favour of the protection of wildlife. The fact is that historical collection sites have not yielded a single specimen of this species despite several explorations by local scientists in recent years (e.g. Campos-Cerqueira and Aide 2017). In this regard, we believe that all effects that systematic research could have had on the population status of *E. karlschmidti* (and *E. eneidae* among others) in its historical locations in PR have already occurred.

It has been suggested that the disappearance of *E. karlschmidti* from its historical collection sites may merely be a matter of extirpation, because 'it is likely that many streams in the Luquillo

Forest were never visited by collectors' (Hedges and Thomas 1991: 59) and 'many potential habitats remain uninvestigated' (Rivero 1963: 1). For a species that previously had been collected in the lowlands and at mid-elevations throughout PR, however, the number of refugia currently would be limited since '99.4% of the island's area is located within 1 km of the nearest road [and] all forest types should be similarly accessible' (Grau *et al.* 2004: 1173; recall that in PR, road total network was 1,300 km in 1922 [Miller 1922] and had increased to 26,862 km by 2012 [The World Factbook 2022]). Nonetheless, despite the changes in land-use in PR, the island retains an extensive network of rivers, creeks, and ravines (see Figure 1 from Lugo *et al.* 2001), many of which remain unexplored and constitute potential havens for remnant populations of this species.

4. Eleutherodactylus schwartzi

Eleutherodactylus schwartzi (VI coqui frog) exists on Tortola, Virgin Gorda, and Great Dog, among other localities in the BVI, but Schwartz and Henderson (1991: 115), reinforce Schwartz and Thomas's (1975), and Schwartz and Henderson's (1988) suggestion of it being 'evidently extinct' by stating 'formerly on U.S. Virgin Islands'. Consequently, the IUCN Red List refers to *E. schwartzi* as having been extirpated from St John in the USVI (Hedges and Thomas 2010b). However, Mayer (2012: 143) referred to *E. schwartzi* from St John as 'PX' (= possibly or probably extirpated). Mostly, because of this extirpation, this species currently is classified as Endangered in the IUCN Red List (Hedges and Thomas 2010b). It was described based on specimens collected in 1964 from Tortola and Virgin Gorda, BVI and was separated from its Puerto Rican sibling species, *E. coqui*, by differences in colouration and morphology, and because an 'apparently real geographic hiatus exists on Culebra, Vieques, St Thomas, and St. John' (Thomas 1966: 389). Consequently, reading that *E. schwartzi* was extirpated from St John (Hedges and Thomas 2010b), and 'possibly or probably extirpated' from St John (Mayer 2012), led us to investigate these disappearances more thoroughly, as they have profound consequences for conservation of amphibians in the area.

We initiated our investigation by recalling passages in older and current literature that referred to four names of species (in order of relevance): *Eleutherodactylus schwartzi, E. portoricensis, E. antillensis,* and *E. coqui*. Initially, distinguishing *E. schwartzi* and *E. coqui* from their sibling species *E. portoricensis* by Thomas (1966) was possible by a combination of variables – morphological and acoustic – that required examination and description of live animals. Consequently, it is highly important to distinguish different calls when there are no other physical means of confirming the identity of cryptic species based on preserved specimens alone (e.g. Thomas 1966). We started with Mayer's (2012) list of all the major islands in the VI that have records of the following species: *E. antillensis, E. coqui* ('I' [Introduced]), and *Eleutherodactylus schwartzi* ('?') in St Thomas; *E. antillensis* ('I?'), *E. coqui* ('I'), and *E. schwartzi* ('I?') in St Croix; *E. antillensis, E. coqui* ('I'), and *E. schwartzi* ('possibly or probably extirpated') in St John; *E. antillensis* and *E. schwartzi* in Tortola (see Table 8.1 for other islands in the BVI). According to Mayer (2012), *E. schwartzi* has been mentioned from all major USVI, along with its closest-relative *E. coqui* (e.g. Hedges *et al.* 2008). Notably, the structure and dominant frequency of the two-note advertisement call of these two species are similar (see Figure 8 in Thomas 1966; Ovaska and Caldbeck 1999). These parameters of their calls are within the values of those of the two-note advertisement call of the sympatric *E. antillensis* in the BVI (Ovaska and Caldbeck 1997a, 1997b; Drewry and Rand 1983). For reasons discussed here, we believe the original record of *E. schwartzi* from St John and subsequent claims of its extirpation are doubtful, or at least unverifiable from the available data. We consider that until the identity of specimens collected from St John can be verified, the record of *E. schwartzi* from St John may have been of another species instead. To clarify, we discuss the historical account of *E. schwartzi* on St John.

Shortly before the paper by Hedges and Thomas (2010b), Henderson and Powell (2009: 77) mentioned the fact that *E. schwartzi* 'may be extirpated' from St John, citing Philibosian and

Yntema (1977) among others. Platenberg and Boulon (2006: 219) considered this species as 'endemic to St. John and the British Virgin Islands' (citing the IUCN from 2004), but that it had been 'extirpated from the U.S. Virgin Islands', again citing Philibosian and Yntema (1977). Schwartz and Henderson (1991: 115), referred to *E. schwartzi* as 'formerly on U.S. Virgin Islands', but they did not provide a reference to sustain this claim. MacLean (1982) did not include St John within the species' geographical distribution but limited it to Tortola and Virgin Gorda instead. Similar to Schwartz and Henderson (1991), MacLean (1982) did not provide a reference to sustain this geographical distribution for the species.

Heatwole *et al.* (1981) did include St John, USVI, and Tortola, Virgin Gorda, and Great Dog in the BVI as being within the species' geographical range. While Heatwole *et al.* (1981) did not make any reference to an extirpation of the species from St John, they did include St Croix for the first time in the recent literature as part of the species' geographical distribution but did so with a tentative 'introduced?'. Contrary to MacLean (1982), MacLean *et al.* (1977: 32) did include St John for *E. schwartzi*, but these authors listed the species with a question mark as '?*Eleutherodactylus schwartzi*'. We then examined Philibosian and Yntema (1977: 5), who listed *E. schwartzi* from St John with an '*E*', which they defined as: 'Extinct or possibly extinct in the area marked. Formerly present as either a breeding species or a non-breeding migrant.' Consistent with Schwartz and Henderson (1991), Schwartz and Thomas (1975: 36) referred to *E. schwartzi* from St John as 'evidently extinct', a statement that might originate from Thomas (1966), who discovered *E. schwartzi*. Thomas (1966) made three particular statements, the first being: 'No *portoricensis* or *coqui* equivalent has been recorded from St Thomas, St John, Vieques, or Culebra, and my own collecting experience substantiates this' (p. 386). Three years before Thomas collected specimens of *E. schwartzi* (i.e. 1964), Seaman (1961) considered '*E. portoricensis*' to be rare on St Thomas. However, he also mentioned that this species is 'more abundant on St. John and Tortola' (Seaman 1961: 9; a phrase not reproduced by Thomas 1966). Consequently, the search for an answer to whether or not *E. schwartzi* existed on St John in the first place seemed to become increasingly complex as we went further back in time.

Thomas (1966: 386) also mentioned having seen (although uncommonly) *E. 'portoricensis'* on St Croix 'where it has presumably been introduced (Grant and Beatty, 1944; Seaman, pers. comm.) along with *E. antillensis*'. In fact, Seaman (1961: 9), considered *E. antillensis* 'Not native to St. Croix, but introduced at least twice, once by the writer in 1951. Now found in Christiansted gardens, and along Canaan gut and Salt River.' More specifically, Grant and Beatty (1944: 110; included in Thomas [1966]) indicated having collected a specimen of frog from Christiansted, St Croix, and that this specimen:

> prove[d] to be a typical, average-sized *E. antillensis* which is probably the species referred to by the writer on p. 505 when *E. portoricensis* Schmidt was listed with a question mark as probably the species liberated. Mr. Beatty reported that in 1937 a Puertorican had liberated a number of "coqui", the Puertorican vernacular for this species. Thus it is possible that *antillensis* has become established and that *portoricensis* was not introduced.

This passage from Grant and Beatty (1944) was based on Grant (1937: 505), who described this species on St Croix as: '*Eleutherodactylus portoricensis*? Schmidt. Mr Beatty wrote on 24 October: "A Puerto Rican liberated a number of Coqui Here". This species cannot be included in the fauna.' Grant and Beatty (1944) did not mention the record of *E. portoricensis* from St John documented by Grant (1937). Schwartz and Thomas (1975: 19), however, listed *E. coqui* (considered specimens of *E. portoricensis* in Grant's time) as 'introduced on St. Thomas and St. Croix, U.S.

Virgin Is'. A salient fact becomes clear at this point: the introduction of species from PR into the USVI was more common than frequently thought at the time of Grant and Beatty (1944).

A major turning point for our investigation came from Grant (1937: 518), who mentioned that 'A few hours were spent at Cruz Bay, St John' (on his way to Tortola), but he also remarked that 'this expedition unfortunately coincided with a very severe drought' (1937: 501). Nonetheless, Grant (1937: 518) collected two specimens of *Eleutherodactylus portoricensis* at Cinnamon Bay, northern St John: 'the frog was not found at Cruz Bay in 1932, but was reported as heard from the general locality of Cinnamon Bay. (Jour. Dept. Agri., P. R., XVI. P. 340.)'. The work Grant (1937) cited as not having found this species in 1932 was Grant (1932e) – *The herpetology of St. John and adjacent keys, U.S. Virgin Islands* – a study based on a collection trip conducted in St John from 13 to 17 April 1932. However, the work Grant (1937) was referring to when he mentioned having 'heard' this species from Cinnamon Bay in St John actually was Grant (1932f) – *Herpetology of Tortola; notes on Anegada and Virgin Gorda, British Virgin Islands* – a study based on a collection trip to Tortola from 10 to 16 August 1932. Grant (1937) did not distinguish the studies from which these claims originated, which made it difficult to track the circumstances under which *E. schwartzi* was first documented from St John. Moreover, it seems that Grant's (1932e) paper was not available to Thomas (1966). Thomas (1966: 386) cited Grant (1932f): '*Eleutherodactylus "portoricensis"* has been reported only from St John and Tortola in the Virgin Islands by Grant (1932 [*sic*; should be Grant 1932b in Thomas' {1966} literature cited], 1937), who commented on the difference between Tortolan specimens and those of Puerto Rico'; and 'On Tortola and on Virgin Gorda, however, we collected specimens of a *coqui* cognate (= *portoricensis* in the sense of Grant 1932b, 1937).' It is important to distinguish and track the origin of these passages because they offer a clue on how *E. schwartzi* was first documented from St John, an event that has profound implications for conservation of this species that last until today.

First, Grant (1932f: 340) mentioned how he documented *E. schwartzi* (known as *E. portoricensis* at that time) from St John:

> Late one evening while tacking up the channel between St. John and Thatch Island of Tortola, the "bo-peep" of *E. portoricensis* could be heard each time that the shores of St. John were neared. **This adds a new record to the herpetological list of St. John**. [emphasis added]

Based on his writing, Grant (1932f: 340) was sailing the channel between Great Thatch Island, BVI, and St John, and the shores he was referring to were those of the Mary Point Peninsula, which faces Great Thatch Island from northern St John. Second, Grant (1937) mentioned having collected two specimens of *E. portoricensis* (*E. schwartzi* of today) at Cinnamon Bay, northern St John. Richard Thomas (pers. comm. to NR-L 2014) clarified that he did not have the specimens collected by Grant from St John for examination in his 1966 work because the location of these specimens was hard to find (see next paragraphs).

The third statement by Thomas (1966: 389) is that made about *E. antillensis*: '*E. antillensis* has an unaccented two-note call, similar in structure but quite distinct from that of *schwartzi* or of *coqui* or *portoricensis*; the clicking call noted by Grant (1932b) is an occasional variant.' Indeed, the data recalled by Thomas (1966) is from Grant's trip to Tortola (Grant 1932f). Grant (1932f: 340) described the call of *E. antillensis* as a 'click-click-click' and mentioned that the animal 'is identical to the Puerto Rico and St John specimens'. This reference to the call of *E. antillensis* from St John corresponded to data from Grant (1932e: 331) – although not explicitly cited by Grant (1932f) – which were based on his trip to St John in April 1932: 'The voice did not seem to differ greatly from the Puerto Rico specimens.' Thomas (1966: 389) stated that 'the clicking call noted by Grant [1932b] is an occasional variant'. From Ovaska and Caldbeck (1997a; 1997b) we now know that this is a courtship call, not the advertisement call frequently made after dusk (e.g. Drewry and Rand 1983). Grant (1932f: 340) commented: 'NOTE: The voice of another species, resembling that of *E. gryllus* was heard, but the frogs were not located.' He (1932f) did not provide additional clues about the source of this sound.

As mentioned, Grant (1937) collected two specimens of '*Eleutherodactylus portoricensis*' from Cinnamon Bay, St John, but a *coqui/portoricensis* cognate was not collected back in April 1932 (Grant 1932e). A query of collections of *Eleutherodactylus* frogs in various databases from several sources yielded no record of *Eleutherodactylus schwartzi* nor *E. portoricensis* from St John, USVI (e.g. Florida Museum of Natural History, Kansas University Biodiversity Institute and Natural History Museum, and the Museum of Comparative Zoology and University of Michigan Museum of Zoology, the latter two institutions holding Grant's herpetological collection, containing about 7,500 specimens [Smith 1986]; also see VertNet© [2020] for additional institutions). However, the Smithsonian's National Museum of Natural History yielded three records of specimens with the name *E. schwartzi*, not collected by Grant but by Harry A. Beatty and in June 1941 (catalogue numbers USNM 115914, USNM 115921, USNM 115922). The record log from the USNM reads 'No Further Locality Data' and that these specimens were initially identified as *E. cochranae* by Beatty, but reidentified 'to *E. schwartzi* by Ronald I. Crombie, date unknown'. One specimen, USNM 115914, had the label 'Collected in bromeliad' under 'Specimen Note' in the record log. Notably, the only reference to bromeliads as a collection microhabitat of a species of anuran made by Grant came from Grant (1932d: 325) for *E. cochranae* and Grant (1932e: 331) for *E. cochranae* and 'a few *E. antillensis*'.

The corroboration of the identity of these specimens and the existence of *E. schwartzi* in St John could be problematic for several reasons. First, similar to Thomas, our efforts to locate the specimens originally collected by Grant (1937) have been unsuccessful so far, and the locality of the three specimens collected by Beatty from St John in 1941 is unknown. Second, nothing named '*E. portoricensis*' exists in collections – to our knowledge – from St John. However, the literature revealed that the (human-mediated) translocation of individuals of at least two species of amphibians between PR and the USVI (*E. antillensis* and '*E. portoricensis/coqui*') is more common than frequently thought (e.g. Grant 1937; Grant and Beatty 1944). In fact, Waddle *et al.* (2006: 487) collected three specimens of *E. coqui* (a sister species of *E. schwartzi*) from Cruz Bay, St John 'near a drainage ditch located in a landscape and irrigated area'. Third, Thomas (1966) illustrated the challenges for the identification of specimens of cryptic species in the '*E. portoricensis* complex' – to which *E. schwartzi* belongs – based on preserved specimens alone (i.e. body and eye colouration faint in old specimens). Finally, the extraction of intact DNA from old specimens presumably preserved in formalin may be difficult for purposes of identification. As an alternative, the specimens collected by Beatty in 1941 could be examined by osteological analyses using X-rayed images to avoid damage of the tissue and compare the osteological data with data from closely related and sympatric species (*E. coqui, E. antillensis, E. cochranae*) by using multivariate analyses (e.g. see Ríos-López and Thomas 2007). We also recommend that the Mary Point Peninsula in northern St John be explored for individuals of species of amphibians in bromeliads growing on the ground and on trees (e.g. Ovaska *et al.* 1998), and that listening at night for calling individuals be employed. References to extirpation and related terms should be made with caution in the IUCN Red List database and only used once the identity of the original specimens from Grant (and Beatty) is verified as recommended. Consequently, we think that history and evidence show that the record of *E. schwartzi* from St John is sufficiently doubtful to warrant its conservation listing and geographical distribution (and presumed existence) as 'unconfirmed'.

5. *Peltophryne lemur*: distinguishing between rarity and decline of an elusive habitat specialist

Several passages from the literature on the amphibians and reptiles of the PRA are worth reproducing here in chronological order. Stejneger (1904: 572–573) mentioned that *Peltophryne lemur* 'seems to be somewhat rare in Porto Rico, possibly because of lack of suitable localities. Dr Stahl informed me that during his forty years of collecting in the island he had never come across it himself, and the six specimens which in that time he had obtained were brought to him by the peasants as curiosities.' Later, Schmidt (1920: 168) mentioned that 'its occurrence at Coamo Springs, nearly at the opposite side of the island, proves that it is widely distributed'. However, Schmidt (1928: 31) mentioned that 'its rarity is perhaps due to the mongoose'. Grant (1932a: 41) commented: '*Bufo lemur* seems to have always been scarce, or its hiding places are not yet understood.' García-Díaz (1967: 1) cited Schmidt (1928) and wrote: 'Schmidt thought the mongoose (*Herpestes a. auropunctatus*) was responsible for scarcity of the toad.' Schwartz and Thomas (1975: 12) refer to this species as 'apparently now uncommon', while Rivero *et al.* (1980) highlighted that *P. lemur* seems to have been scarce since before *R. marina* was imported.

Pregill (1981: 21–22) commented on fossilized remains of this species:

> The abundance of *Peltophryne lemur* as a fossil is striking by contrast to its present scarcity. Schmidt (1928) secured only five specimens for his synopsis, and Grant (1932 [Grant 1932a]) reported that only 28 individuals had been collected since the species was first described by Cope in 1868. As a result, nothing is known of its habits. The toad is not abundant anywhere in its range; Richard Thomas informs me, however, that the animal is probably more common than currently believed because of inappropriate collecting methods. Its presence in five of the fossil localities suggests that formerly *P. lemur* was more abundant than it is today. Probably the animal leads a secretive, semifossorial existence, a lifestyle which could account for its rarity in museum collections. Obviously, it must spend some time above ground exposed to predation (e.g. while breeding) in order to account for the frequency of its occurrence as a fossil. Schmidt (1928) suggested that the introduction of *Bufo marinus* to the island in the early 1920's might account for the rarity of the species today. *Bufo marinus* is well established in disturbed and undisturbed areas over most of the island and, although it is somewhat larger, by Schmidt's reckoning *B. marinus* may be displacing the endemic. This seems unlikely to me.

In 1973, Rivero was concerned about the population status of *Peltophryne lemur* and conducted a search, using flyers distributed in several municipalities within its historical range in PR (see Rivero and Seguí-Crespo 1992: 52). The results of that effort produced the capture of seven females (presumably a few more, including males; E. Estremera, pers. comm. to NR-L 2017) in Quebradillas in 1974 (Rivero and Seguí-Crespo 1992). Initially, an attempt to maintain these toads in captivity in the Experimental Agriculture Station in Isabela failed. Later, two pairs were collected in Quebradillas in 1980 and used for a captive-breeding project at the Jardín Zoológico in Mayagüez. In 1984, a reproductive event was found in the south-western part of PR, and the number of adult individuals participating in the event was estimated between 700 and 1,000 (Rivero and Seguí-Crespo 1992). In 1987, the US Fish and Wildlife Service listed *P. lemur* as 'Threatened' and stated (US Department of the Interior 1987: 28830): 'Disease and predation have not been documented as factors in the decline of this species. However, predation on dispersing toadlets may be heavy, particularly from wading birds (M. Canals, Guánica Commonwealth

Forest, pers. comm. 1986), and could become a significant factor if populations are greatly reduced by other problems.'

Moreno (1991a: 9) recommended that this species 'needs to be listed as endangered (rather than threatened). Recent unpublished mitochondrial DNA studies (A. Goebles, pers. comm.) suggest northern and southern populations in PR are distinct. If so, the northern form may be extirpated in the wild (zoo individuals remain).' In 1991, the species was observed in several municipalities along the north/north-western coast and the south/south-western coasts of PR (Rivero and Seguí-Crespo 1992). Additional anecdotal sightings in PR include Cupey, Trujillo Alto, Carolina, Luquillo, and Fajardo in the north/north-east; Dorado, Vega Baja, Manatí, Arecibo, Quebradillas, Isabela, and Aguadilla in the north/north-west; Aguada, San Germán, and Mayagüez in the west; and Lajas, Guánica, Coamo, Santa Isabel, and Guayama in the south, but most of these records need to be confirmed (Rivero and Seguí-Crespo 1992). As far as we know, there are only three wild populations of *P. lemur*, all located in southern PR. The largest population is found in Guánica State Forest and consists of approximately 2,000 individuals. The second population was discovered in the municipality of Yauco, south-western PR (Moreno and Canals 1985), and it was considered extirpated until its rediscovery in 2010 (USFWS 2016). A third population was found in Punta Ventana in 2008, which is a private property located in the municipality of Guayanilla, southern PR (USFWS 2016).

Why are there so few specimens of *Peltophryne lemur* collected from PR? According to Kunin and Gaston (1993) and Holt (1997), the 'rarity' of specimens in collections – i.e. the low number of specimens collected and deposited – may not be indicative of problems from a population conservation context, and the low frequency of specimens of *P. lemur* detected may reflect its secretive habits instead (e.g. Pacheco and Barber 2013). Moreover, its brief but explosive breeding after heavy rains makes this species difficult to detect in the wild solely with traditional survey techniques (e.g. visual and acoustic [not automated] surveys). *Peltophryne lemur* is an ecological specialist, and the low frequency of specimens detected could be because of equally low availability of suitable areas for its reproduction and survival (Stejneger 1904; Rivero and Seguí-Crespo 1992). Many predators may include it in their diet (e.g. birds, a wide range of aquatic and terrestrial invertebrates, *Anolis* spp. and other reptiles, amphibians [including cannibalism], and mammals [e.g. rats and mongooses]), but these occurrences are anecdotal and we lack published studies addressing the frequency and intensity of predation pressure on this species.

Peltophryne lemur is found in the lowlands, and it must have been affected by agriculture and urban development there. Ross (2007) presented an analysis of the influence of changes in land-cover between 1963 and 2001 on the breeding ponds of *P. lemur* in the municipality of Quebradillas, north-western PR. Her analysis revealed a 21% increase in the development of urban areas adjacent to the breeding ponds during this period, but no development in areas between the ponds and the Quebrada Bellaca to the west (Ross 2007). A substantial change in land-cover for *P. lemur* occurred in an approximately 1,061 ha of wetland in the Valle de Lajas, south-western PR, and that lies within the geographical distribution of this species (Fig. 8.17). This wetland consisted of the largest freshwater lagoon in PR, the Guánica Lagoon, which was connected to the El Anegado Swamp to the west (i.e. the Guánica-Anegado wetland). This wetland is within the subtropical dry forest life zone (Fig. 8.1) (Ewel and Whitmore 1973) and was seasonally flooded (Danforth 1926; DRNA 2011). However, in 1955 it was drained for agriculture, initially for sugarcane plantations. Once the most important area for migratory and resident birds in PR, the Puerto Rican slider (*Trachemys stejnegeri*), and many other vertebrates including *Leptodactylus albilabris* (Danforth 1926), the Guánica-Anegado wetland is no more. The Cartagena Lagoon (the third largest freshwater lagoon behind the Tortuguero Lagoon in the north) is farther to the west and drains into the Rincón Lagoon on the coast, and it is equally influenced by

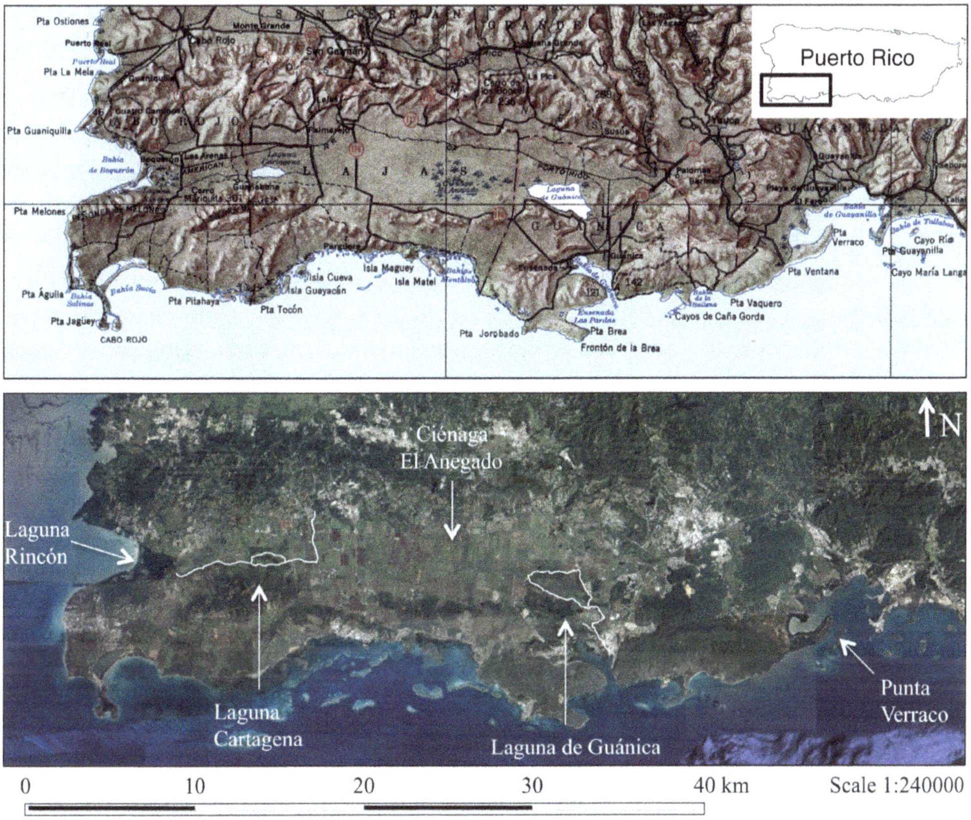

Figure 8.17. Maps showing the largest natural freshwater lagoon, 'Laguna de Guánica', that existed in south-western PR (top panel; modified from a 1952 topographic map, scale 1:240,000, developed by the US Geological Survey, US Department of the Interior), and same region today (bottom panel) after drainage of this lagoon and the associated wetland areas in the Lajas Valley. The 'Laguna Cartagena' to the west of the Lajas Valley is the third-largest natural freshwater-wetland documented in PR. The historical and current geographical distribution of the Puerto Rican crested toad in south-western PR lies within this region and extends to Punta Verraco in the municipality of Guayanilla (see text; section IV. B).

past and current transformation of the landscape by human activities. However, it is still considered one of the most significant areas for waterfowl in south-western PR.

The proximity of the Guánica-Anegado wetland system to nearby localities known to be inhabited by *Peltophryne lemur*, Lajas to the west, Guánica to the south, and Guayanilla to the east, makes it likely that this species may also have occurred there. In fact, *P. lemur* was recently found in numbers in a large plantation of mangos (*Mangifera indica*) adjacent to an old coastal dry forest in Punta Verraco, Guayanilla, just south-east of the Guánica-Anegado wetland (Fig. 8.17) (ARP-R, pers. obs.). This plantation has been under cultivation at least since 1930 (see *Porto Rico 1930 Georeferenciado: Un Mosaico de la Costa* 2022). Currently, the species has been introduced into a karst area in the region of Arecibo, northern PR, where historically it had been collected, and individuals are already calling and breeding there.

Finally, the vast north-western karst region (subtropical moist forest life zone) in PR, is characterized by relatively old forests (Kennaway and Helmer 2007), and seems to be little explored herpetologically; this area could be promising for additional findings (Hedges 2006). These remarks apply not only to *Peltophryne lemur*, but also to other Puerto Rican amphibians. For example, some of the species thought to occur mostly in highland areas can also be found in this karst region (e.g. *Eleutherodactylus richmondi*, at < 350 m asl; ARP-R, pers. obs.). In addition, an unconfirmed species of anuran – or a new geographical record of an extant species of anuran – is present there at < 200 m asl; unpublished data). However, the rate of discovery of native/endemic species of amphibians in PR has declined significantly (Fig. 8.2). Nonetheless, herpetological exploration in the karst region may still yield new taxa, as this region has presumably experienced little transformation from human activities in the recent past compared with other parts of the island (e.g. Picó 1950); this region is still relatively inaccessible by road and difficult to access on foot.

C. Pathogens

Table 8.5 lists all species of amphibians reported from the wild in the PRA and highlights the state of the information on their pathogens (and the microbiota in skin). First, and as far as we know, eight species (~26.6%) have not been assessed for pathogens, most notably two endemic species from the VI (*Eleutherodactylus lentus* and *E. schwartzi*) that are still extant. Second, eight species of amphibians have been sampled only for detection of the chytrid fungus, *Batrachochytrium dendrobatidis* (*Bd*), and studies on parasites are still needed; three species (10%) have been sampled for viruses and *Bd*. Finally, the remaining 11 species of amphibians in this list (~36.6%) have been sampled either for ectoparasites, endoparasites, or both, with pathogens found in them coming from a wide range of taxonomic groups represented by a variety of helminths, nematodes, tongue worms, protozoans, bacteria, annelids, fungi, and ticks, but no blood parasites so far. Some examples are discussed here in detail.

1. Ectoparasites

Ectoparasites were present on the first specimens of *Rhinella marina* intended for introduction into PR from Jamaica (Wolcott 1948a) but were removed before the introduction occurred (Wolcott 1948a); the current status of ticks on *R. marina* in PR is not known. In St Croix, of more than 67 specimens of *R. marina* and 20 specimens of *Eleutherodactylus lentus* examined, none had tick-like organisms on their skins (NF-A, pers. obs.). Unidentified ticks have been found on toadlets of *Peltophryne lemur* at Tamarindo Beach, Guánica State Forest, south-western PR (NR-L, pers. obs.). Unfortunately, these ticks were not collected.

Ticks were first reported for *Eleutherodactylus cooki* by Grant (1932b: 147), who mentioned that 'There was an infestation of small red ticks on the legs and sides of these frogs'. Fox (1947) documented *Ornithodoros* [= *Carius*, see next paragraph] *puertoricensis* as new to science and, based on specimens collected from rats (*Rattus norvegicus* and *R. rattus*) collected in Santurce, PR. Fox (1947) distinguished *O. puertoricensis* from the *O. talaje* group – a little-known and cosmopolitan taxon that may consist of many unknown species. Maldonado-Capriles and Medina-Gaud (1977) mentioned that *O. puertoricensis* was also found in South America, St Croix (USVI), and Trinidad, although the host from which it was found was not specified. Nonetheless, they described specimens of the *O. talaje* group, instead of *O. puertoricensis*, from specimens of *E. cooki* collected by Rivero and Mayorga from the vicinity of Yabucoa, south-eastern PR. Goldberg *et al.* (1996) described the prevalence, mean intensity of infestation, and histopathology caused by a tick, *O. talaje*, in *E. cooki* (the first such report from an *Eleutherodactylus* sp.). However, Maldonado-Capriles and Medina-Gaud (1977) and Goldberg *et al.* (1996) identified the tick found on *E. cooki* tentatively as *O. talaje*. According to Dr J. E. Keirans, who was consulted by both sets of authors, the well-preserved hypostomal denticles were missing in specimens studied by Maldonado-Capriles and Medina-Gaud (1977), and the hypostome was missing in specimens from those studied by Goldberg *et al.* (1996), thereby precluding specific identification of this tick. Several authors referred to the ticks on *E. cooki* as *O. talaje* (e.g. Joglar *et al.* 1996; López-Torres and Longo 2013). Longo *et al.* (2020) examined the influence of ticks and *Bd* on male parental care and sexual selection in *E. cooki*.

Table 8.5. Checklist of documented parasites and pathogens – along with notes and references in parentheses – in individuals of species of amphibians from the PRA. *see Dyer *et al.* (1995) for a discussion – based on Adamson's (1981) reassessment of the Genera *Parapharyngodon* and *Thelandros* – on the taxonomy of nematodes documented by Schmidt and Whittaker (1975) from Puerto Rican frogs. **Chytrid fungus, *Batrachochitrium dendrobatidis*, confirmed by Greenhawk *et al.* (2017) at Casas de la Selva (Patillas, south-eastern PR). Norman Greenhawk, Tony Báez, Iniciativa Herpetológica Inc., and Patricia Burrowes kindly provided unpublished data.

FAMILY AND *SPECIES*	PARASITES/PATHOGENS
Bufonidae	
Peltophryne lemur	Flatworms (Trematoda): *Mesocoelium* sp. and cercariae, a nematode (**Oxyuridae**), other nematodes (more than three species), pentastomid worms (undetermined), ciliate protozoans (undetermined spp.); no blood-parasites found – like the nematodes (**Onchocercidae**) *Foleyella* sp. microfilaria (García-Díaz (1967). Pentastomida (tongue worms: **Raillietiellidae**)[a] *Raillietiella bufonis* (Ali *et al.* 1982). A wide range of infectious pathogens (under captive conditions); most pathogens are opportunistic and include bacteria, fungi, and protozoans associated with sepsis and dermatitis (Lentini 2002). No chytrid fungus, *Batrachochitrium dendrobatidis* (*Bd*), detected at Arecibo (northern PR), Guánica and Guayanilla (south-western PR, several sites sampled), and Coamo (south-central PR) (Barber 2012; Iniciativa Herpetológica, Inc. 2015). Unidentified agents leading to dysecdysis (Brown Skin disease) under captive conditions; no *Bd* detected in captive toads (Crawshaw *et al.* 2014).
Rhinella marina	Acari: ticks (**Ixodidae**) *Amblyomma dissimile* from first specimens brought into PR from Jamaica (Wolcott 1948a). Helmith ova (**Schistosomatidae**): *Schistosoma mansoni*; giant roundworm (**Ascarididae**) *Ascaris lumbricoides*; round worm (**Trichuridae**) *Trichuris trichiura*; nematodes, hookworms, undetermined sp.; bacterial cysts (**Enterobacteriaceae**), '*Escherichia coli*?'; tapeworm ova (**Hymenolepididae**) *Hymenolepis nana*; roundworm larva (**Strongyloididae**), '*Strongyloides stercoralis*?' (Hoffman and Janer 1941). Nematodes (**Onchocercidae**): microfilaria of *Foleyella* sp. (García-Díaz 1967). Bacteria: (following Kauffmann-White classification of serovariants for *Salmonella*),[b] *S. arberdeen* (Maryland, USA) and [b] *S. miami* (Florida, USA); [b] intestinal amoebas (undetermined); [c] Flatworms (Trematoda) (undetermined); [c] Nematodes (*Physocephalus*-like); [c] Myxozoans (undetermined); [c] Protozoans; [c] Polyoma-like virus (Burrowes *et al.* 2004). *Bd* detected at Yauco (south-western PR; Iniciativa Herpetológica, Inc. 2015). No *Bd* detected at Toro Negro (central PR) and [c] Humacao (eastern PR) (Burrowes *et al.* 2008a). No *Bd* detected at Quebradillas (north-western PR), Manatí (northern PR) and Arecibo, nor at Guánica and Guayanilla (several sites sampled), Coamo (southern PR), and Vieques Island (offshore, eastern PR) (Barber 2012; Iniciativa Herpetológica, Inc. 2015). Microbiota: on average, over 3,779 Operational Taxonomic Units on the skin of ten individuals sampled from Bayamón (northern PR) (see tables S2 and S3 in Abarca *et al.* 2017); the finding of the Genus *Niabella* sp. associated with the skin of *R. marina* is 'the first report' for 'a new world amphibian'; also, species of *Niabella* are 'found associated with leeches and macroalgae' (Abarca *et al.* 2017).
Eleutherodactylidae	
Eleutherodactylus antillensis	Nematodes: (**Pharyngodonidae**)* [b] *Parapharyngodon garciai*; (**Strongyloididae**) [b,d] *Strongyloides* sp. (Schmidt and Whittaker 1975); *Cosmocercoides* n. sp. (**Cosmocercidae**) (unidentified locality in PR; Rios-Franceschi *et al.* 2020). *Bd* detected at Toro Negro, Manatí, Guayanilla, and Vieques Island (Burrowes *et al.* 2008a; Iniciativa Herpetológica, Inc. 2015). *Bd* detected from a wetland in Toa Baja (northern PR) ([e] T. Báez, pers. comm. to NR-L). *Bd* detected at Río Encantado (Ciales, north-central PR) and Manatí; no *Bd* detected at Arecibo and Quebradillas, nor at Coamo, Guánica, and Guayanilla (Barber 2012; Iniciativa Herpetológica, Inc. 2015).
E. brittoni	*Bd* detected at Casas de la Selva (Patillas, south-eastern PR), Toro Negro and Adjuntas (west-central PR); no *Bd* detected at Carite (east-central PR; Burrowes *et al.* 2008a) nor at Ciales (Iniciativa Herpetológica, Inc. 2015).
E. cochranae	No *Bd* detected in San Juan (UPRRP Campus, northern PR) (Burrowes *et al.* 2008a).

(*Continued*)

Table 8.5. (Continued)

FAMILY AND *SPECIES*	PARASITES/PATHOGENS
E. cooki	Acari: ticks (**Argasidae**) [f] *Carios* (= *Ornithodorus*) sp. (larval specimens; *talaje* group) (Maldonado-Capriles and Medina-Gaud 1977). Acari, ticks [f] *Carios* (= *Ornithodorus*) *talaje* group or [f] *C. puertoricensis* (Goldberg *et al.* 1996). *Bd* detected at Casas de la Selva (Greenhawk *et al.* 2017). No *Bd* detected in San Lorenzo (east-central PR; Burrowes *et al.* 2008a).
E. coqui	Nematodes: (**Trichostrongylidae**) [b] *Poekilostrongylus puertoricensis*, [g] *Oswaldocruzia lenteixierai*; (**Cosmocercidae**) [b, d] *Aplectana* sp. (two species) (Schmidt and Whittaker 1975). Nematodes: (**Pharyngodonidae**) [h] *Parapharyngodon garciae*; (**Cosmocercidae**) [d, h] *Aplectana* sp. (one species) (Dyer *et al.* 1995). Nematode (**Molineidae**) [i] *Oswaldocruzia lenteixierai* (Moravec and Kaiser 1995). *Bd* detected at EYNF and Río Abajo State Forest (Arecibo-Utuado, north-central PR); [b] Nematodes (**Rhabdiasidae**) *Rhabdias* sp.; [j] Nematodes (*Physocephalus*-like) and pinworms; [j] Acanthocephala; no virus detected (Burrowes *et al.* 2004). *Bd* detected at EYNF, **Casas de la Selva, Carite, Toro Negro, and Adjuntas, and Maricao (west-central PR); no *Bd* detected in San Juan, Bayamón (northern PR), Toa Baja, Mayagüéz (western PR), and Juana Díaz (south-central PR) (Burrowes *et al.* 2008a; Caligari Damestoy 2017). No *Bd* detected at Yauco and Guayanilla, nor at Quebradillas, Arecibo, and Manatí (Iniciativa Herpetológica, Inc. 2015). Nematode (**Ascaridoidea**), *Porrocaecum* sp.; Cestoda (unidentified); Oligochaeta (**Naididae**), *Dero* (*Allodero*) sp.; [k] (other Genera already herein, including *Rhabdias* sp.) (Marr *et al.* 2008). Nematode (**Rhabdiasidae**) *Rhabdias elegans* (Marr *et al.* 2010). *Bd* detected at Arecibo (Barber 2012). Acari: ticks (**Argasidae**) [f] *Carios* (= *Ornithodorus*) sp. (larval specimens; *talaje* group) (Longo and López-Torres 2017).
E. eneidae	[b] Nematodes (*Physocephalus*-like); no *Bd* detected at Toro Negro; no virus detected (Burrowes *et al.* 2004). No *Bd* detected at EYNF (Burrowes *et al.* 2008a).
E. gryllus	No virus, *Bd*, and parasites detected at EYNF (Burrowes *et al.* 2004 and Burrowes *et al.* 2008a). No *Bd* detected at Casas de la Selva (N. Greenhawk, unpublished data).
E. hedricki	No *Bd* detected at EYNF (Burrowes *et al.* 2008a).
E. jasperi	No virus, *Bd*, and parasites detected at Carite (Burrowes *et al.* 2004 and Burrowes *et al.* 2008a).
E. johnstonei	No data published for the PRA.
E. juanariveroi	No *Bd* detected in Toa Baja (Burrowes *et al.* 2008a).
E. karlschmidti	No virus and parasites detected; [b] *Bd* detected at EYNF but not at Guavate Forest (Carite; Burrowes *et al.* 2004 and Burrowes *et al.* 2008a).
E. lentus	No data published for the PRA.
E. locustus	Nematodes: (**Cosmocercidae**) [b, d] *Aplectana* sp. (two species) (Schmidt and Whittaker 1975). No virus, *Bd*, and parasites detected at EYNF (Burrowes *et al.* 2004 and Burrowes *et al.* 2008a), but P. Burrowes (pers. comm.) detected *Bd* at EYNF (unpublished data); *Bd* detected at Carite (Burrowes *et al.* 2008a). *Bd* detected at Casas de la Selva (N. Greenhawk, unpublished data).
E. monensis	Although Burrowes *et al.* (2008a) did not detect *Bd* at Mona Island (offshore, western PR), nor Stabile (2020; $n = 36$ individuals sampled between 2006 and 2019), P. Burrowes (pers. comm.) detected *Bd* (unpublished data).
E. portoricensis	Nematodes: (**Pharyngodonidae**)* [b] *Parapharyngodon garciai* (Schmidt and Whittaker 1975). [b] Nematodes (**Rhabdiasidae**) *Rhabdias* sp.; [b] Nematodes (*Physocephalus*-like) and pinworms; [b] Acanthocephala; no virus detected (Burrowes *et al.* 2004). *Bd* detected at EYNF (in two localities at a different elevation) and Carite (Burrowes *et al.* 2008a). No *Bd* detected at Casas de la Selva (N. Greenhawk, unpublished data).

(Continued)

Table 8.5. (Continued)

FAMILY AND *SPECIES*	PARASITES/PATHOGENS
E. richmondi	Nematodes: (**Cosmocercidae**) [b,d] *Aplectana* sp. (two species) (Schmidt and Whittaker 1975). *Bd* detected at Carite (Burrowes *et al.* 2008a). *Bd* detected at Casas de la Selva (Greenhawk *et al.* 2017). No *Bd* detected at EYNF and Toro Negro; no virus and parasites detected (Burrowes *et al.* 2004).
E. schwartzi	No data published for the PRA.
E. unicolor	*Bd* detected at EYNF (elfin-forest site) (Burrowes *et al.* 2008a).
E. wightmanae	*Bd* detected at EYNF (mid-elevation site), **Casas de la Selva, Carite, and Toro Negro; no *Bd* detected at Maricao (Burrowes *et al.* 2008a; Caligary Damestoy 2017).
Hylidae	
Dryophytes cinereus	No data published for the PRA.
Osteopilus septentrionalis	No *Bd* detected at Quebradillas (Barber 2012).
Scinax ruber	No data published for the PRA.
Leptodactylidae	
Leptodactylus albilabris	Platyhelminthes: (**Diphyllobothridae**) *Diphyllobothrium mansoni* (tapeworm) (Puerto Rico Experiment Station 1937). *Bd* detected at Casas de la Selva, Carite, Toro Negro, Maricao; no *Bd* detected at EYNF, Carite, San Juan (UPRRP Campus), Mayagüez (western PR), and Juana Díaz (Burrowes *et al.* 2008a; Caligari Damestoy 2017). *Bd* detected at Guayanilla and Vieques (Iniciativa Herpetológica, Inc. 2015). No *Bd* detected at Quebradillas, Arecibo, Manatí, Ciales, Guánica, Yauco, and Coamo (Barber 2012; Iniciativa Herpetológica, Inc. 2015).
L. fallax	No data published for the PRA.
Ranidae	
Lithobates catesbeianus	Platyhelminthes (trematode and nematode): undetermined spp. (Pérez 1951a). No *Bd* detected at [c] Humacao (Burrowes *et al.* 2008a), nor at Arecibo (Barber 2012).
L. grylio	No data published for the PRA.
L. pipiens	No data published for the PRA.

[a] New species.
[b] In specimens collected at EYNF (north-eastern PR).
[c] Specimens collected at the Humacao Wildlife Reserve.
[d] Only females of this nematode have been collected, but the specific identification requires the examination of males.
[e] From specimens collected by Báez, herpetofauna keeper at the Abilene Zoo, Texas, USA.
[f] See text for further discussion on the identification of this parasitic tick.
[g] From specimens misidentified as *E. portoricensis* collected at Cupey Alto, San Juan, PR.
[h] These authors collected frogs at Barrio Guamá, San Sebastián (west-central PR).
[i] Moravec and Kaiser (1995) suggested *Poekilostrongylus puertoricensis* and *Oswaldocruzia lenteixierai* could be conspecific.
[j] In specimens collected at Río Abajo State Forest, Arecibo-Utuado municipalities.
[k] Although not parasites, Marr *et al.* (2008) documented external mites, 'likely belonging to the order Mesostigmata'.

Longo and López-Torres (2017) documented ticks similar to those in the *talaje* group on specimens of *E. coqui* located in the same boulders inhabited by *E. cooki*. The ticks found on *E. coqui* were similar to those collected from *E. cooki* in the 1980s (Longo and López-Torres 2017). Longo and López-Torres consulted Dr James W. Mertins, entomologist at the US Department of Agriculture's National Veterinary Services Laboratories, Iowa, USA, who identified the genus of this tick as *Carios*. According to Longo and López (2017: 160), comparisons of the larvae of this tick found on *E. coqui* with *C. talaje* and *C. puertoricensis* suggested that it is 'distinctly different and represent an undescribed species of *Carios* (Mertins, pers. comm.)'. Consequently, the taxonomic problem of the specific identity of the tick on *E. coqui* and also on *E. cooki* still has not been resolved.

2. *Endoparasites*

Endoparasites have been documented from nine species of amphibians (30%) in the PRA. Some parasites were documented as new species from hosts such as *Peltophryne lemur* (*Raillietiella bufonis* by Ali *et al.* 1982). Other parasites await specific description (*Amplectana* spp. in *E. coqui*, Schmidt and Whittaker 1975, and Dyer *et al.* 1995; *Amplectana* sp. in *E. locustus* and *E. richmondi*, Schmidt and Whittaker 1975; *Strongyloides* sp. in *E. antillensis*, Schmidt and Whittaker 1975, and *Cosmocercoides* n. sp. also in *E. antillensis*, Rios-Franceschi *et al.* 2020). Notably, Hoffman and Janer (1941) found helminth ova in *Rhinella marina* and concluded that, on occasions, this species can be coprophagous and may have served as a mechanical vector of parasites frequently found in humans in PR at that time. However, the socioeconomic development experienced by PR since the 1940s included advances in the treatment of used waters, development of community-level sanitary systems, and increased education on human health, with a corresponding awareness of how to prevent common diseases. These advances reduced and eliminated many of the diseases that frequently affected the human population at the time (e.g. malaria), and led to the end of the 'long standing and objectionable habit of promiscuous defecation on the part of the "Jíbaro" (countryman) and his children, to which the toad has adapted itself' (Hoffman and Janer 1941: 504). Today, helminths infections in PR are seldom known.

3. *The chytrid fungus,* Batrachochytrium dendrobatidis *(Bd)*

In the PRA, 22 species of amphibians have been screened for the chytrid fungus, *Batrachochytrium dendrobatidis* or *Bd* (Table 8.5); in PR, so far the oldest record of an anuran positive to *Bd* was an individual (undetermined) collected in 1961 (see section III. B. 3). As documented elsewhere (Berger *et al.* 2016; Lips 2016), *Bd* has been implicated in the population decline and extinction of amphibians in the PRA, specifically on the main island of PR (Burrowes *et al.* 2004, 2008a, 2008b; Longo and Burrowes 2010). In PR, great advances have been made on the identification of some of the risk factors that increase the probability of infection of individual amphibians by *Bd*. For instance, weather-related liability for infection and mortality (Joglar *et al.* 2007; Longo *et al.* 2010, 2013a); effectiveness of different sampling techniques used for the detection of *Bd* in infected individuals (Burrowes *et al.* 2011); the role of bacterial diversity and species composition in the skin of frogs in relation to susceptibility of juveniles and adults to infection (Longo *et al.* 2015); and the role that microhabitat partitioning/use between adults (which are frequently found higher on the surface of vegetation) and juveniles (which are frequently found at ground level) has on the risk of infection (Burrowes *et al.* 2017). Despite these advances, much remains to be learned about the role that *Bd* may have on suspected declines of populations of most species of amphibians in the PRA. For example, physiology, behavioural adjustments, and acclimation under environmental stress vary among species of amphibians, which may generate different patterns of population dynamics among species tested positive for *Bd* – or indeed in relation to other environmental stressors (e.g. Navas *et al.* 2016; see next paragraphs).

Recent studies have shown that infections caused by the chytrid fungus *Bd* in amphibians vary by taxon and reproductive strategy (Tarrant *et al.* 2013); geographical area (Riley *et al.* 2013; for *Rhinella marina* from Costa Rica and PR, compare data for the former in Lips *et al.* [2006] with data for the latter in Burrowes *et al.* [2008a] and Barber [2012]); elevation (Burrowes *et al.* 2008a; Bletz *et al.* 2015); confounding effects due to methodology of experimental design and effectiveness in detectability while sampling (Miller *et al.* 2012; Mutnale *et al.* 2018; also see Arya *et al.* 2012); level of sensitivity of the technique used to collect skin samples (Shin *et al.* 2014); genetic strain of the fungus (Longo *et al.* 2013b; O'Hanlon *et al.* 2018); and size of the sample and power of the statistical tests applied (Stevens *et al.* 2012; Weldon *et al.* 2013). Not surprisingly, myriad factors can influence results based on *Bd* samples testing either positive or negative (e.g. Burrowes *et al.* 2008a), which may lead to equivocal interpretations on prevalence and presumed epidemics. In turn, these interpretations, when detection is imperfect, can have profound consequences for mitigation in an arena characterized by limited resources for conservation (e.g. Riley *et al.* 2013; Kolby and Skerratt 2015).

In South Africa (Tarrant *et al.* 2013), China (Bai *et al.* 2012), India (Mutnale *et al.* 2018), and PR (Burrowes *et al.* 2008a), for example, infections by *Bd* included enzootic (i.e. endemic) lineages, which implies that native and endemic amphibians may have a longer-lasting association with the chytrid fungus than previously acknowledged (Jani *et al.* 2017; see O'Hanlon *et al.* 2018). Unfortunately, we lack the thorough assessment of most of the amphibians in the PRA, particularly of most species collected prior to the 1970s. We believe that such studies are highly important and much needed for a more accurate assessment of this problem in its ecological and historical contexts. Despite significant advances in the state of knowledge of *Bd* on amphibians in the PRA and elsewhere, we still lack adequate sample sizes to assess the current prevalence and extent of pathogenic infection for most species in the area. The recommended sample size is between 50 (Weldon *et al.* 2013) and 60 individuals (per species per locality) in order to reach 95% certainty of detecting one specimen positive for *Bd* when apparent prevalence is medium to high (Hossack *et al.* 2010; Steven *et al.* 2012; Zumbado-Ulate *et al.* 2014; but see Tarrant *et al.* 2013) (Table 8.6).

Inferences about the conservation of the amphibians of the PRA related to the prevalence of *Bd* must be made with caution because of limitations in sampling. For example, the presence of the chytrid fungus, *Bd*, has been detected in 12 species of amphibians in the PRA (possibly 13 species; see Tables 8.5 and 8.6), but adequate data for estimating prevalence exist for only nine species based on guidelines for sample sizes (Table 8.6; Gray *et al.* 2017). In addition, adequate sample sizes for estimating the prevalence of *Bd* in six of these nine species exist only for one geographical region: *Peltophryne lemur* (coastal lowlands/plains region; 0% prevalence); *Rhinella marina* (coastal lowlands/plains region; 0.4% prevalence); *Eleutherodactylus antillensis* (coastal lowlands/plains region; 6.7% prevalence); *E. cooki* (Cayey Mountains; 1.6% prevalence); *E. portoricensis* (Luquillo Mountains; 53.3% prevalence); and *E. richmondi* (Cayey Mountains; 13.3% prevalence). Tarrant *et al.* (2013) considered a medium to high probability threshold of occurrence of 10%, which results in *E. portoricensis* and *E. richmondi* having considerably higher values of prevalence (53.3% and 13.3%, respectively), but only from the localities for which there was the recommended minimum sample size (unfortunately, only one region per species for these cases). The species for which we have adequate data on the prevalence of *Bd* from more than one geographical region are *E. coqui* (common throughout PR), *E. wightmanae* (common throughout mature forests in the highlands), and *Leptodactylus albilabris* (common throughout PR) (Table 8.6). The populations of *E. coqui* have a 47.2% prevalence at the Luquillo Mountains, 30.5% prevalence at the Cayey Mountains, 59.9% prevalence at the Central Mountain Range, and 0.5% prevalence at the coastal lowlands/plains region. The populations of *E. wighmanae* have a 78.5% prevalence at the Luquillo Mountains, 30.8% prevalence at the Cayey Mountains, and 63.9% prevalence at the Central Mountain Range. The populations of *L. albilabris* have a 74.8% prevalence at the Luquillo Mountains, 60.2% prevalence at the Cayey Mountains, 76.9% prevalence at the Central Mountain Range, and 3.8% prevalence at the coastal lowlands/plains region.

Likewise, we can calculate the prevalence of *Bd* at the level of assemblages of amphibians by geographical region considering assemblages whose species comply with the recommended

Table 8.6. The number of specimens per species of anuran sampled for the chytrid fungus, *Batrachochytrium dendrobatidis* (*Bd*), within each of four major geographic regions in Puerto Rico (based on published works for the PR Area and available unpublished data). N/D = no data; '—' = no data from the Virgin Islands. Under major geographic regions, the numerator represents the number of specimens tested positive for *Bd*, and the denominator represents the total number of specimens sampled, grouped for all datasets examined. In **bold** are samples that comply with the recommended range of the number of specimens that should be sampled for statistical validation of prevalence (e.g. between 50 and 60 samples/species/locality; see text and cited works for details on each species). Data from: Burrowes *et al.* 2004, 2008a; Longo and Burrowes 2010; Barber 2012; Caligary Damestoy 2017; Greenhawk *et al.* 2017; Stabile 2020; and unpublished data kindly provided to NR-L by Tony Báez in 2013 (*E. antillensis* from Río Piedras [*n* = 21; 0/21 *Bd* positive] and Toa Baja [*n* = 9; 3/6 *Bd* positive], northern PR), and Norman Greenhawk in 2017 (various species from Cayey). * includes the adjacent highland areas of Carite (Guavate State Forest), and the municipalities of San Lorenzo, Patillas, Yabucoa, and Maunabo (Sierra de Pandura and Guardarraya, south-eastern PR). ** also includes data from the island of Vieques and Mona Island by Burrowes *et al.* 2008a, and coastal plains and Vieques by Iniciativa Herpetologica, Inc. (2015). *** Patricia Burrowes (*Bd* detected, unpublished data to NR-L, pers. comm.).

	MAJOR GEOGRAPHIC REGIONS			
FAMILY AND *SPECIES*	LUQUILLO MOUNTAINS	CAYEY* MOUNTAINS	CORDILLERA CENTRAL	COASTAL** LOWLANDS/PLAINS
Bufonidae				
Peltophryne lemur	N/D	N/D	N/D	[a] **0 / 140**
Rhinella marina	0 / 2	N/D	0 / 1	**1 / 224**
Eleutherodactylidae				
Eleutherodactylus antillensis	N/D	N/D	20 / 20	**7 / 105**
E. brittoni	N/D	0 / 5	9 / 33	0 / 2
E. cochranae	N/D	N/D	N/D	0 / 5
E. cooki	N/D	**1 / 63**	N/D	N/D
E. coqui	**> 688 / > 1,457**	**86 / 282**	**127 / 212**	**1 / 190**
E. eneidae	0 / 11	N/D	0 / 6	N/D
E. gryllus	0 / 11	0 / 6	N/D	N/D
E. hedricki	0 / 1	N/D	N/D	N/D
E. jasperi	N/D	0 / 8	N/D	N/D
E. johnstonei	—	—	—	—
E. juanariveroi	N/D	N/D	N/D	0 / 5
E. karlschmidti	2 /6	0 / 5	N/D	N/D
E. lentus	—	—	—	—
E. locustus	0 / 15***	18 / 33	N/D	N/D
E. monensis	N/D	N/D	N/D	0 / 44***
E. portoricensis	**> 212 / > 398**	1 / 4	N/D	N/D
E. richmondi	0 / 2	**8 / 60**	0 / 3	N/D
E. schwartzi	—	—	—	—
E. unicolor	3 / 6	N/D	N/D	N/D
E. wightmanae	**102 / 130**	**120 / 316**	**76 / 119**	N/D

(*Continued*)

Table 8.6. (Continued)

FAMILY AND *SPECIES*	MAJOR GEOGRAPHIC REGIONS			
	LUQUILLO MOUNTAINS	CAYEY* MOUNTAINS	CORDILLERA CENTRAL	COASTAL** LOWLANDS/PLAINS
Hylidae				
Dryophytes cinereus	N/D	N/D	N/D	N/D
Osteopilus septentrionalis	N/D	N/D	N/D	0 / 5
Scinax ruber	N/D	N/D	N/D	N/D
Leptodactylidae				
Leptodactylus albilabris	**95 / 127**	**77 / 128**	**100 / 130**	**6 / 160**
L. fallax	N/D	N/D	N/D	N/D
Ranidae				
Lithobates catesbeianus	N/D	N/D	N/D	0 / 6
L. grylio	N/D	N/D	N/D	N/D
L. pipiens	—	—	—	—

[a] We did not include results from 18 (possibly 21) captive-bred specimens affected by brown-skin disease and examined by Crawshaw *et al.* (2014): given the clinical external-similarities between the brown-skin disease and chytrid infection (p. 562), both conditions differ, and these specimens were tested negative for *Bd* by way of histological standard and specific staining techniques, and by PCR analyses.

minimum sample size of 50 or more individuals sampled: 1.8% prevalence of *Bd* in the coastal lowlands/plains region (among five species: 15/819 individuals *Bd* positive); 51.9.1% prevalence in the Luquillo Mountains (among four species: 1,097/2,112 individuals *Bd* positive); 65.7% prevalence in the Central Mountain Range (among three species: 303/461 individuals *Bd* positive); 34.4.3% prevalence in the Cayey Mountains (among five species: 292/849 individuals *Bd* positive) (Table 8.6); and 4.8% prevalence in St Thomas (at selected localities; Platenberg 2012) (see next paragraph). However, note that if *E. cooki* is treated as a lowland plains species for computing prevalence at the level of assemblages, the prevalence of *Bd* in the coastal lowlands/plains region remains low at 1.8% (16/882 individuals *Bd* positive) but rises to 37.0% (291/786 individuals *Bd* positive) in the Cayey Mountains. Noteworthy is the marked influence that some species have on the high prevalence of *Bd* in the assemblages of amphibians in given geographical regions: in the Cayey Mountains and the Central Mountain Range, *E. coqui*, *E. wightmanae*, and *L. albilabris* are responsible for the high prevalence. In the Luquillo Mountains, these same three species and *E. portoricensis* are responsible for the high prevalence of *Bd*.

The spread of *Bd* across the VI is unlikely. The optimal temperatures (3–25°C) for the growth of *Bd* is lower than the average ambient air temperature found across the VI (24–32°C). The majority of species are direct developers and foam nesters, and the lack of freshwater across many of the cays and islands is a deterrent to the water-facilitated spread and establishment of this disease. Sampling by the Fort Worth Zoo on St John indicated no occurrence of *Bd* (D. Barber, unpublished data). No records are known of *Bd* in St Croix. However, in 2011, *Bd* was found at three of nine localities in the north-central mountains of St Thomas at low prevalence (3 out of 63 frogs, or 4.8%). The infected frogs included *Eleutherodactylus antillensis* and *E. coqui* (Platenberg 2012).

These data contrast with claims of epidemic infections (i.e. the rapid and wide-scale spread of a disease) leading to extinction of amphibian species in PR (e.g. *E. karlschmidti* and *E. jasperi*); suggested extirpation of species such as *E. wightmanae, E. richmondi,* and *E. locustus* at EYNF; and declines in abundances of other species of amphibians elsewhere in the region. The data do fit the hypothesis that *Bd* is implicated, in part, in the population dynamics of *E. coqui* and *E. portoricensis* at EYNF (Longo and Burrowes 2010), but causality with presumed declines in population size is difficult to prove. For instance, the fact that level of infection by *Bd* in individuals of *E. coqui, E. wightmanae,* and *L. albilabris* varies between recaptures throughout the year (Caligari Damestoy 2017: 15) – with outcomes like the lowering of infection level and clearing of the infection between successive captures, and the gradual decrease in infection level between successive recaptures – is revealing. However, plausible explanations have been provided by Longo and Burrowes (2010) on ontogeny-mediated susceptibility to infection in synergy with seasonality of weather variables, and by Burrowes *et al.* (2017) on the microhabitat-mediated susceptibility of infection for the closely related *E. coqui* and *E. portoricensis.* We need similar data from other species that differ in their ecological requirements and their physiological responses to environmental conditions across their geographical distributions.

Two related questions are relevant for conservation applications given the available data on infection and prevalence of *Bd* in the PRA. First, how much of the variation in the prevalence of *Bd* among constituent species in anuran assemblages is explained by geographical region? Second, how much influence does sample size between highly abundant species (most data available) and low abundance ones (minimal data available) have on the prevalence of *Bd* among constituent species of relatively high conservation concern (i.e. species deemed rare and ecological specialists)? From the perspective of conservation, we have a lot to learn about the population dynamics and distribution of the chytrid fungus itself, as it is challenging to manage without understanding its biology and ecology. Nonetheless, recent techniques may help generate part of the much-needed data.

Techniques for sampling environmental DNA of a target of interest are available (e.g. Torresdal *et al.* 2017; see Mutnale *et al.* 2018 for nested PCR versus qPCR), which supplement methods for the rapid detection of *Bd* from a wide range of environmental conditions, other than from frogs' skin, and at any time (Phalen *et al.* 2011; Kolby *et al.* 2015). These techniques may help to understand and monitor the population dynamics of the chytrid fungus at different seasons on a multi-year basis, and at several localities that differ in environmental conditions (e.g. lowlands versus highlands; grass/shrub ecosystems versus forested ecosystems). We believe that this information is of prime importance for conservation as it would help evaluate areas for the re-establishment of populations of amphibians produced *ex-situ* under captive-breeding programmes. Findings from Harris *et al.* (2009) are equally valuable as they provide a potential non-invasive treatment for species (and areas) tested positive for *Bd*. For example, several species of bacteria, such as *Janthinobacterium lividum,* produce the anti-chytrid metabolite violacein, which has been proven to prevent morbidity and mortality in the endangered frog *Rana muscosa* from California, USA, but not in all systems (Harris *et al.* 2009; Becker *et al.* 2011). Similar prospects may also apply to PR.

In 2010, Ríos-López and undergraduate students detected a bacterium, *Chromobacterium piscinae,* in water samples from a wetland inhabited by the endangered *Eleutherodactylus juanariveroi* and *E. coqui* in Toa Baja, northern PR (NR-L, unpublished data). This bacterium is closely related to *C. violaceum* and *C. pseudoviolaceum* (Kämpfer *et al.* 2009), which are known to produce violacein. In this wetland, individuals of *E. juanariveroi* frequently submerge in the water that covers the base of the bull tongue arrowhead, *Sagittaria lancifolia,* and coincidentally, *E. juanariveroi* and *E. coqui* from the same wetland have tested negative for the chytrid fungus (e.g. Burrowes *et al.* 2008a). Violacein-producing bacteria may have profound implications for environmental remediation and probiotic therapy, and their study should be pursued further (Woodhams *et al.* 2015).

D. Harvesting amphibians for various uses

Horticulturist Dr Thomas Brown McClelland mentioned that in July 1929 one lot of 12 individuals of *Leptodactylus fallax* from Dominica was released in Cartagena Lagoon, south-western PR, because this lagoon (McClelland 1932: 419) was 'highly favorable for the development of the toad' *Rhinella marina*; a second lot of 90 individuals was imported in June 1932 and 50 individuals 'were released on the bank of the reservoir located on the station farm just north of Mayagüez, and forty were released on the river bank near Las Vegas' in the vicinity of Maricao, western PR. McClelland (1932: 419) ended his report with: 'This species is shown in colors in the National Geographic Magazine, May, 1932, and under the plate is the legend: An Edible Species In Some Danger of Extermination. Let us hope that its introduction into Puerto Rico may not reduce the danger of extermination, but that it may also supply the Island with an epicurean dish not previously procurable.' In November 1932, 'a young frog of this species was found' on the station grounds at Mayagüez (McClelland 1932: 419), but the species was never seen again (Rivero 1998).

In 1935, 40 individuals of the common bullfrog *Lithobates catesbeianus* (*Rana catesbeiana* at that time) collected in Florida, USA, were introduced in a specially constructed pond at Río Piedras, PR (Pérez 1951a; 1951b). By 1936, the Annual Report to the Insular Department of Agriculture (author anonymous) stated that this introduction was very successful as frogs were breeding readily (Pérez 1951a; 1951b). By 1951, it was noted that the species 'may be expanding rapidly in the more humid coastal areas' and sightings of an area other than Río Piedras and neighbouring towns included Humacao (eastern PR), Barceloneta (northern PR), and Mayagüez (western PR) (Pérez 1951a; 1951b). It did not become a food source (Rivero 1998) and today, other than records of sightings along coastal areas in PR and the presumed failed introduction into St Croix (Mayer 2012), estimates of population sizes of *L. catesbeianus* are lacking (see section III. E. 2.c).

Also in 1935, the PR Experiment Station at Mayagüez collaborated with the School of Medicine of the Johns Hopkins University at Baltimore, USA, to extract secretions from the paratoid glands of approximately 100 *Rhinella marina* for pharmacological and chemical studies of their venom. These samples were sent to Herman H. Jensen for his studies during 1930s at the Laboratory of Pharmacology of the Johns Hopkins University (Puerto Rico Experiment Station 1937).

In 1974, the US Bureau of Sport Fisheries and Wildlife revised the Lacey Act of 1900 (which regulates the importation of injurious wildlife) to 'establish a list of "low risk" species which may be imported without permit' to the USA and its territories (PR and the USVI) (US Department of the Interior 1974). The US Department of the Interior (1974: 11) further informed: 'At the present time only *Rana pipiens* and *Rana catesbeiana* are included in the proposed "low risk" list of amphibians and reptiles.' Erdman (1967) mentioned that bullfrogs have not become abundant in PR because of the limited areas of swamp suitable for this species. He did not recommend the establishment of a bullfrog culture in PR because, he reasoned, attempts to do so in the USA and Japan failed. However, according to Rivero (1998), the only title referring to the development of a frog-culture industry in PR (and the VI) comes from Riollano (1976). Agronomist Mr Arturo Riollano wrote a detailed prospectus (47 pages) to the Agriculture Experiment Station of the University of Puerto Rico for the development of an industry based on *Lithobates catesbeianus*, similar to initiatives now under way in the Dominican Republic, Costa Rica, and Mexico, among other countries. Rivero (1998) mentioned that this prospectus never progressed in PR (for reasons not explained) and no frog industry was established; however, two species of frogs,

Leptodactylus fallax (McClelland 1932) and *L. catesbeianus* (Pérez 1951a; 1951b), were introduced into the wild to provide the 'Jíbaro' with a source of protein (Rivero 1998).

In April 2014, a US citizen claimed to have collected 16 individuals of two species of coqui frogs from Vieques and to have tried to send them via private postal mail and alleged to have sent 'wallets' to a friend in Ohio. That person was never apprehended to face charges in PR for violating a local law (NotiCel 2014). Being an illegal activity in PR, USVI, and BVI, reports of such attempts to export native and endemic species of amphibians from the PRA are hard to document. Nonetheless, many species of amphibians (anurans, caecilians, and salamanders) were (and are) imported and sold by the pet-trading business – either legally or illegally – but we are unaware of the establishment of any of these species in PR or the VI.

E. Naturalized species

1. Freshwater fishes

The origin of most freshwater fishes currently in the islands is not natural, which justifies a brief discussion of the history (natural and human) of fishes and fisheries. In doing so, we identify areas of needed research that may provide for a broader understanding of the potential impact of a human phenomenon (introduction and maintenance of freshwater fishes) on amphibians in the PRA.

Few studies address the effects of naturalized species of fishes on the abundance and distributional patterns of native freshwater fishes (e.g. Bacheler *et al.* 2004), much less on the amphibians. Two species of native anurans from the PRA need standing bodies of water for reproduction: *Leptodactylus albilabris* and *Peltophryne lemur.* As far as we know, *P. lemur* breeds in intermittent seasonally flooded ponds and creeks throughout its historical and current distribution. However, *L. albilabris* produce larvae that can develop in the presence of freshwater fishes in creeks elsewhere, in palustrine and brackish-water wetlands (herbaceous and forested) in coastal areas, and in virtually all water reservoirs on the island. Consequently, the native (and naturalized) anurans overlap the aquatic habitat that many species of native and inland fishes (*sensu* Erdman 1967, 1972) inhabit in the PRA or can inhabit if introduced.

Recently, López-Hernández *et al.* (2020) conducted a laboratory experiment using aquaria to examine the behavioural responses of larvae of *Leptodactylus albilabris* to the presence of the naturalized cichlid fish *Astronotus ocellatus* ('*Oscar*'; see Neal *et al.* 2009 for data on distribution). Once added to the aquaria, larvae were aware of the predatory fish and responded to it by using areas covered with leaf litter more frequently than expected, and overall, larvae became less active than expected, when the fish was present. The implications of these findings under natural conditions need further study; and also to determine whether similar responses result in the presence of other naturalized or native fishes – including differences in developmental time and size and/or body condition at metamorphosis.

The native fish fauna in Puerto Rican palustrine-ecosystems consists of seven diadromous species, one of which, *Sicydium plumieri*, has been split into four species that are difficult to distinguish externally, which raises this number to ten (Neal *et al.* 2009). Initially, fishes frequently caught in fresh waters and consumed by locals in PR were mostly the native species *Agnostomus monticola* ('*dajao*' in Spanish), with a contribution from '*guavinas*' (a Spanish common name used for several species; Evermann 1900), but the vast majority of the commercial fish-trade was based on importations prior to the transition between Spain and the USA in 1898. Importations dictated the local fish industry afterwards, and mainly from the USA, until 1915 when Hugh McCormick Smith, Commissioner of the US Bureau of Commercial Fisheries, reported to the Secretary of Commerce of the USA that (Smith 1917: 12–13) 'In compliance with official requests received from Porto Rico, Cuba, India, and Japan, consignments of rainbow-trout eggs and of various species of pond fishes were made as follows… (fingerlings, yearlings, and adults introduced to PR) 600 Black bass (= Largemouth [*Micropterus salmoides*]; Peterson *et al.* 2017), 600 Catfish, 1,200 Rock bass, and 600 Sunfish'. The origin of these stocks was not given, nor was the location of release on the island, but this action established a live-fish culture in PR by the government of the USA. The local government in PR continued these efforts in 1934 to strengthen this culture, and many species of sport-fishes were imported from the USA during the next two decades (Erdman 1967, 1972). At EYNF, efforts to establish a trout culture was initiated between 1934 and 1937, and by the end of the 1930s, eggs of rainbow and brown trout were imported from the USA and introduced into a pool constructed for this purpose near one of the trails in EYNF (Gerhart 1936; Erdman 1967, 1972; Weaver 2012). According to Gerhart (1936: 297; translated freely from Spanish), this introduction intended that 'All streams in the forests will be stocked with trout.' However, adults failed to spawn because the temperature of the water was still too high for them and the efforts were discontinued in 1942 (Erdman 1967, 1972). Weaver (2012: 90) summarized the attempts of an introduction into EYNF:

> Rainbow trout eggs (*Oncorhynchus mykiss*) were introduced into the La Mina area nursery in 1934 and 1935 in an attempt to provide sport fishing. The effort was unsuccessful because the streams were too warm for reproduction (Gerhart 1936, see endnotes). During the 1990s, koi (or carp, *Carpinus carpio*) were placed into a shallow, 75 m^2 depression scraped out of impermeable soil along FS-27 (Pico del Este road). Rainfall, cloud moisture deposition, and possibly runoff from the road provide enough water for the koi to survive.

On occasions, larvae of *Leptodactylus albilabris* also can be seen in water that accumulates along the edges of this, and other, roads at high elevations in EYNF (NR-L, pers. obs.).

The establishment of the largemouth bass (*Micropterus salmoides*) was successful throughout PR as new stocks of this species were introduced in 1946 and 1957 to further promote sport-fishing in freshwater reservoirs and provide the rural people on the island with additional protein (Erdman 1967, 1972; Peterson *et al.* 2017). Where introduced, besides PR (it is also one of the nine non-native inland species of fish known or reported from St Croix; Smith-Vaniz and Jelks 2014), the diet of the largemouth bass includes larval and adult anurans (Hickley *et al.* 1994; Lee *et al.* 2009; Purdom *et al.* 2015). The Department of Natural and Environmental Resources of the Commonwealth of Puerto Rico (PRDNER) continues to breed this and other species of fishes to stock reservoirs for sport-fishing, and further introductions from Floridian (USA) stocks are still recommended for fishing purposes (reviewed by Peterson *et al.* 2017). Consequently, the introduction of this, and other, species of fishes into the PRA added more than five times the number of native fishes; the introduced species complete their reproductive cycles entirely in the palustrine environment (Neal *et al.* 2009). Most of these introductions have been intentional for sport-fishing, but also as additional sources of food for the people, for mosquito control, and fortuitously via the pet trade or other means (Hildebrand 1935; Gerhart 1936; Erdman 1967, 1972; Neal *et al.* 2009; see Loftus 2004 for St John, and Smith-Vaniz and Jelks 2014 for St Croix). As a result, introduced freshwater fishes frequently are found throughout river systems on PR and in artificial bodies of water elsewhere in the PRA. Nonetheless, we know little about the influence of freshwater fishes on native (and non-native) amphibians.

2. *Amphibians*

In the PRA, species such as *Rhinella marina, Lithobates catesbeianus, L. grylio, Osteopilus septentrionalis*, and *Scinax ruber* can frequently be found in the lowlands. These five species have in common a dependency on bodies of freshwater for reproduction. In contrast, native and endemic anurans are underrepresented in this habitat, except for two species: *Leptodactylus albilabris* (terrestrial foam-nester with free-swimming larvae) and *Peltophryne lemur* (lays eggs in water). As mentioned, these two species require standing and slow-moving water to complete, either partially (*L. albilabris*) or completely (*P. lemur*), the development of their larvae. The rest of the native anurans in the region consist of direct developers (*Eleutherodactylus*) that lay terrestrial eggs that hatch into miniature replicas of adults (one species, *E. jasperi*, was ovoviviparous). As an evolutionary group of taxa in these islands, native anurans are underrepresented in the freshwater habitat, which coincidentally is exploited by all but one introduced species of anuran (*E. johnstonei* on Tortola; Table 8.1).

(a) *Rhinella marina*

In this section, the reader should be aware that the level of detail of the information provided for *Rhinella marina* in the PRA is necessary for three reasons. First, this information (mostly available from PR) contrasts significantly with that from continental areas, and most dramatically, from Australia, where *R. marina* has been directly implicated in population declines and extirpations of a wide range of species of animals. Second, ecological processes do not necessarily occur similarly in continental areas compared with islands, and that spatiotemporal variation in ecological requirements, along with local species diversity, species abundance, and human-related variables, influence colonization and establishment of introduced species differently, amphibians included (for insights about this topic see Stachowicz *et al.* 1999, Rodda and Dean-Bradley 2002, Kraus 2009, and Capinha *et al.* 2017). Therefore, generalities drawn upon the effects of *R. marina* on native/endemic species must be taken with caution as they may not apply under all biotic and abiotic conditions and in all geographical areas. Moreover, this species was present in the PRA at least 85 years earlier than previously acknowledged, and ecological studies that include this species are lacking, which has profound implications for ecological dynamics and conservation contexts. Finally, the evidence presented herein reveals that dynamics of populations and assemblages of amphibians (and reptiles) are influenced by forest succession (e.g. Ríos-López and Aide 2007; Herrera-Montes and Brokaw 2010), which also may influence perceptions of the effects of *R. marina* (and other naturalized species) on other species of amphibians in this archipelago.

Initially, evidence of an intentional introduction revealed that 12 individuals of *R. marina* (formerly *Bufo marinus*) were brought (see section II. A.; but see Reinhardt and Lütken 1863) to the main island of PR from Barbados in 1920 by David W. May, Director of the US Department of Agriculture-Mayagüez Experimental Station, western PR. May released these specimens into the Experiment Station grounds (Danforth 1925; Dexter 1932; McClelland 1932; Seín 1937a; Wolcott 1937; Barlett 1949). Its introduction was intentional as a biological control for native white grubs that were threatening the sugarcane industry of PR (Dexter 1932; Wolcott 1937; Barlett 1949). In 1923, the second group of 40 *R. marina* was brought into PR (to US Department of Agriculture-Río Piedras Experimental Station, north-eastern PR) from Jamaica (Seín 1937a; Pérez 1951a,b). In 1926, the third group of 1,114 individuals was introduced into Aguirre, south-eastern PR (Dexter 1932; Leonard 1933). As their numbers increased, the toads were sent to various parts of PR in lots of two to 2,000 (Leonard 1933), and the species became well established throughout the island.

The diet of *R. marina* in PR was first assessed methodically by Mrs Raquel R. Dexter, at the time a member of the Department of Biology of the University of Puerto Rico, between early autumn of 1931 and winter of 1932 (Leonard 1933). The main objective of her study was to evaluate the effectiveness of *R. marina* as a biological control in the agricultural landscape (moreover, on p. 1 of her publication, she attributed a description of a foam nest, eggs, and larvae to *R. marina*, but these were of *Leptodactylus albilabris* instead; Rivero 1978). Dexter collected 301 specimens of *R. marina* from 18 localities from around PR; 100% of the prey items were invertebrates. Wolcott (1937) documented the diet of *R. marina* at Río Piedras (and one sample from Loíza Aldea, northern PR) and found no remains of vertebrates in 91 pellets of excrement. However, Rivero (1978, 1998) asserts that *R. marina* is capable of feeding upon mice (*Mus musculus*), small birds, and baby chicks (*Gallus gallus*), while its tadpoles feed on vegetable matter.

Until the mid-1960s, information on the breeding biology of *R. marina* in PR was mostly anecdotal and from chance observations (e.g. Seín 1937b; Mercado-Burgos 1966). The populations of *R. marina* in PR consisted of only 12–1,114 individuals during the 1920s (see above paragraphs and this section). According to Seín (1937a), Wolcott (1937), Pérez (1951a,b), and Rivero (1998) the species reached a peak in abundance between 1932 and 1933, only to decline a few years later as favourable conditions for its main food source (white grubs) changed, among other climatic and human phenomena that influenced the landscape in PR (see below). Wolcott (1948b: 261) reviewed what happened since the introduction of this species in the early 1940s into PR, and its relation to the sugarcane industry and its biological control of the white grub:

> The solution of the white grub problem by the introduction of the toad was indeed a major triumph for biological control, at least during the years when it was most effective. Its fundamental weakness was in its very effectiveness, however, for despite the supposed omnivorous food habits of the toad, the major elements, in Puerto Rico are May beetles and other scarabaeids, "vaquita" leaf beetles and millipedes. When these are gone, and we have close to total control of white grubs in cane fields, there is little else for the toads to eat. Vast numbers of them must have perished of starvation. As was explained in detail in "What has happened to the giant Surinam Toad, *Bufo marinus* L., in Puerto Rico" (*Revista de Agricultura y Comercio de P. II.*, **38** (1): 25-29, fig. 7. San Juan, January–April 1947), "scarcity of food is only one factor which increasingly tends to limit the abundance of the Surinam toad. In the West Indies at least, the weakest point in its life-history is the tadpole or pollywog stage, which must be passed in fresh water. In addition to all the large and fierce dragon fly nymphs which attack the helpless tadpoles, the even larger and more voracious larvae of the endemic dytiscid beetle, *Megadytes gigantea* Castelnau, appear to be the deciding factor in whether any tadpoles will survive to become adults. A single one of these large beetles may lay dozens of eggs in a pool, from which emerge small worm-like larvae with enormous jaws. At first they feed on insects and very small animals, but before they are more than half-grown, they attack tadpoles almost ready to leave the water, and often are forced to complete their growth by cannibalistically feeding on each other when every tadpole in the pool has been destroyed.

Wolcott (1948b: 261–2) continued:

> Reproduction also became more difficult because after three exceptionally dry years, Cartagena and Guánica Lagoons completely dried up and thus the most favourable environments for the development of the tadpoles on the south coast disappeared. Rapidly the balance for the toad began to swing in the opposite direction. Cane growers of a new generation, inexperienced in the plagues of white grubs that had infected the fields of their fathers, again began to complain about

> the abundance of white grubs. When advised to bring toads to infected fields, the problem was where to find the toads.

Finally, Wolcott (1948b: 262) added:

> The rapid increase in abundance in recent years of the common bullfrog of the southeastern United States, *Rana catesbeiana* Shaw, introduced in 1935 into Puerto Rico, at the same time that the toad was becoming so scarce was thought to indicate a possible incompatibility between the two introduced amphibians. Their pollywogs inhabit the same pools, altho [*sic*] the adults occupy entirely different ecological niches. In a study of the food habits of the frog, Mr. Mario Pérez found that altho [*sic*] the frog may be cannibalistic in devouring its own smaller adults and larvae, it did not eat either the pollywogs or the small adults of *Bufo marinus*. The tadpoles of both feed primarily on aquatic algae, of which an amount ample to feed them rapidly grows on the stones and vegetation of every pool. There is thus no serious competition between frog and toad, either as larva or adult, for the frog adult feeds extensively on aquatic insects and other items, and but rarely on the Maybeetles that are the mainstay of the toad.

However, interspecific interactions among anurans and other taxa are more complex and influenced by different natural and human phenomena (see section III. E. 2.a).

For the rest of the twentieth century, estimates of population density of *R. marina* were lacking, and information on its feeding and breeding ecology continued to be anecdotal. For example, Rivero and Seguí-Crespo (1992) mentioned that Mr Estremera (then a graduate student of Rivero; now a retired high-school teacher from Quebradillas) observed one specimen of *R. marina* feeding upon recently metamorphosed *Osteopilus septentrionalis* as these emerged from the surface of the water in a pond at Quebradillas, north-western PR. Estremera dissected the toad and found that 52 metamorphosed *O. septentrionalis* had been ingested (E. Estremera, pers. comm. to NR-L 2018). Quebradillas lies in the karst belt on north-western PR and is one of the localities where the endemic *Peltophryne lemur* used to occur in the north. Scientists long suspected *R. marina* of preying upon *P. lemur* (Ross 2007, Rivera-Sostre 2008), which contributed, by association, to strengthen the justification of listing *R. marina* as 'harmful' to the native and endemic wildlife (e.g. DRNA 2004a; Ross 2007; Rivera-Sostre 2008; USFWS 2016). However, Rivero *et al.* (1980) noted that scientists had been aware of the rarity of *P. lemur* since the mid-1880s, well before the introduction of *R. marina*. Presumably, *P. lemur* has always been a species of low abundance that occurs in groups around suitable breeding sites within the karst belts in southern and northern PR, and the VI. Nonetheless, studies of *R. marina* in different ecosystems throughout the PRA were lacking, which may have contributed to the increase in interest, in the 2000s, to study this species for conservation purposes (see next paragraphs, this section).

Initially, studies on *R. marina* continued to be mostly anecdotal or from fortuitous observations in the early 2000s (e.g. Lever 2003; Vargas-Salinas 2005; Solís *et al.* 2009). For example, at El Tallonal Private Nature Reserve – a nature reserve in a karst forest in Arecibo, northern PR – Vega-Castillo and Puente-Rolón (unpublished data) include individuals of *R. marina* as preying upon rats, bats, and chicks of the bananaquit, *Coereba flaveolata*. This reserve was selected for the release of larvae (see sections V. B. and VI.) of the threatened *Peltophryne lemur* on a frequent basis in artificial ponds with the objective of establishing a self-sustained population of this species in the northern karst region where it was no longer detected at its historical collection sites. The initiative for conservation of *P. lemur* at El Tallonal also led to a thorough study by Rivera-Sostre (2008) that examined 300 individuals of *R. marina* collected from this reserve during 2006 and 2007. She documented that vertebrates (three snakes, one anole, and one freshwater turtle) accounted for 1.7% of the stomachs (5/300) examined and 0.05% (5/9,338) of all prey items. Furthermore, approximately 2,000 individuals of *R. marina* were collected during an hour-long period each month between 2006 and 2017 in an area of ~2 ha cultivated with plantains and oranges, and adjacent to a large artificial pond at El Tallonal. The removal of this toad was also part of the initiative for conservation of *P. lemur* in the presence of *R. marina*. Recently, however, the number of collected individuals of *R. marina* at this area dropped to approximately 20 per monthly visit, ~2 individuals ha^{-1}. Finally, Flores-Vallejo (2011) documented a juvenile *R. marina* preying upon a metamorphic *P. lemur* at El Tallonal.

Other studies generated new knowledge of the reproductive biology, interspecific interactions (including microbial associations with the skin of *R. marina* in PR), and population ecology (we are unaware of any studies from the VI). These studies offer information on its breeding behaviour (Vargas-Salinas 2007); larval-interactions with *Leptodactylus albilabris* (Flores-Nieves *et al.* 2014); characterization of 'Operational Taxonomic Units' of microbiota found in the skin of *R. marina* in northern PR (Abarca *et al.* 2017; Table 8.5); and its distribution, abundance, and larval development along a salinity gradient in a coastal forested wetland, northern PR (Ríos-López 2008a). In addition, estimates of its population size from several locations within a lowland coastal area in northern PR were 5.6 individuals ha^{-1} in a deforested karst valley (Ríos-López and Aide 2007), 5.6 individuals ha^{-1} in mangroves, 2.8 individuals ha^{-1} in a *Pterocarpus officinalis* brackish-water wetland forest, and 11.1 individuals ha^{-1} in a pasture with a long history of agriculture (Ríos-López 2007). However, these estimates were based on diurnal and nocturnal censuses along census-transects distributed throughout the landscape, which contrast with the estimate provided above at El Tallonal. As far as we know, estimates of population size of *R. marina* at the ponds used by *Peltophryne lemur* in south-western PR (Guánica forest) are not available.

Rhinella marina, having been around for at least 186 years in the PRA, is already integrated as a prey into local food webs. Initially, Wolcott (1948b) noticed the role that predatory aquatic beetles have on the survival of larvae of *R. marina*. Rivero (1978, 1998) mentioned that on occasions, rats and mongooses feed on *R. marina*, apparently without ill effects. Consequently, he reasoned that *R. marina* is uncommon today because of predators, parasites, and diseases; unfortunately, Rivero did not describe these events. Yet we do know that monkeys may ingest *R. marina* in PR (see section III. E. 5.e). Furthermore, the larvae of *Leptodactylus albilabris* are facultatively carnivorous (carrion feeders; Lebrón *et al.* 1995) and feed on dead *R. marina* (Joglar *et al.* 2005). Specifically, Joglar and Ríos-López witnessed the larvae of *L. albilabris* feeding on the flesh of a dead adult *R. marina*; they stripped the body clean to the bone, but left the area of the paratoid glands intact. In St Croix, USVI, Gassett *et al.* (2000) documented the cattle egret (*Bubulcus ibis*) as a predator of *R. marina*. Other predators of *R. marina* were listed by Henderson and Powell (2009: 31) and included the Puerto Rican boa, *Chilabotrus inornatus* (under captive conditions), red-tailed hawks, giant centipede (*Scolopendra alternans*), and the '*bruquena*' (or '*buruquena*'; indigenous name for the river crab *Epilobocera sinuatifrons* [Hernández Aquino 1977]) as predators upon *R. marina* in PR. The Puerto Rican racer, *Borikenophis portoricensis*, captures and attempts to prey upon *R. marina* in the wild (ingestion was not observed; A. Cubiñá, pers. comm. and picture provided to NR-L), and eats large-bodied juveniles of *R. marina* without ill effects (NR-L, unpublished data). Among birds, the Smooth-billed Ani or '*judío*' in Spanish (*Crotophaga ani*) and the cattle egret (*Bubulcus ibis*) in PR prey upon large-bodied juveniles of *R. marina* (NR-L, pers. obs.).

As mentioned, the number of *R. marina* in PR increased rapidly as food resources and widely open breeding areas (not exploited by native/endemic species of amphibians) were plentiful relative to its population size during colonization of the lowlands and sugarcane plantations (hence one of its common names 'cane toad'). Initially, its population dynamics is described by the density-independent model of population growth with overlapping generations under unlimited resources. The following words from Leonard (1933: 69) are reproduced here for comparative purposes: 'During rains the toads often become active in considerable numbers in the daytime. Many are killed by automobiles on the highways by night.' However, the rate of population growth of *R. marina* decreased with forest succession after humans ceased its introduction into sugarcane plantations and resources became limited with the abandonment of agriculture in PR. Climatic phenomena (e.g. severe droughts throughout PR) may have contributed to further declines in its population size and distribution since its introduction and colonization (Wolcott 1948b).

On other islands, such as the Philippines, the pattern of the rise and fall of *R. marina* as seen in PR in just 20 years after its introduction to agricultural land was also noted over a similar timespan. Dr Angel Chua Alcala (1957: 91) from the Silliman University (Philippines), examined several factors that led him to conclude that on Negros Island in the Philippines '*Bufo marinus* occupies a habitat and an ecological niche not occupied by any local amphibian', which contributes to its 'rapid increase in number'. However, Alcala (1957: 94) highlighted that 'it is obvious that a species cannot indefinitely increase' and added (p. 95):

> My field observations tend to point to the existence of an equilibrium between the toad population and food on Negros. In the southwestern part of Negros Occidental the toad population has not increased materially in numbers since the time it reached this area 15 years ago. In fact, there are reasons to believe that the population on Negros is declining. While no attempt to make actual count has been made, I have observed that in the Dumaguete area there are now fewer toads than there were four or five years ago. Whereas, five years ago many specimens crushed by vehicles at night littered the streets, now only a few can be seen.

Finally, Alcala (1957: 96) summarized his observations and analyses:

1. *Bufo marinus* in the first fifteen years or so of its existence following its introduction in the Philippines did increase in great number in many areas.
2. *It* has now reached its peak in population density [~1950s].
3. It may actually be declining in numbers due to unknown causes, one of which is probably a disease characterized by an extreme emaciation.
4. This species is probably here to stay as one of the permanent components of our amphibian fauna, but there are at present no evidences to show that it is competing with the local amphibians, resulting in the driving out or replacement of any of the local species.

The similarities in the population dynamics of *R. marina* during colonization of these two different groups of islands – the Philippines and the PRA – are notable and deserve a final thought here. In PR, the images of 'considerable numbers' of toads active during the day after rains and of roadkills of toads are now rare if at all. Currently, its population size may fluctuate around lower values of density – compared with values early during colonization – and is dependent upon geographical location, microhabitat conditions, and type of land-cover (e.g. urban versus forests in succession) as also observed by Alcala (1957). Both instances reflect the fundamental process of ecological succession after colonization, and we believe that all significant effects that *R. marina* could have inflicted on the amphibians in the PRA had already occurred by the mid-twentieth century; it is now a naturalized species that is under population regulation by natural phenomena (including humans). Nonetheless, we acknowledge that there is a need for planned studies aimed at examining the role of *R. marina* in assemblages of amphibians on these and other islands and that we must take generalities on the status and effects of this species drawn from different biogeographical (e.g. continents) and historical (e.g. socioeconomic and cultural) contexts with caution.

(b) *Osteopilus septentrionalis*

Genetic evidence suggests that the introductions of *Osteopilus septentrionalis* into PR originated in Florida, USA, most likely mediated by humans who brought them from Cuba (along with a possible contribution from natural dispersion) (Heinicke *et al.* 2011). This species was introduced into PR before 1959 (Rivero 1978, 1998; Joglar and Ríos 1995) and spread throughout the region between Aguadilla and Isabela in the north-west (Schwartz and Thomas 1975; Schwartz and Henderson 1991; Rivero 1998). Rivero (1978, 1998) suggested that individuals of this species were first brought as pets by children of military personnel serving at US Ramey Air Force Base in Aguadilla (north-western coastal PR). Since then, the species has been found in coastal (e.g. Estrada 2010) and various inland areas, mostly associated with human dwellings, e.g. Río Piedras (Alejandro Hernández and Joglar 2020), Trujillo Alto, Caguas, Gurabo, Morovis, Jayuya, Río Grande, Fajardo, and Humacao. Morales (2011a) also documented this species in the Cabo Rojo municipality, coastal south-western PR. The spread of this species in PR is associated, in large part, with the ornamental plant business, but transportation in military cargo to Vieques also may have occurred (Herrera-Giraldo 2010).

In the VI, *O. septentrionalis* was first sighted on St Croix in 1974, and breeding populations were found in 1975 across the island, presumably originating from individuals that escaped after being brought from PR (Philibosian and Yntema 1976, 1978; Lever 2003; Heinicke *et al.* 2011). This species was also introduced accidentally into the BVI, primarily in ornamental plants and construction materials. Currently, in the VI it is associated with human dwellings, particularly residential water reservoirs under drier conditions (Philibosian and Yntema 1978; Perry *et al.* 2006).

The fecundity of *O. septentrionalis* is high, with females producing clutches of between 3,961 and 4,831 eggs, and an average size of ova of 1.2 mm diameter (Meshaka 2011). This species is considered voracious and harmful (Meshaka 2011) and a threat to other species of amphibians in the PRA (DRNA 2004a; Owen 2005; Perry and Gerber 2006; Platenberg 2006; Platenberg and Boulon 2006).

We examined the available data on the influence of *O. septentrionalis* on other amphibians in the PRA and identified gaps in information that justify further studies. Individuals of this species prey upon adults of *Leptodactylus albilabris* (F. Vargas-Salinas, pers. comm. to NR-L; based on fortuitous observations in the vicinity of a shallow pool at a cattle ranch in north-western PR). In contrast, in 2011, Vega-Castillo *et al.* (unpublished data) examined the stomach contents of 139 adult *O. septentrionalis* (77 females and 32 males) from a coastal lagoon in the municipality of Cabo Rojo, south-western PR, and found that *L. albilabris* was represented in 7.8% of the stomachs examined. However, based on 127 prey items among 16 prey categories found, the fraction of anurans used as a food resource by *O. septentrionalis* (3.9%) was lower than the minimum cut-off of the number of frequently exploited prey items (6.25%, after Krebs' [2014] definition; Ríos-López 2021). Similarly, Ríos-López (2021) found an adult male of *Eleutherodactylus coqui* (SVL 26.1 mm) in a stomach of a female *O. septentrionalis* (SVL 73.4 mm; catalogue number UPRRP-5774) out of 27 stomachs with prey items, examined from an urban area in the municipality of Trujillo Alto, northern PR. In this study, he found that anurans (the only vertebrate prey item found in the diet of *O. septentrionalis*) represented 3.7% among stomachs, but anurans represented 0.18% of prey used as a food resource, which did not reach the minimum cutoff number of frequently used prey items (= 3.3%; Ríos-López 2021). In lowland areas where *O. septentrionalis* is frequently found throughout PR, Joglar *et al.* (2005) suggested that the population status of *L. albilabris* is stable; in these areas, *E. coqui* is the most abundant species of frog. It is also the anuran with 'the highest published density of non-aggregated frogs' (Rodda *et al.* (2001: 334).

Contrary to what has been described from continental areas (e.g. Florida, USA; Meshaka 2011), predation of anurans (and vertebrates at large) by *O. septentrionalis* in PR, and throughout the geographic distribution of its introduction, is very low, and even lower in the BVI (for a review, see Ríos-López 2021). However, inferences about the influence of *O. septentrionalis* on native anurans could be equivocal in several ways. For example, on St Croix, *Leptodactylus albilabris* is less frequently found than in the past, and its potential rarity today is attributed to the prevalence of *O. septentrionalis* and *Rhinella marina* (NF-A, unpublished data). Perry and Gerber (2006) and Platenberg (2007) suggested that tadpoles of *O. septentrionalis* could be a threat to tadpoles of the native frog *L. albilabris* in the VI. Nonetheless, if the feeding ecology of *O. septentrionalis* in the BVI and PR mirrors that in St Croix, then its threat as a predator on native amphibians there may be less important than other aspects of the interactions within the anuran assemblage: *L. albilabris-O. septentrionalis-R. marina* in the PRA (see next paragraphs, this section).

Larvae of *Leptodactylus albilabris* hatch from clutches that consist of 75–200 large eggs (diameter 2–3 mm). After hatching, the larvae stay for several days within a foam nest in terrestrial chambers under logs, rocks, or debris. In fact, larvae can survive for up to 40 days when maintained on moist cotton under laboratory conditions (Candelas *et al.* 1961). In the field, incoming rains wash the nest into temporary pools where larvae complete their development; at this stage, larvae may be between 12 mm and 13 mm in body length (Schmidt 1920; Rivero 1998; Joglar *et al.* 2005). Therefore, based on their reproductive biology, it is possible that by the time temporary pools form after heavy rains, larvae of *L. albilabris* can be fully formed in their foam nests and may have an ecological advantage over larvae of *O. septentrionalis* once they are together in pools (e.g. Heatwole *et al.* 1968 for *L. albilabris* and *Rhinella marina*).

The interaction between *Leptodactylus albilabris* and *O. septentrionalis* may become more complex because of another naturalized species, *Rhinella marina*, frequently found in sympatry throughout the landscape. For example, the eggs of *R. marina* affect the survival of larval *O. septentrionalis*; Punzo and Lindstrom (2001) documented a 30% mortality of larvae of *O. septentrionalis* from feeding upon eggs of *R. marina*. In the BVI, *Peltophryne lemur*, a bufonid-like *R. marina*, co-occurred with *L. albilabris*. Presumably, adaptive responses by *L. albilabris* to bufonids may have been retained because of its past interactions with *P. lemur* (Holt and Bonsall 2017), thereby pre-adapting it to coexistence with *R. marina*. Under laboratory conditions, Flores-Nieves *et al.* (2014) showed that larvae of *L. albilabris* from PR outcompete larvae of *R. marina* when at similar densities. However, larvae of *R. marina* have a higher tolerance to increased temperature in ponds compared with *L. albilabris*, which has been interpreted as adaptive for development and survival in temporary ponds (Heatwole *et al.* 1968; see Beltrán *et al.* 2019 for *Leptodactylus lithonaetes* from northern South America). We are not aware of studies on the physiology of larvae of *O. septentrionalis* under different thermal conditions or other microhabitat-related variables. Consequently, the apparent 'rarity' of *L. albilabris* in the VI may not be attributed solely to these naturalized species of anurans because other phenomena acting synergistically (e.g. changes in land-cover, availability of pools throughout the landscape, seasonality in rain, and temperature) may influence the population dynamics of amphibians, both naturalized and native, independently. For instance, projected increases in air temperature as a result of climatic change may operate on the physiology of larvae of *L. albilabris* in temporary pools, thereby resulting in changes in population size, independently of the effect of naturalized species.

The species *O. septentrionalis, Leptodactylus albilabris*, and *Rhinella marina* are sympatric in many localities throughout the PRA (e.g. Tortola, St Croix, and PR; Table 8.1) and their coexistence provides fruitful avenues for research, such as controlled experiments with larvae of all three species to examine their interspecific interaction under varying environmental conditions. Also needed are long-term studies in different types of land-cover to elucidate the effects of naturalized species of amphibians on the native (and endemic) species of amphibians in the PRA. Other studies could examine the feeding ecology and niche-overlap among species, and conduct systematic monitoring of species in different assemblages throughout the landscape. Consequently, and similar to the discussion on *R. marina* in section III. E. 2.a, these and other studies are needed to elucidate whether the generalities drawn from studies with *O. septentrionalis* and native species of amphibians from continental areas also apply under the insular settings of the PRA (Ríos-López 2021).

Finally, in March 2017, NR-L collected a juvenile (~25 mm SVL) of *O. septentrionalis* in Toa Baja, northern PR, from herbaceous plants at the edge of the critical habitat of the endangered *Eleutherodactylus juanariveroi*. This is the first time that *O. septentrionalis* has been detected in this wetland (Santiago-Suliveras *et al.* 2019). The implications for conservation of this finding need to be studied. Interspecific interactions with *O. septentrionalis* and species other than amphibians include snakes and birds (see sections III. E. 3. and III. E. 4., respectively). Adult individuals of this species fall prey to the Puerto Rican racer, *Borikenophis portoricensis*, which seems not to be affected after ingestion of them either in the wild or under laboratory conditions (NR-L, unpublished data), and to the cattle egret, *Bubulcus ibis* (Gasset *et al.* 2000). These observations suggest that local food webs in the PRA have integrated *O. septentrionalis*, much as have food webs from continental areas (e.g. Meshaka 2011). However, we do not know how common these predation events are throughout the geographical distribution of *O. septentrionalis* in the PIRA.

(c) *Lithobates catesbeianus*

In section III. D., we introduced *Lithobates catesbeianus* and described its history in the PRA, along with the purpose for its introduction. It was introduced in 1935 and then spread over the coastal areas and, according to Rivero (1998), up to ~150 m asl at Patillas and San Lorenzo, eastern PR. Not surprisingly, Ríos-López (2007; 163, his table 5), as part of a year-long monitoring study, detected *L. catesbeianus* in a canal that connects the karst hills with a herbaceous wetland in Toa Baja, northern PR (see Fig. 8.15). However, he only documented the presence of this species because insufficient individuals were detected to generate a reliable estimate of abundance. Burrowes *et al.* (2008a) did not detect the chytrid fungus (*Bd*) in the Humacao Wildlife Reserve (south-eastern coastal PR), but was only able to capture three individuals.

Barber (2012) collected amphibians opportunistically between 2006 and 2011, and on three occasions captured an individual of *L. catesbeianus* in a pond located in Arecibo, northern PR. Similar to results from Burrowes *et al.* (2008a), she did not detect *Bd* in these samples (Barber 2012). In March 2017, Ríos-López and a group of ten senior students from the Wildlife Management Bachelor of Science (BS) at the University of Puerto Rico in Humacao, spent two nights actively searching for *L. catesbeianus* along canals in pastures and shrublands in the valley of Yabucoa (eastern PR), an area with a long tradition in agriculture and cattle-ranching. They collected 13 individuals for an estimated efficiency of one individual of *L. catesbeianus* per two person-hours of effort. As far as we know, the instances described herein are the only estimates of abundance of *L. catesbeianus* in PR, but the data are localized and were not collected with an intention of estimating the population size of this species.

In section III. E. 2.a, we highlighted Wolcott's (1948b) comment that *L. catesbeianus* was 'abundant' at that time. Similar to other examples described in this chapter, estimates of abundance for this species were anecdotal. However, Wolcott (1948b) highlighted a study of the diet of *L. catesbeianus* carried out by Mr Mario E. Pérez, at the time Assistant Entomologist at the Agriculture Experiment Station of the University of Puerto Rico. Pérez's (1951a,b) study is the only account of the diet of *L. catesbeianus* in PR so far published. He examined stomach contents of *L. catesbeianus* from two areas, Mayagüez (coastal western PR), sent to him by Mr Harold C. Plank (Entomologist for the Federal Experiment Station in PR) and from 'a nearby brook' (Pérez 1951a: 137), probably from the vicinity of the facilities of the Agriculture Experiment Station in Río Piedras and presumably the progeny of 40 individuals of *L. catesbeianus* from Florida, USA, originally introduced in 1935 (Pérez 1951a: 136). His information came mostly from one locality in Río Piedras (38 specimens), with a minor contribution from a sample from Mayagüez ($n = 12$). He grouped specimens from both localities for his analysis, however, which precludes discovering whether there were differences in the composition of prey between geographical areas. Pérez (1951a) found that 3.9% of the stomach contents of his combined sample of *L. catesbeianus* consisted of 'pollywogs and small adults' of its species. Finally, Thomas and Joglar (1996: 188) mentioned that 'in part, it [*L. catesbeianus*] preys on native species such as *Leptodactylus albilabris*, which occurs in the same habitats (Thomas, personal observation)'.

Unpublished data from 13 specimens of *Lithobates catesbeianus* collected by NR-L and his students in 2017 included one individual of an adult *Eleutherodactylus coqui* in the stomach of one *L. catesbeianus* and an adult *E. antillensis* in the stomach of another *L. catesbeianus*. Nonetheless, the low sample size of specimens of *L. catesbeianus* and lack of identification of the rest of its prey items precludes an adequate analysis of the relative contribution of anurans in its diet.

These accounts provide important insights into the abundance of *L. catesbeianus* and its effects on other species of amphibians in PR. First, the claims frequently made of the wide distribution of this species in the PRA (past and present) must not be taken as indicative of abundance. Other than records of sightings at various localities in PR, and the presumed failed introduction into St Croix (Mayer 2012), estimates of population size of *L. catesbeianus* are lacking (Table 8.1). Second, fortuitous observations obtained from unplanned and unstructured field studies can be misleading if used as generalities in assessments for conservation; there are limited data on the effects of this species on native/endemic species of amphibians in PR. Consequently, about 85 years since its introduction, we still have a lot to learn about the basic ecology of this species on an island such as PR, and on its effects on the native/endemic species of amphibians, compared with published works from continental regions. Similar to the cases described earlier for *Rhinella marina* and *Osteopilus septentrionalis*, we believe that more studies are still needed on this and other naturalized species of amphibians and their effects on native and endemic amphibians in the PRA. Presumably, the generalization of claims about naturalized amphibians affecting populations of native and endemic species of amphibians may not fit the evidence available in the PRA; ecological interactions may differ among different environmental, biogeographical, and historical contexts.

(d) *Dryophytes cinereus*, *Scinax ruber*, and *Lithobates grylio*

Dryophytes cinereus was first collected from PR in 1954 in north-western PR. The species was seen during the mid-1980s in Quebradillas municipality (E. Rivera-Ocasio, pers. comm. to NR-L), and was frequently brought in by the pet trade during the mid-1990s (NR-L, pers. obs.). Thomas and Joglar (1996) and Rivero (1998) mentioned that the distribution of this species in the wild was limited to a pool in the (Agriculture) Experimental Substation in Isabela, north-western PR, but that no specimen has been seen for years (still the case today). As a result, nothing is known about its population in PR.

Scinax ruber was first collected from Carolina, northern PR, in 1988 (Thomas and Joglar 1996) and Dorado. Ríos-López (1999) found the species in Toa Baja, which is midway between Dorado and Carolina municipalities in northern PR, and Ríos-López (2007) provided an estimate of its abundance in three areas differing in the compositions of their vegetation. The estimates were 33.3 individuals ha^{-1} in a pastureland; 5.6 individuals ha^{-1} in a *Pterocarpus officinalis* wetland forest; and 1,272.2 individuals ha^{-1} in a herbaceous wetland dominated by cattail, *Typha domingensis* (Ríos-López 2007). Not surprisingly, Duellman (1978) mentioned that in its native distribution *S. ruber* is found in low abundance in tropical forests throughout Central and South America, but the species becomes abundant in disturbed situations such as pasturelands after forests have been cleared. In PR, this species has also been located in a coastal herbaceous swamp at Cabo Rojo, south-western PR (Morales 2011b).

Ríos-López (2007) used mating pairs of *Eleutherodactylus coqui* and *E. cochranae* to examine the influence of *Scinax ruber* on their calling behaviour and reproduction under captive conditions. During the 91-day duration of his experiment, he did not detect any differences in either the reproductive output or the calling of either native species from PR that could be attributed to the presence of *S. ruber*. He did detect, though, a significant reduction in the production of egg clutches by *E. cochranae* in the presence of *E. coqui*; the number of egg clutches laid by *E. cochranae* in enclosures with *E. coqui* was one-third the number of egg clutches produced by *E. cochranae* when kept alone. Yet he could not determine whether this reduction was attributable to oophagy or to modification in the reproductive behaviour of female *E. cochranae* in the presence of *E. coqui*. Likewise, even though food was supplied *ad libidum* and all enclosures were frequently misted, *S. ruber* was not provided with standing water that could have influenced its calling, mating, or territorial behaviour, and consequently could have affected the results.

Lithobates grylio is now found along the coastal belt of wetlands that runs from Bayamón to Arecibo, northern PR. Estimates of abundance were provided by Ríos-López (2007) for three areas that differed in composition of vegetation. The estimates were 38.9 individuals ha^{-1} in a *Pterocarpus officinalis* wetland forest; 66.7 individuals ha^{-1} in a herbaceous wetland dominated by cattail, *Typha domingensis*; and 44.4 individuals ha^{-1} in a palustrine herbaceous wetland. The number of individuals detected along a 1.1 km road, which runs parallel to a salinity gradient, decreased with distance from the coast, similar to the pattern of decrease in abundance shown by *L. albilabris* (Ríos-López 2008a). Virtually nothing has been published about its population status and breeding biology, or its interactions with other amphibians in PR.

3. Reptiles

Wolcott (1924) examined the stomach contents of native and endemic lizards in PR but did not find remains of amphibians. However, other studies (Reagan 1996; Stewart and Woolbright 1996; Thomas and Gaa Kessler 1996; Henderson and Powell 2009; Ríos-López *et al.* 2015c) have indicated that a wide range of lizards and snakes from the PRA sparingly include amphibians in their diet. Notably, the crested anole, *Anolis cristatellus*, frequently preys on metamorphs of *Peltophryne lemur* that come out of the water and stay at the edges of artificial ponds (for the reintroduction of *P. lemur* at El Tallonal) in order to reabsorb the tail completely (SIV-C and ARP-R, pers. obs.).

Two introduced reptilian species known to feed on amphibians in the PRA are: the spectacled caiman, *Caiman crocodilus*, and the Indonesian semi-aquatic snake *Xenochrophis vittatus*. In the Tortuguero Lagoon, northern PR, Bontemps *et al.* (2016: 291, their Table 1) documented remains of anurans in two of 138 stomachs of the spectacled caiman analysed for content to be 1.4% frequency by prey category among three size classes of caimans – hatchlings, juveniles, and adults. The remains of anurans were found only in stomachs of juvenile caimans, which represented

4.2% of items by prey category in their diet. Bontemps *et al.* (2016: 293, their Table 2) identified half of these remains as belonging to *Lithobates catesbeianus*, while the rest were classified as 'Unidentified Anuran'. The pig frog *Lithobates grylio*, however, also occurs at the Tortuguero Lagoon (Ríos-López and Joglar 1999), which may represent a significant contribution to the caiman's diet.

The Indonesian semi-aquatic snake *Xenochrophis vittatus* preys upon fishes and small amphibians in its natural geographical range, and now has become naturalized in PR (Herrera-Montes *et al.* 2015). Four adult specimens of this snake were collected between February 2017 (n = 1) and February 2018 (n = 3) in the critical habitat of the endangered *Eleutherodactylus juanariveroi* (Santiago-Suliveras *et al.* 2019). The remains of an *Eleutherodactylus* (presumably *E. coqui* judging by the large size of the specimen) was found in one stomach and the skin from a *Leptodactylus albilabris* was found in the intestines of a different snake. Thus, this semi-aquatic snake is a predator upon the native amphibians in the region, and its diet should be examined in detail. Tentatively, it may be a threat to the endangered *E. juanariveroi*, particularly if no other prey is easily exploited in this wetland. This snake may occur only in low numbers, however, and efforts to control its population may prove effective.

4. *Birds*

The maximum species richness of birds documented from PRA has fluctuated between 350 and more than 400 species at various recent times, with relevant observations of birds having preyed upon amphibians. However, a landmark study based on quantitative analyses of content of stomachs of birds from the region is *Birds of Porto Rico* published by Alexander Wetmore in 1916. His results are summarized in Table 8.7 and are expanded by other contributions from PR and the USVI. In this table, we include quantitative data on the abundance of amphibians as prey for a given species of bird, rather than fortuitous observations of predation. We also include available data published in the format of a per cent of the diet that includes amphibians (i.e. Snyder *et al.* 1987; Gassett *et al.* 2000). Therefore, this summary represents a useful first approximation of the importance of amphibians in the diet of birds – naturalized and native/endemic – from the PRA. However, our review is somewhat incomplete, because studies that lumped amphibians and reptiles as a single category of prey were not included.

Wetmore (1916) found 18 species of birds – out of 178 species treated in his text – that included remains of amphibians in their stomach in sufficient amount to permit quantification (i.e. 10.1% of birds known at the time that had amphibians as part of their diet). As far as we know, the contributions from Snyder *et al.* (1987) and Gasset *et al.* (2000) augment the list in Table 8.7 by 21 species (out of more than 350 species currently known for the area). Notably, Gassett *et al.* (2000) combined data on the abundance and volume of amphibians that were preyed upon and generated an index of the relative importance of amphibians in the diet of the cattle egret on St Croix. The index of relative importance was zero, and thus amphibians are unimportant in the diet of the cattle egret (Gassett *et al.* 2000).

There are no comparable data on the relative contribution of amphibians to the diet of species of birds introduced into the PRA. The region lacks large predatory species that characterize food webs in continental areas; consequently, amphibians and reptiles (rather than birds) represent the most abundant, conspicuous, and ecologically important group of vertebrates that regulate the flow of nutrients, carbon, and energy by way of predator–prey interactions in the insular ecosystem (Ríos-López *et al.* 2015b; Smith *et al.* 2018). However, the topic of the diet and feeding ecology of birds should be pursued in the PRA, as such studies will put into perspective the relative influence of amphibians (and reptiles) in structuring species assemblages in insular ecosystems such as occur here in particular, and in the Caribbean region in general.

Table 8.7. A summary of data from studies on the diet of birds collected from the PRA and that included amphibians in the birds' stomachs (also included are a few related notes). '?' = undetermined number of stomachs. '—' = other items not specified. (**Ex**) = extinct. (**E**) = endemic; (**U**) = category unknown. Taxonomy of birds follows Lepage (2019) and Chesser *et al.* (2019).

FAMILY AND *SPECIES*	COMMON NAME	FRACTION OF STOMACHS WITH REMAINS OF AMPHIBIANS	RELATIVE FREQUENCY BY ABUNDANCE (%) OF AMPHIBIANS/PREYED ITEMS	SPECIES OF AMPHIBIANS (AND NOTES FROM CITED REFERENCES)
Accipitridae				
Buteo platypterus brunnescens	[a] broad-winged hawk	12 / 84	14.3 ([d] may include an additional 9.5% from unidentified small prey).	'frogs'
Ardeidae				
Ardea alba egretta	[b] great egret	1 / 1	69.0	*Leptodactylus albilabris* ($n = 7$), 'with fragments of others'
Bubulcus ibis	[c] cattle egret	3 / 44 5 / 44 1 / 44 1 / 44	0.2 0.2 < 0.0 < 0.0	*L. albilabris* *Osteopilus septentrionalis* *E. antillensis* *Rhinella marina*
Butorides striatus	[b] green-backed heron	3 / 51	0.61	*L. albilabris*
Egretta caerulea	[b] little blue heron	? / 15	0.59 (at most)	miscellaneous animal matter, including remains from two *L. albilabris*
E. thula	[b] snowy egret	1 / 2	—	'a small frog'
Ixobrychus exilis	[b] least bittern	1 / 3	0.33	undetermined
Corvidae				
(**Ex**; **U**) *Corvus leucognaphalus*	[b] white-necked crow	2 / 9	7.04	*Eleutherodactylus* sp. (bones)
Icteridae				
(**E**) *Icterus portoricensis*	[b] Puerto Rican oriole	3 / 71	1.06	*Eleutherodactylus* sp.; an *Anolis* lizard
Quiscalus niger brachypterus	[b] Greater Antillean grackle	12 / 98	5.74	Mostly *L. albilabris* with a minor contribution from *Eleutherodactylus* sp.
Mimidae				

(Continued)

Table 8.7. (Continued)

FAMILY AND *SPECIES*	COMMON NAME	FRACTION OF STOMACHS WITH REMAINS OF AMPHIBIANS	RELATIVE FREQUENCY BY ABUNDANCE (%) OF AMPHIBIANS/PREYED ITEMS	SPECIES OF AMPHIBIANS (AND NOTES FROM CITED REFERENCES)
Margarops fuscatus	[b] pearly-eyed thrasher	? / 38	1.45	*Eleutherodactylus* sp. ($n = 1$; two *Anolis* lizards)
	[d] adults	6 / 913	0.66	*Eleutherodactylus* spp. (one event of eggs)
	[d] fed to nestlings	19 / 491	3.9	*Eleutherodactylus* spp. (three events of eggs)
Parulidae				
Parkesia motacilla	[b] Louisiana waterthrush	1 / 5	—	*Eleutherodactylus* sp. ($n = 1$)
(**E**) *Setophaga adelaidae*	[b] Adelaide's warbler	1 / 41	0.63	*Eleutherodactylus* sp. ($n = 1$; miscellaneous matter)
Seiurus aurocapilla	[b] ovenbird	1 / 13	0.93	*Eleutherodactylus* sp. ($n = 1$)
Picinae				
(**E**) *Melanerpes portoricensis*	[b] Puerto Rican woodpecker	7 / 59	6.17	*Eleutherodactylus* sp. (remains of small *Anolis* spp.)
Strigidae				
(**E**) *Megascops nudipes*	[e] Puerto Rican screech-owl	49 / 125	39.2	'frogs'
Thraupidae				
(**E**) *Nesospingus speculiferus*	[b] Puerto Rican tanager	2 / 37	6.12	*Eleutherodactylus* sp. (a small *Anolis* lizard)
Turdidae				
Turdus plumbeus ardosiaceus	[b] red-legged thrush	? / 32	1.68**	*Eleutherodactylus* sp. ($n = 2$)
		[f] 2 / 22	2.1	*L. albilabris* ($n = 1$; 'bone fragments of lizards [*Anolis* sp.] and treefrogs')

Tyrannidae					
(**E**) *Myiarchus antillarum*	[b] Puerto Rican flycatcher	1 / 40	1.07		*Eleutherodactylus* sp. ($n = 1$; one *Anolis* lizard)
Tyrannus caudifasciatus taylori	[b] loggerhead kingbird	10 / 61	8.81		*Eleutherodactylus* sp.
Vireonidae					
Vireo altiloquus	[b] black-whiskered vireo	2 / 84	0.61		*Eleutherodactylus* sp. (along with miscellaneous matter)

*This summary does not include published chance observations on bird anurophagy or data that do not specify the number of prey items and category of amphibian (or species of amphibian).

**Wetmore (1916) documented 8.41% of lizards and 'tree toads', but detailed the individuals found as two *Eleutherodactylus* spp., one *Pholidoscelis* (*Ameiva*) *exsul*, and seven *Anolis* spp., which yields 2/10 *Eleutherodactylus* spp. from the percent documented herein for the red-legged thrush (e.g. 1.68% *Eleutherodactylus* spp.).

[a] Snyder *et al.* (1987) based on a large number of observations of food brought into two nests of the broad-winged hawk; also, they include unidentified prey 'very small items, probably mostly frogs', which adds 9.5% of frogs to the diet.

[b] Wetmore (1916), based on 2,200 stomachs and 162 species of birds (and 16 species of birds included as hypothetical) as discussed in his text; all birds from PR and adjacent islands; note that none of the non-native amphibians in the PRA documented in this chapter were recorded at the time of Wetmore's publication.

[c] Gassett *et al.* (2000), based on samples from St Croix, USVI.

[d] Snyder *et al.* (1987) based on observations of adults and items brought into nests of the pearly-eyed thrasher; observations recorded from PR, Mona Island, and the USVI combined.

[e] Snyder *et al.* (1987) based on a large number of observations of food brought into two nests of the Puerto Rican screech-owl.

[f] Rolle (1963: 17–18).

5. Mammals

Records of land mammals in the PRA include 46 (or 47) species of which 16 (or 17) are either extinct or extirpated; excluded are domestic mammals that have not developed feral populations. The alternate number of land mammals in this record originates because of an unidentified agouti (*Dasyprocta* sp.), which was brought into Vieques (Woods 1996), St Thomas (Seaman 1961), and Tortola (Island Resources Foundation 2015), but became extirpated from those places (see section III. E. 5.a). In contrast, two species are identified in the archipelago with certainty, *D. leporina* in the USVI (originally from South America) and *D. punctata* in PR (originally from Central America). The genus *Dasyprocta* includes several species of agoutis from Central and South America, and we were unable to distinguish it from the other two species of agoutis identified. Consequently, its record also causes a difference in the number of extant species versus extinct or extirpated species. Currently, *D. punctata* is extant in north-eastern PR, but we could not find records about whether *D. leporina* still occurs in the USVI.

(a) Extinct or extirpated mammals

The species of mammals known from fossil remains include 12 species (presumably endemics; nine rodents, one shrew, one giant sloth, and one bat) (Gannon *et al.* 2005; Vélez-Juarbe *et al.* 2014; Marivaux *et al.* 2020; Marivaux *et al.* 2021). In addition, extirpated populations of two species of bats that still live elsewhere were found on PR (*Macrotus waterhousii* and *Monophyllus plethodon*; Gannon *et al.* 2005). Moreover, four species of mammals were brought here by the Amerindians: one species of rodent, now extinct (*Isolobodon portoricensis* or West Indian hutia; Flemming and MacPhee 1999), which was brought into PR, St Thomas, and St Croix from Hispaniola as a source of food (Miller 1918; Woods 1996; deFrance *et al.* 2010); the guinea pig (*Cavia porcellus*), which was brought into PR from South America through the Lesser Antilles and also as a source of food (deFrance *et al.* 2010; Giovas *et al.* 2012); an undescribed agouti (*Dasyprocta* sp.); and the dog (*Canis lupus familiaris*), with numerous remains in various locations in PR (Woods 1996; deFrance *et al.* 2010; Cooke *et al.* 2017). Except for the dog and at least one species of agouti (see section III. E. 5.b., none of the other species is known to exist today on the archipelago.

(b) Extant mammals

The species of extant mammals in the archipelago include 13 species of bats; 12 species are native (Gannon *et al.* 2005) and one, *Pteronotus portoricensis*, is endemic to PR (Pavan and Marroig 2016). The rest of the species are naturalized. These species include the dog, which was also intentionally introduced by early European settlers, along with cats, *Felis catus*, later on (Gannon *et al.* 2005). Currently, dogs are found mostly in close association with human settlements and no fully feral populations occur in the archipelago (but see Fray Iñigo Abbad y Lasierra [1782] for PR and Iverson [1978] for the BVI). Cats frequently occur in close association with humans, but a few are feral in secondary forests and natural reserves (e.g. Mona Island; García *et al.* 2001). The introduction of dogs into the Caribbean generated different responses on different islands. For example, dogs have remained mostly dependent on humans for feeding and survival on PR, particularly in the twentieth century and to the present as local production of livestock declined. By contrast, dogs are entirely feral in Central Cuba today, and cattle ranchers conduct systematic hunts, using firearms, to control their number of dogs and minimize their predation pressure upon livestock (Borroto-Páez 2009). Currently, with the secondary succession of forests in its prime, predation of livestock by dogs in rural areas is not so rare in PR, while packs of dogs are now seen capturing and preying upon the common green iguana, *Iguana iguana*, in coastal areas (NR-L, pers. obs.). As far as we know, there are no records of feral dogs and cats preying on amphibians in the PRA.

Unintentional introduction of at least three species of rodents also occurred during European settlement: the black rat (*Rattus rattus*, presumably introduced into PR sometime after 1508; Snyder *et al.* 1987); the brown rat (*R. norvegicus*; Fox 1947; Gannon *et al.* 2005); and the house mouse (*Mus musculus*). Also introduced at various post-European times were feral populations of at least two other naturalized extant species: the agoutis *Dasyprocta leporina* and *D. punctata*.

Other mammals, introduced at a much later time, include four species of monkeys. These are the squirrel monkey, *Saimiri sciureus*, in northern PR (Johnsen and Whitehair 1986; NR-L and ARP-R, pers. obs.); the rhesus monkey (*Macaca mulatta*, northern, eastern, and south-western PR, introduced for research in about 1938 into Cayo Santiago, eastern PR [Fridman 2002] and becoming feral sometime between the 1960s and the 1970s, initially in south-eastern PR [Jensen *et al.* 2004]); the patas monkey (*Erythrocebus patas*), feral since the 1970s, initially in south-western PR (Jensen *et al.* 2004); and the vervet (*Chlorocebus pygerythrus*), introduced in the mid-1980s or mid-1990s into the BVI (Island Resources Foundation 2015).

Five large-bodied species of mammals were also introduced and are now feral. These are the Eurasian wild pig (*Sus scrofa*) into Mona Island about 1516 (García *et al.* 2000), into north-western and northern PR between 1505 and 1508, and into St John in about 1718 (US National Park Service 2003); the white-tailed deer (*Odocoileus virginianus*), into Culebra about 1966, into St Croix in late 1790 (Woods 1996), and Carolina municipality in PR (NR-L, pers. obs.); the wild donkey (*Equus asinus*) into St John, USVI; the wild goat (*Capra hircus*) into Mona Island and onto several cays from Culebra to the VI; and *Homo sapiens* completes this list (see Global Mammal Assessment Team 2008).

Finally, the small Indian mongoose (*Herpestes auropunctatus*) is the only species of mammal intentionally introduced as a biological control. It was introduced into Jamaica in 1872 (Espeut 1882), then into PR in 1877 (Wadsworth 1949), and finally into St Croix between 1882 and 1884 to control rats in sugarcane plantations (Woods 1996; Morgan *et al.* 2017). According to Philibosian and Yntema (1977), it is now feral in the PRA, except on Mona Island and Anegada (BVI). NF-A (unpublished data) has not found it on Desecheo, Culebra, or other small islands and cays. In the following sections III. E. 5.c–h., we discuss the effects on the fauna and flora attributed to some of these species, along with recommendations for conservation of native biota.

(c) Wild goats and pigs

The wild goat (*Capra hircus*) and wild pig (*Sus scrofa*) arrest the recovery of vegetation through secondary succession. They do so mostly because they, along with small animals breeding in the ground, remove seedlings, saplings, and young shrubs and trees (e.g. Heatwole *et al.* 1981; García *et al.* 2000). As a result, the forest understorey remains cleared of most vegetation, which facilitates erosion of soils, simplifies the architecture and species richness of the vegetation, reduces the cover of stems valuable for the establishment of bromeliads, and probably reduces humidity and the retention of water in reservoirs. According to the US National Park Service (2003) in the Virgin Islands National Park, St John, three species of frogs are preyed upon by the wild pig. The US National Park Service (2003: 67) presented a summary of environmental consequences that were analysed for 'the wild hog-control programme within the Virgin Islands National Park'. One of the preferred alternatives for wild-hog control is removal by trapping, shooting, and fencing, as these actions 'would improve riparian habitat for frogs, salamanders and aquatic invertebrates' (US National Park Service 2003: 71). Currently, salamanders do not exist in PR and other Antillean islands (note, however, that Gundlach

[1881: 308], Stahl [1882: 70], and Del Valle Atiles [1887: 29] used 'salamandra' as a common name in Spanish for the Puerto Rican dwarf gecko *Sphaerodactylus macrolepis* [now *S. grandisquamis* in Daza *et al.* 2019]). However, a fossil salamander, *Palaeoplethodon hispaniolae,* is known from the Dominican Republic (Poinar and Wake 2015). Furthermore, we could not find an original reference to sustain these statements concerning frogs and hogs in the document of the US National Park Service (2003). Nonetheless, we emphasize the need to quantify rigorously and methodically the effects of wild hogs (direct and indirect) on vegetation, microclimates, and fauna so as to implement effective conservation measures for the benefit of amphibians in particular, and the biota in general.

Categorizing wild pigs and goats as stand-alone threats can be somewhat misleading because the main issue may not be their effects on fauna and flora, but rather concerning how local agencies and the general public sustain the occurrence of these species (e.g. Heatwole *et al.* 1981; García *et al.* 2000). To illustrate, Mona Island is home to eight endemic species of reptiles and one endemic species of amphibian (*Eleutherodactylus monensis*). Several local and federal governmental agencies recognized Mona Island as special by using terms such as 'Unique Natural Area', 'National Natural Landmark', 'Critical Habitat for several terrestrial and marine vertebrates', 'Natural Reserve', 'National Historic Landmark', and 'Historic Site' (e.g. Ventosa-Febles *et al.* 2005). Despite these accolades, the island harbours populations of goats and pigs (along with feral cats), which interact with the native and endemic biota (e.g. Wiewandt 1977; García *et al.* 2001; Meléndez-Ackerman *et al.* 2008; Pérez-Buitrago *et al.* 2010), and which have required efforts for their eradication. For example, the Federal Aid in Wildlife Restoration Act of 1937, as amended (or the Pittman-Robertson Act), provides funding for management and restoration of wildlife, and the US Fish and Wildlife Service administer it. By this federal law, the Government of PR receives funding to implement sport-hunting as the best strategy for the control of the populations of these feral animals on Mona Island (García *et al.* 2001) and all US territories. However, sport-hunting could be in itself controversial as implemented.

Sport-hunting is locally regulated in PR by the Nueva Ley de Vida Silvestre (1999). It is also deemed a valuable societal, recreational, cultural, and economic activity by the government (Government of Puerto Rico 1999; García *et al.* 2000); currently, there are about 5,500 licensed hunters in PR alone. However, the 'Nueva Ley de Vida Silvestre' also provides for non-sporting hunting, defined as a non-recreational 'hunting activity for the purpose of [...] the control of populations' (translated from the Spanish). Non-sporting, non-recreational hunting requires training and the use of professionals to eradicate these and other feral species, but this provision of the local law is rarely implemented on Mona Island. Presumably, sport-hunting is preferred over non-sporting hunting because it provides recurrent access to funding from the federal government, and eradication of these feral species would reduce federal funds available to the local government. In contrast, on other islands with historical sport-hunting, such as St John in the USVI, federal mandates exclude much of the lands from sport-hunting (i.e. Virgin Islands National Park) and require alternative conservation and community-inclusive management plans; sport-hunting is prohibited in the BVI (Raffaele 1990), even though these islands also harbour populations of feral animals. Furthermore, in PR – and presumably elsewhere in the archipelago – local people use some of the small islands and cays under its jurisdiction as wildlife ranches for introducing and harvesting feral goats (e.g. Heatwole *et al.* 1981). Consequently, sport-hunting on Mona Island and ranching activities (on this and other islands in the PRA) do not eradicate feral populations of these species but perpetuate their effects on the native and endemic biota instead (e.g. Wiewandt 1977; Meléndez-Ackerman *et al.* 2008).

The way the local governments deal with feral species in the PRA has profound implications for conservation. On the one hand, some scientists believe that non-effective regulatory mechanisms and artisanal ranching threaten native amphibians and other terrestrial fauna and flora in the PRA, and call for the eradication of naturalized species on Mona Island (and other islands). On the other hand, other scientists believe that it is possible that after more than 500 years the introduction of these naturalized species onto Mona Island, for example, its native and endemic biota – particularly vegetation – might have adapted to their influence already. We believe that there are still signficant gaps in the information, and so before adopting one perspective over the other there is a need for long-term research on the influence of these and other naturalized species on the biota throughout the landscape (see Pérez-Buitago *et al.* 2010).

An additional measure of paramount importance for an effective conservation agenda in these islands is the development of educational programmes for the general public on issues aimed at reducing (and hopefully ending) the practice of local support of feral animals whenever this is shown to affect the native wildlife. Humans must also be explicitly cited as an additional threat in databases relevant to conservation agendas whenever goats and pigs are listed as problems because we (i.e. 'the ultimate cause') maintain and perpetuate the occurrence of these naturalized species. As a result, the effects that naturalized species have on native and endemic species (i.e. 'the proximal consequence') are linked directly to our actions (or lack thereof) as managers, policymakers, law enforcers, and conservationists.

(d) White-tailed deer, feral donkeys, and agoutis

The effects that white-tailed deer, feral donkeys, and agoutis may have on amphibians is unknown. Consequently, the study of the impact of these mammals on the vegetation and other resources that serve as retreat and breeding sites for amphibians (i.e. indirect effects), constitutes an important area for research in herpetofaunal ecology and conservation agendas. For example, it is known that donkeys are valued as attractions for tourists in St John, but for this and other species of mammals listed here, estimates of abundance, distribution, habitat use, feeding ecology, and effects on amphibians are unknown. As such, local governments and research organizations can sponsor ecological studies more frequently than has been done; for undergraduate and post-graduate students, such studies represent opportunities for their academic and professional development.

(e) Feral monkeys

We do not know the effects of feral monkeys on amphibians, particularly in the northern part of PR where the more humid conditions are appropriate for a large number of species of amphibians compared with the southern side of this island. In EYNF, north-eastern PR, a small colony of squirrel monkeys (*Saimiri sciureus*) was removed in 1973 (Snyder *et al.* 1987). However, colonies of feral individuals of this species exist in Toa Baja, northern PR, at least since the early 1980s (Johnsen and Whitehair 1986) along with colonies of the rhesus monkey (*Macaca mulatta*) (NR-L, pers. obs.). Rhesus monkeys also occur in south-eastern PR, while the patas monkey (*Erythrocebus patas*) occurs mostly in south-western PR. Although these species of monkeys may occasionally include small vertebrates in their diets (*S. sciureus* feeds on *Anolis cristatellus*; NR-L and ARP-R, pers. obs.), their impact on amphibians in the PRA is still unexplored and this provides fruitful opportunities for research. However, a fortuitous observation is worth including here because it occurred within the geographical distribution of the threatened Puerto Rican crested toad, *Peltophryne lemur*. Dr Steven Alexander Sloan, from the University of Puerto Rico-Bayamón Campus, kindly provided the following observation (Fig. 8.18):

Figure 8.18. The anterior part of an adult individual of *Rhinella marina* thrown at researchers after its anterior portion of the body was torn apart presumably by the monkey *Erhythrocebus patas* in Lajas, south-western PR. According to the account (see text), the toad was still moving its forelegs when being photographed. Picture courtesy of Dr Steven A. Sloan.

> On February 17, 2007, two students and I were out conducting fieldwork in the USFWS [Fish and Wildlife Service] "La Tinaja reserve" in the Sierra Bermeja, Cabo Rojo, Puerto Rico. We were researching seed dispersal by introduced *Erythrocebus patas* and *Macaca mulatta* monkeys, hence collecting fecal samples. We had seen a troop of *M. mulatta*, earlier that morning, on our hike along a trail that runs north/south along a stream, in the valley between two hills. Upon our return, later that afternoon, about 3 pm, we looked across the stream to the eastern hill where we had seen the troop, and suddenly heard a thump on the ground. I searched the ground where I had heard the thump and nothing caught my attention. I turned and asked the student next to me if they had heard the sound. They said yes. We looked across the valley and up above our heads in the trees and saw nothing. We began to continue our exit from the forest, when my student said, "profe que eso?" [professor, what is that?] I looked to where she was pointing and hesitated to answer as my mind raced to understand what my eyes were seeing. On the ground lay the upper half of a *Rhinella marina*. We searched around for what had tossed the still moving head and torso to our direction, but nothing was evident. February is the driest time of the year for this area and I would image food resources are scarce. I was informed by the, then director of the UPR Caribbean Primate Center, Dr Janis Gonzalez that it was probably an *Erythrocebus patas* that launched the toxic grenade in our direction, as they are more omnivorous than the *M. mulatta*. While this could have been the work of a large bird, none were seen. Whatever the animal that ate the lower half of this amphibian was familiar enough with this prey to avoid the most toxic portion of the body, the large glands on either side of the head.

(f) *Mus musculus*

To our knowledge, there are no ecological studies of the effects of the house mouse (*Mus musculus*) on amphibians, either in natural areas or around human dwellings in the PRA. According to

Musser *et al.* (2016), this mouse rarely ventures into forests, and it is mostly associated with artificial environments such as human dwellings, agricultural areas, pasturelands, edges of roads, and occasionally coastal dunes and saltmarshes. However, Witmer *et al.* (2007) documented that each year after the eradication of roof, or black, rats (*Rattus rattus*) on Buck Island, offshore of St Croix, the population of *M. musculus* increased across the island, indicating a competitive release on populations of house mice. While no amphibians exist on Buck Island, the conservation implications of this study imply that house mice should be considered during the eradication of rats since the effects of house mice on amphibians are unknown in the Caribbean.

Shields and Pitt (2014) reviewed reports on the diet of *M. musculus* (and also of *R. rattus* and *R. norvegicus*) from the Pacific Islands and did not find records of species of amphibians represented in the diet of this species. However, Shields and Pitt (2014) mentioned that, proportionally, the diet of *M. musculus* (and *R. norvegicus*, see section III. E. 5.g) consists of more animal than plant material compared with the proportion in the material in the diet of *R. rattus*. On these islands, herpetofaunal species were not represented in the stomachs from *M. musculus* and *Rattus* spp. examined by Shields and Pitt (2014).

(g) *Rattus norvegicus*

The earliest history of rats in the PRA is not known, but given the proclivity of rats to have been spread by early sailing ships, it is likely that they have been present since early in Spanish settlement.

Ruedas (2016) mentioned that the brown rat, *Rattus norvegicus*, 'is restricted to habitats modified by humans (sewers, buildings, ports, etc.)' in warmer and tropical regions; in the USVI, it seems restricted mostly to towns (Seaman 1961). In a review of the global data on the effects of rats, Towns *et al.* (2006) analysed those cases of species of vertebrates in which rats reduced recruitment or caused extirpation. They inferred extinction based on distributional data of other species following observed invasions of rats. In this review, Towns *et al.* (2006) found no reports of affected species of amphibians because of *R. norvegicus*, or of *R. rattus* (see section III. E. 5.h). Likewise, Shields and Pitt (2014) found no records of amphibians represented in the diet of *R. norvegicus* from the Pacific Islands. Dietary analyses provide evidence of recent predation, whereas possible long-term effects on populations of amphibians by other means are still needed at least for the PRA.

(h) *Rattus rattus*

Willig and Gannon (1996) summarized the scattered data on the feeding ecology of the black rat, *Rattus rattus*, from the PRA. These authors presented a diet of *R. rattus* at El Verde, mostly from studies under conditions of captivity and showed that it consumes a variety of fruits from at least 12 species of plants. They also mentioned its estimated population density at the tabonuco (*Dacryodes excelsa*) forest (< 610 m asl) at EYNF to be approximately 40 individuals ha^{-1}, whereas at the palo colorado (*Cyrilla racemiflora*) forest (> 610 m asl) it was approximately 281 individuals ha^{-1}. At least since the late 1980s, several authors have been calling attention to the need for studies that address the feeding ecology and dietary composition of prey for the black rat from the PRA (e.g. Snyder *et al.* 1987; Willig and Gannon 1996) and elsewhere (e.g. Towns *et al.* 2006; Shields and Pitt 2014). Unfortunately, no such study seems to have taken place. Moreover, studies that involve capturing rats and analysing the stomach contents are not only readily feasible but are highly relevant, as they may provide much-needed data of consequence for the conservation for amphibians. Hopefully, a study of this nature will be conducted for the PRA.

(i) *Herpestes auropunctatus*

The available data on the small Indian mongoose, *Herpestes auropunctatus*, and its effects on the amphibians in the PRA are summarized here and in Appendix 8.I (note that for the currently accepted taxonomy of the small Indian mongoose, we use *Herpestes* because a recent analysis of records in Google Scholar between 2009 [first proposal for the use of *Urva*] and 2019 about the use of *Herpestes* versus *Urva* yielded 1,010 titles referring to *Herpestes* and 23 titles referring to *Urva*; also see Jennings and Veron 2016). Seaman (1952), as cited by Seaman and Randall (1962), documented the prey found in 42 stomachs of mongooses trapped in St Croix. According to Seaman and Randall (1962) and Wolcott (1953), Seaman (1952) did not find any amphibians in those stomachs; instead, the diet consisted of invertebrates, specifically insects (mostly orthopterans) and centipedes (Seaman and Randall 1962). However, Nellis and Everard (1983) mentioned that Seaman's (1952) results were based on 36 stomachs, not 42, and indicated that 13.9% of the prey were toads. A closer examination of the percentages of predated items listed by Nellis and Everard (1983) based on Seaman's (1952) data adds 158.5%, which may be an error and precludes accurate analyses of Seaman's (1952) data.

Wolcott (1953) documented most of the prey that he could identify from the 42 stomachs of mongooses shot by Seaman in St Croix in 1952 and also those from 56 mongooses trapped by Dr Pimentel from pastures at Roosevelt Roads, eastern coastal PR, and sent to Wolcott in 1952. Wolcott (1953) identified one bullfrog (*Lithobates catesbeianus*) and two individuals of *Eleutherodactylus coqui* among these 98 stomachs, and listed the predated items grouped among all stomachs. However, Pimentel (1955) mentioned that amphibians were found in his sample of 315 predated items among all stomachs of his mongooses. Consequently, all individuals of frogs (*L. catesbeianus*, which was first documented from St Croix by Mayer in 2012, and the two *E. coqui*) were from his sample, and not from the samples sent to Wolcott by Seaman. Wolcott (1953) did emphasize the rather large number of lizards (n = 46) from his list of prey from this sample of mongooses. The second and third most abundant prey in stomachs were *Scolopendra subpinipes* (centipedes, n = 38) and *Cyrtopholis portoricae* (tarantulas, n = 14). In agreement with Wolcott (1953), Pimentel (1955) documented that amphibians accounted for 1.1% (three frogs from 315 total prey) from his 56 mongooses and also estimated the density of mongooses from a coastal lowland dominated by tall pasture to be 2.5 individuals ha^{-1}, with an average area ('circular') of the home range of 97.5 m^2 (males: 125.6 m^2, range 91.4–152.4 m^2; females: 81.4 m^2, range 30.5–213.4 m^2).

Seaman and Randall (1962) described attempts by the mongoose to prey upon a fawn of the introduced white-tailed deer at St Croix, as well as upon the consumption of eggs of sea turtles and some birds; no mention was made of amphibians as prey. Their accounts, however, were based on fortuitous observations rather than on methodical studies.

Nellis and Everard (1983) estimated population sizes of mongooses in St Croix to vary from 3.2–3.4 individuals ha^{-1} to 6.4 individuals ha^{-1} (mean = 4.3 individuals ha^{-1}, n = 3 plots) depending on environmental heterogeneity and microclimatic conditions. They mentioned (p. 9) a behavioural modification that results from the presence of mongooses on this island: '*Rattus rattus* is now almost completely arboreal on St Croix.' In contrast, this species occurred equally on the ground and in trees in the presence of the mongoose at EYNF (Endeman *et al.* 2006). The possibility that *R. rattus* experiences ecological release from the *R. norvegicus* in PR should be examined. Hoagland *et al.* (1989) suggested that *R. norvegicus* and *R. rattus* partition their habitats (presumably because of competition) and that the abundance of the former is reduced by predation pressure from mongooses.

It was not until more than 30 years after the study by Pimentel (1955) and Seaman and Randall (1962) that the ecology of the mongoose in the PRA was described in more detail by

Vilella and Zwank (1993) and Vilella (1998). Specifically, Vilella and Zwank (1993) conducted the first study on mongooses' ecology in the Guánica Forest, south-western PR (< 200 m asl). They sampled mongooses along four 450 m transects; along each transect, ten traps were set every 50 m (90 traps/transect). During 24 days of trapping, 34 mongooses were caught and euthanized for analysis of stomach contents. Even though all stomachs were empty, intestinal faecal content revealed 85% by volume of animal material (arthropods 70%, lizards 10%, and rats 5%, but no amphibians). Finally, Vilella and Zwank (1993) concluded that mongooses were not abundant in the Guánica Forest, but the analysis of their data did not allow precise estimates of the population density.

The ecology of the mongoose was also studied in the EYNF: close to the US Forest Service's El Yunque Ranger District facilities (200 m asl) and adjacent to trails and at the edge of the river (two localities, each at > 400 m asl) at El Verde Field Station (western EYNF) (Vilella 1998). Eighteen specimens of mongooses were trapped for analysis of stomach contents. The most frequently predated items were *Anolis* lizards, centipedes (*Scolopendra* spp.), and Dictyoptera (presumably mostly cockroaches). Unfortunately, all predated items were grouped into ten broad categories, and the estimated number of items in each stomach was not given. As a result, estimates of preference for a given prey and the abundance of predated items within a stomach were not obtained. Nonetheless, two stomachs (11% of dissected mongooses) contained the prey type 'Treefrog'. While the relative abundance of treefrogs in these stomachs could not be extracted from the data as published, we can estimate the percent prey-categories for each of these stomachs. 'Treefrogs' accounted for 25% of the prey-categories listed from one of the stomachs, while 20% was the value for a second stomach. These percentages are not equivalent to those presented by Pimentel (1955) and cannot be interpreted as a diet consisting of 20–25% amphibians.

Vilella (1998) provided estimates of the density of the mongoose at EYNF: a mean of 1.8 individuals ha^{-1} (± 0.3 individuals ha^{-1}, $n = 3$ localities). Vilella's (1998) estimates of population size of mongooses from El Yunque were lower than estimates provided by Pimentel (1955) and Nellis and Everard (1983) for lowland PR and St Croix, respectively. Two more studies of the population of mongooses were conducted in the PRA during the mid-2010s. Guzmán-Colón and Roloff (2014) estimated abundance (mean daily capture of mongooses) along a > 1,000 m elevational gradient, from coastal areas to the high-elevation elfin forests at EYNF. The highest frequency of captures of mongooses occurred in coastal forests (0.17 captures day^{-1}), but the frequency of captures decreased along an elevational gradient from 0.17 captures day^{-1} in coastal forests to 0.07–0.06 captures day^{-1} at localities between 600 m and 850 m asl (i.e. respectively a 58.8% reduction in the frequency of captures of mongooses at the palo colorado forest and a 64.7% reduction at stands of the sierra palm compared with the frequency of captures of mongooses at coastal forests), to 0.02 captures day^{-1} at > 850 m asl (i.e. a 88.2% reduction at the elfin forest compared with the frequency of captures in coastal forests). However, as the authors acknowledged, in their sampling design, several important variables that might have influenced the probability of captures of mongooses in traps were not taken into account (e.g. trap response, sampling localities, and home range), and the interior of forests – instead of edges of forests – remains to be studied.

Johnson *et al.* (2016) provided the first published study that explicitly addressed estimates of population density of mongooses from mark-recapture sampling techniques in PR; they also included several modifications of the traditional trap layout and sampling design. A further improvement was that simultaneous comparisons were made in Guánica Forest, which lies in the dry south-western coastal area, and EYNF, which lies in the wet Luquillo Mountains, north-eastern PR. Their estimates of the density of mongooses were 0.34–0.55 mongooses ha^{-1} at Guánica Forest and 0.33–0.49 ha^{-1} at EYNF.

Across the PRA, Nellis and Everard (1983) documented a mean of 4.3 individuals ha^{-1} from St Croix, a density that is approximately 58% greater than the highest density documented from PR (2.5 individuals ha^{-1}; Pimentel 1955). Later, Horst *et al.* (2001) documented 6.4 mongooses ha^{-1} across St Croix, greater than on any other Caribbean island. However, a study of mark-recapture of mongooses conducted between 2003 and 2015 in a 149-ha site at the Sandy Point National Wildlife Refuge in St Croix yielded a mean population estimate of 0.88 individuals ha^{-1} (NF-A, unpublished data).

Many of the above estimates of population size of mongooses in the PRA were fragmented, had problems in their experimental design, and the first of them was conducted approximately 70 years after the mongoose was introduced into the archipelago. Several conclusions, however, can be drawn. First, amphibians comprised an insignificant component of the diet of mongooses (0% to 1.1% of prey-items). Second, there seems to be an inverse relationship between abundance/density of mongooses and elevation (Guzmán-Colón and Roloff 2014; Johnson *et al.* 2016). Notably, the frequency of mammals in the diet of the red-tailed hawk (*Buteo j. jamaicensis*) also decreases with increasing elevation up to the cloud forests at 870 m asl in EYNF (Santana and Temple 1988). Third, there seem to be differences in density of mongooses attributable to geographic location: Nellis and Everard (1983) found a mean of 4.3 individuals ha^{-1} at St Croix, a density approximately 58% greater than the highest density documented from PR, e.g. 2.5 individuals ha^{-1} (Pimentel 1955). Later, Horst *et al.* (2001) documented 6.4 mongooses ha^{-1} across St Croix, greater than any other Caribbean island. However, a study of mark-recapture of mongooses conducted between 2003 and 2015 in a 149-ha site at the Sandy Point National Wildlife Refuge in St Croix yielded a mean population-estimate of 0.88 individuals ha^{-1} (NF-A, unpublished data).

Finally, while we are tempted to suggest that the density of mongooses could have decreased with time after their introduction into the PRA because of changes in land-cover, we still lack long-term studies that address this question. A more fundamental question related to conservation of amphibians is what it is about the mongoose that leads authors to keep including it as a significant threat for amphibians in the PRA. The available data suggest that mongooses are not a threat to amphibians in this archipelago (see Appendix 8.I, for a review of historical accounts that may provide additional insights).

F. Responses of amphibians to climatic variables: temperature, humidity, and salinity

In this section, we examine data on the responses of amphibians – from physiology to assemblages of species – to the climatic variables of temperature and humidity, with a minor contribution from salinity. These data come from studies in PR. We also identify areas in need of research specific to ecophysiology of amphibians in the PRA. In the next section (III. G. 1–4) we discuss responses of amphibians to natural phenomena such as hurricanes and droughts.

1. Case histories of Puerto Rican amphibians' responses to climate

Several studies show that the fluctuation in population dynamics and distribution of amphibians in PR (and elsewhere) are mostly under the direct influence of climatic variables, with phenomena such as droughts and hurricanes playing a significant role (Stewart 1995; Joglar and Burrowes 1996; Woolbright 1996, 1997; Burrowes *et al.* 2004; Vilella and Fogarty 2005; Ríos-López and Aide 2007; Ospina *et al.* 2013; Ríos-López and Dávila-Casanova 2014; Ríos-López *et al.* 2014; Campos-Cerqueira and Aide 2017; Monroe *et al.* 2017; but see Stallard 2001). However, four topics have dominated most of the discussion about changes in population size and species' distributions for roughly 50 years: habitat destruction, introduced species, pathogenic outbreak (i.e. *Bd*), and climatic disturbances (e.g. droughts and hurricanes) (for an overview, see Blaustein *et al.* 2011 and

Heatwole 2013). We argue that combinations of natural and human-origin phenomena influence the responses of amphibians leading to stages that include declines and extirpations of amphibians also in the PRA (see section III. B. 1–5). First, we review other case studies of patterns of responses by amphibians – from individuals to populations – to changes in abiotic variables emanating from different phenomena.

(a) *Rhinella marina* and *Leptodactylus albilabris*

Heatwole *et al.* (1968) examined the heat tolerance of two species of anurans from PR: *Rhinella marina* and *Leptodactylus albilabris*. Tadpoles of *R. marina* were more heat tolerant than were those of *L. albilabris*, which suggests that the former may be better suited for using temporary ponds for reproduction than is the latter, and that climatic warming may benefit *R. marina* over *L. albilabris*. Moreover, Heatwole and Newby (1972) found that under dehydration stress the activity pattern of juvenile *R. marina* is mostly diurnal, reaching an activity peak at very low body moisture (30–39% water content), which they interpreted as an adaptation to xeric conditions. Although *L. albilabris* remains active during the day, it does so only if moist conditions prevail (Heatwole and Newby 1972). Thus, life-history characteristics may have a large influence in determining the relative contribution of heat tolerance to reproductive success between species. Specifically, what *L. albilabris* lacks in heat tolerance, compared with *R. marina*, is compensated for by its reproductive mode as a foam nester and its capacity for normal physiology under dehydration stress (Candelas *et al.* 1961; Heatwole *et al.* 1968).

In coastal areas, *R. marina* and *L. albilabris* occur along a salinity gradient, but adults of *R. marina* are relatively more abundant in the segments of the gradient within mangrove areas than is *L. albilabris* (Ríos-López 2008a). Under laboratory conditions, tadpoles of *R. marina* withstand higher salinity before physiological impairment and death than do those of *L. albilabris* (Ríos-López 2008a), which may explain the difference in the relative abundance of *R. marina* over *L. albilabris* in the field. The abundance of another species, *Lithobates grylio* appears to decrease with increased salinity, similar to *L. albilabris* (Ríos-López 2007), but more data are required for verification.

(b) *Eleutherodactylus* spp.

Heatwole and Newby (1972) examined activity patterns under dehydration stress among nine species of *Eleutherodactylus* (eight from the PRA and *E. martinicensis* from the Lesser Antilles). They found that species whose activity patterns are considered strictly nocturnal showed a lower level of activity during the day (e.g. *E. antillensis, E. coqui*). The reverse was true of *E. cochranae* and *E. locustus*, and their activity patterns were higher during the day than at night. A species that occurs in the leaf litter, *E. richmondi*, showed an activity level that may fit both patterns. A third pattern was found in which activity levels did not differ between day and night, but did decrease with progressive desiccation (*E. eneidae, E. karlschmidti, E. wightmanae*, and *E. martinicensis*). In addition, all species of *Eleutherodactylus* ceased their movements at a moisture content of their body between 40 and 49%, except for *E. coqui*, which remained active until moisture content lowered to between 30% and 39%. Heatwole and Newby (1972) reasoned, based on the curve of the number of movements per hour plotted against moisture content of the body, that these activity patterns were related to the capacity of individuals to behaviourally minimize dehydration. While this capacity differed among species, the activity level of all species increased initially with increased dehydration of the body, which probably reflected an escape or 'water-searching' behaviour as an early response to desiccation, followed by a later reduction in activity, accompanied by a crouching or 'water-conserving posture' that reduced exposure of surface area to air (Heatwole *et al.* 1969; Heatwole and Newby 1972) (Fig. 8.19).

Figure 8.19. Adult individuals of *Eleutherodactylus portoricensis* in the water conservation posture (top left) and hidden beneath the leaf litter (top right) in response to dehydration stress during the 2015 drought in our forest study site in the Cayey Mountains, PR; an adult individual of *E. wightmanae* using a hole in the ground as a retreat site also during the 2015-drought in our study site (bottom left and right). We do not know if the individual of *E. wightmanae* dug this hole. Pictures by Neftalí Ríos.

Ambient humidity is also a fundamental variable influencing the activity and survival of amphibians with direct development. For instance, Pough *et al.* (1983) found that individuals of *E. coqui* cease their nocturnal activity entirely after approximately a week of dry nights at El Verde in the Luquillo Mountains. However, Pough *et al.* (1983) did not quantify what constitutes a 'dry night'. Nonetheless, at El Verde, Odum *et al.* (1970) documented that more than 1.5 mm of water is needed to keep the leaf litter wet in this forest, and Scatena (1990) provided data suggesting that the interception of rain by the forest canopy at this site may reach 58% (percentages of interception vary with the architecture and species composition of the vegetation). Based on these estimates, Stewart (1995) concluded that approximately 3 mm of rain would be needed to keep the forest floor humid enough to support the activity, and presumably survival, of *Eleutherodactylus* frogs. In fact, Stewart found that the abundance of *E. coqui* declined sharply after five or more days with < 3 mm of rain recorded from rain gauges in the open, a quantity that represents the minimum rainfall needed to keep *E. coqui* active at this site.

Heatwole and Newby (1972) found that *E. coqui* was the only species examined to remain active at the lowest recorded level of moisture content of the body (30–39%), which is a level consistent with more diurnal species such as *R. marina* (juveniles) and *L. albilabris*. Hence, we might expect that other species of *Eleutherodactylus* could respond similarly to low levels of rainfall. Yet individuals could also respond adaptively to changes in environmental conditions (Fig. 8.19) (Heatwole *et al.* 1968 for *L. albilabris*; Pough *et al.* 1983 for *E. coqui*), and a linear rationale may not be entirely accurate as presented.

Van Berkum *et al.* (1982: 135) found that 'a substrate moist enough to permit rehydration is one important feature of a retreat site' for *Eleutherodactylus* frogs. For example, individuals of *E. wightmanae* (a species restricted to humid forested areas, mostly in the mountains) were found in holes in the ground during the severe drought of 2015 in a forest at > 850 m asl in the Cayey Mountains (Fig. 8.19) (NR-L, unpublished data; see section III. F. 4). While limited, this observation may represent a behavioural response to dehydration, which could explain the undifferentiated pattern in diel activity under dehydration stress found by Heatwole and Newby (1972) for this species.

Joglar *et al.* (1996) documented a presumed behavioural modification of another species that, albeit not from montane forests, has specialized ecological requirements: *E. cooki*. According to Rogowitz and Sánchez-Rivoleda (1999: 45) individuals of this species have 'negligible cutaneous resistance to water loss', presumably because 'the high humidity and availability of water in caves provides a mesic environment that decreases chances of desiccation'. Although individuals of *E. cooki* have not been seen submerging in streams or the water that flows at the bottom of the formations of boulders, at night frogs can be seen active on the humid surfaces of rocks in the streambed. Joglar *et al.* (1996) observed individuals leaving the caves at night to perch on the humid vegetation adjacent to the openings among the boulders. They suggested that this behaviour relates to required rehydration at night when relative humidity decreases in the caves, but dew (and rain) outside provides moisture not readily available to the frogs inside the caves.

Beuchat *et al.* (1984) found that the tolerance to dehydration of *E. portoricensis* (a species restricted to 'cool, forested montane habitats above 300 m' of elevation [1984: 579]) decreases at lower temperature, thus concluding that this species is sensitive to relatively small changes in ambient temperature in the wild. However, individuals of *E. portoricensis* have been found deep in the leaf litter in response to the severe drought of 2015 in the forest at the Cayey Mountains (Fig. 8.19) (NR-L, unpublished data), which implies that these frogs also respond to desiccation behaviourally. Consequently, the notion that under rainforest conditions species would not be subject to strong selection for adaptive behavioural responses to desiccation need to be re-evaluated for some of the species found in PR (for a further discussion but solely on differences in thermal adaptation between *E. coqui* and *E. portoricensis* and among populations within *E. coqui* and *L. albilabris*, see Christian *et al.* 1988). At the population level, however, the responses of a ground-dwelling species such as *E. portoricensis* to thermal climatic change could be different from those of a canopy-dwelling species such as *E. hedricki* or a eurytopic species such as *E. antillensis*. Unfortunately, we know little about the responses of *E. hedricki* to microclimatic conditions throughout its geographical distribution. Nonetheless, Scheffers and Williams (2018) found that the thermal regimes in the canopies overlap significantly between lowland and highland tropical-montane forests, compared with the degree of thermal overlap between the understorey and ground levels. These authors concluded that this overlap potentially bypasses the geographical differences in thermal constraints and humidity constraints imposed by climatic change on many species of vertebrates inhabiting the understorey of forests along elevational gradients.

Beuchat *et al.* (1984: 579) provided valuable information for *E. antillensis*, a species they associated with 'hot, arid grasslands at low altitudes', and that when stressed by dehydration show activity patterns similar to those of a strictly nocturnal species (Heatwole and Newby 1972). Beuchat *et al.* (1984) also found that tolerance of dehydration by *E. antillensis* increases with decreases in ambient temperature, which opposes the pattern found for *E. portoricensis*. As such, their study expanded the study of Heatwole and Newby (1972) by adding the influence of another environmental variable acting in synergism: temperature. In fact, Navas *et al.* (2016) emphasized the importance of studies that combine the influence of water and temperature for proper mechanistic modelling of responses of amphibians to projected climatic change. Furthermore, results from Beuchat *et al.* (1984) on *E. antillensis* may help us understand its pattern of dispersion upland to 850 m asl in PR (Joglar 1998) as a function of two important variables for the activity of amphibians (i.e. humidity and temperature). So far, the dispersion of *E. antillensis* occurs mostly along the margins of roads in the highest mountain peaks and not deep into the interior of forests in the highlands (NR-L, pers. obs.). Roads are thus migration routes with abiotic conditions, particularly in terms of temperature, suitable for the upland migration of lowland species (i.e. a situation that resembles the potential for upland migration, through the canopy, of tree-dwelling species) (Scheffers and Williams 2018). Consequently, this topic also provides fruitful avenues for future research because of its profound implications for the conservation of upland species of amphibians under the conditions of the projected changes in climate.

Determining the relative contribution of each abiotic variable and weather phenomenon to changes in the abundance and distribution of the amphibians in the PRA is not straightforward. As discussed earlier (section III. B.) the reduction of the geographic range of species of *Eleutherodactylus*, their abundance, and their extirpation from areas used historically for scientific collection, may be explained in part by the influence of weather. Likewise, in tropical montane forests their responses to climatic change may differ between ground-dwelling and canopy-dwelling species. In EYNF in the Luquillo mountains of north-eastern PR, Narins and Meenderink (2014) provided evidence of a change in parameters of the advertisement call of *E. coqui* in areas where the calls were examined more than two decades earlier at the same locality. These changes, from a lower-pitched and longer duration of call to a higher-pitched and shorter duration of call, were interpreted as a proxy for a reduction in body-size in response to increased temperature at each locality along an elevational gradient (e.g. an *in-situ* response). As a result, these authors suggested that the predicted increase in ambient temperature in PR by 2100 'may result in a culling of the low-elevation populations as animals become increasingly temperature stressed, reducing their total biomass' (Narins and Meenderink 2014: 4).

Other studies have shown alternate scenarios for *E. coqui* in response to increased ambient temperature. It is known, for instance, that a higher temperature may result in a decrease in the developmental time in ectotherms (for *E. coqui* see Townsend and Stewart 1985). In turn, the increase in temperature also results in increased metabolism, decreased time to reach maturity, smaller individuals, and an increase in the frequency of reproductive events (Duellman and Trueb 1994; Pigliucci 1996; Van der Have and De Jong 1996; Atkinson and Sibly 1997; Partridge and Coyne 1997; Van Voorhies 1997; Angilletta *et al.* 2004; Hillman *et al.* 2009). Moreover, these biological processes can be detected at a population level (Morrison and Hero, 2003; Hillman *et al.* 2009). For example, the increase in temperature results in a decrease in the intraoval developmental time of *E. coqui* (Townsend and Stewart 1986), which may explain the differences in population size of *E. coqui* at different elevations in EYNF. In this forest, the population size decreases with elevation: 45,000 individuals ha^{-1} at 30 m asl; 24,800 individuals ha^{-1} at 370 m asl; and 2,500 individuals ha^{-1} at 750 m asl (Stewart and Woolbright 1996). Likewise, ambient temperature decreases linearly with elevation (see Figure 3 in Narins and Smith 1986), but rainfall increases (although non-linearly; see Figure 4 in García-Martinó *et al.* 1996). Taken together, these pieces of evidence suggest that a higher frequency of reproduction by lowland females of *E. coqui* may result in larger population sizes in lowland areas compared with highland areas (but see Woolbright 1991). So, the projected increase in ambient temperature in PR may, in fact, increase the biomass of populations of *E. coqui* along the elevational gradient as long as the projected increase in temperature remains within the levels of physiological tolerance of this species.

The increase in population size of *E. coqui* could have consequences at the ecosystem level, and does not occur solely because of increased temperature. For instance, being a highly abundant local predator (or prey), the scenario of population increase with temperature in *E. coqui* may result because of the influence of a myriad of ecosystem processes related to decomposition, whose rate also increases with temperature (and humidity). Moreover, the increase in population size of *E. coqui* also contributes to the local increases in the flow of nutrients, carbon, and energy through predator–prey interactions along the gradient in elevation. However, these are not linear processes subjected to simple interpretations. In fact, the projected reduction in precipitation and orographic humidity, of which decomposition rate also depends, may leave the suggested scenario of population growth and elevational distribution of *E. coqui* tentative (e.g. Navas *et al.* 2016).

The elevational changes in abundance and distribution of other species of *Eleutherodactylus* (*E. wightmanae* and *E. brittoni*) result in a unimodal pattern, with humidity and precipitation (along with some characteristics of vegetation cover) having a more significant influence on abundance and distribution compared with air temperature (Monroe *et al.* 2017). In these species, Monroe *et al.* (2017) found that the highest relative abundance occurs at mid-elevations, between 400 m and 700 m asl for *E. wightmanae*, and between 500 m and 700 m asl for *E. brittoni* in Maricao (western Central Mountain Range). However, factors leading to the decrease in abundance at higher elevations were less clear (Monroe *et al.* 2017).

Finally, at the assemblage level, Ríos-López and Aide (2007) found that the abundance of species of amphibians and reptiles varies markedly in time (during secondary succession) and space (between valleys and hills), and with differences in relative humidity and ambient temperature. They found that during early stages of succession, when ambient temperature is high and relative humidity is low, the abundance of reptiles is higher than that of amphibians. However, as succession proceeds, air temperature decreases and relative humidity increases with a corresponding initial increase in the abundance of amphibians relative to that of reptiles. Furthermore, the herpetofaunal assemblage is characterized by a higher abundance of species of amphibians in valleys – where temperature is lower, relative humidity is higher, and wind speed is lower (which further reduces losses of humidity) – compared with hilltops. On hilltops, in contrast, the abundance of most reptiles is higher, presumably as microclimatic conditions reverse (see Heatwole 1982 for a review on the structuring of herpetofaunal assemblages at larger geographic scales).

2. Relationship between physiology and ecology in amphibians' responses to predicted changes in climate: a needed research agenda

Navas *et al.* (2016) examined the science citation index between 2000 and 2013 (> 3,300 papers) and indicated the number of papers that bridged the topics of physiology, conservation, and population declines of amphibians. On average, the number of papers published solely about population declines was more than 30 times the number of papers specifically addressing the relationship between population declines and physiology (see Figure 7.1 in Navas *et al.* 2016). Moreover, they highlighted many questions that have profound consequences for the conservation of amphibians, and that could be addressed from the perspective of physiological ecology. These include (among others): examining 'whether individuals of a given species would be able to adjust to a given dimension' of climatic and environmental transformations, particularly those of human origin (2016: 162); how physiological ecology and ethology may elucidate patterns of 'orientation and resource- searching behavior' (2016: 163); how 'chronic stressors on physiological function connects directly to population dynamics' (2016: 164), including inhibition or limitation of reproductive output and susceptibility to pathogenic infections (i.e. the fields of stress biology and eco-immunology); and how trophic dynamics and nutrition may be affected by climatic change, thus altering the quality of the diet in a changing environment. There is still a need for broader integration of ecophysiology and climate to reach a better understanding about the mechanisms that mediate responses of amphibians to projected changes in climate for purposes of conservation (see Appendix 8.II, background information based on reports of the Intergovernmental Panel on Climate Change [IPCC] and related documents).

G. The relationship between amphibians and natural phenomena in the Puerto Rican Archipelago: hurricanes and droughts

Wadsworth and Englerth (1959) provided the first detailed analysis of the influence of a hurricane (named Santa Clara in Spanish or Betsy in English) on forests in PR in 1956, and we hypothesized on the influence of this hurricane on the population status of *Eleutherodactylus eneidae*. However, those interested in the topic of hurricanes from the ecological perspective should consult the book *A Caribbean Forest Tapestry – The Multidimensional Nature of Disturbance and Response*, edited by Brokaw *et al.* (2012a), which provides the most thorough review of the effects of hurricanes (and droughts) as natural phenomena influencing responses of fauna, flora, and ecosystems in PR.

On 18 September 1989, Hurricane Hugo made landfall in St Croix as a strong category 4 tropical storm and caused extensive changes to the island's infrastructure and ecosystems (also the nearby islands of Vieques and Culebra). A few hours later, Hugo struck PR as a strong category 3 tropical cyclone, leading to similar transformations, mainly on the eastern part of the island (National Weather Service 2022a). Hurricane Hugo generated an extensive assessment of the influence of hurricanes on flora and fauna in the PRA, in particular, and on forest ecosystems in the Caribbean in general (Walker *et al.* 1991; Lugo *et al.* 2012). Studies on the population dynamics of amphibians in PR were under way when Hugo struck this area, which also ignited analyses on the role of this hurricane on amphibians. In this section, we will limit discussion mostly to the available data on the influence of hurricanes Hugo in 1989 and Georges in 1998 (category 3) on amphibians in the PRA (see Table 8.8), and briefly, the influence of the severe 2015 drought and Hurricane María in 2017 on amphibians in PR.

1. Hurricane Hugo (1989)

Hurricane Hugo changed the architecture, structure, and composition of species in assemblages, and cover of vegetation (particularly of woody vegetation in forests; Reagan 1991; Tanner and Kapos 1991; Woolbright 1997; for the USVI see Reilly 1991). In addition, Hugo generated cascading effects on abiotic and biotic factors and processes (see reviews by Brokaw *et al.* 2012b; McDowell *et al.* 2012; Scatena *et al.* 2012), including changes in ambient and soil temperature and humidity (Tanner and Kapos 1991), rainfall (Scatena and Larsen 1991), nutrients and toxic fluxes of ammonium (Steudler *et al.* 1991), and interspecific interactions (Waide 1991; Willig and Camilo 1991) that include amphibians in PR (Woolbright 1991, 1996, 1997).

Several studies provided various explanations for the changes in the abundance of different species of amphibians after a hurricane. Explanations included changes in predator–prey interactions; recruitment dynamics; changes in vegetation cover and ecosystem processes related to the availability of nutrients; changes in microclimate; and changes in the availability of retreat/breeding sites and the behaviour of amphibians in their use of these sites (e.g. Joglar and Burrowes 1991, 1996; Woolbright 1991, 1996, 1997; Burrowes *et al.* 2004; Vilella and Fogarty 2005; Klawinski *et al.* 2014). Presumably, the most extensive changes in microclimate and biotic conditions occur rapidly during and immediately after a hurricane (Brokaw *et al.* 2012b; McDowell *et al.* 2012; Scatena *et al.* 2012). The influence of these changes on populations of amphibians may persist for months to several years (Table 8.8; also see Figure 8.1 in Willig *et al.* [2012] for responses of the leaf litter and soils where amphibians forage, retreat, and breed). However, the path of ecosystem dynamics after a hurricane and its role on the status of

Table 8.8. Review of patterns of responses of amphibians to hurricanes in Puerto Rico in the herpetological literature, including pre-hurricane and post-hurricane sampling intervals and duration (years), type of data, geographic region, and elevation. M-R = mark-recapture; VS = visual survey; AS = acoustic survey.

HURRICANE (DATE) AND *SPECIES*	RESPONSE	PRE-SAMPLING	POST-SAMPLING	DATA	REGION
[a] **Hugo (18 September 1989)**					
Eleutherodactylus coqui					
Adults (males and females)	By October 1990: fourfold increase; body size decreased; no difference in survivorship.	Twice yearly; winter (January) and summer (June and/or July); 1987–89.	Every other month (October 1989–October 1990).	M-R	EYNF-El Verde Field Station (350 m asl)
Juveniles	Decreased initially; smallest individuals less likely to survive; trend of increase by October 1990.			VS	
Leptodactylus albilabris	Number of calling males remained the same: 3 individuals heard before and after the hurricane.	July 1989 (13 nights).	June 1990 (11 nights).	AS	EYNF-Bisley Research Area (250 m–266 m asl)
E. hedricki	Increased 14%, from 63 to 72 individuals heard.			AS	
E. wightmanae	Decreased 31%, from 39 to 27 individuals heard.			AS	
E. portoricensis	Decreased 45%, from 20 to 11 individuals heard.			AS	
E. richmondi	Decreased 83%, from 12 to 2 individuals heard.			AS	
[b] **Hugo (18 September 1989)**					
E. coqui	Population density of individuals identified as *E. coqui* neither decreasing nor increasing (juveniles were not identified by species).	Twice monthly since February 1989.	Twice monthly between September 1989 and September 1990.	VS	EYNF-Palo Colorado Area (661 m asl)
E. portoricensis	Population density of individuals identified as *E. portoricensis* decreased (juveniles were not identified by species).			VS	

(*Continued*)

Table 8.8. (Continued)

HURRICANE (DATE) AND *SPECIES*	RESPONSE	PRE-SAMPLING	POST-SAMPLING	DATA	REGION
[c] **Follow up, Post-Hugo**					
E. coqui	Population density of individuals identified as *E. coqui* increased initially and fluctuated within pre-hurricane levels 5 years later (i.e. 'stable'). Mean population estimate of juveniles identified as *E. coqui* fluctuated and continued to vary greatly (see Figure 1 in Woolbright 1996 and Figure 5.8 in Brokaw *et al.* 2012b).	1988–90	1991–94	% nights (counts variable); calling males heard	EYNF-El Verde Field Station
[d] **Follow up, Post-Hugo**					
E. coqui *L. albilabris* *E. hedricki* *E. wightmanae* *E. portoricensis*	(stable; Woolbright 1996) mean (± SD) = 70.0% ± 32.5%, variable. mean (± SD) = 65.9% ± 22.3%, variable. mean (± SD) = 93.1% ± 9.7%, variable. mean (± SD) = 1.6% ± 2.7%, decreased, last heard in 1989.	1988–90	1991–94	% nights (counts variable) with calling males heard	EYNF-El Verde Field Station
E. richmondi	mean (± SD) = 5.1% ± 11.2%, decreased, last heard in 1989.				
[e] **Follow up, Post-Hugo**					
E. coqui	Population density of individuals identified as *E. coqui* increased and fluctuated within pre-hurricane levels (see Figure 2D in Joglar and Burrowes (1996); juveniles were not identified by species).	Twice monthly since February 1989.	Twice monthly between September 1989 and 1991: once a month in the period 1992–94.	VS	EYNF-Palo Colorado Area
E. portoricensis	Population density of individuals identified as *E. portoricensis* fluctuated within pre-hurricane levels, stable not declining by summer 1994 (see Figure 2F in Joglar and Burrowes (1996); juveniles were not identified by species).			VS	

Table 8.8. (Continued)

HURRICANE (DATE) AND *SPECIES*	RESPONSE	PRE-SAMPLING	POST-SAMPLING	DATA	REGION
[f] **Follow up, Post-Hugo**					
E. coqui	Mean density of adults of *E. coqui* decreased steadily after Hugo (see Figure 1 in Burrowes *et al.* 2004).	Data from 1989 dropped from the analyses.	Twice monthly between 1990 and 1992; once a month thereafter until 2001.	AS and VS	EYNF-Palo Colorado Area
E. portoricensis	Mean density of adults of *E. portoricensis* decreased steadily after Hugo (see Figure 1 in Burrowes *et al.* 2004).			AS and VS	
[g] **Georges (21 September 1989)**					
E. coqui	Average abundance increased 66.1%, from 52.2 to 86.7 calling individuals heard.	April–May 1998	May 1999	AS	Cordillera Central-Guilarte State Forest (elevation of census plots not specified)
E. wightmanae	Average abundance increased 35.6%, from 11.8 to 16.0 calling individuals heard.			AS	
L. albilabris	Average abundance increased 61.2%, from 6.7 to 10.8 calling individuals heard.			AS	
E. brittoni	Average abundance decreased 20.8%, from 4.8 to 3.8 calling individuals heard.			AS	
E. coqui	Average abundance increased 10.2%, from 32.5 to 35.8 calling individuals heard.	April–May 1998	May 1999	AS	Cordillera Central-Maricao State Forest (elevation of census plots not specified)
E. wightmanae	Average abundance decreased 20.5%, from 22.4 to 17.8 calling individuals heard.			AS	
L. albilabris	Average abundance decreased 72.7%, from 2.2 to 0.6 calling individuals heard.			AS	
E. brittoni	Average abundance decreased 70.5%, from 4.4 to 1.3 calling individuals heard.			AS	
E. richmondi	Average abundance decreased 85.0%, from 6.7 to 1.0 calling individuals heard.			AS	
E. antillensis	Average abundance decreased 81.0%, from 2.1 to 0.4 calling individuals heard.			AS	

(Continued)

Table 8.8. (Continued)

HURRICANE (DATE) AND *SPECIES*	RESPONSE	PRE-SAMPLING	POST-SAMPLING	DATA	REGION
[h] **María (20 September 2017)**					
E. coqui	Average calling activity decreased 6.0%, but occurrence among sampling sites remained the same.	April–August 2017	April–September 2018	AS	Cayey Mountains -Carite State Forest (elevation between ~500 and ~900 m asl)
E. wightmanae	Average calling activity decreased 11.5%, but occurrence among sampling sites remained the same.			AS	
E. richmondi	Average calling activity decreased 7.5%, and occurrence among sampling sites decreased 23.5%.			AS	
E. cochranae	Average calling activity decreased 43.4%, but occurrence among sampling sites remained the same.			AS	
E. gryllus	This species was rarely detected at sampling sites, yet average calling activity and occurrence among sites decreased 100%.			AS	
E. antilensis	Average calling activity increased 34.2%, and occurrence among sampling sites increased 23.5%.			AS	
E. brittoni	Average calling activity increased 22.0%, and occurrence among sampling sites increased 14.0%.			AS	
E. portoricensis	Average calling activity varied 2.2%, but the change was not significant (2021: 1668); occurrence among sampling sites remained the same.			AS	

[a] Woolbright (1991); [b] Joglar and Burrowes (1991); [c] Woolbright (1996); [d] Woolbright (1997); [e] Joglar and Burrowes (1996); [f] Burrowes *et al.* (2004); [g] Vilella and Fogarty (2005); [h] López-Hernández and Puente-Rolón (2021).

populations of amphibians differ depending on the intensity and frequency of the hurricane, geographic locality, topography, and biotic and abiotic characteristics of the area (Table 8.8).

Table 8.8 shows that responses of amphibians to hurricanes in the PRA vary with geographic location, species (and their ecological requirements when known), life stage, and confounding effects brought on by the duration of the period of observation and methodology of the study. Most of these data pertain to one of the most generalist and vagile species in PR, *Eleutherodactylus coqui*, but a broader understanding of responses of other species of *Eleutherodactylus* in the archipelago remains elusive. Moreover, the understanding of the long-lasting effects of hurricanes on changes in the abundance and distribution of amphibians could be obscured by other phenomena, such as droughts, which also influence populations of amphibians in the long term. For example, Christensen *et al.* (2013) documented two strong events of El Niño, in 1986 and 1987, and three consecutive years of El Niño, between 1990 and 1992, which covered the after-years of Woolbright's assessments on the effects of Hurricane Hugo on amphibians in PR. For *E. coqui*, Woolbright (1996) documented that the number of adults of this species increased a year after Hurricane Hugo, but the number of its adults and juveniles decreased to numbers within the observed variation before Hugo in approximately five years (also see Figure 5.8 in Brokaw *et al.* 2012b). Yet Woolbright (1997) also documented changes in the number of adults of other *Eleutherodactylus*, and that included the extirpation of populations of at least two species, *E. portoricensis* and *E. richmondi*, from low-elevation sites in the Luquillo Mountains, northern PR, after Hurricane Hugo (Table 8.8). Nonetheless, whether these extirpations resulted from the immediate effects of Hurricane Hugo, the severe droughts during the late 1980s and early 1990s (see Stewart 1995), or the cumulative effect of both phenomena on the population dynamics of these species remains elusive and requires long-term studies for clarification.

2. Hurricane Georges (1998)

Hurricane Georges generated different patterns of responses of the population of amphibians in PR (Vilella and Fogarty 2005; Table 8.8). For example, the abundance of *Eleutherodactylus wightmanae* in the Guilarte State Forest, Cordillera Central and central PR, increased after this hurricane, but it decreased in the Maricao State Forest, Cordillera Central, and western PR (Table 8.8). The eye of Hurricane Georges affected the Cordillera Central directly, but the decrease in the abundance of *E. wightmanae* in Maricao – in the short term –was not as notable as that after Hurricane Hugo (which barely struck north-eastern PR; Table 8.8; National Weather Service 2022b). The population dynamics of amphibians in the rest of the island of PR have not been studied as intensely or thoroughly as it has been in the Luquillo Mountains (Vilella and Fogarty 2005). Consequently, the interpretation of the magnitude and diversity of responses of amphibians to hurricanes throughout PR (and the PRA) remains limited.

Regardless of the geographical area in PR, the recovery of populations of species would be influenced by distance to nearest sources of species for recolonization (i.e. metapopulation dynamics leading to a rescue effect), and by ecological processes that correlate physiology and breeding biology of species that influence recolonization (see section III. G. 3. next). These are two areas of research that remain unexplored from the perspectives of the population ecology of amphibians and their responses to human and natural phenomena. Furthermore, several species of amphibians in PR are known to occur in patches in any given forest (i.e. *Eleutherodactylus richmondi, E. locustus, E. hedricki*), which suggests that their populations are structured as metapopulations, and source-sink dynamics could confound statements of extirpations deemed unexplained in the literature.

3. *The 2015 Drought*

From 2014 to the present, Ríos-López *et al.* (previously unpublished) have monitored the population dynamics of the herpetofauna in a subtropical wet forest in the Cayey Mountains, south-eastern PR. Using visual and acoustic-survey techniques, they examined the influence of climatic variables and vegetation on the reproduction, abundance, and distribution of four species of *Eleutherodactylus* and two species of reptiles (*Anolis*). Briefly, we highlight the influence of the severe drought of 2015 (Mote *et al.* 2017) on the most-abundant highland-specialist amphibian at our study site in the Cayey Mountains, *E. portoricensis* (Fig. 8.6b, Fig. 8.19), between March 2014 and March 2018.

As mentioned, Odum *et al.* (1970) estimated that > 1.5 mm of rainfall must reach the leaf litter to keep it wet and Scatena (1990) estimated that the rainfall that is intercepted by the forest canopy depending on the composition of species of trees ranges from 48% to 68%. We estimated the interception of rainfall by the forest canopy at our study site (~823 m asl) in the Cayey Mountains as 58% (the middle value between 48% and 68% from Scatena [1990]). Finally, we followed Stewart's (1995) methodology to estimate the amount of rainfall that must reach the leaf litter to avoid physiological stress because of dehydration in *Eleutherodactylus* frogs at El Verde in the Luquillo Mountains (those interested in the variability of climate in the Luquillo Mountains may review Wade *et al.* 2013 and Murphy *et al.* 2017). Stewart (1995) suggested that five or more days with a deficit in humidity in the understorey resulted in dehydration stress in *Eleutherodactylus* frogs and a pronounced reduction in their activity overall. Therefore, the information from these three studies is fundamental for a description of the influence of rainfall and humidity on the activity pattern of *Eleutherodactylus* frogs at our study site in Cayey.

First, we used the methodology of Stewart (1995) and 50 years of rainfall data from Cayey (1960 to 2010; Jájome Alto Station [720 m asl], Climatological Data PR and the VI, Weather Bureau, US Department of Commerce) to detect periods of drought at our study site. We then examined the deviation of rainfall data collected in the understorey from data in historical records and after correcting for the estimated interception of rainfall provided by Scatena (1990) for a forest similar to that in the Cayey Mountains. We estimated that ~2.6 mm of rainfall from historical records is the minimum needed to keep the leaf litter wet in the study site (i.e. ~1.5 mm of rainfall documented in the understorey at ground level after factoring out the estimated 58% of interception of rainfall by the canopy). Finally, we counted the number of days with < 1.5 mm of rainfall documented in the understorey (data from our rain gauge) for the duration of our study and contrasted it with the historical record of rainfall < 2.6 mm documented in the open by the weather station in Jájome, Cayey. The result illustrates the duration and magnitude of drought as the number of days with less than the minimum rainfall required to avoid physiological stress because of dehydration in *Eleutherodactylus portoricensis*.

Compared with historical deficits, at least 13 months exceeded the total number of days resulting in hydric stress at our study site between 2014 and 2015 (Fig. 8.20a). Between 2014 and 2015, 91.5% of the time (~425 days of drought) consisted of days with rainfall below the threshold to avoid hydric stress in this (and other) species. If five or more days without rain lead to dehydration stress and a reduction in activity (including calling for reproduction) in these frogs (Stewart 1995), consider next what would occur in a four-month period characterized by 61 consecutive days with frogs experiencing hydric stress (between December and January 2015) followed by 56 similar consecutive days between February and March 2015 (Fig. 8.20a).

The abundance of *Eleutherodactylus portoricensis* (adults and large juveniles > 10–12 mm in SVL) increases during the rainy and warmer months (e.g. May to November), similar to most *Eleutherodactylus* in the PRA (Fig. 8.20b). However, there is a lag in response of individuals of

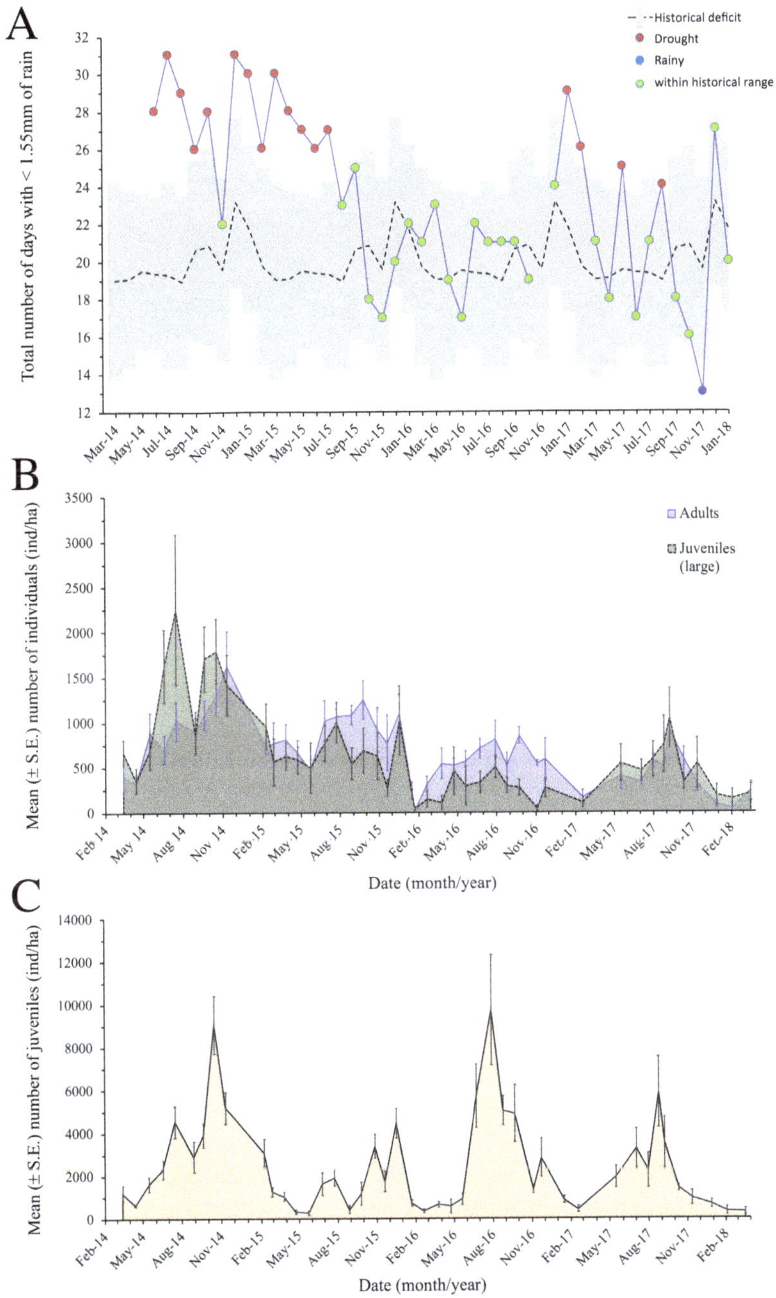

Figure 8.20. Variation in the number of days with hydric deficit (between March 2014 and January 2018; coloured dots and blue line) compared with historical values (grey bars and black dash-lines) between 1960 and 2010 in Jájome Alto, Cayey municipality, PR (A; see text for definition); population fluctuation (mean ± standard error; between March 2014 and March 2018) of adults and large-bodied juveniles of *Eleutherodactylus portoricensis* in Cayey (B); and fluctuation in the number of small-bodied and recently hatched juveniles among all *Eleutherodactylus* frogs (but mostly *E. portoricensis*) in our forest study site in the Cayey Mountains (C). Previously unpublished data of Neftalí Ríos.

Figure 8.21. Changes in vegetation and cover of the canopy one month before Hurricane María (A), a few days after María (B), and one year after María (C) in our forest study site in the Cayey Mountains, PR.

E. portoricensis to the drought in which a reduction in the number of individuals detected became evident a year after the peak of the drought (Fig. 8.20b). A more dramatic reduction was experienced by smaller juveniles (most of them recently hatched), also at approximately a year afterwards (Fig. 8.20c). Initially, this reduction in the number of small juveniles possibly resulted from a combination of a reduction in the reproductive activity of adults (inactive because of stress from dehydration, see above) and increased mortality of juveniles (see next paragraph).

Woolbright (1991) and Stewart (1995) examined the responses of juveniles of *Eleutherodactylus coqui* after climatic phenomena and reasoned that increased mortality of hatchlings resulted from extreme desiccation and dehydration stress in the short term. In the Cayey Mountains, the number of small-sized juveniles rebounded to pre-drought values a year after the 2015-drought (Fig. 8.20c), but its influence as recruits on categories of larger individuals (e.g. older individuals) in this population had not become evident two years later; the number of adults and large juveniles remained low within post-drought values. We agree with Stewart (1995) in that droughts in the Caribbean region have a more significant influence on the population dynamics of *Eleutherodactylus* frogs than is frequently acknowledged (see Stewart 1995; Mote *et al.* 2017; also see Figure 8.16 for *E. karlschmidti*).

4. Hurricane María (2017)

Hurricane María was a high-end category 4 cyclone that caused extensive effects to the PRA (Pasch *et al.* 2018; National Weather Service 2022c; Fig. 8.21a–b). Hurricane María reached PR on 20 September 2017, and at the time of its landfall near Yabucoa (south-eastern PR) defoliated all trees across the island. The eye of Hurricane María crossed over our study site in the Cayey Mountains by late morning, and produced > 889 mm of rainfall (Pasch *et al.* 2018), while its winds (maximum ~155 km; ~130 km in the Cayey Mountains) defoliated all trees in its path (Fig. 8.21b). The canopy of this forest was still open a year after this hurricane (Fig. 8.21c), with trees producing leaves only along the vertical axis (Fig. 8.22); the production of branches would follow in the longer term. In the short term, the opening of the canopy facilitates higher levels of evaporation from air and soils, and results in higher and more variable temperature at ground level (Fig. 8.23a). These conditions affect ecological specialists such as *Eleutherodactylus portoricensis* that are adapted to lower temperature and higher humidity in the lower strata of vegetation in forests in the highlands (see next paragraphs).

The changes in population size of *E. portoricensis* continue with their natural dynamics: the size of the population increases during months of higher temperature and humidity (i.e. May to November), and decreases during months of lower temperature and less humidity (i.e. ~January to March). However, Hurricane María accentuated the decrease in the number of individuals (adults and large juveniles) detected in our study site, which added to the influence that the 2015 drought had on this species (Fig. 8.23b). In contrast, microclimatic conditions in our forest study site in the Cayey Mountains are of higher and more variable ambient temperature, and lower and more variable humidity, which are conditions that characterize forested areas in the lowlands and presumably benefited species considered generalists such as *E. coqui* (Fig. 8.23b). The decrease in the number of recently hatched juveniles (among all species) mirrors that of *E. portoricensis* (Fig. 8.23b–c), the species contributing the greatest number of hatchlings at the study site. Similar to results from Woolbright (1996, 1997), what may happen next is a continued decrease in the abundance of ecological specialists such as *E. portoricensis* and an increase of ecological generalists such as *E. coqui*, which now finds environmental conditions favourable for its reproduction and capitalizes on available resources for retreat during the following years

Figure 8.22. Trees that contributed to a closed canopy in the study site in the Cayey Mountains, PR, lost most of their branches, which opened up the cover of the canopy. Trees produced new leaves but only along the vertical axis of the main trunk one year after Hurricane María.

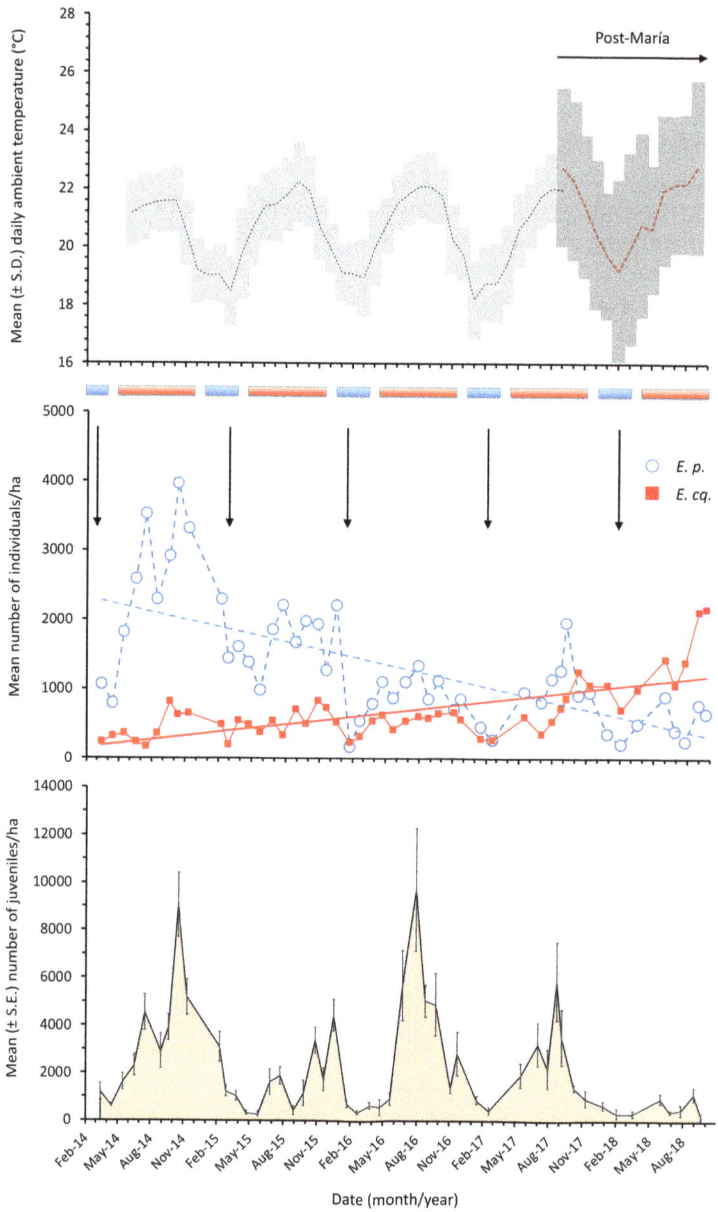

Figure 8.23. (Top) Fluctuation (dashed lines) and variability (grey bars) in ambient temperature in the understorey between May 2014 and September 2018 in the forest study site in the Cayey Mountains, PR. The ambient temperature increased and varied widely because of Hurricane María (darker grey bars and dashed red line) to date. (Middle) Fluctuation in the number of individuals of two species of *Eleutherodactylus*, *E. portoricensis* (open circles and dashed line) and *E. coqui* (filled squares and solid line), between March 2014 and September 2018. The shorter solid bars at the top of the graph highlight dry months with lower temperature (January–April, or dry season) and longer solid bars represent rainy months with higher temperature (May–November, or rainy season); arrows highlight decreases in population size that repeated cyclically on each dry season. Hurricane María resulted in a shift in the pattern of abundance for both species before this hurricane: *E. portoricensis* decreased sharply while *E. coqui* increased sharply after this hurricane. (Bottom) Fluctuation in the number of recently hatched and small juveniles among all species of *Eleutherodactylus* in our forest study site in the Cayey Mountains before and after Hurricane María; the sharp decline in abundance in 2015–16 resulted from the 2015 drought and a more dramatic decline in 2018 resulted from Hurricane María in 2017.

of forest succession (NR-L, unpublished data). Still undetermined is the stage of this assemblage of amphibians that will result from their responses to the combined effects of these natural phenomena longer term (i.e. the next five years as in Woolbright 1996), or if this pattern of change in population size among species repeats in different contexts of the landscape (e.g. aspect, topography, type of land-cover). Notably, López-Hernández and Puente-Rolón (2021: 1664) used automated acoustic monitoring of *Eleutherodactylus* spp. from April 2017 to September 2018, throughout 31 sampling sites at different locations in the Cayey Mountains. These authors found that there were changes in the acoustic activity of species due to Hurricane María varied and that these were site-dependent, but *E. portoricensis* 'did not change its activity or occurrence post-hurricane' by 2018 (2021: 1664) (Table 8.8).

As shown, we think that the rate of transformation of the landscape (because of deforestation, fragmentation, and other anthropogenic activities) decreased in recent decades. We also think that the human footprint remains and influences environmental conditions that are suitable for most ecologically specialist amphibians on the islands, and also that the environmental transformations engendered by climatic phenomena such as droughts and hurricanes are only exacerbated by this anthropogenic influence. Consequently, current environmental conditions in PR facilitate increases in population size and distribution of eurytopic species of *Eleutherodactylus* such as *E. coqui*, *E. antillensis*, *E. cochranae*, and *Leptodactylus albilabris* (the latter species was first detected in our forest study site in the Cayey Mountains several months after Hurricane María). On the other hand, these conditions decrease the population size and distribution of stenotopic species such as *E. portoricensis* and others. In this way, the extinction of species such as *E. eneidae*, *E. jasperi*, and *E. karlschmidti* (see section II. C. 4) becomes evident because of a combination of changes in land-cover, scientific activities of the past, and droughts and hurricanes.

In combination, the effects of anthropogenic and natural phenomena on a single species and the composition of assemblages of amphibians would be more significant than frequently thought; projected changes in climate may also contribute to the structure of assemblages of amphibians whose species composition has no reference in the recent past.

IV. Regulation and conservation status of amphibians in the Puerto Rican Archipelago

A. Current framework of regulation

In PR, two governmental levels of regulatory approaches relevant to the discussion on the conservation of amphibians can be identified. The first is the local government in which PRDNER enforces the regulatory and conservation legislation of the natural resources and wildlife of the island. Specifically, PRDNER implements the law for wildlife in PR (Ley 241 del 15 de Agosto de 1999 – Nueva Ley de Vida Silvestre de Puerto Rico), which provides the guidelines for protection of essential/critical natural habitats for vulnerable and endangered species. Two internal regulations are based on this law. First, Regulation 6766 (entitled *Reglamento para Regir las Especies Vulnerables y en Peligro de Extinción en el Estado Libre Asociado de Puerto Rico*; DRNA 2004b) focuses on the protection of vulnerable and endangered species of wildlife. Second, Regulation 6765 (entitled *Reglamento para Regir la Conservación y Manejo de Vida Silvestre, las Especies Exóticas y la Caza en el Estado Libre Asociado de Puerto Rico*; DRNA 2004a) includes the guidelines for characterizing habitats for conservation purposes, regulates the importation of non-native species, and regulates hunting.

The second governmental level of regulatory approaches is the federal government in which the US Fish and Wildlife Service (USFWS) and the US National Oceanic and Atmospheric Administration Fisheries Service are responsible for enforcing the regulations and conservation legislation for the natural resources and wildlife on land and sea in PR and the USVI. The

USFWS implements the Endangered Species Act of 1973, which provides the guidelines for conservation and protection of endangered and threatened species and their habitats. This act also offers financial assistance and incentives to develop and maintain conservation programmes, and facilitates many of the responsibilities of the USA to comply with international treaties and conventions, such as the Convention on International Trade of Endangered Species of Wild Fauna and Flora and the Western Hemisphere Convention.

In PR, the government announced the results from recently passed legislation, which increased the coverage of protected terrestrial areas from approximately 7% to 16% (1,446.6 km^2) in 2016 (19 April 2016; Gobierno de Puerto Rico 2016) (Fig. 8.14). Two months later, the President of the USA signed into law the Puerto Rico Oversight, Management, and Economic Stability Act of 2016 (Public Law 114–187, 30 June 2016, 130 Stat. 549; also known as PROMESA), which establishes 'an Oversight Board to assist the Government of Puerto Rico, including instrumentalities, in managing its public finances, and for other purposes'. Several passages from Snow (2017) about the consequences of PROMESA on laws and regulations enacted by the local government of PR are worth citing.

First, Snow (2017: 3) argued that PROMESA 'grants powers to the Oversight Board that could be used to affect existing government-owned natural resource areas and agricultural lands in Puerto Rico and affect the applicability and implementation of existing environmental laws'. Specifically, Snow (2017: 11) highlighted that 'the potential reach of PROMESA extends to all lands owned by the Commonwealth and managed by the Department of Natural and Environmental Resources (DNER). This would include Commonwealth Forests [12 L.P.R.A. '193 – Public lands for Commonwealth Forests] which by law also serve as wildlife refuges [12 L.P.R.A. '202 – Refuges]; natural reserves designated by DNER and the PR Planning Board; and estuarine sanctuaries (such as Humacao and Jobanes).'

Snow (2017: 9) hypothesized that under PROMESA 'any attempt by the Legislature to include more areas under such designation may run afoul of the Oversight Board's preemption powers' and that 'assertions of supremacy in PROMESA strongly suggests that one cannot rely on the laws of Puerto Rico to protect environmental resources. To the maximum extent possible, reliance should be on Federal laws and regulations.' Snow (2017: 13) concluded that 'while there is no indication that real property assets of the Commonwealth [PR] will be sold, there is nonetheless that possibility which requires continual monitoring by those concerned with the law's impacts on the natural resources of the island'.

In PR, two local acts passed in 2012 reached centre-stage status under the new legal framework of PROMESA, and have particular relevance for the conservation of amphibians. The PR government passed Act No. 20 of 17 January 2012, as amended, 'Act to Promote the Export of Services' (Gobierno de Puerto Rico 2012a), and Act No. 22 of 17 January 2012, as amended, 'Act to Promote the Relocation of Individual Investors to Puerto Rico' (Gobierno de Puerto Rico 2012b). The objective of Act No. 20 is 'to provide the adequate environment and opportunities to develop Puerto Rico as an international service center, encourage local professionals to stay and return, and attract foreign capital, thus promoting the economic development and social betterment of Puerto Rico'. The objective of Act No. 22 is 'to provide tax exemption for the investment income earned by individuals who become residents of Puerto Rico not later than the year ending on December 31st, 2035'. Coincidentally, in July 2018, the Governor of PR (executive branch in a republican system of government) reverted > 28 km^2 of designated areas (Primera Hora 2018), which currently decreased the coverage of protected areas to 14% on PR; the Supreme Court of PR (judicial branch) revoked the executive branch on 15 November 2019. Also in 2019, the local government reviewed its maps of land-use and classifications of PR – including those areas deemed of ecological importance – by virtue of its 'Ley Núm. 19 de 4 de abril de 2017' (Gobierno de Puerto Rico 2017), in order (among other justifications)

> to simplify and transform the permitting process in the jurisdiction of Puerto Rico, to give certainty, reliability, efficiency and stability to it; continue taking the concerted actions aimed at making Puerto Rico an open place for business by improving the permitting processes, as indicated by the Fiscal Oversight Board, created under the Puerto Rico Oversight, Management and Economic Stability Act, known as 'PROMESA', Pub. L. 114-187.

The period for public comments on this review for a new classification of public lands on PR ended on 24 September 2019 – under significant participation and criticisms from academia, lawyers, environmental agencies, and other sectors of the public interest – while on 27 September 2019, the governor of PR announced the adherence of a revised fiscal plan imposed by the Fiscal Oversight Board under PROMESA on PR.

It was suggested that Acts 20 and 22 would benefit local and foreign speculators on lands within PROMESA's new legal framework over PR. At the time of writing this chapter, however, we can only speculate about other consequences of this new state of legal framework – PROMESA – on the conservation of Puerto Rican amphibians and natural resources in general. Nonetheless, the following example helps understand the extent of this new legal framework in PR. Lugo *et al.* (2004) found that between 1985 and 1995, 85% of lands adjacent to the largest and best-protected forest area in PR, EYNF (north-eastern PR), were transformed to uses not in accordance with the original 1983 zoning regulation by the PR Planning Board (i.e. Junta de Planificación de Puerto Rico 1983); these lands include eight municipalities, and their forests harbour 79% (15/19 species) of native and endemic Puerto Rican amphibians. López-Marrero and Hermansen-Báez (2011) found that from 1998 to 2010, urban land-cover within these lands on average increased 21%, and 71% of urban growth occurred in the areas destined for uses other than urban, also not in accordance with the 1983 zoning regulation. Presumably, a lack of effective regulatory mechanisms generates what has been termed 'conservation *on paper*' (emphasis added) in PR, which since 2012 only worsened with other local acts passed. For example, under a new planning regulation framework that in 2016 changed the 1983 guidelines, the PR Planning Board (see 'Reglamento de Lotificación y Urbanización, Reglamento de Planificación Núm. 34, Reglamento 8695', of 25 February 2016) allowed for flexibility in uses of lands regardless of their zoning classification. It also allowed for the division of large areas of land into lots (or plots), so many see it as bypassing external environmental evaluations and other policies for environmental compliance. Coincidentally, in 2021, 103.2 ha of land within the buffer zone boundaries of EYNF (Naguabo, eastern PR) were advertised for \$1.875 million US dollars (WGS84; centroid at 18° 15.376' N, 65° 48.576' W; National Association of REALTORS® 2022). The 103.2 ha for sale were divided into 51 lots of 2.02 hectares each, but the advertisement for sale went unnoticed by the general public in PR. However, the 103.2 ha were sold as one unit on 31 August 2021 for \$2.2 million US dollars (National Association of REALTORS® 2022).

This discussion reveals that the idea that the development of an economic agenda supersedes the conservation of natural resources of which we humans are also part (in PR and elsewhere) is where the illusion lies. In PR, most changes in land-use categories in favour of urban cover have increased to date, facilitated by these and other acts and a regulation policy that threatens coastal and highland scenic Puerto Rican lands. Currently, only three anurans are listed with a designation of critical habitat under federal law (*Eleutherodactylus cooki*, *E. jasperi*, and *E. juanariveroi*), and one species is also listed but without designation of critical habitat under federal law

(*Peltophryne lemur*). Two of these species occur in the lowlands (*E. juanariveroi* and *P. lemur*), and one has distribution records in the scenic highland areas (*E. jasperi*). Anurans listed under local laws include these four species 'as is' under federal law and five more that lack designation of critical habitat throughout their distribution (*E. eneidae, E. karlschmidti, E. locustus, E. portoricensis,* and *E. richmondi*; see section IV. B). All these species have records from EYNF at an elevation where realtors sold the 103.2 ha in Naguabo, eastern PR.

Finally, Castro-Prieto *et al.* (2019) documented 16.1% of terrestrial area protected in PR (but see Primera Hora 2018). For the rest of the PRA, the USVI currently protects 13.79% (52 km^2) of its terrestrial area according to the portal Protected Planet (https://www.protectedplanet.net), an initiative managed by 'the United Nations Environment World Conservation Monitoring Centre (UNEP-WCMC) with support from IUCN and its World Commission on Protected Areas' (UNEP-WCMC 2018). In the BVI, the Conservation and Fisheries Department of the Ministry of Natural Resources and Labour is responsible for the operationalization and enforcement of regulations and bylaws oriented at managing the natural resources in these islands (Government of the British Virgin Islands, 2019b). Likewise, Protected Planet documents that the BVI currently protects 9.11% (16 km^2) of its terrestrial area (UNEP-WCMC 2018; Fig. 8.14). For comparison, the coverage of protected terrestrial areas in the Greater Antilles is 16.55% (18,481 km^2) for Cuba; 1.95% (534 km^2) for Haiti; 26.24% (12,727 km^2) for the Dominican Republic; and 15.92% (1,760 km^2) for Jamaica (UNEP-WCMC 2018).

B. Conservation status of amphibians

Recently, some agencies in the PRA have included the classifications established by the IUCN in their analyses for the conservation of wildlife, but categories are not necessarily in agreement among agencies or with the opinions of professionals (Table 8.9). The IUCN (2022a) generated its Red List of Threatened Species, which is being used worldwide to guide conservation activities of governments, non-governmental organizations (NGOs), and scientific institutions. As such, the IUCN does not have legal effects on the conservation of amphibians in the PRA for two reasons. First, the IUCN (2022b) Red List has a disclaimer that: 'You further expressly acknowledge and agree that information, text, graphics, and hyperlinks provided to you through The IUCN Red List [...] *are provided solely as a resource and a convenience to you*' (emphasis added). Second, the IUCN (2022b) Red List mentions that 'NO WARRANTY IS MADE AS TO THE RESULTS THAT MAY BE OBTAINED FROM USE OF THE IUCN RED LIST, OR AS TO THE ACCURACY, RELIABILITY, OR CONTENT OF ANY INFORMATION OR SERVICE PROVIDED THROUGH THE IUCN RED LIST'. Consequently, the reader must be aware of the limitations of the data from the IUCN and their level of accuracy and reliability for conservation efforts. In the following paragraphs, we present an overview of the conservation of the amphibians under the applicable regulations and recommendations made by experts in the field.

Since 1991 the conservation status of two-thirds of the species of amphibians in the PRA has changed, usually reaching a higher level of necessity for conservation in later years (Table 8.9). The classification of species in these listings, however, vary with the source of the listing – governmental versus non-governmental origin – and sometimes the classifications have altered dramatically for species such as *Eleutherodactylus cooki, E. gryllus, E. hedricki, E. lentus, E. monensis, E. unicolor,* and *E. wightmanae* (Table 8.9).

PRDNER (2015) lists 26.3% (five species) of the 19 native and endemic species present in PR as Critically Endangered (CR), 21.1% (four species) as Vulnerable (V), 31.6% (six species) as Data Deficient (DD), with 21.4% (four species) not listed. In contrast, the USFWS (2014) lists 4.6% (one species) of the 21 native and endemic species under its jurisdiction as Endangered (E), 14.3% (three species) as Threatened (T), with 81.0% (17 species) not listed. The local government (i.e. PRDNER) usually adopts the listings from the federal government (e.g. USFWS) (Table 8.9).

Table 8.9. Historical (since 1991) and recently published (2021) conservation status of native/endemic amphibians in the Puerto Rican Archipelago (PRA). Designation of species comes from summaries generated by Moreno (1991b), Joglar (1998), and Hedges and Díaz (2011); by state and federal governmental agencies PRDNER (2015) and USFWS (2014), respectively; and by the international organization IUCN (2016, 2021 assessments). The most recent assessments in the IUCN's (2022a) database includes an assessment from 2010 (one species), but has its classification scheme updated for 2021. Not included in this table is Joglar's (2005) listing of all species from Puerto Rico because it is similar to that in Joglar (1998). However, Joglar (2005) assessed two species not assessed in his previous list and these are included herein: *Eleutherodactylus unicolor* as Vulnerable and *Leptodactylus albilabris* as Stable. Population trend ('Pop. Trend'): D = decreasing; I = increasing; S = stable; U = unknown). Categories of population status as follows: CR = Critically Endangered; E or EN = Endangered; T = Threatened; V = Vulnerable; LC = Least Concern; S = Stable; DSA = Decreasing in Some Areas; NSS = Need of Status Surveys; DD = Data Deficient; — = Not Accessed/Specified. CR + EN = designations of disjunct populations, northern and southern Puerto Rico respectively. N/A = not applicable (the species was discovered afterwards). 'Population Status' includes notes from the IUCN database. * refers to species for which a Critical Habitat has been designated. ** refers to superscripts that indicate the most recent year of the published assessment. Also included is a non-exhaustive list of references about the population ecology (e.g. examining seasonality in abundance/acoustics, habitat–abundance relationships, metapopulation dynamics, genetic structure, interspecific interactions) and breeding biology of amphibians (e.g. examining reproductive phenology and habitat-related requirements for reproduction) generated from 1990 to the present for each species; they represent sources highly relevant for the assessment of conservation status of species in the PRA.

FAMILY AND *SPECIES*	MORENO (1991B) PRA	JOGLAR (1998) PR	HEDGES AND DÍAZ (2011) PRA	PRDNER (2015) PR	USFWS (2014) PRA	IUCN (2016)** PRA	IUCN (2021)** PRA	POPULATION STATUS (IUCN 2016 AND 2021)	POP. TREND (2021)
Bufonidae									
Peltophryne lemur	E	E, NSS	CR	CR + EN	T	CR (2010)[a]	EN[v]	Known from only a few localities along the northern and southern coasts of Puerto Rico; considered extirpated from Virgin Gorda Island and St John.	D
	Beauclerc *et al.* (2010); see main text.								
Eleutherodactylidae									
Eleutherodactylus antillensis	S	S, NSS	LC	–	–	LC (2010)[b]	LC[w]	Not declining in the lowlands; unknown in the highlands.	S
	Stewart and Woolbright (1996); Michael (1997a); Ovaska and Caldbeck (1997a, b); see review by Joglar (1998); Ovaska *et al.* (2000); Ovaska (2005); Vilella and Fogarty (2005); Ríos-López (2007); Ríos-López and Aide (2007).								
E. brittoni	S	S, NSS	LC	DD	–	LC (2008)[c]	LC[x]	Abundant in suitable habitat.	S
	Vilella and Fogarty (2005); Ospina *et al.* (2013); Ríos-López and Villanueva-Rivera (2013).								
E. cochranae	S	S, NSS	LC	–	–	LC (2010)[d]	LC[y]	Common.	S
	Michael (1995); Michael (1997b); Ovaska and Caldbeck (1997a); see review by Joglar (1998); Ovaska *et al.* (2000); Ríos-López (2007; Ríos-López and Aide (2007); Ospina *et al.* (2013); Ríos-López and Villanueva-Rivera (2013); Ríos-López *et al.* (2014); Ríos-Franceschi *et al.* (2016).								

(Continued)

Table 8.9. (Continued)

FAMILY AND *SPECIES*	MORENO (1991B) PRA	JOGLAR (1998) PR	HEDGES AND DÍAZ (2011) PRA	PRDNER (2015) PR	USFWS (2014) PRA	IUCN (2016)** PRA	IUCN (2021)** PRA	POPULATION STATUS (IUCN 2016 AND 2021)	POP. TREND (2021)
E. cooki	NSS	T	E	V	T*	V (2008)[e]	EN[z]	Not believed to be declining in 2016; declining in 2021.	D
	Joglar *et al.* (1996); see review by Joglar (1998); Burrowes and Joglar (1999); Burrowes (2000); Rogowitz *et al.* (2001); López-Torres (2008).								
E. coqui	S	V	LC	–	–	LC (2009)[f]	LC[aa]	Lowland populations are common; observed decline in the upland population (2016); increasing throughout PR in 2021.	I
	The most studied *Eleutherodactylus* sp. and one of the most studied anurans in the world; the reader should consult reviews by Joglar (1998), Henderson and Powell (2009), main text, and references within.								
E. eneidae	E	E	CR	CR	–	CR (2010)[g]	CR[bb]	Believed to be most probably extinct; population trend decreasing in 2016, unknown in 2021.	U
	Woolbright (1997); Joglar (1998); Burrowes *et al.* (2004); see main text.								
E. gryllus	NSS	T, NSS	E	DD	–	E (2008)[h]	CR[cc]	Believed to be declining.	D
	See review of Joglar (1998); Burrowes *et al.* (2004); Villanueva-Rivera (2006).								
E. hedricki	NSS	V,NSS	E	DD	–	E (2008)[i]	EN[dd]	There are indications that populations have declined over the past three decades, even in suitable habitat.	D
	Woolbright (1991); Woolbright (1997); see review of Joglar (1998).								
E. jasperi	E	E	CR	CR	T*	CR (2010)[j]	CR[ee]	Believed to be extinct; population trend decreasing in 2016, unknown in 2021.	U
	Drewry and Jones (1976); Wake (1978); USFWS (1984); see main text.								
E. juanariveroi	N/A	N/A	–	CR	E*	CR (2008)[k]	CR[ff]	Common in the type locality; population trend unknown in 2016, decreasing in 2021.	D
	Ríos-López and Thomas (2007); United States Department of the Interior (2012); Ríos-López and Villanueva-Rivera (2013); Ríos-López *et al.* (2014); see main text.								

E. karlschmidti	E	E	CR	CR	–	CR (2010)[l]	CR[gg]	Believed to be extinct; population trend decreasing in 2016, unknown in 2021.	U
	See main text.								
E. lentus	S	–	E	–	–	E (2010)[m]	EN[hh]	This species is rarely encountered, but it has been recorded recently.	D
	Jones (1982); Perry (2009b).								
E. locustus	DSA	E, NSS	CR	V	–	CR (2008)[n]	EN[ii]	Unexplained major decline in abundance in the past two decades; some populations have been extirpated.	D
	See review of Joglar (1998).								
E. monensis	S	V,NSS	V	DD	–	V (2008)[o]	V[jj]	It is an abundant species on the island.	S
	Numazawa and Sengoku (2011); Stabile *et al.* (2014).								
E. portoricensis	DSA	V,NSS	E	V	–	E (2008)[p]	EN[kk]	Studies have shown that populations have been declining; presence in the Cordillera Central should be re-evaluated.	D
	See review of Joglar (1998); Burrowes *et al.* (2004); Ríos-López *et al.* (2015a); Part III-Section C-subsection 3 (The chytrid fungus, *Batrachochitrium dendrobatidis* [*Bd*]) above, and references within.								
E. richmondi	DSA	E, NSS	CR	V	–	CR (2008)[q]	EN[ll]	Unexplained decline of this species during the past two decades.	D
	Woolbright (1991); Stewart and Woolbright (1996); see review of Joglar (1998); Vilella and Fogarty (2005); Villanueva-Rivera (2006); Zegarra-Vilá (2007); López-Hernández *et al.* (2021); Part III-Section C-subsection 3 (The chytrid fungus, *Batrachochitrium dendrobatidis* [*Bd*]) above, and references within.								
E. schwartzi	DSA	–	E	–	–	E (2010)[r]	EN (2010)[r]	Appears to be declining wherever *Osteopilus septentrionalis* spreads; the Great Dog population (BVI) is somewhat distinct, extremely limited in range, and very susceptible to development there.	D
	Jones (1982); Ovaska *et al.* (1998); Ovaska and Caldbeck (1999); Ovaska *et al.* (2000); see main text.								
E. unicolor	S	–	CR	DD	–	V (2008)[s]	CR[mm]	Believed to be stable in 2016, decreasing in 2021.	D
	See review of Joglar (1998); Burrowes *et al.* (2004).								

(*Continued*)

Table 8.9. (Continued)

FAMILY AND *SPECIES*	MORENO (1991B) PRA	JOGLAR (1998) PR	HEDGES AND DÍAZ (2011) PRA	PRDNER (2015) PR	USFWS (2014) PRA	IUCN (2016)** PRA	IUCN (2021)** PRA	POPULATION STATUS (IUCN 2016 AND 2021)	POP. TREND (2021)
E. wightmanae	S	V,NSS	E	DD	–	E (2008)[t]	EN[nn]	Studies show that populations of this species are declining in both the higher and lower elevations of its range.	D
	Woolbright (1991); Vilella and Fogarty (2005); Ríos-López and Dávila-Casanova (2014); Part III-Section C-subsection 3 (The chytrid fungus, *Batrachochitrium dendrobatidis* [*Bd*]) above, and references within.								
Leptodactylidae									
Leptodactylus albilabris	S	–, NSS	LC	–	–	LC (2010)[u]	LC[oo]	Abundant and widespread in the USVI, it appears to be in decline in the BVI.	S
	Woolbright (1991); Vilella and Fogarty (2005); Joglar *et al.* (2005); Ríos-López (2008a).								

[a] Angulo (2010a); [b] Angulo (2010b); [c] Ríos-López (2008b); [d] Ríos-López (2010a); [e] Ríos-López (2008c); [f] Hedges *et al.* (2009); [g] Hedges and Ríos-López (2010); [h] Hedges and Ríos-López (2008a); [i] Angulo (2008a); [j] Angulo (2010c); [k] Ríos-López (2008d); [l] Hedges *et al.* (2010); [m] Hedges and Thomas (2010a); [n] Hedges and Rios-López (2008b); [o] Angulo (2008b); [p] Angulo (2008c); [q] Angulo (2008d); [r] Hedges and Thomas (2010b); [s] Ríos-López (2008e); [t] Angulo (2008e); [u] Ríos-Lopez (2010b); [v] IUCN SSC Amphibian Specialist Group (2021a); [w] IUCN SSC Amphibian Specialist Group (2021b); [x] IUCN SSC Amphibian Specialist Group (2021c); [y] IUCN SSC Amphibian Specialist Group (2021d); [z] IUCN SSC Amphibian Specialist Group (2021e); [aa] IUCN SSC Amphibian Specialist Group (2021f); [bb] IUCN SSC Amphibian Specialist Group (2021g); [cc] IUCN SSC Amphibian Specialist Group (2021h); [dd] IUCN SSC Amphibian Specialist Group (2021i); [ee] IUCN SSC Amphibian Specialist Group (2021j); [ff] IUCN SSC Amphibian Specialist Group (2021k); [gg] IUCN SSC Amphibian Specialist Group (2021l); [hh] IUCN SSC Amphibian Specialist Group (2021m); [ii] IUCN SSC Amphibian Specialist Group (2021n); [jj] IUCN SSC Amphibian Specialist Group (2021o); [kk] IUCN SSC Amphibian Specialist Group (2021p); [ll] IUCN SSC Amphibian Specialist Group (2021q); [mm] IUCN SSC Amphibian Specialist Group (2021r); [nn] IUCN SSC Amphibian Specialist Group (2021s); [oo] IUCN SSC Amphibian Specialist Group (2021t).

NOTE: Following Henderson and Powell (2009), we do not include in this row studies on systematics, phylogeny/taxonomy, and omitted studies documenting fortuitous observations (e.g. non-predesigned studies that document reproductive and predation events, and new geographic records within the geographic range of the species).

We recommend readers to consult studies on the feeding ecology of amphibians in the PRA by Lavigne and Drewry (1970), Jones (1982), and Stewart and Woolbright (1996), among others, because they are particularly relevant for *ex-situ* conservation (e.g. captive care, maintenance, and breeding of amphibians).

Exceptions to studies published prior to 1990 that are included herein are all three studies from *Eleutherodactylus jasperi*, a species that has been neither seen nor heard since the early 1980s, and that of Jones (1982) for *E. lentus* and *E. schwartzi*, which is the only study known on the diet of these two species in the wild.

Non-governmental sources of listings are mostly by individuals either serving as assessors for non-governmental, non-profit conservation organizations or acting in a personal capacity. Of the former, the most recently published assessment (2021 IUCN Red List's assessments cited in Table 20.9; IUCN 2022a) of the 21 native and endemic species of amphibians in the PRA was 4.8% of the populations increasing, 23.8% stable, 57.1% decreasing, and 14.3% unknown. In agreement with the population trend of species in the 2021 IUCN Red List's assessments in Table 8.9 (IUCN 2022a), 71.4% (15 species) were classified as CR (6 species) and EN (9 species). Moreover, 4.8% of species (one species) are listed as V, and 23.8% (five species) as Least Concern (LC). As a result, species whose population trends are considered to be decreasing are also listed EN or CR. Hedges and Díaz (2011) adopted these categories and listed 70% of species as CR (seven species) and E (seven species), 5.0% (one species) as V, and 25.0% (five species) as LC. Consequently, according to these recent listings by IUCN (Table 8.9), most Puerto Rican species of amphibians are currently in need of conservation; however, proactive actions from governmental agencies are significantly lagging.

An examination of classification categories and population status since 1991 revealed two major facts (Table 8.9). First, nine species (out of 20) classified Stable by Moreno (1991b) were reclassified in 2021 either as LC (five species), V (*Eleutherodactylus monensis*), EN (*E. lentus* and *E. wightmanae*), and CR (*E. unicolor*); 33.3% considered Stable initially are currently considered Endangered and Critically Endangered. This reclassification scheme revealed a significant change in the population status of several species in three decades. Second, Moreno (1991b) indicated that 33.3% of species in the area (seven species) are in 'Need of Status Survey' (NSS; three species) or are Declining in Some Areas (DSA but not in other areas; four species). The species mentioned as NSS by Moreno (1991b) are listed EN (two species) or CR (one species) by 2021 (Table 20.9), whereas species considered DSA currently are listed as EN (four species). The listings supplied by Joglar (1998, 2005b) were 63.2% NSS (12 species out of the 19 native Puerto Rican species assessed; he did not assess *E. lentus* and *E. schwartzi* from the VI), almost doubling the percentage of species as NSS by Moreno (1991b) in approximately a decade. Most of the species considered as NSS by Joglar (1998, 2005b) included species currently classified by the IUCN in various other categories (Table 8.9), usually of higher conservation concern. In contrast, PRDNER (2015) assessed 15 species from PR (species from the VI excluded) and listed 40.0% of them (6 species) as DD; three as EN, two as V, and one as LC. The species with inconsistencies in their listings of status need across assessments most likely reflect the weight of the authority – or other related considerations – over data from published studies (see next paragraphs, this section).

Hedges and Díaz (2011: 38, 40) did not discuss whether these species were in need of further survey of their status, but indicated instead that:

> This unusual attention [of the many reviews of conservation status of amphibians mentioned in their work] has made the West Indies one of the best known regions of the world in terms of knowledge of the conservation status of its amphibian fauna. This does not mean that all of the data needed are in hand, only that the available information has been summarized and the gaps in our knowledge have been identified. One of us (S.B.H) has searched extensively – without success – for all five "missing" species **in habitats and localities where they once occurred in abundance** [emphasis added]. In addition, others have searched for *E. karlschmidti* extensively (Joglar *et al.* 2007), also without success.

The acknowledgement by Hedges and Díaz (2011), however, that not all data needed for a robust assessment of the status of any given species listed are in hand is revealing, and has consequences for the species listed by IUCN in 2016 and 2021 assessments (Table 8.9). First, we acknowledge that the listing of the population status of species of amphibians, based on a system of classification using predefined criteria, has been one of the most valuable tools for summarizing knowledge. In IUCN's 2016 listing, however, the population status of 71.4% of 15 species were tentatively categorized as 'believed to', 'not currently believed', 'currently believed', 'appears to be', 'there are indications that', 'considered', 'observed', 'rarely encountered', 'unexplained', or 'unknown'. These vague terms suggest that the robust data vital for an accurate assessment of population status of any given species in the PRA are still missing. Nonetheless, by 2021, the IUCN Red List has three species that changed status for the worse (*Eleutherodactylus cooki, E. gryllus, E. unicolor*), three for relatively the better (*Peltophryne lemur, E. locustus, E. richmondi*). Second, we highlight that the lack of data (and the use of collection data as a proxy for abundance; see section III. B. 3) constitutes a conundrum so complex that it cannot be resolved solely by the available classification system. In the above example, with one exception (*E. richmondi*), we lack published studies on the population dynamics of these species that might justify the changes in categories of population status listed in the IUCN between 2016 and 2021. The gaps in information are still there, most of them have been identified, and fortunately, some assessments of the vulnerability of species to climatic change are being revised (e.g. Foden and Young 2016). Consequently, we still need to bridge the gap between empirical data and the listing of a species' status in order for conservation efforts to be instrumental for the decision-making process by the government; the weight of an 'opinion' is not equivalent to the weight of 'hard evidence', and information gaps should be explicit and repeatedly stated in various passages in these assessments. After all, the governments in the region are the ultimate enforcers of conservation policies and need the best objective truth (over personal truths based on a system of beliefs) available to implement them as effectively as intended.

For the vast majority of amphibian species in the PRA, only a few studies have provided vital information for the assessment of status (see again Table 8.9). We believe that this fact has profound consequences for the effective conservation of amphibians in the area (e.g. listing by governmental agencies, the designation of critical habitat, regulation and implementation of policy, to name a few). Acknowledging this gap between the available data and the assessments of status is not trivial; listings of status frequently are used as leverage in research proposals, justification of research projects, manuscripts submitted for publication, fundraising, and guidance for decision-making processes by governmental agencies. Furthermore, this gap could explain why governmental agencies do not necessarily adopt the assessments proposed by individuals/organizations. They frequently reserve their right to adopt a given recommendation for the conservation of a species of amphibian. One reason for this institutional behaviour may be that the agencies are budget-limited and their policies may change after every election. In turn, the limitation in their budgets frequently has an impact on the sponsoring of research that would otherwise benefit their decision-making process about the listings of species and the designations of critical habitats (Table 8.9).

The listings in Table 8.9 also show that, for most species, at least two decades – for some species three decades – have passed without generation of much new fundamental information on natural history, reproductive ecology, physiology, demography, the dynamics of assemblages, or relation to climate over a range of different locations. Certainly, dissemination of the research on natural history during the past several decades has not been as fruitful as desirable; this is unfortunate because information on the natural history of species is fundamental for effective conservation and accurate classification of their population status. Presumably, the deficiency of such studies is due in large part to limited funding, a lack of release-time for research, either from governmental agencies or institutions of higher education, and limited teaching and training about natural history at public and private educational institutions. The urgent need for more studies on the natural history of amphibians is not exclusive to the PRA, but also needs to be addressed elsewhere (Noss 1996; Joglar 1998; Pyle 2001; Fleischner 2005; Schmidly 2005; Tewksbury *et al.* 2014; Tschinkel and Wilson 2014; LoPresti *et al.* 2016).

Several outlets for the publication of studies of natural history of amphibians are still available for the PRA, including the *Caribbean Journal of Science* (BioOne 2022), *Acta Científica–una revista transdisciplinaria de Puerto Rico y el Caribe* (USDAFS 2022), *Reptiles & Amphibians: Conservation and Natural History* (International Reptile Conservation Foundation 2022), *Caribbean Naturalist* (Eagle Hill Institute 2022), and *LIFE: The Excitement of Biology* (Blay Publishers LLC 2022), among others. There are also many fruitful areas for research on amphibians from the PRA available to undergraduate and graduate students, and to established herpetologists in the region, that will generate the information needed for effective conservation and accurate listing of species' status at local and international levels.

V. Captive breeding

Initiatives for the conservation of amphibians include the successful *ex-situ* captive-breeding of four species (other than breeding *Eleutherodactylus cochranae* and *E. coqui* in captivity and for other purposes; e.g. Ríos-López 2007, among other authors): *E. antillensis* (Báez *et al.* 2014; Colón-Piñeiro *et al.* 2017); *E. portoricensis* (Ríos-López *et al.* 2015a); *E. monensis* (Numazawa and Sengoku 2011; Stabile *et al.* 2014); and *Peltophryne lemur*. Here, we briefly mention projects that resulted in captive-care manuals/protocols and troubleshooting valuable for conservation *ex-situ*.

A. *Eleutherodacylus* spp.

Báez *et al.* (2014) – Herpetology division, Abilene Zoo, Texas, USA – bred *Eleutherodactylus antillensis* successfully and developed a detailed captive-care manual for this species (unpublished); Colón-Piñeiro *et al.* (2017) also bred *E. antillensis*, although their project mostly yielded egg clutches of *E. coqui*. Colón-Piñeiro *et al.* (2017), however, provided detailed laboratory settings for captive-care projects for *Eleutherodactylus* frogs. Ríos-López *et al.* (2015a) bred *E. portoricensis* successfully, and first-generation individuals produced in captivity were all released into the wild; this project was terminated by the end of 2015. Stabile and Joglar (2017) bred *E. monensis* successfully and produced a second generation in captivity. They modified previous *ex-situ* protocols (i.e. Stabile *et al.* 2014) and attributed their success to modifications in diet and feeding regime, UV lighting, environmental enrichment, and management of recently hatched individuals. As far as we know, captive-bred *E. antillensis* and *E. monensis* have not been released into the wild.

B. *Peltophryne lemur*

Rivero initiated efforts to find extant populations of *Peltophryne lemur* in north-western PR in 1973. Over the next seven years, he collected approximately two dozen individuals that were transferred to the University of Puerto Rico, Mayagüez Campus, to produce a captive-bred population, but unfortunately, this effort was unsuccessful (Rivero and Seguí 1992). However, in 1980 Rivero collected two mating pairs of *P. lemur*, transferred them to the Mayagüez Zoo, and a year later, two successful breedings allowed for the expansion of the conservation effort to the USA. The progeny of these individuals from Mayagüez were sent to the Buffalo Zoological Garden, New York, as founders for a captive-breeding programme (Rivero and Seguí 1992). In 1982, the American Zoological Association (AZA) collected and sent individuals from the northern population (USFWS 2016) to its participating zoological institutions. One year later, 75 toadlets from a captive-breeding programme established by the Buffalo Zoological Garden were released in Quebradillas (USFWS 2016). During the same year, the progeny of a second breeding event from the Mayagüez Zoo were sent to Buffalo and Indianapolis, and their progeny were released at Quebradillas.

The multi-institutional captive-breeding and reintroduction programme for the species was formalized in 1984 when *P. lemur* was selected by the AZA for their Species Survival Plan programme (SSPP; USFWS 2016). Today, more than 30 zoos and aquaria from the USA and Canada participate in this programme and collaborate with PRDNER and USFWS in the conservation of this species. Over the years, non-profit organizations, private landowners, community groups, and volunteers have joined the conservation efforts initiated by AZA, USFWS, and PRDNER to collectively constitute the Puerto Rican Crested Toad Working Group (PRCTWG). This group is led by the USFWS and PRDNER, and since its beginning, in 2003, has worked to secure the long-term survival of the species. After a series of trials and errors, a population of captive-bred individuals was established at one location in south-eastern PR and has resulted in successful breeding (USFWS 2016; Mendelson 2018).

Two additional populations of captive-bred individuals of *P. lemur* were established in southern PR: in the municipalities of Coamo in 2006 and in Guayanilla, 10.9 km from the natural population at Ventana, in 2012. In Coamo, PRDNER built a release pond less than a kilometre from '*Los Baños de Coamo*', the historical site for *P. lemur* where Chapman Grant last collected it in 1931 (e.g. Grant 1932a). Adults currently can be observed at Coamo, but reproduction has not been documented. The project in Guayanilla is part of a cooperative agreement between USFWS, PRDNER, AZA, and the Puerto Rico Conservation Trust (USFWS 2016).

Study of the preferences of *P. lemur* for particular microhabitats (Matos-Torres 2006), and of their uses by adults and juveniles (Pacheco and Barber 2013), led to the successful introduction of captive-bred tadpoles into the wild, where they metamorphosed, reached sexual maturity, and began to breed (USFWS 2016). In addition, a reintroduction programme managed by the non-profit organizations Ciudadanos del Karso and Iniciativa Herpetológica in the municipality of Arecibo in northern PR was initiated in 2005 at El Tallonal Private Nature Reserve. Since 2006, tens of thousands of tadpoles have been released there in three artificial ponds (USFWS 2016; Fig. 8.24a–b). Currently, adults can be seen foraging there at night, and they use the ponds for calling; several reproductive events have been documented since 2010. Two additional areas for reintroductions are now established on land managed by the Puerto Rico Conservation Trust in the municipalities of Manatí and Florida [PR], northern PR, where reproductive events also have been documented.

On 21 November 2019, an announcement was made by Fort Worth Zoo (Texas, USA; Fort Worth Zoo 2022) about their first successful effort for *in vitro* fertilization of *P. lemur* (and cryopreservation of sperm and ovules for future procedures). The results obtained were based on a combination of techniques (called 'assisted reproductive technologies') developed by Mississippi State University, and that includes 'hormone therapies, sperm cryopreservation, and *in-vitro* fertilization' (Mississippi State University 2022). This combination also allows for the catch-and-release of adults in the wild (instead of euthanizing individuals for the extraction of sperm from their testes). According to a press release, more than 300 tadpoles metamorphosed from this effort and about two-thirds of them will be released into the wild. For the Puerto Rican Crested Toad, this is a breakthrough that has profound implications for its conservation.

In summary, the PRCTWG works actively with eight of the nine known populations of *P. lemur*: two of the three natural populations (Guánica and Guayanilla) and the six reintroduced populations (Manglillo Grande in Guánica; Gabia in Coamo; El Convento in Guayanilla; El Tallonal in Arecibo; La Esperanza in Manatí, and Río Encantado in Florida [PR]). This endeavour has been recognized as the most extended and successful amphibian-breeding programme in the United States, with over 32 years of continuous effort (USFWS 2016). In 2019, a multisectoral meeting

Figure 8.24. Artificial ponds constructed for the release of captive-bred tadpoles of the Puerto Rican crested toad, *Peltophryne lemur*. These ponds are also used for reproduction by adults that had grown from tadpoles released in previous reintroduction events at El Tallonal, Arecibo, northern PR. The ponds are covered by mesh screens with narrow apertures near the ground that allow adults of *P. lemur* to reach the pond but preclude larger-bodied amphibians in this locality (*R. marina* and *L. catesbeianus*) from reaching the pond (A). Tadpoles produced in captivity are acclimated while still in the plastic bags before they are ready for release in these ponds (B). Pictures by Alberto R. Puente.

about the state of this conservation programme was held by the PRCTWG in the headquarters of the USFWS in Boquerón, south-western PR, which contributed new information from previous meetings (e.g. CBSG 2006). This 2019 meeting set a common ground for future conservation, management, and education efforts among stakeholders, including the presentation and discussion of the working document entitled 'Puerto Rican Crested Toad (*Peltophryne lemur*) Reintroduction and Field Management: Standard Operating Procedures' (Barber *et al.* 2019). Although much work still needs to be done, the long-term establishment of *P. lemur* now seems likely.

VI. Education

Efforts to educate about amphibians have led to a wide range of communication formats, including talks at local and international forums for the general public and professional audiences; publication of flyers for governmental organizations and NGOs; publications in journals of popular science and in peer-reviewed professional journals; field trips and initiatives of citizen science; and dissemination of information via local and international newspapers, radio, television, and electronic media; and through oversight by reserve managers and law enforcers.

Currently, courses in herpetology are being offered at three campuses of the University of Puerto Rico (UPR-Río Piedras, UPR-Mayagüez, UPR-Humacao), along with courses in wildlife management (including amphibians; UPR-Humacao). The University of the Virgin Islands (UVI; with campuses in St Croix and St Thomas) offers a Master of Science (MS) in Marine & Environmental Science and accepts graduate students interested in terrestrial wildlife, including amphibians.

In PR, herpetological activities that include the participation of people of all ages are being conducted by Iniciativa Herpetológica. This organization has coordinated the removal of *Rhinella marina* from release-sites for *Peltophryne lemur* for over ten years. It also founded and coordinated the Herpetology Symposium, held every two years at the UPR-Arecibo campus since 2006.

The St Croix Environmental Association has organized 'frog walks' for over 30 years on the island of St Croix. The 'Great VI Frog Count' at UVI collected thousands of recordings of frogs calling at night since 2015 from at least six islands: St Croix, St Thomas, Water Island, and St John in the USVI, and Jost Van Dyke and Virgin Gorda in the BVI.

Finally, a Caribbean chapter of the organization Partners for Amphibian and Reptile Conservation (PARC; supported by the US government) was founded in 2013 under the mission statement 'to conserve amphibians, reptiles and their habitats as integral parts of our ecosystem and culture through proactive and coordinated public–private partnerships'. The 'Caribbean PARC' chapter is part of Southeast Partners in Amphibian and Reptile Conservation; it has held symposia at its meeting in Atlanta (Georgia, USA; 2013), Cartagena (Colombia; 2014), at the Museo de Historia Natural in Santo Domingo (Dominican Republic; 2015), and in Santo Domingo (Dominican Republic; 2017) (J. Stabile and C. Martínez, pers. comm.), with the collaboration of herpetologists from the PRA. The Caribbean PARC held its fourth conference in 2019 in Kingston, Jamaica; PR will host the fifth conference in 2022.

VII. Resources and reserves

Allocation of resources for governmental personnel responsible for law enforcement, managers in reserves, and institutional support are never enough, and often there are acute shortages. As a result, there is an increasing concern that protection of reserves in the PRA may occur mostly on paper (similar to claims from the Dominican Republic in Pasachnik *et al.* 2015). Consequently, many aspects of the conservation of amphibians are controversial in various ways, and governmental agencies, NGOs, and scientists in PR and abroad must address them adequately from a transdisciplinary perspective.

VIII. Conclusions and future directions

In general, destruction of habitat is agreed to be the main factor influencing the distribution and abundance of amphibians. However, we have demonstrated that collecting for scientific purposes as well as changes in land-use and land-cover each played a more significant role in explaining patterns of distribution, abundance, and declines of amphibians than frequently acknowledged. Likewise, we have shown that evidence of the influence of non-native species on native amphibians is mostly anecdotal to barely existent for the PRA. Frequently, the effects of non-native taxa, while real for some local species (e.g. of reptiles and birds in the presence of mongooses, among others), have been extrapolated uncritically to include native amphibians as victims of the same non-native taxa. However, amphibians differ significantly from other vertebrates in ecological and physiological requirements. Consequently, there is a huge opportunity for research on the ecology of non-native amphibians and other non-native taxa to assess their influence on patterns of distribution and abundance of native amphibians.

We have also examined the evidence for the role of pathogens in population declines of amphibians in the area (Table 8.6). Other than having detected *Bd* among some samples from some geographic regions (Table 8.6), robust evidence for presence is available mainly from only two closely related species (*Eleutherodactylus coqui* and *E. portoricensis*), which show equivocal results in their population dynamics (contrast data in Tables 8.6 and 8.9). All species of *Eleutherodactylus* in the area differ in their ecological requirements, and presumably in their physiology. Future studies should examine the status of populations and the relative sensitivity to *Bd* of other species compared with *E. coqui* and its closest relative, *E. portoricensis*. Bacterially mediated acquired immunity (Longo *et al.* 2015; Hughey *et al.* 2017) and fungicidal secretions of the skin (e.g. Rollins-Smith *et al.* 2015) may also occur and should be examined along with regular monitoring of population dynamics in the wild. The hypothesis of climate-mediated susceptibility of amphibians to *Bd* and other pathogens (Pounds and Crump 1994; for the PRA see Burrowes *et al.* 2004 and Longo *et al.* 2010) should also be tested. Moreover, more species, other than *E. coqui*, need to be examined from an ecophysiological perspective.

Studies that explicitly link patterns of distribution, abundance, and population dynamics with behaviour and physiology are urgently needed to understand, from a conservation perspective, the mechanisms that influence responses of amphibians to projected changes in climate and the landscape in the PRA. Such studies must be conducted with different species and geographic regions, to draw generalizations that help prioritize conservation actions within the realm of limited financial resources and policies. Bridging the communication gap between scientists, policymakers, managers, and the general public is therefore of paramount importance for the conservation of amphibians in the PRA.

We believe that the short-term conservation of amphibians in the PRA currently rests on four principal sources of efforts: (1) outreach and formal education, (2) research, particularly in natural history, (3) law enforcement, (4) sufficient resources being allocated to wildlife management for efficient delivery and oversight of conservation efforts. As mentioned, the transformation of lands for human uses continues to be, without a doubt, the number one threat to amphibians (and other species) in the area; anthropogenic changes in climate and the environment only exacerbate this threat. Understanding the consequences of this threat to our amphibians, however, is still compounded by gaps in the knowledge of the natural history of the vast majority of species in the area. Particularly needed are studies that elucidate patterns of population dynamics from the perspective of climatic change and landscape-related variables in synergism.

Of equal importance, particularly for younger generations of herpetologists and scientists at large, is the examination of the 'old literature' (i.e. contributions published a long time ago in various formats and frequently held in exclusive collections in libraries and elsewhere; these are mostly now readily available in electronic form on the internet). The examination of such literature will help to reconstruct the forces of the past that led to the current state of vulnerability of the amphibians in the PRA (Foster *et al.* 1999). However, the incomplete state of knowledge of the natural history, physiology, behaviour, and population dynamics of Puerto Rican amphibians within the context of the whole ecosystem limits the ability of the government and other institutions to achieve effective conservation and management efforts of native amphibians. Consequently, further research on the natural history of the amphibians and review of the 'old literature' from a transdisciplinary approach represents the best trajectory for a more comprehensive understanding of the conservation needs of the amphibians in the PRA. Such studies will help us to understand how amphibians may respond to projected changes in climate acting in synergy with changes in the landscape; whether individuals can adapt swiftly to the rapid changes in climate currently documented as in progress; and whether conservation of our amphibians will happen in time to circumvent the worst effects of climatic change in the Anthropocene (e.g. Hansen *et al.* 2016).

We consider that the worst effects that the transformation of lands could cause to the distribution and abundance of amphibians in the PRA have already occurred. This transformation includes the extirpation and extinction of several species from their historical collection sites, but several promising localities in the archipelago also remain unexplored (e.g. in PR, the northern karst region). In fact, the '*coquí llanero*' was recently found in a small patch (~1.5 ha) of coastal herbaceous-palustrine wetland in 2018; this locality is in the northern karst region, part of the Río Grande de Arecibo basin, within public lands administered by the Land Authority of Puerto Rico (Morales-Pérez *et al.* 2022). We hope that other unexplored areas in the PRA may harbour species currently deemed extirpated or extinct.

In this chapter, we have highlighted many opportunities for transdisciplinary research to strengthen conservation efforts for the amphibians in the PRA; we have also highlighted how much work is still pending for current and future generations of herpetologists, ecologists, and other scientists interested in amphibians. Therefore, we hope that our work may help these professionals bridge the information gap that limits the activities of other stakeholders engaged in the conservation and study of amphibians and their ecosystems in the Anthropocene.

IX. Acknowledgements

I (NR-L) am grateful to Harold Heatwole and Ariel E. Lugo for critical feedback and editorial recommendations on various drafts of this manuscript during its preparation; I am also indebted to my undergraduate students for their support in the field and laboratory at the University of Puerto Rico in Humacao. I am also grateful to José A. Colón for providing the original photographs of *Eleutherodactylus jasperi*; José R. Almodóvar Rivera and Gloria J. Yukavetsky Colón for their assistance and advice in digitizing them; Bent Erik Kramer Lindow and Daniel Klingberg Johansson from the National History Museum of Denmark for their assistance with records of *Rhinella marina* in the PRA. Finally, I am most grateful to my family (my wife Elsie Rivera and our two daughters, Laura and Sofia) for their unconditional support of me and their loving understanding of the long hours devoted to this project during the past five years, including during the harshest period of Hurricane María (2017) and COVID-19 pandemia (2020–present).

We (SIV-C and ARP-R) appreciate the commitment and hard work that the PRCTWG has maintained for so many years. We also thank the countless students of the UPR-Arecibo and other institutions who since 2006 have helped to control and investigate the impact of *Rhinella marina* on *Peltophryne lemur* and other wildlife at the El Tallonal Nature Reserve at Arecibo, PR.

I (NF-A) would like to acknowledge my academic mentors and friends, including Karen Lips, Roy McDiarmid, Kevin de Queiroz, Ana Longo, Grace DiRenzo, and Lee Fitzgerald for teaching me how to survey amphibians. My work was partially supported by a Smithsonian Institution fellowship to explore the Puerto Rican Bank herpetofauna.

I (DD-C) am grateful to Dr Yogani Govender for her support and ideas.

X. References

Abarca, J. G., Zuniga, I., Ortiz-Morales, G., Lugo, A., Viquez-Cervilla, M., Rodriguez-Hernandez, N., Vázquez-Sánchez, F., Murillo-Cruz, C., Torres-Rivera, E. A., Pinto-Tomás, A. A., and Godoy-Vitorino, F., 2017. Characterization of the Skin Microbiota of the Cane Toad *Rhinella cf. marina* in Puerto Rico and Costa Rica. *Frontiers in Microbiology* 8: 2624. Published online 2018 Jan 5. https://doi.org/10.3389/fmicb.2017.02624.

Abbad y Lasierra, I., 1782. *Historia geográfica, civil y natural de la Isla de San Juan Bautista de Puerto Rico.* Editorial de la Universidad de Puerto Rico, Río Piedras, Puerto Rico, USA.

Abelleira, O. J. and Lugo, A. E., 2008. Post sugar cane succession in moist alluvial sites in Puerto Rico. Pp. 73–92 in *Post-agricultural succession in the Neotropics,* edited by R. Myster. Springer, New York, USA.

Adamson, M. L., 1981. *Parapharyngodon osteopili* n.sp. (Pharyngodonidae: Oxyuroidea) and a revision of *Parapharyngodon* and *Thelandros. Systematic Parasitology* 3:105–117. https://doi.org/10.1007/BF00012216

Adler, K., 2014a. Herpetologists of the past, Part 1. *Contributions to the History of Herpetology, Volume 1 (revised and expanded),* edited by K. Adler. Society for the Study of Amphibians and Reptiles, Ithaca, New York, USA.

Adler, K., 2014b. Herpetologists of the past, Part 2. *Contributions to the History of Herpetology, Volume 2,* edited by K. Adler. Society for the Study of Amphibians and Reptiles, Ithaca, New York, USA.

Alcala, A. C., 1957. Philippines notes on the ecology of the marine toad. *Silliman Journal* 4: 90–96.

Alejandro Hernández, S. and Joglar, R. L., 2020. *Osteopilus septentrionalis. Herpetological Review* 51: 534.

Ali, J. H., Riley, J., and Self, J. T., 1982. Amphibians as definite hosts for pentastomids: *Raillietiella bufonis* n.sp. from *Bufo lemur* in Puerto Rico and a reassessment of *Raillietiella indica* Gedoelst, 1921. *Systematic Parasitology* 4: 279–284. https://doi.org/10.1007/BF00009630

Alonso, R., Crawford, A. J., and Bermingham, E., 2011. Molecular phylogeny of an endemic radiation of Cuban toads (Bufonidae: *Peltophryne*) based on mitochondrial and nuclear genes. *Journal of Biogeography* 39: 434–451. https://doi.org/10.1111/j.1365-2699.2011.02594.x

AmphibiaWeb, 2018. <https://amphibiaweb.org> University of California, Berkeley, California, USA. Accessed 12 February 2018.

Angeli, N. F., 2017. Identifying mechanisms of persistence for ground lizards (Reptilia: *Ameiva*) in the Caribbean. PhD Dissertation. Texas A&M University, College Station, Texas, USA.

Angilletta, M. J., Steury, T. D., and Sears, M. W., 2004. Temperature, growth rate, and body size in ectotherms: fitting pieces of a life-history puzzle. *Integrative and Comparative Biology* 44: 498–509. https://doi.org/10.1093/icb/44.6.498

Angulo, A., 2008a. *Eleutherodactylus hedricki.* The IUCN Red List of Threatened Species 2008: e.T56648A11513578. http://dx.doi.org/10.2305/IUCN.UK.2008.RLTS.T56648A11513578.en. Accessed on 26 February 2017.

Angulo, A., 2008b. *Eleutherodactylus monensis.* The IUCN Red List of Threatened Species 2008: e.T56774A11519092. http://dx.doi.org/10.2305/IUCN.UK.2008.RLTS.T56774A11519092.en. Accessed on 26 February 2017.

Angulo, A., 2008c. *Eleutherodactylus portoricensis.* The IUCN Red List of Threatened Species 2008: e.T56875A11547757. http://dx.doi.org/10.2305/IUCN.UK.2008.RLTS.T56875A11547757.en. Accessed on 26 February 2017.

Angulo, A., 2008d. *Eleutherodactylus richmondi.* The IUCN Red List of Threatened Species 2008: e.T56914A11551678. http://dx.doi.org/10.2305/IUCN.UK.2008.RLTS.T56914A11551678.en. Accessed on 26 February 2017.

Angulo, A., 2008e. *Eleutherodactylus wightmanae.* The IUCN Red List of Threatened Species 2008: e.T57056A11575451. http://dx.doi.org/10.2305/IUCN.UK.2008.RLTS.T57056A11575451.en. Accessed on 26 February 2017.

Angulo, A., 2010a. *Peltophryne lemur.* The IUCN Red List of Threatened Species 2010: e.T54345A11127094. http://dx.doi.org/10.2305/IUCN.UK.2010-2.RLTS.T54345A11127094.en. Accessed on 25 February 2017.

Angulo, A., 2010b. *Eleutherodactylus antillensis.* The IUCN Red List of Threatened Species 2010: e.T56425A11475987. http://dx.doi.org/10.2305/IUCN.UK.2010-2.RLTS.T56425A11475987.en. Accessed on 25 February 2017.

Angulo, A., 2010c. *Eleutherodactylus jasperi.* The IUCN Red List of Threatened Species 2010: e.T7142A12829636. http://dx.doi.org/10.2305/IUCN.UK.2010-2.RLTS.T7142A12829636.en. Accessed on 26 February 2017.

Anonymous, 1877. The development of batrachians without metamorphosis. *Nature* 15 (388): 491–492. https://doi.org/10.1038/015491a0

Arendt, W. J., 2006. Adaptations of an Avian Supertramp: Distribution, Ecology, and Life History of the Pearly-Eyed Thrasher (*Margarops fuscatus*). General Technical Report 27. San Juan, PR: United

States Department of Agriculture, Forest Service, International Institute of Tropical Forestry. https://doi.org/10.2737/IITF-GTR-27

Arya, R., Antonisamy, B., and Kumar, S., 2012. Sample size estimation in prevalence studies. *The Indian Journal of Pediatrics* 79: 1482–1488. https://doi.org/10.1007/s12098-012-0763-3

Asprey, G. F. and Robbins, R. G., 1953. The vegetation of Jamaica. *Ecological Monographs* 23: 359–412. https://doi.org/10.2307/1948625

Atkinson, D. and Sibly, R. M., 1997. Why are organisms usually bigger in colder environments? Making sense of a life history puzzle. *Trends in Ecology and Evolution* 12: 235–239. https://doi.org/10.1016/S0169-5347(97)01058-6

Atkinson, E. E. and Marín-Spiotta, E., 2015. Land use legacy effects on structure and composition of subtropical dry forests in St. Croix, U.S. Virgin Islands. *Forest Ecology and Management* 335: 270–280. https://doi.org/10.1016/j.foreco.2014.09.033

Bacheler, N. M., Neal, J. W., and Noble, R. L., 2004. Diet overlap between native bigmouth sleepers (*Gobiomorus dormitor*) and introduced predatory fishes in a Puerto Rico reservoir. *Ecology of Freshwater Fish* 13: 111–118. https://doi.org/10.1111/j.1600-0633.2004.00040.x

Backstrom, A. C., Garrad, G. E., Hobbs, R. J., and Bekessy, S. A., 2018. Grappling with the social dimensions of novel ecosystems. *Frontiers in Ecology and the Environment* 16: 109–117. https://doi.org/10.1002/fee.1769

Báez, T, Strong, M., and Salinas, K., 2014. Red Eyed Coqui *Eleutherodactylus antillensis* Care Manual. Herpetology division, Abilene Zoo, Texas, USA.

Bai, C., Liu, X., Fisher, M. C., Garner, T. W. J., and Li, Y., 2012. Global and endemic Asian lineages of the emerging pathogenic fungus *Batrachochytrium dendrobatidis* widely infect amphibians in China. *Diversity and Distributions* 18: 307–318. https://doi.org/10.1111/j.1472-4642.2011.00878.x

Barber, D. M., 2012. Chytrid fungus, *Batrachochytrium dendrobatidis* (*Bd*), detected at lower elevations in Puerto Rico: implications for conservation of Puerto Rican Crested Toad (*Peltophryne lemur*). *Herpetological Review* 43: 73–75.

Barber, D., Pacheco, C., Smith, D., Jiménez, N. M., and Rivera, R. L., 2019. Puerto Rican Crested Toad (*Peltophryne lemur*) Reintroduction and Field Management: Standard Operating Procedures. *Puerto Rican Crested Toad Working Group.* Fort Worth, Texas, USA.

Barbour, T., 1914. A contribution to the zoögeography of the West Indies, with especial reference to amphibians and reptiles. *Memoirs of the Museum of Comparative Zoölogy* 44: 205–363. https://doi.org/10.5962/bhl.title.49187

Barbour, T., 1915. Recent notes regarding West Indian reptiles and amphibians. *Proceedings of the Biological Society of Washington* 28: 71–78.

Barbour, T., 1917. Notes on the herpetology of the Virgin Islands. *Proceedings of the Biological Society of Washington* 30: 97–104.

Barbour, T., 1930a. Some faunistic changes in the Lesser Antilles. *Proceedings of the New England Zoological Club* 11: 73–85.

Barbour, T., 1930b. A list of Antillean reptiles and amphibians. *Zoologica* 11: 61–116. https://doi.org/10.5962/p.203735

Barbour, T., 1935. A second list of Antillean reptiles and amphibians. *Zoologica* 19: 77–141. https://doi.org/10.5962/p.203717

Barbour, T., 1937. Third list of Antillean reptiles and amphibians. *Bulletin of the Museum of Comparative Zoölogy* 82: 77–166.

Barker, B. S. and Rodríguez-Robles, J. A., 2017. Origins and genetic diversity of introduced populations of the Puerto Rican Red-Eyed Coquí, *Eleutherodactylus antillensis*, in Saint Croix (U.S. Virgin Islands) and Panamá. *Copeia* 105: 220–228. https://doi.org/10.1643/CG-16-501

Barlett, K. A., 1949. A toad serves the sugar industry of Puerto Rico. Unites States Department of Agriculture, Agricultural Research Administration, Research Achievement Sheet 118.

Baskin, J. N. and Williams, E. E., 1966. The Lesser Antillean *Ameiva* (Sauria, Teiidae). Re-evaluation, zoogeography and the effects of predation. *Studies on the Fauna of Curaçao and other Caribbean Islands,* No. 89: 144–176 + 2 plates.

Beauclerc, K. B., Johnson, B., and White, B. N., 2010. Genetic rescue of an inbred captive population of the critically endangered Puerto Rican crested toad (*Peltophryne lemur*) by mixing lineages. *Conservation Genetics* 11(1): 21–32. https://doi.org/10.1007/s10592-008-9782-z

Becker, M. H., Harris, R. N., Minbiole, K. P., Schwantes, C. R., Rollins-Smith, L. A., Reinert, L. K., Brucker, R. M., Domangue, R. J., and Gratwicke., B., 2011. Towards a better understanding of the use of probiotics for preventing chytridiomycosis in Panamanian golden frogs. *EcoHealth* 8: 501–506. https://doi.org/10.1007/s10393-012-0743-0

Bello y Espinosa, D., 1871. Zoologische Notizen aus Puerto Rico. Nach dem Spanischen frei bearbeitet von Herrn E. von Martens in Berlin. *Zoologischer Garten* (Frankfurt a. M.) 12: 348–351.

Beltrán, I., Ramírez-Castañeda, V., Rodríguez-López, C., Lasso, E., and Amézquita, A., 2019. Dealing with hot rocky environments: Critical thermal maxima and locomotor performance in *Leptodactylus lithonaetes* (Anura: Leptodactylidae). *The Herpetological Journal* 29: 155–161. https://doi.org/10.33256/hj29.3.155161

Berger, L., Roberts, A. A., Voyles, J., Longcore, J. E., Murray, K. A., and Skerratt, L. F., 2016. History and recent progress on chytridiomycosis in amphibians. *Fungal Ecology* 19: 89–99. https://doi.org/10.1016/j.funeco.2015.09.007

Beuchat, C. A., Pough, F. H., and Stewart, M. M., 1984. Response to simultaneous dehydration and thermal stress in three species of Puerto Rican frogs. *Journal of Comparative Physiology B* 154: 579–585. https://doi.org/10.1007/BF00684411

Bevan, A., 1940. Possibilities for forestry in the Virgin Islands: St. Thomas, St. John, St. Croix. *Caribbean Forester* 2: 8–12.

BioOne, 2022. https://bioone.org/journals/caribbean-journal-of-science/issues. Accessed on 11 January 2022.

Birdsey, R. A. and Weaver, P. L., 1987. *Forest area trends in Puerto Rico.* United States Department of Agriculture Forest Service, Southern Forest Experiment Station, New Orleans, LA. Research Note SO-331. https://doi.org/10.2737/SO-RN-331

Blackburn, D. C., Keeffe, R. M., Vallejo-Pareja, M. C., and Vélez-Juarbe, J., 2020. The earliest record of Caribbean frogs: a fossil coquí from Puerto Rico. *Biology Letters* 16: 20190947. https://doi.org/10.1098/rsbl.2019.0947

Blaustein, A. R. and Wake, D. B., 1995. The puzzle of declining amphibian populations. *Scientific American* 272: 52–57. https://doi.org/10.1038/scientificamerican0495-52

Blaustein, A. R. and Kiesecker, J. M., 2002. Complexity in conservation: lessons from the global decline of amphibian populations. *Ecology Letters* 5: 597–608. https://doi.org/10.1046/j.1461-0248.2002.00352.x

Blaustein, A. R., Han, B. A., Relyea, R. A., Johnson, P. T. J., Buck, J. C., Gervasi, S. S., and Kats, L. B., 2011. The complexity of amphibian population declines: understanding the role of cofactors in driving amphibian losses. *Annals of the New York Academy of Sciences* 1223: 108–119. https://doi.org/10.1111/j.1749-6632.2010.05909.x

Blay Publishers LLC, 2022. https://blaypublishers.com/. Accessed on 11 January 2022.

Bletz, M. C., Rosa, G. M., Andreone, F., Courtois, E. A., Schmeller, D. S., Rabibisoa, N. H. C., Rabemananjara, F. C. E., Raharivololoniaina, L., Vences, M., Weldon, C., Edmonds, D., Raxworthy, C. J., Harris, R. N., Fisher, M. C., and Crottini, A., 2015. Widespread presence of the pathogenic fungus *Batrachochytrium dendrobatidis* in wild amphibian communities in Madagascar. *Scientific Reports* 5: 8633. https://doi.org/10.1038/srep08633

Bontemps, D. R., Cuevas, E., Ortiz, E., Wunderle, J. M., Jr., and Joglar, R. L., 2016. Diet of the non-native spectacled caiman (*Caiman crocodilus*) in Puerto Rico. *Management of Biological Invasions* 7: 287–296. https://doi.org/10.3391/mbi.2016.7.3.08

Borroto-Páez, R., 2009. Invasive mammals in Cuba: an overview. *Biological Invasions* 11: 2279–2290. https://doi.org/10.1007/s10530-008-9414-z

Borroto-Páez, R. and Woods, C. A., 2012. Status and impact of introduced mammals in the West Indies. Pp. 241–258 in *Terrestrial Mammals of the West Indies,* edited by R. Borroto-Páez, C. A. Woods, and F. E. Sergile. Contributions. Wocahoota Press and Florida Museum of Natural History, Florida, USA.

Boulenger, G. A., 1881. *Leptodactylus caliginosus* Girard et *L. albilabris* Günther. *Bulletin de la Société Zoologique de France* 6: 30–35.

Bowden, M. J., 1968. Water balance of a dry island–The hydroclimatology of St. Croix, Virgin Islands and potential for agriculture and urban growth. Worcester, MA, Clark University. *Geography Publications at Dartmouth* 6: 1–89.

Bowdish, B. S., 1902. Birds of Porto Rico. *The Auk* 19: 356–366. https://doi.org/10.2307/4069597

Brandeis, T. J. and Oswalt, S. N., 2007. The status of U.S. Virgin Islands' forests, 2004. US Department of Agriculture Forest Service, Southern Research Station, Resource Bulletin SRS-122. Asheville, North Carolina, USA. https://doi.org/10.2737/SRS-RB-122

Brokaw, N., Crowl, T. A., Lugo, A. E., McDowell, W. H., Scatena, F. N., Waide, R. B. and Willig, M. R. (eds), 2012a. *A Caribbean Forest Tapestry. The Multidimensional Nature of Disturbance and Response.* Oxford University Press, New York, USA. https://doi.org/10.1093/acprof:osobl/9780195334692.001.0001

Brokaw, N., Zimmerman, J. K., Willig, M. R., Camilo, G. R., Covich, A. P., Crowl, T. A., Fetcher, N., Haines, B. L., Lodge, D. J., Lugo, A. E., Myster, R. W., Pringle, C. M., Sharpe, J. M., Scatena, F. N., Schowalter, T. D., Silver, W. L., Thompson, J., Vogt, D. J., Vogt, K. A., Waide, R. B., Walker, L. R., Woolbright, L. L., Wunderle, J. M., Jr., and Zou, X., 2012b. Response to Disturbance. Pp. 201–271 in *A Caribbean Forest Tapestry: The multidimensional Nature of Disturbance and Response,* edited by N. Brokaw, T. A. Crowl, A. E. Lugo, W. H. McDowell, F. N. Scatena, R. B. Waide, and M. R. Willig. Oxford University Press, New York, USA. https://doi.org/10.1093/acprof:osobl/9780195334692.003.0005

Bucciarelli, G. M., Blaustein, A. R., Garcia, T. S., and Kats, L. B., 2014. Invasion complexities: the diverse impacts of nonnative species on amphibians. *Copeia* 14: 611–632. https://doi.org/10.1643/OT-14-014

Burrowes, P. A., 2000. Parental care and sexual selection in the Puerto Rican cave-dwelling frog, *Eleutherodactylus cooki*. *Herpetologica* 56: 375–386.

Burrowes, P. A. and Joglar, R. L., 1991. A survey of the population status and an ecological evaluation of three Puerto Rican frogs. Pp. 42–46 in *Status y Distribución de los Reptiles y Anfibios de la Región de Puerto Rico*, edited by J. A. Moreno. Departamento de Recursos Naturales, Publicación Científica Miscelánea No. 1, San Juan, Puerto Rico, USA.

Burrowes, P. A. and Joglar, R. L., 1999. Population genetics of the Puerto Rican cave-dwelling frog, *Eleutherodactylus cooki*. *Journal of Herpetology* 33: 706–711. https://doi.org/10.2307/1565591

Burrowes, P. A., Joglar, R. L., and Green, D. E., 2004. Potential causes for amphibian declines in Puerto Rico. *Herpetologica* 60: 141–154. https://doi.org/10.1655/03-50

Burrowes, P. A., Longo, A. V., and Joglar, R. L., 2008a. Geographic distribution of *Batrachochytrium dendrobatidis* in Puerto Rico. *Herpetological Review* 39: 321–324.

Burrowes, P. A., Longo, A. V., and Rodríguez, C. A., 2008b. Potential fitness cost of *Batrachochytrium dendrobatidis* in *Eleutherodactylus coqui*, and comments on environment-related risk of infection. *Herpetotropicos* 4: 51–57.

Burrowes, P. A., Alicea, A., Longo, A. V., and Joglar, R. L., 2011. Toes versus swabs? evaluation of the best tissue source for detection of *Batrachochytrium dendrobatidis* in field-caught amphibians. *Herpetological Review* 42: 359–362.

Burrowes, P. A., Martes, M. C., Torres-Ríos, M., and Longo, A. V., 2017. Arboreality predicts *Batrachochytrium dendrobatidis* infection level in tropical direct-developing frogs. *Journal of Natural History* 51(11–12): 643–656. https://doi.org/10.1080/00222933.2017.1297504

Caligari Damestoy, P. G., 2017. Status of the pathogenic chytrid fungus (*Batrachochytrium dendrobatidis*) in three species of endemic amphibians from Puerto Rico, and impact of local climate on host response. MS Thesis. University of Puerto Rico, Río Piedras Campus, San Juan, Puerto Rico, USA.

Campos-Cerqueira, M. and Aide, T. M., 2016. Improving distribution data of threatened species by combining acoustic monitoring and occupancy modelling. *Methods in Ecology and Evolution* 2016: 1–9.

Campos-Cerqueira, M. and Aide, T. M., 2017. Lowland extirpation of anuran populations on a tropical mountain. *PeerJ* 5:e4059. https://doi.org/10.7717/peerj.4059

Candelas, G. C., Ortíz, E., Vásquez, C., and Felix, L., 1961. Respiratory metabolism in tadpoles of *Leptodactylus albilabris*. *American Zoologist* 1: 348. (abstract).

Capinha, C., Seebens, H., Cassey, P., García-Díaz, P., Lenzner, B., Mang, T., Moser, D., Pyšek, P., Rödder, D., Scalera, R., Winter, M., Dullinger, S., and Essl, F., 2017. Diversity, biogeography and the global flows of alien amphibians and reptiles. *Diversity and Distributions* 23: 1313–1322. https://doi.org/10.1111/ddi.12617

Carey, W. M., 1972. The herpetology of Anegada, British Virgin Islands. *Caribbean Journal of Science* 12: 79–89.

Carlo, T. A. and Morales, J. M., 2016. Generalist birds promote tropical forest regeneration and increased plant diversity via rare-biased seed dispersal. *Ecology* 97: 1819–1831. https://doi.org/10.1890/15-2147.1

Castro-Prieto, J., Gould, W. A., Ortiz-Maldonado, C., Soto-Bayó, S., Llerandi-Román, I., Gaztambide-Arandes, S., Quiñones, M., Cañón, M., and Jacobs, K. R., 2019. A Comprehensive Inventory of Protected Areas and other Land Conservation Mechanisms in Puerto Rico. Gen. Tech. Report IITF-GTR-50. San Juan, PR: US Department of Agriculture Forest Service, International Institute of Tropical Forestry. https://doi.org/10.2737/IITF-GTR-50

CBSG (Conservation Breeding Specialist Group), 2006. Population and Habitat Viability Analysis for the Puerto Rican Crested Toad FINAL REPORT. IUCN/SSC Conservation Breeding Specialist Group: Apple Valley, Minnesota, USA.

Chesser, R. T., Burns, K. J., Cicero, C., Dunn, J. L., Kratter, A. W., Lovette, I. J., Rasmussen, P. C., Remsen, J. V., Jr., Stotz, D. F., and Winker, K., 2019. Check-list of North American Birds (online). American Ornithological Society. http://checklist.aou.org/taxa. Accessed on 06 November 2019.

Christensen, J. H., Krishna Kumar, K., Aldrian, E., An, S.-I., Cavalcanti, I. F. A., de Castro, M., Dong, W., Goswami, P., Hall, A., Kanyanga, J. K., Kitoh, A., Kossin, J., Lau, N.-C., Renwick, J., Stephenson, D. B., Xie, S.-P., and Zhou, T., 2013. Climate Phenomena and their Relevance for Future Regional Climate Change. Pp. 1217–1308 in *Climate Change 2013: The Physical Science Basis. Contribution of Working Group I to the Fifth Assessment Report of the Intergovernmental Panel on Climate Change*, edited by T. F. Stocker, D. Qin, G.-K. Plattner, M. Tignor, S. K. Allen, J. Boschung, A. Nauels, Y. Xia, V. Bex, and P. M. Midgley. Cambridge University Press, Cambridge, UK, and New York, USA.

Christian, K. A., Nunez, F., Clos, L., and Diaz, L., 1988. Thermal relations of some tropical frogs along an altitudinal gradient. *Biotropica* 20: 236–239. https://doi.org/10.2307/2388239

Cochran, D. M., 1923. A new frog of the genus *Leptodactylus*. *Journal of the Washington Academy of Sciences* 13: 184–185. https://doi.org/10.5962/bhl.part.5293

Collins, M., Knutti, R., Arblaster, J., Dufresne, J.-L., Fichefet, T., Friedlingstein, P., Gao, X., Gutowski, W. J., Johns, T., Krinner, G., Shongwe, M., Tebaldi, C., Weaver, A. J., and Wehner, M., 2013. Long-term Climate Change: Projections, Commitments and Irreversibility. Pp. 1029–1136 in *Climate Change 2013: The Physical Science Basis. Contribution of Working Group I to the Fifth Assessment Report of the Intergovernmental Panel on Climate Change*, edited by T. F. Stocker, D. Qin, G.-K. Plattner, M. Tignor, S. K. Allen, J. Boschung, A. Nauels, Y. Xia, V. Bex, and P. M. Midgley. Cambridge University Press, Cambridge, UK, and New York, USA.

Colón-Piñeiro, Z., Rosario, L., and Restrepo, C., 2017. Retrofitting rodent housing for captive breeding of the direct developing frogs *Eleutherodactylus coqui* and *E. antillensis*: from troubleshooting to monitoring. *Herpetological Review* 48: 368–375.

Cooke, S. B., Dávalos, L. M., Mychajliw, A. M., Turvey, S. T., and Upham, N. S., 2017. Anthropogenic extinction dominates Holocene declines of West Indian mammals. *Annual Review of Ecology, Evolution, and Systematics* 48: 301–327. https://doi.org/10.1146/annurev-ecolsys-110316-022754

Cope, E. D., 1862. On some new and little known American ANURA. *Proceedings of the Academy of Natural Sciences of Philadelphia* 14: 151–159+594

Cope, E. D., 1868. Sixth contribution to the herpetology of tropical America. *Proceedings of the Academy of Natural Sciences of Philadelphia* 20: 305–313.

Crawshaw, G., Pienkowski, M., Lentini, A., Dutton, C., Delnatte, P., Russell, D., Berkvens, C., Barker, I., and Smith, D., 2014. Brown Skin disease: A syndrome of dysecdysis in Puerto Rican Crested Toads (*Peltophryne lemur*). *Zoo Biology* 33: 558–564. https://doi.org/10.1002/zoo.21168

Cubasch, U., Wuebbles, D., Chen, D., Facchini, M. C., Frame, D., Mahowald, N., and Winther, J.-G., 2013. Introduction. Pp. 119–158, in *Climate Change 2013: The Physical Science Basis. Contribution of Working Group I to the Fifth Assessment Report of the Intergovernmental Panel on Climate Change*, edited by T. F. Stocker, D. Qin, G.-K. Plattner, M. Tignor, S. K. Allen, J. Boschung, A. Nauels, Y. Xia, V. Bex, and P. M. Midgley. Cambridge University Press, Cambridge, UK, and New York, USA.

Danforth, S. T., 1925. Porto Rican herpetological notes. *Copeia* 143: 76–79. https://doi.org/10.2307/1437281

Danforth, S. T., 1926. Birds of the Cartagena Lagoon, Porto Rico. *The Journal of the Department of Agriculture of Porto Rico* 10: 1–136.

Davis, M. A., Thompson, K., and Grime, J.P., 2001. Charles S. Elton and the dissociation of invasion ecology from the rest of ecology. *Diversity and Distribution* 7: 97–102. https://doi.org/10.1046/j.1472-4642.2001.00099.x

Davis, M. A., Chew, M. K., Hobbs, R. J., Lugo, A. E., Ewel, J. J., Vermeij, G. J., Brown, J. H., Rosenzweig, M. L., Gardener, M. R., Carroll, S. P., Thompson, K., Pickett, S. T. A., Stromberg, J. C., Del Tredici, P., Suding, K. N., Ehrenfeld, J. G., Grime, J. P., Mascaro, J., and Briggs, J. C., 2011. Don't judge species on their origins. *Nature* 474: 153–154. https://doi.org/10.1038/474153a

Daza, J. D., Pinto, B. J., Thomas, R., Herrera-Martínez, A., Scantlebury, D. P., Padilla García, L. F., Balaraman, R. P., Perry, G., and Gamble, T., 2019. The sprightly little sphaerodactyl: Systematics and biogeography of the Puerto Rican dwarf geckos *Sphaerodactylus* (Gekkota, Sphaerodactylidae). *Zootaxa* 4712: 151–201. https://doi.org/10.11646/zootaxa.4712.2.1

deFrance, S. D., Hadden, C. S., LeFebvre, M. J., and DuChemin, G., 2010. Animal use at the Tibes Ceremonial Center. Pp. 115–151 in *Tibes: People, Power, and Ritual at the Center of the Cosmos*, edited by L. A. Curet, and L. M. Stringer. The University of Alabama Press, Tuscaloosa, Alabama, USA.

del Valle Atiles, F., 1887. *Los animales vertebrados útiles y los dañinos a la Agricultura del pais*. Imprenta del Boletín Mercantil, San Juan, Puerto Rico.

Dexter, R. R., 1932. The food habits of the imported toad *Bufo marinus* in the sugar cane sections of Porto Rico. *International Society of Sugar Cane Technologists*, 4th Congress, San Juan, P. P., Bulletin No. 74: 1–6.

Diamond, J. M., 1974. Colonization of exploded volcanic islands by birds: the supertramp strategy. *Science* 184: 803–806. https://doi.org/10.1126/science.184.4138.803

Dietz, J. L., 1989. *Historia Económica de Puerto Rico*. Ediciones Huracán. Río Piedras, San Juan, Puerto Rico, USA.

Dml'el, R., Perry, G., and Lazell, J., 1997. Evaporative water loss in nine insular populations of the lizard *Anolis cristatellus* group in the British Virgin Islands. *Biotropica* 29: 111–116. https://doi.org/10.1111/j.1744-7429.1997.tb00012.x

Domínguez-Cristóbal, C. M., 2000. *Panorama Histórico Forestal de Puerto Rico*. Editorial de la Universidad de Puerto Rico. San Juan, Puerto Rico, USA.

Domínguez-Cristóbal, C. M., 2006. Presencia del café (*Coffea arabica*) en el acontecer histórico de Puerto Rico. *Acta Científica* 20: 43–46.

Drewry, G. E., 1970. The role of amphibians in the ecology of a Puerto Rican rain forest. Pp. 16–54 in *The Rain Forest Project Annual Report* (June 1970), edited by R. G. Clements, G. E. Drewry and R. J. Lavigne. Puerto Rico Nuclear Center, PRNC–147, US Atomic Energy Commission, USA.

Drewry, G. E. and Jones, K. L., 1976. A new ovoviviparous frog, *Eleutherodactylus jasperi* (Amphibia, Anura, Leptodactylidae), from Puerto Rico. *Journal of Herpetology* 10: 161–165. https://doi.org/10.2307/1562976

Drewry, G. E. and Rand, A. S., 1983. Characteristics of an acoustic community: Puerto Rican frogs of the genus *Eleutherodactylus*. *Copeia* 1983: 941–953. https://doi.org/10.2307/1445095

DRNA (Departamento de Recursos Naturales y Ambientales), 2004a. Reglamento para regir la conservación y el manejo de la vida silvestre, las especies exóticas y la caza en el Estado Libre Asociado de Puerto Rico. Reglamento Número 6765, Departamento de Recursos Naturales y Ambientales, Estado Libre Asociado de Puerto Rico, USA.

DRNA (Departamento de Recursos Naturales y Ambientales), 2004b. Reglamento para regir las especies vulnerables y en peligro de extinción en el Estado Libre Asociado de Puerto Rico. Reglamento Número 6766, Departamento de Recursos Naturales y Ambientales, Estado Libre Asociado de Puerto Rico, USA.

DRNA (Departamento de Recursos Naturales y Ambientales), 2011. Deposición ante la Comisión de Recursos Naturales Ambiente y Energía de la Cámara de Representantes sobre la Resolución de la Cámara Número 277 (*R. de la C. 277*) para investigar el estado en que se encuentran los terrenos donde se encontraba la Laguna Guánica con el fin de retomar los trabajos de restauración de dicha área. 3 de marzo de 2011. Estado Libre Asociado de Puerto Rico, USA.

Duellman, W. E., 1978. The biology of an Equatorial herpetofauna in Amazonian Ecuador. *Miscellaneous Publication, Museum of Natural History, University of Kansas* 65: 1–352.

Duellman, W. E. and Crombie, R. I., 1970. *Hyla septentrionalis* Duméril and Bibron, Cuban treefrog. *Catalogue of American Amphibians and Reptiles* 11 September 1970: 92.1–92.4.

Duellman, W. E. and Trueb, L., 1994. *Biology of Amphibians*, 2nd edn. The Johns Hopkins University Press, Baltimore, Maryland, USA.

Duméril, A. M. C. and Bibron, G., 1841. Erpétologie Générale ou Histore Naturelle complète Des Reptiles. Tome Huitième. Paris. Librairie Encyclopédique de Roret.

Dundee, H. A., 1991. When is an introduction not an introduction? *Herpetological Review* 22: 122.

Dunn, G. E., Davis, W. R., and Moore, P. L. Hurricane season of 1956. *Monthly Weather Review* 84, No. 12, December 1956: 436–443. https://doi.org/10.1175/1520-0493(1956)084<0436:HSO>2.0.CO;2

Dyer, W. G., Williams, E. H., Jr., and Bunkley-Williams, L., 1995. Nematode parasites of a Puerto Rican tree frog, *Eleutherodactylus coqui*. *Transactions of the Illinois State Academy of Science* 88: 39–41.

Eagle Hill Institute, 2022. https://www.eaglehill.us/programs/journals/cana/caribbean-naturalist.shtml. Accessed on 11 January 2022.

Endeman, R., Whisson, D., Quinn, J., Cano, F., Quiñonez, P., and White, T. H., Jr., 2006. Monitoring invasive mammalian predator populations sharing habitat with the Critically Endangered Puerto Rican parrot *Amazona vittata*. *Oryx* 40: 95–102. https://doi.org/10.1017/S0030605305001286

Erdman, D. S., 1967. *Inland game fishes of Puerto Rico*. Puerto Rico Federal Aid Projects F-1-R, Job No. 7. Commonwealth of Puerto Rico, Department of Agriculture, San Juan, Puerto Rico, USA.

Erdman, D. S., 1972. *Los peces de las aguas interiores de Puerto Rico*, segunda edición (traducido al español por Víctor E. Márquez). Proyecto de Ayuda Federal F-1-20, volumen 4, número 2 (96 p.). Servicios Auxiliares y Operaciones Centralizadas, Departamento de Agricultura, Estado Libre Asociado de Puerto Rico.

Espeut, W. B., 1882. On the acclimatization of the Indian mungoos in Jamaica. *Proceedings of the Zoological Society of London* 1882: 712–714. https://doi.org/10.1111/j.1096-3642.1883.tb02783.x

Estrada, A. R., 2010. *Osteopilus septentrionalis* (Cuban Treefrog). Distribution. *Caribbean Herpetology* 18: 1. https://doi.org/10.31611/ch.18

Evermann, B. W., 1900. General report on the investigations in Porto Rico of the United States Fish Commission Steamer Fish Hawk in 1899. *Bulletin of the United States Fish Commission* XX (First Part; published by the Government Printing Office, Washington, 1902.): 1–350 + 52 Plates.

Ewel, J. J. and Whitmore, J. L., 1973. *The ecological life zones of Puerto Rico and the U. S. Virgin Islands*. Forest Service, US Department of Agriculture. Institute of Tropical Forestry, Río Piedras, Puerto Rico, USA.

Field, C. B., Barros, V. R., Mach, K. J., Mastrandrea, M. D., van Aalst, M., Adger, W. N., Arent, D. J., Barnett, J., Betts, R., Bilir, T. E., Birkmann, J., Carmin, J., Chadee, D. D., Challinor, A. J., Chatterjee, M., Cramer, W., Davidson, D. J., Estrada, Y. O., Gattuso, J.-P., Hijioka, Y., Hoegh-Guldberg, O., Huang, H. Q., Insarov, G. E., Jones, R. N., Kovats, R. S., Romero-Lankao, P., Larsen, J. N., Losada, I. J., Marengo, J. A., McLean, R. F., Mearns, L. O., Mechler, R., Morton,

J. F., Niang, I., Oki, T., Olwoch, J. M., Opondo, M., Poloczanska, E. S., Pörtner, H.-O., Redsteer, M. H., Reisinger, A., Revi, A., Schmidt, D. N., Shaw, M. R., Solecki, W., Stone, D. A., Stone, J. M. R., Strzepek, K. M., Suarez, A. G., Tschakert, P., Valentini, R., Vicuña, S., Villamizar, A., Vincent, K. E., Warren, R., White, L. L., Wilbanks, T. J., Wong, P. P., and Yohe, G. W., 2014. Technical summary. Pp. 35–94 in Climate Change 2014: Impacts, Adaptation, and Vulnerability, ed by C. B. Field, V. R. Barros, D. J. Dokken, K. J. Mach, M. D. Mastrandrea, T. E. Bilir, M. Chatterjee, K. L. Ebi, Y. O. Estrada, R. C. Genova, B. Girma, E. S. Kissel, A. N. Levy, S. MacCracken, P. R. Mastrandrea, and L. L. White. Part A: Global and Sectoral Aspects. Contribution of Working Group II to the Fifth Assessment Report of the Intergovernmental Panel on Climate Change, Cambridge University Press, Cambridge, United Kingdom and New York, New York, USA. https://doi.org/10.1017/CBO9781107415379.004

Figueroa-Colón, J. C., 1996. Phytogeographical trends, centers of high species richness and endemism, and the question of extinctions in the native flora of Puerto Rico. Pp. 89–102 in *The scientific survey of Puerto Rico and the Virgin Islands–an eighty-year reassessment of the island's natural history*, edited by J. C. Figueroa-Colón. Annals of the New York Academy of Sciences, 776, New York, USA. https://doi.org/10.1111/j.1749-6632.1996.tb17414.x

Fitzinger, L. I., 1826. Neue classification der reptilien nach ihren natürlichen verwandtschaften: nebst einer verwandtschafts-tafel und einem verzeichnisse der reptilien-sammlung des K. K. zoologischen museum's zu Wien. Wien: J. G. Heubner. https://doi.org/10.5962/bhl.title.4683

Fitzinger, L. J. F. J., 1843. Systema Reptilium. Fasciculus Primus. Amblyglossae. Apud Braumüller et Seidel Bibliopolas, Vindobonae, Austriae.

Fleischner, T. L., 2005. Natural history and the deep roots of resource management. *Natural Resources Journal* 45: 1–13.

Flemming, C. and MacPhee, R. D. E., 1999. Redetermination of holotype of *Isolobodon portoricensis* (Rodentia, Capromyidae), with notes on recent mammalian extinctions in Puerto Rico. *American Museum Novitates* No. 3278: 1–11.

Flinter, G. D., 1834. An account of the present state of the island of Puerto Rico. Samuel Bagster, Jun., Printer, 14, Bartholomew Close, London, UK.

Flores-Nieves, C. Z., Logue, D. M., and Santos-Flores, C. J., 2014. Native White-Lipped Frog larvae (*Leptodactylus albilabris*) outcompete introduced Cane Toad larvae (*Rhinella marina*) under laboratory conditions. *Herpetological Conservation and Biology* 9: 378–386.

Flores-Vallejo, I., 2011. Recomendaciones para la reintroducción del sapo concho puertorriqueño (*Peltophryne lemur*) en la Reserva Natural El Tallonal in Arecibo, Puerto Rico. MS thesis, Universidad Metropolitana, San Juan, Puerto Rico, USA.

Foden, W. B. and Young, B. E., 2016. IUCN SSC Guidelines for Assessing Species' Vulnerability to Climate Change. Version 1.0. Occasional Paper of the IUCN Species Survival Commission No. 59. Cambridge, UK and Gland, Switzerland: IUCN Species Survival Commission.

Fort Worth Zoo, 2022. https://www.fortworthzoo.org/conservation-blog/posts/meet-olaf-the-ivf-toad#dis//phys.org/news/2019-11-1st-vitro-puerto-rico-crested.amp. Accessed on 11 January 2022.

Foster, P., 2001. The potential negative impacts of global climate change on Tropical Montane Cloud Forests. *Earth Science Reviews* 55: 73–106. https://doi.org/10.1016/S0012-8252(01)00056-3

Foster, D. R., Fluet, M., and Boose, E. R., 1999. Human or natural disturbance: landscape-scale dynamics of the tropical forests of Puerto Rico. *Ecological Applications* 9: 555–572. https://doi.org/10.1890/1051-0761(1999)009[0555:HONDLS]2.0.CO;2

Fournier, A. M. V., White, E. R., and Heard, S. B., 2019. Site-selection bias and apparent population declines in long-term studies. *Conservation Biology* 33: 1370–1379. https://doi.org/10.1111/cobi.13371

Fox, I., 1947. *Ornithodoros puertoricensis*, a new tick from rats in Puerto Rico. *The Journal of Parasitology* 33: 253–259. https://doi.org/10.2307/3273558

Fridman, E. P., 2002. History of medical primatology. Pp 12–56 in *Medical Primatology: History, biological foundations and applications*, edited by R. D. Nadler. Taylor & Francis Inc., New York, New York, USA.

Frishkoff, L. O., Gabot, E., Sandler, G., Marte, C., and Mahler, D. L., 2019. Elevation shapes the reassembly of Anthropocene lizard communities. *Nature Ecology & Evolution* 3: 638–646. https://doi.org/10.1038/s41559-019-0819-0

Frost, D. R., 2020. Amphibian Species of the World: an online reference. Version 6.1 (3 March 2020). Electronic Database accessible at https://amphibiansoftheworld.amnh.org/index.php. American Museum of Natural History, New York, USA.

Gannon, M. R., Kurta, A., Rodríguez-Durán, A., and Willig, M. R., 2005. *Bats of Puerto Rico: an island focus and a Caribbean perspective*. Texas Tech University Press, Lubbock, Texas, USA.

García, M. A., Diez, C. E., and Álvarez, A. O., 2000. El control de los cabros y cerdos en la Isla de Mona, Puerto Rico. *Acta Científica* 14: 83–87.

García, M. A., Diez, C. E., and Álvarez, A. O., 2001. The impact of feral cats on Mona Island wildlife and recommendations for their control. *Caribbean Journal of Science* 37: 107–108.

García-Díaz, J., 1967. Rediscovery of *Bufo lemur* (Cope) and additional records of reptiles from Puerto Rico. *Stahlia* 10: 1–6.

García-Martinó, A. R., Warner, G. S., Scatena, F. N., and Civco, D. L., 1996. Rainfall, runoff and elevation relationships in the Luquillo mountains of Puerto Rico. *Caribbean Journal of Science* 32: 413–424. https://doi.org/10.1111/j.1752-1688.1996.tb03495.x

Garman, S., 1888 (1887). West Indian Batrachia in the Museum of Comparative Zoology. *Bulletin of the Essex Institute* 19: 13–16.

Gassett, J. W., Folk, T. H., Alexy, K. J., Miller, K. V., Chapman, B. R., Boyd, F. L., and Hall, D. I., 2000. Food habits of Cattle Egrets on St. Croix, U.S. Virgin Islands. *The Wilson Bulletin* 112: 268–271. https://doi.org/10.1676/0043-5643(2000)112[0268:FHOCEO]2.0.CO;2

Gerhart, G. A., 1936. El Cuerpo Civil de Conservación en Puerto Rico. *Revista de Agricultura de Puerto Rico* 28: 292–299.

Gervasi, S. S., Stephens, P. R., Hua, J., Searle, C. L., Yang Xie, G., Urbina, J., Olson, D. H., Bancroft, B. A., Weis, V., Hammond, J. I., Relyea, R.A., and Blaustein, A. R., 2017. Linking ecology and epidemiology to understanding predictors of multi-host responses to an emerging pathogen, the amphibian chytrid fungus. *PloS ONE* 12: E0167882. https://doi.org/10.1371/journal.pone.0167882

Gilormini, J. A., 1936. Consideraciones generales sobre los bosques en Puerto Rico. *Revista de Agricultura de Puerto Rico* 28: 288–291.

Giovas, C. M., LeFebvre, M., and Fitzpatrick, S. M., 2012. New records for prehistoric introduction of Neotropical mammals to the West Indies: evidence from Carriacou, Lesser Antilles. *Journal of Biogeography* 39: 476–487. https://doi.org/10.1111/j.1365-2699.2011.02630.x

Global Mammal Assessment Team, 2008. *Homo sapiens*. The IUCN Red List of Threatened Species 2008. e.T136584A4313662. http://dx.doi.org/10.2305/IUCN.UK.2008.RLTS.T136584A4313662.en. Accessed on 30 August 2017.

Goldberg, S. R., Holshuh, J., and Joglar, R. L., 1996. *Eleutherodactylus cooki* ("Guajón" or "Demonio de Puerto Rico"). *Herpetological Review* 27: 18.

Gobierno de Puerto Rico, 2012a. Available at https://www.lexjuris.com/lexlex/Leyes2012/lexl2012020.htm. Accessed on 11 January 2021.

Gobierno de Puerto Rico, 2012b. Available at https://www.lexjuris.com/lexlex/Leyes2012/lexl2012022.htm. Accessed on 11 January 2022.

Gobierno de Puerto Rico, 2016. http://www.fortaleza.pr.gov/content/gobierno-logra-proteger-el-16-del-territorio-de-puerto-rico. Accesible at https://www.metro.pr/pr/noticias/2016/04/19/gobierno-logra-proteger-16-territorio-p-r.html. Accessed on 11 January 2022.

Gobierno de Puerto Rico, 2017. Available at http://www.lexjuris.com/lexlex/Leyes2017/lexl2017019.htm. Accessed on 11 January 2022.

Government of Puerto Rico, 1999. Ley Número 241 del 15 de agosto de 1999: Nueva Ley de Vida Silvestre de Puerto Rico. http://www.lexjuris.com/LEXLEX/Ley1999/lex99241.htm. Accessed on 10 September 2017.

Government of the British Virgin Islands, 2019a. http://www.bvi.gov.vg/content/geography. Accessed on 04 January 2022.

Government of the British Virgin Islands, 2019b. http://www.bvi.gov.vg/content/ministry-natural-resources-and-labour. Accessed on 11 January 2022.

Grant, C., 1931a. A new frog from Porto Rico. *Copeia* 1931: 55–56. https://doi.org/10.2307/1435749

Grant, C., 1931b. Notes on *Bufo marinus* (Linnaeus). *Copeia* 1931: 62. https://doi.org/10.2307/1435755

Grant, C., 1932a. *Bufo lemur*, a rare Porto Rican toad. *The Journal of the Department of Agriculture of Puerto Rico* 16: 41–43. https://doi.org/10.46429/jaupr.v16i1.15024

Grant, C., 1932b. A new frog from Puerto Rico. *The Journal of the Department of Agriculture of Puerto Rico* 16: 145–148. https://doi.org/10.46429/jaupr.v16i2.15038

Grant, C., 1932c. Herpetological notes from the Puerto Rico Area. *The Journal of the Department of Agriculture of Puerto Rico* 16: 161–165. https://doi.org/10.46429/jaupr.v16i2.15042

Grant, C., 1932d. A new frog from the Virgin Islands. *The Journal of the Department of Agriculture of Puerto Rico* 16: 325–326. https://doi.org/10.46429/jaupr.v16i3.15050

Grant, C., 1932e. The herpetology of St John and adjacent keys, U. S. Virgin Islands. *The Journal of the Department of Agriculture of Puerto Rico* 16: 331–334. https://doi.org/10.46429/jaupr.v16i3.15052

Grant, C., 1932f. Herpetology of Tortola; notes on Anegada and Virgin Gorda, British Virgin Islands. *The Journal of the Department of Agriculture of Puerto Rico* 16: 339–349. https://doi.org/10.46429/jaupr.v16i3.15054

Grant, C., 1932g. The growth of herpetology in the Puerto Rico and Virgin Island Area. *The Journal of the Department of Agriculture of Puerto Rico* 16: 401–404. https://doi.org/10.46429/jaupr.v16i4.15057

Grant, C., 1937. Herpetological notes with new species from the American and British Virgin Islands, 1936. *The Journal of Agriculture of the University of Puerto*

Rico 21: 503–526. https://doi.org/10.46429/jaupr.v21i4.14316

Grant, C. and Beatty, H. A., 1944. Herpetological notes on St. Croix, Virgin Islands. *Herpetologica* 2: 110–113.

Grau, H. R., Aide, T. M., Zimmerman, J. K., Thomlinson, J. R., Helmer, E., and Zou, X., 2003. The ecological consequences of socioeconomic and land-use changes in postagriculture Puerto Rico. *Bioscience* 53: 1159–1168. https://doi.org/10.1641/0006-3568(2003)053[1159:TECOSA]2.0.CO;2

Grau, H. R., Aide, T. M., Zimmerman, J. K., and Thomlinson, J. R., 2004. Trends and scenarios of the carbon budget in postagricultural Puerto Rico (1936–2060). *Global Change Biology* 10: 1163–1179. https://doi.org/10.1111/j.1529-8817.2003.00792.x

Grau, H. R. and Aide, T. M., 2007. Are rural-urban migration and sustainable development compatible in mountain systems? *Mountain Research and Development* 27: 119–123. https://doi.org/10.1659/mrd.0906

Gray, M. J., Duffus, A. L., Haman, K. H., Harris, R., Allender, M., Thompson, T., Christmas, M., Sacerdote-Velat, A., Sprague, L., Williams, J., and Miller, D., 2017. Pathogen surveillance in herpetofaunal populations: guidance on study design, sample collection, biosecurity, and intervention strategies. *Herpetological Review* 48: 334–351.

Greenhawk, N., 2013. Range extension of *Eleutherodactylus cooki*, the "Coqui Guajon", Grant, 1932 (Amphibian: Eleutherodactylidae). *Check List* 95: 1050–1053. https://doi.org/10.15560/9.5.1050

Greenhawk, N., Zlotnik, S., Billy, L. M., Boas, S., and Gabel, S., 2017. Baseline amphibians survey and sampling of *Batrachochytrium dendrobatidis* in the Icaco and Hormiga valleys, Patillas, Puerto Rico. *Phyllomedusa* 16: 63–69. https://doi.org/10.11606/issn.2316-9079.v16i1p63-69

Gundlach, J. C., 1881. Apuntes para la fauna Puerto-Riqueña, Tercera Parte, Anfibios. *Anales de la Sociedad Española de Historia Natural* (Madrid) 10: 305–317.

Günther, A. C. L. G., 1859. On the reptiles from St. Croix, West Indies, collected by Messrs. A. and E. Newton. *Annals and Magazine of Natural History* 4, Third Series, No. XXI, part XXII: 209–217. https://doi.org/10.1080/00222935908697110

Guzmán-Colón, D. and Roloff, G. J., 2014. Small Indian mongoose (*Herpestes auropunctatus*) population abundance and effects of habitat features on trapping success in protected areas of eastern Puerto Rico. *Caribbean Naturalist* 2014(19): 1–12.

Hammerson, G., Solís, F., Ibáñez, R., Jaramillo, C., and Fuenmayor, Q., 2004. *Lithobates pipiens*. The IUCN Red List of Threatened Species 2004: e.T58695A11814172. http://dx.doi.org/10.2305/IUCN.UK.2004.RLTS.T58695A11814172.en. Accessed on 17 September 2017.

Hansen, J., Sato, M., Hearty, P., Ruedy, R., Kelley, M., Masson-Delmotte, V., Russell, G., Tselioudis, G., Cao, J., Rignot, E., Velicogna, I., Tormey, B., Donovan, B., Kandiano, E., von Schuckmann, K., Kharecha, P., Legrande, A. N., Bauer, M., and Lo, K-W. 2016. Ice melt, sea level rise and superstorms: evidence from paleoclimate data, climate modeling, and modern observations that 2°C global warming could be dangerous. *Atmospheric Chemistry and Physics* 16: 3761–3812. https://doi.org/10.5194/acp-16-3761-2016

Harris, R. N., Brucker, R. M., Walke, J. B., Becker, M. H., Schwantes, C. R., Flaherty, D. C., Lam, B. A., Woodhams, D. C., Briggs, C. J., Vredenburg, V. T., and Minbiole, K. P. C., 2009. Skin microbes on frogs prevent morbidity and mortality caused by a lethal skin fungus. *International Society for Microbial Ecology* 3: 818–824. https://doi.org/10.1038/ismej.2009.27

Hartmann, D. L., Klein Tank, A. M. G., Rusticucci, M., Alexander, L. V., Brönnimann, S., Charabi, Y., Dentener, F. J., Dlugokencky, E. J., Easterling, D. R., Kaplan, A., Soden, B. J., Thorne, P. W., Wild, M., and Zhai, P. M., 2013. Observations: Atmosphere and Surface. Pp. 159–254 in *Climate Change 2013: The Physical Science Basis. Contribution of Working Group I to the Fifth Assessment Report of the Intergovernmental Panel on Climate Change*, edited by T. F. Stocker, D. Qin, G.-K. Plattner, M. Tignor, S. K. Allen, J. Boschung, A. Nauels, Y. Xia, V. Bex, and P. M. Midgley. Cambridge University Press, Cambridge, UK, and New York, USA.

Heatwole, H., 1982. Herpetological communities. Pp. 1–19 in *Herpetological communities: A symposium of the Society for the Study of Amphibians and Reptiles and the Herpetologists League, August 1977*, edited by N. J. Scott, Jr. United States Department of the Interior, US Fish and Wildlife Service. Wildlife research Report 13. Washington, District of Columbia, USA.

Heatwole, H., 1983. Physiological responses of animals to moisture and temperature. Pp. 239–265 in *Tropical Forest Ecosystems, A. Structure and Function*, ed. by F. B. Golley. Elsevier Scientific Publishing Company, Amsterdam, The Netherlands.

Heatwole, H., 2013. Worldwide decline and extinction of amphibians. Pp. 259–278 in *The Balance of Nature and Human Impact*, edited by K. Rohde. Cambridge University Press, Cambridge, UK. https://doi.org/10.1017/CBO9781139095075.025

Heatwole, H. and MacKenzie, F., 1967. Herpetogeography of Puerto Rico. IV. Paleogeography, faunal similarity and endemism. *Evolution* 21: 429–438. https://doi.org/10.1111/j.1558-5646.1967.tb03400.x

Heatwole, H. and Newby, R. C., 1972. Interactions of internal rhythm and loss of body water in influencing activity levels of amphibians. *Herpetologica* 28: 156–162.

Heatwole, H. and Torres, F., 1967. Distribution and geographic variation of the ameivas of Puerto Rico and the Virgin Islands. Herpetogeography of Puerto Rico III. *Studies on the Fauna of Curaçao and other Caribbean Islands*, 24(92): 63–111.

Heatwole, H. and Wilkinson, J. W. (eds), 2009. Amphibian Decline: Diseases, Parasites, Maladies and Pollution. Volume 8 in *Amphibian Biology*, edited by H. Heatwole. Surrey Beatty & Sons, Baulkham Hills, New South Wales, AU.

Heatwole, H. and Wilkinson, J. W. (eds), 2011. Conservation and Decline of Amphibians: Ecological Aspects, Effect of Humans, and Management. Volume 10 in *Amphibian Biology*, edited by H. Heatwole. Surrey Beatty & Sons, Baulkham Hills, New South Wales, AU.

Heatwole, H., Blasini De Austin, S., and Herrero, R., 1968. Heat tolerances of tadpoles of two species of tropical anurans. *Comparative Biochemistry and Physiology* 27: 807–815. https://doi.org/10.1016/0010-406X(68)90620-8

Heatwole, H., Levins, R., and Byer, M. D., 1981. Biogeography of the Puerto Rican Bank. *Atoll Research Bulletin* 251: 1–63. https://doi.org/10.5479/si.00775630.251.1

Heatwole, H., Sade, D. S., and Hildreth, R., 1963a. Herpetogeography of Puerto Rico, I. Herpetofauna of Cayo Santiago and Cayo Batata. *Caribbean Journal of Science* 3: 1–5.

Heatwole, H., Kelts, L., Levins, R., and Torres, F., 1963b. Faunal notes on Culebra Island, Puerto Rico. *Caribbean Journal of Science* 8: 29–30.

Heatwole, H., Torres, F., de Austin, S. B., and Heatwole, A., 1969. Studies on anuran water balance—I. Dynamics of evaporative water loss by the coquí, *Eleutherodactylus portoricensis*. *Comparative Biochemistry and Physiology* 28: 254–269. https://doi.org/10.1016/0010-406X(69)91342-5

Hedges, S. B., 1993. Global amphibian declines: a perspective from the Caribbean. *Biodiversity and Conservation* 2: 290–303. https://doi.org/10.1007/BF00056674

Hedges, S. B., 1996. The origin of West Indian amphibians and reptiles. Pp. 95–128 in *Contributions to West Indian Herpetology: A Tribute to Albert Schwartz*, edited by R. Powell and R. W. Henderson. Contributions to Herpetology 12, Society for the Study of Amphibians and Reptiles, Ithaca, New York, USA.

Hedges, S. B., 2006. An overview of the evolution and conservation of West Indian amphibians and reptiles. *Applied Herpetology* 3: 281–292. https://doi.org/10.1163/157075406778905072

Hedges, S. B. and Díaz, L. M., 2011. The conservation status of amphibians in the West Indies. Pp. 31–47 in *Conservation of Caribbean Island Herpetofaunas Volume 1: Conservation Biology and the Wider Caribbean*, edited by A. Hailey, B. S. Wilson, and J. A. Horrocks. Brill, Leiden, The Netherlands. https://doi.org/10.1163/ej.9789004183957.i-228.23

Hedges, S. B. and Heinicke, M. P., 2007. Molecular phylogeny and biogeography of West Indian frogs of the genus *Leptodactylus* (Anura, Leptodactylidae). *Molecular Phylogenetics and Evolution* 44: 308–314. https://doi.org/10.1016/j.ympev.2006.11.011

Hedges, B. and Ríos-López, N., 2008a. *Eleutherodactylus gryllus*. The IUCN Red List of Threatened Species 2008: e.T56634A11510937. http://dx.doi.org/10.2305/IUCN.UK.2008.RLTS.T56634A11510937.en. Accessed on 26 February 2017.

Hedges, B. and Rios-López, N., 2008b. *Eleutherodactylus locustus*. The IUCN Red List of Threatened Species 2008: e.T56725A11524637. http://dx.doi.org/10.2305/IUCN.UK.2008.RLTS.T56725A11524637.en. Accessed on 26 February 2017.

Hedges, B. and Ríos-López, N., 2010. *Eleutherodactylus eneidae*. The IUCN Red List of Threatened Species 2010: e.T7150A12831858. http://dx.doi.org/10.2305/IUCN.UK.2010-2.RLTS.T7150A12831858.en. Accessed on 26 February 2017.

Hedges, S. B. and Thomas, R., 1989. Supplement to West Indian amphibians and reptiles: a check-list. Milwaukee Public Museum, *Contribution in Biology and Geology* No. 77: 1–11.

Hedges, S. B. and Thomas, R., 1991. The importance of systematic research in the conservation of amphibian and reptile populations. Pp. 56–61 in *Status y Distribución de los Reptiles y Anfibios de la Región de Puerto Rico*, edited by J. A. Moreno. Departamento de Recursos Naturales, Publicación Científica Miscelánea No. 1, San Juan, Puerto Rico, USA.

Hedges, B. and Thomas, R., 2010a. *Eleutherodactylus lentus*. The IUCN Red List of Threatened Species 2010: e.T56711A11520415. http://dx.doi.org/10.2305/IUCN.UK.2010-2.RLTS.T56711A11520415.en. Accessed on 26 February 2017.

Hedges, B. and Thomas, R., 2010b. *Eleutherodactylus schwartzi*. The IUCN Red List of Threatened Species 2010: e.T56954A11560198. http://dx.doi.org/10.2305/IUCN.UK.2010-2.RLTS.T56954A11560198.en. Accessed on 26 February 2017.

Hedges, S. B., Duellman, W. E., and Heinicke, M. P., 2008. New World direct-developing frogs (Anura:

Terrarana): Molecular phylogeny, classification, biogeography, and conservation. *Zootaxa* 1737: 1–182. https://doi.org/10.11646/zootaxa.1737.1.1

Hedges, S. B., Hass, C. A., and Maxson, L. R., 1992. Caribbean biogeography: molecular evidence for dispersal in West Indian terrestrial vertebrates. *Proceedings of the National Academy of Sciences* 89: 1909–1913. https://doi.org/10.1073/pnas.89.5.1909

Hedges, B., Joglar, R., and Thomas, R., 2010. *Eleutherodactylus karlschmidti*. The IUCN Red List of Threatened Species 2010: e.T7146A12830790. http://dx.doi.org/10.2305/IUCN.UK.2010-2.RLTS.T7146A12830790.en. Accessed on 26 February 2017.

Hedges, B., Joglar, R., Thomas, R., Powell, R., and Ríos-López, N., 2009. *Eleutherodactylus coqui*. The IUCN Red List of Threatened Species 2009: e.T56522A11491306. http://dx.doi.org/10.2305/IUCN.UK.2008.RLTS.T56522A11491306.en. Accessed on 26 February 2017.

Hedges, S. B., Powell, R., Henderson, R. W., Hanson, S., and Murphy, J. C., 2019. Definition of the Caribbean Islands biogeographic region, with checklist and recommendations for standardized common names of amphibians and reptiles. *Caribbean Herpetology* 67: 1–53. https://doi.org/10.31611/ch.67

Heinicke, M. P., Díaz, L. M., and Hedges, S. B., 2011. Origin of invasive Florida frogs traced to Cuba. *Biology Letters* 7: 407–410. https://doi.org/10.1098/rsbl.2010.1131

Helmer, E. H., 2004. Forest conservation and land development in Puerto Rico. *Landscape Ecology* 19: 29–40. https://doi.org/10.1023/B:LAND.0000018364.68514.fb

Henderson, R. W., 1992. Consequences of predator introductions and habitat destruction on amphibians and reptiles in the post-Columbus West Indies. *Caribbean Journal of Science* 28: 1–10.

Henderson, R. W. and Powell, R., 2009. *Natural History of West Indian Reptiles and Amphibians*. University Press of Florida, Gainesville, Florida, USA.

Henry, T. R., 1958. *The Strangest Things in the World: A Book About Extraordinary Manifestations of Nature*. Public Affairs Press, Washington, DC, USA.

Hernández Aquino, L., 1977. Diccionario de voces indígenas de Puerto Rico, Segunda Edición: ampliamente aumentada y corregida. Editorial Cultural, Río Piedras, Puerto Rico, USA.

Herrera-Giraldo, J. L., 2010. Herpetofaunal species composition on the Vieques National Wildlife Refuge, Vieques, Puerto Rico. MS thesis, University of Puerto Rico, Mayagüez Campus, Mayagüez, Puerto Rico, USA.

Herrera-Montes, A. and Brokaw, N. 2010. Conservation value of tropical secondary forest: a herpetofaunal perspective. *Biological Conservation* 143: 1414–1422. https://doi.org/10.1016/j.biocon.2010.03.016

Herrera-Montes, A., Ríos-Droz, B., Puente-Rolón, A. R., Dávila-Casanova, D., and Ríos-López, N., 2015. *Xenochrophis vittatus* (Striped Keelback). *Herpetological Review* 46: 64.

Herrera, D. A., Ault, T. R., Fasullo, J. T., Coats, S. J., Carrillo, C. M., Cook, B. I., and Williams, A. P., 2018. Exacerbation of the 2013–2016 Pan-Caribbean drought by anthropogenic warming. *Geophysical Research Letters* 45: 10619–10626. https://doi.org/10.1029/2018GL079408

Heyer, W. R., 1978. Systematics of the *fuscus* group of the frog genus *Leptodactylus* (Amphibia, Leptodactylidae). *Natural History Museum of Los Angeles County Science Bulletin* 29: 1–85. https://doi.org/10.5479/si.00810282.301

Hickley, P., North, S. M., and Harper, D. M., 1994. The diet of largemouth bass, *Micropterus salmoides*, in Lake Naivasha, Kenya. *Journal of Fish Biology* 44: 607–619. https://doi.org/10.1111/j.1095-8649.1994.tb01237.x

Hildebrand, S. F., 1935. An annotated list of fishes of the fresh waters of Puerto Rico. *Copeia* 1935: 49–56. https://doi.org/10.2307/1436106

Hill, R. T., 1898. *Cuba and Porto Rico with the other islands of the West Indies, their topography, climate, flora, products, industries, cities, people, political conditions, etc.* The De Vinne Press, The Century Co., New York, USA. https://doi.org/10.5962/bhl.title.4428

Hill, R. T., 1899. *Notes on the Forest Conditions of Porto Rico*. Bulletin No. 25. Division of Forestry, United States Department of Agriculture. Washington, District of Columbia, USA. https://doi.org/10.5962/bhl.title.66228

Hillman, S. S., Withers, P. C., Drewes, R. C., and Hillyard, S. D., 2009. Ecological and Environmental Physiology of Amphibians. Oxford University Press Inc., New York, USA. https://doi.org/10.1093/acprof:oso/9780198570325.001.0001

Hoagland, D. B., Horst, G. R., and Kilpatrick, C. W., 1989. Biogeography and population biology of the mongoose in the West Indies. Pp. 611–633 in *Biogeography of the West Indies: Past, Present, and Future*, edited by C. A. Woods. Sandhill Crane Press, Gainesville, Florida, USA.

Hoffman, W. M. and Janer, J. L., 1941. *Bufo marinus* as a vector of helminth ova in Puerto Rico. *Puerto Rico Journal of Public Health and Tropical Medicine* 16:501–504.

Hoffmann, M., Hilton-Taylor, C., Angulo, A., Böhm, M., Brooks, T. M., Butchart, S. H. M., Carpenter, K. E., Chanson, J., Collen, B., Cox, N. A., Darwall, W. R. T., Dulvy, N. K., Harrison, L. R., Katariya, V., Pollock,

C. M., Quader, S., Richman, N. I., Rodrigues, A. S. L., Tognelli, M. F., Vié, J.-C., Aguiar, J. M., Allen, D. J., Allen, G .R., Amori, G., Ananjeva, N. B., Andreone, F., Andrew, P., Aquino-Ortiz, A. L., Baillie, J. E. M., Baldi, R., Bell, B. D., Biju, S. D., Bird, J. B., Black-Decima, P., Blanc, J. J., Bolaños, F., Bolivar-G., W., Burfield, I. J., Burton, J. A., Capper, D. R., Castro, F., Catullo, G., Cavanagh, R. D., Channing, A., Chao, N. L., Chenery, A. M., Chiozza, F., Clausnitzer, V., Collar, N. J., Collett, L. C., Collette, B. B., Cortez-Fernandez, C. F., Craig, M. T., Crosby, M. J., Cumberlidge, N., Cuttelod, A., Derocher, A. E., Diesmos, A. C., Donaldson, J. S., Duckworth, J. W., Dutson, G., Dutta, S. K., Emslie, R. H., Farjon, A., Fowler, S., Freyhof, J., Garshelis, D. L., Gerlach, J., Gower, D. J., Grant, T. D., Hammerson, G. A., Harris, R. B., Heaney, L. R., Hedges, S. B., Hero, J.-M., Hughes, B., Hussain, S. A., Icochea M., J., Inger, R. F., Ishii, N., Iskandar, D. T., Jenkins, R. K. B., Kaneko, Y., Kottelat, M., Kovacs, K. M., Kuzmin, S. L., La Marca, E., Lamoreux, J. F., Lau, M. W. N., Lavilla, E. O., Leus, K., Lewison, R. L., Lichtenstein, G., Livingstone, S. R., Lukoschek, V., Mallon, D. P., McGowan, P. J. K., McIvor, A., Moehlman, P. D., Molur, S., Muñoz-Alonso, A., Musick, J. A., Nowell, K., Nussbaum, R. A., Olech, W., Orlov, N. L., Papenfuss, T. J., Parra-Olea, G., Perrin, W. F., Polidoro, B. A., Pourkazemi, M., Racey, P. A., Ragle, J. S., Ram, M., Rathbun, G., Reynolds, R. P., Rhodin, A. G. J., Richards, S. J., Rodríguez, L. O., Ron, S. R., Rondinini, C., Rylands, A. B., de Mitcheson, Y. S., Sanciangco, J. C., Sanders, K. L., Santos-Barrera, G., Schipper, J., Self-Sullivan, C., Shi, Y., Shoemaker, A., Short, F. T., Sillero-Zubiri, C., Silvano, D. L., Smith, K. G., Smith, A. T., Snoeks, J., Stattersfield, A. J., Symes, A. J., Taber, A. B., Talukdar, B. K., Temple, H. J., Timmins, R., Tobias, J. A., Tsytsulina, K., Tweddle, D., Ubeda, C., Valenti, S. V., van Dijk, P. P., Veiga, L. M., Veloso, A., Wege, D. C., Wilkinson, M., Williamson, E. A., Xie, F., Young, B. E., Akçakaya, H. R., Bennun, L., Blackburn, T. M., Boitani, L., Dublin, H. T., da Fonseca, G. A. B., Gascon, C., Lacher, T. E., Jr., Mace, G. M., Mainka, S. A., McNeely, J. A., Mittermeier, R. A., Reid, G. M., Rodriguez, J. P., Rosenberg, A. A., Samways, M. J., Smart, J., Stein, B. A., and Stuart, S. N., 2010. The impact of conservation on the status of the World's vertebrates. *Science* 330: 1503–1509. https://doi.org/10.1126/science.1194442

Holdridge, L. R., 1939. Forestry in Puerto Rico. *Caribbean Forester* 1: 7–11.

Holt, R. D., 1997. Rarity and evolution: some theoretical considerations. Pp. 209–234 in *The Biology of Rarity*, edited by W. E. Kunin and K. J. Gaston. Chapman & Hall, London, UK. https://doi.org/10.1007/978-94-011-5874-9_12

Holt, R. D. and Bonsall, M. B., 2017. Apparent competition. *Annual Review of Ecology, Evolution, and Systematics* 48: 447–471. https://doi.org/10.1146/annurev-ecolsys-110316-022628

Honegger, R. E., 1981. List of amphibians and reptiles either known or thought to have become extinct since 1600. *Biological Conservation* 19: 141–158. https://doi.org/10.1016/0006-3207(81)90049-5

Hopkins, D., 1989. An extraordinary eighteenth-century map of the Danish sugar-plantation island St. Croix. *Imago Mundi* 41:44–58. https://doi.org/10.1080/03085698908592667

Horst, G. R., Hoagland, D. B., and Kilpatrick, C. W., 2001. The mongoose in the West Indies: the biogeography and population biology of an introduced species. Pp. 409–424 in *Biogeography of the West Indies: Patterns and Perspectives*, 2nd edn, edited by C. A. Woods and F. E. Sergile. CRC Press, Boca Raton, Florida, USA. https://doi.org/10.1201/9781420039481-21

Hossack, B. R., Adams, M. J., Campbell Grant, E. H., Pearl, C. A., Bettaso, J. B., Barichivich, W. J., Lowe, W. H., True, K., Ware, J. L., and Corn, P. S., 2010. Low prevalence of chytrid fungus (*Batrachochytrium dendrobatidis*) in amphibians of U.S. headwater streams. *Journal of Herpetology* 44: 253–260. https://doi.org/10.1670/09-058.1

Hughey, M. C., Pena, J. A., Reyes, R., Medina, D., Belden, L. K., and Burrowes, P. A., 2017. Skin bacterial microbiome of a generalist Puerto Rican frog varies along elevation and land use gradients. *PeerJ* 5:e3688. https://doi.org/10.7717/peerj.3688

Iniciativa Herpetológica, Inc., 2015. Final Report. Cooperative Agreement between U.S. Department of the Interior, Fish and Wildlife Service, and Iniciativa Herpetológica, Inc. Agreement # 40181AG112. 53 pp.

International Reptile Conservation Foundation, 2022. Available at https://journals.ku.edu/reptilesandamphibians/about. Accessed on 11 January 2022.

IPCC (Intergovernmental Panel on Climate Change), 2002. Climate Change and Biodiversity. Technical Paper V prepared by IPCC Working Group II Technical Support Unit for the United Nations Convention on Biological Diversity, edited by H. Gitay, A. Suárez, R. T. Watson, and D. J. Dokken.

IPCC, 2013: Summary for Policymakers. Pp. 1–29, in *Climate Change 2013: The Physical Science Basis. Contribution of Working Group I to the Fifth Assessment Report of the Intergovernmental Panel on Climate Change*, edited by T. F. Stocker, D. Qin, G.-K. Plattner, M. Tignor, S. K. Allen, J. Boschung, A. Nauels, Y. Xia, V. Bex, and P. M. Midgley. Cambridge University Press, Cambridge, UK, and New York, USA.

Island Resources Foundation. 2013. *An Environmental Profile of the Island of Anegada, British Virgin Islands.*

Island Resources Foundation. Tortola, British Virgin Islands and Washington, DC.

Island Resources Foundation. 2015. *An Environmental Profile of the Island of Tortola, British Virgin Islands.* Island Resources Foundation. Tortola, BVI and Washington, DC.

IUCN, 2016. The IUCN Red List of Threatened Species. Version 2016-3. <http://www.iucnredlist.org>. Accessed on 25 February 2017.

IUCN, 2022a. The IUCN Red List of Threatened Species. Version 2021-3. <https://www.iucnredlist.org>. Accessed on 12 January 2022.

IUCN, 2022b. https://www.iucnredlist.org/terms/terms-of-use#10.%20Disclaimer%20of%20Warranty. The IUCN Red List of Threatened Species. Version 2021-3. https://www.iucnredlist.org. Accessed on 04 January 2022.

IUCN SSC Amphibian Specialist Group, 2021a. *Peltophryne lemur.* The IUCN Red List of Threatened Species 2021: e.T54345A172692429. https://dx.doi.org/10.2305/IUCN.UK.2021-1.RLTS.T54345A172692429.en. Accessed on 14 June 2021.

IUCN SSC Amphibian Specialist Group, 2021b. *Eleutherodactylus antillensis.* The IUCN Red List of Threatened Species 2021: e.T56425A3038940. https://dx.doi.org/10.2305/IUCN.UK.2021-1.RLTS.T56425A3038940.en. Accessed on 14 June 2021.

IUCN SSC Amphibian Specialist Group, 2021c. *Eleutherodactylus brittoni.* The IUCN Red List of Threatened Species 2021: e.T56477A172739340. https://dx.doi.org/10.2305/IUCN.UK.2021-1.RLTS.T56477A172739340.en. Accessed on 14 June 2021.

IUCN SSC Amphibian Specialist Group, 2021d. *Eleutherodactylus cochranae.* The IUCN Red List of Threatened Species 2021: e.T56515A3041363. https://dx.doi.org/10.2305/IUCN.UK.2021-1.RLTS.T56515A3041363.en. Accessed on 14 June 2021.

IUCN SSC Amphibian Specialist Group, 2021e. *Eleutherodactylus cooki.* The IUCN Red List of Threatened Species 2021: e.T7149A172755657. https://dx.doi.org/10.2305/IUCN.UK.2021-1.RLTS.T7149A172755657.en. Accessed on 14 June 2021.

IUCN SSC Amphibian Specialist Group, 2021f. *Eleutherodactylus coqui.* The IUCN Red List of Threatened Species 2021: e.T56522A3041672. https://dx.doi.org/10.2305/IUCN.UK.2021-1.RLTS.T56522A3041672.en. Accessed on 14 June 2021.

IUCN SSC Amphibian Specialist Group, 2021g. *Eleutherodactylus eneidae.* The IUCN Red List of Threatened Species 2021: e.T7150A172793860. https://dx.doi.org/10.2305/IUCN.UK.2021-1.RLTS.T7150A172793860.en. Accessed on 14 June 2021.

IUCN SSC Amphibian Specialist Group, 2021h. *Eleutherodactylus gryllus.* The IUCN Red List of Threatened Species 2021: e.T56634A172795370. https://dx.doi.org/10.2305/IUCN.UK.2021-1.RLTS.T56634A172795370.en. Accessed on 14 June 2021.

IUCN SSC Amphibian Specialist Group, 2021i. *Eleutherodactylus hedricki.* The IUCN Red List of Threatened Species 2021: e.T56648A172795557. https://dx.doi.org/10.2305/IUCN.UK.2021-1.RLTS.T56648A172795557.en. Accessed on 14 June 2021.

IUCN SSC Amphibian Specialist Group, 2021j. *Eleutherodactylus jasperi.* The IUCN Red List of Threatened Species 2021: e.T7142A172793582. https://dx.doi.org/10.2305/IUCN.UK.2021-1.RLTS.T7142A172793582.en. Accessed on 14 June 2021.

IUCN SSC Amphibian Specialist Group, 2021k. *Eleutherodactylus juanariveroi.* The IUCN Red List of Threatened Species 2021: e.T135828A172796720. https://dx.doi.org/10.2305/IUCN.UK.2021-1.RLTS.T135828A172796720.en. Accessed on 14 June 2021.

IUCN SSC Amphibian Specialist Group, 2021l. *Eleutherodactylus karlschmidti.* The IUCN Red List of Threatened Species 2021: e.T7146A172793731. https://dx.doi.org/10.2305/IUCN.UK.2021-1.RLTS.T7146A172793731.en. Accessed on 14 June 2021.

IUCN SSC Amphibian Specialist Group, 2021m. *Eleutherodactylus lentus.* The IUCN Red List of Threatened Species 2021: e.T56711A3047786. https://dx.doi.org/10.2305/IUCN.UK.2021-1.RLTS.T56711A3047786.en. Accessed on 14 June 2021.

IUCN SSC Amphibian Specialist Group, 2021n. *Eleutherodactylus locustus.* The IUCN Red List of Threatened Species 2021: e.T56725A172795683. https://dx.doi.org/10.2305/IUCN.UK.2021-1.RLTS.T56725A172795683.en. Accessed on 14 June 2021.

IUCN SSC Amphibian Specialist Group, 2021o. *Eleutherodactylus monensis.* The IUCN Red List of Threatened Species 2021: e.T56774A172795883. https://dx.doi.org/10.2305/IUCN.UK.2021-1.RLTS.T56774A172795883.en. Accessed on 14 June 2021.

IUCN SSC Amphibian Specialist Group, 2021p. *Eleutherodactylus portoricensis.* The IUCN Red List of Threatened Species 2021: e.T56875A110924108. https://dx.doi.org/10.2305/IUCN.UK.2021-1.RLTS.T56875A110924108.en. Accessed on 14 June 2021.

IUCN SSC Amphibian Specialist Group, 2021q. *Eleutherodactylus richmondi.* The IUCN Red List of Threatened Species 2021: e.T56914A172795982. https://dx.doi.org/10.2305/IUCN.UK.2021-1.RLTS.T56914A172795982.en. Accessed on 14 June 2021.

IUCN SSC Amphibian Specialist Group, 2021r. *Eleutherodactylus unicolor.* The IUCN Red List of Threatened Species 2021: e.T57023A172796109. https://dx.doi.org/10.2305/IUCN.UK.2021-1.RLTS.T57023A172796109.en. Accessed on 14 June 2021.

IUCN SSC Amphibian Specialist Group, 2021s. *Eleutherodactylus wightmanae*. The IUCN Red List of Threatened Species 2021: e.T57056A172796246. https://dx.doi.org/10.2305/IUCN.UK.2021-1.RLTS.T57056A172796246.en. Accessed on 14 June 2021.

IUCN SSC Amphibian Specialist Group, 2021t. *Leptodactylus albilabris*. The IUCN Red List of Threatened Species 2021: e.T57113A3055043. https://dx.doi.org/10.2305/IUCN.UK.2021-1.RLTS.T57113A3055043.en. Accessed on 14 June 2021.

IUCN SSC Amphibian Specialist Group, 2021u. *Latonia nigriventer*. The IUCN Red List of Threatened Species 2021: e.T6715A177694004. https://dx.doi.org/10.2305/IUCN.UK.2021-3.RLTS.T6715A177694004.en. Accessed on 04 January 2022.

Iverson, J. B., 1978. The impact of feral cats and dogs on populations of the West Indian Rock Iguana, *Cyclura carinata*. *Biological Conservation* 14: 63–73. https://doi.org/10.1016/0006-3207(78)90006-X

Jani, A. J., Knapp, R. A., and Briggs, C. J., 2017. Epidemic and endemic pathogen dynamics correspond to distinct host population microbiomes at a landscape scale. *Proceedings of the Royal Society B* 284: 20170944. https://doi.org/10.1098/rspb.2017.0944

Jensen, K., Alvarado-Ramy, F., González-Martínez, J., Kraiselburd, E., and Rullán, J., 2004. B-Virus and free-ranging macaques, Puerto Rico. *Emerging Infectious Diseases* 10: 494–496. https://doi.org/10.3201/eid1003.030257

Jennings, A. and Veron, G., 2016. *Herpestes auropunctatus*. The IUCN Red List of Threatened Species 2016: e.T70204120A70204139. http://dx.doi.org/10.2305/IUCN.UK.2016-1.RLTS.T70204120A70204139.en. Accessed on 24 September 2018.

Joglar, R. L., 1998. *Los Coquíes de Puerto Rico: Su Historia Natural y Conservación*. Editorial de la Universidad de Puerto Rico, San Juan, Puerto Rico.

Joglar, R. L., 2005a. Anfibios. P. 53 in *Biodiversidad de Puerto Rico: Vertebrados Terrestres y Ecosistemas*, edited by R. L. Joglar. Editorial del Instituto de Cultura Puertorriqueña, San Juan, Puerto Rico, USA.

Joglar, R. L. 2005b. Conservación. Pp. 56–59 in *Biodiversidad de Puerto Rico: Vertebrados Terrestres y Ecosistemas*, edited by R. L. Joglar. Editorial del Instituto de Cultura Puertorriqueña, San Juan, Puerto Rico, USA.

Joglar, R. L. and Burrowes, P. A., 1991. El efecto del huracan Hugo sobre una comunidad de anfibios en El Yunque, Puerto Rico y algunas recomendaciones para la protección de las especies del género *Eleutherodactylus*. Pp. 47–53 in *Status y Distribución de los Reptiles y Anfibios de la Región de Puerto Rico*, edited by J. A. Moreno. Departamento de Recursos Naturales, Publicación Científica Miscelánea No. 1, San Juan, Puerto Rico, USA.

Joglar, R. L. and Burrowes, P. A., 1996. Declining amphibian populations in Puerto Rico. Pp. 371–380 in *Contributions to West Indian Herpetology: A Tribute to Albert Schwartz*, edited by R. Powell and R. W. Henderson. Contributions to Herpetology 12, Society for the Study of Amphibians and Reptiles, Ithaca, New York, USA.

Joglar, R. L. and Rios-López, N., 1995. *Osteopilus septentrionalis* (Cuban Treefrog, Rana Platanera). *Herpetological Review* 26: 105–106.

Joglar, R. L. and Rivero, J. A., 1986. Sinonimia de *Eleutherodactylus ramosi* Rivero. *Caribbean Journal of Science* 22: 123.

Joglar, R. L., Burrowes, P. A., and Ríos, N., 1996. Biology of the Puerto Rican cave-dwelling frog, *Eleutherodactylus cooki*, and some recommendations for conservation. Pp. 251–258 in *Contributions to West Indian Herpetology: A Tribute to Albert Schwartz*, edited by R. Powell and R. W. Henderson. Contributions to Herpetology 12, Society for the Study of Amphibians and Reptiles, Ithaca, New York, USA.

Joglar, R. L., Rentas, V., and Li, F. C., 2005. Ranita de labio blanco (*Leptodactylus albilabris*). Pp. 74–88 in *Biodiversidad de Puerto Rico: Vertebrados Terrestres y Ecosistemas*, edited by R. L. Joglar. Editorial del Instituto de Cultura Puertorriqueña, San Juan, Puerto Rico, USA.

Joglar, R. L., Ríos-López, N., and Cardona, M., 1998. *Osteopilus septentrionalis* (Cuban Treefrog, Rana Platanera). *Herpetological Review* 29: 107.

Joglar, R. L., Álvarez, A. O., Aide, T. M., Barber, D., Burrowes, P. A., García, M. A., León-Cardona, A., Longo, A. V., Pérez-Buitrago, N., Puente, A., Ríos-López, N., and Tolson, P. J., 2007 (reprinted 2011). Conserving the Puerto Rican herpetofauna. *Applied Herpetology* 4: 327–345. https://doi.org/10.1163/157075407782424566

Johnsen, D. O. and Whitehair, L. A., 1986. Research facility breeding. Pp. 499–512 in *Primates: The Road to Self-Sustaining Populations*, edited by K. Benirschke. Springer-Verlag New York Inc., New York, USA. https://doi.org/10.1007/978-1-4612-4918-4_40

Johnson, S. R., Berentsen, A. R., Ellis, C., Davis, A., and Vercauteren, K. C., 2016. Estimates of small Indian mongoose densities: implications for rabies management. *The Journal of Wildlife Management* 80: 37–47. https://doi.org/10.1002/jwmg.998

Jones, K. L., 1982. Prey patterns and trophic niche overlap in four species of Caribbean frogs. Pp. 49–55 in *Herpetological communities: A symposium of the Society for the Study of Amphibians and Reptiles and the Herpetologists League, August 1977*, edited by N. J. Scott, Jr. United States Department of the Interior, US Fish and Wildlife Service. Wildlife research Report 13. Washington, DC.

Junta de Planificación de Puerto Rico. 1983. Reglamento de zonificación especial para las zonas no urbanas de los municipios circundantes al Bosque Nacional del Caribe (El Yunque). Estado Libre Asociado de Puerto Rico, Oficina del Gobernador, San Juan, PR, USA.

Kämpfer, P., Busse, H., and Scholz, H. C., 2009. *Chromobacterium piscinae* sp. nov. and *Chromobacterium pseudoviolaceum* sp. nov., from environmental samples. *International Journal of Systematic and Evolutionary Microbiology* 59: 2486–2490. https://doi.org/10.1099/ijs.0.008888-0

Kennaway, T. and Helmer, E. H., 2007. The forest types and ages cleared for land development in Puerto Rico. *GIScience & Remote Sensing* 44: 356–382. https://doi.org/10.2747/1548-1603.44.4.356

Kennaway, T. A., Helmer, E. H., Lefsky, M. A., Brandeis, T. A., and Sherrill, K. R. 2008. Mapping land cover and estimating forest structure using satellite imagery and coarse resolution lidar in the Virgin Islands. *Journal of Applied Remote Sensing* 2: 1–27. https://doi.org/10.1117/1.3063939

Khalyani, A. H., Gould, W. A., Harmsen, E., Terando, A., Quinones, M., and Collazo, J. A., 2016. Climate change implications for tropical islands: interpolating and interpreting statistically downscaled GCM projections for management and planning. *Journal of Applied Meteorology and Climatology* 55: 265–282. https://doi.org/10.1175/JAMC-D-15-0182.1

Klawinski, P. D., Dalton, B., and Shields, A. B., 2014. Coqui frog populations are negatively affected by canopy opening but not detritus deposition following an experimental hurricane in a tropical rainforest. *Forest Ecology and Management* 332: 118–123. https://doi.org/10.1016/j.foreco.2014.02.010

Kolby, J. E. and Skerratt, L. F., 2015. Amphibian chytrid fungus in Madagascar neither shows widespread presence nor signs of certain establishment. *PLoS ONE* 10: e0139172. https://doi.org/10.1371/journal.pone.0139172

Kolby, J. E., Smith, K. M., Ramirez, S. D., Rabemananjara, F., Pessier, A. P., Brunner, J. L., Golberg, C. S., Berger, L., and Skerratt, L. F., 2015. Rapid response to evaluate the presence of amphibian chytrid Fungus (*Batrachochytrium dendrobatidis*) and ranavirus in wild amphibian populations in Madagascar. *PLoS ONE* 10: E0125330. https://doi.org/10.1371/journal.pone.0125330

Kraus, F., 2009. Alien reptiles and amphibians: a scientific compendium and analysis. Volume 4 in *Invading Nature: Springer Series in Invasion Ecology*, edited by J. A. Drake. Springer & Business Media, New York, USA.

Krebs, C. J., 2014. *Ecological Methodology*. Third edition (in preparation). Available at https://www.zoology.ubc.ca/~krebs/books.html.

Kunin, W. E. and Gaston, K. J., 1993. The biology of rarity: patterns, causes and consequences. *Trends in Ecology and Evolution* 8: 298–301. https://doi.org/10.1016/0169-5347(93)90259-R

LaBastille, A. and Richmond, M., 1973. Birds and mammals of Anegada island, British Virgin Islands. *Caribbean Journal of Science* 13: 91–109.

Larsen, M. C., 2000. Analysis of 20th century rainfall and streamflow to characterize drought and water resources in Puerto Rico. *Physical Geography* 21: 494–521. https://doi.org/10.1080/02723646.2000.10642723

Larue, D. K., 1994. Puerto Rico and the Virgin Islands. Pp. 151–165 in *Caribbean Geology: An Introduction*, edited by S. K. Donovan and T. A. Jackson. The University of the West Indies Publisher's Association, Kingston, Jamaica.

Lavigne, R. J. and Drewry, G. E., 1970. Feeding behavior of the frogs and lizards in the Tropical Wet Forest – preliminary report. Pp. 64–73 in *The Rain Forest Project Annual Report* (June 1970), edited by R. G. Clements, G. E. Drewry, and R. J. Lavigne. Puerto Rico Nuclear Center, PRNC–147, US Atomic Energy Commission, USA.

Lazell, J., 1991. The herpetofauna of Guana Island: diversity, abundance, rarity, and conservation. Pp. 28–33 in *Status y Distribución de los Reptiles y Anfibios de la Región de Puerto Rico*, edited by J. A. Moreno. Departamento de Recursos Naturales, Publicación Científica Miscelánea No. 1, San Juan, Puerto Rico, USA.

Lebrón, R., Batra, A. T., Bontempo, J., Buckley, C., Cron, M., Fenstermacher, L., Mahoney, C., Schmitt, L., and Bauer, A. M., 1995. *Leptodactylus albilabris* (Caribbean White-lipped Frog). *Herpetological Review* 26: 31.

Ledru, A. P. and Sonnini de Manoncourt, C. N. S., 1810. *Voyage aux îles de Ténériffe, la Trinité, Saint Thomas, Sainte-Croix et Porto-Ricco, exécuté par ordre du gouvernement français, depuis le 30 septembre 1796 jusqu'au 7 juin 1798, sous la direction du capitaine Baudin, pour faire des recherches et des collections relatives à l'histoire naturelle*. Editeur: A. Bertrand. Paris. https://doi.org/10.5962/bhl.title.162382

Lee, W. O., Yang, H., Yoon, S. W., and Park, J. Y., 2009. Study on the feeding habits of *Micropterus salmoides* in Lake Okjeong and Lake Yongdam, Korea. *Korean Journal of Ichthyology* 21: 200–2007.

Lentini, A., 2002. Husbandry manual, Puerto Rican crested toad (*Peltophryne lemur*), 2002 update. Keeper and Curator Edition. Toronto Zoo, Scarborough, Ontario, Canada. 48 p.

Leonard, M. D., 1933. Notes on the giant toad, *Bufo marinus* (L.), in Puerto Rico. *Journal of Economic Entomology* 26: 67–72. https://doi.org/10.1093/jee/26.1.67

Lepage, D., 2019. Checklist of the birds of Puerto Rico. Avibase, the world bird database. Retrieved from https://avibase.bsceoc.org/checklist.jsp?lang=EN&region=pr&list=clements&format=1 (Accessed on 11 November 2019).

Lever, C., 2003. *Naturalized Reptiles and Amphibians of the World.* Oxford University Press Inc., New York, USA.

Levins, R. and Heatwole, H., 1973. Biogeography of the Puerto Rican Bank: introduction of species onto Palominitos Island. *Ecology* 54: 1056–1064. https://doi.org/10.2307/1935571

Lewis, E. R. and Narins, P. M., 1985. Do frogs communicate with seismic signals? *Science* 227: 187–189. https://doi.org/10.1126/science.227.4683.187

Lillywhite, H. B., 2008. *Dictionary of Herpetology.* Krieger Publishing Company, Malabar, Florida.

Lips, K. R., 2016. Overview of chytrid emergence and impacts on amphibians. *Philosophical Transactions of the Royal Society B* 371: 20150465. https://doi.org/10.1098/rstb.2015.0465

Lips, K. R., Brem, F., Brenes, R., Reeve, J. D., Ross, A. A., Voyles, J., Carey, C., Livo, L., Pessier, A. P., and Collins, J. P., 2006. Emerging infectious disease and the loss of biodiversity in a neotropical amphibian community. *Proceedings of the National Academy of Sciences of the United States of America* 103: 3165–3170. https://doi.org/10.1073/pnas.0506889103

Loftus, W. F. 2004. Assessment of heavy rainfall event on the inland habitats and fishes of Virgin Islands National Park, St. John, USVI. US Geological Survey Florida Integrated Science Center, Everglades National Park Field Station, Homestead, Florida, USA.

López-Hernández, D. and Puente-Rolón, A., 2021. The effect of Hurricane María on the richness and calling activity of the *Eleutherodactylus* assemblages of the Carite State Forest, Puerto Rico. *Biotropica* 53: 1664–1672. https://doi.org/10.1111/btp.13015

López-Hernández, D. and Puente-Rolón, A. R., and Bird-Picó, F., 2021. Identification of habitat features associated with site occupancy of *Eleutherodactylus richmondi* at the Carite State Forest in Puerto Rico. *Caribbean Naturalist* 84: 1–11.

López-Marrero, T. and Hermansen-Báez, L. A., 2011. Expansion of urban land cover around El Yunque National Forest. Fact Sheet. USDA Forest Service, Southern Research Station, Gainesville, FL, USA.

Longo, A. V. and Burrowes, P. A., 2010. Persistence with chytridiomycosis does not assure survival of direct-developing frogs. *EcoHealth* 7: 185–195. https://doi.org/10.1007/s10393-010-0327-9

Longo, A. V. and López-Torres, A. L., 2017. *Eleutherodactylus coqui* (Common Coquí). *Herpetological Review* 48: 160.

Longo, A. V., Burrowes, P. A., and Joglar, R. L., 2010. Seasonality of *Batrachochytrium dendrobatidis* infection in direct-developing frogs suggests a mechanism for persistence. *Diseases of Aquatic Organisms* 92: 253–260. https://doi.org/10.3354/dao02054

Longo, A. V., Ossiboff, R. J., Zamudio, K. R., and Burrowes, P. A., 2013a. Lability in host defenses: terrestrial frogs die from chytridiomycosis under enzootic conditions. *Journal of Wildlife Diseases* 49: 197–199. https://doi.org/10.7589/2012-05-129

Longo, A. V., Rodríguez-Gómez, C. A., Zegarra, J. P., Monzón, O., Claudio-Hernández, H. J., Joglar, R. L., Zamudio, K. R., Burrowes, P. A., and López-Torres, A. L., 2020. Tick parasitism as a cost of sexual selection and male parental care in a Neotropical frog. *Ecosphere* 11(1): e03010. https://doi.org/10.1002/ecs2.3010

Longo, A. V., Savage, A. E., Hewson, I., and Zamudio, K. R., 2015. Seasonal and ontogenetic variation of skin microbial communities and relationships to natural disease dynamics in declining amphibians. *Royal Society Open Science* 2: 140377. https://doi.org/10.1098/rsos.140377

Longo, A. V., Rodríguez, D., da Silva-Leite, D., Toledo, L. F., Mendoza-Almeralla, C., Burrowes, P. A., and Zamundio, K. R., 2013b. ITS1 copy number varies among *Batrachochytrium dendrobatidis* strains: implications for qPCR estimates of infection intensity from field-collected amphibian skin swabs. *PLoS ONE* 8: e59499. https://doi.org/10.1371/annotation/e4c527e6-4b6f-4c12-970c-a9e2e6ed8f79

López-Hernández, D. C., Muñiz-Torres, A. Y., Rodríguez-Rivera, K. X., Puente-Rolón, A. R., 2020. Behavioral changes in tadpoles of the Antillean white lipped frog (*Leptodactylus albilabris*) in the presence of a predator: habitat use preference and activity levels. *Caribbean Journal of Science* 50: 132–138. https://doi.org/10.18475/cjos.v50i1.a14

López-Marrero, T. M. and Villanueva-Colón, N., 2006. *Atlas ambiental de Puerto Rico.* Editorial de la Universidad de Puerto Rico, San Juan, Puerto Rico, USA.

López-Marrero, T. and Hermansen-Báez, L. A., 2011. Expansión urbana y zonificación de los terrenos circundantes al Bosque Nacional El Yunque. [Hoja informativa]. Gainesville, FL: USDA Forest Service, Southern Research Station. USA.

López-Torres, A. L., 2008. Distribution and habitat status of the coquí guajón *Eleutherodactylus cooki*. MS thesis, Universidad Metropolitana, San Juan, Puerto Rico, USA.

López-Torres, A. L. and Longo, A. V., 2013. Implementing recovery actions for the threatened Coquí Guajón (*Eleutherodactylus cooki*): where to start? *FrogLog* 21: 41–42.

LoPresti, E. F., Karban, R., Robinson, M., Grof-Tisza, P., and Wetzel, W., 2016. The natural history supplement: furthering natural history amongst ecologists and evolutionary biologists. *Bulletin of the Ecological Society of America* 97: 305–310. https://doi.org/10.1002/bes2.1239

Lugo, A. E., 1996. Ninety years of plant ecology research in Puerto Rico. Pp. 73–87 in *The scientific survey of Puerto Rico and the Virgin Islands–an eighty-year reassessment of the island's natural history*, edited by J. C. Figueroa-Colón. Annals of the New York Academy of Sciences, 776, New York, USA. https://doi.org/10.1111/j.1749-6632.1996.tb17413.x

Lugo, A. E., 2004. The outcome of alien tree invasions in Puerto Rico. *Frontiers in Ecology and the Environment* 2: 265–273. https://doi.org/10.1890/1540-9295(2004)002[0265:TOOATI]2.0.CO;2

Lugo, A. E., 2008. Visible and invisible effects of hurricanes on forest ecosystems: an international review. *Austral Ecology* 33: 368–398. https://doi.org/10.1111/j.1442-9993.2008.01894.x

Lugo, A. E., 2014. Evolving conservation paradigms for the Anthropocene. Pp. 47–59 in *Forest conservation and management in the Anthropocene: Conference proceedings*, edited by A. V. Sample and R. P. Bixler. United States Department of Agriculture Forest Service, Rocky Mountain Research Station, Fort Collins, Colorado, USA. RMRS-P-71.

Lugo, A. E., 2015. Forestry in the Anthropocene. *Science* 349: 771. https://doi.org/10.1126/science.aad2208

Lugo, A. E., 2018. Characterization of ten extreme disturbance events in the context of social and ecological systems. *Biogeochemistry* 141: 385–400. https://doi.org/10.1007/s10533-018-0453-y

Lugo, A. E., 2019. Correction to: Characterization of ten extreme disturbance events in the context of social and ecological systems. *Biogeochemistry* 144: 117–118. https://doi.org/10.1007/s10533-019-00563-8

Lugo, A. E., Carlo, T. A., and Wunderle, J. M., Jr., 2012. Natural mixing of species: novel plant-animal communities on Caribbean Islands. *Animal Conservation* 15: 233–241. https://doi.org/10.1111/j.1469-1795.2012.00523.x

Lugo, A. E., López, T del M., and Ramos, O. M., 2004. Urbanización de los terrenos en la periferia de El Yunque. General Technical Report IITF-16. United States Department of Agriculture Forest Service, International Institute of Tropical Forestry, Río Piedras, Puerto Rico, USA.

Lugo, A. E., Miranda-Castro, L., Vale, A., del M. López, T., Hernández-Prieto, E., García-Martinó, A., Puente-Rolón, A. R., Tossas, A. G., McFarlane, D. A, Miller, T., Rodríguez, A., Lundberg, J., Thomlinson, J., Colón, J., Schellekens, J. H., Ramos, O., and Helmer, E., 2001. Puerto Rican Karst: A Vital Resource. US Forest Service General Technical Report WO-65, Río Piedras, Puerto Rico, USA. https://doi.org/10.2737/WO-GTR-65

Lugo-López, M. A., 1953. *Moisture Relationships of Puerto Rican Soils*. Technical Paper 9, University of Puerto Rico, Agricultural Experiment Station, Río Piedras, Puerto Rico.

MacLean, W. P., Kellner, R., and Dennis, H., 1977. *Island Lists of West Indian Amphibians and Reptiles*. Smithsonian Herpetological Information Service No. 40. https://doi.org/10.5479/si.23317515.40.1

MacLean, W. P., 1982. *Reptiles and Amphibians of the Virgin Islands*. Macmillan Education Limited, London and Basingstoke, UK.

Maldonado-Capriles, J. and Medina-Gaud, S., 1977. The ticks in Puerto Rico (Arachnida: Acarina). *Journal of Agriculture of the University of Puerto Rico* 61: 402–404. https://doi.org/10.46429/jaupr.v61i3.10443

Marcano-Vega, H., 2017. Forests of Puerto Rico, 2014. Resource Update FS–121. United States Department of Agriculture Forest Service, Southern Research Station, Asheville, North Carolina, USA. https://doi.org/10.2737/FS-RU-121

Marcano-Vega, H., Brandeis, T. J., and Turner, J. A., 2015. *Los bosques de Puerto Rico*. Resource Bulletin SRS-202. United States Department of Agriculture Forest Service, Southern Research Station, Asheville, North Carolina, USA. https://doi.org/10.2737/SRS-RB-202ES

Marivaux, L., Vélez-Juarbe, J., Merzeraud, G., Pujos, F., Viñola López, L. W., Boivin, M., Santos-Mercado, H., Cruz, E. J., Grajales, A., Padilla, J., Vélez-Rosado, K. I., Philippon, M., Léticée, J.-L., Münch, P., and Antoine, P.-O., 2020. Early Oligocene chinchilloid caviomorphs from Puerto Rico and the initial rodent colonization of the West Indies. *Proceedings of the Royal Society B: Biological Sciences* 287: 20192806. https://doi.org/10.1098/rspb.2019.2806

Marivaux, L., Vélez-Juarbe, J., Viñola López, L. W., Fabre, P.-H., Pujos, F., Santos-Mercado, H., Cruz, E. J., Grajales Pérez, A. M., Padilla, J., Vélez-Rosado, K. I., Corné, J.-J., Philippon, M., Münch, P., and Antoine, P.-O., 2021. An unpredicted ancient colonization of the West Indies by North American rodents: dental evidence of a geomorph from the early Oligocene of

Puerto Rico. *Papers in Palaeontology*: 1–19. https://doi.org/10.1002/spp2.1388

Marr, S. R., Mautz, W. J., and Hara, A. H., 2008. Parasite loss and introduced species: a comparison of the parasites of the Puerto Rican tree frog (*Eleutherodactylus coqui*), in its native and introduced ranges. *Biological Invasions* 10: 1289–1298. https://doi.org/10.1007/s10530-007-9203-0

Marr, S. R., Johnson, S. A., Hara, A. H., and McGarrity, M. E., 2010. Preliminary evaluation of the potential of the helminth parasite *Rhabdias elegans* as a biological control agent for invasive Puerto Rican coquís (*Eleutherodactylus coqui*) in Hawaii. *Biological Control* 54: 69–74. https://doi.org/10.1016/j.biocontrol.2010.04.012

Marrero, J., 1950. Reforestation of degraded lands in Puerto Rico. *Caribbean Forester* 11: 3–15.

Martin, P. H., Sherman, R. H., and Fahey, T. J., 2004. Forty years of tropical forest recovery from agriculture: structure and floristic of secondary and old-growth riparian forests in the Dominican Republic. *Biotropica* 36: 297–317. https://doi.org/10.1111/j.1744-7429.2004.tb00322.x

Martinuzzi, S., Gould, W. A., Lugo, A. E., and Medina, E., 2009. Conversion and recovery of Puerto Rican mangroves: 200 years of change. *Forest Ecology and Management* 257: 75–84. https://doi.org/10.1016/j.foreco.2008.08.037

Matos-Torre, J. J., 2006. Habitat characterization for the Puerto Rican Crested Toad (*Peltophryne* [*Bufo*] *lemur*) at Guánica State Forest, Puerto Rico. MS thesis. University of Puerto Rico, Mayagüez Campus, Mayagüez, Puerto Rico, USA.

Mayer, G. C. and Lazell, J. D., Jr., 1988. Distributional records for reptiles and amphibians from the Puerto Rican Bank. *Herpetological Review* 19: 23–24.

Mayer, G. C., 2012. Puerto Rico and the Virgin Islands. Pp. 136–147 in *Island Lists of West Indian Amphibians and Reptiles*, edited by R. Powell and R. W. Henderson. *Bulletin of the Florida Museum of Natural History* 51: 85–166.

McClelland, T. B., 1932. Introduction of *Leptodactylus fallax* in Puerto Rico. *The Journal of the Department of Agriculture of Puerto Rico* 16: 419. https://doi.org/10.46429/jaupr.v16i4.15060

McDowell, W. H., Scatena, F. N., Waide, R. B., Brokaw, N., Camilo, G. R., Covich, A. P., Crowl, T. A., González, G., Greathouse, E. A., Klawinski, P., Lodge, D. J., Lugo, A. E., Pringle, C. M., Richardson, B. A., Richardson, M. J., Schaefer, D. A., Silver, W. L., Thompson, J., Vogt, D. J., Vogt, K. A., Willig, M. R., Woolbright, L. L., Zou, X., and Zimmerman, J. K., 2012. Geographic and Ecological Setting of the Luquillo Mountains. Pp. 72–163 in *A Caribbean Forest Tapestry: The Multidimensional Nature of Disturbance and Response*, edited by N. Brokaw, T. A. Crowl, A. E. Lugo, W. H. McDowell, F. N. Scatena, R. B. Waide, and M. R. Willig. Oxford University Press, New York, USA. https://doi.org/10.1093/acprof:osobl/9780195334692.003.0003

Meerwarth, H., 1901. Die Westindischen Reptilien und Batrachier des Naturhistorischen Museums in Hamburg. *Mitteilungen aus dem Naturhistorischen Museum in Hamburg* 18: 1–43.

Meléndez-Ackerman, E. J., Cortés, C., Sustache, J., Aragón, S., Morales-Várgas, M., García-Bermúdez, M., and Fernández, D. S., 2008. Diet of feral goats in Mona Island Reserve, Puerto Rico. *Caribbean Journal of Science* 44: 199–205. https://doi.org/10.18475/cjos.v44i2.a7

Mendelson, J. R., III, 2018. Frogs in glass boxes: responses of zoos to global amphibian extinctions. Pp. 298–310 in *The Ark and Beyond: The Evolution of Zoo and Aquarium Conservation*, edited by B. A. Minteer, J. Maienschein, and J. P. Collins. The University of Chicago Press, Chicago, Illinois, USA. https://doi.org/10.7208/chicago/9780226538631.003.0025

Mercado-Burgos, N., 1966. The influence of external factors on the spermatogenic cycle of *Bufo marinus*. MS thesis. University of Puerto Rico, Río Piedras Campus, San Juan, Puerto Rico, USA.

Meshaka, W. E., Jr., 2011. A Runaway Train in the Making: The Exotic Amphibians, Reptiles, Turtles, and Crocodilians of Florida. Monograph 1. *Herpetological Conservation and Biology* 6: 1–101.

Meyerhoff, H. A., 1927. Tertiary physiographic development of Porto Rico and the Virgin Islands. *Bulletin of the Geological Society of America* 38: 557–576. https://doi.org/10.1130/GSAB-38-557

Michael, S. F., 1995. Captive breeding of two species of *Eleutherodactylus* (Anura: Leptodactylidae) from Puerto Rico, with notes on behavior in captivity. *Herpetological Review* 26: 27–29.

Michael, S. F., 1997a. Captive breeding of *Eleutherodactylus antillensis* (Anura: Leptodactylidae) from Puerto Rico with notes on behavior in captivity. *Herpetological Review* 28: 141–143.

Michael, S .F., 1997b. Vocalization and diurnal retreat defense in the Puerto Rican frog *Eleutherodactylus cochranae*. *Journal of Herpetology* 31: 453–456. https://doi.org/10.2307/1565681

Miller, G. S., Jr., 1918. Mammals and reptiles collected by Theodore de Booy in the Virgin Islands. *Proceedings of the United States National Museum* 54: 507–511, with one plate. https://doi.org/10.5479/si.00963801.54-2244.507

Miller, P. G., 1922. *Historia de Puerto Rico*. Rand McNally & Company, Chicago, Ill.

Miller, D. A. W., Talley, B. L., Lips, K. R., and Grant, E. H. C., 2012. Estimating patterns and drivers of infection prevalence and intensity when detection is imperfect and sampling error occurs. *Methods in Ecology and Evolution* 3: 850–859. https://doi.org/10.1111/j.2041-210X.2012.00216.x

Mississippi State University, 2022. https://www.msstate.edu/newsroom/article/2019/12/one-kind-toad-born-through-msu-pioneering-technology-thats-saving. Accessed on 11 January 2022.

Molina-Colón, S. and Lugo, A. E., 2006. Recovery of a subtropical dry forest after abandonment of different land uses. *Biotropica* 38: 354–364. https://doi.org/10.1111/j.1744-7429.2006.00159.x

Monroe, K. D., Collazo, J. A., Pacifici, K., Reich, B. J., Puente-Rolón, A. R., and Terando, A. J., 2017. Occupancy and indes of abundance of *Eleutherodactylus wightmanae* and *E. brittoni* along elevations gradients in West-Central Puerto Rico. *Caribbean Naturalist* 40: 1–18.

Morales, A. L., 2011a. *Osteopilus septentrionalis* (Cuban Treefrog). Distribution. *Caribbean Herpetology* 20: 1. https://doi.org/10.31611/ch.20

Morales, A. L., 2011b. *Scinax ruber* (Red-snouted Treefrog). Distribution. *Caribbean Herpetology* 21: 1. https://doi.org/10.31611/ch.21

Morales-Pérez, A. L., Rodríguez-Gómez, C. A., Monzón-Carmona, O., Santiago-Breván, L., and Montalvo-Serrano, D., 2022. *Eleutherodactylus juanariveroi* (Plains Coqui, Coquí Llanero). *Herpetological Review* 53(3): 438–439.

Moravec, F. and Kaiser, H., 1995. Helminth parasites from West Indian frogs, with descriptions of two new species. *Caribbean Journal of Science* 31: 252–268.

Moreno, J. A., 1991a. Accounts of those species considered to be of concern. Pp. 9–10 in *Status y Distribución de los Reptiles y Anfibios de la Región de Puerto Rico*, edited by J. A. Moreno. Departamento de Recursos Naturales, Publicación Científica Miscelánea No. 1, San Juan, Puerto Rico, USA.

Moreno, J. A., 1991b. Appendix II. Status recommendations/recomendaciones de designación. Pp. 65–67 in *Status y Distribución de los Reptiles y Anfibios de la Región de Puerto Rico*, edited by J. A. Moreno. Departamento de Recursos Naturales, Publicación Científica Miscelánea No. 1, San Juan, Puerto Rico, USA.

Moreno, J. A. and Canals, M., 1985. Reproductive ecology of *Peltophryne lemur* in Guánica. Unpublished Report to the Puerto Rico Department of Natural Resources, San Juan, Puerto Rico, USA.

Morgan, M., Daley, B., Hilgemann, L., and Zimmerman, T. W., 2017. A reforestation profile of the U.S. Virgin Islands. *Tree Planters' Notes* 60: 4–18.

Morrison, C. and Hero, J., 2003. Geographic variation in life-history characteristics of amphibians: a review. *Journal of Animal Ecology* 72: 270–279. https://doi.org/10.1046/j.1365-2656.2003.00696.x

Mote, T. L., Ramseyer, C. A., and Miller, P. W., 2017. The Saharan Air Layer as an early rainfall season suppressant in the eastern Caribbean: The 2015 Puerto Rico drought. *Journal of Geophysical Research: Atmospheres* 122. https://doi.org/10.1002/2017JD026911

Muñiz-Castro, M. A., Williams-Linera, G., and Martínez-Ramos, M., 2012. Dispersal mode, shade tolerance, and phytogeographical affinity of tree species during secondary succession in tropical montane cloud forest. *Plant Ecology* 213: 339–352. https://doi.org/10.1007/s11258-011-9980-5

Murphy, L. S., 1916. *Forests of Porto Rico, past present, and future, and their physical and economic development.* United States Department of Agriculture Forest Service, Washington, District of Columbia, USA, Bulletin No. 354. https://doi.org/10.5962/bhl.title.65087

Murphy, S. F., Stallard, R. F., Scholl, M. A., González, G., and Torres-Sánchez, A. J., 2017. Reassessing rainfall in the Luquillo Mountains, Puerto Rico: Local and global ecohydrological implications. *PLoS ONE* 12: e0180987. https://doi.org/10.1371/journal.pone.0180987

Musser, G., Hutterer, R., Kryštufek, B., Yigit, N., and Mitsain, G., 2016. *Mus musculus*. (errata version published in 2017) The IUCN Red List of Threatened Species 2016: e.T13972A115117618. http://dx.doi.org/10.2305/IUCN.UK.2016-3.RLTS.T13972A22405706.en. Accessed on 24 August 2017.

Mutnale, M. C., Anand, S., Eluvathingal, L. M., Roy, J. K., Reddy, G. S., and Vasudevan, K., 2018. Enzootic frog pathogen *Batrachochytrium dendrobatidis* in Asian tropics reveals high ITS haplotype diversity and low prevalence. *Scientific Reports* 8: 10125. https://doi.org/10.1038/s41598-018-28304-1

Narins, P. M. and Smith, S. L., 1986. Clinal variation in anuran advertisement calls: basis for acoustic isolation? *Behavioral Ecology and Sociobiology* 19: 135–141. https://doi.org/10.1007/BF00299948

Narins, P. M. and Meenderink, S. W., 2014. Climate change and frog calls: long-term correlations along a tropical altitudinal gradient. *Proceedings of the Royal Society B* 281: 20140401. https://doi.org/10.1098/rspb.2014.0401

National Association of REALTORS®, 2022. https://www.realtor.com/realestateandhomes-detail/191-Bo-Rio-Blanco-Km-97_Naguabo_PR_00718_M97303-48533. Accessed on 11 January 2022.

National Weather Service, 2022a. https://www.weather.gov/ilm/HurricaneHugo. Accessed on 11 January 2022.

National Weather Service, 2022b. https://www.weather.gov/mob/georges. Accessed on 11 January 2022.

National Weather Service, 2022c. https://www.weather.gov/sju/maria2017. Accessed on 11 January 2022.

Navas, C. A., Gomes, F. R., and Domenico, E. A., 2016. Physiological ecology and conservation of anuran amphibians. Pp. 155–188 in *Amphibian and Reptiles Adaptations to the Environment: Interplay Between Physiology and Behavior,* edited by D. V. de Andrade, C. R. Bevier, and J. E. de Carvalho. CRC Press, Taylor & Francis Group, Boca Raton, Florida, USA. https://doi.org/10.1201/b20420-8

Neal, J. W., Lilyestrom, C. G., and Kwak, T. J., 2009. Factors influencing tropical island freshwater fishes: species, status, and management implications in Puerto Rico. *Fisheries* 34: 546–554. https://doi.org/10.1577/1548-8446-34.11.546

Nellis, D. W. and Everard, C. O. R., 1983. The Biology of the Mongoose in the Caribbean. Pp. 1–162 in *Studies on the Fauna of Curaçao and other Caribbean Islands,* No. 195, edited by P. Wagenaar Hummelinck and L. J. van der Steen. Foundation for Scientific Research in Surinam and the Netherlands Antilles, Utrecht, The Netherlands.

NOAA (National Oceanic and Atmospheric Administration), 2021. https://www.ncei.noaa.gov/access/services/data/v1?dataset=normals-monthly-1991-2020&startDate=0001-01-01&endDate=9996-12-31&stations=RQC00666258&format=pdf. Archived from the original on 2021-07-13. Accesed on 04 January 2022.

NotiCel, 2014. El reto del DRNA ante el tráfico ilegal de las especies exóticas. An online newspaper report by Laura M. Quintero, from 13 April 2014. Available at http://www.noticel.com/noticia/158572/el-reto-del-drna-ante-el-trafico-ilegal-de-las-especies-exoticas.html

Noss, R. F., 1996. The naturalists are dying off. *Conservation Biology* 10: 1–3. https://doi.org/10.1046/j.1523-1739.1996.10010001.x

Numazawa, M. and Sengoku, S., 2011. Observations on the reproduction in captivity of the Mona Coqui *Eleutherodactylus monensis* (Anura: Leptodactylidae) from Puerto Rico. *Current Herpetology* 30: 181–186. https://doi.org/10.5358/hsj.30.181

Nytch, C. J., Hunter, W. C., Núñez-García, F., Fury, C., and Quiñones, M., 2015. *Avian conservation planning priorities for Puerto Rico and the U.S. Virgin Islands (BCR 69).* Report submitted to the US Fish and Wildlife Service in compliance to contract number F12PX02389.

Odum, H. T., Drewry, G., and Kline, J. R., 1970. Climate at El Verde, 1963–1966. Pp. B347–B418 in *A Tropical Rain Forest: A Study of Irradiation and Ecology at El Verde, Puerto Rico,* edited by H. T. Odum and R. F. Pigeon. Division of Technical Information, US Atomic Energy Commission, Oak Ridge, Tennessee, USA.

O'Hanlon S. J., Rieux, A., Farrer, R. A., Rosa, G. M., Waldman, B., Bataille, A., Kosch, T. A., Murray, K. A., Brankovics, B., Fumagalli, M., Martin, M. D., Wales, N., Alvarado-Rybak, M., Bates, K. A., Berger, L., Böll, S., Brookes, L., Clare, F., Courtois E. A., Cunningham, A. A., Doherty-Bone, T. M., Ghosh, P., Gower, D. J., Hintz, W. E., Höglund, J., Jenkinson, T. S., Lin, C. F., Laurila, A., Loyau, A., Martel, A., Meurling, S., Miaud, C., Minting, P., Pasmans, F., Schmeller, D. S., Schmidt, B. R., Shelton, J. M., Skerratt, L. F., Smith, F., Soto-Azat, C., Spagnoletti, M., Tessa, G., Toledo, L. F., Valenzuela-Sánchez, A., Verster, R., Vörös, J., Webb, R. J., Wierzbicki, C., Wombwell, E., Zamudio, K. R., Aanensen, D. M., James, T. Y., Gilbert, M. T. P., Weldon, C., Bosch, J., Balloux, F., Garner, T. W. J., and Fisher, M. C., 2018. Recent Asian origin of chytrid fungi causing global amphibian declines. *Science* 360: 621–627. https://doi.org/10.1126/science.aar1965

Ospina, O. E., Villanueva-Rivera, L. J., Corrada-Bravo, C. J., and Aide, T. M., 2013. Variable response of anuran calling activity to daily precipitation and temperature: implications for climate change. *Ecosphere* 4(4): 1–12. https://doi.org/10.1890/ES12-00258.1

Ovaska, K., 2005. The frog (*Eleutherodactylus antillensis*). Pp. 175–180 in *Island: Fact and Theory in Nature,* edited by J. Lazell. University of California Press, Berkeley and Los Angeles, California, USA.

Ovaska, K. E. and Caldbeck, J., 1997a. Courtship behavior and vocalizations of the frogs *Eleutherodactylus antillensis* and *E. cochranae* on the British Virgin Islands. *Journal of Herpetology* 31: 149–155. https://doi.org/10.2307/1565348

Ovaska, K. E. and Caldbeck, J., 1997b. Vocal behavior of the frog *Eleutherodactylus antillensis* on the British Virgin Islands. *Animal Behaviour* 54: 181–188. https://doi.org/10.1006/anbe.1996.0414

Ovaska, K. E. and Caldbeck, J., 1999. Courtship call of the frog *Eleutherodactylus schwartzi* from the British Virgin Islands. *Journal of Herpetology* 33: 501–504. https://doi.org/10.2307/1565653

Ovaska, K., Caldbeck, J., and Lazell, J., 1998. *Eleutherodactylus schwartzi* (NCN). *Herpetological Review* 29: 97.

Ovaska, K. E., Caldbeck, J., and Lazell, J., Jr., 2000. Ecological notes of leptodactylid frogs, *Leptodactylus* and *Eleutherodactylus,* from the British Virgin Islands. *Breviora* 508: 1–25.

Owen, J. L., 2005. The Cuban Tree Frog (*Osteopilus septentrionalis*): Distribution, Diet, and Reproduction of an Invasive Species in the British Virgin Islands. MS thesis, Department of Range, Wildlife and Fisheries Management, Texas Tech University, Texas, USA.

Owen, J., Perry, G., Lazell, J., Petrovic, C., and Egelhoff, J., 2005. *Osteopilus septentrionalis* (Cuban Tree Frog). *Herpetological Review* 36: 76.

Owen, J., Perry, G., Lazell, J., Petrovic, C., and Egelhoff, J., 2006. *Osteopilus septentrionalis* (Cuban Tree Frog). *Herpetological Review* 37: 74–75.

Pacheco, C. G. and Barber, D. M., 2013. *Peltophryne lemur* (Puerto Rican Crested Toad). *Herpetological Review* 44: 124–126.

Partridge, L. and Coyne, J. A., 1997. Bergmann's rule in ectotherms: is it adaptive? *Evolution* 51: 632–635. https://doi.org/10.1111/j.1558-5646.1997.tb02454.x

Pasch, R. J., Penny, A. B., and Berg, R., 2018. Hurricane María, 16–30 September 2017. National Hurricane Center Tropical Cyclone Report, number AL152017, dated 10 April 2018. National Hurricane Center, Miami, Florida, USA.

Pasachnik, S. A., Carreras De León, C., and León, Y. M., 2015. Protected only On paper? three case studies from protected areas in the Dominican Republic. *Caribbean Naturalist* 2016(30): 1–19.

Pascarella, J. B., Aide, T. M., Serrano, M. I., and Zimmerman, J. K., 2000. Land-use history and forest regeneration in the Cayey Mountains, Puerto Rico. *Ecosystems* 3: 217–228. https://doi.org/10.1007/s100210000021

Pavan, A. C. and Marroig, G., 2016. Integrating multiple evidences in taxonomy: species diversity and phylogeny of mustached bats (Mormoopidae: *Pteronotus*). *Molecular Phylogenetics and Evolution* 103: 184–198. https://doi.org/10.1016/j.ympev.2016.07.011

Pechmann, J. H. K., Scott, D. E., Semlitsch, R. D., Caldwell, J. P., Vitt, L. J., and Gibbons, W., 1991. Declining amphibians populations: the problem of separating human impacts from natural fluctuations. *Science* 253: 892–895. https://doi.org/10.1126/science.253.5022.892

Pérez, M. E., 1951a. The food of *Rana catesbeiana* Shaw in Puerto Rico. *The Journal of Agriculture of the University of Puerto Rico* 35: 136–142. https://doi.org/10.46429/jaupr.v35i4.12805

Pérez, M. E., 1951b. The food of *Rana catesbeiana* Show in Puerto Rico. *Herpetologica* 7: 102–104.

Pérez-Buitrago, N., Sabat, A. M., and McMillan, W. O., 2010. Spatial ecology of the endangered Mona Island iguana *Cyclura cornuta stejnegeri*: does territorial behavior regulate density? *Herpetological Monographs* 24: 86–110. https://doi.org/10.1655/HERPMONOGRAPHS-D-09-039.1

Perry, G., 2009a. First record of the frog *Eleutherodactylus lentus* in the British Virgin Islands: Conservation implications of native or introduced status. *Applied Herpetology* 6: 185–187. https://doi.org/10.1163/157075409X427171

Perry, G., 2009b. First record of *Osteopilus septentrionalis* on Guana Island, British Virgin Islands. *Applied Herpetology* 6: 191–192. https://doi.org/10.1163/157075408X397518

Perry, G. and Gerber, G. P., 2006 (reprinted 2011, with addendum). Conservation of amphibians and reptiles in the British Virgin Islands: Status and patterns. *Applied Herpetology* 3: 237–256. https://doi.org/10.1163/157075406778116203

Perry, G. and Platenberg, R., 2007. Recent additions to the herpetofauna of Little St. James, US Virgin Islands. *Applied Herpetology* 4: 387–389. https://doi.org/10.1163/157075407782424593

Perry, G., Powell, R., and Watson, H., 2006. Keeping invasive species off Guana Island, British Virgin Islands. *Iguana* 13: 273–277.

Peters, R. H. and Wassenberg, K., 1983. The effect of body size on animal abundance. *Oecologia* 60: 89–96. https://doi.org/10.1007/BF00379325

Peters, W. C. H., 1876. Hr. W. Peters las über eine von Hrn. Viceconsul L. Krug und Dr. J. Gundlach auf der Insel Puertorico gemachte Sammlung von Säugethieren und Amphibien, so wie über die Entwickelung eines Batrachiers, *Hylodes martinicensis* Dum. Bibr., ohne Metamorphose. *Monatsberichte der Königlichen Preussische Akademie de Wissenschaften zu Berlin*, vol 13. November 1876: 703–714 + 1 Plate, 914.

Peterson, N. R., Lutz-Carrillo, D., and Neal, J. W., 2017. Genetic composition of largemouth bass in Puerto Rico reservoirs. *North American Journal of Fisheries Management* 37: 1199–1207. https://doi.org/10.1080/02755947.2017.1353560

Phalen, D., Slapeta, J., King, J., and Rose, K., 2011. Development and validation of a rapid field test to detect the chytrid fungus *Batrachochytrium dendrobatidis* at a high specificity and sensitivity, for use in surveys to determine the distribution of Chytridiomycosis. Australian Registry of Wildlife Health, Taronga Conservation Society Australia, Mosman New South Wales. A progress report submitted for the Australian Government Department of Sustainability, Environment, Water, Population and Communities, downloadable at http://www.environment.gov.au/biodiversity/invasive-species/publications/rapid-field-test-detect-chytrid-fungus-batrachochytrium

Philibosian, R. and Ruibal, R., 1971. Conservation of the lizards *Ameiva polops* in the Virgin Islands. *Herpetologica* 27: 450–454.

Philibosian, R. and Yntema, J. A., 1976. Records and status of some reptiles and amphibians in the Virgin Islands, I. 1968–1975. *Herpetologica* 32: 81–85.

Philibosian, R. and Yntema, J. A., 1977. *Annotated checklist of the birds, mammals, reptiles, and amphibians of the Virgin Islands and Puerto Rico.* Published privately by Information Services, Frederiksted, St. Croix, Virgin Islands, USA.

Philibosian, R. and Yntema, J. A., 1978. Records and status of some reptiles and amphibians in the Virgin Islands, II. 1975–1976. *Herpetologica* 34: 47–51.

Picó, R., 1950. *The Geographic Regions of Puerto Rico.* University of Puerto Rico Press, Río Piedras, Puerto Rico, USA.

Pigliucci, M., 1996. How organisms respond to environmental changes: from phenotypes to molecules (and vice versa). *Trends in Ecology and Evolution* 11: 168–173. https://doi.org/10.1016/0169-5347(96)10008-2

Pimenta, B. V. S., Haddad, C. F. B., Nascimento, L. B., Gonçalves Cruz, C. A., and Pombal, J. P., Jr., 2005. Comment on "status and trends of amphibian declines and extinctions worldwide." *Science* 309: 1999b. https://doi.org/10.1126/science.1112996

Pimentel, D., 1955. Biology of the Indian mongoose in Puerto Rico. *Journal of Mammalogy* 36: 62–68. https://doi.org/10.2307/1375723

Platenberg, R. J., 2006. Wetland Conservation Plan for St. Thomas and St. John, U.S. Virgin Islands. US Virgin Islands Division of Fish and Wildlife, Department of Planning and Natural Resources, St Thomas, Virgin Islands, USA.

Platenberg, R. J., 2007. Impacts of introduced species on an island ecosystem: non-native reptiles and amphibians in the US Virgin Islands. Pp. 168–174 in *Managing Vertebrate Invasive Species: Proceedings of an International Symposium*, edited by G. W. Witmer, W. C. Pitt, and K. A. Fagerstone. USDA/APHIS/WS, National Wildlife Research Center, Fort Collins, Colorado, USA.

Platenberg, R. J., 2012. Species and Habitats of Concern. Annual Report, State Wildlife Grant VI-T-9-R-2. Division of Fish and Wildlife. St. Thomas, Virgin Islands, USA.

Platenberg, R. J. and Boulon, R. H., Jr., 2006 (reprinted 2011). Conservation status of reptiles and amphibians in the U.S. Virgin Islands. *Applied Herpetology* 3: 215–235. https://doi.org/10.1163/157075406778116159

Poinar, G., Jr., and Wake, D. B., 2015. *Palaeoplethodon hispaniolae* gen. n., sp. n. (Amphibia: Caudata), a fossil salamander from the Caribbean. *Palaeodiversity* 8: 21–29.

Porto Rico 1930 Georeferenciado: un mosaico de la costa, 2022. http://prgeoref.weebly.com/. Accessed on 10 January 2022.

Pough, F. H., Taigen, T. L., Stewart, M. M., and Brussard, P. F., 1983. Behavioral modification of evaporative water loss by a Puerto Rican frog. *Ecology* 64: 244–252. https://doi.org/10.2307/1937072

Pounds, J. A., 2001. Climate and amphibian declines. *Nature* 410: 639–640. https://doi.org/10.1038/35070683

Pounds, J. A. and Crump, M. L., 1994. Amphibian declines and climate disturbance: the case of the golden toad and the harlequin frog. *Conservation Biology* 8: 72–85. https://doi.org/10.1046/j.1523-1739.1994.08010072.x

Pounds, J. A., Bustamante, M. R., Coloma, L. A., Consuegra, J. A., Fogden, M. P. L., Foster, P. N., La Marca, E., Masters, K. L., Merino-Viteri, A., Puschendorf, R., Ron, S. R., Sánchez-Azofeifa, G. A., Still, C. J., and Young, B. E., 2006. Widespread amphibian extinctions from epidemic disease driven by global warming. *Nature* 439: 161–167. https://doi.org/10.1038/nature04246

Powell, R. and Henderson, R. W., 1996. A brief history of West Indian herpetology. Pp. 29–50 in *Contributions to West Indian Herpetology: A Tribute to Albert Schwartz*, edited by R. Powell and R. W. Henderson. Contributions to Herpetology 12, Society for the Study of Amphibians and Reptiles, Ithaca, New York, USA.

Powell, R. and Henderson, R. W., 1999. Addenda to the checklist of West Indian amphibians and reptiles. *Herpetological Review* 30: 137–139.

Powell, R. and Henderson, R. W., 2003. A second set of addenda to the checklist of West Indian amphibians and reptiles. *Herpetological Review* 34: 341–345.

Powell, R. and Henderson, R. W., 2012. Introduction. Pp. 86–90 in *Island Lists of West Indian Amphibians and Reptiles*, edited by R. Powell and R. W. Henderson. *Bulletin of the Florida Museum of Natural History* 51: 85–166.

Powell, R., Henderson, R. W., Adler, K., and Dundee, H. A., 1996. An annotated checklist of West Indian amphibians and reptiles. Pp. 51–93 in *Contributions to West Indian Herpetology: A Tribute to Albert Schwartz*, edited by R. Powell and R. W. Henderson. Contributions to Herpetology 12, Society for the Study of Amphibians and Reptiles, Ithaca, New York, USA.

Powell, R., Henderson, R. W., Farmer, M. C., Breuil, M., Echternacht, A. C., van Buurt, G., Romagosa, C. M., and Perry, G., 2011. Introduced amphibians and reptiles in the greater Caribbean: Patterns and conservation implications. Pp. 63–143 in *Conservation of Caribbean Island Herpetofaunas, Volume 1: Conservation Biology and the Wider Caribbean*, edited by A. Hailey, B. S. Wilson, and J. A. Horrocks. Koninklijke Brill NV, Leiden, The Netherlands. https://doi.org/10.1163/ej.9789004183957.i-228.38

Pramuk, J. B., 2000. Prenasal bones and snout morphology in West Indian bufonids and the *Bufo granulosus* species group. *Journal of Herpetology* 34: 334–340. https://doi.org/10.2307/1565439

Pramuk, J. B., 2006. Phylogeny of South American *Bufo* (Anura: Bufonidae) inferred from combined evidence. *Zoological Journal of the Linnean Society* 146: 407–452. https://doi.org/10.1111/j.1096-3642.2006.00212.x

Pramuk, J. B., Hass, C. A., and Hedges, S. B., 2001. Molecular phylogeny and biogeography of West Indian toads (Anura: Bufonidae). *Molecular Phylogenetics and Evolution* 20: 294–301. https://doi.org/10.1006/mpev.2001.0985

PRCCC (Puerto Rico Climate Change Council) Working Group 1, 2013. Geophysical and Chemical Scientific Knowledge. Pp. 21–84 in *Puerto Rico's State of the Climate 2010-2013: Assessing Puerto Rico's Social-Ecological Vulnerabilities in a Changing Climate*, edited by K. R. Jacobs, A. Terando, and E. Díaz. Puerto Rico Coastal Zone Management Program, Department of Natural and Environmental Resources, NOAA Office of Ocean and Coastal Resource Management. San Juan, Puerto Rico, USA.

PRDNER (Puerto Rico Department of Natural and Environmental Resources), 2015. Puerto Rico state wildlife action plan: ten year review. Department of Natural and Environmental Resources, San Juan, Puerto Rico, USA.

Pregill, G., 1981. Late Pleistocene herpetofaunas from Puerto Rico. *University of Kansas Museum of Natural History Miscellaneous Publication No.* 71: 1–72. https://doi.org/10.5962/bhl.title.63988

Pregill, G. K. and Olson, S. L., 1981. Zoogeography of West Indian vertebrates in relation to Pleistocene climatic cycles. *Annual Review of Ecology and Systematics* 12: 75–98. https://doi.org/10.1146/annurev.es.12.110181.000451

Primera Hora, 2018. https://www.primerahora.com/noticias/ciencia-ambiente/nota/criticandecisionderossellodedesarticular7reservasnaturales-1293820/. Accessed on 11 January 2022.

Protected Areas Conservation Action Team, 2018. Puerto Rico Protected Areas Database [version of December, 2018]. GIS data. San Juan, Puerto Rico, USA.

Public Law 114–187, 30 June 2016, 130 Stat. 549. Puerto Rico Oversight, Management, and Economic Stability Act. Available at https://www.congress.gov/114/bills/hr5278/BILLS-114hr5278rfs.pdf

Puerto Rico Climate Change Council (PRCCC) Working Group 1, 2013. Geophysical and Chemical Scientific Knowledge. Pp. 21–84 in *Puerto Rico's State of the Climate 2010-2013: Assessing Puerto Rico's Social-Ecological Vulnerabilities in a Changing Climate*, ed by K. R. Jacobs, A. Terando, and E. Díaz. Puerto Rico Coastal Zone Management Program, Department of Natural and Environmental Resources, NOAA Office of Ocean and Coastal Resource Management. San Juan, Puerto Rico, USA.

Puerto Rico Experiment Station, 1937. Report of the Puerto Rico Experiment Station, 1936. Puerto Rico Experiment Station of the United States Department of Agriculture, Mayagüez, Puerto Rico. Office of Experiment Stations, United States Department of Agriculture, Washington, DC, USA.

Punzo, F. and Lindstrom, L., 2001. The toxicity of eggs of the Giant Toad, *Bufo marinus* to aquatic predators in a Florida retention pond. *Journal of Herpetology* 35: 693–697. https://doi.org/10.2307/1565916

Purdom, T., Schoofs, A., Hodgson, J., Maki, T., and Hodgson, J., 2015. A thirty year diet record of largemouth bass (*Micropterus salmoides*) from a small north temperate lake. *Bios* 86: 20–30. https://doi.org/10.1893/0005-3155-86.1.20

Pyle, R. M., 2001. The rise and fall of natural history. *Orion Autumn* 2001: 16–23.

Raffaele, H. A., 1990. *Una Guía a las Aves de Puerto Rico y las Islas Vírgenes*, Edición Revisada. Princeton University Press, Princeton, New Jersey, USA.

Reagan, D. P., 1991. The response of Anolis lizards to hurricane-induced habitat changes in a Puerto Rican rain forest. *Biotropica* 23: 468–474. https://doi.org/10.2307/2388268

Reagan, D. P., 1996. Anoline Lizards. Pp. 321–345 in *The Food Web of a Tropical Rain Forest*, edited by D. P. Reagan and R. B. Wade. University of Chicago Press, Chicago, USA.

Reilly, A. E., 1991. The effects of hurricane Hugo in three tropical forests in the U.S. Virgin Islands. *Biotropica* 23: 414–419. https://doi.org/10.2307/2388260

Reinhardt, J. and Lütken, C. F., 1863 (1862). Bidrag til det vestindiske Öriges og navnligen til de dansk-vestindiske Öers Herpetologie. *Videnskabelige Meddelelser fra den Naturhistoriskl Forening i Kjöbenhavn* 10–18: 153–291.

Rice, K. G., Waddle, J. H., Crockett, M. E., Carthy, R. R., and Percival, H. F., 2005. Herpetofaunal Inventories of the National Parks of South Florida and the Caribbean: Volume II. Virgin Islands National Park. Open-File Report 2005-1301. US Geological Survey, Fort Lauderdale, Florida, USA. https://doi.org/10.3133/ofr20051301

Riley, K., Berry, O. F., and Roberts, J. D., 2013. Do global models predicting environmental suitability for the amphibian fungus, *Batrachochytrium dendrobatidis*, have local value to conservation managers? *Journal of Applied Ecology* 50: 713–720. https://doi.org/10.1111/1365-2664.12091

Riollano, A., 1976. Development of a frog culture industry in Puerto Rico. Mimeographed prospect submitted to the Agriculture Experiment Station of the University of Puerto Rico, USA.

Ríos-Franceschi, A., Medina, S., and Joglar, R. L., 2020. *Eleutherodactylus antillensis* (Puerto Rican Red-eyed Frog). *Herpetological Review* 51: 817–818.

Ríos-Franceschi, A., García-Cancel, J. G., Bird-Picó, F. J., and Carrasquillo, L. D., 2016. Spatiotemporal changes of the herpetofaunal community in Mount Resaca and Luis Peña Cay, Culebra National Wildlife Refuge, Culebra, Puerto Rico. *Life: The Excitement of Biology* 3: 254–289.

Ríos-López, N., 1999 (reprinted 2000). *Scinax rubra*. *Herpetological Review* 30: 232.

Ríos-López, N., 2007. The structuring of herpetofaunal assemblages in human-altered coastal ecosystems. PhD Dissertation. University of Puerto Rico, Río Piedras Campus, San Juan, Puerto Rico, USA.

Ríos-López, N., 2008a. Effects of increased salinity on tadpoles of two anurans from a Caribbean coastal wetland in relation to their natural abundance. *Amphibia-Reptilia* 29: 7–18. https://doi.org/10.1163/156853808783431451

Ríos-López, N., 2008b. *Eleutherodactylus brittoni*. The IUCN Red List of Threatened Species 2008: e.T56477A11470157. http://dx.doi.org/10.2305/IUCN.UK.2008.RLTS.T56477A11470157.en. Accessed on 25 February 2017.

Ríos-López, N., 2008c. *Eleutherodactylus cooki*. The IUCN Red List of Threatened Species 2008: e.T7149A12831612. http://dx.doi.org/10.2305/IUCN.UK.2008.RLTS.T7149A12831612.en. Accessed on 26 February 2017.

Ríos-López, N., 2008d. *Eleutherodactylus juanariveroi*. The IUCN Red List of Threatened Species 2008: e.T135828A4207317. http://dx.doi.org/10.2305/IUCN.UK.2008.RLTS.T135828A4207317.en. Accessed on 26 February 2017.

Ríos-López, N., 2008e. *Eleutherodactylus unicolor*. The IUCN Red List of Threatened Species 2008: e.T57023A11569269. http://dx.doi.org/10.2305/IUCN.UK.2008.RLTS.T57023A11569269.en. Accessed on 26 February 2017.

Ríos-López, N., 2010a. *Eleutherodactylus cochranae*. The IUCN Red List of Threatened Species 2010: e.T56515A11488661. http://dx.doi.org/10.2305/IUCN.UK.2010-2.RLTS.T56515A11488661.en. Accessed on 26 February 2017.

Ríos-Lopez, N., 2010b. *Leptodactylus albilabris*. The IUCN Red List of Threatened Species 2010: e.T57113A11581929. http://dx.doi.org/10.2305/IUCN.UK.2010-2.RLTS.T57113A11581929.en. Accessed on 26 February 2017.

Ríos-López, N., 2021. The Cuban treefrog (Amphibia: Anura: Hylidae: *Osteopilus septentrionalis*), an exotic anurophagous on the loose: a case study and review. *Acta Científica: Una revista transdisciplinaria de Puerto Rico y el Caribe* 32(2): 33–51.

Ríos-López, N. and Joglar, R. L., 1999 (reprinted 2000). *Rana grylio* (Pig Frog). *Herpetological Review* 30: 231–232.

Ríos-López, N. and Aide, T. M., 2007. Herpetofaunal dynamics during secondary succession. *Herpetologica* 63: 35–50. https://doi.org/10.1655/0018-0831(2007)63[35:HDDSS]2.0.CO;2

Ríos-López, N. and Thomas, R., 2007. A new species of palustrine *Eleutherodactylus* (Anura: Leptodactylidae) from Puerto Rico. *Zootaxa* 1512: 51–64. https://doi.org/10.11646/zootaxa.1512.1.3

Ríos-López, N. and Villanueva-Rivera, L. J., 2013. Acoustic characteristics of a native anuran (Amphibia) assemblage in a palustrine herbaceous wetland from Puerto Rico. *Life: The Excitement of Biology* 1: 118–135. https://doi.org/10.9784/LEB1(2)Rios.04

Ríos-López, N. and Dávila-Casanova, D., 2014. *Eleutherodactylus wightmanae* (Melodious Coqui). *Herpetological Review* 45: 678–679.

Ríos-López, N., Alicea-Ortíz, J. D., and Burgos-Aponte, E., 2016. Natural history observations in the Puerto Rican Dwarf Anole, *Anolis occultus* Williams and Rivero 1965 (Squamata: Dactyloidae): nectar-feeding, bird predation, and a review of the species' natural history with implications for its conservation. *Life: The Excitement of Biology* 4: 100–113. https://doi.org/10.9784/LEB4(2)Rios.01

Ríos-López, N., Reyes-Díaz, M., Ortíz-Rivas, L., Negrón-Del Valle, J. E., and de Jesús-Villanueva, C. N., 2014. Natural history and ecology of the critically endangered Puerto Rican plains coquí, *Eleutherodactylus juanariveroi* (Amphibia: Anura: Eleutherodactylidae). *Life: The Excitement of Biology* 2: 69–93. https://doi.org/10.9784/LEB2(2)Rios.01

Ríos-López, N., Agosto-Torres, E., Hernández-Muñíz, R. M., Vicéns-López, C., Bernardi-Salinas, A., Tirado-Casillas, W. N., and Flores-Rodríguez, Y. M., 2015a. Conservation efforts for the Puerto Rican Mountain Coqui (Anura: Eleutherodactylidae: *Eleutherodactylus portoricensis* Schmidt, 1927): reproductive biology in captivity. *Life: The Excitement of Biology* 3: 61–82. https://doi.org/10.9784/LEB3(2)Rios.01

Ríos-López, N., Joglar, R. L., Rodríguez-Gómez, C. A., Díaz-Vázquez, C. J., and Rivera, I., 2015b. Natural history notes of saurophagy: an update from the

Puerto Rican vertebrate fauna. *Life: The Excitement of Biology* 3: 118–136. https://doi.org/10.9784/LEB3(2)Rios.02

Ríos-López, N., Flores-Rodríguez, Y. M., Agosto-Torres, E., Vicéns-López, C., and Hernández-Muñíz, R. M., 2015c. Life History Observations on the Melodious Coqui, *Eleutherodactylus wightmanae* (Anura: Eleutherodactylidae), from Puerto Rico: Double Clutches and Adult Predation by the Yellow-Chinned Anole, *Anolis gundlachi* (Squamata: Dactyloidae). *Life: The Excitement of Biology* 3: 137–148. https://doi.org/10.9784/LEB3(2)Rios.03

Rivera-Lebrón, R. L., 2014. Viabilidad ecológica de un habitat artificial para disminuir la pérdida del Coquí Guajón *Eleutherodactylus cooki* durante la construcción de la represa y embalse Valenciano en Juncos, Puerto Rico. MS thesis, Universidad Metropolitana, San Juan, Puerto Rico, USA.

Rivera-Sostre, M. L., 2008. Variación en la dieta del sapo de la caña (*Chaunus* [*Bufo*] *marinus*) en la Reserva Natural Privada El Tallonal en Arecibo, Puerto Rico. MS thesis, Universidad de Puerto Rico, Mayagüez Campus, Mayagüez, Puerto Rico, USA.

Rivero, J. A., 1959. Two new species of *Eleutherodactylus* from Puerto Rico. *Breviora* 103: 1–7.

Rivero, J. A., 1963. *Eleutherodactylus hedricki*, a new species of frog from Puerto Rico (Salientia, Leptodactylidae). *Breviora* 185: 1–7.

Rivero, J. A., 1978. *The Amphibians and Reptiles of Puerto Rico*. Editorial de la Universidad de Puerto Rico, San Juan, Puerto Rico, USA.

Rivero, J. A., 1991. Divagaciones sobre las especies de coquíes en peligro de extinción y las causas posibles de esa situación. Pp. 54–55 in *Status y Distribución de los Reptiles y Anfibios de la Región de Puerto Rico*, edited by J. A. Moreno. Departamento de Recursos Naturales, Publicación Científica Miscelánea No. 1, San Juan, Puerto Rico, USA.

Rivero, J. A., 1998. *The Amphibians and Reptiles of Puerto Rico*· 2nd edn. Editorial de la Universidad de Puerto Rico, San Juan, Puerto Rico, USA.

Rivero, J. A. and Joglar, R., 1979. *Eleutherodactylus cochranae* (Whistling *Eleutherodactylus*; Coqui Pitito). *Herpetological Review* 10: 101.

Rivero, J. A. and Mayorga, H., 1963. Notes on the distribution of some Puertorican frogs with a discussion on the possible origin of *Eleutherodactylus locustus*. *Caribbean Journal of Science* 3: 81–85.

Rivero, J. A. and Seguí Crespo, D., 1992. *Anfibios y reptiles en nuestro folklore*. Imprenta San Rafael, Quebradillas, Puerto Rico, USA.

Rivero, J. A. and Seguí Crespo, D., 1997. *Los Coquíes de Puerto Rico*. Imprenta San Rafael, Quebradillas, Puerto Rico, USA.

Rivero, J. A., Maldonado, J., and Mayorga, H., 1963. On the habits and food of *Eleutherodactylus karlschmidti* Grant. *Caribbean Journal of Science* 3: 25–27.

Rivero, J. A., Mayorga, H., Estremera, E., and Izquierdo, I., 1980. Sobre el *Bufo lemur* (Amphibia, Bufonidae). *Caribbean Journal of Science* 15: 33–40.

Robertson, W. B., Jr., 1962. Observations on the birds of St. John, Virgin Islands. *The Auk* 79: 44–76. https://doi.org/10.2307/4082450

Rodda, G. H. and Dean-Bradley, K., 2002. Excess density compensation of island herpetofaunal assemblages. *Journal of Biogeography* 29: 623–632. https://doi.org/10.1046/j.1365-2699.2002.00711.x

Rodda, G. H., Perry, G., Rondeau, R. J., and Lazell, J., 2001. The densest terrestrial vertebrate. *Journal of Tropical Ecology* 17: 331–338. https://doi.org/10.1017/S0266467401001225

Rogers, S. D. and Peacock, M. M., 2012. The disappearing northern leopard frog (*Lithobates pipiens*): conservation genetics and implications for remnant populations in western Nevada. *Ecology and Evolution* 2: 2040–2056. https://doi.org/10.1002/ece3.308

Rogowitz, G. L. and Sánchez-Rivoleda, J., 1999. Locomotor performance and aerobic capacity of the Cave Coquí, *Eleutherodactylus cooki*. *Copeia* 1999: 40–48. https://doi.org/10.2307/1447383

Rogowitz, G. L., Candelaria, C. L., Denizard, L. E., and Meléndez, L. J., 2001. Seasonal reproduction of a Neotropical frog, the cave coquí (*Eleutherodactylus cooki*). *Copeia* 2001: 542–547. https://doi.org/10.1643/0045-8511(2001)001[0542:SROANF]2.0.CO;2

Rohde, K., Ford, H., Andrew, N. R., and Heatwole, H., 2013. How to conserve biodiversity in a nonequilibrium world. Pp. 393–406 in *The Balance of Nature and Human Impact*, edited by K. Rohde. Cambridge University Press, Cambridge, UK. https://doi.org/10.1017/CBO9781139095075.036

Rolle, F. J., 1963. Life history of the Red-Legged Thrush (*Mimocichla plumbea ardosiacea*) in Puerto Rico. *Studies on the Fauna of Curaçao and Other Caribbean Islands* 60. 1–40 + Plate I.

Rollins-Smith, L. A., Reinert, L. K., and Burrowes, P.A., 2015. Coqui frogs persist with the deadly chytrid fungus despite a lack of defensive antimicrobial peptides. *Diseases of Aquatic Organisms* 113: 81–83. https://doi.org/10.3354/dao02823

Rosenzweig, M. L., 2003. *Win-win ecology: how the earth's species can survive in the midst of human enterprise*. Oxford University Press, New York, USA.

Ross, G. S., 2007. Biotic and abiotic factors affecting calling activity at traditional breeding ponds of Puerto Rican crested toads (*Bufo* [*Peltophryne*] *lemur*) in Quebradillas, Puerto Rico. MS thesis, University of Puerto Rico, Mayagüez Campus, Mayagüez, Puerto Rico, USA.

Ruedas, A. R., 2016. *Rattus norvegicus*. The IUCN Red List of Threatened Species 2016: e.T19353A22441833. http://dx.doi.org/10.2305/IUCN.UK.2016-3.RLTS.T19353A22441833.en. Accessed on 24 August 2017.

Sabaj M. H., 2016. Standard symbolic codes for institutional resource collections in herpetology and ichthyology: an Online Reference. Version 6.5 (16 August 2016). Accessible at http://www.asih.org/. American Society of Ichthyologists and Herpetologists, Washington, DC, USA.

Sampson, L. V., 1904. A contribution to the embryology of *Hylodes martinicensis*. *American Journal of Anatomy* 3: 473–504. https://doi.org/10.1002/aja.1000030404

Sánchez Muñoz, A. J., 2011. First record of the Lesser Antillean frog (*Eleutherodactylus johnstonei*) in Puerto Rico. *IRCF Reptiles & Amphibians* 18: 186.

Santana C., E. and Temple, S. A., 1988. Breeding biology and diet of Red-Tailed Hawks in Puerto Rico. *Biotropica* 20: 151–160. https://doi.org/10.2307/2388188

Santiago-Suliveras, G., Crespo-Zapata, J. D., Ríos-López, N., and Puente-Rolón, A. R., 2019. *Xenochrophis vittatus* (Striped Keelback). *Herpetological Review* 50: 604.

Scatena, F. N., 1990. Watershed scale rainfall interception on two forested watersheds in the Luquillo Mountains of Puerto Rico. *Journal of Hydrology* 113: 89–102. https://doi.org/10.1016/0022-1694(90)90168-W

Scatena, F. N., 1993. The management of Luquillo Elfin Cloud Forest ecosystem: irreversible decisions in a nonsubstitutable ecosystem. Pp.191–198 in *Tropical Montane Cloud Forests*, edited by L. S. Hamilton, J. O. Juvik, and F. N. Scatena. Proceedings of an International Symposium at San Juan, Puerto Rico, East-West Center, USA.

Scatena, F. N. and Larsen, M. C., 1991. Physical aspects of hurricane Hugo in Puerto Rico. *Biotropica* 23: 317–323. https://doi.org/10.2307/2388247

Scatena, F. N., Blanco, J. F., Beard, K. H., Waide, R. B., Lugo, A. E., Brokaw, N., Silver, W. L., Haines, B. L., and Zimmerman, J. K., 2012. Disturbance Regime. Pp. 164–200 in *A Caribbean Forest Tapestry: The multidimensional Nature of Disturbance and Response*, edited by N. Brokaw, T. A. Crowl, A. E. Lugo, W. H. McDowell, F. N. Scatena, R. B. Waide, and M. R. Willig. Oxford University Press, New York, USA. https://doi.org/10.1093/acprof:osobl/9780195334692.003.0004

Scheffers, B. R. and Williams, S. E., 2018. Tropical mountain passes are out of reach – but not for arboreal species. *Frontiers in Ecology and the Environment* 2018: 1–8. https://doi.org/10.1002/fee.1764

Schmidly, D. J., 2005. What means to be a naturalist and the future of natural history at American universities. *Journal of Mammalogy* 86: 449–456. https://doi.org/10.1644/1545-1542(2005)86[449:WIMTBA]2.0.CO;2

Schmidt, G. D. and Whittaker, F. H., 1975. Nematode parasites of Puerto Rican tree frogs, *Eleutherodactylus* spp: two new species and a proposal of *Poekilostrongylus* gen.nov. (Trichostrongylidae). *Parasitology* 70: 287–294. https://doi.org/10.1017/S003118200004974X

Schmidt, K. P., 1920. Contributions to the herpetology of Porto Rico. *Annals of the New York Academy of Science* 28: 167–200. https://doi.org/10.1111/j.1749-6632.1918.tb55351.x

Schmidt, K. P., 1926. The amphibians and reptiles of Mona Island, West Indies. *Field Museum of Natural History, Zoological Series* 12: 149–163. https://doi.org/10.5962/bhl.title.2782

Schmidt, K. P., 1927. A new tree-frog from Porto Rico. *American Museum Novitates* 279: 1–3.

Schmidt, K. P., 1928 (reprinted 1930). Amphibians and land reptiles of Porto Rico, with a list of those reported from the Virgin Islands, Part 1. *Annals of the New York Academy of Sciences* 10: 1–160.

Schuchert, C., 1935. *Historical geology of the Antillean-Caribbean region or the lands bordering the Gulf of Mexico and the Caribbean sea*. J. Wiley & Sons, Inc., New York, USA.

Schwartz, A. and Thomas, R., 1975. A check-list of West Indian amphibians and reptiles. *Carnegie Museum of Natural History, Special Publication* 1: 1–126. https://doi.org/10.5962/bhl.title.123681

Schwartz, A. and Henderson, R. W., 1985. *A Guide to the Identification of the Amphibians and Reptiles of the West Indies Exclusive of Hispaniola*. Milwaukee Public Museum, Milwaukee, Wisconsin, USA.

Schwartz, A. and Henderson, R. W., 1988. West Indian amphibians and reptiles: a check-list. Milwaukee Public Museum, *Contribution in Biology and Geology* 74: 1–274.

Schwartz, A. and Henderson, R. W., 1991. *Amphibians and Reptiles of the West Indies: Descriptions, Distributions, and Natural History*. University of Florida Press, Gainesville, Florida, USA.

Schwartz, A., Thomas, R., and Ober, L., 1978. First supplement to a check-list of West Indian amphibians and reptiles. *Carnegie Museum of Natural History, Special Publication* 5: 1–35.

Seaman, G. A., 1952. The mongoose and Caribbean wildlife. *Transactions of the North American Wildlife Conference* 17: 188–197.

Seaman, G. A., 1961. *Mammals, reptiles and amphibians of the Virgin Islands*. Privately printed.

Seaman, G. A., 1989. *Ay-Ay: An Island Almanac Interspersed with Reminiscences*. Macmillan Caribbean, London, United Kingdom.

Seaman, G. A. and Randall, J. E., 1962. The mongoose as a predator in the Virgin Islands. *Journal of Mammalogy* 43: 544–546. https://doi.org/10.2307/1376922

Seín, S., Jr., 1937a. El muy calumniado sapo. *Revista de Agricultura de Puerto Rico* 28: 505–509.

Seín, S., Jr., 1937b. The development of the giant Surinam toad, *Bufo marinus* L. *The Journal of the Department of Agriculture of Puerto Rico* 21: 77–78. https://doi.org/10.46429/jaupr.v21i1.14344

Settele, J., Scholes, R., Betts, R., Bunn, S., Leadley, P., Nepstad, D., Overpeck, J. T., and Taboada, M. A., 2014. Terrestrial and inland water systems. Pp 271–359 in *Climate Change 2014: Impacts, Adaptation, and Vulnerability. Part A: Global and Sectoral Aspects*, edited by C. B. Field, V. R. Barros, D. J. Dokken, K. J. Mach, M. D. Mastrandrea, T. E. Bilir, M. Chatterjee, K. L. Ebi, Y. O. Estrada, R. C. Genova, B. Girma, E. S. Kissel, A. N. Levy, S. MacCracken, P. R. Mastrandrea, and L. L. White. Contribution of Working Group II to the Fifth Assessment Report of the Intergovernmental Panel on Climate Change, Cambridge University Press, Cambridge, UK, and New York, New York, USA.

Shields, A. B. and Pitt, W. C., 2014. A review of invasive rodent (*Rattus* spp. and *Mus musculus*) diets on Pacific Islands. *Proceedings of the 26th Vertebrate Pest Conference* 26: 161–165. https://doi.org/10.5070/V426110336

Shin, J., Bataille, A., Kosch, T. A., and Waldan, B., 2014. Swabbing often fails to detect amphibian chytridiomycosis under conditions of low infection load. *PLoS ONE* 9: e111091. https://doi.org/10.1371/journal.pone.0111091

Smith, H. M., 1917. Report of the United States Commissioner of Fisheries for the Fiscal Year 1915 with Appendixes. Washington Government Printing Office, Bureau of Fisheries, Washington, District of Columbia, USA.

Smith, H. M., 1986. Chapman Grant, *Herpetologica*, and the Herpetologists' League. *Herpetologica* 42: 1–32.

Smith, L., 1920. Report of the Virgin Islands Agricultural Experiment Station. US Department of Agriculture, St. Croix, Virgin Islands, USA.

Smith, R. L., Beard, K. H., and Koons, D. N., 2018. Invasive coqui frogs are associated with greater abundances on nonnative birds in Hawaii, U.S.A. *The Condor: Ornithological Applications* 120: 16–29. https://doi.org/10.1650/CONDOR-17-109.1

Smith-Vaniz, W. F. and Jelks, H. L., 2014. Marine and inland fishes of St. Croix, U.S. Virgin Islands: an annotated checklist. *Zootaxa* 3803: 1–120. https://doi.org/10.11646/zootaxa.3803.1.1

Snow, J. B., 2017. Implications for Puerto Rico's Natural Heritage under PROMESA. Pinchot Institute for Conservation, Washington, DC, USA.

Snyder, N. F. R., Wiley, J. W., and Kepler, C. B. 1987. *The parrots of Luquillo: natural history and conservation of the Puerto Rican parrot*. The Western Foundation of Vertebrate Zoology, Los Angeles, California, USA.

Solís, F., Ibáñez, R., Hammerson, G, Hedges, B., Diesmos, A., Matsui, M., Hero, J-M, Richards, S., Coloma, L., Ron, S., La Marca, E., Hardy, J., Powell, R., Bolaños, F., Chaves, G., and Ponce, P., 2009. *Rhinella marina*. The IUCN Red List of Threatened Species 2009: e.T41065A10382424. http://dx.doi.org/10.2305/IUCN.UK.2009- 2.RLTS.T41065A10382424.en. Accessed on 17 September 2019.

Stabile, J. L., 2020. Ecology, behavior, and conservation of the Mona Coqui (*Eleutherodactylus monensis*) in Puerto Rico. MS thesis, University of San Francisco, San Francisco, California, USA.

Stabile, J. L. and Joglar, R. L., 2017. *Eleutherodactylus monensis* (Mona Coqui). *Herpetological Review* 48: 792.

Stabile, J. L., Santiago, L. A., and Joglar, R. L., 2014. Reproductive biology of the Mona Coqui (*Eleutherodactylus monensis*) in Captivity. *Herpetological Review* 45: 248–250.

Stachowicz, J. J., Whitlatch, R. B., and Osman, R. W., 1999. Species diversity and invasion resistance in a marine ecosystems. *Science* 286: 1577–1579. https://doi.org/10.1126/science.286.5444.1577

Stahl, A., 1882. *Catálogo del Gabinete Zoológico del Dr. A. Stahl, en Bayamon (Pto.-Rico.) Precedido de una Clasificacion sistemática de los animales que corresponden á esta fauna*. Imprenta del Boletín Mercantil, San Juan, Puerto Rico.

Stallard, R. F., 2001. Possible environmental factors underlying amphibian decline in eastern Puerto Rico: analysis of U.S. government data archives. *Conservation Biology* 15: 943–953. https://doi.org/10.1046/j.1523-1739.2001.015004943.x

Steffen, W., Crutzen, P. J., and McNeill, J. R., 2007. The Anthropocene: are humans now overwhelming the great forces of nature? *Ambio* 36: 614–621. https://doi.org/10.1579/0044-7447(2007)36[614:TAAHNO]2.0.CO;2

Stejneger, L., 1904 (1902). The herpetology of Porto Rico. Government Printing Office, Washington, Report of the United States National Museum for 1902, *Smithsonian Institution United States National Museum* 129: 548–724.

Stejneger, L., 1913. A new lizard from Porto Rico. *Proceedings of the Biological Society of Washington* 26: 69–72.

Steudler, P. A., Melillo, J. M., Bowden, R. D., Castro, M. S., and Lugo, A. E., 1991. The effects of natural and human disturbances on soil nitrogen dynamics and trace gas fluxes in a Puerto Rican wet forest. *Biotropica* 23: 356–363. https://doi.org/10.2307/2388252

Stevens, S. D., Prescott, D. R. C., and Whiteside, D. P., 2012. Occurrence and prevalence of chytrid fungus (*Batrachochytrium dendrobatidis*) in amphibian species of Alberta. Alberta Sustainable Resource Development, Fish and Wildlife Division, Alberta Species at Risk Report No. 143, Edmonton, Alberta, Canada.

Stewart, M. M., 1995. Climate driven population fluctuations in rain forest frogs. *Journal of Herpetology* 29: 437–446. https://doi.org/10.2307/1564995

Stewart, M. M. and Pough, F. H., 1983. Population density of tropical forest frogs: relation to retreat sites. *Science* 221: 570–572. https://doi.org/10.1126/science.221.4610.570

Stewart, M. M. and Woolbright, L. L., 1996. Amphibians. Pp. 273–320 in *The Food Web of a Tropical Rain Forest*, edited by D. P. Reagan, and R. B. Wade. University of Chicago Press, Chicago, USA.

Stuart, S. N., Chanson, J. S., Cox, N. A., Young, B. E., Rodrigues, A. S. L., Fischman, D. L., and Waller, R. W., 2005. Response to comment on "status and trends of amphibian declines and extinctions worldwide." *Science* 309: 1999c. https://doi.org/10.1126/science.1113265

Tanner, E. V. J. and Kapos, V., 1991. Hurricane effects on forest ecosystems in the Caribbean. *Biotropica* 23: 513–521. https://doi.org/10.2307/2388274

Tarrant, J., Cilliers, D., du Preez, L. H., and Weldon, C., 2013. Spatial assessment of amphibian chytrid fungus (*Batrachochytrium dendrobaditis*) in South Africa confirms endemic and widespread infection. *PLoS ONE* 8: e69591. https://doi.org/10.1371/journal.pone.0069591

Tewksbury, J. J., Anderson, J. G. T., Bakker, J. D., Billo, T. J., Dunwiddie, P. W., Groom, M. J., Hampton, S. E., Herman, S. G., Levey, D. J., Machnicki, N. J., Cartínez del Rio, C., Power, M. E., Rowell, K., Salomon, A. K., Stacey, L., Trombulak, S. C., and Wheeler, T. A., 2014. Natural history's place in science and society. *BioScience* 64: 300–310. https://doi.org/10.1093/biosci/biu032

The World Fackbook, 2022. https://www.cia.gov/the-world-factbook/countries/puerto-rico/#transportation. Accessed on 10 January 2022.

Thieler, R. E., Rodriguez, R. W., and Himmselstoss, E.A., 2007. Historical Shoreline Changes at Rincón, Puerto Rico 1936-2006. US Geological Survey, US Department of the Interior, USA. https://doi.org/10.3133/ofr20071017

Thomas, R., 1966. New species of Antillean *Eleutherodactylus*. *Quarterly Journal of the Florida Academy of Sciences* 28: 375–391.

Thomas, R., 1999. The Puerto Rico area. Pp. 169–179 in *Caribbean Amphibians and Reptiles*, edited by B. I. Crother. Academic Press, San Diego, California, USA. https://doi.org/10.1016/B978-012197955-3/50017-3

Thomas, R. and Gaa Kessler, A., 1996. Nonanoline Reptiles. Pp. 347–362 in *The Food Web of a Tropical Rain Forest*, edited by D. P. Reagan and R. B. Wade. University of Chicago Press, Chicago, USA.

Thomas, R. and Joglar, R., 1996. The herpetology of Puerto Rico: past, present, and future. Pp. 181–200 in *The scientific survey of Puerto Rico and the Virgin Islands–an eighty-year reassessment of the island's natural history*, edited by J. C. Figueroa-Colón. *Annals of the New York Academy of Sciences*, 776, New York, USA. https://doi.org/10.1111/j.1749-6632.1996.tb17420.x

Thomson, A. M., Calvin, K. V., Smith, S. J., Kyle, G. P., Volke, A., Patel, P., Delgado-Arias, S., Bond-Lamberty, B., Wise, M. A., Clarke, L. E., and Edmonds, J.A., 2011. RCP4.5: a pathway for stabilization of radiative forcing by 2100. *Climatic Change* 109: 1–25. https://doi.org/10.1007/s10584-011-0151-4

Thompson, J., Brokaw, N., Zimmerman, J. K., Waide, R. B., Everham, E. M., III, Lodge, D. J., Taylor, C. M., García-Montiel, D., and Fluet, M., 2002. Land use history, environment, and tree composition in a tropical forest. *Ecological Applications* 12: 1344–1363. https://doi.org/10.1890/1051-0761(2002)012[1344:LUHEAT]2.0.CO;2

Torresdal, J. D., Farrell, A. D., and Goldberg, C. S., 2017. Environmental DNA detection of the Golden Tree Frog (*Phytotriades auratus*) in bromeliads. *PLoS ONE* 12: e0168787. https://doi.org/10.1371/journal.pone.0168787

Towns, D. R., Atkinson, A. E., and Daugherty, C. H., 2006. Have the harmful effects of introduced rats on islands been exaggerated? *Biological Invasions* 8: 863–891. https://doi.org/10.1007/s10530-005-0421-z

Townsend, D. S. and Stewart, M. M., 1985. Direct development in *Eleutherodactylus coqui* (Anura: Leptodactylidae): a staging table. *Copeia* 1985: 423–436. https://doi.org/10.2307/1444854

Townsend, D. S. and Stewart, M. M., 1986. The effect of temperature on direct development in a terrestrial-breeding, neotropical frog. *Copeia* 1986: 520–523. https://doi.org/10.2307/1445011

Townsend, D. S., Stewart, M. M., and Pough, F. H., 1984. Male parental care and its adaptive significance in a Neotropical frog. *Animal Behaviour* 32: 421–431. https://doi.org/10.1016/S0003-3472(84)80278-X

Tschinkel, W. R. and Wilson, E. O., 2014. Scientific natural history: telling the epics of nature. *BioScience* 64: 438–443. https://doi.org/10.1093/biosci/biu033

Tucker, R. W. and Wolcott, G. W., 1935. Parasite introductions: Barbados and Puerto Rico. *Proceedings of the 5th Congress of the International Society of Sugar Cane Technologists, Brisbane*: 398–404.

Turner, F. B. and Gist, C. S., 1970. Observation of lizards and frogs in an irradiated Puerto Rican forest. Pp. E25–E49, Chapter E-2 in, *A Tropical Rain Forest: A Study of Irradiation and Ecology at El Verde, Puerto Rico,* edited by H. T. Odum and R. F. Pigeon. United States Atomic Energy Commission. Oak Ridge, Tennessee, USA.

Turvey, N., 2013. *Cane toads: a tale of sugar, politics and flawed science.* Sydney University Press, University of Sydney, New South Wales, Australia. https://doi.org/10.2307/j.ctv1sjwpjh

Tschudi, J. J., 1838. Classification der Batrachier mit Berücksichtigung der fossilen Thiere dieser Abtheilung der Reptilien. In der buchdruckerei von Petitpierre, Neuchâtel, Switzerland. https://doi.org/10.5962/bhl.title.4883

Tyson, G. F., 1992. On the periphery of the peripheries: the cotton plantations of St. Croix, Danish West Indies, 1735-1815. *The Journal of Caribbean History* 26: 1–36.

Uetz, P., 2010. The original descriptions of reptiles. *Zootaxa* 2334: 59–68. https://doi.org/10.11646/zootaxa.2334.1.3

United States Census Bureau, 2012. *2010 Census of Population and Housing: Summary Population and Housing Characteristics, CPH-1-53, Puerto Rico.* United States Government Printing Office. Washington, District of Columbia, U.S.A.

UNEP-WCMC, 2018. Protected Area Profile for British Virgin Islands from the World Database of Protected Areas, June 2018. Available at: www.protectedplanet.net.

USDAFS (United States Department of Agriculture Forest Service), 2022. https://www.fs.usda.gov/treesearch/pubs/54663. Accessed on 11 January 2022.

US Department of the Interior, 1974. Summary–Lacy Act Revision: Importation of Injurious Wildlife. *Herpetological Review* 5: 11.

US Department of the Interior, 1987. Endangered and threatened wildlife and plants; determination of threatened status for the Puerto Rican Crested Toad. US Fish and Wildlife Service, *Federal Register* 52: 28828–28831.

US Department of the Interior, 1997. Endangered and threatened wildlife and plants; Threatened status for the Guajón. US Fish and Wildlife Service, *Federal Register* 62: 31757–31762.

US Department of the Interior, 2012. Endangered and threatened wildlife and plants; determination of endangered species status for Coquí Llanero throughout its range and designation of critical habitat. US Fish and Wildlife Service, *Federal Register* 77: 60778–60802.

USFWS (United States Fish and Wildlife Service), 1984. Golden Coqui Recovery Plan. US Fish and Wildlife Service, Atlanta, Georgia, USA.

USFWS (United States Fish and Wildlife Service), 2014. Caribbean Endangered and Threatened Animals. https://www.fws.gov/caribbean/es/Endangered-Animals.html accessed February 25, 2017.

USFWS (United States Fish and Wildlife Service), 2016. Puerto Rican Crested Toad (*Peltophryne lemur*) 5-Year Review: Summary and Evaluation. Boquerón, Puerto Rico: Caribbean Ecological Services Field Office. US Fish and Wildlife Service, Atlanta, Georgia, USA.

US National Park Service, 2003. Sustained reduction plan for non-native wild hogs within Virgin Islands National Park. Final Environmental Assessment, St. John, Virgin Islands National Park, US Department of the Interior, Southeast Region, USA.

US Naval Security Group Activity Sabana Seca, 1999. Cultural Resources Management-Installation. US Navy, Department of Defense, USA.

van Berkum, F., Pough, F. H., Stewart, M. M., and Brussard, P. F., 1982. Altitudinal and interspecific differences in the rehydration abilities of Puerto Rican frogs (*Eleutherodactylus*). *Physiological Zoology* 55: 130–136. https://doi.org/10.1086/physzool.55.2.30155847

van der Have, T. M. and De Jong, G., 1996. Adult size in ectotherms: temperature effects on growth and differentiation. *Journal of Theoretical Biology* 183: 329–340. https://doi.org/10.1006/jtbi.1996.0224

van der Molen, M. K., 2002. Meteorological impacts of land use change in the maritime tropics. PhD thesis, VU University Amsterdam, Amsterdam, The Netherlands.

van der Molen, M. K., Dolman, A. J., Waterloo, M. J., and Bruijnzeel, L. A., 2006. Climate is affected more by maritime than by continental land use change: a multiple-scale analysis. *Global and Planetary Change* 54: 128–149. https://doi.org/10.1016/j.gloplacha.2006.05.005

van Voorhies, W. A., 1997. On the adaptive nature of Bergmann size clines: a reply to Mousseau, Partridge and Coyne. *Evolution* 51: 635–640. https://doi.org/10.1111/j.1558-5646.1997.tb02455.x

Vargas-Salinas, F., 2005. *Bufo marinus* (Cane Toad). *Herpetological Review* 36: 431–432.

Vargas-Salinas, F., 2007. Breeding behavior of the Cane Toad *Bufo marinus* (Bufonidae): a successfully invasive species. *Herpetological Review* 38: 12–17.

Vega-Castillo, S. I., 2000. Habitat description and comparison of the Puerto Rican Demon or Guajón (*Eleutherodactylus cooki*) at two localities on the Eastern part of Puerto Rico. MS thesis, University of Puerto Rico, Mayagüez Campus, Puerto Rico, USA.

Vélez-Juarbe, J., Martin, T., MacPhee, R. D. E., and Ortega-Ariza, D., 2014. The earliest Caribbean rodents: Oligocene caviomorphs from Puerto Rico. *Journal of Vertebrate Paleontology* 34: 157–163. https://doi.org/10.1080/02724634.2013.789039

Ventosa-Febles, E. A., Camacho-Rodríguez, M., Chabert-Llompart, J. L., Sustache-Sustache, J., and Dávila-Casanova, D., 2005. Puerto Rico Critical Wildlife Areas. Terrestrial Resources Division, Bureau of Fish and Wildlife, Department of Natural and Environmental Resources, Commonwealth of Puerto Rico. San Juan, Puerto Rico, USA.

VertNet©, 2020. http://portal.vertnet.org/publishers. VertNet© webapp version 2020-07-04T11:51-03:00. VertNet© version 2016-09-29. Accessed on 10 January 2022.

Vilella, F. J., 1998. Biology of the mongoose (*Herpestes javanicus*) in a Rain Forest of Puerto Rico. *Biotropica* 30: 120–125. https://doi.org/10.1111/j.1744-7429.1998.tb00374.x

Vilella, F. J. and Zwank, P. J., 1993. Ecology of the small Indian mongoose in a coastal Dry Forest of Puerto Rico where sympatric with the Puerto Rican nightjar. *Caribbean Journal of Science* 29: 24–29.

Vilella, F. J. and Fogarty, J. H., 2005. Diversity and abundance of forest frogs (Anura: Leptodactylidae) before and after hurricane Georges in the Cordillera Central of Puerto Rico. *Caribbean Journal of Science* 41: 157–162.

Villanueva-Rivera, L. J., 2006. Calling activity of *Eleutherodactylus* frogs of Puerto Rico and habitat distribution of *E. richmondi.* MS thesis. University of Puerto Rico, Río Piedras Campus, San Juan, Puerto Rico, USA.

Vitousek, P. M., Mooney, H. A., Lubchenco, J., and Melillo, J. M. 1997. Human domination of Earth's ecosystems. *Science* 277: 494–499. https://doi.org/10.1126/science.277.5325.494

Waddle, J. H., Crockett, M. E., and Rice, K.G., 2005. *Osteopilus septentrionalis* (Cuban Treefrog). *Herpetological Review* 36: 333.

Waddle, J. H., Crockett, M. E., and Rice, K. G., 2006. *Eleutherodactylus coqui* (Coqui). *Herpetological Review* 37: 487.

Wade, R. B., Comarazamy, D. E., González, J. E., Hall, C. A. S., Lugo, A. E., Luvall, J. C., Murphy, D. J., Ortiz-Zayas, J. R., Ramírez-Beltran, N. D., Scatena, F. N., and Silver, W. L., 2013. Climate variability at multiple spatial and temporal scales in the Luquillo Mountains, Puerto Rico. *Ecological Bulletins* 54: 21–41.

Wadsworth, F. H., 1949. The development of the forest land resources of the Luquillo Mountains, Puerto Rico. PhD Dissertation, University of Michigan, Ann Arbor, Michigan, USA.

Wadsworth, F. H., 1950. Notes on the climax forests of Puerto Rico and their destruction and conservation prior to 1900. *Caribbean Forester* 11: 38–47.

Wadsworth, F. H., 2008. La reforestación gubernamental en Puerto Rico. *Acta Científica* 22: 105–113.

Wadsworth, F. H., 2012. Los bosques y el uso de madera en Puerto Rico. Printed online only as an open access paper in: http://www.edicionesdigitales.info/biblioteca/Wadsworth2012.pdf.

Wadsworth, F. H., 2014. A forestry assignment to Puerto Rico. Forestry Memoirs of Frank Wadsworth. Impresos Emmanuelli, Inc., Aguas Buenas, Puerto Rico, USA.

Wadsworth, F. H. and Englerth, G. H., 1959. Effects of the 1956 hurricane on forests in Puerto Rico. *Caribbean Forester* 20: 38–51.

Waide, R. B., 1991. Summary of the response of animal populations to hurricanes in the Caribbean. *Biotropica* 23: 508–512. https://doi.org/10.2307/2388273

Wake, M. H., 1978. The reproductive biology of *Eleutherodactylus jasperi* (Amphibia, Anura, Leptodactylidae), with comments on the evolution of live-bearing systems. *Journal of Herpetology* 12: 121–133. https://doi.org/10.2307/1563398

Wake, D. B., 1991. Declining amphibian populations. *Science* 253: 860. https://doi.org/10.1126/science.253.5022.860

Walker, E. P., 1975. Mammals of the World. Johns Hopkins University Press, Baltimore, Maryland, USA.

Walker, L. R., Lodge, D. J., Brokaw, N. V. L., and Waide, R. B., 1991. An introduction to hurricanes in the Caribbean. *Biotropica* 23: 313–316. https://doi.org/10.2307/2388246

Ward, S., Rodriguez, A., and Morgan, R., 2000. Estate Thomas experimental forest: past, present, and future. Pp 105–114 in *Proceedings of the 5th Annual Caribbean Urban Forestry Conference*, edited by T. W. Zimmerman. University of the Virgin Islands, Cooperative Extension Service, St. Croix, Virgin Islands, USA.

Weather Underground, 2022. https://www.wunderground.com/weather/vg/virgin-gorda. Accesed on 04 January 2022.

Weaver, P. L., 2000. Elfin woodland recovery 30 years after a plane wreck in Puerto Rico's Luquillo Mountains. *Caribbean Journal of Science* 36: 1–9.

Weaver, P. L., 2006. *A summary of 20 years of forest monitoring in Cinnamon Bay watershed, St. John, U.S. Virgin Islands.* General Technical Report IITF-34, United States Department of Agriculture Forest Service, International Institute of Tropical Forestry, Río Piedras, Puerto Rico, USA. https://doi.org/10.2737/IITF-GTR-34

Weaver, P. L., 2008. Dwarf forest recovery after disturbances in the Luquillo Mountains of Puerto Rico. *Caribbean Journal of Science* 44: 150–163. https://doi.org/10.18475/cjos.v44i2.a4

Weaver, P. L., 2012. *The Luquillo Mountains: forest resources and their history.* General Technical Report IITF-44, United States Department of Agriculture Forest Service, International Institute of Tropical Forestry, Río Piedras, Puerto Rico, USA.

Weldon, C., Crottini, A., Bollen, A., Rabamananjara, F. C. E., Copsey, J., Garcia, G., and Andreone, F., 2013. Pre-emptive national monitoring plan for detecting the amphibian chytrid fungus in Madagascar. *EcoHealth* 10: 234–240. https://doi.org/10.1007/s10393-013-0869-8

Wells, K. D., 2007. *The ecology and behavior of amphibians.* The University of Chicago Press. Chicago, Illinois, USA.

Westergaard, W., 1938. A St. Croix map of 1766: with a note on its significance in West Indian plantation economy. *The Journal of Negro History* 23: 216–228. https://doi.org/10.2307/2714346

Wetmore, A., 1916. Birds of Porto Rico. Bulletin No. 326, United States Department of Agriculture Forest Service, Washington, District of Columbia, USA. https://doi.org/10.5962/bhl.title.64703

Wetmore, A., 1917a. The Birds of Vieques Island, Porto Rico. *The Auk* 33: 403–419. https://doi.org/10.2307/4072636

Wetmore, A., 1917b. The Birds of Culebra Island, Porto Rico. *The Auk* 34: 51–62. https://doi.org/10.2307/4072538

Wetmore, A., 1918. The Birds of Desecheo Island, Porto Rico. *The Auk* 35: 333–340. https://doi.org/10.2307/4072911

Whitbeck, R. H., 1932. The agricultural geography of Jamaica. *Annals of the Association of American Geographers* 22: 13–27. https://doi.org/10.1080/00045603209357096

Wiewandt, T. A., 1977. Ecology, behavior, and management of the Mona Island ground iguana, *Cyclura stejnegeri*. PhD Dissertation. Cornell University, Ithaca, New York, USA.

Wilcox, W. A., 1902. The fisheries and fish trade of Porto Rico in 1902. Report of the United States Commissioner of Fish and Fisheries: 367–395.

Wiley, J. W., 1985. Bird Conservation in the United States Caribbean. Pp. 107–159 in *Bird Conservation 2*, edited by S. A. Temple. The University of Wisconsin Press, Wisconsin, USA.

Wiley, J. W., 1986. Habitat change and its effects on Puerto Rican raptors. *Birds of Prey Bulletin* No. 3: 51–56.

Willig, M. R. and Camilo, G. R., 1991. The effect of hurricane Hugo on six invertebrate species in the Luquillo Experimental Forest of Puerto Rico. *Biotropica* 23: 455–461. https://doi.org/10.2307/2388266

Willig, M. R. and Gannon, M. R., 1996. Mammals. Pp. 399–431 in *The Food Web of a Tropical Rain Forest*, edited by D. P. Reagan, and R. B. Wade. University of Chicago Press, Chicago, USA.

Willig, M. R., Bloch, C. P., Covich, A. P., Hall, C. A. S., Lodge, D. L., Lugo, A. E., Silver, W. L., Waide, R. B., Walker, L. R., and Zimmerman, J. K., 2012. Long-term research in the Luquillo mountains. Synthesis and foundations for the future. Pp. 361–441 in *A Caribbean Forest Tapestry: The multidimensional Nature of Disturbance and Response*, edited by N. Brokaw, T. A. Crowl, A. E. Lugo, W. H. McDowell, F. N. Scatena, R. B. Waide, and M. R. Willig. Oxford University Press, New York, USA.

Witmer, G. W., Boyd, F., and Hillis-Starr, Z., 2007. The successful eradication of introduced roof rats (*Rattus rattus*) from Buck Island using diphacinone, followed by an irruption of house mice (*Mus musculus*). *Wildlife Research* 34: 108–115. https://doi.org/10.1071/WR06006

Wolcott, G. N., 1924. The food of Porto Rican lizards. *The Journal of the Department of Agriculture of Porto Rico* 7: 5–37.

Wolcott, G. N., 1937. What the Giant Surinam Toad, *Bufo marinus* L., is eating now in Puerto Rico. *The Journal of Agriculture of the University of Puerto Rico* 21: 79–84. https://doi.org/10.46429/jaupr.v21i1.14345

Wolcott, G. N., 1948a. The insects of Puerto Rico. *The Journal of Agriculture of the University of Puerto Rico* 32 (1): 1 224. https://doi.org/10.46429/jaupr.v32i1.13611

Wolcott, G. N., 1948b. The insects of Puerto Rico. *The Journal of Agriculture of the University of Puerto Rico* 32 (2): 225–416. https://doi.org/10.46429/jaupr.v32i2.13615

Wolcott, G. N., 1953. The food of the mongoose (*Herpestes javanicus auropunctatus* Hodgson) in St. Croix and Puerto Rico. *The Journal of Agriculture of the University of Puerto Rico* 37 (3): 241–247. https://doi.org/10.46429/jaupr.v37i3.12754

Woodhams, D. C., Alford, R. A., Antwis, R. E., Archer, H., Becker, M. H., Belden, L. K., Bell, S. C, Bletz,

M., Daskin, J. H., Davis, L. R., Flechas, S. V., Lauer, A., Gonzalez, A., Harris, R. N., Holden, W. M., Hughey, M. C., Ibáñez, R., Knight, R., Kueneman, J., Rabemananjara, F., Reinert, L. K., Rollins-Smith, L. A., Roman-Rodriguez, F., Shaw, S. D., Walke, J. B., and McKenzie, V., 2015. Antifungal isolates database of amphibian skin-associated bacteria and function against emerging fungal pathogens. *Ecology* 96: 595. https://doi.org/10.1890/14-1837.1

Woods, C. A., 1996. The Land Mammals of Puerto Rico and the Virgin Islands. Pp. 131–148 in *The scientific survey of Puerto Rico and the Virgin Islands–an eighty-year reassessment of the island's natural history*, edited by J. C. Figueroa-Colón. *Annals of the New York Academy of Sciences*, 776, New York, USA. https://doi.org/10.1111/j.1749-6632.1996.tb17418.x

Woolbright, L. L., 1985. Patterns of nocturnal movement and calling by the tropical frog *Eleutherodactylus coqui*. *Herpetologica* 41: 1–9.

Woolbright, L. L., 1991. The impact of hurricane Hugo on forest frogs in Puerto Rico. *Biotropica* 23: 462–467. https://doi.org/10.2307/2388267

Woolbright, L. L., 1996. Disturbance influences long-term population patterns in the Puerto Rican frog, *Eleutherodactylus coqui* (Anura: Leptodactylidae). *Biotropica* 28: 493–501. https://doi.org/10.2307/2389091

Woolbright, L. L., 1997. Local extinctions of anuran amphibians in the Luquillo Experimental Forest of northeastern Puerto Rico. *Journal of Herpetology* 31: 572–576. https://doi.org/10.2307/1565612

Wu, W., Hall, C., and Zhang, L., 2006. Predicting the temporal and spatial probability of orographic cloud cover in the Luquillo experimental forest in Puerto Rico using generalized linear (mixed) models. *Ecological Modelling* 192: 473–498. https://doi.org/10.1016/j.ecolmodel.2005.07.031

Yu, M., Rivera-Ocasio, E., Heartsill-Scalley, T., Davila-Casanova, D., Rios-López, N., and Gao, Q., 2019. Landscape-level consequences of rising sea-level on coastal wetlands: saltwater intrusion drives displacement and mortality in the twenty-first century. *Wetlands* 39: 1343–1355. https://doi.org/10.1007/s13157-019-01138-x

Zegarra-Vila, J. P., 2007. Assessment of *Eleutherodactylus richmondi* populations at the Maricao State Forest, Puerto Rico. MS thesis. University of Puerto Rico, Mayagüéz Campus, Mayagüéz, Puerto Rico, USA.

Zumbado-Ulate, H., Bolaños, F., Gutiérez-Espeleta, G., and Puschendorf, R., 2014. Extremely low prevalence of *Batrachochytrium dendrobatidis* in frog populations from Neotropical Dry Forest of Costa Rica supports the existence of a climatic refuge from disease. *EcoHealth*. Available at https://doi.org/10.1007/s10393-014-0967-2

APPENDIX 8.I: The case of the mongoose and amphibians from the Puerto Rican Archipelago

Anecdotal accounts since the early 1900s place the mongoose as a highly detrimental predator on reptiles and birds (e.g. Bowdish 1902; Stejneger 1904; Barbour 1914; Wetmore [1916, 1917a,b, 1918]; Schmidt 1928; Grant 1932e), and although some mammals are also included as prey, amphibians are rarely or not included. How did the population status of amphibians from the Puerto Rican Archipelago (PRA) become tied to the mongoose in the scientific literature? In this appendix, we argue that two phenomena relate to perceptions about the influence of the mongoose on the population status of amphibians from the PRA. First, ecological processes such as secondary succession might have influenced the perception of early scientists in the PRA, which required the examination of this topic – the mongoose and its effects on amphibians – from an historical context (see main text about the history of uses of the land). Second, we believe that the weight of authority of Thomas Barbour's (1930a) publication 'Some faunistic changes in the Lesser Antilles', *Proceedings of the New England Zoological Club* **11**: 73–85, had a significant influence on this perception that still holds today. The available data – besides anecdotal accounts since the arrival of the mongoose in the Puerto Rico Archipelago (PRA) – suggest that the mongoose is not a threat to our amphibians. Next we examine several passages that provide information seemingly equivocal or subjective – or in the words of Powell and Henderson (1996: 32–33) about lizards and snakes, 'based largely on circumstantial evidence' – that may explain how the mongoose also is attributed to the rarity of our amphibians until today. Stejneger (1904: 555) mentioned in reference to *Peltophryne lemur* that 'the apparent rarity of the toad (*Bufo lemur*) is probably due less to the mongoose than to its very retired habits and the scarcity of suitable localities'; however, he mentioned 'mongoose' four more times, all in relation to its influence on the rarity of reptiles from Puerto Rico. Barbour (1914) mentioned the word 'mongoose' 21 times, all references related exclusively to the population status of reptiles from the West Indies (WI). However, Schmidt (1928: 31) said in reference to *Peltophryne lemur* that 'Its rarity is perhaps due to the mongoose'; this is the only reference that links the mongoose to an amphibian from Puerto Rico and the US Virgin Islands. Barbour (1930b: 76) mentioned in reference to *Eleutherodactylus lentus* that 'its subterranean habits protect it against capture by the mongoose', and he referred to the mongoose in terms of an additional amphibian (from Dominica). Barbour (1937: 99) omitted any reference to the influence of the mongoose such as mentioned in his previous list (Barbour 1930b) and reassessed the population status of *E. lentus* as 'this seems not to be a common species, according to notes kindly furnished by Major Chapman Grant'. Grant (1932a: 41) indicated that '*Bufo* [= *Peltophryne*] *lemur* seems to have always been scarce, or its hiding places are not yet understood'.

Some 60 years after the last previous mention of the mongoose in the herpetological literature for the PRA, Hedges (1993: 297) brought the topic up again in reference to *Eleutherodactylus karlschmidti*:

> However, rats and mongooses, which were introduced, are abundant in Puerto Rico and occur in undisturbed forest. Black Rats (*Rattus rattus*) especially are a problem in Caribbean National Forest where they are very common, even in the dwarf forest on El Yunque Peak. It is possible that these arboreal nocturnal omnivores prey on *Eleutherodactylus* eggs or the frogs themselves. The mongoose (*Herpestes auropunctatus*), although primarily diurnal, is known to prey on frogs (Walker, 1975; Nellis and Everard, 1983) and this particular species of frog would be especially vulnerable because it characteristically sits on exposed rocks in and around streams (Rivero, 1978).

Later, Henderson and Powell (2009: 9) mentioned that mongooses 'are known to prey on frogs (*Eleutherodactylus* spp.) in forested areas in eastern Puerto Rico (Vilella 1998)'. Powell and Henderson (2012: 87) also stated:

> Humans bring with them cats and dogs that prey on reptiles, and goats, burros, and cattle that crop vegetation to bare soil, eliminating critical ground cover and food for a variety of amphibians and reptiles. The mongoose was introduced to many islands in the 19th century, putatively to control rodent populations that were having a deleterious effect on sugar production. The impact on ground-dwelling lizards, snakes, and birds has frequently been catastrophic.

Powell and Henderson (2012), however, did not mention mongooses (or rats) as threats specifically to amphibians from the WI. Nonetheless, the references mentioned in this appendix so far are the primary sources of literature that might have cemented the reputation of the mongoose concerning the population status of amphibians from a conservation perspective. We probably never will know, however, what scientists witnessed in the PRA during the early 1900s that led them to refer to the mongoose in the scientific literature with terms such as 'villain', 'destroyer', 'voracious', 'nuisance', and 'pest'. These terms were frequently used alongside others, as well as phrases such as 'ravages', 'overwhelming damage', 'drastic', and 'evil results of its introduction'. Notably, we frequently found these passages also accompanied by terms and phrases such as 'possibly', 'presumably', 'most likely', 'may', 'perception', 'is believed to', 'been implicated in', 'certainly', 'undeniable', 'are a problem', and similar statements in the early literature just reviewed. These subjective terms – which are not supported by quantitative data (Powell and Henderson 1996) – are frequent and vague enough as to suggest that data for the accurate assessment of the influence of the mongoose on the amphibians in the PRA were and are still missing (see the main text). Most likely, visiting scientists to the PRA witnessed the exponential population growth of the mongoose – much as others witnessed that of *Rhinella marina* in the Philippines and PR (see main text) – that frequently occurs during colonization and early stages of secondary succession of introduced species in new environments when unlimited resources (e.g. naïve prey, breeding sites, among others) relative to the mongoose's initial density, low predation pressure, a simplified (homogeneous) landscape (see section III. in the main text), and protection by people occurred (e.g. the mongoose used as a biological control of rats in agriculture lands). As a result, predation events of species other than amphibians (e.g. reptiles) frequently could have been witnessed during the early stages of population expansion of the mongoose and faunal succession because of the transformations of the land; amphibians could have been included as 'victims' by reasoning alone. Other species, in contrast, may have contributed in part to the reputation attributed to the mongoose in the PRA about the scarcity of reptiles, and by extension in the scientific literature, of amphibians. For example, Arendt (2006: 273) cited Seaman (1952) as suggesting 'that much of the blame for the depleted faunas of numerous Caribbean islands, so often attributed to the mongoose, should be accredited to the thrasher and two species of *Rattus* (black rat, *R. rattus*, and brown rat, *R. norvegicus*)'. The thrasher or pearly-eyed thrasher, *Margarops fuscatus*, has been described as a 'Supertramp' (*sensu* Diamond 1974; also see Chapter 1 of the present volume), which refers to a species with outstanding vagility, and that incurs in density-compensation on small islands. 'Supertramp' also refers to a species that has the ability to change its reproductive strategy in situations of disturbance (natural and man-made), and is a prolific breeder with raptor-like predatory behaviour (it also preys on a wide range of small mammals, adult birds and their nests, and herpetofauna from the PRA and the Caribbean islands at large). As for mammals, for example, the PRA had species whose extant relatives from other Caribbean islands are known to include small vertebrates in their diet, including lizards (e.g. Capromyidae; Borroto-Páez and Woods 2012; Snyder *et al.* 1987). More often than frequently acknowledged, it is possible that interspecific responses also can be imprinted and persistent in extant reptiles (and amphibians) to these non-native species such as the mongoose (see review by Holt and Bonsall 2017). Consequently, we may infer the possibility that the mongoose could be a scapegoat for the scarcity of reptiles and amphibians from the PRA when early herpetologists did not fully know the geographic distribution and ecological requirements of most species of vertebrates (see next paragraph).

Later stages of succession in the PRA would lead to environmental conditions under which populations of the mongoose must have fluctuated at lower levels of population size and density, which would accompany the presumed adaptation to this predator by the native fauna at large. For example, Raffaele (1990: 16; translated freely from Spanish) mentioned – about vulnerable, threatened, and endangered birds – that apparently 'these birds had reached a condition of certain equilibrium with the mongoose, after suffering great losses when they had reached their numerical peak 50 years ago'. Heatwole *et al.* (1981: 12) mentioned that 'The role of the mongoose in the decline and extinction of West Indian native fauna may not be as great as is sometimes attributed to it, but it is almost certainly responsible for local extinction of some lizards, snakes and ground-nesting birds (see Barbour 1930a; Heatwole and Torres 1967; Philibosian and Ruibal 1971).' Finally, Honegger (1981) presented a list of amphibians (and reptiles) either known or thought to have become extinct since 1600 and for which the mongoose amounted to the critical threat faced by many of these and other species. Two species of amphibians were listed by Honegger (1981): *Latonia* (*Discoglossus*) *nigriventer* from Israel, a species last collected in 1955 until it was rediscovered in 2011 (IUCN SSC Amphibian Specialist Group 2021u), and *Lithobates* (*Rana*) *pipiens* from Las Vegas, Nevada, USA. However, the mongoose is not mentioned in 'Threats in detail' for *L. nigriventer*, and the IUCN SSC Amphibian Specialist Group (2021u) specifies:

> The Huleh marshes were drained in the 1950s in an attempt to both eradicate malaria and to make the land suitable for agricultural use. Of the original 6,000 ha of marshland, only 300 ha (3 km^2) remained after drainage. The remaining wetland was set aside as a nature reserve - the Hula Nature Reserve - in 1964. While the reserve is well-managed, it functions as a refuge for many water birds, possibly more than those that used to be in the original swamp, and predation pressure is thought to be very high and a real threat to a small amphibian population comprised of few individuals (S. Gafny pers. comm. March 2012). This appears to be supported by high rates of injury recorded among adults in Yesod HaMa'ala, and may reflect predation attempts by both native species and invasive species such as mosquitofish (Perl *et al.* 2017). In addition, most of the area surrounding the reserve is cultivated. (S. Gafny pers. comm. March 2012).

Rogers and Peacock (2012: 2040) in referring to the losses of *Lithobates pipiens* mentioned that these 'are largely due to habitat fragmentation and the introduction of non-native fish, amphibian, and plant species' instead. Currently, *L. pipiens* is classified as Least Concern (Hammerson *et al.* 2004), with the caveat that it is part of a species complex that needs further taxonomic resolution.

In summary, the mongoose is present in many localities in the PRA (Fig. A8.1), but we found many deficiencies that need to be addressed before further generalizing about its influence on our amphibians. Fortunately, rather than being a problem, these deficiencies provide fruitful opportunities for future research and projects by graduate students and scientists in general, to distinguish between apparent and real effects brought about by the mongoose and other naturalized species. Until then, a reassessment of the population status of the amphibians in the PRA (e.g. on the IUCN Red List) must omit the mongoose as a threat.

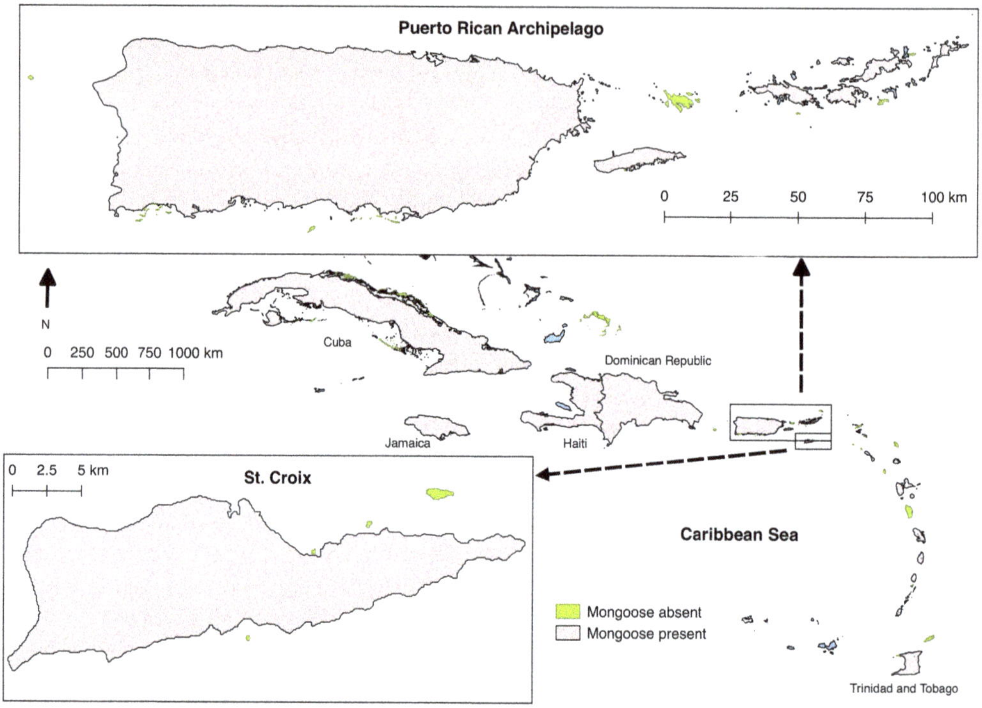

Figure A8.1 Geographic distribution of *Herpestes auropunctatus* within most Caribbean islands in general, and PRA in particular. In blue, unknown presence/introduction status.

APPENDIX 8.II: Generalities and scenarios of projected changes in climate in the Puerto Rican Archipelago (PRA)

In this appendix, we discuss the changes in climate expected in the immediate future in the Caribbean and the PRA, briefly focusing on interactions among temperature, precipitation, sea level, hurricanes, and droughts. We acknowledge that predictions and estimates of changes in climate in reports and other studies on the topic continue to vary as parameters are added to existing models, with increases in the precision of scientific equipment and analyses, and as models are changed and updated accordingly (Cubasch *et al.* 2013). Even with these limitations, we wanted to provide a brief background of technical definitions of models that allows for projections of climatic change relevant to the amphibians in the PRA. These definitions and subsequent discussion of the topic may help us guide the decisionmaking process for the conservation of amphibians and their ecosystems in the PRA.

A. Definition of technical terms and models of the IPCC: representative concentration pathways (RCPs) and radiative forcing (RF)

According to the IPCC (2013: 19), 'a new set of scenarios, the Representative Concentration Pathways (RCPs), was used for the new climate model simulations carried out under the framework of the Coupled Model Intercomparison Project Phase 5 (CMIP5) of the World Climate Research Programme'. The model CMIP5 produced a 'multi-data set', based on the best information available, that allowed for 'long-term simulations of 20th century climate and projections for the 21st century and beyond' (Cubasch *et al.* 2013: 146). This model (and older models) also allowed for comprehensive analyses of projections in climatic change based on four Representative Concentration Pathways scenarios (RCPs, namely RCP2.6, RCP4.5, RCP6.0, and RCP8.5) that updated scenarios in the most recent report of the IPCC. Each of these pathways is 'designed to explore a wide range of future climate characterized by representative trajectories of well-mixed greenhouse gas (WMGHG) concentrations and other anthropogenic forcing agents' relative to pre-industrial levels (Collins *et al.* 2013: 1044).

The RCPs represent scenarios about the 'concentration and emission of gas and aerosols, land use changes', and other emissions from natural and human sources that 'can perturb the Earth's radiation budget, producing a radiative forcing [RF] that affects climate' (Cubasch *et al.* 2013: 127). According to Cubasch *et al.* (2013: 127), 'RF is a measure of the net change in the energy balance in response to an external perturbation.' RF is measured in $W\,m^{-2}$, and each of the RCPs represents a scenario of radiative forcing that allows us to examine the relative contribution of different sources of emissions to changes in climate at various spatiotemporal scales and make projections of climatic change for effective mitigation and conservation actions of our climatic system and biodiversity at large. Consequently, the RCP2.6 (the lowest of the four scenarios) represents the projected endpoint in radiative forcing by the year 2100: 2.6 $W\,m^{-2}$ (Cubasch *et al.* 2013). Likewise, RCP4.5 (medium-low), RCP6.0 (medium-high), and RCP8.5 (highest) are scenarios that under a specific set of circumstances reflect projected endpoints in radiative forcing stabilized by the year 2100. However, the higher the value of RCP, the higher the endpoint in radiative forcing and the more dramatic the changes in phenomena affecting the atmospheric, marine, and terrestrial interphases. As a result, Thomson *et al.* (2011) suggested that these scenarios represent mitigation ones in the sense that they guide the decisionmaking process whereby the changes needed to achieve a preferred endpoint can be reached but with explicit actions that mitigate the emissions of greenhouse gases by a given period. As such, RCP4.5 is frequently designated as a preferred scenario because it 'invokes all available technology options that can cost-effectively contribute to mitigation' (Thomson *et al.* 2011: 12).

B. Projected changes in variables and phenomena relevant to populations of amphibians in the PRA

1. Temperature

Global temperature has increased on land and on the surface of seas between the late 19th century and the beginning of the 20th century respectively, as supported by many lines of evidence (see Frequently Asked Questions [FAQ 2.1] and Figure 1 in Hartmann *et al.* 2013: 198; Cubasch *et al.* 2013). Projections of changes in global mean temperature under the conservative scenario of the RCP4.5 (see section B., will 'likely' [i.e. 66–100% probability]) be an increase between '1.1°C and 2.6°C' by the year 2100 (Collins *et al.* 2013: 1031).

In the Caribbean, there is 'high confidence' (Field *et al.* 2014) that current and future projections of changes in temperature will be for an increase (projections based on CMIP5 under RCP4.5), with virtually no difference between high-temperature seasons and low-temperature ones by the year 2100 (Christensen *et al.* 2013; see Khalyani *et al.* 2016 for Puerto Rico). A fundamental question is whether amphibians have the capacity to adapt rapidly to the rapid changes in climate expected in the immediate future. Yet we noted in the text that we need studies that allow for broader integration of ecophysiology and climate to reach a better understanding about the mechanisms that mediate responses of amphibians to projected changes in climate for conservation purposes. However, it is known that climatic variables interact in complex ways, which make predictions that fit observations a more difficult task (see next section).

2. Precipitation

Projections for precipitation on land and sea in the Caribbean region (see table 14.1 in Christensen *et al.* 2013) indicated a decrease in annual precipitation of 9% by the year 2100 in the Caribbean; there will also be a decrease in precipitation between 12% and 15% by 2100 during the early rainy season (between June and August) compared with the reference period 1986–2005. In Puerto Rico, McDowell *et al.* (2012) suggested that annual precipitation decreased over most of Puerto Rico during the past century. However, these authors also noted that precipitation increased between November and April, which covers the months of the dry season in PR.

In the Caribbean and in Central America, the warming phase of El Niño Southern Oscillation (ENSO), (the opposite of El Niño being La Niña), is frequently related to dry conditions (Christensen *et al.* 2013). However, Mote *et al.* (2017) showed that the Saharan Air Layer (SAL), known in the region as the incoming Saharan dust, is linked directly to the reduction of rainfall in Puerto Rico that occurred during the early season in 2015. This reduction in rainfall was manifest as a historic drought; Mote *et al.* (2017) noted a similar scenario in 1994 in Puerto Rico when a drought followed an intense ENSO-El Niño period (see Christensen *et al.* 2013). Moreover, Herrera *et al.* (2018: 7) estimated that increases in temperature attributed to human activities explained ~15–17% of this historic drought, and predicted increases in the severity of future droughts in the Caribbean 'because of higher temperatures alone, regardless of changes in precipitation.'

Increased temperature raises the height of orographic formation of clouds in tropical montane forests elsewhere, which also reduces the frequency and daily regime of cloud immersion and humidity (Foster 2001; Pounds *et al.* 2006; van der Molen *et al.* 2006; Wu *et al.* 2006). In Puerto Rico, this phenomenon has been documented by van der Molen (2002) after Hurricane Hugo defoliated most elfin forests at El Yunque National Forest in the Luquillo Mountains in 1989. The cloud-base was uplifted beyond the highest peak in this forest (1,074 m at El Toro Peak; Weaver 2012; Wu *et al.* 2006), which reduced the frequency of daily mist and relative humidity with local warming. Assuming an increase in warming in the immediate future, the projected deficit in precipitation in the Caribbean – including the PRA – will be accompanied by a reduction in the local orographic humidity, which may increase the uncertainty of the expected range in the decrease in precipitation.

Finally, the expected decrease in precipitation for the Caribbean, including PR, may also depend on the expected rainfall associated with high-intensity hurricanes whenever these occur (see Figure 14.17 in Christensen *et al.* 2013). Therefore, even though the agreement is in the decrease in annual precipitation accompanied by a redistribution of precipitation throughout the year, the relative contribution of phenomena such as hurricanes and ENSO (Christensen *et al.* 2013), SAL (e.g. Mote *et al.* 2017), and rate of warming from human activities (Herrera *et al.* 2018) remains unknown. However, Khalyani *et al.* (2016) suggested that a decrease in precipitation – either annual or seasonal – will change the distribution and extent of ecosystems at high-elevation mountain ranges, which are the areas where the vast majority of endemic and ecologically specialized species of amphibians occur in the PRA.

3. Rise in sea level

There is also 'high confidence' that the increase in the global mean surface temperature during the past century was accompanied by an increase in the rate of rise in sea level compared with the estimate for the past two millennia (IPCC 2013: 11). According to the Puerto Rico Climate Change Council Working Group 1 (PRCCC 2013: 61), the current rate of rise in sea level in the Caribbean is 10 cm per century. In Puerto Rico, the increased rate of change in sea level – along with human activities on the shoreline – is expected to continue or increase current rates of erosion of the coast in some areas (e.g. Rincón at the western tip of Puerto Rico; Thieler *et al.* 2007; PRCCC 2013). The combined effects of natural and human phenomena in the coastal and inland landscapes, however, makes the dynamics of erosion and deposition of sediments on the coasts complex, but there is agreement that the rate of coastal erosion caused by increased rate of rise in sea level in PR will be higher by the year 2100 compared with the past century. Two native and endemic species of amphibians in the PRA occur only in coastal ecosystems: the threatened *Peltophryne lemur,* which depends on water for reproduction, and the endangered *Eleutherodactylus juanariveroi,* which is a palustrine-herbaceous-wetland specialist.

In conclusion, even though natural and human phenomena interact in complex ways, there are some generalities in the literature about climatic change that are relevant to the PRA and the Caribbean in general. First, ecosystems most vulnerable to climatic change include, but are not limited to, those high in mountains, those on lowlands adjacent to coastal areas, and oceanic islands with low topography (Intergovernmental Panel on Climate Change [IPCC] 2002). Second, species most vulnerable to climatic change include, but are not limited to, those with low capacity for dispersal, those confined to small and isolated areas such as mountaintops, and those encroached between inland urban areas and coastal plains subjected to the intrusion of saltwater because of increased sea level (Settele *et al.* 2014).

As noted, the PRA has ecosystems and species of amphibians that can be considered vulnerable in light of these generalities (see main text, sections III. F. and G.), but for some species and ecosystems we still lack information on their capacity to respond and adapt to the changes in climate at the pace that these changes are occurring in the Anthropocene.

9 Amphibians of the Lesser Antilles

Robert Powell and Robert W. Henderson

Abbreviations and acronyms

asl	*above sea level*
Bd	Batrachochytrium dendrobatidis *(chytrid fungus)*
BP	*Before Present*
CERMES	*Centre for Resource Management and Environmental Studies (University of the West Indies)*
CITES	*Convention on International Trade in Endangered Species of Wild Fauna and Flora*
CR	*Critically Endangered*
DOE	*Department of Environment*

Robert Powell and Robert W. Henderson, 'Chapter 9: Amphibians of the Lesser Antilles' in: *The Conservation and Biogeography of Amphibians in the Caribbean*. Pelagic Publishing (2023). © Robert Powell and Robert W. Henderson. DOI: 10.53061/GKFK9725

EN	*Endangered*
FAO	*Food and Agriculture Organization (of the United Nations)*
h	*hour*
IRCF	*International Reptile Conservation Foundation*
IUCN	*International Union for the Conservation of Nature*
LC	*Least Concern*
MCZ	*Museum of Comparative Zoology, Harvard University*
m	*metres*
MCRP	*Mountain Chicken Recovery Programme*
MNHN	*Muséum national d'Histoire naturelle (Paris)*
mya	*million years ago*
NT	*Near Threatened*
p-h	*person-hour*
RP	*Robert Powell*
RWH	*Robert W. Henderson*
SSC	*The Species Survival Commission (of the IUCN)*
USAID	*United States Agency for International Development*
VU	*Vulnerable*
ZSL	*Zoological Society of London*

I. Introduction and physiography of the islands

The Lesser Antilles, as traditionally defined, comprise an arc-like archipelago of oceanic islands that extends some 700 km from Sombrero in the north to Grenada in the south (Fig. 9.1). Collectively, these islands form the eastern boundary of the Caribbean Sea. Sixteen island banks include from one each (Sombrero, Saba, Redonda, Montserrat, Marie-Galante, Barbados) to more than 30 (Grenada) named islands, ranging in size from barely emergent rocks to > 1,400 km^2 (Guadeloupe, when Basse-Terre and Grand-Terre are considered together).

These islands developed along the border of the Caribbean and South American tectonic plates where the latter is being subducted beneath the former (Mann 1999; Potter *et al.* 2004). The present Lesser Antillean Arc represents Eocene to Holocene eastward migration of the resultant volcanic activity (Malfait and Dinkelman 1972). Seventeen volcanoes remain active (Bouysse *et al.* 1990).

The first islands to form along this abutment arose over 36 mya and formed the 'outer-arc' islands along the eastern (Atlantic) side of the archipelago (Fig. 9.1). This 'outer arc' includes the present-day islands of the Sombrero, Anguilla, Antigua, Marie-Galante, and Barbados Banks, plus the Grande-Terre portion of Guadeloupe. The volcanoes responsible for forming these islands are extinct. Since their undersea birth some 36 to > 50 mya, the outer-arc islands have been uplifted by tectonic activities, rising above sea level – only to be submerged again as a consequence of erosion and local subsidence of Earth's crust. During submerged periods, corals formed limestone caps over the previously exposed volcanic rock. Subsequent uplifts added areas to the islands that had never before been above sea level. Because the original volcanic surfaces of the islands in this outer, much older, arc have become capped with limestone, they are sometimes called the 'Limestone Caribees' (Martin-Kaye 1969).

The 'inner-arc' islands (Fig. 9.1) or 'Volcanic Caribees' are of much more recent volcanic origins (Martin-Kaye 1969). These islands form the Saba, St Christopher, Redonda, Montserrat, Île des Saintes, Dominica, Martinique, St Lucia, St Vincent, and Grenada Banks plus the Basse-Terre portion of Guadeloupe. These islands continue to show varying degrees of volcanic activity, and

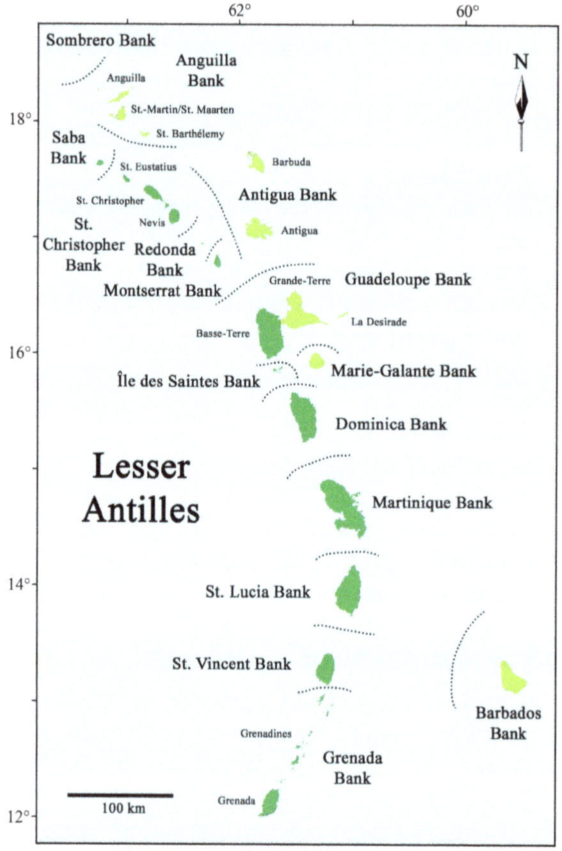

Figure 9.1. Map of the Lesser Antilles showing the banks and some of the major islands. 'Outer-arc' islands (Limestone Caribees) are coloured light green and 'inner-arc' islands (Volcanic Caribees) are in darker green. Map by John S. Parmerlee, Jr.

threats of eruption exist. Every major island has a maximum elevation at least 50% higher than the highest peak in the Limestone Caribees (Table 9.1).

During the coldest periods of the ice ages, when sea levels were about 130 m lower than today (Murray-Wallace and Woodroffe 2014), the present islands of each inner-arc and outer-arc bank were connected to form single large islands. When the glaciers melted and sea levels rose, only the areas at higher elevations remained above water, resulting in the separated islands on most banks today.

Inner-arc islands have peaks that are sufficiently high (above 600 m) to generate orographic lift. Referred to in the English-speaking nations of the region as 'snag islands' (because they 'snag' clouds), the peaks often are covered by clouds and receive as much as twice the rainfall as drier sites at lower elevations (Powell *et al.* 2015). For example, inner-arc Saba receives an island-wide average of about 1,100 mm of rain per year, whereas the higher slopes of Mount Scenery receive over 2,000 mm. Similarly, areas on the Quill (St Eustatius) with elevations above 400 m receive from 1,500 to 2,000 mm of rain per year, significantly more than the island-wide average of about 1,070 mm. In sharp contrast, the highest peaks of the outer-arc islands lack the elevation required for clouds to form as a result of an orographic effect. Consequently, those islands show less disparity in the amount of precipitation.

Table 9.1. Lesser Antillean banks and major islands. Sizes and maximum elevations rounded to the nearest km² or m, respectively (except for Sombrero's size). Human population sizes and densities estimated as of January 2018 (except where noted). Islands that do not support amphibian populations are marked with an asterisk (not all islands with established populations are listed – i.e. some small satellite islands are omitted).

BANK/ISLAND	ISLAND SIZE (KM²)	MAXIMUM ELEVATION (M)	POPULATION (INDIVIDUALS/KM²)
Outer-arc Islands ('Limestone Caribees')			
Sombrero Bank			
Sombrero*	0.4	12	0 (0)
Anguilla Bank			
Anguilla	91	64	15,174 (166)
Dog Island*	2	30	0 (0)
Scrub Island*	3	25	0 (0)
St.-Martin/St Maarten	85	424	72,677 (855)
St.-Barthélemy	22	286	7,160 (325)
Antigua Bank			
Barbuda	160	38	**
Antigua	280	402	95,882 (342)
Guadeloupe Bank			
Grand Terre	590	150	209,089 (354)
La Désirade	22	273	1,481 (67)
Marie-Galante Bank			
Marie-Galante	160	204	11,561 (72)
Barbados Bank			
Barbados	430	340	286,388 (666)
Inner-arc Islands ('Volcanic Caribees')			
Saba Bank			
Saba	13	870	2,155 (166)
St Christopher Bank			
St Eustatius (Statia)	20	600	3,348 (167)
St Christopher (St Kitts)	170	1,156	34,983 (206)***
Nevis	130	985	11,108 (85)***
Redonda Bank			
Redonda*	1	296	0 (0)
Montserrat Bank			
Montserrat	100	914****	5,203 (52)
Guadeloupe Bank			
Basse-Terre	848	1,467	178,244 (210)

(Continued)

Table 9.1. (Continued)

BANK/ISLAND	ISLAND SIZE (KM²)	MAXIMUM ELEVATION (M)	POPULATION (INDIVIDUALS/KM²)
Île des Saintes Bank			
Terre-de-Bas	10	293	1,030 (103)
Terre-de-Haut	5	306	1,838 (367)
Dominica Bank			
Dominica	751	1,447	74,308 (99)
Martinique Bank			
Martinique	1,100	1,397	368,640 (335)
St Lucia Bank			
St Lucia	616	950	179,667 (292)
St Vincent Bank			
St Vincent	350	1,234	99,429 (284)
Grenada Bank			
Grenada	311	840	102,258 (329)
Bequia	18	118	5,300 (294)
Canouan	7	267	1,700 (243)
Carriacou	32	294	6,081 (190)
Mayreau*	3	58	271 (90)
Mustique	5	152	500 (100)
Union	8	305	3,000 (375)

* Islands without populations of amphibians.
** Estimated population of 1,638 was evacuated after Hurricane Irma in September 2017.
*** Estimates from 2011, the most recent year for which data for each island were available.
**** 1,236 m prior to the eruption of Soufrière Hills Volcano.

Generally, the inner-arc islands, with greater topographic heterogeneity and variable levels of precipitation, support more diverse biotic communities. Coastal and other low-elevation communities, which dominate the outer-arc islands, tend to be the most biologically uniform. For instance, although outer-arc St Martin/St Maarten is nearly seven times the size of tiny inner-arc Saba, the two islands harbour nearly the same number of native plant and animal species.

Areas of higher elevation on inner-arc islands can sustain perpetually moist forests. However, equally important are the transitional vegetative communities that vary according to elevation and orientation in regard to prevailing winds and consequently variable levels of humidity and precipitation (generally speaking, windward slopes are moister and subject to greater erosive forces than leeward slopes). So, inner-arc islands can support eight or more distinct vegetative communities. Such biotic diversity, combined with dramatic seasonal fluctuations in rainfall (approximately twice as much rain falls in August–November than in January–April), results in conditions that force terrestrial animals, particularly moisture-dependent amphibians, to be either ecologically versatile or able to sustain distributions restricted to the tiny fragments of the islands that provide suitable habitats.

A. Lesser Antillean amphibians

The Lesser Antillean amphibian fauna – all anurans – is relatively depauperate. Only ten native species occur in the entire archipelago, two of which have become established on islands on which they are not native. Another six non-native species from outside the region have established populations on one or more islands. Yet, of the hundreds of islands in the archipelago, anurans have been recorded from only 47 (Henderson and Breuil 2012), and 21 of those support only non-native species (either from within or outside the region). One obvious factor limiting the number of species on these oceanic islands is the difficulty faced by amphibians in crossing expanses of salt water without human mediation. Another is the paucity of extensive suitable (i.e. sufficiently moist) habitat on many islands, especially in the coastal areas that would be encountered initially by prospective colonists – while the remaining moister areas continue to disappear at alarming rates.

All of the established non-native and a few of the native anurans do demonstrate the considerable ecological versatility (e.g. Henderson and Powell 2001; Powell *et al.* 2011; 2013) necessary to reach, survive, and thrive on largely inhospitable islands. At the same time, the native species that lack that attribute are thus restricted to increasingly smaller areas of suitable habitat and are suffering population declines that render them progressively susceptible to extinction.

The ecologically versatile species can benefit from human activities beyond transport to islands they could not have reached on their own. For example, imported plants (often from Florida, USA) not only facilitate new introductions, they also provide familiar habitat for the associated species. In addition, artificial sources of moisture (e.g. cisterns and other water-storage containers, ditches and drainages, ornamental pools and ponds, watering gardens, golf courses, and lawns) provide habitats that would otherwise not be available, especially in arid lowlands (e.g. Hodge *et al.* 2011). On xeric Anguilla, for instance, introduced *Eleutherodactylus johnstonei* was initially limited to the Agriculture Station (Censky 1989) and subsequently to a few gardens of residences and hotels until an unusually wet year (1999), culminating with the landfall of two major hurricanes, allowed them to expand to many other artificially moist habitats. Similarly, *Osteopilus septentrionalis,* which had been seen on the island as early as the late 1980s, took advantage of those conditions in 1999 to move to other areas of the island; subsequently, when drier conditions resumed, they retreated to moist refugia, including household cisterns in which they were breeding (Townsend *et al.* 2000). Similarly, on St Barthélemy, *O. septentrionalis* arrived in 1995 but did not spread to other parts of the island until the very wet years of 1998–2000 (Breuil 2002; Questel 2018).

In sharp contrast, native species closely associated with moist upland forests are dramatically and negatively affected by the deforestation that accompanies human activity (Table 9.2). For instance, *Pristimantis euphronides* occurs on Grenada in mesic forests at elevations above 300 m (Kaiser and Henderson 1994) but has not been encountered in recent years in areas with elevations below 400 m (Harrison *et al.* 2011; Berg *et al.* 2013). Consequently, populations surviving at sites with higher elevations are effectively isolated from one another by intervening areas of unsuitable, largely deforested lower-elevation habitats. Clearing of forests for agriculture, which peaked during the nineteenth century, affected entire islands, although the steepest slopes of the topographically complex Volcanic Caribees were sometimes spared, providing ever-shrinking refuges for these species (e.g. Henderson and Powell 2018). Primary forests remain only in Dominica, Grenada, and Guadeloupe (Butler 2006).

Table 9.2. Forest cover and changes on some Lesser Antillean islands (adapted from Butler 2006 and FAO 2015; major discrepancies are attributable to the use of different criteria to define forests on some islands). The Dutch Islands (Saba, St Eustatius, St Maarten) are not listed, as they were combined with Bonaire and Curaçao. St Martin and St Barthélemy were not included in the 2006 study.

	2006		2015		
COUNTRY	TOTAL FOREST AREA (HA)	PERCENT FOREST COVER (%)	TOTAL FOREST AREA (HA)	PERCENT FOREST COVER (%)	ANNUAL CHANGE 1990–2015 (%)
Anguilla	6,000	58.8	6,000	61.1	0.0
Antigua and Barbuda	9,000	20.5	10,000	22.3	- 0.2
Barbados	2,000	4.7	6,000	14.7	0.0
Dominica	46,000	61.3	43,000	57.8	- 0.6
Grenada	4,000	11.8	17,000	50.0	0.0
Guadeloupe	80,000	47.3	71,000	42.5	- 0.1
Martinique	46,000	43.4	49,000	45.8	0.0
Montserrat	4,000	40.0	3,000	25.0	- 1.3
St Barthélemy	—	—	0	0.0	—
St Kitts and Nevis	5,000	13.9	11,000	42.3	0.0
St Lucia	17,000	27.9	20,000	33.3	- 0.3
St Martin	—	—	1,000	21.2	0.0
St Vincent and the Grenadines	11,000	28.2	27,000	69.2	0.3

Another potentially existential threat, also triggered by human actions, is global climatic change. This generates shifts in hydric conditions, which are capable of inundating low-lying coastal areas and increasing the frequency and intensity of storms (e.g. Lewsey *et al.* 2004; Mercer *et al.* 2012) but also of causing relatively cool and moist uplands to become drier and warmer and ultimately uninhabitable by amphibians intolerant of such conditions (e.g. Pounds *et al.* 1999).

B. International and national environmental legislation

Complicating regional trade and environmental legislation is the reality that the region in question is a political hodgepodge of 16 distinct entities. Anguilla (including Sombrero) and Montserrat are British Overseas Territories. St Martin/St Maarten is the smallest landmass comprised of two nations – Saint-Martin is a French Overseas Collectivity, whereas Sint Maarten is a constituent nation of the Kingdom of the Netherlands. St Barthélemy, like St Martin, is a French Overseas Collectivity. Saba and St Eustatius (Statia) are Special Municipalities of the Netherlands. The sovereign nations of Antigua and Barbuda (including Redonda), Saint Kitts and Nevis, Dominica, St Lucia, Barbados, St Vincent and the Grenadines, and Grenada are in the British Commonwealth of Nations. Guadeloupe (including La Desirade, Marie-Galante, and the Îles des Saintes) and Martinique are French Overseas Departments.

Some regional unity is provided by membership in the Caribbean Community (CARICOM), which includes nations other than those in the Lesser Antilles. Antigua and Barbuda, Barbados, Dominica, Grenada, Montserrat, St Kitts and Nevis, St Lucia, and St Vincent and the Grenadines are full members. Anguilla is an associate member and St Maarten is an observer. Designed primarily to facilitate economic integration and cooperation and coordinate foreign policy, the organization has not effectively addressed many environmental issues, such as monitoring and eradicating invasive species (e.g. Powell *et al.* 2011).

All native and established non-native species of amphibian in the Lesser Antilles have been assessed for the IUCN Red List (Table 9.3). Of the ten native species, one (*Leptodactylus fallax*) is CR, five (*Eleutherodactylus amplinympha, E. barlagnei, E. pinchoni, Pristimantis euphronides, P. shrevei*) are EN, one (*Allobates chalcopis*) is VU (but see the species account below), one (*Eleutherodactylus martinicensis*) is NT, and two (*Eleutherodactylus johnstonei, Leptodactylus validus*) are LC. All six of the established non-native species (*Rhinella marina, Eleutherodactylus planitrostris, Osteopilus septentrionalis, Scinax ruber, S. x-signatus, Pleurodema brachyops*) are LC.

Some Lesser Antillean nations have signed the Protocol Concerning Specially Protected Areas and Wildlife (SPAW Protocol), which seeks to take the necessary measures to protect, preserve, and manage in a sustainable way areas of special value and threatened or endangered species. Signatories in the region are Antigua and Barbuda, Republic of France (St Martin, St Barthélemy, Martinique, Guadeloupe), Kingdom of the Netherlands (St Maarten, Saba, St Eustatius), St Lucia,

Table 9.3. IUCN Red List Assessments for frogs in the Lesser Antilles (only native and established non-native species are listed, i.e. no waifs).

FAMILY AND *SPECIES*	NATIVE OR INTRODUCED	IUCN ASSESSMENT
Aromobatidae		
Allobates chalcopis	Native	Vulnerable
Bufonidae		
Rhinella marina	Introduced	Least Concern
Eleutherodactylidae		
Eleutherodactylus amplinympha	Native	Endangered
Eleutherodactylus barlagnei	Native	Endangered
Eleutherodactylus johnstonei	Native/Introduced	Least Concern
Eleutherodactylus martinicensis	Native/Introduced	Near Threatened
Eleutherodactylus pinchoni	Native	Endangered
Eleutherodactylus planirostris	Introduced	Least Concern
Hylidae		
Osteopilus septentrionalis	Introduced	Least Concern
Scinax ruber	Introduced	Least Concern
Scinax x-signatus	Introduced	Least Concern
Leptodactylidae		
Leptodactylus fallax	Native	Critically Endangered
Leptodactylus validus	Native	Least Concern
Pleurodema brachyops	Introduced	Least Concern
Strabomantidae		
Pristimantis euphronides	Native	Endangered
Pristimantis shrevei	Native	Endangered

St Vincent and the Grenadines, and the United Kingdom of Great Britain and Northern Ireland (Anguilla, Montserrat). Interestingly, the only Lesser Antillean amphibians listed in Annex II of the protocol – indicating that species merit strict protection by all parties – are in the genus *Eleutherodactylus*. *Eleutherodactylus barlagnei* and *E. pinchoni* are listed as EN on the IUCN Red List, but *E. johnstonei* (LC) and *E. martinicensis* (NT) are abundant and have been widely introduced onto islands where they are not native. As surprising as the inclusion of the latter two species is the omission of *Allobates chalcopis* (VU), *Eleutherodactylus amplinympha* (EN), *Leptodactylus fallax* (CR), *Pristimantis euphronides* (EN), and *P. shrevei* (EN). Some nations also are signatories of the Convention on International Trade in Endangered Species of Wild Fauna and Flora (CITES). However, none of the CITES appendices list any species of regional frogs.

On islands with native species, amphibians do benefit from national legislation and protected areas in some countries. For example, in Dominica (Malhotra *et al.* 2007; 2011), the National Parks and Protected Areas Act of 1975 (available at: https://elaw.org/sites/default/files/content_type_law_attachment/Dominica%20National%20Parks%20and%20Protected%20Areas%20Act.pdf) provided for the establishment of national parks and a protected area system. Although *Eleutherodactylus amplinympha* and *E. martinicensis* (and introduced *E. johnstonei*) inhabit protected areas, ironically the species of greatest concern, *Leptodactylus fallax*, does not occur in any currently protected area. The Forestry and Wildlife Act of 1976 (http://extwprlegs1.fao.org/docs/pdf/dmi14187.pdf) made provisions for the protection and management of animals and forested habitats. It also established a closed hunting season for *L. fallax* from 1 March to 31 August. The season was extended to nine months (January to September) in 2000, and further legislation in 2003 not only gave full protection to *L. fallax* but also banned the import of froglegs and other amphibian products.

In the French West Indies (Lorvelec *et al.* 2007; 2011), legislation has protected some terrestrial and freshwater species since 1989. *Eleutherodactylus barlagnei*, *E. martinicensis*, *E. pinchoni*, and the introduced *E. johnstonei* are included, but *Allobates chalcopis*, which was not described until after 1989, is not listed. National parks and nature reserves created between 1976 and 1998 provide protection for some habitats occupied by frogs.

In Antigua and Barbuda, the Environmental Protection and Management Act of 2015 (http://laws.gov.ag/wp-content/uploads/2019/02/a2015-11.pdf), states only that amphibians are wildlife but lists no taxa among protected species. Similarly, in St Lucia, the Wildlife Protection Act of 1980 (https://elaw.org/sites/default/files/content_type_law_attachment/Wildlife%20Protection%20Act%201980.pdf) makes no mention of amphibians. In St Vincent and the Grenadines, the only mention of amphibians in the Wildlife Protection Act of 1987 as amended (https://elaw.org/sites/default/files/content_type_law_attachment/Wildlife%20Protection%20Act%201987.pdf) is that the definition of wildlife includes 'frogs, and the eggs thereof'. In Grenada, the Birds and Other Wild Life (Protection of) Ordinance of 1957 (http://extwprlegs1.fao.org/docs/pdf/grn7859.pdf) does not mention any amphibians. In Montserrat, the Endangered Animals and Plants Act of 1976 as amended (http://extwprlegs1.fao.org/docs/pdf/msr95474.pdf) mysteriously restricts the importation of four species of salamanders and five anurans, with no mention of native species. However, the Forestry, Wildlife, National Parks and Protected Areas Act of 1996 (http://extwprlegs1.fao.org/docs/pdf/msr97811.pdf) includes a section stating 'Wildlife, whether resident, migratory or transient, native or introduced, including captive wild species having their origin in Montserrat are property of the Crown and may not be hunted, taken or possessed except at such times and in such manner as may be provided in this Act or by regulations made thereunder' (pp. 18–19). The Second Schedule of that law lists *Leptodactylus fallax* among species that are partially protected, which allows an open hunting season. The Centre Hills, site of the only known remaining population, is protected and managed by the Department of Environment (DOE 2008).

II. Threats to amphibians

Henderson and Powell (2009) described threats to West Indian amphibians and reptiles using examples from the literature and personal experience. Herein we list most of the same causes and cite some of the same examples but focus on the amphibians of the Lesser Antilles.

The impact of Amerindians on island environments in the pre-Columbian Antilles should not be underestimated (e.g. Newsom and Wing 2004; Fitzpatrick and Keegan 2007; Rick *et al.* 2013). Although the extent to which forests were cleared was negligible when compared to the devastation wrought by later-arriving Europeans, ranges of forest-adapted species such as *Allobates chalcopis*, *Pristimantis euphronides*, and *P. shrevei* contracted. With the arrival of Europeans, the transformation of habitats accelerated, and many alien plants and animals were introduced. The eighteenth and nineteenth centuries saw the clear-felling of forests to make way for agriculture (especially sugarcane), further reducing habitats available for those species. Today, the burgeoning growth of human populations coupled with rampant development (often driven by tourism) on many islands does not bode well for many native amphibians. Henderson and Powell (2001) estimated that human-introduced resources benefited at best 5–10% of the herpetofauna (almost entirely reptiles), whereas at least 50% (including most amphibians) have been negatively affected.

Interventions (*sensu* Smith and Sutherland 2014) intended to ameliorate the effects of threats on amphibian populations in the Lesser Antilles have been limited mostly to efforts to protect *Leptodactylus fallax* (see the species account below). However, efforts have failed to establish (or re-establish) populations on islands other than where the species is currently extant, and the eventual success or failure of breeding these frogs in captivity and releasing them to augment or establish new populations on Montserrat and Dominica remains to be determined.

A. Introduced invasive predators and competitors

Invasive alien species are disproportionately prevalent on islands, where they also generate disproportionate impacts compared to continental areas (e.g. Russell *et al.* 2017). During the past few centuries, alien species have entered Antillean islands with human assistance – intentional or inadvertent – and, as a group, have had a devastating effect on native taxa. Introduced plants alter the very nature of the habitats used by amphibians, non-native carnivores consume them, large mammalian herbivores severely alter habitats, and even rats (*Rattus norvegicus* and *R. rattus*) can modify habitats by exploiting seeds and seedlings. Invasive amphibians also may compete with and/or consume native species.

The small Indian mongoose (*Urva auropunctata*) was introduced initially to Jamaica in 1872 (Espeut 1882) and subsequently became established on more than 20 additional islands, including most of the major Lesser Antillean islands, some as late as the early twentieth century (Hoagland *et al.* 1989). Mongooses have been implicated in the extirpation of populations and even extinctions of several species of squamate reptiles (e.g. Powell and Henderson 2005; Hedges and Conn 2012); they are known to eat frogs (e.g. Gorman 1975; Hays and Conant 2011) and have been implicated in the extirpation of *Leptodactylus fallax* on three Lesser Antillean islands (Barbour 1930). Among Caribbean UK Overseas Territories and Bermuda, Oppel *et al.* (2019) identified Montserrat, one of only two islands with extant populations of *L. fallax*, as a high priority for biosecurity to reduce the risk of invasion of black rats (*R. rattus*) and the small Indian mongoose.

Introduced common opossums (*Didelphis marsupialis*) are abundant on some islands, where they occupy a variety of arboreal and terrestrial habitats. They are known to consume snakes and presumably eat lizards and frogs. Domestic or feral cats (*Felis catus*) and dogs (*Canis lupus familiaris*) are known predators of reptiles and have preyed on *Leptodactylus fallax* on some Lesser

Antillean islands (Faaborg and Arendt 1985). Adams *et al.* (2014) noted that 5–27% of *L. fallax* found on Montserrat during the 2005 dry season had evidence of bites from black rats.

Grazers such as goats (*Capra hircus*), cattle (*Bos taurus*), sheep (*Ovis aries*), and donkeys (*Equus asinus*) can quickly eliminate palatable vegetation from an area, degrading habitats that once provided food and cover for native frogs. Predation by feral pigs (*Sus scrofa*) is threatening one of the last remaining populations of *Leptodactylus fallax* on Montserrat (Peh *et al.* 2015).

Three West Indian species of anurans (*Rhinella marina, Osteopilus septentrionalis,* and *Eleutherodactylus johnstonei*) are widely distributed outside of their native ranges. Although Lever (2003) considered all West Indian populations of *R. marina* to be introduced, some may have arrived in the islands by natural means (Henderson and Powell 2009; Powell *et al.* 2011; 2013). This toad occurs in a wide range of habitats, from nearly intact rainforest to urban settings composed largely of concrete (e.g. Mallery *et al.* 2007). The range of *O. septentrionalis* has increased in recent years, mainly due to the transport of ornamental plants and building materials from the southeastern United States to resort sites (e.g. Townsend *et al.* 2000; Somma and Graham 2015). The frogs either are hidden in the plants themselves or are hitchhikers in containers carrying plants and building materials. These large frogs prey on smaller anurans or outcompete them for food, especially in altered habitats around human habitations, including outbuildings, swimming pools, decorative gardens, and other sources of fresh water (e.g. cisterns for drinking water). *Eleutherodactylus johnstonei,* like *O. septentrionalis,* has been introduced to many islands via accidental transport (Kaiser 1992). Although St George's, Grenada, is the type locality for the species, it is an introduced species on that island. Once established, these frogs quickly become ubiquitous in even ephemerally moist areas at low to moderate elevations, with populations increasing rapidly in a variety of habitats but especially in disturbed situations. However, these frogs also occur in relatively undisturbed habitats and may have the potential to outcompete endemic species such as *Pristimantis shrevei* on St Vincent (Powell and Henderson 2007; 2011) and *P. euphronides* on Grenada (Henderson and Berg 2006; 2011). On Grenada, *Eleutherodactylus johnstonei* and *Pristimantis euphronides* once appeared to have essentially parapatric distributions based on elevational boundaries (Kaiser and Henderson 1994), but the two species now are syntopic in some forested sites at elevations over 500 m (Henderson and Berg 2006; 2011; Berg *et al.* 2013).

Non-native plants dominate many habitats in the West Indies, and invasive species of plants can negatively impact frogs (e.g. Powell *et al.* 2005; 2011; 2015). Wilson *et al.* (2006) noted that invasive plants effectively alter habitats and prey bases in much the same way as introduced herbivores.

B. Loss and alteration of habitats

Destruction of habitats is the major factor contributing to the extinction of species (e.g. Tilman *et al.* 2002). Gallant *et al.* (2007) demonstrated that many regions of the world with the richest assemblages of amphibians also were undergoing the highest rates of landscapes being modified. The Lesser Antilles have a long history of altered habitats, beginning with the first human inhabitants about 4,000 years ago. Some modifications to habitats probably occurred with the arrival of the earliest colonizers. Newsom and Wing (2004) suggested that Archaic hunter-gatherers might have introduced food crops and engaged in some low-level food production, and Pagán-Jiménez *et al.* (2019) argued that the use of domestic plants by the earliest settlers of the Antilles was a major feature of the initial peopling of the region. Between 2,500 and 2,250 BP, the transformation of habitats likely accelerated with the arrival of people into the region from the Orinoco Drainage and other river systems of South America's northeastern coast. These newcomers made ceramics and grew food crops. They brought with them dogs and domesticated plants from the mainland that included manioc or cassava (*Manihot esculenta*), the primary food crop (Wilson 1997). Ceramic Age cultivation of corn (*Zea mays*), sweet potato (*Ipomoea batatas*), and manioc (among other cultivated food crops) certainly would have required extensive expanses of cleared ground (although corn may not have been as widely cultivated in the Antilles as on the Neotropical mainland, being overshadowed by the primary system of root-cropping) (Newsom and Wing 2004; but see Lane *et al.* 2008). Destruction of habitats increased exponentially with the arrival of European settlers in the late fifteenth century. Forest giants were converted to lumber for building houses and boats, and vast expanses of forest were cleared for agriculture, most notably for sugarcane and later for bananas. In addition, the use of woody plants for fuel (wood or charcoal) occurred wherever people settled, and that exploitation continues to this day. The landscapes of many islands became dissected by plantations, creating a patchwork of habitat-islands that dictated intra-island distributions capable of inhibiting gene flow among anuran populations.

Myers *et al.* (2000) and Smith *et al.* (2005) considered the West Indies a biodiversity hotspot. Only 11.3% of the original extent of primary vegetation remains in the region (Myers *et al.* 2000), and a more realistic estimate of intact forests might be 10% (Hedges 2006). Despite the exploitation by native frogs of introduced coconut palms, banana, cacao, and coffee trees, as well as human edifices and associated debris, relatively few species have benefited from human-influenced resources (Henderson and Powell 2001). Introduced orchard and forest trees (e.g. mahogany) may simulate natural patterns of vegetation, but complex habitats are used more frequently by more species, and habitats associated with clear-cuts are rarely used by amphibians (Henderson and Powell 2001).

For example, 8,000 years ago, about 63% of Grenada was covered in forest; in 2000, forests covered only an estimated 15% of the island (EarthTrends 2003). On a more positive note, satellite image-based mapping on Grenada indicated that cultivated land areas declined 65% from about 1945 to 2000, whereas dry scrub woodlands increased 716% and evergreen forest and forest/shrub increased by 83% (Helmer *et al.* 2008). Between 2005 and 2010, total forested areas declined not at all on Grenada (Table 9.2), and the area of 'primary forest' on Grenada did not change (if, indeed, any primary forest remains) (FAO 2010).

Caribbean islands are popular tourist destinations, and tourism is important to the economies of many insular nations. Tourism also places pressure on the islands to provide housing and vacation-related activities for visitors. Expansive resorts and golf courses can affect a variety of habitats, especially those in coastal areas. For instance, Helmer *et al.* (2008) indicated that changes in land-cover on Grenada from 1945 to about 2000 showed an increase of 1,458% in urban or developed land and areas with no vegetation and a 65% decrease in cultivated land. Helmer *et al.* (2008) found similar patterns on other islands (i.e. St Christopher, Nevis, St Eustatius, Barbados), but the extant anurans on these islands were introduced.

Some islands promote and feature nature-related tourism, but such 'nature' tours often are nothing more than a short ride in a minibus and an abbreviated hike in a national park or reserve – nevertheless, they do provide incentives for conserving the few remaining natural areas.

C. Accidental and intentional killing

For terrestrial wildlife, the major threat from accidental killing probably occurs on roads. Anurans (e.g. *Rhinella marina*) often are victims of automobiles on rainy nights. As the resident and transient human populations increase, so too will paved roads, automobile traffic, and wildlife mortality. The exploitation of *Leptodactylus fallax* has a long history (Martin *et al.* 2007), but prior to the population crash in 2009–10, as many as fifty individuals of the critically endangered frog were killed daily for consumption by local people and tourists on Montserrat (Daltry and Gray 1999). See further below.

D. Commercial exploitation

Schlaepfer *et al.* (2005) noted that 'species with restricted ranges [and] high levels of endemism [...] could be detrimentally affected by even a small number of individuals being removed from the wild'. Legal, commercial exploitation of West Indian frogs appears to be of minimal concern, although active animal vendors are based in Barbados and St Maarten (Powell *et al.* 2011). Most regional governments do not allow commercial collection and exportation of native wildlife. However, *Leptodactylus fallax* has been unsustainably hunted for domestic consumption as a traditional dish and for sale to restaurants, where it is offered to tourists (Martin *et al.* 2007). Prior to the 1996 eruption of the Sufrière Hills Volcano on Montserrat, roughly 60–80 hunters were active, and 15–20 restaurants were serving 'mountain chicken'. By 2007, only about eight hunters were active, although most hunted only occasionally. However, one person hunted frequently and was not selective, taking males, females, adults, and juveniles throughout the year (McCauley and Mendes 2006). On Dominica, hunting has been regulated since 1976, and legislation in 2003 gave full protection to *L. fallax*.

E. Restricted population sizes and ranges

Many West Indian endemics are geographically limited to either a single small island or to a small area on a larger island and consequently are vulnerable to stochastic events, including inadvertent or intentional alterations of already limited habitats. For example, *Pristimantis euphronides* has a range of at most 16 km^2 on Grenada and the range of *P. shrevei* on St Vincent is at most 90 km^2 (Hedges 1999).

F. Climatic changes

The earth is undergoing human-mediated climatic changes that are adversely affecting biodiversity on a global scale (e.g. Vinnikov *et al.* 1999; Cahill *et al.* 2013; Heatwole 2013). Effects can be direct (e.g. Stewart 1995; Burrowes *et al.* 2004) or indirect, mainly by synergistically enhancing the impact of invasive species (e.g. Hellmann *et al.* 2008; Mainka and Howard 2010) or alien diseases and their vectors (see below). Fernandez Amaya and Mahon (2011) documented the vulnerability of Union Island (St Vincent Grenadines) to climatic change, although they did not directly assess the potential effects on biodiversity. Subtle but insidious responses by frogs to a warming climate include lowland species moving to areas at higher elevations, where they might displace already stressed high-elevation endemics (e.g. Huey *et al.* 2012).

Long-term and short-term effects of altered temperature and rainfall regimes have been studied in Puerto Rico (Stewart 1995; Burrowes *et al.* 2004) and have been implicated in declines in many frog populations. Although we cannot unequivocally state that climatic change is responsible, declining population numbers of *P. euphronides* could, at least in part, be attributable to several consecutive dry years on Grenada in concert with other contributing factors that include the continuing spread of *Eleutherodactylus johnstonei*, Hurricane Ivan, and the fungus *Batrachochytrium dendrobatidis* (Henderson and Berg 2011; Berg *et al.* 2013).

G. Hurricanes and volcanic eruptions

The threat of hurricanes is an annual event in the Lesser Antilles, and hurricanes can have substantial effects on vegetation, the structure of forests, and biotic communities (e.g. Walker *et al.* 1991). Because the islands have been subjected to hurricanes for thousands of years, however, they probably have caused few extinctions or even extirpations. On the other hand, because of the intermittent nature and unpredictability of hurricanes, native species probably have not evolved behaviours adaptive for surviving these events.

On Grenada, Henderson and Berg (2005) documented short-term effects of Hurricane Ivan in September 2004. At that time (November 2004), the frog *Pristimantis euphronides* was encountered at pre-hurricane levels at Grand Etang despite the absence of crown vegetation providing shaded daytime retreats. However, surveys at the same site in December 2007 and March 2008 indicated a substantive decrease in the numbers of *P. euphronides*. What we do not know is how many animals perish during a hurricane; the numbers could be substantial – and effects on populations already small for various other reasons could be catastrophic. Some populations of *P. euphronides* (e.g. Grand Etang) have not recovered from Hurricane Ivan.

Volcanic eruptions, although less frequent than hurricanes, can be more devastating. Each eruption has the potential of killing virtually everything in its path and is an important and natural source of catastrophic disturbance to ecological communities. The 1979 eruption of La Soufrière on St Vincent was considered less extensive than that of 1902 (Fiske and Sigurdsson 1982), and the endemic frog *Pristimantis shrevei* was collected near the summit in 1990 (Kaiser *et al.* 1994c). Kaiser and Henderson (1994) reported that ground vegetation was beginning to cover the bare volcanic debris in 1992, and *Eleutherodactylus johnstonei* was found syntopically with *P. shrevei* at that time. More recently, on Montserrat during the mid-1990s, the Sufrière Hills Volcano erupted and caused widespread damage to the environment and native wildlife (Daltry and Gray 1999; Dalsgaard *et al.* 2007; Martin *et al.* 2007; Hilton *et al.* 2008; Muñoz and Hewlett 2011; McTighe 2015). Martin *et al.* (2007) noted complete destruction of forests in the South Sufrière Hills by pyroclastic flows and degradation by ash falls in the Centre Hills.

H. Alien pathogens and parasites

The fungal parasite *Batrachochytrium dendrobatidis* (*Bd*) causes chytridiomycosis (or chytrid), which has been implicated in amphibian extinctions and population declines throughout the world but especially in the Neotropics (e.g. Whitfield *et al.* 2012), including the West Indies (e.g. Díaz *et al.* 2007). It has been documented in *Leptodactylus fallax* on Dominica (e.g. Magin 2004; Malhotra *et al.* 2007; 2011) and Montserrat (e.g. Garcia *et al.* 2009; Adams *et al.* 2014); in *E. johnstonei* and *Pristimantis shrevei* on St Vincent (Sweeney 2016); *E. johnstonei, Leptodactylus validus*, and *P. euphronides* on Grenada (Henderson and Berg 2011). To date, no infections have been found in *Rhinella marina* on St Vincent (Sweeney 2016) or Grenada (e.g. Drake *et al.* 2014). Rödder *et al.* (2010) presented models that showed many areas of the world less suitable for *Bd* as a consequence of projected changes in climate; however, they also noted that the 'tropics are least suitable for *Bd*, **with the exception of highland regions**' (bold added for emphasis) – exactly where the most threatened species of Lesser Antillean frogs occur.

Many parasites infecting native amphibians also are native and probably evolved in conjunction with their hosts. Others undoubtedly have been introduced with the many alien species that have become established. Effects of such invaders on native hosts are unknown.

I. Environmental pollution

Although we have no documented evidence demonstrating the impact of pesticides on any species of the West Indian herpetofauna, anecdotal accounts suggest that they have had a deleterious and sometimes substantial effect on some species of Antillean frogs. Sugarcane dominated insular economies until the late nineteenth century. Today, cultivation of bananas is the principal wide-scale cash crop on larger islands, and plantations cover large expanses on many Lesser Antillean islands, including formerly forested hillsides. This industry has introduced the subsidized use of chemical fertilizers and extensive applications of pesticides and herbicides. Powell and Henderson (2007; 2011) noted that these chemicals might not have directly affected the frogs but surely had a negative impact on their arthropodan prey – and, for all we know, a 'direct' impact may have been felt as anurans absorbed chemicals through their skin. Short of finding animals dead in the field, we may never be able to assess the full impact of pesticides (Henderson and Powell 2009).

J. Over-population

The West Indies already include several of the most densely populated countries in the world (e.g. Barbados, St Martin/St Maarten) (Table 9.2), and the human populations on some Lesser Antillean islands are expected to double between mid-1994 and 2025 (Jenkins and Harcourt 1996). The growth of the human population on Antillean islands (as elsewhere) aggravates many of the factors already negatively affecting the herpetofauna. More people need additional housing and infrastructure, which in turn require more development – and more urban (and agricultural and industrial) development means that more habitats will be transformed to accommodate and sustain those additional people. Development also may require that building materials be imported from other islands or, more likely, from the United States. Those materials may harbour non-native species, thereby introducing aliens that may eat, compete with, or infect with equally alien parasites and pathogens native to a particular island. Such alterations of habitats render species already stressed by the contraction of habitats even more vulnerable to extirpation or extinction (Henderson and Powell 2009).

III. The native amphibians

The native amphibians of the Lesser Antilles comprise ten (maybe 11 or 12) species in four (or five) families. Nine (maybe 10) are regional endemics and six (maybe 7) occur on single islands (see the following paragraph). Of the regional endemics, *Eleutherodactylus johnstonei* has been widely introduced within and beyond the region and *E. martinicensis* is established on other islands within the region, whereas multiple attempts to introduce *Leptodactylus fallax* on other islands have failed. In addition to St Vincent, Grenada, and at least one of the Grenadines, *Leptodactylus validus* occurs on Trinidad and Tobago and northeastern South America. Although generally considered an introduced species throughout the Antilles (and so treated herein), *Rhinella marina*, native to northern South America east of the Andes, might well have reached some of the more southerly Lesser Antilles by natural overwater dispersal.

L'Herminier (1817, in Breuil 2009) mentioned the presence of a frog on St Barthélemy under the name *Rana*. In 1821, A. Plée collected a frog on St Barthélemy that he described as being similar to those of Puerto Rico and Martinique and lacking webbed feet (A. Plée in Breuil 2002). This frog appears to be present in the collection of the Muséum national d'Histoire naturelle in Paris as MNHN 4881 (Breuil 2002). This species, listed by Questel (2018) as *Eleutherodactylus* sp., if endemic to St Barthélemy, is almost certainly extinct. We have not included this record among the enumerated native species nor have we included an account in the following section.

Because many species of Lesser Antillean frogs have multi-island distributions and, therefore, in many cases, occur in more than one country, a country-by-country assessment of these species would include a great deal of redundancy. Instead, in each of the following species accounts, we list countries of occurrence, Appendix 9.I lists the species of frogs that occur on each island, and Appendix 9.II lists all of the islands on which each species is known to occur.

A. *Allobates chalcopis* Kaiser, Coloma, and Gray, 1994 (Fig. 9.2A)

1. Distribution

This species is endemic to Martinique. The type locality is on the southeastern slope of Montagne Pelée at ~500 m asl, but fieldwork in 2011 failed to find *A. chalcopsis* at that site. It was, however, found on the summit of Montagne Pelée from ~800 to ~1,400 m asl (Fouquet *et al.* 2013).

2. Relationships

'*Allobates chalcopsis* is unambiguously embedded within *Allobates*' (Fouquet *et al.* 2013) and is nested within the *A. trilineatus* clade. This clade comprises species largely from northern South America (Grant *et al.* 2006; Fouquet *et al.* 2013).

Figure 9.2. Photographs of amphibians from the Lesser Antilles. (A) *Allobates chalcopis* from Martinique. (photograph by Maël Dewynter); (B) *Eleutherodactylus amplinympha* from Dominica (photograph by Robert Powell); (C) *Eleutherodactylus barlagnei* from Basse-Terre (Guadeloupe) (photograph by Renaud Boistel): (D) *Eleutherodactylus johnstonei* from St Vincent (photograph by Robert Powell); (E) *Eleutherodactylus martinicensis* from Dominica (photograph by Robert Powell); (F) *Eleutherodactylus pinchoni* from Basse-Terre (Guadeloupe) (photograph by Renaud Boistel).

3. Habitat and natural history

The type series was collected in a montane rainforest at ~500 m asl consisting of shrub thickets, palm brakes, ferns, and stands of bamboo (Kaiser *et al.* 1994d). More recently (2011; Fouquet 2013), it was taken in a 'transitional' situation (at ~800 m asl) between forest and a savannah dominated by ferns (*Dicranopteris* spp.) and Poaceae; calling males were found under 1.0–1.5 m of dense herbaceous vegetation. At ~1,200 m asl the density of calling males maximized in a habitat of savannahs consisting of Poaceae, Bromeliaceae, lycopods, and scattered mats of *Sphagnum* spp. and shrubs (*Charianthus nodosus*). The species demonstrated a crepuscular activity pattern based on peaks in calling (Kaiser *et al.* 1994d), and it was heard calling up to the summit of Montagne Pelée (Fouquet *et al.* 2013). Eggs are laid on the ground in a nest under leaves and guarded by the adults (Hedges *et al.* 2010a).

4. Conservation status, threats, and actions

Kaiser and Henderson (1994) described this species as locally abundant at the type locality. They did, however, note that the site was used for unlawful disposal of garbage and that several free-flowing water pipes with uncontrolled runoff into ravines attracted people who wanted to wash their cars and do laundry at the site. Because of this degradation of the habitat, they recommended the establishment of a sanctuary to preserve an area in its pristine condition as a means of protecting populations of this frog.

This species was assessed as VU on the IUCN Red List (Hedges *et al.* 2010a). The assessment determined that the current population trend was stable and no current major threats affected the species; however, its small range made it susceptible to stochastic events, especially volcanic eruptions. Fouquet *et al.* (2013) argued convincingly that it should be reassessed and designated CR. The species has experienced either a shift in its range or a severe reduction thereof, possibly associated with a population decline since about the turn of the millennium. Its range has been reduced dramatically, from 33 km^2 to 3.7 km^2, which could be associated with climatic change, and 'there is simply no higher ground available for the species' (Fouquet *et al.* 2013). The current geographic range of *A. chalcopis* lies entirely within the Réserve Biologique Intégrale de la Montagne Pelée, a protected area that is managed by the *Office National des Forêts – Direction régionale de la Martinique*. The species is not, however, protected by any national conservation law. In addition to an IUCN Red List reassessment, Fouquet *et al.* (2013) recommended a national conservation programme to protect *A. chalcopis*.

B. *Eleutherodactylus amplinympha* Kaiser, Green, and Schmid, 1994 (Fig. 9.2B)

1. Distribution

This species is endemic to Dominica, where it has an elevational range between 300 m and 1,200 m.

2. Relationships

Eleutherodactylus amplinympha is a member of the *E. martinicensis* species group (Hedges *et al.* 2008), and *E. amplinympha* and *E. martinicensis* are sister species (Kaiser *et al.* 1994b). Note, however, that Yuan *et al.* (2022) found no evidence supporting this sister-group relationship.

3. Habitat and natural history

This species occurs in a transition zone between montane rainforest (few trees, height < 20 m; shrub thickets, palm brakes, and ferns) and elfin woodland. Kaiser *et al.* (1994b) considered the low, thicket-like forest 'an ideal refugium for anurans and nearly impenetrable to humans'. *Eleutherodactylus amplinympha* was most abundant in the area of Morne Macaque in Morne Trois Pitons National Park, where a density of ~18 frogs/m^2 was observed in a pile of flat rocks. Frogs were found perched on the ground and low vegetation at heights < 1.0 m at Boeri in Lake Morne Trois Pitons National Park (Carter *et al.* 2009). A single terrestrial egg mass was found in a rock crevice near Freshwater Lake; it was on a thin mat of ground moss and was attended by a male (Kaiser *et al.* 1994b). This species is broadly sympatric with *E. martinicensis* and syntopic at some sites (Malhotra and Thorpe 1999).

4. Conservation status, threats, and actions

Although Kaiser and Henderson (1994) described this species as abundant in localities surveyed, they expressed concern about a hydroelectric development in the southern and eastern parts of Morne Trois Pitons National Park (at Freshwater Lake) that had damaged the previously pristine elfin woodland. They considered the species 'well protected, barring further disturbances of the protected zones'. This species has subsequently been assessed as EN on the IUCN Red List (Hedges and Powell 2010a) due to habitat degradation (small-scale farming and selective logging), especially in the southern portion of its range. Those authors recommended improved habitat protection 'at lower elevations, particularly in the south'.

C. *Eleutherodactylus barlagnei* Lynch, 1965 (Fig. 9.2C)

1. Distribution

This species is restricted to the Basse-Terre portion of Guadeloupe, where it inhabits streams and valleys from sea level to ~1,300 m.

2. Relationships

Eleutherodactylus barlagnei is a member of the *E. martinicensis* species group (Hedges *et al.* 2008) and, according to Kaiser *et al.* (1994b), is considered the sister species of *E. pinchoni.*

3. Habitat and natural history

This semi-aquatic species occurs in the humid forest as well as in elfin woodlands on the upper reaches of Soufrière, where it is associated with boulder-strewn mountain torrents (Breuil 2002) and grassy savannahs (Hedges *et al.* 2010b). It also tolerates disturbed habitats, but only when trees occur along streams (Stuart *et al.* 2008; Hedges *et al.* 2010b). In streams at night, it perches on boulders (often moss-covered) with diameters of 0.3–1.8 m (Schwartz 1967). During the day, it may be encountered under rocks embedded in coarse sand on small bars in streams, and males call from rocks in slack water along the edges (Schwartz 1967). It occurs syntopically with *E. johnstonei* and *E. martinicensis* at some sites (Breuil 2002).

4. Conservation status, threats, and actions

As this species is dependent on fast-flowing water, Kaiser and Henderson (1994) expressed concern regarding its sensitivity to the effects of upstream pollution. However, they described the species as 'common' in sites at a lower elevation and 'abundant at higher elevations'. Furthermore, based on protected natural areas on the island, these authors considered 'the species sufficiently protected at this time'. More recently, however, and based on IUCN guidelines, *E. barlagnei* has been assessed as EN (Hedges *et al.* 2010b). Threats to the species include loss of forest habitat to agriculture (especially banana plantations) and for construction of human settlements, as well as pollution from pesticides used on bananas and from domestic sources. In addition, introduced predators (rats, mongooses, cats) undoubtedly take frogs, and the introduced and ubiquitous *E. johnstonei* might provide serious competition (Hedges *et al.* 2010b). Lorvelec *et al.* (2011) noted that Guadeloupe is under French legislation and, under that legislation, *E. barlagnei* is protected internationally. However, Hedges *et al.* (2010b) remarked that Parc National de la Guadeloupe, where *E. barlagnei* occurs, 'is not well managed for biodiversity conservation' and '[t]here is a need to expand and strengthen the existing protected area coverage of Guadeloupe'. The species certainly warrants careful monitoring, especially in regard to introduced predators and the threat of chytrid.

D. *Eleutherodactylus johnstonei* Barbour, 1914 (Fig. 9.2D)

1. Distribution

Eleutherodactylus johnstonei ranges throughout the Lesser Antilles (in large part mediated by humans), where the elevational distribution extends from sea level to at least 1,300 m. A high degree of uncertainty prevailed until recently regarding the origin of the species, which had been considered native to the Antigua and Barbuda Bank (Pregill *et al.* 1994) or possibly Montserrat, Nevis, St Kitts, or St Lucia (Kaiser 1997), all islands with only one native frog, until Yuan *et al.* (2022) demonstrated that one of two major clades within the species almost certainly originated on Montserrat. In addition to the islands previously considered parts of the native range, although they may now possibly be extirpated on Barbuda, these frogs are known from Anguilla (where the population might be extirpated), St Barthélemy, St Maarten/St Martin,

Saba, St Eustatius, St Christopher, Guadeloupe, Terre-de-Haut (Les Saintes), La Désirade, Marie-Galante, Martinique, St Vincent, the Grenadines (Bequia, Mustique, Canouan, Union Island, Petit St Vincent, Carriacou), Grenada (including Hog Island), and Barbados (e.g. Kaiser and Hardy 1994a; Malhotra and Thorpe 1999; Henderson and Powell 2009; Powell *et al.* 2011; Henderson and Breuil 2012; Frost 2019); on Dominica, this species was restricted to a narrow strip of the Caribbean coast and probably is extirpated (Kaiser 1992). Introductions outside the Lesser Antilles are known from Bermuda, Jamaica, Puerto Rico (a single individual at a local hotel with no evidence of any further introductions), Tortola (British Virgin Islands), Aruba, Bonaire, Curaçao, Trinidad, San José (Costa Rica), Panama, Venezuela, Barranquilla and Tolima (Colombia), Georgetown (Guyana), coastal French Guiana, probably coastal Surinam, and possibly Brazil (Henderson and Powell 2009; Powell *et al.* 2011; Sánchez Muñoz 2011; Melo *et al.* 2014; Frost 2019). Although the type locality for *E. johnstonei* is on Grenada, it was introduced to that island from Barbados (where it also was introduced) in about 1885 (Barbour 1914).

2. Relationships

Based on evidence from electrophoresis, morphology, chromosomes, and vocalizations, Kaiser *et al.* (1994b) found *E. johnstonei* to be sister-taxon to a clade comprising *E. amplinympha* and *E. martinicensis*. *Eleutherodactylus johnstonei* is in the *E. martinicensis* species group of Hedges *et al.* (2008), along with other Lesser Antillean congeners. This species often was confused with *E. martinicensis* in the older literature (Schwartz 1967; Kaiser and Hardy 1994a; Kaiser *et al.* 1994a, 1994b). However, Yuan *et al.* (2022) found no support for the assumption that *E. johnstonei* is sister to a clade comprised of *E. amplinympha* and *E. martinicensis*.

Yuan *et al.* (2022) documented two major mitochondrial clades within *E. johnstonei*. The western clade, in which males are significantly smaller than those of the eastern clade, occurs on Montserrat, Nevis, Saba, St Kitts, St Eustatius, St Maarten/St Martin, and Bermuda, suggesting that all but the one on Bermuda might have dispersed naturally (rather than via human mediation). The eastern clade, which is not derived from Montserratian populations and the origin of which is unknown, accounts for frogs on Antigua, Guadeloupe, Martinique, St Lucia, St Vincent, Grenada, Barbados, and for all of those sampled beyond the Lesser Antilles (Jamaica, Curaçao, Trinidad, and the mainland) except Bermuda.

3. Habitat and natural history

Eleutherodactylus johnstonei is the anuran version of a weed. Individuals of this species can live in a variety of habitats, become established in many places, thrive in disturbed situations in close proximity to human activity, reproduce quickly, and apparently resist eradication. Of the 47 islands in the Lesser Antilles that harbour at least one species of frog, *E. johnstonei* occurs on 42, although it is considered native only on Montserrat and one other unknown island (Yuan *et al.* 2022). This species occurs mainly in disturbed habitats, including mowed fields, yards, gardens, sugarcane fields, towns, houses, plantations (Henderson and Powell 2009), and, when in forests, usually in clearings or gaps (Hedges *et al.* 2010d). A colony was found initially on Anguilla at the Agriculture Station (Censky 1989), one of the wettest places on the island, due in part to the irrigation of plant seedlings; subsequently, it has been encountered in watered yards and gardens (Hodge *et al.* 2003). On St Martin/St Maarten, it is found in watered gardens and plant nurseries, and it readily takes advantage of suitable artificial conditions around human dwellings. It is abundant in generally moist habitats at a higher elevation on Saba, where it occurs in leaf litter, under rocks and logs, in bromeliads, and water tanks, and is often edificarian (Henderson and Powell 2009). On St Eustatius, *E. johnstonei* is abundant only on the Quill (especially in the crater), but also occurs in moist gardens (Powell *et al.* 2015). It has been encountered under rocks and in a coastal *Terminalia* grove on St Christopher (Schwartz 1967). It occupies forest remnants on

Antigua, particularly along streams, but also watered yards and gardens and is often edificarian (Henderson and Powell 2009). On Marie-Galante, it was associated with sugarcane fields and gardens (Henderson *et al.* 1992). Residential gardens and roadside and beachside vegetation were favoured on Dominica (Kaiser and Wagenseil 1995). On Martinique, Breuil (2009) found it in dry and wet forests and often in bromeliads. On St Lucia, it readily enters homes (Long 1974). On St Vincent, Mallery *et al.* (2007) found the species to be 'ubiquitous' in urban (cement ditches, houses, gardens), agricultural, and more or less natural areas, both xeric and mesic. Díaz-Lameiro *et al.* (2007) and Sweeney (2015) found these frogs in dense forests, even at a high elevation otherwise dominated by *Pristimantis shrevei*, in streambeds, leaf litter, and urban alleys. Germano *et al.* (2003) also found *E. johnstonei* to be 'ubiquitous' on Grenada in urban situations (homes, cement ditches, small parks, roadside ditches), agricultural sites, and in more or less natural areas at elevations below 600 m. Diurnal retreats on Grenada included the interior of an indoor toilet (Henderson 2010) and the water tank of a toilet on Carriacou, which contained 50–75 subadults during the dry season (RWH), cisterns (Harrison *et al.* 2011), and drainage pipes in a retaining wall (RP, RWH). On Barbados, Fields and Horrocks (2011) found *E. johnstonei* to be distributed island-wide and especially abundant in disturbed habitats (cane fields, gullies, gardens). Refugia on Barbados, which might be used by 1–9 frogs, included bromeliads and cracks and cavities under and within rocks (Ovaska 1991a).

Daltry (2011) described *E. johnstonei* as 'abundant' on Antigua, one of the islands on which it had been considered native (Kaiser 1997; Yuan *et al.* 2022). Harrison *et al.* (2011) and Berg *et al.* (2013) noted that numbers on Grenada had remained relatively stable even during recent (2007; 2008; 2010) dry periods. Sweeney (2015), despite selecting high-elevation sites on St Vincent that strongly favoured *Pristimantis shrevei*, found *E. johnstonei* during transect surveys at all but one (Montreal) of seven locations. On Barbados, Ovaska (1991a) recorded densities of 0.32–2.73 frogs/m^2 in various habitats. Diet on Barbados included ants (in 79.6% of 113 stomachs), springtails (36.3%), leafhoppers (35.4%), adult dipterans (23.9%), earwigs (22.1%), and mites (20.4%); ants composed 51.4% of the total number of prey items ($n = 939$; Ovaska 1991b). de Silva and Wilson (2006) found eggs on Mustique in damp areas behind tree bark, under stones, and in the leaf bracts of spreading airplants (*Tillandsia utriculata*). On Grenada, Harrison *et al.* (2011) found females laying clutches of 5–30 eggs, which typically are attended by the male until hatching, although Townsend (1996) indicated that both sexes attend nests.

4. Conservation status, threats, and actions

The species was assessed as LC on the IUCN Red List (Hedges *et al.* 2010e). This widely distributed and often invasive species is common and adaptable, with both native and introduced populations occurring in both intensely altered and natural habitats. It is described as abundant on Antigua, where the chytrid fungus is the greatest potential threat to the species (Daltry 2011). Concerns pertaining to this species revolve primarily around its impact on populations of native frogs. On Basse-Terre (Guadeloupe), Kaiser (1997) suspected *E. johnstonei* to be a competitor of the two endemic species of *Eleutherodactylus*; on Martinique, it has replaced *E. martinicensis* in terrestrial habitats (Breuil 1997). Although restricted to a narrow strip of disturbed habitat on Dominica, if established, *E. johnstonei* has the potential to compete with *E. martinicensis* (Kaiser and Wagenseil 1995) and, if not now, eventually with the endemic *E. amplinympha*. This species has certainly impacted the endemic strabomantid frogs on St Vincent (*Pristimantis shrevei*) and Grenada (*P. euphronides*), with which it competes and apparently is displacing in some areas (Kaiser *et al.* 1994c; Henderson and Berg 2006; 2011; Powell and Henderson 2007; 2011; Harrison *et al.* 2011; Berg *et al.* 2013; Sweeney 2015; Henderson and Powell 2018). Henderson and Berg (2011) provided statistical evidence indicating that over time at Grand Etang (Grenada), *E. johnstonei* has contributed to the decline of *P. euphronides*. In addition, the chytrid fungus has been documented in several populations of *E. johnstonei* (Henderson and Berg 2011; Berg *et al.* 2013; Sweeney 2015; 2016) and the species might serve as a vector for this disease.

E. *Eleutherodactylus martinicensis* (Tschudi, 1838) (Fig. 9.2E)

1. Distribution

The natural range of this species is in question, although it is most likely endemic to Martinique and/or Dominica. It is known from Antigua, Guadeloupe, Îlet à Kahouanne, Îlet-à-Cochons, Marie-Galante, La Désirade, the Île des Saintes, Dominica, and Martinique. It has been encountered in areas at elevations ranging from sea level to at least 1,250 m asl (on Guadeloupe). Lescure (2000) proposed Dominica, Martinique, and St Lucia as the native range for the species, but it apparently is absent from St Lucia (Kaiser 1992; Kaiser and Hardy 1994b; Malhotra and Thorpe 1999), and populations on Antigua and the Guadeloupean Archipelago likely are introduced. The populations on St Martin/St Maarten and St Barthélemy also are introduced, likely in the early 1980s (Magras 1992; Breuil 2002). Although its occurrence on Antigua has been questioned (Pregill *et al.* 1988; Kaiser and Hardy 1994b), recent publications do place it on that island (e.g. Hedges *et al.* 2010d; Daltry 2011).

Besides islands on which it does occur, Barbour (1914: 248), placed *E. martinicensis* on Jamaica, St Christopher, Saba, Montserrat, and St Eustatius. Those records almost certainly refer to *E. johnstonei*, which Barbour (1914: 249) described as a new species on the page following his account of *E. martinicensis*.

2. Relationships

Based on evidence from electrophoresis, morphology, chromosomes, and vocalizations, Kaiser *et al.* (1994b) found *E. martinicensis* and *E. amplinympha* in a sister-group relationship, with *E. johnstonei* as the sister-taxon to that clade. *Eleutherodactylus martinicensis* is in the *E. martinicensis* species group of Hedges *et al.* (2008), along with other Lesser Antillean congeners. Note, however, that Yuan *et al.* (2022) found no evidence to support this sister-group relationship.

3. Habitat and natural history

This species is primarily a denizen of mesic forests but has been collected in dry forests as well. In addition, it has been encountered in disturbed situations, including *Musa* plantations, sugarcane fields, gardens, and human edifices (Hedges *et al.* 2010d). On Antigua, it was taken from beneath stones and in beachside coconut trash (Lynn 1957). It occurs in mesic and xeric forests on Guadeloupe, as well as in savannahs, orchards, greenhouses, washrooms, in *Cocos* trash, and associated with *Clusia* sp. and *Cyathea* sp. high on Soufrière (Breuil 2002). On Dominica, it has been encountered in rainforest, dry woodlands, cacao and banana groves, in a *Coccoloba* stand, and on rocks in montane streams (Schwartz 1967; Ackley *et al.* 2009).

Eleutherodactylus martinicensis 'is a very common species', especially in undisturbed habitats (Hedges *et al.* 2010d). It is nocturnal and has been observed active in the rainforest at the top of Pic Napoleon (Soufrière, Guadeloupe) at an ambient temperature of 14.5°C (Schwartz and Henderson 1991). It will hunt insects attracted to lights at night. Males call from small mounds of dirt, leaves, and branches (rarely > 2.0 m above the ground; Breuil 2002). Egg clutches have been found in a bromeliad, under palm fibre (29 eggs) in a *Cocos* grove, and in a rolled palm petiole (40 eggs; Schwartz and Henderson 1991).

4. Conservation status, threats, and actions

Eleutherodactylus martinicensis has been assessed as NT on the IUCN Red List (Hedges *et al.* 2010d) based on its Extent of Occurrence (likely < 5,000 km^2) and declining habitat. Although it is 'not seriously threatened', the species is affected by introduced predators (rats, cats, mongooses), by loss of forests to agricultural, residential, and commercial development, and possibly by pesticides. Furthermore, *E. johnstonei* 'is perhaps a potential competitor, and appears to be replacing this species in open areas' (Hedges *et al.* 2010d). The species does occur in protected areas in portions of its range.

F. *Eleutherodactylus pinchoni* Schwartz, 1967 (Fig. 9.2F)

1. Distribution

Eleutherodactylus pinchoni is restricted to the Basse-Terre portion of Guadeloupe, where it is widely distributed in habitats at elevations ranging from 180 m to ~1,470 m (Breuil 2002).

2. Relationships

Eleutherodactylus pinchoni is a member of the *E. martinicensis* species group (Hedges *et al.* 2008) and is considered the sister species of *E. barlagnei* (Kaiser *et al.* 1994b).

3. Habitat and natural history

This species is specialized for life in large moss mats found in forests and elfin woodlands; it also occurs in creeping ground vegetation (Kaiser and Henderson 1994). Schwartz (1967) encountered it in rainforest, cacao groves (including in wet husks), forest litter, under wet banana trash, beneath palm husks and fronds at the edges of forests, and under logs and rocks in densely shaded forests. It does occur in secondary forests, but not outside forested habitats (Stuart *et al.* 2008). According to Hedges *et al.* (2010c), '[t]he forest habitat on Guadeloupe is fragile, small in extent, and habitat quality continues to decline'.

Males call from the ground, low herbaceous growth (~15 cm above the ground), and from deep grass (<1.0 m above the ground) (Schwartz 1967). Eggs are laid on the ground and in bromeliads; they develop directly, without a larval stage (Hedges *et al.* 2010c).

4. Conservation status, threats, and actions

Kaiser and Henderson (1994) considered *E. pinchoni* 'common' throughout Basse-Terre, and 'abundant' in elfin woodlands on the slopes of La Soufrière, and suggested that potential external pressures on *E. pinchoni* were minimal. More recently, however, and based on IUCN Red List guidelines, *E. pinchoni* has been assessed as EN (Hedges *et al.* 2010c). This assessment noted a general decrease in numbers in the northern portion of the species' range, possibly due to alterations of its habitat. Numbers also have decreased around Volcán Soufrière in the southern part of its range since 1991, probably due to volcanic eruptions. Major threats to the species include loss of forest habitat to agriculture (especially banana plantations) and for construction of human settlements, as well as pollution from pesticides used on bananas. Additional threats to *E. pinchoni* include introduced predators (rats, mongooses, cats) and the introduced frog *Eleutherodactylus johnstonei* 'appears to be displacing this species' (Hedges *et al.* 2010c). Lorvelec *et al.* (2011) noted that Guadeloupe is under French legislation, and *E. pinchoni* is therefore protected internationally. Hedges *et al.* (2010c) noted that it occurs in Parc National de la Guadeloupe, but that additional protection of the forest habitat is required in addition to controlling introduced predators within existing protected areas.

G. *Leptodactylus fallax* Müller, 1926 (Fig. 9.3G)

1. Distribution

This species presumably is extant only on the leeward coasts of Dominica and Montserrat in mesic habitats at elevations ranging from sea level to about 400 m on Dominica and about 1,000 m on Montserrat (Henderson and Powell 2009) – although the IUCN SSC Amphibian Specialist Group (2017) listed the upper elevational limit as 430 m for both islands. The species might be extirpated in the wild on Montserrat but reintroduction efforts are underway (see below).

Figure 9.3. Photographs of amphibians from the Lesser Antilles. (A) *Leptodactylus fallax* from Montserrat (photograph by Gerardo García); (B) *Leptodactylus validus* from St Vincent (photograph by Robert Powell); (C) *Pristimantis euphronides* from Grenada (photograph by Billie Harrison); (D) *Pristimantis shrevei* from St Vincent (photograph by Robert Powell); (E) *Rhinella marina* from Grenada (photograph by Richard A. Sajdak); (F) *Eleutherodactylus planirostris* from St Martin (photograph by Mark Yokoyama).

Preserved specimens purportedly from St Kitts are in the Museum of Comparative Zoology at Harvard, but no frogs have been seen on that island since 1881 (Kaiser 1994). Reports of this species on Antigua, Guadeloupe, Martinique, and St Lucia are not documented by specimens (Kaiser 1994), but historical information in Lescure (1979a, 1979b) suggests that these frogs did occur on Guadeloupe and Martinique. Attempts to introduce *L. fallax* on Grenada (Groome 1970), Martinique (Lescure 1983), and Puerto Rico (Barbour 1937) failed. Based on a thorough review of the literature and searches of museum specimens, King and Ashmore (2014) concluded that the species is unlikely to have ever occurred on Jamaica, Hispaniola (Dominican Republic), Puerto Rico, and Trinidad (although some frogs might have existed for a short period on Puerto Rico after an attempted introduction); might or might not have occurred on Nevis and Guadeloupe; probably occurred on Antigua and St Lucia; and did occur on St Christopher (St Kitts) and Martinique (but populations no longer exist).

2. Relationships

Leptodactylus fallax belongs to the *L. pentadactylus* group (Heyer 1979; de Sá *et al.* 2004; Hedges and Heinicke 2007), which is otherwise distributed in mainland South America and from which it diverged 27–29 mya (Hedges and Heinicke 2007). The distribution of *L. fallax* in the Lesser Antilles is consistent with a South American origin and the flow of oceanic currents facilitating natural over-water dispersal (Hedges 1996; 2001), although King and Ashmore (2014) concluded that the most likely mechanism for the colonization of the Lesser Antilles was the introduction of the ancestor of *L. fallax* by pre-Columbian Amerindians. Hedges and Heinicke (2007) further noted that the cytochrome b tree revealed that frogs from Montserrat and Dominica are identical. Since those islands have never been joined, the species likely evolved on one island and moved recently (possibly even with human mediation) to the others where populations exist or existed. One potential benefit of that situation is that reintroductions of frogs could be accomplished without concern for mixing genetically unique populations.

3. Habitat and natural history

Optimal habitat for these very large frogs (females to 280 mm and males to 190 mm SVL; Guarino *et al.* 2014) appears to be undisturbed primary forest with permanently flowing water, with which frogs are closely associated during the dry season but from which they disperse widely during the rainy season (Daltry and Gray 1998; García 2008). However, *L. fallax* exploits other moist habitats, including those altered by human activity, such as fruit and vegetable plantations in various stages of succession (e.g. Lescure 1979a; Davis *et al.* 2000). Frogs apparently associate with soil types that allow digging of nests (e.g. Stuart *et al.* 2008).

These frogs hide in burrows or moist leaf litter by day, leaving retreats at dusk to forage or search for mates during the breeding season (García 2008). Prey consists of just about anything an individual can handle and includes a wide variety of invertebrates and small vertebrates, including frogs, lizards, snakes, bats, and one record of a small rodent (Lescure 1979a; Brooks 1982; Buley 2003).

Males compete for territories with battles involving elevation onto the hind limbs in an effort to force the competitor to the ground (García 2008). Peptide skin secretions produced only by males apparently promote aggressive behaviour (King *et al.* 2005). Males call on rainy nights from at least mid-February through September, most intensely in May and June (Lescure 1979a, Davis *et al.* 2000; García 2008). More than 2,000 eggs are deposited in a foam nest at the bottom of a burrow (Lescure 1979a). Both males and females guard nests and females deposit 10,000–25,000 unfertilized eggs that serve as the only food source for 26–43 developing larvae (Gibson and Buley 2004).

4. Conservation status, threats, and actions

Leptodactylus fallax is listed as CR on the IUCN Red List (IUCN SSC Amphibian Specialist Group 2017). These relatively long-lived frogs (to 7 years; Guarino *et al.* 2014) are experiencing substantial and ongoing population declines on both Dominica and Montserrat (e.g. Adams *et al.* 2014; IUCN SSC Amphibian Specialist Group 2017). The assessment details in IUCN SSC Amphibian Specialist Group (2017) cited monthly monitoring on Dominica since 2014 that led to a population estimate of 132 individuals and noted that the population crash on Dominica from 2002 to 2004 resulted in near-extirpation, with population levels so low that the species was not detected from 2006 to 2010 (M. Hudson in IUCN SSC Amphibian Specialist Group 2017). The population crash on Montserrat in 2009 to 2010 also resulted in near-extirpation, with an initial count of four individuals which was reduced to only two individuals that survived exposure to *Bd* at one site after 2011 (M. Hudson in IUCN SSC Amphibian Specialist Group 2017). The current population is severely fragmented into seven or eight subpopulations, the two largest of which (both on Dominica) contain 40–45% of all frogs in the wild (M. Hudson in IUCN SSC Amphibian Specialist Group 2017). Moreover, fewer than 50 mature individuals remain in each subpopulation, and a continuing decline in the number of mature individuals is inferred from ongoing threats.

The threats facing all amphibians in the Lesser Antilles have affected *Leptodactylus fallax*. Mongoose predation (Barbour 1930, 1937; Groome 1970) and competition with invasive *Rhinella marina* (Groome 1970; Lescure 1983) have been implicated in the apparent extirpation of populations of *L. fallax* on Martinique and Guadeloupe. However, the most recent records of the species on Martinique predates the introduction of the mongoose in the 1890s (Lescure 1979b) and these frogs have coexisted with *R. marina* in some areas of Montserrat (Kaiser 1994). However, domestic or feral cats and dogs are known to take *L. fallax* (Faaborg and Arendt 1985), and many frogs on Montserrat during the dry season of 2005 had suffered rat bites (Adams *et al.* 2014).

Loss and alteration of habitats as a consequence of human activity continues on Dominica (IUCN SSC Amphibian Specialist Group 2017), where substantial habitat loss is taking place over much of the species' range due to agriculture, human settlement, and pressure from development (primarily for tourism but also to meet the needs of a growing population seeking higher standards of living). The continuing loss of habitats on Montserrat was interrupted by the massive eruption of the Soufrière Hills Volcano (see below).

Humans heavily and unsustainably harvested *L. fallax* for both subsistence and commercial use for tourists (e.g. Martin *et al.* 2007). After the major population declines caused by chytridiomycosis in the early 2000s, a range-wide ban was put in place to stop the hunting of this species – but some illegal hunting almost certainly continues (IUCN SSC Amphibian Specialist Group 2017).

Even prior to the major population crashes in the early 2000s, *L. fallax* had ranges restricted to relatively small areas on both islands. On Dominica, where the frogs occupy no protected areas (Daltry 2002), human activities encroach on ever-larger areas, including some remaining anuran habitat. On Montserrat, where the Centre Hills protected area encompasses a portion of the species' historical range, subpopulations in the south of the island were lost to lava flows from volcanic eruptions and subsequent ash-fall (Daltry and Gray 1999; Adams *et al.* 2014; IUCN SSC Amphibian Specialist Group 2017). Today, most remaining subpopulations on both islands are separated by considerable distances and inhospitable habitats, resulting in genetic isolation and even greater vulnerability to stochastic events.

Natural disasters, including volcanic eruptions (see above) and hurricanes (Daltry and Gray 1998; Daltry 2002), have had major effects on *L. fallax*. Although hurricanes occur regularly in the Antilles, the episodic nature of such storms suggests that animal populations are unlikely to adapt to the immediate and longer-term effects on habitats. For example, Kaiser and Henderson (1994), in regard to Montserrat, stated: 'The devastation of that island by Hurricane Hugo in 1989 has probably caused sufficient habitat disruption to seriously imperil the remaining populations of the species'.

García (2008) addressed environmental pollution, focusing on the impact of agrichemicals on frogs directly or on their prey-base. He concluded that such effects are likely, but their extent on Montserrat is unknown.

Climatic change can adversely affect biodiversity by altering temperature and rainfall regimes (e.g. Burrowes *et al.* 2004), with such changes having been implicated in declining frog populations on Puerto Rico. Although comparable studies do not exist for Dominica and Montserrat, an evident dependence on moist forests and flowing streams suggests that *L. fallax* is particularly vulnerable to such effects. A more insidious impact might be that altered climatic parameters could further stress populations already affected by other threats, rendering them more vulnerable to *Bd* (Li *et al.* 2013).

The subpopulation on Dominica, where the species was once most abundant, declined catastrophically (at least 80%) between 2002 and 2004, following a major confirmed outbreak of the fungal disease chytridiomycosis (e.g. Magin 2004; Malhotra *et al.* 2007; 2011; Adams *et al.* 2014). Although the subpopulation on Montserrat was thought to be chytrid-free as late as 2008 (García *et al.* 2007; Martin *et al.* 2007; García 2008), it declined from thousands of individuals to just four following an outbreak of *Bd* in 2009 and 2010 (e.g. García *et al.* 2009; Adams *et al.* 2014; IUCN SSC Amphibian Specialist Group 2017). Two of the individuals have subsequently disappeared, although *Bd* was not confirmed as the cause; the two remaining individuals on Montserrat both have histories of surviving *Bd* infections (M. Hudson in IUCN Amphibian Specialist Group 2017). As *Leptodactylus fallax* has a known lifespan of 7–9 years in nature (Guarino *et al.* 2014) and 12 years in captivity (Jayson *et al.* 2018), the subpopulation on Montserrat is almost certainly extirpated.

The Mountain Chicken Recovery Programme (MCRP) was started in 2008 and currently consists of Durrell Wildlife Conservation Trust, Zoological Society of London (ZSL), Chester Zoo, Norden's Ark, Paignton Zoo, and the governments of Montserrat and the Commonwealth of Dominica. The MCRP undertakes all conservation actions, guided by a 20-year long-term recovery plan (Adams *et al.* 2014). Captive, biosecure populations are housed at the Durrell Wildlife Conservation Trust on Jersey (Montserrat origin), ZSL, Chester Zoo, and Norden's Ark in Sweden. Captive-breeding programmes are ongoing, and some breeding attempts have been successful (IUCN SSC Amphibian Specialist Group 2017; Donaldson 2018). Jayson *et al.* (2018a) developed a body-condition score metric for the species, and Jayson *et al.* (2018b) compared the nutritional content of diets in captivity and in nature.

In the world's first trial on a wild population, Hudson *et al.* (2016) demonstrated that *in-situ* treatment of frogs by immersion in an aqueous solution of the antifungal agent itraconazole can be an effective means of reducing chytridiomycosis-induced mortality in the short term but was not feasible in the long term.

No hunting of animals on Dominica has been allowed since 2003, and public-awareness actions have taken place to inform the Dominican public of the magnitude of threats facing this species and to discourage illegal hunting (IUCN SSC Amphibian Specialist Group 2017). Between 2011 and 2014, a series of experimental reintroductions were conducted in order to understand the continued impact and seasonal dynamics of disease; that analysis is ongoing (M. Hudson in IUCN SSC Amphibian Specialist Group 2017). Remnant subpopulations on both islands are being monitored and public-awareness activities such as 'International Mountain Chicken Day' and educational outreach in schools continue to raise awareness of the species (M. Hudson and J. Dawson in IUCN SSC Amphibian Specialist Group 2017).

H. *Leptodactylus validus* Garman, 1888 (1887) (Fig. 9.3H)

1. Distribution

As currently defined, this species ranges from the lowlands of the Guiana Shield, north through the continental islands of Trinidad and Tobago to the Grenada and St Vincent Banks (de Sá *et al.* 2014; Frost 2019). Murphy (1997), Hedges (1999), Hardy *et al.* (2010), and Henderson and Powell (2009) indicated that this species reaches an elevation of ~730 m (presumably on Trinidad). The highest elevation for which we have documentation on either St Vincent or Grenada is 530 m at Grand Etang (Grenada); the species has not been encountered at a frequently visited site at 705 m asl on Grenada (B. C. Harrison, *in litt.*, 14/04/2016).

On the St Vincent Bank, *L. validus* is known only from St Vincent; on the Grenada Bank, records exist for Grenada and Bequia (e.g. Garman 1888; Barbour 1914; Kaiser and Henderson 1994; Henderson and Berg 2006; 2011; Daudin and de Silva 2007; 2011; Powell and Henderson 2007; 2011; Henderson and Powell 2018). The status of the population on Bequia is unknown (Daudin and de Silva 2007; 2011), but the species is locally abundant in the lowlands of both St Vincent and Grenada and sporadically present in areas at higher elevations.

Evidence presented by Camargo *et al.* (2009) suggested that *L. validus* originated on the mainland and first colonized Trinidad with subsequent independent colonizations of Tobago and the Lesser Antilles from Trinidad. The colonization of Trinidad occurred about one mya and the colonization of the Lesser Antillean islands about 0.4 mya. The likeliest scenario suggested that *L. validus* dispersed from Trinidad to St Vincent and from there to Grenada, the latter a dispersal event that could have been mediated by humans as recently as 1,600 years ago. Interestingly, Kaiser and Henderson (1994) suggested that these frogs fortuitously 'invaded' the southern Lesser Antilles by rafting but noted that the distribution might be an example of human agency because it 'closely follows the lines of habitat disturbance'.

2. *Relationships*

Although Yanek *et al.* (2006) demonstrated that *L. validus* and *L. pallidirostris* are conspecific, Pyron and Wiens (2011) included both in their phylogeny of the Amphibia as sister taxa to *L. wagneri*. de Sá *et al.* (2004) assigned *L. validus* to the *L. melanonotus* species group; *L. colombiensis* and *L. wagneri* are sister taxa in a clade with *L. validus*, and this clade is sister to all remaining taxa in the *L. melanonotus* species-clade. All other species included in the *L. melanonotus* species-group have mainland South American distributions.

3. *Habitat and natural history*

Although G. Allen (in Barbour 1914) and Barbour (1916) suggested that this is a species of the 'primeval forest', we consider it more often an inhabitant of disturbed situations, including roadside ditches. Schwartz and Henderson (1991) indicated that these ground-dwelling frogs were in mesic habitats and in shallow, roadside ditches on St Vincent and Grenada. Hardy *et al.* (2010) stated that they are generally terrestrial along forest edges, but often close to streams, shaded gullies, and caves, but that they also occur in meadows and roadsides in parks and rural yards, suggesting that they adapt well to altered habitats. Kaiser and Henderson (1994) indicated that the species was resistant to a certain amount of pollution and was locally abundant in urban ditches, gardens, and plantations, as well as more natural settings such as ponds, lakes, and small, slow-flowing streams. At Grand Etang (St Andrew Parish, Grenada, elevation 530 m), A. Schwartz (field notes, 20/11/1961) encountered them in ditches, on muddy paths, and on stumps up to 0.45 m above the ground. Also on Grenada, the species was not uncommon in roadside ditches along a road dominated by upscale hotels at Grand Anse on the outskirts of St George's. Henderson and Powell (2009) found them in leaf litter near streams in the Botanic Garden on St Vincent.

Leptodactylus validus is mostly nocturnal but can be heard calling by day after rains. A. Schwartz (field notes, 20/11/1961) collected 26 during one night at Grand Etang (Grenada). Also on Grenada, Germano *et al.* (2003) found single individuals at four of eight sites surveyed in June. Mallery *et al.* (2007) encountered the species at only one of seven surveyed sites on St Vincent (the Botanic Garden), also in June, but noted that it was locally common at that site. The species breeds in streams and drainage ditches, and the larvae develop in such wetlands; eggs are laid in foam nests in temporary ponds and ditches, and the larvae develop in water (Hardy *et al.* 2010). They are known to fall prey to the snake *Mastigodryas bruesi* (Schwartz and Henderson 1991). Powell and Henderson (2007; 2011) suggested that *Rhinella marina* might prey on these frogs.

4. *Conservation status, threats, and actions*

Leptodactylus validus is on the IUCN Red List as LC (Hardy *et al.* 2010). The species is common and adaptable, and the sizes of its populations are presumed to be large and unlikely to be declining quickly enough to qualify for listing in a more threatened category. However, Kaiser and Henderson (1994) indicated that several 'previously productive' roadside and plantation habitats on Grenada and St Vincent with improved drainage systems lacked frogs in October 1993. Despite those observations, they did not consider this species to be threatened due to 'its abundance and its ability to resist a variety of disturbances or pollutants'. Germano *et al.* (2003), despite the species' apparent ability to tolerate 'marginal habitats', posited that the 'small, fragmented distribution' indicated a need for protecting existing populations on Grenada. Powell and Henderson (2007; 2011) noted potential vulnerability to alterations in habitats and suggested monitoring of extant populations on St Vincent. Possibly most alarming, Henderson and Berg (2011) indicated that an individual from Grand Anse (Grenada) tested positive for the chytrid fungus. Hardy *et al.* (2010) considered the main threat to *L. validus* to be the destruction of habitats due to urbanization, development for tourism, and agriculture; they also felt the species was able to adapt well to altered habitats. More recently (February 2019), we (RP, RWH) heard the species calling from a number of sites in disturbed situations at low elevation on Grenada.

Although not affecting the status of the species as a whole, Daudin and de Silva (2007; 2011) suggested that the population on Bequia might be a remnant of the pre-colonial era, when the island was heavily wooded and water more available; they also indicated that the present status of the Bequia population is unknown.

l. *Pristimantis euphronides* (Schwartz, 1967) (Fig. 9.3l)

1. *Distribution*

Pristimantis euphronides is endemic to Grenada, where it occurs in mesic forests at elevations > 300 m (Kaiser and Henderson 1994); Harrison *et al.* (2011) and Berg *et al.* (2013), however, never encountered the species at sites with elevations below 400 m. A population from a coastal site mentioned by Schwartz (1967) presumably led to Groome's (1970) mistaken assumption that both *Pristimantis urichi* and *P. euphronides* were present on Grenada. We are confident that records of this species from a coastal area (collected at St George's in 1910) are based on erroneous locality data.

Garman, who presumably never left the vicinity of St George's, failed to collect *P. euphronides* during his visit to Grenada in 1879. Thirty-one years later, G. M. Allen and C. T. Brues did travel to Grand Etang and collected a series of 35 *P. euphronides* on 28 August 1910 (MCZ 2910–30, 2932–35, 51760–69), but they are catalogued as having been collected at St George's (near sea level). Schwartz (1967) subsequently reported those specimens as having been collected at St George's, but no records for the species occur anywhere between St George's and sites with elevations of about 305 m asl. Furthermore, according to Allen and Brues, 'not a single specimen [of *P. euphronides*] was taken in the lowlands' (Barbour 1916). We suspect that the Allen–Brues locality was a shipping point rather than a collection locality. In 1961, Schwartz and associates collected the species at several localities; again, never at sites with elevations below 300 m asl.

Hedges (1999) and Hedges and Powell (2010b) listed an elevational range of 300–840 m and a total range of 16 km^2, although the latter is likely less than that today (Harrison *et al.* 2011).

2. *Relationships*

Kaiser *et al.* (1994a, 1994b) considered *P. euphronides* and *P. shrevei* sister species. According to Hedges *et al.* (2008), *P. euphronides* belongs to the *P. unistrigatus* species group and 191 of 193 species (as of 2008) in the group are distributed over northern South America and Trinidad and Tobago.

Pristimantis euphronides is one of two species (*P. shrevei* is the other) that occurs outside of that range, and those two species form a clade distinct from all others in the subgenus *Pristimantis* (Hedges *et al.* 2008).

3. *Habitat and natural history*

Although largely restricted to cool, damp forests (Kaiser *et al.* 1994c), these frogs occur on the ground and on vegetation in rainforests, forest edges, montane meadows surrounded by agriculture (Stuart *et al.* 2008; Hedges and Powell 2010b; Henderson and Powell 2018), wet forested uplands (Henderson and Berg 2006; 2011), in grass and low herbaceous growth in a small montane meadow surrounded by forest and nutmeg plantations, non-calling individuals on bromeliads growing low on trees, under a plank at Grand Etang (Schwartz 1967), and on low (0.3 m) to moderately high (1.8 m) perches, usually ferns or leaves of broadleaf vegetation, along trails in forests at Grand Etang (B. C. Harrison, unpublished data). Frogs retreat beneath leaf litter and enter crevices in the substrate to avoid desiccation during the day and in dry periods (Harrison *et al.* 2011; Berg *et al.* 2013). In captivity, individuals dig holes into which they retreat or deposit eggs (Berg *et al.* 2013). This species appears to be less tolerant of dry forest-floor conditions than is *Eleutherodactylus johnstonei* (Henderson and Berg 2011).

The species is moderately abundant in suitable habitat (Stuart *et al.* 2008; Hedges and Powell 2010). Germano *et al.* (2003) found frogs along only one of eight sampled 100-m transects (at Grand Etang) but encountered 27 frogs at that site. More recent work (Harrison *et al.* 2011; Berg *et al.* 2013) documented steady declines in numbers at several localities.

4. *Conservation status, threats, and actions*

The species is designated EN on the IUCN Red List (Hedges and Powell 2010b; also in Stuart *et al.* 2008) because its Extent of Occurrence is less than 5,000 km^2, individuals occur in few locations, and the extent and quality of its habitat continue to decline. Major threats include habitat loss and destruction, primarily due to urbanization, development (largely for tourism), and agriculture. The species also is threatened by competition with invasive *Eleutherodactylus johnstonei* (e.g. Kaiser and Henderson 1994; Kaiser *et al.* 1994c; Henderson and Berg 2006; 2011; Henderson 2008; Harrison *et al.* 2011) and the chytrid fungus (Harrison *et al.* 2011; Henderson and Berg 2011). Whatever the cause(s), populations are declining (Harrison *et al.* 2011).

Alarm for the status of *P. euphronides* was not sounded until the early 1990s (Kaiser *et al.* 1994a, 1994c; Kaiser and Henderson 1994), 30 years after Schwartz's fieldwork in 1961. During visits to Grand Etang in 1990–92, Kaiser *et al.* (1994c) failed to find or hear *P. euphronides*, but *Eleutherodactylus johnstonei* was 'very common'. In 1990, many *P. euphronides* and few *E. johnstonei* were calling at the Cable & Wireless station (at Kubalal; near Mt St Catherine). In 1991 and 1992, the calls of *E. johnstonei* drowned out the calls of *P. euphronides* at that site (Kaiser *et al.* 1994c). In 2004–2006, Henderson and Berg (2006; 2011) reported encounter rates of 2.2–30.7 frogs/person-hour of searching depending on the site, time of year, and prevalence of *Eleutherodactylus johnstonei*. Between 2007 and 2010, encounter rates for *P. euphronides* were routinely < 1.0/p-h at Grand Etang, whereas those for *E. johnstonei* were 6.3–22.8/p-h. At the Cable & Wireless station, the encounter rate for *P. euphronides* was 24/p-h in 2007, but by 2010, it had dropped to 3.6/p-h (Henderson and Berg 2011). The persistence of the invasive species at Grand Etang is likely, over time, to be contributing to the decline in the numbers of *P. euphronides*. More recently (24 May 2019) at Grand Etang, B. C. Harrison (*in litt.*, 09/04/2019) encountered 47 *E. johnstonei* and no *P. euphronides* (but heard five) during three hours of searching. At Les Avocats (St Andrew Parish; 400 m asl) over two nights (23 and 26 May 2019; six hours of searching), Harrison observed 17 *P. euphronides* and 41 *E. johnstonei*; she heard an additional 26 *P. euphronides* (B. C. Harrison, *in litt.*, 09/04/2019). The numbers at Grand Etang are discouraging; the Avocats numbers are hopeful.

Harrison *et al.* (2011) found frogs infected with the chytrid (*Bd*) fungus at four sites (Grand Etang National Park, Cable & Wireless station, Les Avocats water works, Mt. Stanhope) on Grenada. *Pristimantis euphronides* and *E. johnstonei* have been observed perching on the same plants and even on the same leaf (e.g. Germano *et al.* 2003), which could facilitate the transmission of the chytrid fungus (Harrison *et al.* 2011). B. C. Harrison (*in litt.*, 01/09/2016) noted that many of the individual *P. euphronides* found in close proximity to *E. johnstonei*, especially at the Cable & Wireless station, were 'gaunt' and emaciated.

J. *Pristimantis shrevei* (Schwartz, 1967) (Fig. 9.3J)

1. Distribution

This species is endemic to St Vincent. Hedges (1999), Hedges and Powell (2004), and Stuart *et al.* (2008) listed an elevational range of 275–922 m; Henderson and Powell (2009a) gave the upper limit of the elevational range as 925 m. Hedges (1999) reported a total range of 90 km^2, although it is probably less than that today.

2. Relationships

Pristimantis shrevei was originally described as a subspecies of *Eleutherodactylus urichi* by Schwartz (1967), elevated to species by Kaiser *et al.* (1994c), and was until recently assigned to the genus *Eleutherodactylus* (e.g. Heinicke *et al.* 2007). Kaiser *et al.* (1994a, 1994b) considered *P. shrevei* and *P. euphronides* sister species. According to Hedges *et al.* (2008), *P. shrevei* belongs to the *P. unistrigatus* species group, and 191 of 193 species (as of 2008) in the group are distributed over northern South America and Trinidad and Tobago. *Pristimantis shrevei* is one of the two species (*P. euphronides* is the other) that occur outside of that range, and those two species form a clade distinct from all others in the subgenus *Pristimantis* (Hedges *et al.* 2008).

3. Habitat and natural history

These largely arboreal frogs occur in upland (> 300 m asl) rainforests, along forest edges, and in montane meadows surrounded by agriculture, where they are primarily on vegetation but occasionally on the ground (Schwartz 1967; Hedges and Powell 2004; Diaz Lameiro *et al.* 2007; Stuart *et al.* 2008; Henderson and Powell 2018). Surveys by Rodríguez *et al.* (2017) indicated that attached leaves were the favoured microhabitat at both upper Vermont (394–458 m asl) and lower Vermont (274–386 m asl); perch heights (rarely >1.0 m) did not vary significantly among sex and age classes, but *P. shrevei* perched significantly higher than *Eleutherodactylus johnstonei* where both species were present.

Kaiser *et al.* (1994c) collected *P. shrevei* near the summit of the volcano La Soufrière in 1990, only 11 years after it erupted. Kaiser and Henderson (1994) reported that ground vegetation was beginning to cover the bare volcanic debris in 1992 and that *Eleutherodactylus johnstonei* was found syntopically with *P. shrevei* at even the higher elevation.

Goldberg *et al.* (1998) observed plerocercoid cysts (*Rhabdias elegans*) and polymorphid cystacanths in these frogs. Moravec and Kaiser (1995) found nematodes (*Oswaldocruzia brevispicula* and *Strongyloides* sp.) and an acanthocephalan (*Centrorhynchus* sp.). The snake *Chironius vincenti* is known to prey on this species (Henderson and Powell 2009).

4. Conservation status, threats, and actions

Pristimantis shrevei has been assessed as EN on the IUCN Red List (Hedges and Powell 2004; also in Stuart *et al.* 2008) based on Extent of Occurrence (less than 5,000 km^2), all individuals in fewer than five locations, and the extent and quality of its habitat continuing to decline. The national list (USAID 2013) also included the species as EN. The latter (USAID 2013) indicated that populations were declining, a trend that might have begun with the clearing of forests by the first Amerindian settlers (Henderson and Powell 2009; 2018). Major threats include loss and destruction of habitats, primarily due to urbanization, development for tourism, and agriculture. The species also could be threatened by competition with invasive *Eleutherodactylus johnstonei* (e.g. Kaiser and Henderson 1994; Kaiser *et al.* 1994c; Treglia 2006; Powell and Henderson 2007; 2011; Sweeney 2015) and the threat of the chytrid fungus. Sweeney (2015; 2016) first detected chytrid on St Vincent in *P. shrevei* from La Soufrière (and in *Eleutherodactylus johnstonei* in the Fenton Forest Reserve, where *P. shrevei* also was present).

This species is moderately abundant in suitable habitat (Hedges and Powell 2004; Diaz-Lameiro *et al.* 2007; Mallery *et al.* 2007; Stuart *et al.* 2008). Treglia (2006) stated that the species remains essentially ubiquitous in relatively pristine habitats at elevations above those where *Eleutherodactylus johnstonei* is abundant; however, Sweeney (2015) found 6–30 frogs at seven of seven sampled transects, all at elevations > 250 m and all but one also inhabited by *E. johnstonei*. Mallery *et al.* (2007) found frogs at the only two of seven sampled sites with elevations > 250 m. During frog-dedicated surveys, Rodríguez *et al.* (2017) recorded encounter rates (frogs encountered/h) of 11.5 (at upper Vermont, 394–458 m asl), 6.5 (lower Vermont, 274–386 m asl), 70 (near the summit of La Soufrière), and 33.6 (~halfway up or down La Soufrière).

IV. The introduced amphibians

We herein define 'introduced taxa' as those that reached islands where they are not native by means of dispersal mediated by human activity. In some instances (e.g. *Rhinella marina*), the introductions were intentional, with the stated purposes of controlling pests in sugarcane plantations (Easteal 1981). Most introductions, however, were inadvertent and in most instances involved transport with goods (especially ornamental plants; e.g. Goin 1944; Powell *et al.* 2011; 2013) that sometimes originated in locations (e.g. Florida, USA) where the species in question were not native. Greater detail is provided in the individual species accounts and the section on patterns of invasions (below).

Amphibians introduced to Lesser Antillean islands include two species native to the region (*Eleutherodactylus johnstonei, E. martinicensis*) that have become established on other islands, six species from outside the region (*Rhinella marina, Eleutherodactylus planirostris, Osteopilus septentrionalis, Scinax ruber, S. x-signatus, Pleurodema brachyops*) that have become established on one or more islands, and four species (*R. marina, E. planirostris, Hyla squirella* [= *Dryophytes squirellus*], *O. septentrionalis*) from outside the regions that have been documented on islands but are not known to have established populations. Note that *Rhinella marina,* which is firmly established on many islands and for which a waif has been documented on another, has traditionally been considered – and is treated herein – as introduced. Nonetheless, we cannot rule out the possibility that the ancestors of some toads might well have reached some of the more southerly Lesser Antilles by natural overwater dispersal (e.g. Henderson and Powell 2009; Powell *et al.* 2011; 2013).

A. Patterns of introductions

Kraus (2009) noted that amphibians (and reptiles) are unusual among introduced species in that they are transported via a diversity of both intentional and unintentional pathways. Unintentional introductions tend to involve individuals inadvertently transported with goods or the vehicles and/or containers used in transport. Intentional introductions occur primarily because a species is perceived to provide some value to humans.

The vast majority of amphibian introductions into the Lesser Antilles for which we have pertinent data were unintentional, with the most common denominator being transport with ornamental plants or the shipping containers in which they are packaged (e.g. Powell *et al.* 2011, 2013, and references therein). Such events are ongoing; for example, as recently as 2017, Questel

(2018) documented six *Osteopilus septentrionalis* in containers of plants shipped from Florida (USA) that were intercepted in St Barthélemy. Obvious advantages of such pathways include an environment during transport that is more conducive to survival and an association with familiar plants upon arrival. Equally unintentional but – based on apparent frequency – less effective is transport with other goods. For instance, R. I. Crombie (*in litt.* in Kaiser 1992) suggested that *Eleutherodactylus johnstonei* might have been transported to Dominica with emergency supplies after the island was struck by Hurricane David in 1979. Also in the unintentional category was the hapless *Salamandra salamandra* trapped in a plastic salad container shipped from France to St Barthélemy (Breuil *et al.* 2012) – although similar events are apparently quite common (e.g. Hughes *et al.* 2019).

Intentional introductions include the unsuccessful attempts to relocate *Leptodactylus fallax* from either Montserrat or Dominica to Grenada (Groome 1970), Martinique (Lescure 1983), and Puerto Rico (Barbour 1937), presumably as a delicacy for both local consumption and as a tourist attraction. More successful have been the efforts to introduce *Rhinella marina*, primarily during and immediately after the colonial period and expressly for the purpose of controlling insect pests on sugarcane plantations (Easteal 1981).

In addition to arriving at a new destination, a successful invasion requires that the introduced alien species become established or 'naturalized' (e.g. Kraus 2009). This entails the new location providing a suitable habitat. As mentioned previously, low-lying coastal areas in the Lesser Antilles are rarely favourable for many amphibians – but notable exceptions are those extremely versatile species that include all of the non-native taxa established in the region. Powell and Henderson (2008), characterizing species that invaded urban habitats in the West Indies, listed six features, noting that the successful species inevitably exhibit one or more of them. Invaders (1) are ecologically versatile and capable of tolerating a broad range of sometimes rapidly and dramatically changing conditions; (2) exhibit edificarian tendencies outside the urban environment; (3) tend to be edge species or, at least, species that are euryecious, not habitat specialists; (4) feed on invertebrates; (5) are heliotherms if diurnal, and (6) are often scansorial. That list closely parallels descriptions by Henderson and Powell (1999) of a West Indian herpetofauna that is derived largely from successful colonizers and whose ancestors might be expected to exhibit colonizing characteristics. Also similar are descriptions by Henderson and Powell (2001) when discussing the 5–10% of West Indian species that have responded positively to human-mediated changes in their environments. Although all of those studies included reptiles, the amphibians associated with the relevant traits inevitably included the non-native species that are successfully established on islands of the Lesser Antilles. In addition, Meshaka (2001) cited four factors likely responsible for success by *Osteopilus septentrionalis* (all broadly applicable to other introduced species in the region) in establishing populations: (1) High fecundity allows the recruitment of sufficient numbers of offspring to offset the inevitably high mortality during the dry season; (2) a short larval period permits the exploitation of transient conditions when standing surface water enables reproduction and dispersal; (3) a broad dietary niche facilitates a quick adjustment to new and potentially different prey; and (4) open habitat and vacant dietary niches and the resultant lack of competition.

B. Introductions of native amphibians to new islands within the area

1. Eleutherodactylus johnstonei *Barbour, 1914*

Most likely native to Monterrat and a second unknown island (Yuan *et al.* 2022), populations – presumably introduced – have been recorded in Anguilla (where it might now be extirpated), St Barthélemy, St Maarten/St Martin, Saba, St Eustatius, St Christopher, Nevis, Guadeloupe, Terre-de-Haut (Les Saintes), La Désirade, Marie-Galante, St Lucia, Martinique, St Vincent, the

Grenadines (Bequia, Mustique, Canouan, Union Island, Petit St Vincent, Carriacou), Grenada (including Hog Island), and Barbados (e.g. Kaiser and Hardy 1994a; Malhotra and Thorpe 1999; Powell *et al.* 2005, 2011, 2015; Henderson and Powell 2009; Henderson and Breuil 2012; Yokoyama 2010, 2012, 2013; Frost 2019). An apparently short-lived population on Dominica probably emanated from frogs arriving with emergency supplies from a variety of islands in the aftermath of Hurricane David in 1979 (R. I. Crombie, *in litt.* in Kaiser 1992). At best, the range never extended beyond a narrow strip of the Caribbean coast and is almost certainly extirpated (Kaiser 1992). Noteworthy is that, although the type locality for the species is on Grenada, these frogs were introduced to that island from Barbados (where they also were introduced) in about 1885 (Barbour 1914).

2. Eleutherodactylus martinicensis *(Tschudi, 1838)*

The natural range of this species is in question but probably is Martinique, Dominica, or both. It has been recorded from Antigua, St Martin/St Maarten, St Barthélemy, Guadeloupe, Îlet à Kahouanne, Îlet-à-Cochons, Île des Saintes Bank, Marie-Galante, La Désirade, Dominica, and Martinique (Kaiser 1992; Magras 1992; Kaiser and Hardy 1994b; Breuil 2002; Powell *et al.* 2005, 2015; Henderson and Breuil 2012; Yokoyama 2012, 2013). Although Lescure (1983) suggested that the species might have been introduced to Dominica by refugees fleeing Martinique and Guadeloupe during the French Revolution, Lescure (2000) proposed Dominica, Martinique, and St Lucia as the native range; however, the species apparently is absent from St Lucia (Kaiser 1992; Kaiser and Hardy 1994b; Malhotra and Thorpe 1999).

Populations on Antigua (Schwartz 1967; Schwartz and Thomas 1975) and the Guadeloupean Archipelago probably are introduced, although Schwartz (1967) suggested that Guadeloupe was most likely the type locality. Populations on St Martin/St Maarten and St Barthélemy certainly are introduced, likely in the early 1980s (Magras 1992; Breuil 2002). The population on St Barthélemy is now widely distributed, mainly around buildings but also in non-urban areas; the density appears to be low (Questel 2018).

In addition to the islands on which it does occur, Barbour (1914) placed *E. martinicensis* on Jamaica, St Christopher, Saba, Montserrat, and St Eustatius, but those records almost certainly refer to *E. johnstonei*. Moreover, although Pregill *et al.* (1988) questioned the species' occurrence on Antigua (attributing those records to confusion with *E. johnstonei*), recent publications (e.g. Hedges *et al.* 2010d; Daltry 2011) do confirm the presence of *E. martinicensis* on that island.

C. Established non-native amphibians

1. Rhinella marina *(Linnaeus, 1758) (Fig. 9.3K)*

Distribution: As currently defined, this species ranges east of the Andes throughout Amazonian and Guianan South America (the Guianas to central Brazil and Amazonian Colombia, Peru, and Bolivia) (Frost 2019). Introduced populations occur in Florida (USA), many West Indian islands, Hawaii, Fiji, the Philippines, Taiwan, Ryukyu Islands (Japan), New Guinea, the Solomon Islands, Australia, and many other Pacific islands (e.g. Lever 2003; Kraus 2009; Frost 2019). Solis *et al.* (2009) indicated that the elevational range extends from sea level to 3,000 m (the latter certainly on the mainland).

In the Lesser Antilles, *Rhinella marina* occurs on at least 18 islands (Appendices 9.I and 9.II); it is established on all major island banks except the Sombrero, Saba, Redonda, Îles des Saintes, Marie Galante, and Dominica Banks, albeit not on every island on every colonized bank (e.g. Malhotra and Thorpe 1999; Powell *et al.* 2011, 2013; Henderson and Breuil 2012). The population on Dennery Island, a satellite of St Lucia, might have been extirpated after its breeding pond was destroyed (J. C. Daltry, *in litt.*, 15/05/2019). On the Grenada Bank, the introductions to Canouan, Carriacou, and Mustique are very recent (Daudin and de Silva 2007; 2011). The means of arrival on Carriacou in about 1998 and, more recently, Canouan are unknown. However, toads were accidentally transported to Mustique as recently as 2001 in a shipment of topsoil and plant material from St Vincent (Paice 2005).

The first introductions of this species to the Lesser Antilles were from French Guiana to Martinique and from Martinique to Barbados before 1844 (Waite 1901), although Schomburg (1971) reported that *R. marina* was introduced to Barbados from Guyana in approximately 1833. Barbour (1914) first documented the presence of the species on Grenada, and Clark (1916) reported that the species had become 'extinct' on St Vincent (the next record of its presence there was by Schwartz and Thomas 1975). Easteal (1981), in his history of introductions, stated that 'almost no information is available about any of these introductions'; however, he did make notes of museum specimens from Grenada collected in the 1870s.

Although we have designated *R. marina* an introduced species on every island in the Lesser Antilles on which it occurs, we would be remiss if we did not acknowledge that, because the species has demonstrated tolerance to salinity and high temperatures (e.g. Malhotra and Thorpe 1999; Rios-López 2008; Wijethunga *et al.* 2016), it had the potential to reach the eastern Caribbean (e.g. Grenada or St Vincent) by way of overwater dispersal on flotsam originating on the Neotropical mainland rather than by human mediation.

Relationships: Using molecular, morphological, and skin-secretion data, Maciel *et al.* (2010) employed Baysian and maximum-parsimony inferences to recover a monophyletic *Rhinella marina* group that included ten species comprising two major clades that have a wide collective distribution on the South American mainland east of the Andes. *Rhinella marina* belongs to the 'north-central clade'.

As defined until recently, *Rhinella marina* was a complex of several species (e.g. Solis *et al.* 2009). Historical accounts refer to this species under a variety of names; for example, until 2006 it was assigned to the genus *Bufo*, then to the genus *Chaunus* (Frost *et al.* 2006), and subsequently by implication to the genus *Rhinella* (Chaparro *et al.* 2007). Vallinoto *et al.* (2010) provided molecular evidence that the Amazonian population to which the name *R. marina* was originally assigned is closely related to other members of the complex (the *R. marina* group of Maciel *et al.* 2010) but distant from clades in Central America (and possibly north into Mexico) and western Ecuador. Acevedo *et al.* (2016) clearly demonstrated the existence of at least two genetically and morphologically distinct species (see also Frost 2019). Consequently, introduced populations currently assigned to *R. marina* with at least some ancestors that originated from areas west of the Andes might comprise mixed populations or be misnamed. Although no detailed studies have examined populations of these toads on Lesser Antillean islands, due to the proximity of these islands to the continental range of *R. marina* and the very real possibility that the ancestors of the toads on at least the most southerly islands could have reached them by natural overwater dispersal, we continue to recognize the Lesser Antillean populations as *R. marina*.

Habitat and natural history: *Rhinella marina* occurs in a variety of habitats, but usually those that are altered to some degree (agricultural, urban, suburban). On Antigua, it occurs in most gardens and pastures (Lever 2003), but also the leaf litter in remaining patches of forest (Henderson and Powell 2009). It is often associated with human-altered habitats (urban and rural) such as agricultural areas, gardens, and roads on Montserrat (Henderson and Powell 2009). On Guadeloupe, it is encountered in sugarcane fields, *Musa* plantations, gardens, pastures, beaches, ravines, and culverts (Breuil 2002). It occurs in cultivated areas (primarily *Musa* groves) on Martinique, as well as in littoral dry forest (Breuil 2009). According to Long (1974), on St Lucia it 'can be seen on any night doing its best to get killed on the road'. On St Vincent, *R. marina* favours agricultural areas, roadside ditches, vacant lots, and it was common in the Botanical Garden (Mallery *et al.* 2007; Henderson and Powell 2018). In the Grenadines, loud, enthusiastic breeding was observed in a rural pond on Carriacou. Livestock watering-troughs, mudflats, roadside ditches, sugarcane fields, and residential lawns and gardens are exploited on Grenada (Henderson and Powell 2009; 2018). It has been encountered in plant nurseries on Barbados (Lever 2003).

The large paratoid glands contain secretions that are cardiotoxic, vasoconstricting, convulsant, and hallucinogenic, and that can sicken or kill animals that try to ingest it (Rivero 1998). These nocturnal anurans will sometimes forage for light-attracted insects. The diet of *R. marina* is taxonomically broad, and these toads will take insects, spiders, scorpions, centipedes, and millipedes, as well as some inanimate items (e.g. cigarette butts, small stones) (Long 1974; Breuil 2002; Henderson and Powell 2009). Males will call from almost any standing water (ponds, rivers, creeks, cisterns); multiple exuberant males will sometimes attempt to mount a single female (amplexus is axillary). Eggs (up to 35,000 per female) are deposited in long gelatinous strings and hatch after 36 hours (Rivero 1998) to four days (Schwartz and Henderson 1991).

Conservation status, threats, and actions: The species is designated on the IUCN Red List as LC; it is widely distributed, even more widely introduced, and tolerant of a broad range of habitats (Solis *et al.* 2009). The principal conservation concern where non-native populations have been established is competition with and predation on native species.

Although firmly established on many islands, we strongly recommend careful monitoring of cargo from areas known to be inhabited by *R. marina* to prevent the species from becoming established on those islands currently free of these toads. An officially organized eradication programme on Mustique (Daudin and de Silva 2007; 2011) is now concentrating on finding and destroying eggs (M. de Silva, pers. comm., 19/04/2016).

The effects of *R. marina* on native fauna has been most intensively studied in Australia. Shine (2010) stated that the major impact was the lethal toxic ingestion of toads by frog-eating predators. Phillips *et al.* (2005), citing Freeland (1984), listed four potential effects: (1) predation on native fauna, (2) competition with native fauna, (3) as toxic prey of native predators, and (4) vectors for disease. Mayer *et al.* (2015) demonstrated that *R. marina* inhibits the activity of native anurans by inducing avoidance or by reducing activity, and consequently might curtail activities such as feeding and breeding.

2. Eleutherodactylus planirostris *(Cope, 1862) (Fig. 9.3L)*

Distribution: This species is native to the Bahama Islands, Cayman Islands, and Cuba, where its elevational distribution ranges from sea level to ~725 m. It has been introduced to the Caicos Bank (Providenciales, North Caicos, Middle Caicos, and Big Ambergris Cay) (Buckner *et al.* 2012), Jamaica, Grenada, southeastern USA, Mexico, Honduras, Nicaragua (Miskito Cays), Panama, Suriname, China (Hong Kong), the Hawaiian Islands, the Philippines, and Guam (Christy *et al.* 2007; McCranie *et al.* 2008; Kraus 2009; Crawford *et al.* 2011; Heinicke *et al.* 2011; Ouboter and Jairam 2012; Cedeño-Vázquez *et al.* 2014; Lee *et al.* 2016; Frost 2019). In the Lesser Antilles, it is established only on St Martin/St Maarten (Yokoyama 2012, 2013; Powell *et al.* 2015; Frederick *et al.* 2021), where it was first documented in 2012. Another introduced population had been established on Grenada, likely introduced during the Cuban occupation in the early 1980s (although an introduction via shipments of ornamental plants, possibly from Florida, USA, cannot be ruled out) (Henderson and Powell 2018). Specimens were collected in 1991 and 1992 (A. Wynn, *in litt.*, 06/012016) but recent fieldwork has failed to locate *E. planirostris* at the site where it was previously known to occur (Henderson and Powell 2018). We assume that this population is now extirpated. Although several waifs of this species entered St Vincent with ornamental plants shipped from Miami, Florida, USA, they were intercepted by quarantine officers (Glasgow 2011). Eight individuals in plant containers shipped from Florida, USA, were intercepted on St Barthélemy in November 2017, but no evidence suggests that a population is established on that island (Questel 2018).

Relationships: *Eleutherodactylus planirostris* belongs to the *E. planirostris* series and the *E. planirostris* group; the species group is distributed in Cuba, the Cayman Islands, and the Bahamas (Heinicke *et al.* 2008).

Habitat and natural history: *Eleutherodactylus planirostris* is largely ground-dwelling and is commonly found in mesic and xeric habitats, including forests, caves, beaches, gardens, and plant nurseries, as well as urban areas (Hedges *et al.* 2009). In St Maarten, it was initially encountered in Emilio Wilson Park (Yokoyama 2012, 2013), but it has dispersed at least as far as St Dominic High School in South Reward. Both sites are along L. B. Scott Road and about 1.7 km apart (M. Yokoyama, *in litt.*, 29/04/2019). Daytime hours are spent in leaf litter and under stones (Yokoyama 2013). The ability of this species to invade new areas might be attributable at least in part to its relatively high resistance to desiccation (Stewart and Martin 1980).

Conservation status, threats, and actions: This species has received an IUCN Red List assessment of LC based on its wide distribution, ecological plasticity, presumed large population, and that it is unlikely to be declining (Hedges *et al.* 2009). The only concern this species could raise in the Lesser Antilles is if it were in competition with native species. Since all extant anurans on St Martin/St Maarten are introduced, that concern currently is moot.

3. Osteopilus septentrionalis *(Duméril and Bibron, 1841) (Fig. 9.4M)*

Distribution: This species is native to Cuba (including many satellites), the Cayman Islands, and the Bahamas (e.g. Henderson and Powell 2009; Powell and Henderson 2012), where the elevational range extends from sea level to ~1,110 m. In the West Indies, introduced populations are known from the Turks & Caicos, the Puerto Rico Bank (including the American and British Virgin Islands), and many Lesser Antillean Islands (Powell *et al.* 2011; 2013; Powell and Henderson 2012). Introductions outside the West Indies include Curaçao, Oahu (Hawaii), Florida and coastal Georgia, and Costa Rica (e.g. Kraus 2009; Frost 2019). Although populations in the Florida Keys might have arrived naturally via overwater dispersal from Cuba (Meshaka 2001 and references therein), long-established populations in southern Florida are the most common origin of frogs establishing introduced populations outside the species' native range (e.g. Meshaka 2001; Townsend *et al.* 2000; Kraus 2009; Powell *et al.* 2011; 2013).

In the Lesser Antilles, introduced populations are established on Anguilla, St Maarten/St Martin, St Barthélemy, Nevis, Antigua, St Lucia, and Mustique in the St Vincent Grenadines; waifs are known from Saba and Dominica (Powell 2007; Lindsay and Cooper 2008; Henderson and Breuil 2012). Frogs identified as *Ololygon rubra* (now *Scinax ruber*) on St Maarten/Martin (Schwartz and Henderson 1991; Malhotra and Thorpe 1999; Yokoyama 2010) were *Osteopilus septentrionalis* (Powell *et al.* 1992; Yokoyama 2012, 2013). Two individuals were interdicted in Grenada in 2013 (Somma and Graham 2015). The inclusion of Guadeloupe in the species' range (Hedges *et al.* 2010f) was an error.

Relationships: *Osteopilus septentrionalis* is sister to the clade containing all other congeners (Moen and Wiens 2009). The most likely origin of this Greater Antillean genus is South America via the Lesser Antilles during the early Cenozoic (Hedges 1996).

Habitat and natural history: *Osteopilus septentrionalis* is most abundant in mesic habitats but can occur in xeric situations; it lives in all kinds of disturbed habitats, including towns and houses, but also is present in forests, mangroves, and coastal areas, where it can tolerate brackish water (Hedges *et al.* 2009; Henderson and Powell 2009). It frequently functions as a human commensal and is closely associated with cisterns, drainage ditches, basements, axils of banana trees and palms, flowerpots, and ornamental plantings (Henderson and Powell 2009).

On Anguilla, frogs were in residential areas (gardens, outbuildings, cisterns), on hotel grounds, and in an abandoned well (Townsend *et al.* 2000). On St Martin/St Maarten, the species

Figure 9.4. Photographs of amphibians from the Lesser Antilles. (A) *Osteopilus septentrionalis* from Mustique (St Vincent Grenadines) (photograph by Mark de Silva); (B) *Scinax ruber* from Martinique (photograph by Michel Breuil); (C) *Scinax x-signatus* from Guadeloupe (photograph by Michel Breuil); (D) *Pleurodema brachyops* from St Martin (photograph by Mark Yokoyama); (E) *Hyla squirella* (= *Dryophytes squirellus*) from St Barthélemy (photograph by Karl Questel); (F–G) the salamander, *Salamandra salamandra* from St Barthélemy (photographs by Karl Questel).

has been encountered in scrub situations with some standing water, in wall crevices, gutters, open canals, behind window shutters, on mango and *Ficus* trees, and in axils of palm, coconut, and banana trees (Breuil 2002; Powell *et al.* 2005; 2015; Henderson and Powell 2009; Yokoyama 2010, 2012, 2013). Tadpoles are rarely found in roadside ditches where guppies are abundant, presumably because the latter will eat eggs and small tadpoles (Yokoyama 2012, 2013). On St Barthélemy, frogs were on roof beams, in cisterns, under planks, and in banana trees (Breuil 2002). After spreading throughout much of the island during the wet years of 1998–2000 (Breuil 2002), the population decreased following several years of drought and is now restricted to a few neighbourhoods where it often is associated with old cisterns (Questel 2018). Frogs sought natural crevices in tree trunks on Mustique in the St Vincent Grenadines (Henderson and Powell 2018). Waifs on Saba were perched on a house (J. Magor, pers. comm.; Powell 2007).

These exceedingly adaptable frogs occur wherever high humidity and fresh water (necessary for breeding) are available, even in situations dramatically altered by human activity (Breuil 2002; Powell *et al.* 2005; 2015; Henderson and Powell 2009).

Osteopilus septentrionalis is primarily nocturnal and hides by day in natural or artificial crevices or hangs from the undersides of large leaves (Townsend *et al.* 2000; Breuil 2002;

Powell *et al.* 2005; 2015; Henderson and Powell 2009; 2018). Frogs often hunt around insect-attracting lights (Henderson and Powell 2009 and references therein), where multiple individuals frequently wait along the edge of the most brightly lit area. Prey includes a variety of arthropods, snails, other frogs (including conspecifics), and lizards (Meshaka 2001; Henderson and Powell 2009 and references therein). Tadpoles eat larvae and eggs of other anurans (e.g. Meshaka 2001; Dodd 2013 and references therein). Meshaka (2001) also noted that he never observed more than one cohort of tadpoles at closed sites, attributing this to the fact that tadpoles readily eat conspecific eggs and larvae. In turn, various snakes, birds, and arboreal mammals throughout the natural and introduced range of the species consume these frogs (e.g. Meshaka 2001; Henderson and Powell 2009 and references therein).

Males call throughout the year from vegetation or other surfaces near water (e.g. Hedges *et al.* 2010f), but intensity increases dramatically during rainy periods (Henderson and Powell 2009 and references therein). Almost any ephemeral or fish-free permanent body of standing or slow-moving water is exploited – including pools, marshes, flooded pastures, small streams, ditches, ornamental jars, cisterns, large metal containers, water storage tanks, and swimming pools (Henderson and Powell 2009 and references therein; Powell *et al.* 2015). Eggs are deposited in a thin surface film as partial clutches of 200–300 and rarely exceeding 1,000 eggs (e.g. Meshaka 2001; Dodd 2013). Eggs frequently hatch in less than 30 hours (although some take much longer); hatchlings are ~7 mm in length, fore and hind limbs develop at 40–50 mm, and the tail is resorbed 2–5 days later (e.g. Meshaka 2001; Henderson and Powell 2009; Dodd 2013). Lindsay *et al.* (2019) reported larval gigantism in the introduced population on Nevis.

Skin secretions are irritants that can produce a burning, itching sensation to mucous membranes for as long as 20 minutes; these presumably dissuade especially mammalian predators (e.g. Meshaka 2001; Dodd 2013).

Conservation status, threats, and actions: This species has received an IUCN Red List assessment of LC based on its wide distribution, tolerance of a broad range of habitats, presumed large population, and because it is unlikely to be declining (Hedges *et al.* 2010f). The principal concerns regarding this species in the Lesser Antilles are competition with or predation on native species and the possibility that it might serve as a vector for pathogens. On Puerto Rico, N. Ríos, M. Cardona, and R. Joglar (in Burrowes and Joglar 2005) documented *Osteopilus septentrionalis* affecting native species by: (1) displacing them from diurnal refuges, (2) reducing their calling frequencies, (3) causing them to capture fewer prey, (4) reducing attendance of clutches, (5) reducing hatching success, and (6) preying on both adults and clutches.

4. Scinax ruber *(Laurenti, 1768) (Fig. 9.4N)*

Distribution: This species occurs naturally in eastern Panama, Colombia, Venezuela, Trinidad and Tobago, the Guianas, and is widespread in Amazonia; the elevational range is from sea level to ~2,600 m. It has been introduced into Puerto Rico, St Lucia, and Martinique; reports of the species in St Martin/St Maarten (Schwartz and Henderson 1991; Malhotra and Thorpe 1999; Yokoyama 2010) are in error (Powell *et al.* 1992; Yokoyama 2012, 2013).

Relationships: Fouquet *et al.* (2007) found *S. ruber* to be a paraphyletic taxon. Their *S. ruber* A emerged as a sister taxon to *S. x-signatus* with a high level of support.

Habitat and natural history: On the South American mainland, this arboreal, nocturnal frog occurs in a diverse array of habitats, including open environments to moist forests. It is considered a 'pest' species that inhabits cleared areas in rainforests. It breeds in roadside ditches and shallow temporary pools; males call from the ground and low vegetation around temporary ponds in the forest (Solis *et al.* 2010). On Martinique, it occurs in pastures, meadows, orchards, *Musa* groves, residential gardens, and in edificarian situations (Breuil and Ibéné 2008). On

St Lucia, it is usually found in disturbed (including agricultural) areas, and Malhotra and Thorpe (1999) considered it a human commensal. Schwartz and Henderson (1991) reported it climbing on the wall of a hotel dining room one evening at 1900 h.

Conservation status, threats, and actions: Although Kaiser and Henderson (1994) considered *Scinax ruber* 'almost certainly a recent introduction on St Lucia', it had been reported on that island by Boulenger in 1891. It was first discovered in Martinique in 1997 but probably arrived in the mid-1980s; it is now widespread and abundant in the south of the island (Breuil and Ibéné 2008). The latter authors believed that an introduction from French Guiana was more likely than from St Lucia. *Scinax ruber* has an IUCN Red List designation of LC (Solis *et al.* 2010). Breuil and Ibéné (2008) considered this species (along with *S. x-signatus* and *Osteopilus septentrionalis*) 'a threat for all herpetological communities in the Lesser Antilles' and they lamented that 'it is a pity that nearly nothing was [done] in French Islands to fight these invasive species'. The concern, apparently, is that *S. ruber* will compete with native anurans. All conservation concerns relative to this species in the Lesser Antilles should focus on limiting its distribution and preventing additional colonizations.

5. **Scinax x-signatus** *(Spix, 1824) (Fig. 9.4O)*

Distribution: This species has a wide continental distribution on the South American mainland, including Colombia, Venezuela (and Isla Margarita), the Guianas, Amazonia, and into southern Brazil; elevational distribution is from sea level to 900 m (Rodrigues *et al.* 2010). In the Lesser Antilles, it has been introduced to Guadeloupe (Grande-Terre and Basse-Terre), Marie-Galante, and La Désirade.

Relationships: *Scinax x-signatus* is apparently sister to one of the *S. ruber* lineages of Fouquet *et al.* (2007).

Habitat and natural history: On the South American mainland, *S. x-signatus* occurs in tropical savannahs and along forest edges with open areas; it is apparently 'very adaptable to human disturbance' (Rodrigues *et al.* 2010). Males were observed calling from *Heliconia* leaves at 1.5 m above the ground and 2.0 m from the edge of a forest pond. On Guadeloupe, it occurs in freshwater lakes, brackish lagoons with mangroves, edges of forest swamps, and pastures devoid of trees (Breuil and Ibéné 2008).

Conservation status, threats, and actions: *Scinax x-signatus* has been assessed as LC on the IUCN Red List due to its wide distribution, tolerance of a broad range of habitats, and presumed large population (Rodrigues *et al.* 2010). Breuil and Ibéné (2008) considered this species (along with *S. ruber* and *Osteopilus septentrionalis*) 'a threat for all herpetological communities in the Lesser Antilles and [they lamented that] it is a pity that nearly nothing was down [*sic*] in French Islands to fight these invasive species'. Its range continues to expand on Guadeloupe and may pose a threat to native anurans. All conservation concerns relative to this species in the Lesser Antilles should focus on limiting its distribution and preventing additional colonizations.

6. **Pleurodema brachyops** *(Cope, 1869) (Fig. 9.4P)*

Distribution: *Pleurodema brachyops* is known from the Pacific lowlands of Panama from Azuero to central Panama, eastern and northern Colombia, northern and western Venezuela (including Isla Margarita), Guyana, northern Brazil (state of Roraima), and Aruba. It has been introduced to Curaçao, Klein Bonaire, and Bonaire (Hummelinck 1940; van Buurt 2005). In the Lesser Antilles, *P. brachyops* is known only from St Martin/St Maarten, where it was introduced, possibly in the early 2010s (M. Yokoyama; *in litt.*, 08/04/2019), although the first frogs were discovered on 24 October 2015 (Janse *et al.* 2020). LaMarca *et al.* (2010) listed Saba and St Eustatius as additional Lesser Antillean sites of introduced populations, but initial queries to knowledgeable residents on those islands provided contrary information, suggesting that the species does not occur on either island. We tentatively attribute these inadvertent additions to the herpetofaunas of Saba and St Eustatius to possible confusion over which islands comprise the Dutch Caribbean. However, during a herpetological inventory on St Eustatius in 2015, a resident mentioned toads coming out of the ground after heavy rainfall. Because of the considerable traffic in goods between St Martin/St Maarten and neighboring islands, Janse *et al.* (2020) recommended ongoing searches for *P. brachyops* and other introduced species on both St Eustatius and Saba.

Relationships: This species belongs to the *Pleurodema brachyops* clade, but interpretations differ as to its position within the clade. It is the sister taxon of *P. alium* + *P. diplolister*, or sister to *P. tucumanum* + *P. borellii* + *P. cinereum*, or it is the sister taxon to all other species (Faivovich *et al.* 2012).

Habitat and natural history: In its native range, *P. brachyops* occurs in open grasslands and savannahs. Staton and Dixon (1977) described it as ubiquitous in the central Llanos of Venezuela. At the onset of the first rains, these frogs emerge from sandy soils where they hide during the dry season. Breeding frogs make foam nests in ephemeral or permanent small pools of water. This species adapts well to human-induced changes in its habitat (LaMarca *et al.* 2010). On Aruba, Curaçao (onto which it was introduced from Aruba in 1910), and Bonaire (introduced from Curaçao in 1928), *P. brachyops* is appreciated because it consumes mosquito larvae in stagnant pools (van Buurt and Debrot 2012). It apparently now occupies a previously empty ecological niche on those islands (van Buurt 2006). On St Martin/St Maarten, *P. brachyops* has been encountered at four known sites on both the Dutch and French sides of the island, including individuals in an abandoned swimming pool, in drainage ditches, and roadside puddles; it has become more conspicuous since Hurricane Irma hit the island in 2017 (M. Yokoyama; *in litt.*, 08/04/2019; Janse *et al.* 2020).

Conservation status, threats, and actions: Although an introduced species in the Lesser Antilles, it is unlikely to pose a conservation threat to anurans on St Martin/St Maarten as all extant frogs on that island are introduced.

D. Unestablished non-native amphibians

1. Rhinella marina *(Linnaeus, 1758)*

Henderson and Breuil (2012) documented a single individual from Union Island (St Vincent Grenadines). No evidence exists for any further introductions.

2. Eleutherodactylus planirostris *(Cope, 1862)*

An introduced population apparently was established on Grenada, likely introduced during the Cuban occupation in the early 1980s (although an introduction via shipments of ornamental plants, possibly from Florida, USA, cannot be ruled out) (Henderson and Powell 2018). Specimens were collected in 1991 and 1992 (A. Wynn, *in litt.*, 06/01/2016), but recent fieldwork has failed to locate the species at the site where it was previously known to occur (Henderson and Powell 2018). We assume that this population is now extirpated. However, Harrison and Cheng (2022) documented a new island record for *E. planirostris* on Carriacou (Grenada, Grenadines), reporting the discovery of two individuals at sites some 1,400 m apart from one another, which suggests that an established population might exist on the island; the origins of the frogs are unknown. In addition, several waifs shipped with ornamental plants from Miami, Florida, USA were intercepted by quarantine officers in St Vincent (Glasgow 2011). Eight individuals in plant containers shipped from Florida, USA, were intercepted on St Barthélemy in November 2017, but no evidence suggests that a population became established (Questel 2018).

3. Hyla squirella (= Dryophytes squirellus) *(Daudin, 1800) (Fig. 9.4Q)*

Hyla squirella (recently assigned to the genus *Dryophytes* by Duellman *et al.* 2016) is native to the southeastern USA (Powell *et al.* 2016). An introduced population is known from Grand Bahama

Island (Crombie 1972) and additional populations might occur on San Salvador (Hillbrand *et al.* 2011) and North Bimini (Lee 2004). Hammerson and Hedges (2017) stated that it was established on Grand Bahama Islands and the Little Bahama Bank. In the Lesser Antilles, a single waif in a plant container shipped from Florida, USA, was intercepted on St Barthélemy (Questel 2018).

4.* Osteopilus septentrionalis *(Duméril and Bibron, 1841)

Waifs are known from Saba and Dominica (Powell 2007; Henderson and Breuil 2012), and two individuals in separate shipments from Florida (USA) were interdicted into Grenada in 2013 (Somma and Graham 2015).

5.* Salamandra salamandra *(Linnaeus, 1758) (Fig. 9.4R)

This species is native to Europe, where it is widely distributed from Portugal to southeastern Ukraine (Frost 2019). On 7 January 2012, a live individual was found in a supermarket in St Jean on St Barthélemy in a plastic box of fresh salad shipped from Loire-Atlantique, France; it died two days after its discovery (Breuil *et al.* 2012).

V. Acknowledgements

We thank Ronald Altig, Renaud Boistel, Michel Breuil, Jenny Daltry, Mark de Silva, Maël Dewynter, Darrel Frost, Gerardo García, Billie Harrison, Kevel Lindsay, Olivier Lorvelec, Kate Orchard, John Parmerlee Jr., Karl Questel, Richard Sajdak, and Mark Yokoyama for insights, use of photographs, and production of the map.

VI. References

Acevedo, A. A., Lampo, M., and Cipriani, R., 2016. The cane or marine toad, *Rhinella marina* (Anura, Bufonidae): two genetically and morphologically distinct species. *Zootaxa* 4103: 574–586. https://doi.org/10.11646/zootaxa.4103.6.7.

Ackley, J. W., Muellman, P. J., Carter, R. E., Henderson, R. W., and Powell, R., 2009. A rapid assessment of herpetofaunal diversity in variously altered habitats on Dominica. *Applied Herpetology* 6: 171–184. https://doi.org/10.1163/157075408X394124.

Adams, S. L., Morton, M. N., Terry, A., Young, R. P., Dawson, J., Martin, L., Sulton, M., Hudson, M., Cunningham, A., Garcia, G., Goetz, M., Lopez, J., Tapley, B., Burton, M., and Gray, G., 2014. Long-term recovery strategy for the critically endangered mountain chicken 2014–2034. Mountain Chicken Recovery Programme. http://www.amphibians.org/wp-content/uploads/2015/08/Mountain-Chicken-SAP-2014-working-draft-FINAL.pdf.

Barbour, T., 1914. A contribution to the zoögeography of the West Indies, with especial reference to amphibians and reptiles. *Memoirs of the Museum of Comparative Zoölogy* 44: 205–359. https://doi.org/10.5962/bhl.title.49187.

Barbour, T., 1916. The reptiles and amphibians of Grenada. Pp. 236–243 in *The Grenada Handbook, Directory and Almanac for the Year 1916*, The Colonial Secretary (compiler). Wyman & Sons, Ltd., London.

Barbour, T., 1930. A list of Antillean reptiles and amphibians. *Zoologica* 11: 61–116. https://doi.org/10.5962/p.203735.

Barbour, T., 1937. Third list of Antillean reptiles and amphibians. *Bulletin of the Museum of Comparative Zoölogy* 82: 77–166.

Berg, C. S., Harrison, B. C., and Henderson, R. W., 2013. *Pristimantis euphronides*, observations in the field and captivity: conservation implications. Pp. 109–115 in *Biodiversité insulaire: la flore, la faune et l'homme dans les Petites Antilles. Actes du Colloque international, Schoelcher, 8–10 Novembre 2010*, edited by J.-L. Vernier and M. Burac. Direction de l'Environment, de l'Aménagement et du Logement de Martinique and Université de Antilles et de la Guyana, Schoelcher, Martinique.

Boulenger, G. A., 1891. On reptiles, batrachians and fishes from the Lesser West Indies. *Proceedings of the Zoological Society of London* 1891: 351–357. https://doi.org/10.1111/j.1096-3642.1891.tb01759.x.

Bouysse, P., Westercamp, D., and Andreieff, P., 1990. The Lesser Antilles island arc. *Proceedings of the Ocean Drilling Program, Scientific Results* 110: 29–44. https://doi.org/10.2973/odp.proc.sr.110.166.1990.

Breuil, M., 1997. L'Herpétofaune de la Reserve Biologique de la Montagne Peléé (Martinique). Fort-de-France, Martinique, Report, Office national des Forêts, Paris, Association des Amis du Laboratoire des Reptiles et Amphibiens du MNHN, Paris.

Breuil, M., 2002. Histoire naturelle des Amphibiens et des Reptiles terrestres de l'archipel Guadeloupéen. Guadeloupe, Saint-Martin, Saint-Barthélemy. *Partrimoines Naturels* 54: 1–339.

Breuil, M., 2009. The terrestrial herpetofauna of Martinique: Past, present, future. *Applied Herpetology* 6: 123–149. https://doi.org/10.1163/157075408X386114.

Breuil, M. and Ibéné, B., 2008. Les Hylidés envahissants dans les Antilles françaises et le peuplement batrachologique naturel. *Bulletin de la Société Herpétologique de France* 125: 41–67.

Breuil, M., Guiougou, F., Questel, K., and Ibéné, B., 2009. Modifications du peuplement herpétologique dans les Antilles françaises: disparitions et espèces allochtones. Partie 1: Historique-Amphibiens. *Le Courrier de la Nature* 249: 30–37.

Breuil, M., Maille, J.-C., and Le Quellec, F., 2012. *Salamandra salamandra* (Fire Salamander). Distribution. *Caribbean Herpetology* 25: 1. https://doi.org/10.31611/ch.25.

Brooks, G. R., Jr., 1982. An analysis of prey consumed by the anuran, *Leptodactylus fallax*, from Dominica, West Indies. *Biotropica* 14: 301–309. https://doi.org/10.2307/2388091.

Buckner, S. D., Franz, R., and Reynolds, R. G., 2012. Bahama Islands and Turks & Caicos Islands. Pp. 93–110 in *Island Lists of West Indian Amphibians and Reptiles*, edited by R. Powell and R. W. Henderson. *Bulletin of the Florida Museum of Natural History* 51: 85–166.

Buley, K. R., 2003. *Leptodactylus fallax* (Mountain Chicken). Ophiophagy. *Herpetological Review* 34: 358–359.

Burrowes, P. A. and Joglar, R. L., 2005. El estado de los anfibios de Puerto Rico. Pp. 60–65 in *Biodiversidad de Puerto Rico. Vertebrados Terrestres y Ecosistemas*, edited by R. L. Joglar. Serie de Historia Natural. Editorial Instituto de Cultura Puertorriquena, San Juan, Puerto Rico.

Burrowes, P. A., Joglar, R. L., and Green, D. E., 2004. Potential causes for amphibian declines in Puerto Rico. *Herpetologica* 60: 141–154. https://doi.org/10.1655/03-50.

Butler, R. A., 2006. Caribbean Islands. Mongabay.com/A Place Out of Time: Tropical Rainforests and the Perils They Face. https://rainforests.mongabay.com/20carib.htm.

Cahill, A. E., Aiello-Lammens, M. E., Fisher-Reid, M. C., Hua, X., Karanewsky, C. J., Yeong Ryu, H., Sbeglia, G. C., Spagnolo, F., Waldron, J. B., Warsi, O., and Wiens, J. J., 2013. How does climate change cause extinction? *Proceedings of the Royal Society B* 280: 20121890. http://dx.doi.org/10.1098/rspb.2012.1890.

Camargo, A., Heyer, W. R., and de Sá, R. O., 2009. Phylogeography of the frog *Leptodactylus validus* (Amphibia: Anura): Patterns and timing of colonization events in the Lesser Antilles. *Molecular Phylogenetics and Evolution* 53: 571–579. https://doi.org/10.1016/j.ympev.2009.07.004.

Carter, R. E., Berg, C. S., Ackley, J. W., and Powell, R., 2009. Frogs of Dominica, with notes on habitat use by two species of *Eleutherodactylus*. *Herpetological Bulletin* 108: 14–23.

Cedeño-Vázquez, J. R., González-Vázquez, J., Martínez-Arce, A., and Canseco-Márquez, L., 2014. First record of the invasive greenhouse frog (*Eleutherodactylus planirostris*) in the Mexican Caribbean. *Revista Mexicana de Biodiversidad* 85: 650–653. https://doi.org/10.7550/rmb.43234.

Censky, E. J., 1989. *Eleutherodactylus johnstonei* (Salientia: Leptodactylidae) from Anguilla, West Indies. *Caribbean Journal of Science* 25: 229–230.

Chaparro, J. C., Pramuk, J. B., and Gluesenkamp, A. G., 2007. A new species of arboreal *Rhinella* (Anura: Bufonidae) from cloud forest of southeastern Peru. *Herpetologica* 63: 203–212. https://doi.org/10.1655/0018-0831(2007)63[203:ansoar]2.0.C);2.

Christy, M. T., Clark, C. S., Gee II, D. E., Vice, D., Warner, D. S., Tyrrell, C. L., Rodda, G. H., and Savidge, J. A., 2007. Recent records of alien anurans on the Pacific island of Guam. *Pacific Science* 61: 469–483. https://doi.org/10.2984/1534-6188(2007)61[469:rroaao]2.0.C);2.

Clark, A. H., 1916. The present status and breeding season of the Giant Toad (*Bufo agua*) in Barbados, St. Vincent, Trinidad and Demerara. *Copeia* 1916: 13–14.

Crawford, A. J., Alonso, R., Jaramillo A., C. A., Sucre, S., and Ibañez D., R., 2011. DNA barcoding identifies a third invasive species of *Eleutherodactylus* (Anura: Eleutherodactylidae) in Panama City, Panama. *Zootaxa* 2890: 65–67. https://doi.org/10.11646/zootaxa.2890.1.6.

Crombie, R. I., 1972. The presence of *Hyla squirella* in the Bahamas. *Quarterly Journal of the Florida Academy of Sciences* 35: 49–52.

Dalsgaard, B., Hilton, G. M., Gray, G. A. L., Aymer, L., Boatswain, J., Daley, J., Fenton, C., Martin, J., Martin, L., Murrain, P., Arendt, W. J., Gibbons, D. W., and Oleson, J. M., 2007. Impacts of a volcanic eruption on the forest bird community of Montserrat, Lesser Antilles. *Ibis* 149: 298–312. https://doi.org/10.1111/j.1474-919X.2006.00631.x.

Daltry, J. C., 2002. *Mountain Chicken Monitoring Manual*. Fauna and Flora International, Cambridge, and the Forestry and Wildlife Division, Dominica.

Daltry, J. C., 2007. An introduction to the herpetofauna of Antigua, Barbuda and Redonda, with some conservation recommendations. *Applied Herpetology* 4: 97–130. https://doi.org/10.1163/157075407780681338.

Daltry, J. C., 2011. An introduction to the herpetofauna of Antigua, Barbuda and Redonda, with some conservation recommendations. Pp. 17–51 in *Conservation of Caribbean Island Herpetofaunas. Volume 2: Regional Accounts of the West Indies*, edited by A. Hailey, B. S. Wilson, and J. A. Horrocks. Brill, Leiden, The Netherlands.

Daltry, J. C. and Gray, G., 1998. *Mountain Chicken Emergency Assessment: Findings of Fieldwork in January and February 1998*. Fauna & Flora International, Cambridge.

Daltry, J. C. and Gray, G., 1999. Effects of volcanic activity on the endangered Mountain Chicken Frog (*Leptodactylus fallax*). *Froglog* 32: 1–2.

Daudin, J. and de Silva, M., 2007. An annotated checklist of the amphibians and terrestrial reptiles of the Grenadines with notes on their local natural history and conservation. *Applied Herpetology* 4: 163–175. https://doi.org/10.1163/157075407780681329.

Daudin, J. and de Silva, M., 2011. An annotated checklist of the amphibians and terrestrial reptiles of the Grenadines with notes on their local natural history and conservation. Pp. 259–272 in *Conservation of Caribbean Island Herpetofaunas. Volume 2: Regional Accounts of the West Indies*, edited by A. Hailey, B. S. Wilson, and J. A. Horrocks. Brill, Leiden, The Netherlands.

Davis, S. L., Davis, R. B., James, A., and Talyn, B. C. P., 2000. Reproductive behaviour and larval development of *Leptodactylus fallax* in Dominica, West Indies. *Herpetological Review* 31: 217–220.

de Sá, R. O., Grant, T., Camargo, A., Heyer, W. R., Ponssa, M. L., and Stanley, E., 2014. Systematics of the Neotropical genus *Leptodactylus* Fitzinger, 1826 (Anura: Leptodactylidae): Phylogeny, the relevance of non-molecular evidence, and species accounts. *South American Journal of Herpetology* 30, Special Issue no. 1: S1–S128. https://doi.org/10.2994/sajh-D-13-00022.1.

de Silva, M. and Wilson, D., 2006. *A Natural History of Mustique*. The Mustique Company Ltd., Mustique Island, St Vincent and the Grenadines.

Díaz, L. M., Cádiz, A., Chong, A., and Silva, A., 2007. First report of chytridiomycosis in a dying toad (Anura: Bufonidae) from Cuba: A new conservation challenge for the island. *EcoHealth* 4: 172–175. https://doi.org/10.1007/s10393-007-0094-4.

Díaz-Lameiro, A. M., Powell, R., and Berg, C. S., 2007. Colour and pattern polymorphism in *Pristimantis shrevei* and *Eleutherodactylus johnstonei* (Leptodactylidae) on St. Vincent, West Indies. *Herpetological Bulletin* 101: 18–25.

Dodd, C. K., Jr., 2013. *Frogs of the United States and Canada*. 2 volumes. Johns Hopkins University Press, Baltimore, Maryland.

DOE (The Department of Environment), 2008. *Montserrat Centre Hills Management Plan, 2008–2010: Enabling the Effective Conservation and Management of Natural Resources within Montserrat's Centre Hills*. Department of the Environment, Ministry of Agriculture, Lands, Housing, and the Environment, Brades, Montserrat, and the Royal Society for the Protection of Birds, Sandy, United Kingdom.

Donaldson, C., 2018. Analyzing factors influencing reproductive success of the Mountain Chicken Nordens Ark Captive Breeding Program. Unpublished bachelor degree project in bioscience, University of Skövde, Skövde, Sweden.

Drake, M. C., Zieger, U., Groszkowski, A., Gallardo, B., Sages, P., Reavis, R., Faircloth, L., Jacobson, K., Lonce, N., Pinckney, R., and Cole, R. A., 2014. Survey of helminths, ectoparasites, and chytrid fungus of an introduced population of Cane Toads, *Rhinella marina* (Anura: Bufonidae), from Grenada, West Indies. *Journal of Parasitology* 100: 608–615. https://doi.org/10.1645/13-470.1.

Duellman, W. E., Marion, A. B., and Hedges, S. B., 2016. Phylogenetics, classification, and biogeography of the treefrogs (Amphibia: Anura: Arboranae). *Zootaxa* 4104: 1–109. https://doi.org/10.11646/zootaxa.4104.1.1.

EarthTrends. 2003. Country profiles. Grenada: Forests, grasslands, and drylands. Pp. 1–7. World Resources Institute, Washington, DC (website no longer available).

Easteal, S., 1981. The history of *Bufo marinus* (Amphibia: Anura) introductions; a natural experiment in evolution. *Biological Journal of the Linnaean Society* 16: 93–113. https://doi.org/10.1111/J.1095-8312.1981.TB01645.X.

Espeut, W. B., 1882. On the acclimatization of the Indian mungoos in Jamaica. *Proceedings of the Zoological Society of London* 1882: 712–714. https://doi.org/10.1111/j.1096-3642.1883.tb02783.x.

Faaborg, J. R. and Arendt, W. J., 1985. *Wildlife Assessment in the Caribbean*. Institute of Tropical Forestry, Río Piedras, Puerto Rico.

Faivovich, J., Ferraro, D. P., Basso, N. G., Haddad, C. F. B., Rodrigues, M. T., Wheeler, W. C., and Lavilla, E. O., 2012. A phylogenetic analysis of *Pleurodema* (Anura: Leptodactylidae: Leiuperinae) based on mitochondrial and nuclear gene sequences, with comments on the evolution of anuran foam nests. *Cladistics* 28: 460–482. https://doi.org/10.1111/j.1096-0031.2012.00406.x.

FAO (Food and Agriculture Organization of the United Nations), 2010. Global forest resources assessment 2010. *FAO Forestry Paper* 163: xxxi + 340 pp.

Fernandez Amaya, M. M. and Mahon, R., 2011. Livelihoods vulnerability and capacity for adaptation to climate change in Union Island, St. Vincent and the Grenadines. CERMES Technical Report No. 48. Centre for Resource Management and Environmental Studies, University of the West Indies, Faculty of Pure and Applied Sciences, Cave Hill Campus, Barbados.

Fields, A. and Horrocks, J. A., 2011. The herpetofaunas of Barbados: Anthropogenic impacts and conservation status. Pp. 90–104 in *Conservation of Caribbean Island Herpetofaunas. Volume 2: Regional Accounts of the West Indies*, edited by A. Hailey, B. S. Wilson, and J. A. Horrocks. Brill, Leiden, The Netherlands.

Fiske, R. S. and Sigurdsson, H., 1982. Soufriere Volcano, St. Vincent: Observations of its 1979 eruption from the ground, aircraft, and satellites. *Science* 216: 1105–1126. https://doi.org/10.1126/science.216.4550.1105.

Fitzpatrick, S. M. and Keegan, W. F., 2007. Human impacts and adaptations in the Caribbean islands: an historical ecology approach. *Earth and Environmental Science Transactions of the Royal Society of Edinburgh* 98: 29–45. https://doi.org/10.1017/S1755691007000096.

Fouquet, A., Vences, M., Slducci, M.-D., Meyer, A., Marty, C., Blanc, M., and Gilles, A., 2007. Revealing cryptic diversity using molecular phylogenetics and phylogeography in frogs of the *Scinax ruber* and *Rhinella margaritifera* species groups. *Molecular Phylogenetics and Evolution* 43: 567–582. https://doi.org/10.1016/j.ympev.2006.12.006.

Fouquet, A., Pineau, K., Trefaut Rodrigues, M., Mailles, J., Schneider, J.-B., Ernst, R., and Dewynter, M., 2013. Endemic or exotic: The phylogenetic position of the Martinique Volcano frog *Allobates chalcopis* (Anura: Dendrobatidae) sheds light on its origin and challenges current conservation strategies. *Systematics and Biodiversity* 11: 87–101. https://doi.org/10.1080/14772000.2013.764944.

Frederick, J. H., Yuan, M. L., Jung, C., and Meijer zu Scholochtern, M. 2021. Geographic distribution. *Eleutherodactylus planirostris* (Greenhouse Frog). Herpetological Review 52: 790.

Freeland, W. J., 1984. Cane Toads: a review of their biology and impact on Australia. Technical Report No. 19. Conservation Commission of the Northern Territory, Darwin, Australia.

Frost, D. R., 2019. Amphibian Species of the World: An Online Reference. Version 6.0. Electronic Database accessible at http://research.amnh.org/herpetology/amphibia/index.html. American Museum of Natural History, New York, USA (accessed 1 August 2019). https://doi.org/10.5531/db.vz.0001.

Frost, D. R., Grant, T., Faivovich, J., Bain, R. H., Haas, A., Haddad, C. F. B., de Sá, R. O., Channing, A., Wilkinson, M., Donnellan, S. C., Raxworthy, C. J., Campbell, J. A., Blotto, B. L., Moler, P. E., Drewes, R. C., Nussbaum, R. A., Lynch, J. D., Green, D. M., and Wheeler, W. C., 2006. The amphibian tree of life. *Bulletin of the American Museum of Natural History* 297: 1–370. https://doi.org/10.1206/0003-0090(2006)297[0001:tatol]2.0.C);2.

Gallant, A. L., Klaver, R. W., Casper, G. S., and Lannoo, M. J., 2007. Global rates of habitat loss and implications for amphibian conservation. *Copeia* 2007: 967–979. https://doi.org/10.1643/0045-8511(2007)7[967:grohla]2.0.CO;2.

García, G., 2008. Projekt "Berghuhn" – der Antillen-Ochsenfrosch (*Leptodactylus fallax*). *Draco* 9: 70–79.

García, G., Cunningham, A., Horton, D., Garner, T., Hyatt, A., Hengstberger, S., Lopez, J., Ogrodowczyk, A., Fenton, C., and Fa, J., 2007. Mountain chickens *Leptodactylus fallax* and sympatric amphibians appear to be disease free on Montserrat. *Oryx* 41: 398–401. https://doi.org/10.1017/S0030605307001012.

García, G., Lopez, J., Fa, J. E., and Gray, G. A. L., 2009. Chytrid fungus strikes mountain chickens in Montserrat. *Oryx* 43: 323–324.

Garman, S., 1888 (1887). West Indian Batrachia in the Museum of Comparative Zoology. *Bulletin of the Essex Institute* 19: 13–16.

Germano, J. M., Sander, J. M., Henderson, R. W., and Powell, R., 2003. Herpetofaunal communities in Grenada: a comparison of altered sites, with an annotated checklist of Grenadian amphibians and reptiles. *Caribbean Journal of Science* 39: 68–76.

Gibson, R. C. and Buley, K. R., 2004. Maternal care in oophagy in mountain chickens. *Copeia* 2004: 128–135. https://doi.org/10.1643/CE-02-091R2.

Glasgow, A., 2011. A failed (?) introduction of lizards and frogs on St Vincent. *IRCF Reptiles & Amphibians* 18: 54–55. https://doi.org/10.17161/randa.v18i1.

Goin, C. J., 1944. *Eleutherodactylus ricordii* at Jacksonville, Florida. *Copeia* 1944: 192.

Goldberg, S. R., Bursey, C. R., and Kaiser, H., 1998. Gastrointestinal helminths of five species of *Eleutherodactylus* (Anura: Leptodactylidae) from the West Indies. *Caribbean Journal of Science* 34: 146–149.

Gorman, M. L., 1975. The diet of feral *Herpestes auropunctatus* (Carnivora: Viverridae) in the Fijian Islands. *Journal of Zoology* 175: 273–278. https://doi.org/10.1111/j.1469-7998.1975.tb01401.x.

Grant, T., Frost, D. R., Caldwell, J. P., Gagliardo, R., Haddad, C. F. B., Kok, P. J. R., Means, D. B., Noonan, B. P., Schargel, W. E., and Wheeler, W. C., 2006. Phylogenetic systematics of dart-poison frogs and their relatives (Amphibia: Athesphatanura: Dendrobatidae). *Bulletin of the American Museum of Natural History* 299: 1–262. https://doi.org/10.1206/0003-0090(2006)299[1:psodfa]2.0.CO;2.

Groome, J. R., 1970. *A Natural History of the Island of Grenada, W.I.* Grenada National Trust, St George's, Grenada.

Guarino, F. M., García, G., and Andreone, F., 2014. Huge but moderately long-lived: age structure in the mountain chicken, *Leptodactylus fallax*, from Montserrat, West Indies. *Herpetological Journal* 24: 167–173.

Hammerson, G. A. and Hedges, B., 2017. *Dryophytes squirellus* (amended version of 2004 assessment). *The IUCN Red List of Threatened Species* 2017: e.T55662A112715025. http://dx.doi.org/10.2305/IUCN.UK.2017-1.RLTS.T55662A112715025.en. Downloaded 1 August 2019.

Hardy, J., Heyer, R., Hedges, B., and Powell, R., 2010. *Leptodactylus validus*. *The IUCN Red List of Threatened Species* 2010: e.T57172A11595666. http://dx.doi.org/10.2305/IUCN.UK.2010-2.RLTS.T57172A11595666.en. Downloaded 1 August 2019.

Harrison, B., Berg, C. S., and Henderson, R. W., 2011. The Grenada Frog (*Pristimantis euphronides*): an endemic species in decline and the combined effects of habitat loss, competition, and chytridiomycosis. *IRCF Reptiles & Amphibians* 18: 66–73. https://doi.org/10.17161/randa.v18i2.

Harrison, B. C. and Cheng, S., 2022. Geographic distribution. *Eleutherodactylus planirostris* (Greenhouse Frog). *Herpetological Review* 53: 439.

Hays, W. S. T. and Conant, S., 2011. Biology and impacts of Pacific island invasive species. 1. A worldwide review of effects of the Small Indian Mongoose, *Herpestes javanicus* (Carnivora: Herpestidae). *Pacific Science* 61: 3–16. https://doi.org/10.1353/psc.2007.0006.

Heatwole, H., 2013. Worldwide decline and extinction of amphibians. Pp. 259–278 in *The Balance of Nature and Human Impact*, edited by K. Rohde. Cambridge University Press, Cambridge.

Hedges, S. B., 1996. The origin of West Indian amphibians and reptiles. Pp. 95–127 in *Contributions to West Indian Herpetology: A Tribute to Albert Schwartz*, edited by R. Powell and R. W. Henderson. Society for the Study of Amphibians and Reptiles, Ithaca, New York.

Hedges, S. B., 1999. Distribution patterns of amphibians in the West Indies. Pp. 211–254 in *Patterns of Distribution of Amphibians: A Global Perspective*, edited by W. E. Duellman. The Johns Hopkins University Press, Baltimore.

Hedges, S. B., 2001. Caribbean biogeography: an outline. Pp. 15–33 in *Biogeography of the West Indies: Patterns and Perspectives*, edited by C. A. Woods and F. E. Sergile. CRC Press, Boca Raton, Florida.

Hedges, S. B., 2006. An overview of the evolution and conservation of West Indian amphibians and reptiles. *Applied Herpetology* 3: 281–292. https://doi.org/10.1163/157075406778905072.

Hedges, S. B. and Conn, C. E., 2012. A new skink fauna from Caribbean islands (Squamata, Mabuyidae, Mabuyinae). *Zootaxa* 3288: 1–244. https://doi.org/10.11646/zootaxa.3288.1.1.

Hedges, S. B. and Heinicke, M. P., 2007. Molecular phylogeny and biogeography of West Indian frogs of the genus *Leptodactylus* (Anura, Leptodactylidae). *Molecular Phylogenetics and Evolution* 44: 308–314. https://doi.org/10.1016/j.ympev.2006.11.011.

Hedges, B. and Powell, R., 2004. *Pristimantis shrevei. The IUCN Red List of Threatened Species* 2004: e.T56961A11561177. http://dx.doi.org/10.2305/IUCN.UK.2004.RLTS.T56961A11561177.en. Downloaded 1 August 2019.

Hedges, B. and Powell, R., 2010a. *Eleutherodactylus amplinympha. The IUCN Red List of Threatened Species* 2010: e.T56411A11472323. http://dx.doi.org/10.2305/IUCN.UK.2010-2.RLTS.T56411A11472323.en. Downloaded 1 August 2019.

Hedges, B. and Powell, R., 2010b. *Pristimantis euphronides. The IUCN Red List of Threatened Species* 2010: e.T56593A11491004. http://dx.doi.org/10.2305/IUCN.UK.2010-2.RLTS.T56593A11491004.en. Downloaded 1 August 2019.

Hedges, S. B., Duellman, W. E., and Heinicke, M. P., 2008. New World direct-developing frogs (Anura: Terrarana): Molecular phylogeny, classification, biogeography, and conservation. *Zootaxa* 1737: 1–182. https://doi.org/10.11646/zootaxa.1737.1.1.

Hedges, B., Díaz, L., and Powell, R., 2009. *Eleutherodactylus planirostris. The IUCN Red List of Threatened Species* 2009: e.T56864A11534006. http://dx.doi.org/10.2305/IUCN.UK.2004.RLTS.T56864A11534006.en. Downloaded 1 August 2019.

Hedges, B., Ibéné, B., Breuil, M., and Powell, R., 2010a. *Allobates chalcopis. The IUCN Red List of Threatened Species* 2010: e.T55066A11247344. http://dx.doi.org/10.2305/IUCN.UK.2010-2.RLTS.T55066A11247344.en. Downloaded 1 August 2019.

Hedges, B., Ibéné, B., Breuil, M., and Powell, R., 2010b. *Eleutherodactylus barlagnei. The IUCN Red List of Threatened Species* 2010: e.T56451A11480379. http://dx.doi.org/10.2305/IUCN.UK.2010-2.RLTS.T56451A11480379.en. Downloaded 1 August 2019.

Hedges, B., Ibéné, B., Breuil, M., and Powell, R., 2010c. *Eleutherodactylus pinchoni. The IUCN Red List of Threatened Species* 2010: e.T56860A11546413. http://dx.doi.org/10.2305/IUCN.UK.2010-2.RLTS.T56860A11546413.en. Downloaded 1 August 2019.

Hedges, B., Ibéné, B., Breuil, M., and Powell, R., 2010d. *Eleutherodactylus martinicensis. The IUCN Red List of Threatened Species* 2010: e.T56747A11530334. http://dx.doi.org/10.2305/IUCN.UK.2010-2.RLTS.T56747A11530334.en. Downloaded 1 August 2019.

Hedges, B., Ibéné, B., Koenig, S., la Marca, E., Ibáñez, R., and Hardy, J., 2010e. *Eleutherodactylus johnstonei. The IUCN Red List of Threatened Species* 2010: e.T56684A11505248. http://dx.doi.org/10.2305/IUCN.UK.2010-2.RLTS.T56684A11505248.en. Downloaded 1 August 2019.

Hedges, B., Diaz, L., Ibéné, B., Joglar, R., Powell, R., Bolaños, F., and Chaves, G., 2010f. *Osteopilus septentrionalis. The IUCN Red List of Threatened Species* 2010: e.T55811A11368202. http://dx.doi.org/10.2305/IUCN.UK.2010-2.RLTS.T55811A11368202.en. Downloaded 1 August 2019.

Heinicke, M. P., Duellman, W. E., and Hedges, S. B., 2007. Major Caribbean and Central American frog faunas originated by oceanic dispersal. *Proceedings of the National Academy of Sciences* 104: 10092–10097. https://doi.org/10.1073/pnas.0611051104.

Heinicke, M. P., Diaz, L. M., and Hedges, S. B., 2011. Origin of invasive Florida frogs traced to Cuba. *Biology Letters* 7: 407–410. https://doi.org/10.1098/rsbl.2010.1131.

Hellmann, J. J., Byers, J. E., Bierwagen, B. G., and Dukes, J. S., 2008. Five potential consequences of climate change for invasive species. *Conservation Biology* 22: 534–543. https://doi.org/10.1111/j.1523-1739.2008.00951.x.

Helmer, E. H., Kennaway, T. A., Pedreros, D. H., Clark, M. L., Marcano-Vega, H., Tieszen, L. L., Ruzycki, T. R., Schill, S. R., and Carrington, C. M. S., 2008. Land cover and forest formation distributions for St. Kitts, Nevis, St. Eustatius, Grenada and Barbados from decision tree classification of cloud-cleared satellite imagery. *Caribbean Journal of Science* 44: 175–198. https://doi.org/18475/cjos.v44i2.a6.

Henderson, R. W., 2010. *Eleutherodactylus johnstonei* (NCN). Microhabitat. *Herpetological Review* 41: 195.

Henderson, R. W. and Berg, C. S., 2005. A post-Hurricane Ivan assessment of frog and reptile populations on Grenada, West Indies. *Herpetological Bulletin* 91: 4–9.

Henderson, R. W. and Berg, C. S., 2006. The herpetofauna of Grenada and the Grenada Grenadines: conservation concerns. *Applied Herpetology* 3: 197–213. https://doi.org/10.1163/157075406778116195.

Henderson, R. W. and Berg, C. S., 2011. The herpetofauna of Grenada and the Grenada Grenadines: conservation concerns. Pp. 239–258 in *Conservation of Caribbean Island Herpetofaunas. Volume 2: Regional Accounts of the West Indies*, edited by A. Hailey, B. S. Wilson, and J. A. Horrocks. Brill, Leiden, The Netherlands.

Henderson, R. W. and Breuil, M., 2012. Lesser Antilles. Pp. 148–159 in Island lists of West Indian amphibians and reptiles, edited by R. Powell and R. W. Henderson. *Bulletin of the Florida Museum of Natural History* 51: 85–166.

Henderson, R. W. and Powell, R., 1999. West Indian herpetoecology. Pp. 223–268 in *Caribbean Amphibians and Reptiles*, edited by B. I. Crother. Academic Press, San Diego.

Henderson, R. W. and Powell, R., 2001. Responses by the West Indian herpetofauna to human-influenced resources. *Caribbean Journal of Science* 37: 41–54.

Henderson, R. W. and Powell. R., 2009. *Natural History of West Indian Reptiles and Amphibians*. University Press of Florida, Gainesville, Florida.

Henderson, R. W. and Powell, R., 2018. *Amphibians and Reptiles of the St. Vincent and Grenada Banks, West Indies*. Edition Chimaira, Frankfurt am Main.

Henderson, R. W., Daudin, J., Haas, G. T., and McCarthy, T. J., 1992. Significant distribution records for some amphibians and reptiles in the Lesser Antilles. *Caribbean Journal of Science* 28: 101–103.

Heyer, W. R., 1979. Systematics of the *pentadactylus* pecies group of the frog genus *Leptodactylus* (Amphibia: Leptodactylidae). *Smithsonian Contributions to Zoology* 301: 1–43. https://doi.org/10.5479/si.00810282.301.

Hillbrand, P. A., Sloan, A. T., and Hayes, W. K., 2011. The terrestrial reptiles of San Salvador Island, Bahamas. *IRCF Reptiles & Amphibians* 18: 154–166. https://doi.org/10.17161/randa.v18i3.

Hilton, G., Martin, L., Daly, J., and Allcorn, R., 2008. Montserrat. Pp. 230–36 in *Important Bird Areas in the Caribbean: Key Sites for Conservation*, edited by D. C. Wege and V. Anadon-Irizarry. BirdLife International, Cambridge.

Hoagland, D. B., Horst, G. R., and Kilpatrick, C. W., 1989. Biogeography and population biology of the mongoose in the West Indies. Pp. 611–634 in *Biogeography of the West Indies: Past, Present, and Future*, edited by C. A. Woods. Sandhill Crane Press, Gainesville, Florida.

Hodge, K. V. D., Censky, E. J., and Powell, R., 2003. *The Reptiles and Amphibians of Anguilla, British West Indies*. Anguilla National Trust, The Valley, Anguilla.

Hodge, K. V. D., Powell, R., and Censky, E. J., 2011. Conserving the herpetofauna of Anguilla. Pp. 3–15 in *Conservation of Caribbean Island Herpetofaunas. Volume 2: Regional Accounts of the West Indies*, edited by A. Hailey, B. S. Wilson, and J. A. Horrocks. Brill, Leiden, The Netherlands.

Hudson, M. A., Young, R. P., Lopez, J., Martin, L., Fenton, C., Mccrea, R., Griffiths, R. A., Adams, S.-L., Gray, G., García, G., and Cunningham, A. A., 2016. *In-situ* itraconazole treatment improves survival rate during an amphibian chytridiomycosis epidemic. *Biological Conservation* 195: 37–45. https://doi.org/10.1016/j.biocon.2015.12.041.

Huey, R. B., Kearney, M. R., Krockenberger, A., Holtum, J. A. M., Jess, M., and Williams, S. E., 2012. Predicting organismal vulnerability to climate warming: Roles of behaviour, physiology and adaptation. *Philosophical Transactions of the Royal Society B* 367: 1665–1679. https://doi.org/10.1098/rstb.2012.0005.

Hughes, D. F., Green, M. L., Warner, J. K., and Davidson, P. C., 2019. There's a frog in my salad! A review of online media coverage for wild vertebrates found in prepackaged produce in the United States. *Science of the Total Environment* 675: 1–12. https://doi.org/10.1016/j.scitotenv.2019.03.254.

Hummelinck, P. W., 1940. Studies on the fauna of Curaçao, Aruba, Bonaire, and the Venezuelan Islands. No. 2. A survey of the mammals, lizards and molluscs. *Studies on the Fauna of Curaçao and Other Caribbean Islands* 1: 59–108.

IUCN SSC Amphibian Specialist Group, 2017. *Leptodactylus fallax. The IUCN Red List of Threatened Species* 2017: e.T57125A3055585. http://dx.doi.org/10.2305/IUCN.UK.2017-3.RLTS.T57125A3055585.en. Downloaded 1 August 2019.

Janse, J., Schotsmans, E., Yokoyama, M., Chalifour, J., and van Wagensveld, T. P., 2020. First records of the Stout Four-eyed Frog, *Pleurodema brachyops* (Cope 1869) on St. Martin. *IRCF Reptiles & Amphibians* 27: 250–251. https://doi.org/10.17161/randa.v27i2.14220.

Jayson, S., Harding, L., Michaels, C. J., Tapley, B., Hedley, J., Goetz, M., Barbon, A., Garcia, G., Lopez, J., and Flach, E., 2018a. Development of a body condition score for the mountain chicken frog (*Leptodactylus fallax*). *Zoo Biology* 37: 196–205. https://doi.org/10.1002/zoo.21409.

Jayson, S., Ferguson, A., Goetz, M., Routh, A., Tapley, B., Harding, L., Michaels, C. J., and Dawson, J., 2018b. Comparison of the nutritional content of the captive and wild diets of the critically endangered mountain chicken frog (*Leptodactylus fallax*) to improve its captive husbandry. *Zoo Biology* 37: 332–346. https://doi.org/10.1002/zoo.21442.

Jenkins, M. and Harcourt, C., 1996. Lesser Antilles. Pp. 120–137 in *The Conservation Atlas of Tropical Forests: The Americas*, edited by C. S. Harcourt and J. A. Sayer. Simon & Schuster, New York.

Kaiser, H., 1992. The trade-mediated introduction of *Eleutherodactylus martinicensis* (Anura: Leptodactylidae) on St. Barthélemy, French Antilles, and its implications for Lesser Antillean biogeography. *Journal of Herpetology* 26: 264–273. https://doi.org/10.2307/1564880.

Kaiser, H., 1994. *Leptodactylus fallax. Catalogue of American Amphibians and Reptiles* 583: 1–3.

Kaiser, H., 1997. Origins and introductions of the Caribbean frog, *Eleutherodactylus johnstonei* (Leptodactylidae): management and conservation concerns. *Biodiversity and Conservation* 6: 1391–1407. https://doi.org/10.1023/A:1018341814510.

Kaiser, H. and Hardy, J. D., Jr., 1994a. *Eleutherodactylus johnstonei. Catalogue of American Amphibians and Reptiles* 581: 1–5.

Kaiser, H. and Hardy, J. D., Jr., 1994b. *Eleutherodactylus martinicensis. Catalogue of American Amphibians and Reptiles* 582: 1–4.

Kaiser, H. and Henderson, R. W., 1994. The conservation status of Lesser Antillean frogs. *Herpetological Natural History* 2: 41–56.

Kaiser, H. and Wagenseil, R., 1995. Colonization and distribution of *Eleutherodactylus johnstonei* Barbour (Anura: Leptodactylidae) on Dominica, West Indies. *Caribbean Journal of Science* 31: 341–344.

Kaiser, H., Sharbel, T. F., and Green, D. M., 1994a. Systematics and biogeography of eastern Caribbean *Eleutherodactylus* (Anura: Leptodactylidae): evidence from allozymes. *Amphibia-Reptilia* 15: 375–394. https://doi.org/10.1163/156853894X00416.

Kaiser, H., Green, D. M., and Schmid, M., 1994b. Systematics and biogeography of eastern Caribbean frogs (Leptodactylidae: *Eleutherodactylus*), with the description of a new species from Dominica. *Canadian Journal of Zoology* 72: 2217–2237. https://doi.org/10.1139/z94-297.

Kaiser, H., Hardy, J. D., Jr., and Green, D. M., 1994c. Taxonomic status of Caribbean and South American frogs currently ascribed to *Eleutherodactylus urichi* (Anura: Leptodactylidae). *Copeia* 1994: 780–796. https://doi.org/10.2307/1447195.

Kaiser, H., Coloma, L. A., and Gray, H. M., 1994d. A new species of *Colostethus* (Anura: Dendrobatidae) from Martinique, French Antilles. *Herpetologica* 50: 23–32.

King, J. D. and Ashmore, P. C. 2014. Conservation and historical biogeography: How did the mountain chicken frog get to the Caribbean? *International Journal of Biodiversity and Conservation* 6: 754–764. https://doi.org/10.5897/ijbc2011.075.

King, J. D., Rollins-Smith, L. A., Nielsen, P. F., John, A., and Conlon, J. M., 2005. Characterization of a peptide from skin secretions of male specimens of the frog, *Leptodactylus fallax* that stimulates aggression in male frogs. *Peptides* 26: 597–601. https://doi.org/10.1016/j.peptides.2004.11.004.

Kraus, F., 2009. *Alien Reptiles and Amphibians: A Scientific Compendium, and Analysis.* Invading Nature: Springer Series in Invasion Biology 4. Springer, Dordrecht, The Netherlands.

LaMarca, E., Solís, F., Ibáñez, R., Jaramillo, C., Fuenmayor, Q., Reynolds, R., Rodrigues, M. T., Mijares, A., and MacCulloch, R., 2010. *Pleurodema brachyops. The IUCN Red List of Threatened Species* 2010: e.T57283A11599287. http://dx.doi.org/10.2305/IUCN.UK.2010-2.RLTS.T57283A11599287.en. Downloaded 1 August 2019.

Lane, C. S., Horn, S. P., and Orvis, K. H., 2008. The earliest evidence of Ostionoid maize agriculture from the interior of Hispaniola. *Caribbean Journal of Science* 44: 43–52. https://doi.org/10.18475/cjos.v44i1.a6.

Lee, D. S., 2004. Additional reptiles and amphibians introduced to the Bahamas: a growing conservation concern. *Bulletin of the Chicago Herpetological Society* 39: 161–4.

Lee, W. H., Lau, M. W.-N., Lau, A., Rao, D.-Q., and Sung, Y.-H., 2016. Introduction of *Eleutherodactylus planirostris* (Amphibia, Anura, Eleutherodactylidae) to Hong Kong. *Acta Herpetologica* 11: 85–89. https://doi.org/10.13128/acta_herpetol-16491.

Lescure, J., 1979a. Étude taxinomique et éco-éthologique d'un Amphibien des petites Antilles: *Leptodactylus fallax* Muller, 1926 (Leptodactylidae). *Bulletin du Muséum National d'Histoire Naturelle*, Paris 4, serie 1, section A, no. 3: 757–774.

Lescure, J., 1979b. Singularité et fragilitié de la faune en vertébrés des petites Antilles. *Compté rendu des séances de la Société biogéographie* 482. 93–109.

Lescure, J., 1983. Introductions passives et actives de reptiles et d'amphibiens dans les Antilles et les Guyanes. *Compté rendu des séances de la Société biogéographie* 9: 59–70.

Lescure, J., 2000. Répartition passée de *Leptodactylus fallax* Muller, 1923 et d'*Eleutherodactylus johnstonei* Barbour, 1914 (Anoures, Leptodactylidés). *Bulletin de la Societe Herpetologique de France* 94: 13–23.

Lever, C., 2003. *Naturalized Amphibians and Reptiles of the World.* Oxford University Press, Oxford.

Lewsey, C., Cid, G., and Kruse, E., 2004. Assessing climate change impacts on coastal infrastructure in the Eastern Caribbean. *Marine Policy* 28: 393–409. https://doi.org/10.1016/j.marpol.2003.10.016.

Li, Y., Cohen, J. M., and Rohr, J. R., 2013. Review and synthesis of the effects of climate change on amphibians. *Integrative Zoology* 8: 145–161. https://doi.org/10.1111/1749-4877.12001.

Lindsay, K., and Cooper, B., 2008. *Osteopilus septentrionalis* becomes established on Antigua, West Indies. *Applied Herpetology* 5: 96–98.

Lindsay, K., Orchard, K., and Yokoyama, M., 2019. Giant Cuban Treefrog (*Osteopilus septentrionalis*) tadpoles in an introduced population on Nevis (Lesser Antilles). *IRCF Reptiles & Amphibians* 26: 125–126. https://doi.org/10.17161/randa.v26i2.14380.

Long, E. G., 1974. *The Serpent's Tale. Reptiles and Amphibians of St. Lucia.* Iounaloa Series No. 2. University of West Indies Extra Mural Department, The Morne, St. Lucia.

Lorvelec, O., Pascal, M., Pavis, C., and Feldmann, P., 2007. Amphibians and reptiles of the French West Indies: Inventory, threats and conservation. *Applied Herpetology* 4: 131–161. https://doi.org/10.1163/157075407780681356.

Lorvelec, O., Pascal, M., Pavis, C., and Feldmann, P., 2011. Amphibians and reptiles of the French West Indies: Inventory, threats and conservation. Pp. 205–237 in *Conservation of Caribbean Island Herpetofaunas. Volume 2: Regional Accounts of the West Indies,* edited by A. Hailey, B. S. Wilson, and J. A. Horrocks. Brill, Leiden, The Netherlands.

Lynn, W. G., 1957. Notes on a collection of reptiles and amphibians from Antigua, B.W.I. *Herpetologica* 13: 53–56.

Maciel, N. M., Collevatti, R. G., Colli, G. R., and Schwartz, E. F., 2010. Late Miocene diversification and phylogenetic relationships of the huge toads in the *Rhinella marina* (Linnaeus, 1758) species group (Anura: Bufonidae). *Molecular Phylogenetics and Evolution* 57: 787–797. https://doi.org/10.1016/j.ympev.2010.08.025.

Magin, C., 2004. *Wildlife Survey Report.* Fauna and Flora International, Cambridge, UK, and the Forestry and Wildlife Division, Dominica.

Magras, M., 1992. La faune de Saint-Barthélemy. The fauna of Saint-Barthélemy. *Tropical St. Barthélemy* 2: 22–30.

Mainka, S. A. and Howard, G. W., 2010. Climate change and invasive species: double jeopardy. *Integrative Zoology* 5: 102–111. https://doi.org/10.1111/j.1749-4877.2010.00193.x.

Malfait, B. T. and Dinkelman, M. G., 1972. Circum-Caribbean tectonic and igneous activity and the evolution of the Caribbean Plate. *Geological Society of America Bulletin* 83: 251–271. https://doi.org/10.1130/0016-7606(1972)83[251:ctaiaa]2.0.CO;2.

Malhotra, A. and Thorpe, R. S., 1999. *Reptiles & Amphibians of the Eastern Caribbean.* Macmillan Education Ltd., London.

Malhotra, A., Thorpe, R. S., Hypolite, E., and James, A., 2007. A report on the status of the herpetofauna of the Commonwealth of Dominica, West Indies. *Applied Herpetology* 4: 177–194. https://doi.org/10.1163/ej.9789004194083.i-439.57.

Malhotra, A., Thorpe, R. S., Hypolite, E., and James, A., 2011. A report on the status of the herpetofauna of the Commonwealth of Dominica, West Indies. Pp. 149–166 in *Conservation of Caribbean Island Herpetofaunas. Volume 2: Regional Accounts of the West Indies,* edited by A. Hailey, B. S. Wilson, and J. A. Horrocks. Brill, Leiden.

Mallery, C. S., Jr., Marcum, M. A., Powell, R., Parmerlee, J. S., Jr., and Henderson, R. W., 2007. Herpetofaunal communities of the leeward slopes and coasts of St. Vincent: a comparison of sites variously altered by human activity. *Applied Herpetology* 4: 313–325. https://doi.org/10.1163/157075407782424494.

Mann, P., 1999. Caribbean sedimentary basins: classification and tectonic setting from Jurassic to present. *Sedimentary Basins of the World* 4: 3–31. https://doi.org/10.1016/s1874-5997(99)80035-5.

Martin, L., Morton, M. N., Hilton, G. M., Young, R. P., Garcia, G., Cunningham, A. A., James, A., Gray, G., and Mendes, S. (eds), 2007. A Species Action Plan for the Montserrat Mountain Chicken *Leptodactylus fallax*. Department of Environment, Montserrat.

Martin-Kaye, P. H. A., 1969. A summary of the geology of the Lesser Antilles. *Overseas Geology and Mineral Resources* 10: 172–206.

Mayer, M., Brown, G. P., Zimmermann, B., Greenlees, M. J., and Shine, R., 2015. Habitat use of the introduced Cane Toad (*Rhinella marina*) and native frog species in tropical Australia. *Journal of Tropical Ecology* 31: 531–540. https://doi.org/10.1017/s0266467415000474.

McCauley, C. and Mendes, S., 2006. Socioeconomic Assessment Report. Unpublished report from the Centre Hills Project, Montserrat.

McCranie, J. R., Collart, J. R., Castañeda, F. E., and Solis, J. M., 2008. *Eleutherodactylus (Euhyas) planirostris.* Geographic distribution. *Herpetological Review* 39: 362–363.

McTighe, A., 2015. Volcano at twenty: Montserrat's rise from the ashes. *Contributoria,* July 2015. http://www.contributoria.com/issue/2015-07/55510657853daab15a0000fc.html.

Melo, M. A., Lyra, M. L., Brischi, A. M., Geraldi, V. C., and Haddad, C. F. B., 2014. First record of the invasive frog *Eleutherodactylus johnstonei* (Anura: Eleutherodactylidae) in São Paulo, Brazil. *Salamandra* 50: 177–180.

Mercer, J., Kelman, I., Alfthan, B., and Kurvits, T., 2012. Ecosystem-based adaptation to climate change in Caribbean small island developing states: Integrating local and external knowledge. *Sustainability* 4: 1908–1932. https://doi.org/10.3390/su4081908.

Meshaka, W. E., Jr., 2001. *The Cuban Treefrog in Florida: Life History of a Successful Colonizing Species.* University Press of Florida, Gainesville.

Moen, D. S. and Wiens, J. J., 2009. Phylogenetic evidence for competitively driven divergence: body-size evolution in Caribbean treefrogs (Hylidae: *Osteopilus*). *Evolution* 63: 195–214. https://doi.org/10.1111/j.1558-5646.2008.00539.x.

Moravec, F. and Kaiser, H., 1995. Helminth parasites from West Indian frogs, with description of two new species. *Caribbean Journal of Science* 31: 252–268.

Muñoz, M. M. and Hewlett, J., 2011. Ecological consequences of continual volcanic activity on the lizard, *Anolis lividus*, from Montserrat. *Herpetological Review* 42: 160–165.

Myers, N., Mittermeier, R. A., Mittermeier, C. G., Fonseca, G. A. B., and Kent, J., 2000. Biodiversity hotspots for conservation priorities. *Nature* 403: 853–858. https://doi.org/10.1038/35002501.

Murphy, J. C., 1997. *Amphibians and Reptiles of Trinidad and Tobago*. Krieger Publishing Co., Malabar, Florida.

Murray-Wallace, C. and Woodroffe, C., 2014. *Quaternary Sea-Level Changes: A Global Perspective*. Cambridge University Press, Cambridge.

Newsom, L. A. and Wing, E. S., 2004. *On Land and Sea: Native American Uses of Biological Resources in the West Indies*. University of Alabama Press, Tuscaloosa, Alabama.

Oppel, S., Havery, S. J., John, L., Bambini, L., Varnham, K., Dawson, J., and Radford, E., 2019. Maximising conservation impact by prioritizing islands for biosecurity. Pp. 659–665 in *Island Invasives: Scaling up to Meet the Challenge*, edited by C. R. Veitch, M. N. Clout, A. R. Martin, J. C. Russell, and C. J. West. IUCN, Gland, Switzerland.

Ouboter, P. E. and Jairam, R., 2012. *Amphibians of Suriname*. Brill, Leiden, The Netherlands.

Ovaska, K., 1991a. Reproductive phenology, population structure, and habitat use of the frog *Eleutherodactylus johnstonei* in Barbados, West Indies. *Journal of Herpetology* 25: 424–430. https://doi.org/10.2307/1564764.

Ovaska, K., 1991b. Diet of the frog *Eleutherodactylus johnstonei* (Leptodactylidae) in Barbados, West Indies. *Journal of Herpetology* 25: 486–488. https://doi.org/10.2307/1567445.

Pagán-Jiménez, J. R., Rodríguez Ramos, R., and Hofman, C. L., 2019. On the way to the islands: The role of domestic plants in the initial peopling of the Antilles. Pp. 89–106 in *Early Settlers of the Insular Caribbean*, edited by C. L. Hofman and A. T. Antczak. Sidestone Press, Leiden, The Netherlands.

Paice, M. R., 2005. Geographic distribution. *Bufo marinus*. *Herpetological Review* 36: 331–332.

Peh, K. S.-H., Balmford, A., Birch, J.C., Brown, C., Butchart, S. H. M., Daley, J., Dawson, J., Gray, G., Hughes, F. M. R., Mendes, S., Jillett, J., Stattersfield, A. J., Thomas, D. H. L., Walpole, M., and Bradbury, R. B., 2015. Potential impact of invasive alien species on ecosystem services provided by a tropical forested ecosystem: a case study from Montserrat. *Biological Invasions* 17: 461–475. https://doi.org/10.1007/s10530-014-0743-9.

Phillips, B. L., Smith, J. G., Greellees, M., Brown, G. P., Shine, R., and Webb, J. K., 2005. Predicting the impact of cane toads on native fauna: a mechanistic approach. In *Proceedings of the East Kimberley Cane Toad Forum facilitated and organized by Kimberley Specialists in Research Inc. with assistance from Andrew Storey*, edited by A. W. Storey and S. L. Boulter. Kimberley Specialists in Research Inc., Kununurra, Western Australia. Available at: http://www.canetoads.com.au/forumprocedespp7.htm accessed 1 August 2019.

Potter, R. B., Barker, D., Conway, D., and Klak, T., 2004. *The Contemporary Caribbean*. Pearson Education Ltd., London.

Pounds, J. A., Fogden, M. P. L., and Campbell, J. H., 1999. Biological response to climate change on a tropical mountain. *Nature* 398: 611–615. https://doi.org/10.1038/19297.

Powell, R., 2007. Geographic distribution. *Osteopilus septentrionalis* (Cuban Treefrog). *Herpetological Review* 38: 215.

Powell, R. and Henderson, R.W., 2005. Conservation status of Lesser Antillean reptiles. *Iguana* 12: 62–77.

Powell, R. and Henderson, R. W., 2007. The St. Vincent (Lesser Antilles) herpetofauna: Conservation concerns. *Applied Herpetology* 4: 295–312. https://doi.org/10.1163/157075407782424539.

Powell, R., and Henderson, R. W., 2008. Urban herpetology in the West Indies. Pp. 389–404 in *Urban Herpetology*, edited by J. C. Mitchell, R. E. Jung Brown, and B. Bartholomew. Herpetological Conservation, volume 3, Society for the Study of Amphibians and Reptiles, Salt Lake City, Utah.

Powell, R. and Henderson, R. W., 2011. The St. Vincent (Lesser Antilles) herpetofauna: Conservation concerns. Pp. 359–376 in *Conservation of Caribbean Island Herpetofaunas. Volume 2: Regional Accounts of the West Indies*, edited by A. Hailey, B. S. Wilson, and J. A. Horrocks. Brill, Leiden, The Netherlands.

Powell, R. and Henderson, R. W. (eds), 2012. Island lists of West Indian amphibians and reptiles. *Bulletin of the Florida Museum of Natural History* 51: 85–166.

Powell, R., Passaro, R. J., and Henderson, R. W., 1992. Noteworthy herpetological records from Saint [sic] Maarten, Netherlands Antilles. *Caribbean Journal of Science* 28: 234–235.

Powell, R., Henderson, R. W., and Parmerlee, J. S., Jr., 2005. *Reptiles and Amphibians of the Dutch Caribbean: St. Eustatius, Saba, and St. Maarten*. St. Eustatius National Parks Foundation, Gallows Bay, St. Eustatius, Netherlands Antilles.

Powell, R., Henderson, R. W., Farmer, M. C., Breuil, M., Echternacht, A. C., van Buurt, G., Romagosa, C. M., and Perry, G. 2011. Introduced amphibians and reptiles in the Greater Caribbean: Patterns and conservation implications. Pp. 63–143 in *Conservation of Caribbean Island Herpetofaunas. Volume 1: Conservation Biology and the Wider Caribbean*, edited by A. Hailey, B. S. Wilson, and J. A. Horrocks. Brill, Leiden, The Netherlands.

Powell, R., Henderson, R. W., Perry, G., Breuil, M., and Romagosa, C. M., 2013. Introduced amphibians and reptiles in the Lesser Antilles. Pp. 74–107 in *Biodiversité insulaire: la flore, la faune et l'homme dans les Petites Antilles. Actes du Colloque international, Schoelcher, 8–10 Novembre 2010*, edited by J.-L. Vernier and M. Burac. Direction de l'Environment, de l'Aménagement et du Logement de Martinique and Université de Antilles et de la Guyana, Schoelcher, Martinique.

Powell, R., Henderson, R. W., and Parmerlee, J. S., Jr., 2015. *The Reptiles and Amphibians of the Dutch Caribbean: Saba, St. Eustatius, and St. Maarten*. 2nd edn, revised and expanded. Nature Guide Series No. 004. Dutch Caribbean Nature Alliance, Kralendijk, Bonaire.

Powell, R., Conant, R., and Collins, J. T., 2016. *Peterson Field Guide to Reptiles and Amphibians of Eastern and Central North America*. 4th edn. Houghton Mifflin Harcourt, Boston, Massachusetts.

Pregill, G. K., Steadman, D. W., Olson, S. L., and Grady, F. V., 1988. Late Holocene fossil vertebrates from Burma Quarry, Antigua, Lesser Antilles. *Smithsonian Contributions to Zoology* 463: iv + 27 pp. https://doi.org/10.5479/si.00810282.463.

Pregill, G. K., Steadman, D. W., and Watters, D. R., 1994. Late Quaternary vertebrate faunas of the Lesser Antilles: Historical components of Caribbean biogeography. *Bulletin of Carnegie Museum of Natural History* 30: iv + 51 pp.

Pyron, R. A. and Wiens, J. J., 2011. A large-scale phylogeny of Amphibia including over 2800 species, and a revised classification of advanced frogs, salamanders, and caecilians. *Molecular Phylogenetics and Evolution* 61: 543–583. https://doi.org/10.1016/j.ympev.2011.06.012.

Questel, K., 2018. Proposition d'une méthodologie de classement des espèces indigenes et des espèces exotiques en vue de prioriser les actions de conservation ou leur gestion. Proposal of a methodology to classify native species and exotic species in order to prioritize conservation actions or their management. *Le Bulletin de l'Agence Territoriale de l'Environnement de Saint-Barthélemy (ATE Bulletin)* 2: 2–33.

Rick, T. C., Kirch, P. V., Erlandson, J. M., and Fitzpatrick, S. M., 2013. Archeology, deep history, and the human transformation of island ecosystems. *Anthropocene* 4: 33–45. https://doi.org/10.1026/j.ancene.2013.08.002.

Rios-López, N., 2008. Effects of increased salinity on tadpoles of two anurans from a Caribbean coastal wetland in relation to their normal abundance. *Amphibia-Reptilia* 29: 7–18. https://doi.org/10.1163/156853808783431451.

Rivero, J. A., 1998. *Los Anfibios y Reptiles de Puerto Rico. The Amphibians and Reptiles of Puerto Rico*. 2nd edn, revised. University of Puerto Rico, San Juan.

Rödder, D., Kielgast, J., and Lötters, S., 2010. Future potential distribution of the emerging amphibian chytrid fungus under anthropogenic climate change. *Diseases of Aquatic Organisms* 92: 201–207. https://doi.org/10.3354/dao02197.

Rodrigues, M. T., Caramaschi, U., and Mijares, A., 2010. *Scinax x-signatus. The IUCN Red List of Threatened Species* 2010: e.T56005A11404900. http://dx.doi.org/10.2305/IUCN.UK.2010-2.RLTS.T56005A11404900.en. Downloaded 1 August 2019.

Rodríguez Gómez, C. A., Díaz-Lameiro, A. M., Berg, C. S., Henderson, R. W., and Powell, R., 2017. Relative abundance and habitat use by the frogs *Pristimantis shrevei* (Strabomantidae) and *Eleutherodactylus johnstonei* (Eleutherodactylidae) on St. Vincent. *Caribbean Herpetology* 58: 1–12. https://doi.org/10.31611/ch.58.

Russell, J. C., Meyer, J.-V., Holmes, N. D., and Pagad, S., 2017. Invasive alien species on islands: impacts, distribution, interactions and management. *Environmental Conservation* 44: 359–370. https://doi.org/10.1017/S0376892917000297.

Sánchez Muñoz, A. J., 2011. First record of the Lesser Antillean Frog (*Eleutherodactylus johnstonei*) in Puerto Rico. *IRCF Reptiles & Amphibians* 18: 186. https://doi.org/10.17161/randa.v18i3.

Schlaepfer, M. A., Hoover, C. and Dodd, C. K. Jr., 2005. Challenges in evaluating the impact of the trade in amphibians and reptiles on wild populations. *BioScience* 55: 256–264. https://doi.org/10.1641/0006-3568(2005)055[0256:cieto]2.0.CO;2.

Schomburg, R. H., 1971. *The History of Barbados*. 2nd edn. Frank Cass, London.

Schwartz, A., 1967. Frogs of genus *Eleutherodactylus* in the Lesser Antilles. *Studies on the Fauna of Curaçao and Other Caribbean Islands* 91: 1–62.

Schwartz, A. and Henderson, R. W., 1991. *Amphibians and Reptiles of the West Indies: Descriptions, Distributions, and Natural History*. University of Florida Press, Gainesville, Florida.

Schwartz, A. and Thomas, R., 1975. A checklist of West Indian amphibians and reptiles. *Carnegie Museum of Natural History Special Publication* 1: 1–216. https://doi.org/10.5962/bhl.title.123682.

Shine, R., 2010. The ecological impact of invasive Cane Toads (*Bufo marinus*) in Australia. *The Quarterly Review of Biology* 85: 253–291. https://doi.org/10.1086/655116.

Smith, M. L., Hedges, S. B., Buck, W., Hemphill, A., Incháustegui, S., Ivie, M., Martina, D., Maunder, M., and Ortega, J. F., 2005. Caribbean islands. Pp. 112–18 in *Hotspots Revisited: Earth's Biologically Richest and Most Endangered Terrestrial Ecoregions*, edited by R. A. Mittermeier, P. Robles-Gil, M. Hoffmann, C. G. Mittermeier, J. Pilgrim, T. Brooks, and G. A. B. Fonseca. CEMEX, Mexico City.

Smith, R. K. and Sutherland, W. J., 2014. *Amphibian Conservation: Global Evidence for the Effects of Interventions*. Pelagic Publishing, Exeter.

Solís, F., Ibáñez, R., Hammerson, G., Hedges, B., Diesmos, A., Matsui, M., Hero, J.-M., Richards, S., Coloma, L., Ron, S., La Marca, E., Hardy, J., Powell, R., Bolaños, F., Chaves, G., and Ponce, P., 2009. *Rhinella marina. The IUCN Red List of Threatened Species* 2009: e.T41065A10382424. http://dx.doi.org/10.2305/IUCN.UK.2009-2.RLTS.T41065A10382424.en. Downloaded 1 August 2019.

Solís, F., Ibáñez, R., Jaramillo, C., Fuenmayor, Q., Azevedo-Ramos, C., La Marca, E., Coloma, L.A., Ron, S., Hardy, J., Hedges, B., Ibéné, B., Breuil, M., and Powell, R., 2010. *Scinax ruber. The IUCN Red List of Threatened Species* 2010: e.T55994A11395509. http://dx.doi.org/10.2305/IUCN.UK.2010-2.RLTS.T55994A11395509.en. Downloaded 1 August 2019.

Somma, L. A. and Graham, P. R., 2015. Cuban Treefrogs, *Osteopilus septentrionalis* (Duméril & Bibron, 1841) (Anura: Hylidae), and other nonindigenous herpetofauna interdicted in Grenada, Lesser Antilles. *IRCF Reptiles & Amphibians* 22: 40–42. https://doi.org/10.17161/randa.v22i1.

Staton, M. A., and Dixon, J. R., 1977. The herpetofauna of the central Llanos of Venezuela: Noteworthy records, a tentative checklist and ecological notes. *Journal of Herpetology* 11: 17–24. https://doi.org/10.2307/1563286.

Stewart, M. M., 1995. Climate driven population fluctuations in rain forest frogs. *Journal of Herpetology* 29: 437–446. https://doi.org/10.2307/1564995.

Stewart, M. M. and Martin, G. E., 1980. Coconut husk piles: A unique habitat for Jamaican terrestrial frogs. *Biotropica* 12: 107–116. https://doi.org/10.2307/2387726.

Stuart, S. N., Hoffmann, M., Chanson, J. S., Cox, N. A., Berridge, R. J., Ramani, P., and Young, B. E. (eds), 2008. *Threatened Amphibians of the World*. Lynx Edicions, Barcelona; IUCN, Gland, Switzerland; and Conservation International, Arlington, Virginia.

Sweeney, R., 2015., Project report: Conservation assessment for the St. Vincent Frog June 2015. Unpublished report submitted to the Forestry Department, Saint Vincent and the Grenadines.

Sweeney, R., 2016., First detection of the amphibian chytrid fungus (*Batrachochytrium dendrobatidis*) in St. Vincent and the Grenadines. *Herpetological Review* 47: 212–214.

Tilman, D., May, R. M., Lehman, C. L., and Nowak, M. A., 2002. Habitat destruction and the extinction debt. *Nature* 371: 65–66. https://doi.org/10.1038/271065a0.

Townsend, D. S., 1996. Patterns of parental care in frogs of the genus *Eleutherodactylus*. Pp. 229–239 in *Contributions to West Indian Herpetology: A Tribute to Albert Schwartz*, edited by R. Powell and R. W. Henderson. Contributions to Herpetology 12, Society for the Study of Amphibians and Reptiles, Ithaca, New York.

Townsend, J. H., Eaton, J. M., Powell, R., Parmerlee, J. S., Jr., and Henderson, R. W., 2000. Cuban Treefrogs (*Osteopilus septentrionalis*) in Anguilla, Lesser Antilles. *Caribbean Journal of Science* 36: 326–328.

Treglia, M. L., 2006. An annotated checklist of the amphibians and reptiles of St. Vincent, West Indies. *Iguana* 13: 251–262.

USAID (United States Agency for International Development), 2013. *Tropical Forest and Biological Diversity Analysis. Antigua and Barbuda – Barbados – Dominica – Grenada – Guyana – St. Kitts and Nevis – St. Lucia – St. Vincent and the Grenadines – Suriname – Trinidad and Tobago*. U.S. Agency for International Development, US Embassy, Wildey, Barbados. Available at: http://www.usaidgems.org/Documents/FAA&Regs/FAA118119LAC/BECFAA%20118%20119%20FB%20Assessment%20Sept%202013%20.pdf (accessed 1 August 2019).

Vallinoto, M., Sequeira, F., Sodré, D., Bernardi, J. A. R., Sampaio, I., and Schneider, H., 2010. Phylogeny and biogeography of the *Rhinella marina* species complex (Amphibia, Bufonidae) revisited: Implications for Neotropical diversification hypotheses. *Zoologica Scripta* 39: 128–140. https://doi.org/10.1111/j.1463-6409.00415.x.

van Buurt, G., 2005. *Field Guide to the Amphibians and Reptiles of Aruba, Curaçao and Bonaire*. Edition Chimaira, Frankfurt a/Main, Germany.

van Buurt, G., 2006. Conservation of amphibians and reptiles in Aruba, Curaçao and Bonaire. *Applied*

Herpetology 3: 307–321. https://doi.org/10.1163/157075406778905027.

van Buurt, G. and Debrot, A. O., 2012. Exotic and invasive terrestrial and freshwater animal species in the Dutch Caribbean. Report No. C001/12. The Ministry of Economic Affairs, Agriculture and Innovation. The Hague. Available at: http://edepot.wur.nl/192697 (accessed 1 August 2019).

Vinnikov, K. Y., Robock, A., Stouffer, R. J., Wals, J. E., Parkinson, C. L., Cavalieri, D. J., Mitchell, J. F. B., Garrett, D., and Zakharov, V. F., 1999. Global warming and northern hemisphere sea ice extent. *Science* 286: 1934–1937. https://doi.org/10.1126/science.286.5446.1934.

Waite, F. C., 1901. *Bufo agua* in the Bermudas. *Science* 13: 342–343. https://doi.org/10.1126/science.13.322.342.

Walker, L. R., Lodge, D. J., Brokaw, N. V. L., and Waide, R. B., 1991. An introduction to hurricanes in the Caribbean. *Biotropica* 23: 313–316. https://doi.org/10.2307/2388246.

Whitfield, S. M., Kerby, J., Gentry, L. R., and Donnelly, M. A., 2012. Temporal variation in infection prevalence by the amphibian chytrid fungus in three species of frogs at La Selva, Costa Rica. *Biotropica* 44: 779–784. https://doi.org/10.2307/23359981.

Wijethunga, U., Greenlees, M., and Shine, R., 2016. Living up to its name? The effect of salinity on development, growth, and phenotype of the "marine" toad (*Rhinella marina*). *Journal of Comparative Physiology B* 186: 205–213. https://doi.org/10.1007/s00360-015-0944-2.

Wilson, B. S., Horrocks, J. A., and Hailey, A., 2006. Conservation of insular herpetofaunas in the West Indies. *Applied Herpetology* 3: 181–195. https://doi.org/10.1163/157075406778116168.

Wilson, S. M., 1997. Introduction to the study of the indigenous people of the Caribbean. Pp. 1–8 in *The Indigenous People of the Caribbean*, edited by S. M. Wilson. University Press of Florida, Gainesville, Florida.

Yanek, K., Heyer, W. R., and de Sá, R. O., 2006. Genetic resolution of the enigmatic Lesser Antillean distribution of the frog *Leptodactylus validus* (Anura, Leptodactylidae). *South American Journal of Herpetology* 1: 192–206. https://doi.org/10.2994/1808-9798(2006)1[192:grotel]2.0.CO;2.

Yokoyama, M., 2010. *The Incomplete Guide to the Wildlife of Saint Martin*. Mark Yokoyama, St. Martin.

Yokoyama, M., 2012. Reptiles and amphibians introduced on St. Martin, Lesser Antilles. *IRCF Reptiles & Amphibians* 19: 271–279. https://doi.org/10.17161/randa.v19i4.13924.

Yokoyama, M., 2013. *The Incomplete Guide to the Wildlife of Saint Martin*. 2nd edn. Mark Yokoyama, St. Martin.

Young, B. E., Stuart, S. N., Chanson, J. S., Cox, N. A., and Boucher, T. M., 2004. *Disappearing Jewels: The Status of New World Amphibians*. NatureServe, Arlington, Virginia.

Yuan, M. L., Frederick, J. H., McGuire, J. A., Bell, R. C., Smith, S. R., Fenton, C., Cassius, J., Williams, R., Wang, I. J., Powell, R., and Hedges, S. B., 2022. Endemism, invasion, and overseas dispersal: The phylogeographic history of the Lesser Antillean frog, *Eleutherodactylus johnstonei*. *Biological Invasions*. https://doi.org/10.1007/s10530-022-02803-9.

APPENDIX 9.I: Islands that harbour amphibians in the Lesser Antilles. (I) = Introduced (non-native to the island listed). (W) = Waif (introduced but not established on the island listed). (PE) = Presumably extirpated (once established on the island but no longer present). (PX) = Presumably extinct. Latitude and longitude gleaned from Google Earth.

ANGUILLA BANK

Anguilla (18.23/-63.03) (90 km^2)
Eleutherodactylus johnstonei (I)
Osteopilus septentrionalis (I)
Rhinella marina (I)

Saint Barthélémy (17.89/-62.83) (22 km^2)
Eleutherodactylus johnstonei (I)
Eleutherodactylus martinicensis (I)
Eleutherodactylus planirostris (W)
Eleutherodactylus sp. (PX)
Osteopilus septentrionalis (I)
Salamandra salamandra (W)

Saint Martin/Sint Maarten (18.06/-63.05) (85 km^2)
Eleutherodactylus johnstonei (I)
Eleutherodactylus martinicensis (I)
Eleutherodactylus planirostris (I)
Eleutherodactylus sp. (PX)
Hyla squirrella (W)
Osteopilus septentrionalis (I)
Pleurodema brachyops (I)

SABA BANK

Saba (17.63/-63.23) (13 km^2)
Eleutherodactylus johnstonei (I)
Osteopilus septentrionalis (I)

ST CHRISTOPHER BANK

St Christopher (= St Kitts) (17.36/-62.78) (170 km^2)
Eleutherodactylus johnstonei (I)
Leptodactylus fallax (PE)
Rhinella marina (I)

Nevis (17.16/-62.59) (130 km^2)
Eleutherodactylus johnstonei (I)
Osteopilus septentrionalis (I)
Rhinella marina (I)

St Eustatius (17.49/-62.98) (20 km^2)
Eleutherodactylus johnstonei (I)

ANTIGUA BANK

Antigua (17.08/-61.80) (280 km^2)
Eleutherodactylus johnstonei
Eleutherodactylus martinicensis (I)
Osteopilus septentrionalis (I)
Rhinella marina (I)

Barbuda (17.65/-61.81) (160 km^2)
Eleutherodactylus johnstonei
Green Island (17.07/-61.67)
Eleutherodactylus johnstonei

MONTSERRAT BANK

Montserrat (16.74/-62.19) (100 km^2)
Eleutherodactylus johnstonei (I)
Leptodactylus fallax
Rhinella marina (I)

GUADELOUPE BANK

Guadeloupe (Basse-Terre) (16.18/-61.68) (848 km^2)
Eleutherodactylus barlagnei
Eleutherodactylus johnstonei (I)
Eleutherodactylus martinicensis (I)
Eleutherodactylus pinchoni
Eleutherodactylus cf. *planirostris* (W)
Leptodactylus fallax (I, PX)
Rhinella marina (I)

Guadeloupe (Grande Terre) (16.33/-61.46) (590 km^2)
Eleutherodactylus johnstonei (I)
Eleutherodactylus martinicensis (I)
Leptodactylus fallax (I, PX)
Rhinella marina (I)
Scinax x-signatus (I)

Islets in the Grand Cul de Sac (between Grande-Terre and Basse-Terre, North)
Îlet à Kahouanne (16.37/-61.78) (0.18 km^2)
Eleutherodactylus martinicensis (I)

Islets in the Grand Cul de Sac (between Grande-Terre and Basse-Terre, South)
Îlet-à-Cochons (16.21/-61.54) (0.20 km²)
Eleutherodactylus martinicensis (I)

La Désirade (16.32/-61.06) (22 km²)
Eleutherodactylus johnstonei (I)
Eleutherodactylus martinicensis (I)

Iles de la Petite Terre
Terre-de-Bas (16.17/-61.12)
Eleutherodactylus sp. (W)

ILE DES SAINTES BANK
Terre de Bas (15.86/-61.68) (9.5 km²)
Eleutherodactylus johnstonei (I)
Eleutherodactylus martinicensis (I)

Terre de Haut (15.87/-61.58) (4.5 km²)
Eleutherodactylus johnstonei (I)
Eleutherodactylus martinicensis (I)

Grande Îlet (=Mare Basse) (15.84/-61.59) (0.7 km²)
Eleutherodactylus martinicensis (I)

Îlet-à-Cabrit (15.88/-61.59) (0.4 km²)
Eleutherodactylus martinicensis (I)

MARIE GALANTE BANK
Marie Galante (15.94/-61.26) (160 km²)
Eleutherodactylus johnstonei (I)
Eleutherodactylus martinicensis (I)
Scinax cf *x-signatus* (I)

DOMINICA BANK
Dominica (15.43/-61.35) (751 km²)
Eleutherodactylus amplinympha
Eleutherodactylus johnstonei (I, PX)
Eleutherodactylus martinicensis
Leptodactylus fallax
Osteopilus septentrionalis (I)

MARTINIQUE BANK
Martinique (14.68/-61.01) (1100 km²)
Allobates chalcopis
Eleutherodactylus johnstonei (I)
Eleutherodactylus martinicensis
Leptodactylus fallax (I, PX)
Rhinella marina (I)
Scinax ruber (I)

Gros Islet (14.55/-61.01) (0.12 km²)
Eleutherodactylus johnstonei (I)

Îlet Chancel (14.69/-60.89) (0.8 km²)
Eleutherodactylus johnstonei (I)
Eleutherodactylus martinicensis
Rhinella marina (I)

Îlet Chevalier (14.43/-60.83) (0.2 km²)
Eleutherodactylus johnstonei (I)

Îlet Lavigne (= Gros Îlet) (14.43/-61.89) (0.05 km²)
Eleutherodactylus johnstonei (I)

Îlet Long (14.61/-60.86) (0.12 km²)
Eleutherodactylus johnstonei (I)
Rhinella marina (I)

Îlet Métrente (14.62/-60.86) (0.06 km²)
Eleutherodactylus johnstonei (I)

Îlet Oscar (14.63/-60.85) (0.04 km²)
Eleutherodactylus johnstonei (I)

Îlet Petite Martinique (14.69/-60.91 (0.05 km²)
Eleutherodactylus johnstonei (I)

Îlet Petite Vincent (14.69/-60.92) (0.005 km²)
Eleutherodactylus johnstonei (I)

Îlet Ragot ou Îlet de la Grotte (14.69/-60.88) (0.03 km²)
Eleutherodactylus johnstonei (I)

Îlet Thierry (14.63/-60.85) (0.05 km²)
Eleutherodactylus johnstonei (I)

ST LUCIA BANK
St Lucia (13.92/-60.96) (616 km²)
Eleutherodactylus johnstonei (I)
Leptodactylus fallax (PE)
Osteopilus septentrionalis (I)
Rhinella marina (I)
Scinax ruber (I)

Praslin Island (13.88/-60.89) (0.01 km²)
Eleutherodactylus johnstonei (I)

BARBADOS BANK

Barbados (13.14/-59.56) (430 km²)
Eleutherodactylus johnstonei (I)
Rhinella marina (I)

ST VINCENT BANK

St Vincent (13.26/-61.19) (350 km²)
Eleutherodactylus johnstonei (I)
Pristimantis shrevei
Leptodactylus validus
Rhinella marina (I)

GRENADA BANK

Grenada (12.12/-61.68) (311 km²)
Eleutherodactylus johnstonei (I)
Eleutherodactylus planirostris (I, PX)
Leptodactylus fallax (I, E)
Leptodactylus validus
Pristimantis euphronides
Osteopilus septentrionalis (I)
Rhinella marina (I)

Bequia (13.02/-61.23) (18.0 km²)
Eleutherodactylus johnstonei (I)
Leptodactylus validus
Rhinella marina (I)

Canouan (12.71/-61.32) (7.4 km²)
Eleutherodactylus johnstonei (I)
Rhinella marina (I)

Carriacou (12.48/-61.32) (32.0 km²)
Eleutherodactylus johnstonei (I)
Rhinella marina (I)

Mustique (12.84/-61.18) (5.2 km²)
Eleutherodactylus johnstonei (I)
Osteopilus septentrionalis (I)
Rhinella marina (I)

Petit St Vincent (12.54/-61.38) (0.35 km²)
Eleutherodactylus johnstonei (I)

Union (12.60/-61.44) (8.1 km²)
Eleutherodactylus sp. (I)
Eleutherodactylus johnstonei (I)
Rhinella marina (I)

ISLANDS SATELLITE TO GRENADA

Hog Island (12.00/-61.74) (0.28 km²)
Eleutherodactylus johnstonei (I)

APPENDIX 9.II: Islands on which Lesser Antillean amphibians occur. (I) = Introduced (non-native to the island listed). (W) = Waif (introduced but not established on the island listed). (PE) = Presumably extirpated (once established on the island but presumably no longer present). (PX) = Presumably extinct.

Allobates chalcopis
Martinique

Eleutherodactylus amplinympha
Dominica

Eleutherodactylus barlagnei
Guadeloupe (Basse-Terre)

Eleutherodactylus johnstonei
Anguilla (I)
Antigua
  Green Island
Barbados (I)
Barbuda
Dominica (I, PE)
Grenada (I)
  Bequia (I)
  Canouan (I)
  Hog Island (I)
  Mustique (I)
  Petit St Vincent (I)
  Union (I)
Guadeloupe (Basse-Terre) (I)
Guadeloupe (Grande Terre) (I)
  La Désirade (I)
Ile des Saintes
  Terre-de-Bas (I)
  Terre-de-Haut (I)
Marie Galante (I)
Martinique (I)
  Gros Islet (I)
  Îlet Chancel (I)
  Îlet Chevalier (I)
  Îlet Lavigne (I)
  Îlet Long (I)
  Îlet Meternte (I)
  Îlet Oscar (I)
  Îlet Petite Martinique (I)
  Îlet Petite Vincent (I)
  Îlet Ragot ou Ilet de la Grotte (I)
  Îlet Thierry (I)
Montserrat (I)
Nevis (I)
Saba (I)
St Barthélémy (I)
St Christopher (= St Kitts) (I)
St Eustatius (I)
St Lucia (I)
  Praslin Island (I)
St Martin/St Maarten (I)
St Vincent (I)

Eleutherodactylus martinicensis
Antigua (I)
Dominica
Guadeloupe (Basse-Terre) (I)
Guadeloupe (Grande Terre) (I)
  Îlet à Kahouanne (I)
  Îlet-à-Cochons (I)
  La Désirade (I)
Ile des Saintes (I)
  Grande Îlet (= Mare Basse) (I)
  Îlet-à-Cabrit (I)
  Terre-de-Bas (I)
  Terre-de-Haut (I)
Marie Galante (I)
Martinique
  Îlet Chancel
St Barthélémy (I)
St Lucia (I)
St Martin/St Maarten (I)

Eleutherodactylus pinchoni
Guadeloupe (Basse-Terre)

Eleutherodactylus planirostris
Grenada (I, PX)
St Barthélémy (W)
St Martin/St Maarten (I)

***Eleutherodactylus* sp.**
Grenada Bank
  Union (I)
Iles de la Petite Terre
  Terre-de-Bas (W)
St Barthélémy (PX)
St Martin/St Maarten (PX)

Hyla squirrella (***Dryophytes squirrellus***)
St Barthélémy (W)

Leptodactylus fallax
Dominica
Grenada (I, PE)
Guadeloupe (Basse-Terre) (I, PE)
Guadeloupe (Grande Terre) (I, PE)
Martinique (I, PE)
Montserrat
St Christopher (= St Kitts) (PE)
St Lucia (PE)

Leptodactylus validus
Grenada
  Bequia
St Vincent

Osteopilus septentrionalis
Anguilla (I)
Antigua (I)
Dominica (I)
Grenada (W)
  Mustique (I)
Nevis (I)
Saba (I)
St Barthélémy (I)
St Lucia (I)
St Martin/St Maarten (I)

Pleurodema brachyops
St Martin/St Maarten (I)

Pristimantis euphronides
Grenada

Pristimantis shrevei
St Vincent

Rhinella marina
Anguilla (I)
Antigua (I)
Barbados (I)
Dominica (W)
Grenada (I)
  Bequia (I)
  Canouan (I)
  Carriacou (I)
  Mustique (I)
  Union (W)
Guadeloupe (Basse-Terre) (I)
Guadeloupe (Grande Terre) (I)
Martinique (I)
  Île Chancel (I)
  Îlet Long (I)
Montserrat (I)
Nevis (I)
St Christopher (= St Kitts) (I)
St Lucia (I)
St Vincent (I)

Scinax ruber
Martinique (I)

Scinax x-signatus
Guadeloupe (Grande Terre) (I)
Marie Galante (I)

Salamandra salamandra
St Barthélémy (W)

10 Amphibians of Aruba, Curaçao, and Bonaire

Gerard van Buurt

Abbreviations and acronyms

CITES	*Convention on International Trade in Endangered Species of Wild Fauna and Flora*
CSIRO	*Commonwealth Scientific and Industrial Research Organizations (Australia)*
DNA	*deoxyribonucleic acid*
ECICC	*Eastern Caribbean Islands Culture Conference*
FPI	*Fundashon Planifikashon di Idioma*
RAMSAR	*Convention on Wetlands of International Importance*
SPAW	*Specially Protected Areas and Wildlife*
UNA	*University of the Netherlands Antilles (now re-named University of Curaçao)*

I. Introduction

A. Geography, climate, extended dry periods, *tankis* and earthen dams, and vegetation

Aruba, Curaçao, and Bonaire, together with the Venezuelan islands of Los Monjes, Islas Aves, Los Roques, La Orchila, and La Blanquilla form an archipelago north of the Venezuelan coast. Biogeographically, these islands do not belong to the West Indian region; their flora and fauna are mainly of South American origin. There are many endemic elements in the fauna, largely derived from South American ancestors. This justifies the view that these islands form a small, but distinct, zoogeographical subregion belonging to the South American realm (Wagenaar Hummelinck 1940).

The fauna (Wagenaar Hummelinck 1940) and especially the flora (Stoffers 1956) both contain West Indian elements too. Wagenaar Hummelinck makes a distinction between older and newer West Indian fauna. The West Indian elements are especially strong in the flora. Stoffers states the following (about the Dutch Leeward Islands):

Gerard van Buurt, 'Chapter 10: Amphibians of Aruba, Curaçao, and Bonaire' in: *The Conservation and Biogeography of Amphibians in the Caribbean*. Pelagic Publishing (2023). © Gerard van Buurt. DOI: 10.53061/XGMW2341

> Of the species with a more restricted distribution the West-Indian plants (35) are the most numerous, the South American element being represented by only 22 species. The vegetation of the Leeward group therefore resembles the flora of the Antilles, rather than the flora of South America, at least from a floristic point of view. However, physiognomically the vegetation corresponds more to that of the dry region of northern South America.

Perhaps this area (the Monjes–La Blanquilla archipelago) should be considered an intermediate zone between the West Indies and the South American realm. The Venezuelan islands of Los Monjes, Islas Aves, Los Roques, La Orchila, and La Blanquilla are devoid of amphibians.

Of the islands covered in this chapter, Curaçao is the largest with a surface area of 444 km^2. Aruba has a surface area of 190 km^2 and Bonaire 289 km^2 (including Klein Bonaire with 7 km^2). Their climate is rather arid compared to most Caribbean islands, with a mean annual rainfall of 409 mm, 557 mm, and 463 mm respectively (Meteorological Service of Netherlands Antilles and Aruba). Aruba is the most arid of the Dutch islands, while the western parts of Curaçao receive the most rainfall; the mean temperature across the archipelago is about 28°C. The hilly areas receive slightly more rain, and rainfall can be quite variable from year to year. Dry years have only 200–300 mm of rain, while the maximum is about 1,100 m. The driest year on record for Curaçao is 1914, which saw an average of 207.9 mm. Every century there occur several extended periods of drought with practically no rain (van Buurt 2010) (Figs. 10.1–10.3).

1. Extended dry periods

The extended dry periods experienced by these islands have had an important function in eliminating invasive species that established themselves during wetter periods, such as the African civet (*Civettictis civetta*) and the red-eared or black-rumped waxbill (*Estrilda troglodytes*) (van Buurt 2010). Since most amphibians are adapted to more humid and damp conditions, these extended dry periods also make it difficult for new invasive amphibian species to permanently establish themselves. Nowadays there are many lush, irrigated gardens, which formerly did not exist since water was too scarce and expensive to be wasted in such a way. These are refuges for invasive amphibians, such as the whistling frog (*Eleutherodactylus johnstonei*) the Cuban treefrog (*Osteopilus septentrionalis*), and the cane toad (*Rhinella marina*). Here they are able to survive dry periods, but they can enter the dry bush, locally called *mondi*, only during the rainy season.

Several times each century, there occur extremely dry periods of two or three dry years in a row and even, more rarely, four successive dry years (Table 10.1). For Curaçao, rainfall data are available since 1830, with short interruptions from 1875 to 1883 and from 1892 to 1894; the dry periods have also extended to the other islands of Aruba and Bonaire, whose patterns of rainfall are similar to that of Curaçao.

Data before 1954 are a weighted average of several locations on the island; after 1978 only the data at Hato airport are given. In this table, both values are given for the years 1958 to 1978. The Hato location is usually somewhat drier than the weighted average. Monthly data, which unfortunately are not available for the early years, give a better view of the true extent of the dry periods. The dry period 2001–2002 (so named based on the yearly values), for example, going by monthly data lasted until the end of September 2003; it thus persisted for 2 years and 9 months. The 1982–83 dry period lasted until the end of September 1984, also 2 years and 9 months. A very dry period with one rain shower exceeding 70 mm – enough to cause trees to green and grass to sprout – will bring some relief, and such a drought will be less 'stressful' than a year with the same amount of precipitation spread out more evenly. The 1958–60 dry period properly

Figure 10.1. Footpath going up the Seru di Warawara (Warawara hill) in Santa Cruz, Curaçao, at the end of a long period of drought (24 May 2014). A low dam running across the hill can be seen. Such structures are called *faha*, meaning 'belt', and many were constructed in the past to conserve run-off rainwater, forcing the water into the ground, and also to a lesser extent in order to combat erosion. Of course hillsides are drier than valleys, but even so it is clear that an invasive species like the Cuban treefrog cannot survive a dry period in such an environment.

speaking began at the end of November 1957 and was interrupted by a rain shower of 104.3 mm (Hato Airport) in January 1960; then the dry period continued until August 1960, with more rain following in October 1960. The Hato data, compared with the weighted data for 1960, strongly suggests that a shower of 104.3 mm fell at Hato only. The rest of the island suffered a long dry period of 2 years and 8 months.

2. *Tankis* and earthen dams

Tankis are constructed and maintained on Aruba in order to collect rainwater for the irrigation of plants in the public parks and for agricultural purposes. These *tankis* are earthen dams with a dug-out basin behind them. The dug-out material is usually andesite gravel or sand that is used for construction purposes. On Curaçao there are many earthen impoundments (Fig. 10.2). These usually have masonry overflows and protective structures at their ends and are called 'dams', rather than *tankis*. The term 'dam' refers to both the dyke and its reservoir. Dams on government-owned land are maintained by the government of Curaçao for the management of water. Bonaire has similar dams, but because a larger part of this island is calcareous (and hence highly permeable), there are not as many as on Curaçao and the system is less extensive. An important difference between the Aruban *tankis* and those of Curaçao is that soils are much more impermeable on Aruba. Since they are dug out, most Aruban *tankis* are also deeper than

Figure 10.2. The dam at Soto. (A) 30 November 2008. This dam is the second-largest on Curaçao and it can hold 450,000 tons of water. The dam extends to the right; the image captures approximately two-thirds of the water surface area of the dam reservoir, which had already receded somewhat when this picture was taken. (B) 16 September 2016. During dry seasons the dam will dry out completely.

Figure 10.3. Lago Disparsé, a lake that disappeared in 2006, and in other dry seasons but later reappears in wet years, as this picture shows. Nevertheless, even when holding water it is still called Lago Disparsé, the local name for the dam at Malpaïs. This is the third-largest dam on Curaçao; the picture covers only part of it.

Table 10.1. Extended dry periods in Curaçao, 1830–2004 (Meteorological Service Netherlands Antilles & Aruba). 'Hato' indicates data from the Hato Airport. All other data are a weighted average of several locations. From 1958 to 1978 both datasets are provided.

EXTENDED PERIOD OF DROUGHT	PRECIPITATION IN MM/YR	EXTENDED PERIOD OF DROUGHT	PRECIPITATION IN MM/YR
1841–1843	272.0 290.0 408.0	1947–1948	301.2 421.1
1868–1869	226.2 296.0	1958–1960	287.4; Hato 280.6 355.8; Hato 282.2 363.3; Hato 482.8
1898–1900	479.6 357.7 437.4	1977–1978	355.5; Hato 270.5 389.2; Hato 271.1
1902–1905	313.4 382.3 473.9 413.6	1982–1983	Hato 380.5 Hato 340.1
1919–1920	270.3 301.4	1986–1987	Hato 321.2 Hato 369.3
1929–1930	323.2 269.1	2001–2002	Hato 331.4 Hato 331.8

the dams on Curaçao. Thus, there is almost always some water in the *tankis* and only at the end of very long extended dry periods will they sometimes become completely dry. So, there is almost always some water available for wildlife.

In Curaçao the soils are much more permeable, and water held in the dams penetrates the ground more rapidly. Because of this, these dams only hold water for a few months at most during the rainy season (Figs 10.1, 10.2). Housing developments in Curaçao have led to the destruction of many dams. There are still a few large examples and about 800 smaller ones that continue to function. The *tankis* and dams are used by several species of frogs for oviposition.

Tankis and dams are clearly important for the survival of amphibian species. *Pleurodema* usually lay their eggs in such dams, but sometimes also use much shallower pools.

3. Vegetation

The vegetation of these islands is typically xeric and contains such plants as candelabra cacti, opuntias, dyewood, lignum vitae, and acacias. During the rainy season of wet years, however, Curaçao and Bonaire can be surprisingly green (Fig. 10.4).

II. The amphibian fauna: native and introduced

Due to their permeable skin, most amphibians cannot survive immersion in salt water and can only cross saltwater barriers with great difficulty. Hence, oceanic islands that arose from the sea without ever having had a connection to a mainland tend to lack native amphibians. When amphibians are encountered on such islands, they have usually been introduced by humans. In rare cases, however, big tropical storms can blow frogs or other small animals from one island to another nearby. Some of the amphibians transported in this way may have a chance of surviving the trip. Furthermore, strong hurricanes generate tornados that can suck up living creatures, which eventually fall back to the ground. A large animal cannot survive such an ordeal, but there are documented cases of live fish and frogs raining from the sky (Gudger 1922; Dees 1961;

Figure 10.4. Landhuis Santa Cruz and surroundings (A) after a good rainy season in 2004, and (B) during the 2007 dry season. The air is a bit hazy with Saharan dust particles, which reach the islands during certain periods.

Dennis and Wolff 1993). A frog landing on long, wet grass, or other soft vegetation, or which somehow ends up in a pool of water, could conceivably survive such a trip.

In contrast to oceanic islands, continental islands have had former connections with a mainland before being separated via eustatic changes in sea level or having split off as a fragment from a continent through tectonic movement, or by the severing of a former land bridge. Consequently, such islands may have retained amphibians from the mainland that have been in continuous residence since separation.

The geological history of Aruba differs from that of Curaçao and Bonaire (van Buurt 2005) and this is reflected in their respective present-day batrachofaunas. Aruba originated in the open sea and later drifted close to the continent and may even have been connected to the mainland at an early period of low sea level. Later a channel formed because of erosion from the sea and by the subsidence of a small fault block in the sea floor between Aruba and the continent (Curet 1992).

The Aruban herpetofauna has several endemic elements; it differs from that of the mainland and has a strongly insular character. This indicates that the island has been isolated from the mainland sufficiently long that differentiation of species has occurred. By contrast, Curaçao and Bonaire were never in such proximity to the mainland of South America, thus making dispersal to them of salt-sensitive taxa like amphibians unlikely except by human agency. No amphibian species are endemic to either of these two islands, although some other taxa, such as reptiles, that are more resistant to saltwater, and hence better adapted to transport over the sea, are endemic.

There follow individual accounts of the amphibian species from Aruba, Curaçao and Bonaire; Table 10.2 shows their occurrences on the various islands. Note that two of the five species, *Eleutherodactylus johnstonei* and *Pleurodema brachyops* occur on all three of the islands, one (*Osteopilus*) on Curaçao and Bonaire, and two species on only one each – of which one did

Table 10.2. The occurrence of frogs on the islands of Aruba, Curaçao and Bonaire. I = introduced; N = native; – = absent.

FAMILY/*SPECIES*	ARUBA	CURAÇAO	BONAIRE
BUFONIDAE			
Rhinella marina	I	–	–
ELEUTHERODACTYLIDAE			
Eleutherodactylus johnstonei	I	I	I
Eleutherodactylus sp.	I*	–	–
HYLIDAE			
Osteopilus septentrionalis	–	I	I
LEPTODACTYLIDAE			
Pleurodema brachyops	N	I	I
TOTAL NUMBER OF SPECIES	**4**	**3**	**2**

* No longer present

not become permanently established. All species were introduced from outside the area except *Pleurodema brachyops* and it is native only to Aruba, having been introduced from there to the other two islands.

A. *Rhinella marina* (Sapo, or Cane Toad): Aruba

Rhinella marina has been introduced into Aruba and of the three islands this is the only one on which it occurs. Originally it was thought that these animals were brought in with the importation of sand from rivers in Surinam in the early 1970s when the Hyatt Hotel was built in Aruba. However, letters in the archives of the Veterinary Service in Aruba indicate that these animals have been present since the early 1960s; they were introduced by an Aruban who went on vacation to Colombia and brought some back with him, releasing them in a *tanki*.

There is a major genetic break in the populations of *R. marina* that separates populations west and north of the Venezuelan Andes from those that are east and south (Moritz 1998). Thus, the *R. marina* on Aruba belong to the western race, unless the original stock was collected in the Colombian llanos east of the Andes, which seems unlikely for various reasons. The ancestors of the *R. marina* on the West Indian islands, Australia, and Hawaii originated in French Guiana and belong to the eastern race (Lever 2001).

Toads of the family Bufonidae have a pair of paratoid glands, extending on each side from just behind the eye to the back of the head. The secretions of these glands are poisonous, and except for predator species adapted to feeding on these creatures, the consumption of toads, or even their eggs, may cause death – even to humans (Licht 1967). In Aruba, many dogs died after attacking these toads. For this reason, *R. marina* is widely considered to be a pest. A cruel, but effective, way of killing these animals is now commonly used in Aruba: household cleaning detergent, such as Pine-Sol®, is poured over them.

Rhinella marina eats many insects, earthworms, frogs, lizards, small snakes, carrion, mice and small rats, and even leftover dogfood or catfood – in short almost anything of a size it can get into its mouth. This animal will even eat bees and wasps and can be a problem for beekeepers. In Aruba, Sapos have been observed to prey on the Santanero snake (*Leptodeira bakeri*) (Van Buurt 2005). The prey can also become the predator, as the Santanero is known to eat small *R. marina*. Snakes of the genus *Leptodeira* are known to be immune to the poison of *R. marina* and other toads, and they constitute one of its few predators.

In Aruba, the Sapo consumes many cockroaches and, at least in this respect, is considered beneficial. Nowadays, however, *R. marina* is mainly deemed a pest. However, it should be remembered that in the past, before the advent of modern insecticides, *R. marina* was widely introduced to combat insect pests in sugarcane fields. However, Aruba is very arid and was not a sugar-producing island, and thus the control of pests was not the reason cane toads were introduced.

It is actually quite remarkable that this animal has managed to establish itself on an arid island such as Aruba. One factor of importance is undoubtedly that although Aruba has a drier climate than either Curaçao or Bonaire, its soils are less permeable. The tanks of Aruba hold water much longer than do the dams of Curaçao or Bonaire. While the basins behind the dams of Curaçao or Bonaire will hold water for a few months at most, the Aruban *tankis* dry up only after a very long and severe dry season, and this rarely occurs. The Sapo also finds refuge in gardens, where it seeks moist spots. Often it burrows beneath the drippers of drip-irrigation systems or under the water outlet of air-conditioning systems. The very fact that the Sapo has managed to establish itself on Aruba indicates that if it were to be introduced on Curaçao or Bonaire, it is very likely that it could survive and establish itself there too.

The females of this species lay long strings of small black eggs in the *tankis*. Larval development is extremely rapid, and metamorphosis takes place at a small size. Since the introduction of the Sapo, some people contend that the Dori (see below) has become rare in Aruba. It is likely that the Sapo competes with the Dori, especially in the larval stage. The Dori is probably also suitable prey for the Sapo.

B. *Eleutherodactylus johnstonei* (whistling frog)

Many species of frog lay their eggs in humid soil, and some attach their eggs to plants. In this way, frogs can be transported to other countries with live, potted plants. The whistling frog *Eleutherodactylus johnstonei* originated in the Lesser Antilles but has been introduced by humans into Aruba, Curaçao, and Bonaire, as well as onto many other Caribbean islands and also into Venezuela. On Curaçao, the whistling frog probably came in from Venezuela with plants that were imported for landscaping. The species has been present there since the late 1970s or early 1980s; it has now spread over large areas and can be found in many gardens. It was introduced into Aruba in the early 1990s and is found in the gardens of some of the hotels, such as La Cabaña. In Bonaire, the whistling frog was present in 1996 at the Harbor Village Hotel. In March 2001, none was found at this site (L. Vitt, pers. comm.) – they may have been eradicated by pest control.

C. *Eleutherodactylus* sp.

In January 2015, Richard van der Wal found a few frogs in bags of garden soil on Aruba. It seems likely that the animals had arrived in these bags. There is, however, no absolute proof since they could have arrived with other garden materials and crept into the bags, which had already been open for some time. The person who bought the bags of soil could not establish their origin. These frogs apparently did not survive and establish themselves and are no longer found on the island. Judging from a picture, they were probably of the genus *Eleutherodactylus*, but they could not be identified to species. They were about 2.5 cm in length and look a bit like *Eleutherodactylus cuneatus* from Cuba. Some anole species from Cuba, *Anolis sagrei* and *Anolis porcatus*, were found in the garden of the Radisson hotel on Aruba and had clearly been living there for some time. I discovered them in August 2008, and they were subsequently seen there by others and are probably still present. The garden was landscaped with trees imported from Cuba.

D. *Pleurodema brachyops* (Dori Maco) (Fig. 10.5)

Pleurodema is a genus of South American origin which occurs both in South and Central America but is not present in the West Indies. The species under discussion here probably arrived in Aruba via natural means rather a long time ago. It is also possible, however, that *Pleurodema brachyops* was imported by the Amerindians. This seems less likely, but the possibility cannot be fully excluded. On Aruba, and also on Curaçao where frogs did not occur historically, many artifacts, such as pots and pendants, had frog motifs. These are both imported and locally produced (de Haseth 1993). The frog was a symbol of fertility and of rain, and its call was believed to attract precipitation (Wassén 1934). This could have been a motive for importing such an animal.

The Papiamentu name 'Dori Maco' is of Aruac (Caquetío) origin from Aruba, indicating that this frog already occurred there before colonial times. It is now used on all three islands. The name 'Maco' is also found in Taïno, as a general name for frogs (Tejera 1977). 'Dori' is an onomatopoeia and thus Dori Maco is the frog that calls 'do-ri'.

This animal is mentioned in an old Papiamentu rhyme from Aruba. This rhyme was noted by a German professor named Martin in about 1883, and has persisted to modern times (Van Meeteren 1947):

> Dori, Dori maco, si mi muri, ken ta derami? Ami, ami, ami
> Dori, Dori maco, ora mi muri, ken ta yorami? Ami, ami, ami

which translated into English is:

> Dori, Dori maco, if I die, who will bury me? Me, me, me
> Dori, Dori maco, when I die, who will cry for me? Me, me, me

Figure 10.5. (A) *Pleurodema brachyops* (Dori Maco) and (B) a froth nest of the same species, taken in a shallow pool alongside a road, in Curaçao.

Judging from a photograph, Manzanilla Puppy (pers. comm.) thinks that the colour pattern of *Pleurodema brachyops* from Aruba is somewhat different from the patterns of individuals found in Venezuela. This would argue in favour of the species having arrived on its own. Comparison of the DNA of frogs from Aruba with that of mainland *P. brachyops*, especially those from Paraguaná, conceivably could resolve this matter.

The Dori Maco was introduced into Curaçao from Aruba in about 1910 (Wagenaar Hummelinck 1940). It came to Curaçao with sand dug from the bottom of *tankis* in Aruba. In the past, this hard, diorite sand was imported as grit to sandblast and clean steamboilers, as well as to build ovens, furnaces, and chimneys (the largely calcareous sands from Curaçao could not be used for these purposes). The sand often was moist when shipped in schooners. During the trip the sand would shift, and many frogs could be seen jumping in the hold of the ship – this was told to me as a child by a friend of my parents, the late Carlos Debrot whose father had been captain of a schooner importing the sand.

If there is sufficient rain to fully penetrate and saturate the soil, adult frogs burrow out of the mud. If rainfall has been insufficient, the frogs will not emerge. In the dry season, they remain buried in the bottom of reservoirs and other moist places, where they aestivate in a state of reduced metabolism. Such an animal appears to be dead and partly mummified. When it is immersed in water, however, it gradually swells and starts to hop away.

Pleurodema was introduced into Bonaire in 1928 (Wagenaar Hummelinck 1940). My late uncle, Luis H. Lopez Ramirez, who was from Bonaire, told me long ago how this occurred: a friend of his was spending vacation with an aunt on the island. He brought some larvae (pollywogs) from Curaçao and released them in a dam that was holding water at the time, since it had been raining both on Curaçao and on Bonaire. *Pleurodema brachyops* is also found on the island of Klein Bonaire where it must have been introduced by humans.

E. *Osteopilus septentrionalis* (Cuban treefrog)

This species is known to have been brought into Curaçao inadvertently on several occasions. In 2000, a live individual was intercepted from an open-top container with palm trees imported from Cuba. This container was transshipped via Venezuela and took five weeks to arrive (Van Buurt 2005). The first free-living animal of this species in Curaçao was encountered in a kitchen sink at the SOLTUNA horticultural project, Groot Piscadera, on 12 February 2007 (Van Buurt 2007). Subsequently, many more were found about 290 metres away on the premises of Vivian's Nursery, a company importing ornamental and landscaping plants from Miami. According to workers, the animals were first seen in about September/October 2006. From the company's records, it seems that the frogs probably arrived in a container of wild-harvested *Sabal palmetto* palms from Miami in July 2006.

On Bonaire, a Cuban treefrog was found in 2002 in Kaya Grandi, the main street of Kralendijk. It probably came in with imported plants sold in a nearby shop. On 9 December 2004, Gijs van Hoorn reported that a rather large population of Cuban treefrogs had now established itself in Bonaire, following heavy rains the previous October and November, in an area around a nursery that imports plants from Florida. These frogs then spread rapidly over Kralendijk and its environs.

III. Threats to amphibians

A. Insufficient information on the effects of introduced species on local ecosystems

We do not have much information on how the various species of frogs on these islands interact with each other, or with indigenous species, or what their effects are on the ecosystem in general. Additionally, some of what we think we know may be wrong.

Since the introduction of the Sapo (*Rhinella marina*), many people contend that the Dori (*Pleurodema brachyops*) has become rare on Aruba. Others dispute this, however, and the controversy is an example of how little we know. Often, after a long sequence of dry years people tend to forget that during past wet years these animals were, in fact, abundant. Even had their numbers declined, this would not necessarily have been caused by the introduction of *Rhinella marina* – there could be other reasons. It seems logical that *Rhinella marina* must be a threat to *Pleurodema brachyops; Rhinella* has toxic egg masses that are very numerous in the *tankis*, so almost automatically we assume it is a threat and do not stop to think that in a large part of South America and Panama these animals coexist, apparently without problems. *Pleurodema* sometimes use much shallower pools of water for their froth nests than the waters used by *Rhinella* for breeding; this could be an ecological segregation that permits coexistence.

Another mystery is the fact that the Cuban treefrog disappears almost completely at the end of a long dry season. I suspect that mortality is very high, yet some must survive. Where do they hide? Some secrete themselves in rainwater cisterns in houses. However, many of these cisterns are no longer in use, and new houses rarely have them. Furthermore, how do the frogs get from their hiding places to the dams when these fill up after the first heavy rains? Sometime after the first rains many small treefrogs appear and then later larger ones too. When the dry season returns and the dams dry out, the frogs invade gardens and nearby houses, where they are a nuisance, entering bathrooms and jumping in the kitchen sink. I have never seen one in the bush (*mondi*), away from the inhabited areas, nor has anyone I know. Maybe the frogs are hiding and only come out at night; in gardens; however, they can sometimes be observed during daylight. During the rainy season, they must be penetrating into the bush – or at least it seems impossible to imagine they would not. Neither do we have any idea how much damage they could do to indigenous fauna of Curaçao and Bonaire, such as the endemic geckoes *Gonatodes antillensis* and *Phyllodactylus martini*. This lack of basic information makes it difficult to assess threats to amphibian populations and the threats to other local taxa from these amphibians. There is no monitoring.

B. Loss and fragmentation of habitat, urbanization, pollution, and diseases, especially chytridiomycosis

There has been a loss of habitat on these islands, mostly because of urbanization. On Curaçao, many dams have been destroyed so that areas previously covered by reservoirs can be used for housing or other construction. This increases the flow to the next dam downstream. Moreover, the direction of the flow of water is often changed because of the building of new roads or other construction. Many old dams are in a state of disrepair. By contrast, species such as the whistling frog are dependent on urbanization, since they can only survive in irrigated gardens; they are thus common in the suburbs. Since the soils in Curaçao are more porous than in Aruba, there are many more wells on the former island than on the latter. Most houses have sewage pits, from where contaminated water filters subterraneously into wells. Almost all wells in areas inhabited by humans contain water that is heavily polluted by bacteria, and many wells have high levels of nitrates. Levels of up to 80 ppm of nitrate in wells are not uncommon and I suspect that these high levels make wells unsuitable habitat for Cuban treefrogs. For the irrigation of gardens, this is not a problem, but cannot be used for drinking or bathing – for these purposes, people rely on water mains that provide desalinated water.

Nowadays there are several sewage treatment plants on the islands, and in some areas houses and new developments are connected to these plants. Aruba has two (another one has been built after the manuscript was written and now Aruba has three), Curaçao three and Bonaire one sewage treatment facility. The tertiary ponds of these sewage plants are the only truly permanent bodies of freshwater on the islands. The nitrate levels in the tertiary ponds are usually about 20 ppm, although they sometimes reach 40 ppm during malfunctions. Some of the ponds have *Oreochromis mossambicus* in them that die at nitrate levels of approximately 40 ppm. Frogs do not use tertiary ponds for oviposition. I suspect that the reason is because, unlike *Tilapia*, their larvae cannot tolerate high levels of nitrates.

Chytridiomycosis has not been reported from Aruba, Curaçao, or Bonaire.

C. Pest control of frogs

Many people consider the whistling frog to be a garden pest and its arrival created demand for pest control companies to eradicate it. It is amazing that such a small frog can emit such a loud sound. Unfortunately, many people cannot appreciate these tones, the more so since they can penetrate an air-conditioned room. The frogs were flushed out of their hiding places, usually potted plants, using ammonia. Now, circumstances have changed: with the arrival of the Cuban treefrog this problem was solved. When Cuban treefrogs reach a garden, they promptly eliminate the whistling frogs. There used to be many gardens in the inhabited areas that harboured whistling frogs but now their number has been markedly reduced.

The Cuban treefrog also preys on *Pleurodema brachyops* and has been seen eating individuals of this species in some shallow ponds around Kralendijk. When the former frog arrived in Bonaire many local people were indignant that it was eating 'their frog' the Dori Maco, not realizing that the Dori also is an introduced species on the island. The Cuban treefrog also eliminates large numbers of the invasive gecko, *Hemidactylus mabouia*, and will certainly also eat indigenous geckos. However, the indigenous geckos had already been eliminated from most houses by *Hemidactylus mabouia*. Indigenous geckos can still be found in the bush (mondi), away from artificial lights, where there are no *Hemidactylus mabouia* (Hughes *et al.* 2014). Thus, the bush probably also serves as a refuge from the Cuban treefrog.

When the dam of Noord Zapateer dried out on Curaçao, Cuban treefrogs invaded houses and a small agricultural/horticultural plot nearby. This plot had a masonry irrigation basin, locally called 'baki di awa' (from the Dutch *waterbak*) that was filled with well-water by a windmill. The farmer had an elegant solution: he got rid of all the frogs in the whole neighbourhood by emptying a bottle of Malathion pesticide into his irrigation basin.

D. Potential threat to amphibians from invasive species

Potential threats to local frogs are newly introduced agents carrying diseases, such as chytridiomycosis, or an invasive predator, such as a snake or another species of frog.

IV. Conservation and monitoring

A. Legislation

Treaties that are of significance for the conservation of amphibians, of which Curaçao (C) and/or Aruba (A) and Bonaire (B) are a party through the Kingdom of the Netherlands (N), are:

(1) CITES (N, B, C, A)
(2) Cartagena Convention (A, B, C) and SPAW protocol (C, B)
(3) Convention on Biological Diversity (N, B, C, A)
(4) Convention on Migratory species or Bonn Convention (N, B, C, A)
(5) RAMSAR Convention (N, B, C, A).

All organisms listed under CITES are subject to the CITES regulations by local law. Thus, when CITES listings get updated, it is not necessary to update local law as this occurs automatically.

Species listed under CITES annex I are automatically protected and a permit to keep them is needed. Such permits are usually given under a grandfather clause, and apply to specimens already in one's possession before the local law came into effect. Permits can be granted for scientific purposes. The legal situation is similar in Aruba and Bonaire: the lists of species and other provisions of the SPAW protocol under the Cartagena Convention are incorporated into local legislation about the natural environment for Aruba, Curaçao, and Bonaire.

There is now some regulation of invasive species. The Head of the Veterinary service in Curaçao must approve the importation of any new animal onto the island. Permission is also now required to export any animal from the island. The police and customs are much more cooperative on these matters than in the past, and help to enforce these rules. Everyone now knows about the disaster of the boa introduced into Aruba and is trying to prevent a similar situation arising in Curaçao.

A remaining area of concern is that there are still no adequate controls on incoming plants. Invasive species, including amphibians and reptiles, often hitchhike on plants brought in by nurseries and for use in ornamental landscaping. For example, as discussed, the Cuban treefrog entered via this route. A phytosanitary certificate is required, but this measure is insufficient.

B. Faunal reserves

All three islands have terrestrial national parks: in Aruba, the Parke Nacional Arikok; in Curaçao, the Christoffel National Park and the Shete Boca Park; and in Bonaire, the Washington/Slagbaai National Park and Klein Bonaire National Park.

References

Buurt, G. van, 2005. *Field Guide to the Amphibians and Reptiles of Aruba, Curaçao and Bonaire*. Edition Chimaira, Frankfurt am Main.

Buurt, G. van, 2006. Conservation of Amphibians and Reptiles in Aruba, Curaçao and Bonaire. *Applied Herpetology* 3: 307–321. https://doi.org/10.1163/157075406778905027

Buurt, G. van, 2007. Breeding population of *Osteopilus septentrionalis* in Curaçao. *Applied Herpetology* 4: 390–391. https://doi.org/10.1163/157075407782424520

Buurt, G. van, 2010. A short natural history of Curaçao. Pp. 229–256 in *Crossing Shifting Boundaries, Language and Changing Political status in Aruba, Bonaire and Curaçao*. Vol. 1 of *Proceedings of the ECICC*, Dominica 2009, edited by N. Faraclas, R. Severing, C. Weijer, and E. Echteld. FPI and UNA, Curaçao.

Curet, E. A., 1992. Stratigraphy and evolution of the Tertiary Aruba basin. *Journal of Petroleum Geology* 15: 283–304. https://doi.org/10.1111/j.1747-5457.1992.tb00873.x

Dees, L., 1961. Rains of fishes. United States Department of the Interior, U.S. Fish and Wildlife Service, Bureau of Commercial Fisheries, Fisheries Leaflet 513: 1–5 [reprinted 1967, 1969, and 1970].

Dennis, J. and Wolff, G., 1993. *It's Raining Frogs and Fishes: Four Seasons of Natural Phenomena and Oddities of the Sky*. Harper Perennial, New York.

Gudger, E. W., 1922. Rains of fishes and frogs. *Natural History* 22: 84.

Haseth de, C. P., 1993. Kikkermotieven op Indiaanseartefacten van Curaçao. *Kristòf*, Jaargang. VIII-4: 17–28.

Hughes, D. F., Meshaka, W. E., Jr., and van Buurt, G., 2014. The Superior colonizing gecko *Hemidactylus mabouia* on Curaçao: conservation implications for the native gecko *Phyllodactylus martini*. *Journal of Herpetology* 48: 60–63. https://doi.org/10.1670/13-161

Lever, C., 2001. *The Cane Toad: History and Ecology of a Successful Colonist*. Westbury Academic and Scientific publishing, West Yorkshire.

Licht, L. E., 1967. Death following possible ingestion of toad eggs. *Toxicon* 5: 141–142. https://doi.org/10.1016/0041-0101(67)90168-7

Moritz, C., 1998. Taxonomic status of *Bufo marinus* from South America, Hawaii and Australia. *SIRO Bufo Project: An Overview of Research Outcomes*: 7, not paginated, edited by B. Green. Commonwealth Scientific and Industrial Research Organisation, Canberra.

Stoffers, A. L., 1956. *The Vegetation of the Netherlands Antilles*. Publications of the Foundation for Scientific Research in Surinam and the Netherlands Antilles, Utrecht, The Netherlands15.

Tejera, E., 1977. *Indigenismos*, vols 1 and 2. Editora de Santo Domingo, Santo Domingo, Dominican Republic.

van Meeteren, N., 1947. *Volkskunde van Curaçao*. Reprint 1977, Antilliaanse reeks 2, S. Emmering, Amsterdam.

Wagenaar Hummelinck, P., 1940. *Studies on the Fauna of Curaçao, Aruba, Bonaire and the Venezuelan Islands*. Dissertatie, Utrecht.

Wassén, H., 1934. The frog motive among the South American Indians. Ornamental studies. *Anthropos* 29: 319–370.

11 Amphibians of the Venezuelan Islands

Harold Heatwole

I. Introduction

The Federal Dependencies of Venezuela consists of about 600 islands, cays, and emergent rocks ranging in size from 0.0005 km^2 to 106.6 km^2 (Federal Dependencies of Venezuela 2019) (Fig. 11.1). These occur in three geographic groups (1) Isla Aves remote from Venezuela in the open north-eastern Caribbean Sea; (2) five islands or archipelagos fringing the Venezuelan coast but beyond the South American continental shelf; (3) and six islands on Venezuela's continental shelf. In addition, there are three inshore islands, Margarita, Isla Cubagua, and Isla Coche that do not belong to the Dependencies.

A. Islands of the Federal Dependencies

1. Isla Aves

Isla Aves – not to be confused with Las Aves (see below) – is the most isolated island of the Caribbean Sea (63° 36' 30" W; 15° 40' 30" N). It is located on a coral reef on the Prominencia de Aves, 500 km north of Venezuela with its nearest neighbour being Dominica, 200 km to the east in the Lesser Antilles (Maloney and Schubert 1968). It is composed of coralline sand partly cemented into a soft sandstone, is used by sea turtles for nesting (Hummelinck 1952), and contains dense populations of nesting laughing gulls (*Leucophaeus atricilla*), sooty terns (*Onychoprion fuscatus*), and noddy terns (*Anous stolidus*) (Zuloaga 1955; Rolle 1966). It has a present-day area of approximately 0.045 km^2, but at one time was much larger. In 1647 it was circular in form with a diameter of 13 km, which by 1705 had reduced to 4 km, leaving two smaller cays isolated from it. By the mid-1800s, the island was losing its circular form and becoming elongated (length slightly less than 1 km). By 1939 its length had receded to 815 m, by 1954 to 680 m, by 1966 to 648 m, and by 1968 to 550 m (Maloney and Schubert 1968; Heatwole *et al.* 2022). The vegetation in two independent surveys conducted 12 years apart (1954 and 1966) consisted only of two low species of succulent halophytes, *Sesuvium portulacastrum* and *Portulaca oleracea* (Zuloaga 1955; Byers unpub. data). Most of the species of plants common on strands in the Caribbean were missing. Isla Aves has only recently been inhabited permanently by humans (see Heatwole *et al.* 2022) and it lacks fresh water (Zuloaga 1955).

Harold Heatwole, 'Chapter 11: Amphibians of the Venezuelan Islands' in: *The Conservation and Biogeography of Amphibians in the Caribbean*. Pelagic Publishing (2023). © Harold Heatwole. DOI: 10.53061/JTRO5536

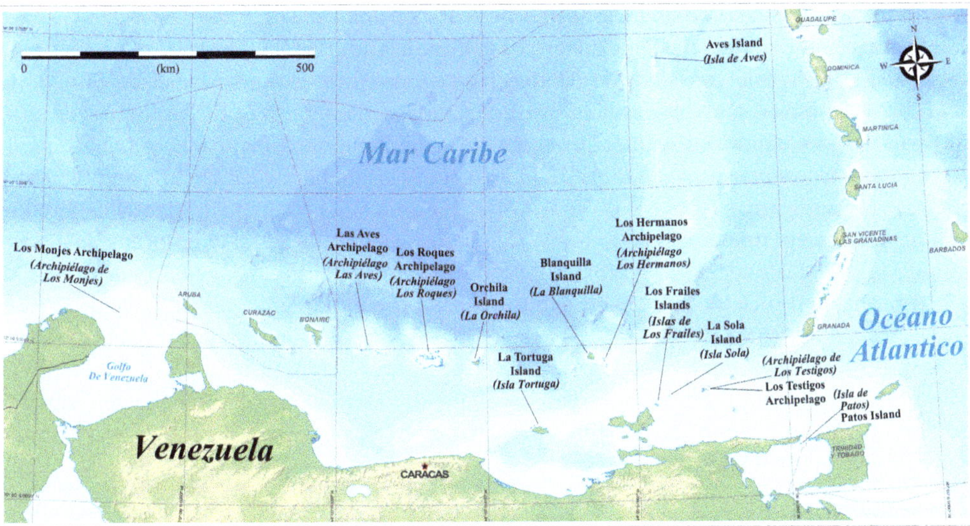

Figure 11.1. Topographic and political map of Federal Dependencies of Venezuela, southern Caribbean ('Mapa de ubicación de las Dependencias Federales de Venezuela' in Spanish). Added elements: Scale bar = 500 km; North compass. English translation of the Work by Original Author. From (title): 'VE-Dependencias Federales ubicacion. png' (https://commons.wikimedia.org/wiki/File:VE-Dependencias_Federales_ubicacion.png). Author: Unukalhai (https://commons.wikimedia.org/wiki/User:Unukalhai) - Own work. Created 10 June 2013. Licensed under CC BY-SA 3.0 (https://creativecommons.org/licenses/by-sa/3.0/deed.en). Created using www.inkscape.org by User:Unukalhai (https://commons.wikimedia.org/wiki/User:Unukalhai).

2. Fringing islands located beyond the South American continental shelf

These islands and archipelagos form a line parallel to the Venezuelan coast beyond the continental shelf (Fig 11.1). They represent an eastward extension of the Dutch islands of Aruba, Curaçao, and Bonaire with which they share similar environments and geologic history, and collectively are considered to constitute a distinct, zoogeographic subregion of the South American realm (see van Buurt, Chapter 10, this issue). From Bonaire progressively eastward lie the Venezuelan islands or archipelagos of Las Aves, Los Roques, La Orchila, La Blanquilla, and Los Hermanos. For characteristics of these islands see Federal Dependencies of Venezuela (2019); some data on coordinates, areas, and other attributes also obtained from Wikipedia and other websites.

Las Aves (not to be confused with Isla Aves, see above) consists of an atoll-like archipelago of 21 coralline islands and mangrove islets, with a collective area of 3.35 km^2 and located at 12° 00' N; 67° 40' W between the Dutch island of Bonaire to the west and the Los Roques archipelago to the east. The main industry is fishing.

Los Roques (11° 51' 16.19" N; 66° 45' 16.19" W) is an archipelago of about 350 islands, islets, and cays with a collective area of 40.6 km^2. The area was declared a (marine) national park in 1972, and the main activity is controlled tourism.

La Orchila is an island of 40 km^2 and located at 11° 47' N; 66° 10' W between Los Roques and La Blanquilla. It contains a Venezuelan military base.

La Blanquilla (11° 51' 12" N; 64° 35' 55" W) is a limestone and sand island with an area of 64.53 km^2. It was named for its white sandy beaches and, although uninhabited, it is visited by tourists, including cruise ships. Feral donkeys are present.

Los Hermanos archipelago consists of a chain of seven barren, rocky islets at 11° 45' N; 64° 25' W and with a collective area of 2.14 km^2.

3. *Fringing islands located on the South American continental shelf*

From west to east and near the Venezuelan coast, these islands (Fig. 11.1) are: (1) Los Monjes archipelago (eight rocky islets or isolated rocks aggregated into three groups, Monjes del Sur, Monjes del Este, and Monjes del Norte, located west of Aruba at the mouth of the Golfo de Venezuela between the La Guajira Peninsula of Colombia and the Paraguaná Peninsula of Venezuela; 12° 21' N; 70° 55' W; collective area 0.2 km^2). (2) La Tortuga (located between Orchila and Blanquilla but farther shoreward; 10° 55' N; 80° 31' W; collective area of its component islands is 156.6 km^2; uninhabited except for occasional tourists). (3) Los Frailes Islands (10 islands; located just northeast of Margarita Island; 11° 12' N; 63° 45' W; collective area of largest island 0.75 km^2). (4) Isla La Sola (uninhabited rock located offshore of the base of the Paria Peninsula of Venezuela; 11° 18' 21" N; 63° 34' 20" W; area 0.0005 km^2). (5) Los Testigos (six major islands and a number of smaller rocky islets located east of La Sola; 11° 22' N; 63° 06' W; area of main island 6.53 km^2). Finally, (6) Patos (10° 38' 18" N; 61° 51' 50" W) is a small uninhabited rocky island just off the eastern tip of the Paria Peninsula in the northern entrance to the Gulf of Paria between Venezuela and Trinidad. It is part of a national park.

B. Margarita Island

Margarita Island (10° 59' 49.56" N; 63° 54' 40.68" W) is not part of the Federal Dependencies of Venezuela, but rather it and its two small satellite islands, Isla Cubagua (10° 49' 12.36" N; 64° 11' 21.12" W; 22.4 km^2) and Isla Coche (10° 46' 58.43" N; 63° 56' 48.49" W; 55 km^2), between it and the Venezuelan mainland, comprise the Venezuelan state of Nueva Esparta. Margarita consists of two mountainous peninsulas that represent an extension of the Venezuelan Coastal Range; it is the largest of the Venezuelan islands (1,020 km^2). This complex of three islands once was connected to the Venezuelan mainland (Rivas-Fuenmayor *et al.* 2005). It is populated and developed and supports a tourism industry.

II. The amphibian fauna

The Venezuelan islands contain few species of amphibians. The present author conducted a faunal survey of Isla Aves on 12–13 May 1966 and no amphibians or other terrestrial vertebrates, other than seabirds, were present (Heatwole, *et al.* 2022). However, given the fact that several hundred years ago the island was much larger (see above), its biota then may have been more diverse.

Bisbal (2008) reviewed the history of 67 surveys of the terrestrial vertebrates of the Federal Dependencies of Venezuela. He found that mammals occurred on all of the island groups, except Los Hermanos and Isla Aves, for a collective total for the Dependencies of ten species, mostly bats and introduced species (but he overlooked the temporary occurrence of mice on Isla Aves; see Hummelinck 1952). Collectively, there were 132 species of birds recorded from the Dependencies, mostly seabirds, shorebirds, and migratory landbirds, but also some terrestrial residents. Every island-group of the Dependencies had birds. The herpetofaunal surveys revealed that every island-group except Isla Aves also had at least one resident species of terrestrial reptile (sea turtles not included) for a collective total of 20 species (1 terrestrial turtle, 3 snakes, 16 lizards).

No species of amphibian was recorded from any of the islands of the Dependencies. Thus, Margarita Island is the only one known to support an amphibian fauna. There are five resident species, all conspecific with forms on mainland Venezuela (Rivas-Fuenmayor *et al.* 2005). They are *Rhinella humboldti*, *Rhinella marina*, *Scinax x-signatus*, *Leptodactylus fuscus*, and *Pleurodema brachyops* (nomenclature updated). Although not recorded from the two satellite islands of Margarita, some or all of these species may occur there as well.

A. *Rhinella humboldti* (Gallardo, 1965)

This species is natively widespread in northern South America, being found in Colombia, Venezuela, Guyana, and Surinam (Narvaes and Rodrigues 2009); its known Caribbean distribution is limited to Margarita Island and Trinidad (Rivas-Fuenmayor *et al.* 2005). It is the most eurytopic species on Margarita, having been recorded from all the major terrestrial habitats on the island: Thorn Scrubland; grasslands; mangroves; deciduous, semi-deciduous, and evergreen forests; dwarf submontane woodlands; and around human habitations.

B. *Rhinella marina* (*sensu stricto*) (Linnaeus, 1758)

Rhinella marina native to northern South America has been widely introduced throughout the world, including a number of Caribbean islands (see other chapters, this volume) and has been the cause of ecological disasters at most destinations (Shine 2018). It is now known that this species in the former broad interpretation of its taxonomy (*sensu lato*) constitutes two different taxonomic entities, one on each side of the Andes (Acevedo *et al.* 2016). The eastern taxon retains the name *Rhinella marina*. Unless a morphological and/or biochemical analysis is made, or the origin of the introduced individuals is known, one cannot be sure which of these two taxa is present on any given island. However, the proximity of Margarita Island to the centre of the range of *R. marina* (*sensu stricto*) makes it highly probable that it is this form that occurs on Margarita, arriving either via natural dispersal, or by an inadvertent or purposeful introduction by humans. Regardless of how it arrived, the event probably was recent as in 2005 its new record from the island was based on only one specimen from deciduous forest (Rivas-Fuenmayor *et al.* 2005).

C. *Scinax x-signatus* (Spix, 1824)

This species is extremely widespread, naturally occurring throughout the Amazon Basin and the northern half of South America, including parts of Colombia, Venezuela, Peru, Bolivia, and Brazil, and all of Guyana, Surinam, and French Guiana (Trefaut-Rodrigues *et al.* 2010) and has been introduced into the Lesser Antillean islands of Guadeloupe (Grande-Terre and Basse-Terre), Marie-Galante, and La Désirade (see Powell and Henderson, Chapter 9 this volume). On Margarita Island, it was recorded only in premontane evergreen forests and dwarf submontane woodlands (Rivas-Fuenmayor *et al.* 2005).

D. *Leptodactylus fuscus* (Schneider, 1799)

This species is another frog with an exceptionally wide natural distribution in South America (Argentina, Bolivia, Brazil, Venezuela, French Guiana, Surinam, Guyana), Central America (Panama) and the Caribbean Islands (Trinidad, Tobago, and Margarita Island) (Heyer and Maxson 1982). On Margarita Island it was found in grassland and around human habitations. Since Margarita was once connected to the Venezuelan mainland, and the speciation of the ancestral *L. fuscus* stock may have begun as early as the Miocene (Heyer 1978), the occurrence of *L. fuscus* on Margarita may date to the time of connection, perhaps reinforced by overwater dispersal since then.

E. *Pleurodema brachyops* (Cope, 1869)

This species has a more restricted distribution than any of the previous species. It occurs naturally only in Aruba (see Chapter 10 of this volume), Panama, northern Colombia, Venezuela, Guyana, and barely into north-western Surinam and northern Brazil (La Marca *et al.* 2010). Its current distribution does not quite reach the Venezuelan coast opposite Margarita Island, although it does slightly farther to the west. On Margarita Island, *P. brachyops* was reported from thorn scrubland, grasslands, deciduous and semi-deciduous forests, and from around human habitation.

Pleurodema brachyops has been introduced into Curaçao, Klein Bonaire, and Bonaire (from Aruba; see van Buurt, Chapter 10 of this volume) and into St Martin/St Maarten of the Lesser Antilles (see Powell and Henderson, Chapter 9 this volume).

III. Conservation

Unlike most islands or archipelagos in the Caribbean, the Venezuelan islands face few challenges in terms of the conservation of amphibians. There are several reasons for this: (1) Most of the individual isolated islands, or those that are part of an archipelago are small and perhaps below the minimum insular area for amphibians (see H. Heatwole and N. Ríos-López, Chapter 1 this volume) or are otherwise unsuitable for amphibians, e.g., lack of standing water; such islands probably never had an established species of frog. (2) Humans seem to have had little impact on the insular populations. Indeed, the total population of the islands of the Federal Dependencies was only slightly more than 2,000 persons (Federal Dependencies of Venezuela 2019). Some small islands that have a significant human presence, such as those now with military bases (Los Monjes, La Blanquilla), lacked amphibians for other reasons anyway so the potential effect of humans was irrelevant. Indeed, the only island large enough and ecologically suitable for amphibians, Margarita Island, was the most populous of humans of all and the most highly developed. All of the species there are widespread on the South American mainland and two were eurytopic on Margarita and three were associated with human habitations and hence adapted to anthropic conditions. All species were of Least Concern according to the IUCN (International Union for the Conservation of Nature) Red Databook: *Rhinella humboldti* (Torres-Suárez and Vargas-Salinas 2014), *Rhinella marina* (Solis *et al.* 2009), *Scinax x-signatus* (Trefaut-Rodrigues *et al.* 2010), *Leptodactylus fuscus* (Reynolds *et al.* 2004), and *Pleurodema brachyops* (La Marca *et al.* 2010).

It seems likely that *R. marina* had recently either dispersed to Margarita naturally or been introduced by humans just prior to 2005 (Rivas-Fuenmayor 2005) and the main conservation measure for amphibians that seems to be needed for the Venezuelan islands is to monitor the influence of this new arrival on the previously established species.

IV. Acknowledgements

I am grateful to my wife, Susan King, for technical assistance with preparation of the manuscript.

V. References

Acevedo, A. A., Lampo, M., and Cipriani, R., 2016. The cane or marine toad, *Rhinella marina* (Anura, Bufonidae): two genetically and morphologically distinct species. *Zootaxa* **4103**: 574–586. https://doi.org/10.11646/zootaxa.4103.6.7

Bisbal E., F. J., 2008. Los vertebrados terrestres de las dependencias federales de Venezuela. Interciencia **33**: 103–111.

Federal Dependencies of Venezuela. Last edited 17 August 2019. https://en.wikipedia.org/wiki/Federal_Dependencies_of_Venezuela

Heatwole, H., Torres, F., King, S. K., Debrot, A. O., and Torres, J., 2022. The history of a remote Caribbean cay, Isla Aves, and its terrestrial biota: A review and new data. (in press). *Atoll Research Bulletin* **630**: 1–38.

Heatwole, H. and Ríos-López, N., this volume. Theories of insular biogeography and ecology. Chaper 14 in Part 5. *The Caribbean*, ed by N. Ríos-López and H. Heatwole: Volume 9. *Status of the Decline and Conservation of Amphibians: Western Hemisphere.* Pelagic Press, Exeter, AUS.

Heyer, W. R., 1978. Systematics of the *fuscus* group of the frog genus *Leptodactylus* (Amphibia, Leptodactylidae). *Science Bulletin. Natural History Museum of Los Angeles County* **29**: 1–85. https://doi.org/10.5479/si.00810282.301

Heyer, W. R. and Maxson, L. R., 1982. Distributions, relationships, and zoogeography of Lowland Frogs: The *Leptodactylus* complex in South America, with special reference to Amazonia. Pp. 375–388 in *Biological Diversification in the Tropics*, ed by G. T. Prance. Columbia University Press, New York, USA.

Hummelinck, P. W., 1952. Islote Aves, een Vogeleiland in de Caraibische Zee. *De West-Indische Gids* **33**: 23–34. https://doi.org/10.1163/22134360-90000129

La Marca, E., Solís, F., Ibáñez, R., Jaramillo, C., Fuenmayor, Q., Reynolds, R., Trefaut-Rodrigues, M., Mijares, A., and MacCulloch, R., 2010. *Pleurodema brachyops. The IUCN Red List of Threatened Species* 2010: e.T57283A11599287. https://doi.org/10.2305/IUCN.UK.2010-2.RLTS.T57283A11599287.en

Maloney, N. J. and Schubert, C., 1968. La Isla de Aves: una isla que desaparecerá? *Acta Cientifica Venezolana* **19**: 2152–2154.

Narvaes, P. and Rodrigues, M. T., 2009. Taxonomic revision of *Rhinella granulosa* species group (Amphibia, Anura, Bufonidae), with a description of a new species. *Arquivos de Zoologia São Paulo* **40**: 1–73. https://doi.org/10.11606/issn.2176-7793.v40i1p1-73

Powell, R. and Henderson, R. W., this volume. Amphibians of the Lesser Antilles. Chapter 9 in Part 5. *The Caribbean*, ed by N. Ríos-López and H. Heatwole: Volume 9. *Status of the Decline and Conservation of Amphibians: Western Hemisphere.* Pelagic Press, Exeter, AUS.

Reynolds, R., Caramaschi, U., Mijares, A., Acosta-Galvis, A., Heyer, R., Lavilla, E., and Hardy, J., 2004. *Leptodactylus fuscus. The IUCN Red List of Threatened Species* 2004: e.T57129A11588348. https://doi.org/10.2305/IUCN.UK.2004.RLTS.T57129A11588348.en

Rivas-Fuenmayor, G., Ugueto, G., Rivero, R., and Miralles, A., 2005. The herpeteofana of Isla de Margarita, Venezuela: New records and Comments. *Caribbean Journal of Science* **41**: 346–351.

Rolle, F. J., 1966. Notes on birds from some West Indian Islands. *Stahlia*, Miscellaneous Papers of the Museum of Biology, University of Puerto Rico Number **7**: 1–3.

Shine, R., 2018. *Cane Toad Wars.* University of California Press, Oakland, California, USA. https://doi.org/10.1525/9780520967984

Solís, F., Ibáñez, R., Hammerson, G., Hedges, B., Diesmos, A., Matsui, M., Hero, J.-M., Richards, S., Coloma, L., Ron, S., La Marca, E., Hardy, J., Powell, R., Bolaños, F., Chaves, G., and Ponce, P., 2009. *Rhinella marina. The IUCN Red List of Threatened Species* 2009: e.T41065A10382424. http://dx.doi.org/10.2305/IUCN.UK.2009-2.RLTS.T41065A10382424.en

Torres-Suárez, O. and Vargas-Salinas, F., 2014. *Rhinella humboldti* (Gallardo 1965) *Catálago de anfibios y reptiles de Colombia* **2**: 19–23.

Trefaut-Rodrigues, M., Caramaschi, U., and Mijares, A., 2010. *Scinax x-signatus. The IUCN Red List of Threatened Species* 2010: e.T56005A11404900. http://dx.doi.org/10.2305/IUCN.UK.2010-2.RLTS.T56005A11404900.en

van Buurt, G., this volume. Amphibians of Aruba, Curaçao, and Bonaire. Chapter 10 in Part 5. *The Caribbean*, ed by N. Ríos-López and H. Heatwole: Volume 9. *Status of the Decline and Conservation of Amphibians: Western Hemisphere.* Pelagic Press, Exeter, AUS.

Zuloaga, G., 1955. The Isla de Aves Story. *Geographical Review* **45**: 172–180. https://doi.org/10.2307/212228

12 Amphibians and their Conservation in Trinidad and Tobago

Renoir J. Auguste, J. Roger Downie, Michael J. Jowers, Richard M. Lehtinen, Ryan S. Mohammed, and John C. Murphy

Abbreviations and acronyms

asl	*above sea level*
CITES	*Convention on International Trades in Endangered Species of Fauna and Flora*
CR	*Critically Endangered*
DD	*Data Deficient*
DNRF	*Department of Natural Resources and Forestry of Tobago*
EN	*Endangered*
eDNA	*environmental deoxyribonucleic acid*
ESA	*Environmentally Sensitive Area*
ESS	*Environmentally Sensitive Species*
IUCN	*International Union for the Conservation of Nature*
JR-D	*J. Roger Downie*
LC	*Least Concern*
MJ-J	*Michael J. Jowers*
RJ-A	*Renoir J. Auguste*
RM-L	*Richard M. Lehtinen*
RS-M	*Ryan S. Mohammed*
UWIZM	*University of the West Indies Zoology Museum*
VU	*Vulnerable*

I. Introduction

The Republic of Trinidad and Tobago (from now on referred to as Trinidad and Tobago) is an island nation, composed of two main islands and associated satellites (Fig. 12.1): they form an archipelago of about 5,000 km^2 of land, which is located off coastal northern Venezuela. Islands within this archipelago are grouped as follows: the Bocas Islands (Chacachacare, Huevos, Monos, Gaspar Grande, Gasparillo), the Five Islands (Caledonia, Craig, Lenagan, Nelson, Pelican, and Rock), San Diego Islands (Cronstadt, Carrera), the islands and satellites of Little Tobago, St Giles,

Renoir J. Auguste, J. Roger Downie, Michael J. Jowers, Richard M. Lehtinen, Ryan S. Mohammed, and John C. Murphy, 'Chapter 12: The Venezuelan Islands' in: *The Conservation and Biogeography of Amphibians in the Caribbean*. Pelagic Publishing (2023). © Renoir J. Auguste et al. DOI: 10.53061/RCJP8789

Figure 12.1. Topographic map of Trinidad and Tobago, and adjacent Venezuela, showing locations mentioned in the text.

Goat, Sisters' Rock, Saut d'Eau, Faralon Rocks, and Soldado Rocks. While this archipelago is a single political unit, its islands and satellites have different geological origins and histories of human exploitation. Based upon the Pindell model of the region's tectonics (Pindell and Kennan 2001, 2009), Trinidad was formed from two terranes; the Northern Range was formed at 250 to 500 km along the west of its current location as part of the South American continent, and Tobago had an oceanic origin on the leading edge of the Caribbean Plate. Currently, all these islands have a herpetofauna, but amphibians appear restricted to the two main islands, probably due to the absence of standing freshwater on the smaller islands.

The first list of the amphibians of Trinidad and Tobago (Court 1858, 1884) contained four names *Pseudis merianae* (= *Pseudis paradoxa*), *Hyla viridis* (= *Phyllomedusa trinitatis*), *Bufo strumosus* (= *Rhinella marina*), and *Pipa americana* (= *Pipa pipa*), but with recognition that other species of hylids and bufonids were present. Other lists of the islands' herpetofauna followed. Mole and Urich (1894) reported 12 species of amphibians. Roux (1926) examined a collection made by Kugler and listed 14 species. Lutz (1927) discussed Venezuelan and Trinidadian anurans and listed 14 species from Trinidad with comments on their distributions on the island. Parker (1933) produced a key to Trinidad's anurans and discussed 23 species based upon collections made by Urich and Vesey-Fitzgerald. A year later, Parker (1934) described *Gastrotheca* (= *Flectonotus*) *fitzgeraldi* from Trinidad. The most in-depth review of Trinidad's anuran fauna came from Kenny (1969, 1977). In 1969, he listed 25 species and provided descriptions of both larval and adult forms, as well as data on life-history and distribution. In 1977, he added one more species and made some corrections to the 1969 paper.

The first amphibian reported from Tobago was by Cope (1879) who listed *Hylodes martincensis* (= *Pristimantis urichi*). Barbour (1916) listed three species of anurans from Tobago ('*Eupemphix trinitatis, Bufo marinus,* and *Hyla crepitans*'), while Mertens (1972) reported nine species. Hardy (1982, 1984) listed 11 species of anurans from Tobago and also discussed some dubious records; one of the species was later identified as a Tobagonian endemic, *Pristimantis* (= *Eleutherodactylus*) *turpinorum* (Hardy, 2001).

Murphy (1997) produced an updated list for Trinidad and Tobago, and included a discussion of some dubious records in museums and changes in nomenclature. His list included 36 species of anurans, two of them known only from single specimens recorded in the late nineteenth and early twentieth centuries, and a dubious caecilian described from one specimen that is lost. No salamander has ever been recorded from the islands. Murphy's list was based on original fieldwork as well as on the study of collections in museums and a review of the literature. Since then, one new hylid has been added to the Trinidad list: *Scarthyla vigilans* (Smith *et al.* 2011).

The development and implementation of effective conservation efforts for amphibians on an archipelago such as Trinidad and Tobago require a good knowledge of which species are present, their distributions and populations, and the threats they face. These efforts also benefit greatly from information about the ecological requirements of amphibians and their life-history strategies, as well as the existing framework and resources available for wildlife conservation. Hailey and Cazabon-Mannette (2011) provided a valuable analysis of these issues both for amphibians and reptiles. Since then, much new information has emerged. This chapter attempts to bring all this together, focusing solely on amphibians and identifying areas for future research and conservation efforts.

II. The species

The general biology of the amphibians of Trinidad and Tobago, accompanied by colour photographs of each species, is treated in the book *A Field Guide to the Amphibians and Reptiles of Trinidad and Tobago* (Murphy *et al.* 2018). Additional photographs appear in Figures 12.2 to 12.5 of the present chapter.

The new amphibian 'tree of life' by Frost *et al.* (2006) set off a revolution in our understanding of amphibian taxonomy and systematics. This process continues and shows no signs of abating (e.g. Duellman *et al.* [2016] in their revision of the family Hylidae). Therefore, the list of species we set out here should be considered as provisional. This tentativeness is largely because previously established boundaries between species are being re-examined, especially by means of molecular tools, and particularly in the case of species currently considered to have very wide distributions (Murphy and Downie 2012). Another factor is the geographic position of Trinidad and Tobago, which renders the country susceptible to invasions of species both from the South American mainland and from the Antilles, thereby making for an unstable fauna. Kenny (1969, 1977, 1979) felt that several species, especially those limited to the south-western part of Trinidad, were recent arrivals from the mainland. Biogeographers tend to consider Trinidad and Tobago as part of the mainland, and accounts of the herpetofauna of the Caribbean often exclude the two islands (e.g. Crother 1999), but the lists of species show influences from both directions. Hedges *et al.* (2019) have included Trinidad and Tobago as Caribbean islands in their account of Caribbean amphibians and reptiles.

The 35 amphibian species known from recent records to occur in Trinidad and Tobago are listed in Table 12.1. This table gives locations, including those of anurans that also occur outside of the two islands, and it provides the IUCN's (2020) Red List categories for each species. We use the nomenclature of families and species given by Frost (2020), unless recent work known to us indicates a justified change. We also draw attention to the occasional differences between Frost (2020) and the other major database, AmphibiaWeb (2020). Common names are listed as suggested by Downie (2013), who discussed the issues concerning these names in Trinidad and Tobago and suggested new names for species when common names are lacking or are currently inappropriate. Hedges *et al.* (2019) have made recommendations on common names, somewhat different from ours.

Figure 12.2. Four species of frogs endemic to Tobago. (A) Tobago Stream Frog, *Mannophryne olmonae* (photograph by John C. Murphy); (B) Tobago Glass Frog, *Hyalinobatrachium orientale tobagoense* (photograph by Renoir J. Auguste). (C) Tobago Litter Frog, *Pristimantis charlottevillensis* (photograph by John C. Murphy); (D) Turpin's Litter Frog, *Pristimantis turpinorum* (photograph by Renoir J. Auguste).

Figure 12.3. A species of frog endemic to Trinidad, Trinidad Stream Frog, *Mannophryne trinitatis* with developing young on its back. Photograph by John C. Murphy.

Figure 12.4. Two Caribbean species that are invasive in Trinidad and Tobago. (A) Johnstone's Rain Frog, *Eleutherodactylus johnstonei*; (B) Maracaibo Treefrog, *Scarthyla vigilans*. Photographs by John C. Murphy.

Some species require individual comments, indicated by superscript '1' in Table 12.1. Manzanilla *et al.* (2007) established that the aromobatid *Mannophryne trinitatis* is endemic to Trinidad. AmphibiaWeb (2020) does not recognize the family Aromobatidae and retains *Mannophryne* within the Dendrobatidae. There has been a dispute over the identity of the smaller bufonid genus *Rhinella*, found in Trinidad (but not in Tobago). Murphy (1997) called this species *beebei*, but Narvaes and Rodrigues (2009) concluded that the Trinidadian species was not distinguishable from *Rhinella humboldti*, a species whose type locality is in western Colombia. Murphy *et al.* (2017) disagreed based on molecular analyses, morphology, biogeography, and differences in call, and restored the Trinidadian population to *beebei*, noting that it likely occurs both in Trinidad and in Venezuela and that there is a need for further assessment of the Venezuelan populations.

Jowers *et al.* (2015) examined DNA from South American and Tobagonian populations of the glass frog, *Hyalinobatrachium orientale*. They concluded that the population on Tobago is not sufficiently different to be designated as a separate species, but agreed with Hardy's (1984) contention that it merited subspecific designation as *H. o. tobagoense* (the grammatically correct form in place of *tobagoensis* as used by Hardy).

The ground-living direct-developing frogs of the Neotropics have undergone an extensive taxonomic revision (Hedges *et al.* 2008), which resulted in the erection of an overarching

Figure 12.5. Four anuran species presumably shared between Trinidad and Venezuela (but see section II). (A) Beebe's Toad, *Rhinella beebei* (formerly *Rhinella humboldti*); (B) Golden Treefrog, *Phytotriades auratus*; (C) Gough's Treefrog, *Dendropsophus minutus*; (D) Miniscule Treefrog, *Dendropsophus minusculus*. Photographs by John C. Murphy.

unranked taxon, the Terrarana, now given a rank as superfamily Brachycephaloidea (Frost 2020), and the reorganization of the genus *Eleutherodactylus* (previously placed within the family Leptodactylidae) into a number of new genera and families. In Trinidad and Tobago, the three native species are placed in the family Craugastoridae and the genus *Pristimantis*. The invasive non-native species *Eleutherodactylus johnstonei* is in the family Eleutherodactylidae.

Dendropsophus minutus is a species under revision. Gehara *et al.* (2014) resurrected the name *D. goughi* and noted that these very widespread small treefrogs might be a complex of species. Frost (2020) now lists *D. goughi* as a Trinidadian endemic. As *D. minutus*, it has been recorded at many locations in Trinidad and Tobago but generally in numbers less than some other small hylids. If it is confirmed as a Trinidad and Tobago endemic, its conservation status needs to be assessed separately from that of *D. minutus*.

There has long been consideration of the need for taxonomic revision of the widely distributed hylid *Hypsiboas* (= *Boana*) *crepitans*. Orrico *et al.* (2017) changed the name to *Boana xerophylla*, followed by Escalona *et al.* (2021) describing the Andean populations (including Trinidad and Tobago) of frogs as a new species, *Boana platanera*.

Phytotriades auratus is still listed in AmphibiaWeb (2020) by its earlier name of *Phyllodytes*, despite the revisionary work of Jowers *et al.* (2008). It has long been believed that this bromeliad-dwelling frog is restricted to a small number of mountaintops in Trinidad. However, Rivas and

Mayke De Freitas (2015) reported what appears to be this species from a mountain in northern Venezuela, so it could be more widespread than frequently recognized. Molecular analysis is needed to establish how closely related the two populations are, and the extent of the population in Venezuela has yet to be determined.

Scarthyla vigilans appeared neither in the papers by Kenny (1969, 1977), nor in his brief note on recent colonizations (1979), nor in the book by Murphy (1997). This species was first noticed in south-western Trinidad in 2006 and was considered limited to that area by Smith *et al.* (2011). However, it is now known to be much more widespread; Mohammed *et al.* (2014) and Auguste (2019) reported it from south-eastern Trinidad; Auguste and Rutherford (2015) found it in the north-west, at Tucker Valley, in the east-central part of the island, at Kernahan, and in the north-east at Wallerfield. Either this species has expanded very rapidly, or its quiet cricket-like call has led to its being previously overlooked. Whatever the explanation, it is now widespread in Trinidad.

Frost (2020) discussed the relationship between the genera *Leptodactylus* and *Adenomera;* we retain *Adenomera.* Like *Scarthyla vigilans,* this species also appears to have expanded its range in Trinidad (Auguste 2017).

The largest leptodactylid in Trinidad was originally listed by Kenny (1969) as *Leptodactylus pentadactylus pentadactylus,* but he corrected this nomenclature (1977) to *Leptodactylus bolivianus,* which refers to a species widespread in South America. However, Heyer and Heyer (2013) revised that species and assigned the Trinidadian population to *Leptodactylus insularum.*

Leptodactylus macrosternum was first noted by Kenny (1977, 1979) and he speculated that it was a new arrival from Venezuela. Since then, the identity of this species has come into question. De Sá *et al.* (2014) concluded that there is much work required to determine the identities of species, and the boundaries between them, in a set of species in the *Leptodactylus macrosternum-latrans-chaquensis* group.

Leptodactylus nesiotus was long considered a Trinidadian endemic with a distribution restricted to the south-west of the island, but populations have recently been identified in the Guianas (Jairam and Fouquet 2018) and it has also been recorded from new localities in Trinidad.

Elachistocleis ovalis and *Elachistocleis surinamensis* are listed by Kenny (1969) and Murphy (1997) as occurring in Trinidad. Some authors have expressed doubt about whether these are really separate species – for instance see comments by Frost (2020) on the nomenclatural status of *Elachistocleis* and the need for a revision of species in this genus. However, genetic analysis of Trinidad specimens (Jowers *et al.* 2021a) uncovered three species of *Elachistocleis* on Trinidad. They (Jowers *et al.* 2021a) renamed the frog that had been referred to as *E. ovalis* as *E. nigrogularis* and they found *Elachistocleis surinamensis* to be genetically distinct from *E. nigrogularis.* The third undescribed species on Trinidad was not represented by molecular data in the study, but its call was distinct from those of *E. nigrogularis* and *E. surinamensis.*

The revision by Frost *et al.* (2006) placed Trinidad's only ranid in a new genus as *Lithobates palmipes.* Yet Yuan *et al.* (2016) criticized the decision to resurrect the genus *Lithobates.*

III. Conservation status

Table 12.1 includes a summary of the global distribution and conservation status (IUCN 2020) of the frogs of Trinidad and Tobago. Thirty species are listed as of Least Concern (LC), mainly because these species have wide distributions beyond Trinidad and Tobago. Three of these species require comment. *Leptodactylus bolivianus* is LC, but there has been no assessment of the population now named *Leptodactylus insularum.* The same is true of *Leptodactylus macrosternum,*

Table 12.1. The frogs of Trinidad and Tobago.

FAMILY/*SPECIES*	LOCATIONS	CONSERVATION STATUS	SUGGESTED COMMON NAMES
AROMOBATIDAE			
Mannophryne olmonae	TO only	VU	Tobago stream frog
Mannophryne trinitatis[1]	TR only	LC	Trinidad stream frog
BUFONIDAE			
Rhinella humboldti (beebei)[1]	TR; Venezuela	LC	Beebe's toad
Rhinella marina	TT; Texas to northern South America[2]	LC	Crapaud; cane toad; marine toad
CENTROLENIDAE			
Hyalinobatrachium orientale tobagoense[1]	TO and Venezuela (as *H. orientale*)	VU	Tobago glass frog
CRAUGASTORIDAE			
Pristimantis charlottevillensis[1]	TO only	VU	Charlotteville litter frog
Pristimantis turpinorum[1]	TO only	DD	Turpin's litter frog
Pristimantis urichi[1]	TT only	LC	Urich's litter frog
ELEUTHERODACTYLIDAE			
Eleutherodactylus johnstonei[2]	TT; widespread in the Caribbean	LC	Johnstone's litter frog
HYLIDAE			
Dendropsophus microcephalus	TR; widespread in South and Central America	LC	Small-headed treefrog
Dendropsophus minusculus	TR; widespread in South America	LC	Minuscule treefrog
Dendropsophus minutus (= *Dendropsophus goughi*)[1]	TT; as *goughi*, may be TT endemic	LC as *minutus*; not assessed as *goughi*	Minute treefrog
Boana boans	TR; widespread in South America	LC	Giant treefrog
Boana geographica	TR; widespread in South America	LC	Map treefrog
Boana platanera[1]	TT; widely distributed north of the Orinoco River in Venezuela also present in Panama and Colombia	LC as *crepitans*; not assessed as *xerophylla* or *platanera*	Rattle-voiced treefrog
Boana punctata	TR; widespread in South America	LC	Lesser green treefrog
Phytotriades auratus[1]	TR; northern Venezuela	EN	Trinidad golden treefrog
Pseudis paradoxa	TR; widespread in South America	LC	Paradox frog
Scarthyla vigilans[1]	TR; widespread in northern South America	LC	Pale grey-green treefrog

(Continued)

Table 12.1. (Continued)

FAMILY/*SPECIES*	LOCATIONS	CONSERVATION STATUS	SUGGESTED COMMON NAMES
Scinax ruber	TT; widespread in South America	LC	Lesser brown treefrog
Sphaenorhynchus lacteus	TR; widespread in South America	LC	Lime treefrog
Trachycephalus typhonius (= *Trachycephalus venulosus*)	TT; widespread in Central and South America	LC	Milky treefrog; Warty treefrog
HEMIPHRACTIDAE			
Flectonotus fitzgeraldi	TT; Paria Peninsula of Venezuela	LC	Marsupial treefrog
LEPTODACTYLIDAE			
Adenomera hylaedactyla[1]	TR; widespread in South America	LC	Lesser dark-spotted thin-toed frog
Engystomops pustulosus	TT; widespread in Central and South America	LC	Túngara frog
Leptodactylus fuscus	TT; widespread in South America	LC	Whistling frog
Leptodactylus insularum (= *Leptodactylus bolivianus*)[1]	TR; widespread in South America as *L. bolivianus*	LC as *L. bolivianus*	Barbour's thin-toed frog
Leptodactylus macrosternum (= *Leptodactylus latrans*)[1]	TR; taxonomic issues cause uncertainty over distribution	LC as *L. latrans*	Greater dark-spotted thin-toed frog
Leptodactylus nesiotus	TR only	LC	Trinidad thin-toed frog
Leptodactylus validus	TT; many Lesser Antillean islands	LC	Garman's thin-toed frog
MICROHYLIDAE			
Elachistocleis nigrogularis[1]	TR; widespread in Central and South America	LC	Common narrow-mouthed frog; Siren frog
Elachistocleis surinamensis[1]	TR; similar to *E. ovalis* and may be the same species	LC	Surinam narrow-mouthed frog; siren frog
PHYLLOMEDUSIDAE			
Phyllomedusa trinitatis	TR; northern Venezuela	LC	Trinidad leaf frog
PIPIDAE			
Pipa pipa	TR; widespread in South America	LC	Pipa toad
RANIDAE			
Lithobates palmipes[1]	TR; widespread in South America	LC	Great olive-green ground frog

TO = Tobago; **TR** = Trinidad; **TT** = Trinidad and Tobago. Conservation status and criteria are according to the criteria of IUCN (2001, 2020): **LC** = Least Concern; **CR** = Critically Endangered; **EN** = Endangered; **VU** = Vulnerable.

[1] For further comment, see text.

[2] Also introduced widely, for example in Barbados, Puerto Rico, Hawaii, Philippines, Australia.

which has no listing, but is LC if correctly identified as *Leptodactylus latrans*. Similarly, as noted earlier, if *Dendropsophus goughi* is confirmed as a Trinidadian endemic, it should be assessed in Trinidad and Tobago.

To have 30/35 species as LC (85%) is considerably better than the global average 41% of species threatened with extinction (IUCN 2020), but this does not mean that there is no conservation interest in these species. For example, as insular populations these may be able to act as reservoirs of populations facing severe continental threats. Later, we consider the health of the populations of all anuran species in Trinidad and Tobago, but first we consider the four species in the threatened categories: Endangered (EN) and Vulnerable (VU), the one species regarded as Data Deficient (DD), and also discuss the rationale behind three endemic species having had their status downgraded to LC in the 2020 update of the IUCN Red List. Table 12.1 includes the criteria used by IUCN (2001) to assess the status of the EN, VU, and DD species.

Mannophryne olmonae is on this list as VU. This species was assessed as Critically Endangered (CR) in 2004 on the basis of noticeable declines of its populations in the previous decade. However, no data were provided to support this claimed decline and more recent evidence (Lehtinen *et al.* 2016) shows a wider distribution than previously reported, successful reproduction over that range, and high occupancy of the suitable habitat within it, including both primary and secondary forest. In addition, although there has been evidence of chytrid infection, the frogs sampled showed no signs of clinical disease (Alemu *et al.* 2008). On the basis of this evidence and the lack of any threats to the species' forested habitat, the IUCN Amphibian Specialist Group (2020d) adjusted its status to VU D2 (D2 = population small, restricted area of occupancy), a substantial downgrade in conservation concern. Thomson *et al.* (2018) enhanced this conclusion by finding no chytrid in the *M. olmonae* population.

Mannophryne trinitatis was originally assessed in 2004. At that time, it was believed to occur both in Trinidad and in the adjacent Paria Peninsula of Venezuela. Yet Manzanilla *et al.* (2007) demonstrated that the Venezuelan population constitutes a distinct species, named *Mannophryne venezuelensis*, making *M. trinitatis* a Trinidadian endemic. The VU listing of B1ab(iii) (= occurrence in less than 20,000 km^2, fragmented distribution, declining population) was based on the species' limited range, on a decline in extent and quality of its habitat, and on its severely fragmented distribution, although as far as we can determine no original fieldwork was conducted to support these descriptions. The assessment was revisited by Angulo (2010), but no change was made and, other than noting that the species was no longer considered to have a population in Venezuela, no new evidence was presented. Greener *et al.* (2017) have now carried out an extensive survey of *M. trinitatis*. Its distribution is fragmented in the sense that it occurs in at least two distinct areas: the Northern Range and the Central Range, with a recent report of its occurrence also in the southern Trinity Hills (Mohammed *et al.* 2014). However, the populations in the two main locations are not fragmented. In the Northern Range, *M. trinitatis* is ubiquitous along the steep mountain streams that drain to the valleys; in the Central Range, Jowers and Downie (2004) found the frog from Mount Harris in the east to Gran Couva in the west, with a major population at Tamana Hill. This species is common both in primary and in secondary forests, and its estimated population size is in the range of millions. Although there are areas of deforestation in the Northern Range, most of the forest is continuous and intact. The chytrid fungus, *Batrachochytrium dendrobatidis*, was reported from *M. trinitatis* in the Northern Range during 2007 and 2009, but with no signs of clinical disease (Alemu *et al.* 2013); Greener *et al.* (2017) encountered no traces of infection in 2014. Based on these results, *M. trinitatis* has been reassessed as LC (IUCN SSC Amphibian Specialist Group 2020e).

The population status of the next seven species treated in this section was assessed in 2004, and the assessment has since been updated in 2020 (IUCN SSC Amphibian Specialist Group 2020a,b,c,f,g,h,i).

The glass frog *Hyalinobatrachium orientale* was listed VU B1ab(iii) (= occurrence in less than 20,000 km^2, fragmented distribution, declining populations). The justification of VU is the limited extent of its occurrence, its fragmented distribution, and the decline in the extent and quality of its habitat (Senaris *et al.* 2004). Jowers *et al.* (2015) carried out a molecular analysis of the Tobagonian and Venezuelan populations of this species and concluded that there was insufficient evidence of divergence to justify separating the two populations into different species. Nonetheless, they found enough divergence to justify Hardy's (1984) designation of the population from Tobago as a distinct subspecies, *Hyalinobatrachium orientale tobagoense. Hyalinobatrachium orientale* inhabits the vegetation (e.g. commonly the leaves of *Heliconia bahai*) frequently found overhanging streams and rivers in forested areas. The area in north-eastern Tobago which is inhabited by *H. orientale* is very similar to that inhabited by *Pristimantis charlottevillensis*, but the latter species can be found anywhere on the forest floor, whereas *H. orientale* is restricted to the banks and edges of streams. It is therefore vulnerable to clearance of the vegetation adjacent to streams. An assessment published in 2004 listed agricultural expansion and human settlement as major threats to this species in Venezuela, but could not identify threats in Tobago. Still, given the potential for clearance of the riverine vegetation by humans and the absence of adequate regulatory mechanisms for protection of habitats, the category VU remains valid for this species (IUCN SSC Amphibian Specialist Group 2020b), and work needs to be done to assess the extent of its distribution, particularly in Tobago.

The golden treefrog *Phytotriades auratus* was listed CR B1ab(iii) and 2ab(iii), because of its small area of occurrence and occupancy, fragmented distribution, and declining populations. The justification of CR is the species' very small range and area of occupancy, extremely fragmented distribution, and decline in the extent and quality of its habitat (Hardy 2004a). The report of *P. auratus* from Venezuela (Rivas and Mayke de Freitas 2015) means that the range of this species is greater than previously thought, depending on how closely related the two populations are. However, it is restricted to a microhabitat, at high elevation, that consists of pools of water enclosed by the leaves of the giant bromeliad *Glomeropitcairnia erectiflora*, which almost inevitably implies a fragmented distribution, although not necessarily a low population density. Clarke *et al.* (1995) estimated the population size of this species as about 20,000 frogs in the vicinity of the summit of Trinidad's El Tucuche alone, although later work (Jowers *et al.* 2008) suggested that this estimate may have been too high. Recent surveys by Torresdal *et al.* (2017) and Brozio *et al.* (2017) have used the method of eDNA, which can detect the presence of amphibians in small bodies of water without disturbing the habitat, and have detected populations on El Tucuche, Cerro del Aripo, and Chaguaramal, the latter being a new distributional record. These surveys found no evidence of any decline in the extent or quality of the habitat. However, it would be inappropriate to reconsider the conservation status of this species until the extent of its occurrence in Venezuela has been established, especially considering the potential effects of climatic change on a species restricted to a microhabitat found only at high elevations. Nevertheless, the IUCN SSC Amphibian Specialist Group (2020f) concluded that the extra records of distribution justify a change to Endangered (EN).

Pristimantis charlottevillensis was listed as LC by the IUCN SSC Amphibians Specialist Group in 2013. Although the species does occur in a wide variety of forested and grassy areas, populations are restricted to a limited extent of occurrence in north-eastern Tobago only, based on recent distributional surveys (Jowers *et al.* 2021b). For this reason, the species has been recommended to have a higher threat status as VU D2 (IUCN SSC Amphibian Specialist Group 2020g).

Pristimantis turpinorum was listed VU D2 (Hardy 2004b; D2 because of small populations and restricted area of occupancy). The species was first described by Hardy (2001) and is therefore absent from Murphy's (1997) list, although it is most likely the species listed by Murphy as *Eleutherodactylus* cf. *rozei* and first identified by Hardy (1982). The assessment as VU, rather than CR, is very surprising, given that the species is known only from fewer than five localities in the forests of northeastern Tobago. Few surveys have attempted to document its distribution in Tobago in detail. In our experience in the field, this species tends to be arboreal and closely associated with *Dypsis* palms. One of us (RJ-A) has seen adults on substrates close to the ground, including an amplexing pair on a plant at 1 m off the ground over a stream. As a direct-developing species, like other *Pristimantis* species, oviposition presumably happens on damp ground or in moist crevices in trees. Vocalizations by males have been heard only for brief periods just before sunset. Therefore, detection of this species is difficult, even when it is present. Consequently, its status has been updated to Data Deficient (DD) since we know so little about either its distribution or natural history (IUCN SSC Amphibian Specialist Group 2020h; Jowers *et al.* 2021).

Pristimantis urichi is probably endemic to Trinidad and Tobago, as Kaiser *et al.* (1994) considered the identification of Venezuelan specimens as *urichi* to be incorrect. *Pristimantis urichi* was listed EN A2ae because of a major population decline claimed by Hardy *et al.* (2004) to have occurred in Tobago (greater than 70% over the previous ten years). However, no data were presented to support this conclusion, and no assessment was made in Trinidad. Finding *P. urichi* throughout Trinidad, Kenny (1969) judged it 'probably the most ubiquitous of Trinidad's amphibians'. Our published surveys in both Trinidad and Tobago over the years 2012–16 support Kenny's view: the frog is found in primary and in secondary forest across both islands, but not in human-disturbed habitats (Jowers *et al.* 2021b). Lacking baseline data on past population size and fluctuations, we cannot determine whether the populations are changing, but we certainly have no evidence of a major decline in numbers, as might have been expected had Hardy *et al.*'s (2004) claim been correct, and had it continued. There is certainly no decline in the frog's range on the two islands. In our view, this is an abundant species in largely stable habitat conditions and its status has been recommended to be downgraded to Least Concern (LC) (IUCN SSC Amphibian Specialist Group 2020i).

The marsupial frog *Flectonotus fitzgeraldi* was listed EN B1ab(iii), the latter based on its small extent of occurrence, fragmented distribution, and declining population. The justification for EN is its small extent of occurrence, severely fragmented distribution, and decline in the extent and quality of its habitat (La Marca *et al.* 2004). Its population is said to be decreasing, but no data have been presented to justify that conclusion. In our experience, this is a difficult species to assess. It lives in trees and shrubs and rarely descends to the ground. The female carries her eggs in a dorsal pouch, and they hatch as late-stage tadpoles, after which she deposits them in small bodies of water, such as those held by bromeliads, to complete development and metamorphose a few days later. The call is soft and easily mistaken for that of an insect, and the males call only for a short period at about dusk. During their work on the Golden Treefrog in Trinidad, Clarke *et al.* (1995) found that 20% of the giant bromeliads (*Glomeropitcairnia erectiflora*) opened on El Tucuche ($n = 20$) contained *F. fitzgeraldi*, as did 11.5% of *Tillandsia* sp. ($n = 25$). More recently, Smith (2007) reported densities of *F. fitzgeraldi* from three forested sites in Trinidad's Northern Range, where disturbance of the habitat ranged from high (Simla: 50 ± 10 frogs per hectare), to intermediate (Springhill Estate; 49 ± 9), to low (Morne Bleu Ridge; 75 ± 23). In addition, Auguste (2019), Auguste *et al.* (2015), and Mohammed *et al.* (2014) confirmed the presence of *F. fitzgeraldi* at Matura forest in northeast Trinidad, Aripo Savanna in central Trinidad, and Edward Trace in the south, respectively, but without attempting any estimate of numbers. In Tobago, a survey of 106 phytotelmata of *Xanthosoma jacquinii* on the Main Ridge found 10% to be occupied by

F. fitzgeraldi. Depth of water was the single statistically significant predictor of the occupancy of these plants; plants with more water were more likely to be occupied (Eyre 2013). Just like *Pristimantis urichi*, which occurs in a similar habitat, the lack of past data on populations makes assessment of trends difficult. However, given these survey results and the extent of suitable forest on both islands, we see no justification for regarding the species as threatened in Trinidad and Tobago. The IUCN status has since been downgraded to LC based on findings of the species' distribution in Trinidad and Tobago, along with data from Venezuela (IUCN SSC Amphibian Specialist Group 2020a; Smith *et al.* 2021).

Leptodactylus nesiotus was listed VU D2 (D2 = population small, restricted area of occupancy). The justification of VU is that this species is known only from a single location, Icacos swamp, in south-western Trinidad and nowhere else in the World (Hardy 2004c). *Leptodactylus nesiotus* was first described by Heyer (1994) from three adult males collected in 1983 and 1986. Ponssa *et al.* (2010) provided natural history notes on this species and analysed the morphology of four additional specimens, including one female. This frog can be heard calling in chorus among dense vegetation in swampy areas. The call is similar to, but subtly different from, that of *Leptodactylus validus*, and Borzée *et al.* (2020) have provided a description of their advertisement and aggressive calls. Adult specimens analysed are similar to, but again distinct from, *L. validus*, being smaller and possessing a labial stripe, which is absent in *L. validus*. Genetic analysis has confirmed that *nesiotus* and *validus* are distinct species, but closely related (de Sá *et al.* 2014). More recently, *L. nesiotus* has been found and specimens collected by Ali, Auguste, Murphy, and Rutherford (specimens in UWIZM) from as far east as Nariva swamp, as well as in localities between Icacos and Nariva. Mohammed *et al.* (2014) and Auguste (2019) did not report this species for their survey of south-eastern Trinidad. Given that this frog has recently been reported to occur in Guyana, Suriname, and French Guiana (Jairam and Fouquet 2018), the status has been downgraded to LC (IUCN SSC Amphibian Specialist Group 2020c).

IV. Distributions

Kenny (1969) visited 562 localities (some of them repeatedly) in Trinidad between 1954 and 1960 in order to determine distributions. Murphy (1997) made six intensive field trips to the two islands between 1982 and 1994, and used the resulting information, plus previous records including Kenny's, to derive distributional maps for each species. The known distributions of anurans on Trinidad and Tobago are indicated in Table 12.2. This table was constructed using the information from these accounts plus data from recent fieldwork, much of it generated by the authors of this chapter and their associates. This table should be read in conjunction with the map in Figure 12.1.

V. Monitoring the populations of frogs in Trinidad and Tobago

As noted earlier, the first efforts to map the distributions of the frogs of Trinidad and Tobago were made by Kenny (1969, 1977) and Murphy (1997), but neither attempted to assess populations or whether they were changing. The Global Amphibian Assessment (Stuart *et al.* 2004) attempted to assess the population status of all the species of amphibians in the world in light of worrying reports of widespread declines. The assessments made for Trinidad and Tobago were based largely on decisions made by committees of experts rather than on recent fieldwork. This section discusses progress since then.

The large variety of frogs on Trinidad and Tobago means that no single monitoring approach can be used; some species are diurnal, but most are nocturnal; some are only easily located

Table 12.2. Distributions of the frogs of Trinidad and Tobago, based on Murphy (1997), Murphy *et al.* (2018), and results of published and unpublished surveys by the present authors.

FAMILY/*SPECIES*	DISTRIBUTION
AROMOBATIDAE	
Mannophryne olmonae	Tobago only: restricted to the margins of rivers and streams in forested areas (both primary and secondary). Lehtinen *et al.* (2016) found that this species extended farther west than previously recorded, essentially occupying streams east of the line from Hillsborough Bay in the south to Castara Bay in the north.
Mannophryne trinitatis	Trinidad only; throughout the Northern Range, occupying the sides of small streams, but not rivers; on the southern slopes, mainly at higher elevations, but on the northern coast, they can be present along streams flowing onto beaches. Also found in the Central Range from Mount Harris in the east to Gran Couva in the west, with a notable population at Tamana Cave (Jowers and Downie 2004; Greener *et al.* 2017). No populations were found in southern Trinidad, despite the occurrence of suitable habitat until Mohammed *et al.* (2014) found this species by a pond on Edward Trace; the extent of this southern population has yet to be determined.
BUFONIDAE	
Rhinella beebei	Trinidad only: common throughout the island in open areas and in woodland, both undisturbed and human-modified, at lower elevations; breeds in shallow pools in the wet season.
Rhinella marina	Trinidad and Tobago: can be found almost anywhere, but mainly at lower elevations; often associated with urban areas, but also along rivers and in forests. Breeds in rivers, temporary pools, and drainage ditches.
CENTROLENIDAE	
Hyalinobatrachium orientale	Central and northern Tobago only: adults found on vegetation overhanging small streams and larger rivers; tadpoles hidden in the gravel of streambeds. Full extent not yet established, but so far detected along many streams to the east of the Roxborough–Parlatuvier Road and in forest on Main Ridge at some streams to the west of the road.
CRAUGASTORIDAE	
Pristimantis charlottevillensis	North-eastern and central Tobago only: Murphy (1997) reported this frog from the Roxborough–Parlatuvier Road and from roadside localities farther east. Surveys by Jowers *et al.* (2021b) confirmed these locations but also found this species farther west along the Northside Road near Castara and in the forest around Hillsborough Dam. This species inhabits forest-floor leaf litter and the sides and beds of streams. It is common along roadsides where there are no human habitations.
Pristimantis turpinorum	North-east Tobago only: adults generally found in palm trees a few metres above the ground. This species has been sighted recently in a few locations in forest along the eastern side of the Roxborough–Parlatuvier Road, but its full distribution still has to be confirmed and appears to be more restricted than reported by Murphy (1997).
Pristimantis urichi	Trinidad and Tobago: in Trinidad; present in primary and secondary forest, essentially throughout the island except for human-inhabited areas. In Tobago, this frog's distribution is very similar to that of *P. charlottevillensis*, i.e. in the north eastern and central parts. This is a species of the forest floor, but often found perched and calling from low bushes; not generally encountered along streams.
ELEUTHERODACTYLIDAE	
Eleutherodactylus johnstonei	This species is a recent arrival in Trinidad (Kenny 1979; Manickchan *et al.* 2011; Downie *et al.* 2017), and even more recently in Tobago (White 2013; Downie *et al.* 2017). In Trinidad, this frog has spread from its original location in the dock area of Port of Spain along the east–west corridor south of the Northern Range and south both along the Southern Main Road in the west and the Cunapo Southern Road in the east. In Tobago, it has been heard in Speyside and Parlatuvier in the north and in Lowlands in the south; it is not known whether these are established populations (Downie *et al.* 2017). This species seems to be predominantly an inhabitant of urban areas and is heard in gardens and parks.

(Continued)

Table 12.2. (Continued)

FAMILY/*SPECIES*	DISTRIBUTION
HYLIDAE	
Dendropsophus microcephalus	Trinidad only: widespread throughout the island at lower elevations, especially amongst grasses and low bushes along roadside ditches, and around ponds and swamps.
Dendropsophus minusculus	Trinidad only: so far, reported only from the Icacos Peninsula. This frog is found amongst grasses and low bushes at the margins of swamps and ditches.
Dendropsophus minutus	Trinidad and Tobago: like *D. microcephalus*, widespread throughout Trinidad at lower elevations, amongst grasses and low bushes along roadside ditches and around ponds and swamps. In Tobago, mainly reported from the Windward Road north-east of Roxborough, but it may be more widespread.
Boana boans	Trinidad only: found in the trees lining riverbanks in the valleys of the Northern and Central Ranges, but also in similar habitat in south-eastern Trinidad (Mohammed *et al.* 2014).
Boana platanera	Trinidad and Tobago: widespread in Trinidad at lower elevations, often in garden ponds, but also in flooded areas in savannahs and at the edges of forests. In Tobago, widespread throughout the mountainous spine of the island.
Boana geographica	Trinidad only: like *Boana boans,* this frog is found along riverbanks in the valleys of the Northern Range, but more often among bushes than high in trees. Possibly now less common in the Northern Range than in the past. It is also found around ponds and at the margins of slow-flowing creeks, for example in Cedros, Nariva, Aripo Savannahs and Edward Trace (southeast) (Auguste *et al.* 2015; Mohammed *et al.* 2014). Tadpoles form dense black shoals.
Boana punctata	Trinidad only: found in low vegetation around swampy areas and ponds; tadpoles hide in very muddy water. Widely distributed in the valley of the Northern Range and across the rest of the island at lower elevations, including Aripo Savannahs and Edward Trace (southeast) (Auguste *et al.* 2015; Mohammed *et al.* 2014).
Phytotriades auratus	Trinidad only: found only in and beside the water reservoirs (phytotelmata) held by the giant bromeliad *Glomeropitcairnia erectiflora* at high elevations; El Tucuche, Cerro del Aripo, Chaguaramal.
Pseudis paradoxa	Trinidad only: a fully aquatic frog, mainly found at the margins of swamps and larger ponds especially amongst emergent vegetation. Very widespread where such habitat exists, including Nariva, Cedros, and Aripo Savannahs (Auguste *et al.* 2015), as well as the Valsayn area, Penal Rock Road and Pitch Lake (RS-M, unpublished). Tadpoles grow to lengths of up to 20 cm.
Scarthyla vigilans	Trinidad only: most likely a recent arrival. Initial location in the southwest at Cedros, but now widespread (Smith *et al.* 2011; Auguste and Rutherford 2015). A savannah frog, found in the low vegetation at the edges of ponds, swamps, and ditches.
Scinax ruber	Trinidad and Tobago; widespread throughout Trinidad at lower elevations; similarly common in Tobago throughout the mountains of the central spine. Essentially a species of savannahs and forest edges, found in vegetation of low to medium height at the margins of ditches and ponds.
Sphaenorhynchus lacteus	Trinidad only: widespread throughout the island at lower elevations. Found amongst low, emergent vegetation at the edges of swamps and overgrown ponds. Recent records include several locations in south-eastern Trinidad and in the Aripo Savannahs (Auguste *et al.* 2015; Mohammed *et al.* 2014).
Trachycephalus typhonius	Trinidad and Tobago: widely distributed on both islands at lower elevations. Adults presumably inhabit trees in forests, but they are most often seen when large numbers congregate to breed around ponds at the edges of forests.
HEMIPHRACTIDAE	
Flectonotus fitzgeraldi	Trinidad and Tobago: probably widely distributed in primary and secondary forest on both islands, but detailed records are sparse. Recently reported in Trinidad from Edward Trace (south-east) and Aripo Savannahs (Auguste *et al.* 2015; Mohammed *et al.* 2014; Smith *et al.* 2021). Found in and around pools of water held by plants (phytotelmata), such as epiphytic bromeliads, *Heliconia* spp. and *Xanthosoma* spp.

(*Continued*)

Table 12.2. (Continued)

FAMILY/*SPECIES*	DISTRIBUTION
LEPTODACTYLIDAE	
Adenomera hylaedactyla	Trinidad only; originally reported only from the southwest, this frog has recently been recorded in Port of Spain (northwest), the Arena Forest and Freeport (central), in the eastern hills of the Northern Range, and Aripo Savannahs (Auguste 2017), suggesting a much wider distribution. This is a species of savannahs, forests, and the edges of swamps.
Engystomops pustulosus	Trinidad and Tobago: widespread throughout both islands, mainly at lower elevations. A species of forest edges and savannahs. It is often a human commensal, found breeding in such temporary pools as tyre-ruts and roadside ditches, frequently alongside *Leptodactylus fuscus*.
Leptodactylus fuscus	Trinidad and Tobago: widespread throughout both islands, mainly at lower elevations. It inhabits forest edges, savannahs, and often is found in human-modified habitats and breeding in the same pools and ditches as *Engystomops pustulosus*.
Leptodactylus insularum	Trinidad only: there are past records from the central area, but the more recent reports are from Cedros/Icacos in the southwest and Rio Claro in the southeast. An inhabitant of savannahs and the edges of forests where it breeds in larger pools and deep, slow-moving ditches.
Leptodactylus macrosternum	Trinidad only: only recorded from Icacos in the southwest; probably occupies savannahs and the edges of forests, but little is known of its habits, identity, or distribution.
Leptodactylus nesiotus	Trinidad only: recorded only from Icacos Swamp in the southwest and Nariva Swamp in the southeast. Probably a denizen of savannahs, and the edges of swamps, but little is known about its habits or distribution.
Leptodactylus validus	Trinidad and Tobago: widespread on both islands mainly at lower elevations. A species of forests and forest-edges, it often is found around pools in dense thickets, but also in densely vegetated ditches along rural roadsides.
MICROHYLIDAE	
Elachistocleis nigrogularis	Trinidad and adjacent Venezuela: most recorded localities are at lower elevations between the Northern and Central Ranges. Mainly a savannah and forest-edge species, including Aripo Savannahs.
Elachistocleis surinamensis	Trinidad only: with a very similar distribution to that of *Elachistocleis nigrogularis* (Jowers *et al.* 2021a).
PHYLLOMEDUSIDAE	
Phyllomedusa trinitatis	Trinidad only: widely distributed at lower elevations. Recent records include Mayaro (southeast) and Aripo Savannahs (Auguste *et al.* 2015; Mohammed *et al.* 2014). Breeds in trees and bushes overhanging pools and along slow-moving ditches at the edges of forests.
PIPIDAE	
Pipa pipa	Trinidad only: a fully aquatic frog, known mainly from permanent swamps such as Nariva, and occasionally recorded from slow-moving streams. Detailed distribution and habits poorly known.
RANIDAE	
Lithobates palmipes	Trinidad only: all records are from the southeast, including a pool on Edward Trace (White and Ali 2009). Probably a forest and forest-margin species, living close to pools, but its habits and distribution are poorly known.

during rainy and humid days, others are active most nights in the wet season; some are more active in the dry season; the presence of most species can be detected by their calls, but others appear to be silent. On a world map, Trinidad and Tobago look small. However, covering 5,000 km^2 of varied terrain in order to monitor populations of amphibians is no easy task. Moreover, the time and personnel with adequate expertise required for monitoring amphibians are considerable.

In developing a programme for monitoring populations of amphibians, some key questions need to be considered. (1) Should the focus fall only on threatened species, or should all be monitored? Gillespie *et al.* (2011) emphasized the need to keep a check beyond the current threatened list in case previously common species went into sudden and unrecorded decline. (2) Should the main assessment be of the extent of occupancy, which is relatively easily monitored, or should population sizes be estimated, which is a much harder task? (3) Should a set of key areas be chosen for regular monitoring, or is this strategy too likely to miss important trends at wider spatial scales? (4) Should threats and opportunities for amphibians be monitored as well as populations? Threats can include adverse changes in habitat condition, including those brought in by several human activities, pollution, and diseases, particularly chytridiomycosis; opportunities can include restoration and rehabilitation of habitats, such as the creation of farm ponds and reforestation. (5) What are the practicalities of implementing a long-term monitoring programme in terms of resources such as time, personnel, and funds? Dodd (2010) included reviews of these issues, and we give them some consideration over the remainder of this chapter.

As discussed in section VII (Conservation: legislation and practice in Trinidad and Tobago; see below), a strategy for monitoring the amphibian populations of Trinidad and Tobago is some way off, but progress has been made. The authors of this chapter are part of a loose grouping of residents of Trinidad and Tobago, as well as interested visitors, mainly from the United States and the United Kingdom, who have worked with members of the Trinidad and Tobago Field Naturalists' Club, and staff and students of The University of the West Indies' St Augustine campus. Visitors have included students from expeditions of the University of Glasgow, Scotland, and from the College of Wooster, Ohio, USA. This work forms part of a worldwide effort in citizen science (Dickinson 2011), which is supplementing the efforts of governmental agencies and other scientists in valuable ways. Some of this work has been carried out during an annual 'bioblitz' (since 2012 to present), during which all species, including amphibians, located in a particular area are recorded over a period of 24 hours, such as at Nariva swamp, Trinidad, in 2014, and around Charlotteville, Tobago, in 2015.

A summary of the monitoring work conducted to date is shown in Table 12.3. Three topics for conservation research agendas arise so far: (1) Recording the spread of new and invasive species; (2) Assessing the status of some of the species on IUCN Red Lists; (3) Monitoring the occurrence and impact of chytridiomycosis. Of the potentially threatened species, the status of *Pristimantis turpinorum* has yet to receive detailed attention.

A government-sponsored programme, funded by the Global Environmental Facility and the Food and Agricultural Organisation, to assess the biodiversity of six pilot protected areas has been in progress since 2016; the areas are Matura National Park, Caroni Swamp, Nariva Swamp, Trinity Hills, Tobago Main Ridge Forest Reserve, and Northeast Tobago Marine Park. Three amphibian species were chosen as bioindicators: *Mannophryne trinitatis* at Matura, and *Mannophryne olmonae* and *Pristimantis turpinorum* at Tobago's Main Ridge. These species were selected because of their endemic status and their potential vulnerability to threats listed next. Based on the baseline survey, these pilot protected areas appear to represent important areas for the amphibians of Trinidad and Tobago, with more than 70% of the country's species recorded (Auguste 2019).

VI. Threats

A. Exploitation

In contrast to the situation on Dominica and Montserrat where the mountain chicken, *Leptodactylus fallax*, has been severely over-harvested as a local delicacy (Tapley *et al.* 2014), we

Table 12.3. Recent monitoring of Trinidad and Tobago's populations of frogs.

AUTHORS	WORK ACCOMPLISHED
Lehtinen *et al.* (2016)	Three years of surveys of audio and visual encounters of *Mannophryne olmonae* in north-eastern Tobago, accompanied by occupancy-modelling analysis.
Manickchan *et al.* (2011); Downie *et al.* (2017)	Several years of audio surveys monitoring the spread of the invasive alien, *Eleutherodactylus johnstonei*.
Alemu *et al.* (2008, 2013); Shepherd *et al.* (2016); Thomson *et al.* (2018)	Recording the occurrence of chytrid infections in various species, including *Mannophryne olmonae* and *M. trinitatis*.
Jowers and Downie (2004); Greener *et al.* (2017)	Assessment of the range of *Mannophryne trinitatis*, followed by a population assessment, partly based on removal sampling.
Mohammed *et al.* (2014)	Intensive surveys during 2009–14, using visual encounters, audio-recording, and dip-netting in 13 core areas in south-eastern Trinidad.
Smith *et al.* (2011); Auguste and Rutherford (2015); Auguste (2017)	Using visual and audio encounters to map the range and spread of *Scarthyla vigilans* and *Adenomera hylaedactyla*.
Auguste *et al.* (2015); Auguste and Hailey (2018); Auguste (2019)	Surveys (2011–16) using visual and audio encounters to record species richness, distribution, and composition across protected areas in Trinidad and Tobago.
Brozio *et al.* (2017); Torresdal *et al.* (2017)	Use of eDNA to assess distribution and to estimate the population size of the golden treefrog.
Gourevitch and Downie (2018); Smith *et al.* (2019)	Methods of evaluating and monitoring populations of *Phyllomedusa trinitatis*.
Smith *et al.* (2021); Jowers *et al.* (2021b)	Extensive surveys of presence/absence data (2006–19) to establish the distribution and conservation status of *Flectonotus fitzgeraldi* in Trinidad and Tobago.

know of no Trinidadian or Tobagonian frogs being used for food or, indeed, for any other purposes. There have been occasional advertisements on the internet for the sale of live specimens of the golden treefrog, but it is not known whether any collection of specimens for commercial purposes was actually carried out.

B. Invasives

Given the geographic positions of Trinidad and Tobago just off the mainland of South America, and their history of colonial exploitation and international trade, it is not surprising that the current fauna and flora are a mixture of native species and those that have arrived by natural processes, such as migration and rafting, or have been brought in either intentionally or accidentally through human agency. Kenny (1969) considered several of his listed species of frogs to be recent arrivals, especially those restricted to the southeast, close to Venezuela. Kenny (1979) later noted the arrival of *Eleutherodactylus johnstonei*, and Smith *et al.* (2011) documented the presence of *Scarthyla vigilans*, unknown from Trinidad as recently as Murphy's publication in 1997 (it is possible that *S. vigilans* was simply overlooked by previous researchers, but we think this unlikely, given the extensiveness of earlier survey work, especially by Kenny). These two recent arrivals have now spread rapidly across Trinidad, with the former recorded also from Tobago (Downie *et al.* 2017). These species can certainly be classed as invasive aliens, but it is not clear that they pose any threat to native species. *Eleutherodactylus johnstonei* is confined (so far) to human settlements, and there is no evidence that *S. vigilans* interacts with, or damages, other species. Many native predators feed on amphibians (e.g. snakes, spiders, waterbirds), but there is no evidence that any invasive predator poses a threat, although the small Indian mongoose (*Urva auropunctata*), introduced to control snakes in the sugar plantations, may take some frogs. In Australia, the cane toad, introduced there in the 1920s, is regarded as a serious threat to native reptiles and amphibians (Shine 2018), but in Trinidad and Tobago it is native and not known to be a threat to other frogs.

C. Pollution

Farmers in Trinidad and Tobago make heavy use of agrichemicals, but there have been few studies so far published on any effects these may have on Trinidadian and Tobagonian amphibians. Sookoo *et al.*'s (2017) study of the effects of insecticides and herbicides on the larvae of two common species of tadpoles suggests that there is scope for further investigation on this topic. As a public-health measure, aimed at reducing the numbers of mosquitoes and therefore reducing the incidence of mosquito-borne diseases (malaria has been eliminated but yellow fever occurs occasionally and outbreaks of dengue fever are rather common), standing water close to human habitation is frequently sprayed with insecticides; the effects of this practice on any amphibian inhabitants of these bodies of water has not been investigated. A common localized pollutant of soil and water in Trinidad is oil released by the petroleum industry, especially in the south. Whether this has adverse effects on amphibians is not clear. We have seen foam nests of *Engystomops pustulosus* on small pools of water contaminated by oil; an oil-industry water reservoir near Guapo (south-western Trinidad), with visible surface films of oil on the water, has been colonized by several species of frogs in large numbers.

D. Disease

As noted earlier, Alemu *et al.* (2008, 2013) reported the presence of the amphibian chytrid fungus in *Mannophryne olmonae* in Tobago and in *M. trinitatis* in Trinidad, but with no signs of clinical disease in either species. Shepherd *et al.* (2016) and Greener *et al.* (2017) re-surveyed *M. trinitatis* for chytrid disease seven years after the work reported by Alemu *et al.* (2013). They found no evidence of infection, nor did they encounter chytrid in 15 other species sampled at nine localities in the Northern Range and in the southwest. A similar survey in Tobago in 2016 (Thomson *et al.* 2018) also found no chytrid in *M. olmonae* or in several other species. So far, it appears that the frogs of Trinidad and Tobago have not been as badly affected by chytrid disease as many species have been elsewhere. In other countries where chytrid-related declines have occurred, few declines are known from lowland areas; upland populations above 800 m asl seem to be the most vulnerable (Kriger and Hero 2008; Lips *et al.* 2003). This may be because the amphibian chytrid fungus does better at lower temperatures (17–25°C) than at higher ones (Piotrowski *et al.* 2004). In Trinidad and Tobago, the only areas where these conditions are met are in the upland forests of the Northern Range. Since this is also the area where the endangered golden treefrog occurs, monitoring for disease in this species should be undertaken. No other amphibian diseases, such as ranaviruses, have been reported from Trinidad and Tobago.

E. Climatic change

Climatic change has the potential to affect Trinidad and Tobago's frogs in several ways. First, significant rise in sea level will adversely affect coastal freshwater habitats such as Nariva Swamp in the east and Los Blanquizales Lagoon and other smaller swamps in southwestern Trinidad, all of which are populated by many species of frogs. These swamps are all at low elevations, with riverine connections to the sea. Incursion of seawater as a result of sea-level rise would be fatal to most species of frogs, most of which have very low tolerance of saline conditions. The southern coastline of Trinidad was categorized as a 'colonising zone' by Kenny (1995). This is mostly from 'waifs' brought by flushing effects of the South American mainland rivers. During these events, species of herpetofauna can be seen washed up on the Trinidad coastline, most of them freshwater animals carried on large floating mats of water hyacinth. Changes to the salinity of Trinidad's freshwater swamps could alter the balance between the survival and mortality of such incomers compared to the native fauna.

Second, spatiotemporal changes in weather variables such as ambient temperature, cloud cover, humidity, and rainfall may have adverse effects on all amphibians, which mainly require waterbodies with suitable conditions for reproduction and survival. At particular risk is the golden treefrog, whose current niche is restricted to one species of epiphytic bromeliad growing in cloud forest on a small number of mountaintops. Climatic warming would be likely to generate an upland shift in the distribution of these bromeliads, but Trinidad lacks mountains higher than El Tucuche and Cerro del Aripo. Another possible threat is that climatic change could reduce rainfall at these higher elevations, generating prolonged dry spells that would increase the risk of bush fires, already a serious problem in the forests of Trinidad's Northern Range.

Third, we noted earlier that chytrid disease has so far not been a serious problem for the frogs of Trinidad and Tobago. However, it is now known that chytrid's dangers to frogs vary with weather and climate, being worse in cooler, damper conditions. Any climatic change in that direction is likely to increase the risk of disease, as happened to the mountain chicken on Montserrat (Piotrowski *et al.* 2004).

F. Loss or change of habitat

Worldwide, anthropogenic change in habitat tends to be the prime threat to wildlife, including amphibians. As Hailey and Cazabon-Mannette (2011) described, the histories of Trinidad and Tobago make these islands unusual in this respect. The two islands had long, but different, experiences of Western colonial management until they were united as a single administrative unit within the British Empire in 1889. They remained together at independence in 1962. Both islands were fully forested when first encountered by Europeans in 1498. From early colonial days, much of the lowland forest of Tobago was cleared for agriculture, mainly for sugar and cacao plantations. Clearance of forests occurred later and was less complete in Trinidad because its original Spanish colonists were little interested in cash crops, but from the early nineteenth century as a British colony, clearance of lowland forest for the production of sugar increased rapidly, along with cultivation of cacao and coffee, which occurred mainly in the valleys and on hills at lower elevations. The sugar industry closed in Tobago in the 1930s and declined in Trinidad in the 1960s, finally closing in 2010. The result, especially in Trinidad, is that extensive tracts of land, previously sugar plantations, now lie fallow awaiting new uses. In some areas, these lands are being developed for housing, but the population growth of humans in Trinidad and Tobago is negligible, with births being offset by emigration. As a result, there is less pressure for new housing projects than in many comparable countries. In Trinidad, large areas have never been deforested, especially in the Northern Range, much of the Central Range, and in the south. This is true also of Tobago, where the extensive Main Ridge has been protected as a forest reserve since 1776, the earliest such protection in the world. This does not mean that these forests are 'pristine'; they are extensively invaded by alien species such as bamboos, cacao, and coffee. Overall, however, the islands are slightly more than 50% forested, with only 5% of this as commercial plantations (teak, Caribbean pine, citrus, and cacao) and much of the timber is native forest cropped on a long rotation technique. The current rate of deforestation is low (less than 1% per annum) and mainly caused by clearance for quarrying (e.g. sand, gravel, and limestone).

For most amphibian species, a key resource is standing water, since relatively few species utilize streams and rivers. Both islands possess freshwater swamps which expand in the rainy season, thereby providing temporary pools at the fringes, but, prior to European colonization, there were no lakes or larger ponds. A series both of large and small reservoirs has been constructed to provide a water supply for the human population. The colonization of these reservoirs by amphibians has been little studied so far; the presence of predatory fish, and reptiles such as anacondas and caiman, may restrict suitable habitat for frogs therein. However, our unpublished data at a small reservoir near Guapo (southwest) show that several species of frogs have established dense populations among the emergent vegetation at the shallower margins. Another anthropogenic feature, easily seen on a flight across Trinidad, is the large number of small farm-ponds scattered over the agricultural areas of the central region and the south. In Europe, research has shown that newly created ponds can provide excellent habitat for amphibians (Rannap *et al.* 2009). In contrast, no systematic work has been carried out on these ponds as amphibian habitat in Trinidad, but our preliminary work has revealed that many species use them, especially once the banks become well vegetated. Another anthropogenic habitat is the ponds that develop in disused quarries. In Europe, these, too, are often colonized by amphibians. No relevant studies have been published for Trinidad and Tobago, but disused sand and gravel quarries are common in some areas and do accumulate water (JR-D, pers. obs.).

Finally, we consider the urban habitat. Towns and cities can offer suitable habitats for some amphibian species, especially in ponds located in parks and gardens, but also in poorly drained areas. When one of us (JR-D) first visited Trinidad in 1982, the university town of St Augustine had numerous areas of undeveloped land where ponds formed seasonally and drainage ditches often clogged up, thereby providing pools for breeding. It was easy to find at least ten species around the town at that time. Since then, a great deal of tidying up has occurred, especially the construction of more effective drains; the result is good for the local residents – less flooding and fewer mosquitoes – but not so good for the frogs.

This is a section on 'threats', but it is important to remember that while changes in habitat may be a threat to some species, they can also present an opportunity for others. The clearance of the lowland forests of Trinidad and Tobago may have restricted the habitat of some frogs, but must have provided more habitat than previously existed for those like *Leptodactylus fuscus* that utilize grassland and forest fringes. In today's Trinidad, the creation of farm and quarry ponds may be supplying habitat for many species. Pot-holed gravel roads also accumulate water that can persist longer than on surrounding land, and some amphibian species rapidly colonize these pools for breeding (JR-D, pers. obs.).

VII. Conservation: legislation and practice in Trinidad and Tobago

Hailey and Cazabon-Mannette (2011) reviewed legislation about conservation pertinent to amphibians and reptiles in Trinidad and Tobago. Here, we summarize that legislation, focusing on amphibians, and give an update on recent changes (Table 12.4).

Two agencies are responsible for conservation of wildlife in Trinidad and Tobago. The Environmental Management Authority drafts legislation and rules, and checks compliance with international treaties that the country has signed. The Forestry Division regulates the use of natural areas, and its Wildlife Section manages permits for hunting and collecting for scientific (and other) purposes. Although both agencies function across the two islands, the Tobago House of Assembly's Department of Natural Resources and Forestry (DNRF) has considerable local authority.

The legislation currently in force provides little legal protection for amphibians, since the Conservation of Wildlife Act (1980) defined animals under its protection as any mammal, bird, or reptile, thereby excluding amphibians (as well as fish and invertebrates). Nevertheless, in practice, the Wildlife Section (in Trinidad) and the DNRF (in Tobago) issue Special Game Licences to researchers who apply to collect amphibians, and Export Licences if any specimens are to be exported (although none of the amphibians of Trinidad and Tobago are in the Appendices of CITES). The Environmentally Sensitive Species (ESS) Rules prevent the removal or disturbance of listed organisms (except for approved research), or pollution of their habitats or any other activity likely to harm them. The list of species protected in this way was originally very short and included no amphibians, but the golden treefrog was added to the ESS list in 2013.

Table 12.4. National environmental legislation and international treaty obligations relevant to amphibian conservation in Trinidad and Tobago. Modified and updated from Hailey and Cazabon-Mannette (2011).

TYPE OF ENACTMENT	YEAR	PURPOSE
NATIONAL LEGISLATION		
Rules for environmentally sensitive areas	2001	To protect defined areas, including prohibition of removal of, or damage to, any animals in those areas.
Define regulations for environmentally sensitive species	2001	To prohibit the taking, possession, or trade of listed organisms or their parts, or any other activities likely to cause them harm.
Environmental Management Act	2000	To establish the Environmental Management Authority, promote understanding and conservation of the environment, and develop a national environmental policy.
Forests Act	1999	To regulate use of forests and forest produce.
Conservation of Wildlife Act	1980	To regulate hunting and scientific collecting, export of animals, and sanctuaries for game.
Animals (Diseases and Importation) Act	1997	To regulate importation of animals and to require notification of animal diseases.
INTERNATIONAL TREATIES		
Convention on International Trade in Endangered Species of Wild Fauna and Flora (CITES)	1984	To regulate importation and exportation of listed wildlife and wildlife products.
Convention on Wetlands of International Importance (Ramsar)	1993	To promote conservation of wetlands.
Convention on Biological Diversity	1996	To protect biodiversity.

Legislation protecting habitats in Trinidad and Tobago has been weak and ineffective. A large number of Wildlife Sanctuaries and Forest Reserves (including the historical Main Ridge Forest in Tobago) exist on paper, but their status provides little protection in practice. For instance, considerable work to develop a set of National Parks has not been implemented. In 2001, a set of Rules for Environmentally Sensitive Areas (ESA) was established. So far, three such areas have been designated: Matura (2004), Nariva Swamp (2005), and Aripo Savannas (2007). Despite the expectation expressed by Hailey and Cazabon-Mannette (2011), no other ESAs have yet been designated. The three ESAs undoubtedly offer some protection for a range of amphibian species, as exemplified by reports on Aripo Savannah and pilot Protected Areas by Auguste *et al.* (2015), Auguste and Hailey (2018), and Auguste (2019). In 2019, a National Protected Areas Systems Plan was accepted by cabinet; this plan proposed to protect more areas of land and water across Trinidad and Tobago. Some of the areas include the ESAs which hopefully will involve more strict management measures being put in place.

The Government of Trinidad and Tobago consulted widely on a National Wildlife Policy in 2013. This process recognized that current systems were 'outdated, inflexible and largely ineffective in addressing the negative impacts on wildlife resources by a rapidly changing, increasingly sophisticated and modern society' (Draft National Wildlife Policy 2013) and aimed to develop a policy that would 'lead to the effective conservation and management of the country's wildlife resources'. This and other work led to the drafting of a Forestry, Protected Areas and Wildlife Conservation Bill in 2014. A change in government postponed the work on this bill and has since been delayed until further notice. One relevant proposed change is that amphibians will be classed as animals that merit protection.

VIII. Conclusions

The overall status of conservation of the frogs of Trinidad and Tobago is more favourable than in many countries. This is partly because so many of the species in question also have wide distributions on mainland South America. In addition, even the endemic species of Trinidad and Tobago generally are not under immediate threat of extinction. This situation is related to the relatively stable human population exerting little negative pressure on wild lands, and to the country's main income being generated by offshore oil and gas, industries that have minimal impact on terrestrial habitats. Furthermore, the residents of Trinidad and Tobago generally ignore frogs, contrasting with their very unfavorable treatment of snakes, which are often killed when encountered. The main folk attitude to frogs is that they are poisonous; although this is only really correct for the cane toad, the result is that most residents of Trinidad and Tobago avoid handling amphibians.

This general assessment, however, gives no grounds for complacency. Much remains to be found out about the frogs of Trinidad and Tobago, especially the detailed distributions and population sizes of the endemic species. We also need to regularly monitor the impact of climatic change and to check for the occurrence and effects of chytrid disease. The use of molecular tools has contributed significantly to the taxonomy and systematics of some of the islands' amphibians in recent years. For example, genetic identification has been particularly useful for discovering cryptic species, as recently shown for *Rhinella beebei* (Murphy *et al.* 2017), and advances in taxonomic knowledge resulted from similar work on *Mannophryne trinitatis* and *Phytotriades auratus*, the latter classified erroneously at generic level (Manzanilla *et al.* 2007; Jowers *et al.* 2008). The advantages of molecular genetics over traditional morphological classification are many; more so when working on endangered species, since sampling need not involve the killing of specimens (Jowers *et al.* 2008; Torresdal *et al.* 2017; Brozio *et al.* 2017), and can generate results quickly and at a relatively low cost. Moreover, besides its contribution to taxonomy, the use of molecular phylogeny and phylogeography on Trinidad and Tobago's amphibians is leading to an understanding of the evolutionary processes occurring in the area, particularly for the assessment of the processes and patterns of amphibian speciation. For instance, the islands have been a stepping-stone for the colonization of many species from South America to the Caribbean (Camargo *et al.* 2009; Manzanilla *et al.* 2009). The islands' geology, close proximity to northern Venezuela, and the existence of well-dated Pliocene–Pleistocene land-bridge formations to the mainland make all three biogeographical hypotheses for the origins of their taxa plausible: vicariance, land-bridge connections, and over-sea dispersal (Manzanilla *et al.* 2007). In addition, two mountain ranges in Trinidad and one in Tobago have served as interglacial refugia (Jowers *et al.* 2011).

Ongoing research by some of the authors of this chapter attempts to assess the taxonomy of most of the islands' amphibian species. However, the distribution of some species in northern South America makes this task challenging, since it requires sampling from the mainland. Another difficulty relates to the uncertainty of taxonomic identification of sequences deposited in GenBank®, often matching geographically distant species from a single locality and the presence of polyphyletic lineages within cryptic species complexes. At the local scale, studies addressing the structure of amphibian populations throughout Trinidad and Tobago are mostly lacking (Jowers *et al.* 2011), but are pivotal to the assessment of populations' gene-pools, levels of genetic isolation, and the presence of lineages based on multiple colonizations. Phylogeographic, phylogenetic, and population-genetic studies focusing on species throughout distributional ranges (the mainland as well as Trinidad and Tobago) should help discover new species and subspecies, detect the presence of evolutionary lineages, and define the patterns of populations' connectivity. These datasets are critical to evidence-based conservation, monitoring, and management efforts (Smith and Sutherland 2014). International databases such as The Global Biodiversity Information Facility (GBIF) provide access to valuable distributional data.

The local lack of appreciation of a wonderful frog fauna is a matter of regret. We hope that the Government's developing policy on the natural world will include a commitment to education about wildlife, including about amphibians. It is our experience through working with local schools, both in Trinidad and in Tobago, that children are very open to developing positive feelings towards amphibians; it is the adults who convey negative attitudes. There has been growing popularity of online resources such as iNaturalist which combines both academic scientific research and citizen-based science. Consequently, education of the youth is of paramount importance for effective conservation of amphibians in the immediate future.

We hope that the ESA network will be expanded to include more of the areas where amphibians are common and diverse: in Trinidad, the Trinity Hills, and in Tobago, the Main Ridge Forest, would be excellent additions. In managing the ESAs, we hope that the opportunity will be taken to create suitable habitat where none currently exists. Experience in temperate regions shows that the creation of ponds can be highly beneficial. Although hardly explored so far, we feel that there is scope for ecotourism to include frogs; in Trinidad and Tobago, ecotourism currently focuses on birds and marine turtles. Therefore, adding frogs should add to the interest of tourists informed by media coverage of the world's biodiversity.

IX. Acknowledgements

We thank Adrian Hailey and Joanna Smith for comments on the manuscript; many students on expeditions of the University of Glasgow and the College of Wooster for assistance with fieldwork; Mike Rutherford for assistance with logistics and museum specimens at UWIZM; and members of the Trinidad and Tobago Field Naturalists' Club for participation in gathering data on biodiversity. JR-D thanks the University of Glasgow, the Glasgow Natural History Society, and the Carnegie Trust for the Universities of Scotland for help with the costs of fieldwork; similarly, RM-L thanks the College of Wooster and the Andrew Mellon Foundation.

X. References

Alemu, J. B., Cazabon, M. N. E., Dempewolf, L., Hailey, A., Lehtinen, R. M., Mannette, R. P., Naranjit, K. T., and Roach, A. C. J., 2008. Presence of the chytrid fungus *Batrachochytrium dendrobatidis* in populations of the critically endangered frog *Mannophryne olmonae* in Tobago, West Indies. *EcoHealth* 5: 34–39. https://doi.org/10.1007/s10393-008-0154-4

Alemu, J. B., Cazabon-Mannette, M. N. E., Cunningham, A. A., Dempewolf, L., Hailey, A., Mannette, R. P., Naranjit, K. T., Perkins, M. W., and Schmidt-Roach, A. C., 2013. Presence of the chytrid fungus *Batrachochytrium dendrobatidis* in a Vulnerable frog in Trinidad, West Indies. *Endangered Species Research* 20: 131–136. https://doi.org/10.3354/esr00485

AmphibiaWeb, 2020. Online database accessible at http://amphibiaweb.org. Accessed 01/10/2020. University of California, Berkeley, CA, USA.

Angulo, A., 2010. *Mannophryne trinitatis*. The IUCN Red List of Threatened Species. Accessed 22/07/2019.

Auguste, R. J., 2017. An updated distribution of the frog *Adenomera* cf *hylaedactyla* (Anura: Leptodactylidae) on Trinidad, West Indies. *World: Journal of the Trinidad and Tobago Field Naturalists' Club* 2017: 61–63.

Auguste, R. J., 2019. Herpetofaunal checklist for six pilot protected areas in Trinidad and Tobago. *Herpetology Notes* 12: 577–585.

Auguste, R. J. and Hailey, A., 2018. Diversity and species composition of amphibians of the Aripo Savannas Scientific Reserve, Trinidad, West Indies. *Journal of Herpetology* 52: 86–93. https://doi.org/10.1670/17-048

Auguste, R. J. and Rutherford, M. G., 2015. Range expansion of *Scarthyla vigilans* (Anura: Hylidae) in Trinidad, West Indies. *Living World: Journal of the Trinidad and Tobago Field Naturalists' Club* 2015: 43.

Auguste, R. J., Charles, S. P., and Murphy, J. C., 2015. An Updated Checklist of the Amphibians and Reptiles of the Aripo Savannas Scientific Reserve, Trinidad, West Indies. *Living World, Journal of the Trinidad and Tobago Field Naturalists' Club* 2015: 37–43.

Barbour, T., 1916. Amphibians and reptiles from Tobago. *Proceedings of the Biological Society of Washington* 29: 221–224.

Borzée, A., Nguyen, H. Q., and Jowers, M. J., 2020. Description of the advertisement and aggressive calls of the enigmatic Trinidad thin-toed frog *Leptodactylus nesiotus*. *South American Journal of Herpetology* 15: 63–67. https://doi.org/10.2994/SAJH-D-18-00011.1

Brozio, S., Manson, C., Gourevitch, E., Burns, T. J., Greener, M. S., Downie, J. R., and Hoskisson, P. A., 2017. Development and application of an eDNA method to detect the critically endangered golden treefrog (*Phytotriades auratus*) in bromeliad phytotelmata. *PLoS ONE* 12(2): e0170619. https://doi.org/10.1371/journal.pone.0170619

Camargo, A., Heyer, R. W., and de Sá, R. O., 2009. Phylogeography of the frog *Leptodactylus validus* (Amphibia: Anura); patterns and timing of colonization events in the Lesser Antilles. *Molecular Phylogenetics and Evolution* 53: 571–579. https://doi.org/10.1016/j.ympev.2009.07.004

Clarke, F. M., Ward, A. I., and Downie, J. R., 1995. Factors affecting the distribution and status of the Golden Tree Frog, *Phyllodytes auratus*, in Trinidad. *Bulletin of the British Herpetological Society* 54: 3–9.

Cope, E. D., 1879. Eleventh contribution to the herpetology of tropical America. *Proceedings of the Academy of Natural Sciences of Philadelphia* 18: 261–277.

Court, J., 1858. Catalogue of Reptiles. Pp. 440–441 in *Trinidad: Its Geography, Natural Resources, Administration, Present Condition, and Prospects*, edited by L. A. A. G. De Verteuil. Ward and Lock, London.

Court, J., 1884. Catalogue of Reptiles. Pp. 381–382 in *Trinidad: Its Geography, Natural Resources, Administration. Present Condition and Prospects*, 2nd edn, edited by L. A. A. G. De Verteuil. Cassell and Company, London.

Crother, B., 1999. *Caribbean Amphibian and Reptiles*. Academic Press, San Diego, California, USA.

de Sá, R. O., Grant, T., Camargo, A., Heyer, R. W. Ponssa, M. L., and Stanley, E., 2014. Systematics of the neotropical genus *Leptodactylus* Fitzinger, 1826 (Anura; Leptodactylidae): phylogeny, the relevance of non-molecular evidence, and species accounts. *South American Journal of herpetology* 9 (special issue 1): 1–123. https://doi.org/10.2994/SAJH-D-13-00022.1

Dickinson, J. L., Zuckerberg, B., and Bonter, D. N., 2011. Citizen science as an ecological research tool: challenges and benefits. *Annual Review of Ecology, Evolution and Systematics* 41: 149–172. https://doi.org/10.1146/annurev-ecolsys-102209-144636

Dodd, C. K., 2010. *Amphibian ecology and conservation: a handbook of techniques*. Oxford University Press, Oxford, UK.

Downie, J. R., 2013. What common names should we use for Trinidad and Tobago's frogs? *Living World: Journal of the Trinidad and Tobago Field Naturalists' Club* 2013: 32–37.

Downie, J. R., Greener, M. S., Auguste, R. J., and Geerah, P., 2017. Further expansion of the range of the frog *Eleutherodactylus johnstonei* (Amphibia: Eleutherodactylidae) in Trinidad and Tobago. *Living World: Journal of the Trinidad and Tobago Field Naturalists' Club* 2017: 58–60.

Draft National Wildlife Policy, 2013. Government of the Republic of Trinidad and Tobago consultation document, available from the Ministry of Environment and Water Resources.

Duellman, W. E., Marion, A. B., and Hedges, S. B., 2016. Phylogenetics, classification, and biogeography of the treefrogs (Amphibia: Anura: Arboranae). *Zootaxa* 4104: 1–109. https://doi.org/10.11646/zootaxa.4104.1.1

Escalona, M., La Marca, E., Castellanos, M., Fouquet, A., Crawford, A. J., Rojas-Runjaic, F. J. M., Giaretta, A. A., Señaris, J. C., and Castroviejo-Fisher, S., 2021. Integrative taxonomy reveals a new but common Neotropical treefrog, hidden under the name *Boana xerophylla*. *Zootaxa*, 4981(1): 401–448. https://doi.org/10.11646/zootaxa.4981.3.1

Eyre, M., 2013. *Exploring the microhabitat of marsupial frogs: a study of the forces driving selection for* Flectonotus fitzgeraldi *within herbaceous* Xanthosoma jacquini *populations of Tobago*. Undergraduate Thesis, The College of Wooster.

Frost, D. R., 2020. *Amphibian Species of the World: an online reference*, version 6.0 (accessed 01/10/2020). Electronic database accessible at http://research.amnh.org/herpetology/amphibia/index.html. American Museum of Natural History, New York, USA.

Frost, D. R., Grant, T., Faivovich, J., Bain, R. H., Haas, A., Haddad, C. F. B., de Sá, R. O., Channing, A., Wilkinson, M., Donnellan, S. C., Raxworthy, C. J., Campbell, J. A., Blotto, B. L., Moler, P., Drewes, R. C., Nussbaum, R. A., Lynch, J. D., Green, D. M., and Wheeler, W. C., 2006. The amphibian tree of life. *Bulletin of the American Museum of Natural History* 297: 1–370. https://doi.org/10.1206/0003-0090(2006)297[0001:TATOL]2.0.CO;2

Gehara, M., Crawford, A. J., Orrico, V. G., Rodríguez, A., Lötters, S., Fouquet, A., Barrientos, L. S., Brusquetti, F., De la Riva, I., Ernst, R., and Urrutia, G. G., 2014. High levels of diversity uncovered in a widespread nominal taxon: continental phylogeography of the Neotropical tree frog *Dendropsophus minutus*. *PloS ONE* 9(9), p.e103958. https://doi.org/10.1371/journal.pone.0103958

Gillespie, G. R., Scroggie, M. P., Roberts, D., Cogger, H. G., Mahony, M. J., and McDonald, K. R., 2011. The influence of uncertainty on conservation assessments; Australian frogs as a case study. *Biological Conservation* 144: 1516–1525. https://doi.org/10.1016/j.biocon.2010.10.031

Gourevitch, E. H. Z. and Downie, J. R., 2018. An evaluation of tree frog tracking methods using *Phyllomedusa trinitatis* (Anura: Phyllomedusidae). *Phyllomedusa: Journal of Herpetology* 17: 233–246. https://doi.org/10.11606/issn.2316-9079.v17i2p233-246

Greener, M. S., Shepherd, R., Hoskisson, P. A., Asmath, H., and Downie, J. R., 2017. How many Trinidad stream frogs (*Mannophryne trinitatis*) are there, and should they be regarded as vulnerable to extinction? *Herpetological Journal* 27: 5–11.

Hailey, A. and Cazabon-Mannette, M., 2011. Conservation of herpetofauna in the Republic of Trinidad and Tobago. *Conservation of Caribbean Island Herpetofaunas* 1: 183–217. https://doi.org/10.1163/ej.9789004183957.i-228.64

Hardy, J. D., 1982. Biogeography of Tobago, West Indies, with special reference to amphibians and reptiles; a review. *Bulletin of the Maryland Herpetological Society* 18: 37–142.

Hardy, J. D., 1984. Herpetology of Tobago; additions, deletions, and taxonomic changes. *Bulletin of the Maryland Herpetological Society* 20: 12–19.

Hardy, J. D., 2001. A new frog of the genus *Eleutherodactylus* from the island of Tobago, West Indies. *Bulletin of the Maryland Herpetological Society* 37: 24–31.

Hardy, J. D., 2004a. *Phytotriades auratus*. The IUCN Red List of Threatened Species. Accessed 22/07/2019.

Hardy, J. D., 2004b. *Pristimantis turpinorum*. The IUCN Red List of Threatened Species. Accessed 22/07/2019.

Hardy, J. D., 2004c. *Leptodactylus nesiotus*. The IUCN Red List of Threatened Species. Accessed 22/07/2019.

Hardy, J. D., Mijares, A., and La Marca, E., 2004. *Pristimantis urichi*. The IUCN Red List of Threatened Species. Accessed 22/07/2019.

Hedges, S. B., Duellman, W. E., and Heinicke, M. P., 2008. New World direct-developing frogs (Anura: Terrarana): molecular phylogeny, classification, biogeography, and conservation. *Zootaxa* 1737: 1–182. https://doi.org/10.11646/zootaxa.1737.1.1

Hedges, S. B., Powell, R., Henderson, R. W., Hanson, S., and Murphy, J. C., 2019. Definition of the Caribbean Islands biogeographic region, with checklist and recommendations for standardized common names of amphibians and reptiles. *Caribbean Herpetology* 67: 1–53. https://doi.org/10.31611/ch.67

Heyer, W. R., 1994. Variation within the *Leptodactylus podicipinus–wagneri* complex of frogs (Amphibia: Leptodactylidae). *Smithsonian Contributions to Zoology* 546: 1–124. https://doi.org/10.5479/si.00810282.546.i

Heyer, W. R. and Heyer, M. M., 2013. Systematics, distribution, and bibliography of the frog *Leptodactylus insularum* Barbour, 1906 (Amphibia: Leptodactylidae). *Proceedings of the Biological Society of Washington* 126: 204–233. https://doi.org/10.2988/0006-324X-126.3.204

IUCN, 2001. The IUCN Red List of Threatened Species; categories and criteria (version 3.1). www.iucnredlist.org/static/categories_criteria3_1. Accessed 11/3/2017.

IUCN, 2020. The IUCN Red List of Threatened Species. Version 2020-2. <http://www.iucnredlist.org>. Accessed 11/10/2020.

IUCN SSC Amphibian Specialist Group. 2020 a. *Flectonotus fitzgeraldi*. The IUCN Red List of Threatened Species.

IUCN SSC Amphibian Specialist Group. 2020 b. *Hyalinobatrachium orientale tobagoense*. The IUCN Red List of Threatened Species.

IUCN SSC Amphibian Specialist Group. 2020 c. *Leptodactylus nesiotus*. The IUCN Red List of Threatened Species.

IUCN SSC Amphibian Specialist Group. 2020 d. *Mannophryne olmonae*. The IUCN Red List of Threatened Species.

IUCN SSC Amphibian Specialist Group. 2020 e. *Mannophryne trinitatis*. The IUCN Red List of Threatened Species.

IUCN SSC Amphibian Specialist Group. 2020 f. *Phytotriades auratus*. The IUCN Red List of Threatened Species.

IUCN SSC Amphibian Specialist Group. 2020 g. *Pristimantis charlottevillensis*. The IUCN Red List of Threatened Species.

IUCN SSC Amphibian Specialist Group. 2020 h. *Pristimantis turpinorum*. The IUCN Red List of Threatened Species.

IUCN SSC Amphibian Specialist Group. 2020 i. *Pristimantis urichi*. The IUCN Red List of Threatened Species.

Jairam, R. and Fouquet, A., 2018. First records of *Leptodactylus nesiotus* (Anura: Leptodactylidae) for Guyana, Suriname and French Guiana. *Herpetology Notes* 11: 997–999.

Jowers, M. J. and Downie, J. R., 2004. Distribution of the frog *Mannophryne trinitatis* (Anura-Dendrobatidae) in Trinidad, West Indies. *Living World: Journal of the Trinidad and Tobago Field Naturalists' Club* 2004: 17–19.

Jowers, M. J., Downie, J. R., and Cohen, B. L., 2008. The Golden Tree Frog of Trinidad, *Phyllodytes auratus* (Anura: Hylidae): systematic and conservation status. *Studies on Neotropical Fauna and Environment* 43: 181–188. https://doi.org/10.1080/01650520801965490

Jowers, M. J., Lehtinen, R. M., Downie, R. J., Georgiadis, A. P., and Murphy, J. C., 2015. Molecular phylogenetics of the glass frog *Hyalinobatrachium orientale* (Anura: Centrolenidae): evidence for Pliocene connections between mainland Venezuela and the island of Tobago. *Mitochondrial DNA* 26: 613–618. https://doi.org/10.3109/19401736.2014.880888

Jowers, M. J., Martinez-Solano, I., Cohen, B. L., Manzanilla, J., and Downie, J. R., 2011. Genetic differentiation in the Trinidad endemic *Mannophryne trinitatis* (Anura: Aromobatidae); Miocene vicariance, *in situ* diversification and lack of geographical structuring across the island. *Journal of Zoological Systematics and Evolutionary Research* 49: 133–140. https://doi.org/10.1111/j.1439-0469.2011.00615.x

Jowers, M. J., Othman, S. N., Borzée, A., Rivas, G. A., Sánchez-Ramírez, S., Auguste, R. J., Downie, J. R., Read, M. and Murphy, J. C., 2021a. Unraveling unique island colonization events in *Elachistocleis* frogs: phylogeography, cryptic divergence, and taxonomical implications. *Organisms Diversity & Evolution* 21:189–206. https://doi.org/10.1007/s13127-021-00487-y

Jowers, M. J., Sánchez-Ramírez, S., Greener, M. S., Harper, L. R., Auguste, R. J., Marshall, T., Thomson, R., Byrne, I., Loughrey, C. F., Graham, L. and McGhee, W. A., 2021b. Ecology, conservation status, and phylogenetic placement of endemic *Pristimantis* frogs (Anura: Craugastoridae) in Trinidad and Tobago and genetic affinities to northern Venezuela. *Population Ecology* 64: 136–149. https://doi.org/10.1002/1438-390X.12107

Kaiser, H., Sharbel, T. F., and Green, D. M., 1994. Systematics and biogeography of eastern Caribbean *Eleutherodactylus* (Anura: Leptodactylidae): evidence from allozymes. *Amphibia-Reptilia* 15: 375–394. https://doi.org/10.1163/156853894X00416

Kenny, J. S., 1969. Amphibia of Trinidad. *Studies on the Fauna of Curaçao and Other Caribbean Islands* 29: 1–78.

Kenny, J. S., 1977. The Amphibia of Trinidad–an addendum. *Studies on the Fauna of Curaçao and Other Caribbean Islands* 51: 91–95.

Kenny, J. S., 1979. Some recent colonizations. *Living World: Journal of the Trinidad and Tobago Field Naturalists' Club* 1978–1979: 27.

Kenny, J. S., 1995. *Views from the bridge; a memoir on the freshwater fishes of Trinidad*. J.S. Kenny Publications, Trinidad.

Kriger, K. M. and Hero, J. M., 2008. Altitudinal distribution of chytrid (*Batrachochytrium dendrobatidis*) infection in subtropical Australian frogs. *Austral Ecology* 33: 1022–1032. https://doi.org/10.1111/j.1442-9993.2008.01872.x

La Marca, E., Manzanilla, J., and Hardy, J. D., 2004. *Flectonotus fitzgeraldi*. The IUCN Red List of Threatened Species. Accessed 22/07/2019.

Lehtinen, R. M., Calkins, T. L., Novick, A. M., and McQuigg, J. L., 2016. Reassessing the conservation status of an island endemic frog. *Journal of Herpetology* 50: 249–255. https://doi.org/10.1670/14-161

Lips, K. R., Reeve, J. D., and Witters, L. R., 2003. Ecological traits predicting amphibian population declines in Central America. *Conservation Biology* 17: 1078–1088. https://doi.org/10.1046/j.1523-1739.2003.01623.x

Lutz, A., 1927. Notes on batrachians from Venezuela and Trinidad. *Memorias do Instituto Oswaldo Cruz* 20: 35–65. https://doi.org/10.1590/S0074-027619270 00100003

Manickchan, S. A., Starr, C. K., Ramjohn, C., Mahabir, S. V., Mahabir, K., and Mohammed, R. S., 2011. The distribution of the frog *Eleutherodactylus johnstonei* (Amphibia: Eleutherodactylidae) in Trinidad, West Indies. *Living World: Journal of the Trinidad and Tobago Field Naturalists' Club* 2011: 46–49.

Manzanilla, J., Jowers, M. J., La Marca, E., and García-París, M., 2007. Taxonomic reassessment of *Mannophryne trinitatis* (Anura: Dendrobatidae) with a description of a new species from Venezuela. *The Herpetological Journal* 17: 31–42.

Manzanilla, J., La Marca, E., and Garcia-Paris, M., 2009. Phylogenetic patterns of diversification in a clade of neotropical frogs (Anura: Aromobatidae: *Mannophryne*). *Biological Journal of the Linnean Society* 97: 185–199. https://doi.org/10.1111/j.1095-8312.2009 .01074.x

Mertens, R., 1972. Herpetofauna tobagana. *Stuttgarter Beitriige zur Naturkunde* 252: 1–22.

Mohammed, R. S., Manickchan, S. A., Charles. S. P., and Murphy, J. C., 2014. The herpetofauna of southeast Trinidad. *Living World: Journal of the Trinidad and Tobago Field Naturalists' Club* 2014: 12–20.

Mole, R. R. and Urich, F. W., 1894. A preliminary list of the reptiles and batrachians of the island of Trinidad. *Journal of the Trinidad Field Naturalists' Club* 2: 77–90.

Murphy, J. C., 1997. *Amphibians and reptiles of Trinidad and Tobago*. Krieger Publishing, Malabar.

Murphy, J. C. and Downie, J. R., 2012. The changing Trinidad and Tobago herpetofauna. *Living World, Journal of The Trinidad and Tobago Field Naturalists' Club* 2012: 87–95.

Murphy, J. C., Downie, J. R., Sierra, T. A., and Jowers, M. J., 2017. Toads, tall mountains and taxonomy, the *Rhinella granulosa* group (Amphibia, Anura, Bufonidae) on both sides of the Andes. *Salamandra* (in press).

Murphy J. C., Downie, J. R., Smith, J. M., Livingstone, S. R., Mohammed, R. S., Auguste, R. J., Lehtinen, R. M., Eyre, M., Sewlal, J. N., Noriega, N., Casper, G. S., Anton, T., Thomas, R. A., Rutherford, M. G., Braswell, A. L., and Jowers, M. J., 2018. A field guide to the amphibians and reptiles of Trinidad and Tobago. Trinidad and Tobago Field Naturalist's Club. 336 pp.

Narvaes, P. and Rodrigues, M. T., 2009. Taxonomic revision of *Rhinella granulosa* species group (Amphibia, Anura, Bufonidae), with a description of a new species. *Arquivos de Zoologia* 40: 1–73. https://doi.org/10.11606/issn.2176-7793.v40i1p1-73

Orrico, V. G. D., Nunes, I., Mattedi, C., Fouquet, A., Lemos, A. W., Rivera-Correa, M., Lyra, M. L., Loebmann, D., Pimenta, B. V. S., Caramaschi, U., Rodrigues, M. T., and Haddad, C. F. B., 2017. Integrative taxonomy supports the existence of two distinct species within *Hypsiboas crepitans* (Anura: Hylidae). *Salamandra* 53: 99–113.

Parker, H. W., 1933. A list of the frogs and toads of Trinidad. *Tropical Agriculture (Trinidad)* 10: 8–12.

Parker, H. W., 1934. Some frogs and toads of Trinidad. *Tropical Agriculture (Trinidad)* 11: 123.

Pindell, J. and Kennan, L., 2001. Processes and Events in the Terrane Assembly of Trinidad and E. Venezuela. *GCSSEPM Foundation 21st Annual Research Conference Transactions, Petroleum Systems of Deep-Water Basins*: 159–192. https://doi.org/10.5724 /gcs.01.21.0159

Pindell, J. L. and Kennan L., 2009. Tectonic evolution of the Gulf of Mexico, Caribbean and northern South America in the mantle reference frame; an update. *Geological Society, London. Special publications* 328: 1–55. https://doi.org/10.1144/SP328.1

Piotrowski, J. S., Annis, S. L., and Longcore, J. E., 2004. Physiology of *Batrachochytrium dendrobatidis*, a chytrid pathogen of amphibians. *Mycologia* 96: 9–15. https://doi.org/10.1080/15572536.2005.11832990

Ponssa, M. L., Jowers, M. J., and de Sá, R. O., 2010. Osteology, natural history notes, and phylogenetic relationships of the poorly known Caribbean frog *Leptodactylus nesiotus* (Anura, Leptodactylidae). *Zootaxa* 2646: 1–25. https://doi.org/10.11646/zootaxa .2646.1.1

Rannap, R., Lohmus, A., and Briggs, L., 2009. Restoring ponds for amphibians: a success story. *Hydrobiologia* 634: 87–95. https://doi.org/10.1007/s10750-009-9884-8

Rivas, G. A. and Mayke de Freitas, S., 2015. Discovery of the Critically Endangered Golden Tree Frog, *Phytotriades auratus* (Boulenger, 1917), in eastern Venezuela, with comments on its distribution, conservation, and biogeography. *Herpetological Review* 46: 153–157.

Roux, J., 1926. Notes d'herpetologie sud-americaine. 1. Sur une collection de Reptiles et d'Amphibiens de l'ile de la Trinite. *Revue Suisse de Zoologie* 33: 291–299.

Shepherd, R., Hoskisson, P. A., and Downie, J. R., 2016. Apparent absence of chytrid infection in Trinidad's frogs. *Living World: Journal of the Trinidad and Tobago Field Naturalists' Club* 2016: 19–22.

Senaris, C., La Marca, E., and Hardy, J. D., 2004. *Hyalinobatrachium orientale*. The IUCN red List of Threatened Species. Accessed 22/07/2019.

Shepherd, R., Hoskisson, P. A., and Downie, J. R., 2016. Apparent absence of chytrid infection in Trinidad's frogs. *Living World: Journal of the Trinidad and Tobago Field Naturalists' Club* 2016: 19–22.

Shine, R., 2018. The impact of an invasive amphibian: The cane toad, *Rhinella marina*. Pp. 107–124 in *Amphibian Biology* Volume 11. *Status of Conservation and Decline of Amphibians: Eastern Hemisphere*. Part 6. *Australia, New Zealand, and the Pacific Islands*, edited by H. Heatwole and J. Rowley.

Smith, J. M., 2007. *University of Glasgow Exploration Society Trinidad Expedition Report, 2007*. University of Glasgow Exploration Society, Glasgow, Scotland.

Smith, R. K. and Sutherland, W. J., 2014. *Amphibian Conservation: Global evidence for the effects of interventions*. Exeter, Pelagic Publishing.

Smith, J. M., Downie, J. R., Dye, R. F., Ogilvy, V., Thornham, D., Rutherford, M. G., Charles, S. P., and Murphy, J. C., 2011. Amphibia: Anura: Hylidae *Scarthyla vigilans* (Solano 1971): Range extension and new country record for Trinidad, WI with notes on tadpoles, habitat, behaviour and biogeographical significance. *Check List* 7: 574–577. https://doi.org/10.15560/7.5.574

Smith, J. M., Bland, A., Gourevitch, E., Hoskisson, P. A., and Downie, J. R., 2019. Stable individual variation in ventral spotting patterns in *Phyllomedusa trinitatis* (Anura: Phyllomedusidae) and other *Phyllomedusa* species: a minimally invasive method for recognizing individuals. *Phyllomedusa: Journal of Herpetology* 18: 13–26. https://doi.org/10.11606/issn.2316-9079.v18i1p13-26

Smith, J. M., Jowers, M. J., Auguste, R. J., Hoskisson, P., Beyts, C., Muir, G., Greener, M. S., Thornham, D., Byrne, I., Lehtinen, R. M., Eyre, M., Rutherford, M. G., Murphy, J. C., de Freitas, M., Rivas, G. A., and Downie, J. R. 2021. The distribution and conservation status of the dwarf marsupial frog (*Flectonotus fitzgeraldi*) in Trinidad, Tobago and Venezuela. *Amphibian and Reptile Conservation*15(1) [General Section]: 36–56 (e273). .

Sookoo, N., Hailey, A., and Mohammed, A., 2017. Toxicity of six commercial pesticide formulations to larvae of two tropical frogs, *Rhinella (Bufo) marina* (Bufonidae) and *Engystomops (Physalaemus) pustulosus* (Leptodactylidae). *Journal of Aquatic Pollution and Toxicology* 1: 2–8.

Stuart, S. N., Chanson, J. S., Cox, N. A., Young, B. E., Rodrigues, A. S., Fischman, D. L., and Waller, R. W., 2004. Status and trends of amphibian declines and extinctions worldwide. *Science* 306: 1783–1786. https://doi.org/10.1126/science.1103538

Tapley, B., Harding, L., Sulton, M., Durand, S., Burton, M., Spences, J., Thomas, R., Douglas, T., Andre, J., Winston, R., George, M., Gaworek-Michalczenia, M., Hudson, M., Blackman, A., Dale, J., and Cunningham, A. A., 2014. An overview of current efforts to conserve the critically endangered mountain chicken (*Leptodactylus fallax*) on Dominica. *Herpetological Bulletin* 128: 9–11.

Thomson, R, Hoskisson, P. A., Brozio, S., and Downie, J. R., 2018. Apparent lack of chytrid infection in northeast Tobago's frogs. *Living World: Journal of the Trinidad and Tobago Field Naturalists' Club* 2018: 82–86.

Torresdal, J. D., Farrell, A. D., and Goldberg, C. S., 2017. Environmental DNA detection in the golden tree frog (*Phytotriades auratus*) in bromeliads. *PLoS ONE* 12: e0168787. https://doi.org/10.1371/journal.pone.0168787

White, G., 2013. Range expansion of the exotic *Eleutherodactylus johnstonei* (Anura: Eleutherodactylidae) to Tobago, Trinidad and Tobago. *Living World: Journal of the Trinidad and Tobago Field Naturalists' Club* 2013: 75.

White, G. and Ali, S., 2009. The status of *Lithobates palmipes* (Spix) in Trinidad and Tobago. *Living World: Journal of The Trinidad and Tobago Field Naturalists' Club* 2009: 44–45.

Yuan, Z. Y., Zhou, W. W., Chen, X., Poyarkov, N. A., Chen, H. M., Jang-Liaw, N. H., Chou, W. H., Matzke, N. J., Iizuka, K., Min, M. S., and Kuzmin, S. L., 2016. Spatiotemporal Diversification of the True Frogs (Genus *Rana*): A Historical Framework for a Widely Studied Group of Model Organisms. *Systematic Biology* 65: 824–842. https://doi.org/10.1093/sysbio/syw055

13 Amphibians of the Bocas del Toro Archipelago, Republic of Panama

Sandra P. Galeano, Layla Freeborn, and Ricardo Cossio

Abbreviations and acronyms

Bd	Batrachochytrium dendrobatidis
CATIE	*Centro Agronómico Tropical de Investigación y Enseñanza*
CITES	*Convention on International Trade in Endangered Species of Wild Fauna and Flora*
CE	*Critically Endangered*
DD	*Data Deficient*
EPYPSA	*Estudios, Proyectos y Planificación Sociedad Anónima*
ESP	*Environmental Science Program*
ITEC	*Institute for Tropical Ecology and Conservation*
IUCN	*International Union for Conservation of Nature*
LC	*Least Concern*
MiAmbiente	*The Ministry of Environment of Panama*
NT	*Near Threatened*
NL	*National Law*
PNUD	*Programa de las Naciones Unidas para el Desarrollo*
SD	*standard deviation*
STRI	*Smithsonian Tropical Research Institute*
VU	*Vulnerable*
ya	*years ago*

I. Introduction

A. Geography

The Bocas del Toro Archipelago is a group of continental islands located in the northwestern Bocas del Toro Province, Panama. It consists of seven main islands, numerous smaller islands, and dozens of mangrove-covered islets. Together, these islands partially delimit two coastal lagoons, Bahía Almirante and Chiriquí Lagoon, from the Caribbean Sea. The largest island, Isla Colón (59 km^2) (Anderson and Handley 2002), lies < 2 km from the adjacent mainland and < 30 km from the Panama–Costa Rica border. In addition to Isla Colón, the relatively large islands in the archipelago include Isla Popa (53 km^2), Isla Bastimentos (52 km^2), Isla San Cristobal (37 km^2),

Sandra P. Galeano, Layla Freeborn, and Ricardo Cossio, 'Chapter 13: Amphibians of the Bocas del Toro Archipelago, Republic of Panama' in: *The Conservation and Biogeography of Amphibians in the Caribbean*. Pelagic Publishing (2023). © Sandra P. Galeano *et al.* DOI: 10.53061/JDCB6361

Figure 13.1. Map of the Bocas del Toro Archipelago (top) and its location in northwestern Panama (within the Bocas del Toro Province) as indicated by the dashed box in the outline of the country (bottom).

Cayo de Agua (16 km^2), Isla Solarte (also referenced in the literature as Cayo Nancy; 7 km^2), Isla Pastores (~2.7 km^2), and Isla Loma Partida (~3.3 km^2). A map depicting the major geographic features of the region is shown in Figure 13.1.

B. Formation and biogeography of the islands

One of the most defining features of the Bocas del Toro Archipelago is its recency. Formation of the islands began approximately 8,900 ya and reached the current configuration during the past 1,000 years (Anderson and Handley 2002). The first island to separate from the mainland was Isla Escudo de Veraguas, but having separated from the southeastern Peninsula Valiente it is unrelated to other islands in the archipelago and will not be considered herein. The remaining islands formed sequentially as rising sea levels separated previously connected landmasses from the Península Tierra Oscura. This process began with the formation of Boca del Drago pass (~6,200 ya) and the separation of Isla Colón from the then J-shaped peninsula (~5,200 ya); it ended during the past 1,000 years with the separation of Isla San Cristobal and Isla Popa from the mainland and the splitting of Isla Solarte from Isla Bastimentos (Anderson and Handley 2002).

Areas of land that are part of a continental shelf and that become partially submerged because of changes in sea level, result in islands referred to as 'land-bridge' or 'continental' – like those of the Bocas del Toro Archipelago. These islands contrast with islands formed by volcanic activity and that lack any previous connection to nearby continents (i.e. 'oceanic' islands). Consequently, both types of islands differ in the predicted assemblage of species, mainly because different processes influence them. For example, the process of extinction dominates species richness on continental islands, but processes related to the rate of immigration and extinction – arriving at some equilibrium in the number of species at a given time – dominate species richness on oceanic ones (Lawlor 1982; Patterson and Atmar 1986; Anderson and Handley 2002).

Biogeographic studies of amphibians from the Bocas del Toro Archipelago mostly have focused on assessing the role of geographic isolation on the evolution of polymorphic colouration in the strawberry poison frog *Oophaga pumilio* (summarized by Gehara *et al.* 2013). Studies that included species of amphibians from this archipelago and from a broader region (e.g. isthmian Central America) have reported the presence of a filter-barrier, termed the 'Bocas break'. This break is located within the area that roughly corresponds to the Panama–Costa Rica border (Crawford *et al.* 2007; Wang *et al.* 2008). Even though there is no barrier recognized formally in this region, this break reduces the probability of dispersal between localities on either side to the border (Wang *et al.* 2008) and results in deep intraspecific-genetic divergence (Crawford *et al.* 2007; Wang *et al.* 2008; Wang and Shaffer 2008). Therefore, it is possible that the Bocas break influenced the genetic structure of species in the region prior to the origin of the Bocas del Toros islands; once formed, the genetic structure could have changed again in these islands. Consequently, future studies could determine the relative contribution of this break, and that of isolation between islands, to present-day genetic structuring of populations of amphibians.

C. Climate

The region that includes Bocas del Toro Province is characterized by tropical rainforest that receives high rainfall without clearly defined dry or wet seasons. Recording of meteorological data for the Province began in 1926 from Changuinola (a continental city located 8 km inland in northwestern Bocas del Toro Province), and the data were collected by the Chiriquí Land Company until 2004. The banana industry, established in the region at the end of the nineteenth century, used rainfall data to predict crop yields (Clyde Stephens, pers. comm., referenced by Kaufmann and Thompson 2005). Recording of rainfall data from northernmost Bocas del Toro Archipelago, began in 1972 from the Bocas del Toro 'Isla Colón' International Airport by the Instituto de Recursos Hidráulicos y Electrificación, and most recently as part of the Smithsonian Tropical Research Institute (STRI) Environmental Science Program (ESP) and Physical Monitoring Program. These programs include (but are not limited to) collection of data using a meteorological tower situated on a floating platform near the Smithsonian station's pier on Isla Colón.

In this region, mean annual rainfall based on airport records from 1973 to 2003 was 3,277 mm (SD = 461 mm; Kaufmann and Thompson 2005), while that from 2003 to 2016 was 3,540 mm (SD = 557 mm) based on data from the STRI. Rainfall is consistently high throughout the year and lacks a pronounced dry period, but in general there is an intra-annual rainfall pattern shared by other Caribbean coastal locations. Specifically, there is a period of relatively higher rainfall and another of lower rainfall; the maximum rainfall occurs in July and the minimum in March (Alfaro 2002; Kaufmann and Thompson 2005).

The annual pattern of rainfall at the Bocas del Toro Archipelago, however, is slightly different from that of the wider region because there are two periods of higher (November/December and July) and of lower (February/March and September/October) rainfall. On the other hand, tipping-bucket records from 2002 to 2004 reveal that 48% of measurable rain comes from approximately 1% of the daytime hours, with most of it occurring at night (e.g. Kaufmann and Thompson 2005). Average relative humidity, recorded since 2010, ranges from 85.5% in January to 89.4% in July.

The monthly average of daily maximum and minimum air temperatures since 2003 ranges from 25.3°C in January to 27.2°C in June and September (STRI). These data, as well as that for solar radiation, sea-surface temperature, and wind speed, show seasonal patterns that are described in detail in other publications (e.g. Kaufmann and Thompson 2005).

D. Types of habitats

A high diversity of terrestrial ecosystems characterizes the Bocas del Toro Archipelago. For example, mangrove forests line the borders between terrestrial habitats and those of the marine environment, while lowland humid tropical forests are present on most islands as primary and secondary forests. Moreover, highly disturbed habitats, such as pastures and abandoned cacao plantations, are found on every island, whereas limestone caves, marshes, and a few streams provide rich environments for many species of plants and animals (Peter Lahanas, pers. comm., ITEC).

Forests inland consist of vegetation that is dominated by evergreen tropical rainforest species (Soto *et al.* 1998). In primary forests, for example, emergent species such as kapok (*Ceiba pentandra*) and prioria (*Prioria* spp.) reach heights of up to 60 m, whereas canopy species such as nutmegs (*Virola* spp.) and figs (*Ficus* spp.) reach heights between 20 m and 40 m (ITEC; Valdespino and Santamaría 1997). The understorey is abundant with ferns, palms, cycads, and aroids. Lianas and a variety of epiphytes, such as bromeliads and orchids, penetrate all subcanopy layers of the forest, from the emergent level to the forest floor (ITEC).

Secondary forests are structurally different from primary forests and generally lack the emergent layer. For instance, secondary forests are abundant with herbaceous species including heliconia (*Heliconia* spp.), pepper (*Piper* spp.), and the Panama hat palm (*Carludovica palmata*). The aroids *Xanthosoma violaceum* (Brust 1993) and *Dieffenbachia* spp. (McKone *et al.* 2014), bromeliads (Duellman and Trueb 1986), and *Heliconia* spp. (Pröhl 2002) – all represented both in primary and secondary forests – provide poison frogs with sites for depositing and rearing their tadpoles. In Bocas del Toro, however, published records of plant–amphibian associations are limited to the dendrobatid *Oophaga pumilio*, which deposits tadpoles in the axils of leaves of *X. violaceum* (Brust 1993).

One protected area, the Bastimentos Island National Marine Park, was established in 1988 to preserve the region's habitat diversity; this area consists of 13.2 ha of terrestrial and marine environments along Isla Bastimentos and the nearby Zapatillas Keys. The designation of this area as a national park provides protection to the habitats of at least 21 amphibian species (see Table 13.1).

II. The species of amphibians of the archipelago

The islands of the Bocas del Toro Archipelago have retained many of the species of amphibians from the adjacent mainland, a feature common to continental islands. At least 37 species of amphibians, all natives, have been reported from the archipelago (Table 13.1 lists the distribution and conservation status of each species). Thirty-one of these species are anurans representing 11 families, whereas four species are salamanders representing one family, and two species are caecilians belonging to one family. Most of the species of anurans are members of three families: Dendrobatidae (7), Strabomantidae (6), and Hylidae (5).

Table 13.1. Amphibian species (under **Order**: **An** = Anura; **Cau** = Caudata; **Gymn** = Gymnophiona) found in the Bocas del Toro Archipelago, Panama. Presence of a species is indicated by +. The conservation status for each species according to the National Law (**NL**), International Union for the Conservation of Nature (**IUCN**), and the Convention on International Trade in Endangered Species of Wild Fauna and Flora (**CITES**) are designated as Critically Endangered (**CE**), Near Threatened (**NT**), Least Concern (**LC**), or Vulnerable (**VU**). Under **CITES**, NT = not listed. References are coded numbers according to the key at the bottom of the table. Asterisks (*) represent species reported only by one reference.

ORDER	FAMILY	SPECIES	ISLAND								CONSERVATION STATUS			REFERENCES
			BASTIMENTOS	CAYO DE AGUA	COLÓN	LOMA PARTIDA	PASTORES	POPA	SAN CRISTOBAL	SOLARTE	NL	IUCN	CITES	
An	Aromobatidae	*Allobates talamancae*	+	+	+	+	+	+		+	LC	LC	–	1, 2, 3, 4, 5
	Bufonidae	*Incilius coniferus*	+	+	+			+	+	+	LC	LC	–	1, 3, 4, 5, 6
		Rhinella horribilis	+	+	+					+	LC	LC	–	4, 5, 6, 7
	Centrolenidae	*Teratohyla spinosa*			+						LC	LC	–	6, 8
	Craugastoridae	*Craugastor polyptychus*	+	+	+	+		+	+	+	LC	LC	–	1, 3, 4, 5, 9
		**Craugastor crassidigitus*		+				+			LC	LC	–	3
		Craugastor megacephalus	+		+			+			LC	LC	–	4, 5
		Craugastor talamancae	+	+	+	+	+	+	+	+	LC	LC	–	1, 3, 4
	Strabomantidae	*Pristimantis altae*						+			VU	LC	–	1, 4
		**Pristimantis caryophyllaceus*			+						LC	LC	–	6
		Pristimantis cerasinus	+		+	+		+		+	LC	LC	–	1, 3, 4, 5, 6, 10
		**Pristimantis cruentus*						+			LC	LC	–	4
		Pristimantis ridens	+		+						LC	LC	–	5, 11
		**Strabomantis bufoniformis*			+						CR	EN	–	4
	Dendrobatidae	*Andinobates claudiae*	+	+	+			+			CR	DD	II	3, 4, 5,12, 14
		Colostethus pratti			+						VU	LC	–	4, 13
		Dendrobates auratus			+	+	+				VU	LC	II	4, 5, 6, 7
		Oophaga pumilio	+	+	+	+	+	+	+	+	VU	LC	II	4,15
		Phyllobates lugubris	+	+	+	+	+	+	+	+	VU	LC	II	1, 4, 5, 12, 14

(*Continued*)

Table 13.1. (Continued)

ORDER	FAMILY	SPECIES	ISLAND								CONSERVATION STATUS			REFERENCES
			BASTIMENTOS	CAYO DE AGUA	COLÓN	LOMA PARTIDA	PASTORES	POPA	SAN CRISTOBAL	SOLARTE	NL	IUCN	CITES	
		Silverstoneia flotator	+		+			+			LC	LC	–	3, 4, 5
		**Silverstoneia nubicola*			+						VU	VU	–	6
	Eleutherodactylidae	*Diasporus diastema*	+	+	+			+	+	+	LC	LC	–	1, 3, 4, 5
	Hylidae	*Agalychnis callidryas*	+	+	+			+	+	+	LC	LC	II	1, 4, 5, 6
		Boana rufitela		+	+			+			LC	LC	–	1, 4, 5
		Dendropsophus ebraccatus	+		+						LC	LC	–	1, 4, 5, 6
		Dendropsophus phlebodes	+	+				+	+	+	LC	LC	–	1, 4, 5
		Scinax elaeochroa	+		+						LC	LC	–	1, 4, 5
		Smilisca phaeota						+	+		LC	LC	–	1, 4
	Leptodactylidae	*Leptodactylus melanonotus*	+	+	+			+			LC	LC	–	4, 5, 10
		Leptodactylus savagei	+	+	+			+	+	+	LC	LC	–	1, 4, 5
	Ranidae	*Lithobates vaillanti*	+	+	+			+	+	+	LC	LC	–	1, 4, 5
Cau	Plethodontidae	*Bolitoglossa biseriata*			+			+			LC	LC	–	1, 4
		Bolitoglossa colonnea	+		+			+			LC	LC	–	4, 16
		Bolitoglossa schizodactyla		+	+			+	+		LC	LC	–	4, 5
		**Oedipina cyclocauda*							+		DD	NT	–	4
Gymn	Dermophiidae	**Dermophis parviceps*						+			DD	LC	–	4
		Gymnopis multiplicata	+		+				+		DD	LC	–	4, 5, 17, 18
Total species number			**22**	**17**	**30**	**7**	**5**	**25**	**14**	**13**				

1 Tejera and Dupuy (2003); **2** Hopkins and Lahanas (2011); **3** Galeano (2016); **4** VertNet, accessed August 2017; **5** Institute for Tropical Ecology and Conservation, accessed August 2017; **6** Valdés (2014); **7** Layla Freeborn (pers. obs.); **8** Ricardo Cossio (pers. obs.); **9** Crawford (2003); **10** César Jaramillo (pers. comm.); **11** Wang *et al.* (2008); **12** Summers *et al.* (1997); **13** Savage (1968); **14** Galeano and Harms (2016); **15** Daly and Myers (1967); **16** Hertz *et al.* (2013); **17** Savage and Wake (2001); **18** IUCN, accessed August 2017.

The vast majority of the amphibians found throughout the archipelago are terrestrial species that live either on the forest floor in the leaf litter or near swamps. Only a few stream-associated breeders (e.g. *Allobates talamancae, Teratohyla spinosa, Silverstoneia flotator*) have been documented from the islands, which also have few streams. Species richness of amphibians is higher on the largest islands (Isla Colón, Isla Popa, and Isla Bastimentos), while it is lower on the smallest islands (Isla Pastores and Loma Partida) (Table 13.1; see section I. A).

One of the anurans of the archipelago, the poison frog, *Oophaga pumilio*, exhibits remarkable variation in skin colouration across islands (Daly and Myers 1967). Individuals of this species exhibit bright colours such as reds, oranges, and yellows, or dull colours such as green, despite being relatively monomorphic (i.e. red body with blue legs) throughout Costa Rica and Nicaragua (Wang and Shaffer 2008; Maan and Cummings 2012). In Bocas del Toro, populations of *O. pumilio* consist mostly of a single colour morph, although a polymorphic population in the northeastern tip of Isla Bastimentos consists of red or yellow individuals (Richards-Zawacki *et al.* 2012). Two polymorphic populations also have been described from the mainland adjacent to the archipelago (at Uyama river [Wang and Shaffer 2008] and at Dolphin Bay [Dugas *et al.* 2015]).

Another notable anuran from the archipelago is *Andinobates claudiae*, the world's smallest poison frog known. It is endemic to the Bocas del Toro province (Jungfer *et al.* 2000). Specifically, this species is found on four islands across the archipelago (Table 13.1) and at least one site on the adjacent mainland (Galeano 2016). Individuals are frequently encountered in the leaf litter in areas where *Heliconia* plants are abundant; breeding adults use these for the deposition of tadpoles (Cossio *et al.* in review). Alteration of habitat is likely to result in a reduced abundance of *A. claudiae*, especially when essential breeding sites such as *Heliconia* spp. are affected by deforestation (R. Cossio, pers. obs.), which requires additional studies on the ecology and conservation of *A. claudiae*.

III. Potential threats to amphibians

The alteration of habitats and their contamination by wastes from human sources are the most obvious threats to amphibian populations across the archipelago. Historically, anthropogenic activities such as urban development and agriculture have contributed significantly to the loss of habitat. Since the mid-2000s, however, the archipelago also has experienced a high rate of immigration and development for tourism (Spalding 2013a), resulting in the rapid development of related infrastructures such as luxury resorts, restaurants, and residences. Unfortunately, such development often has occurred without proper planning (PNUD 2009), thus threatening the extent and quality of areas suitable for amphibians.

The influx of new residents, notably of those expatriated, has increased the demand for inland-forested areas. However, the general impact of this influx on forests, which is the vegetation type with the most extensive coverage in the archipelago, varies among islands (Spalding 2013b). For example, forest cover in the political subdivision that includes Isla Popa and Cayo de Agua decreased 11.2% over a 22-year period (e.g. from 5,307 ha in 1986 to 4,711 ha in 2008). In contrast, forest cover increased 20.5% in the subdivision that includes Isla Bastimentos and Isla Solarte (e.g. from 3,853 ha to 4,848 ha) and 12.6% in the subdivision that includes Isla Colón (e.g. from 3,243 ha to 3,709 ha) during the same period. The dynamic of the coverage of forested areas might have resulted from land purchased for private homes or touristic development, on the one hand, and by the abandonment of lands that resulted in regeneration of forests and/or reforestation, on the other (Spaulding 2013b). However, if the reduction of forest continues at an accelerated rate in some areas, such as in the subdivision that includes Isla Popa and Cayo de Agua, the composition and the structure of communities of amphibians will likely undergo dramatic changes. The impacts of changes in forest cover on the amphibian fauna on the islands are still not known and must be examined in the imminent future for conservation purposes, because several species of amphibians reported from this area strongly depend on these forests (e.g. *Craugastor talamancae, Incilius coniferus,* and *Pristimantis altae*).

Types of vegetative cover that are associated with agriculture for subsistence and as pasture for cattle ranching increased from 14% in 1986 to 90% in 2008 across the archipelago (Spalding 2013b). The cultivation of rice, corn, yucca, cacao, and bananas at small scales have long modified native habitats (EPYPSA 2008). More recently, however, cattle ranching has destroyed natural habitats, mainly through the conversion of forests into pastures, increased erosion and compaction of soils, and contamination of land by the dumping of waste-water from ranches (see next). Unfortunately, the islands most affected by these kinds of modification (e.g. Isla Colón, Isla Bastimentos, and Isla Popa; see Spalding 2013b) also have the highest diversity of species of amphibians (see section II). Consequently, future research needs to address the effects that changes in land-use have on population size, abundance, species richness, and species composition of amphibians across the islands. Of most importance for conservation are those studies focusing on species that depend on well-preserved forests (i.e. *C. talamancae, I. coniferus, P. altae,* and *Strabomantis bufoniformis*) because these species could be disproportionately affected by the modification of habitats. Equally needed in the context of land-use are those studies focused on endemic species such as *A. claudiae* and other amphibians deemed threatened within the archipelago (refer to Table 13.1).

Contamination of water due to insufficient waste-management may represent another threat to Bocas del Toro's amphibians, especially for the larvae and adults of species that breed in aquatic systems. Currently, infrastructure for dealing with trash and sanitation across the archipelago is lacking, leading to littering and the overflow of wastes into the ocean (Bourque 2016). These problems have increased during recent years due to poorly planned tourism (Camargo *et al.* 2013).

Some species of amphibians, such as poison frogs (Dendrobatidae), are threatened by global trade. There is a high demand for these species within the European Union and North America. Auliya *et al.* (2016) placed Panama among the top five exporters of dendrobatids, and Carpenter *et al.* (2014) included *O. pumilio* and *Dendrobates auratus,* species that occur at the Bocas del Toro Archipelago, among the 10 most-traded species of amphibians worldwide. In fact, between 1976 and 2007, the pet-trading business marketed at least 33,429 individuals of *O. pumilio* and 61,839 individuals of *D. auratus* (Carpenter *et al.* 2014). As a result, the unsustainable capture of wild amphibians for the pet trade has the potential to cause local declines of populations (Schlaepfer *et al.* 2005; Nijman and Shepherd 2010). However, there have been no official reports of the extraction and trafficking of amphibians from the Bocas del Toro Archipelago.

Another threat to amphibians comes from the chytrid fungus, *Batrachochytrium dendrobatidis* (*Bd*), which causes chytridiomycosis. Several declines of populations of amphibians throughout Panama are attributed to this disease (Lips *et al.* 2006; Crawford *et al.* 2010). However, this pathogen was not detected in populations of *O. pumilio* on Isla Bastimentos (Richards *et al.* 2008). Consequently, additional research on the distribution of *Bd* across Bocas del Toro, including data from multiple populations and species, is needed to determine whether this pathogen is a threat there.

IV. Conservation status of amphibians

Several species of amphibians present in the Bocas del Toro Archipelago are included on the IUCN's Red List (2017) and have CITES status (see Table 13.1). *Oedipina cyclocauda* is the only species classified as Near Threatened (NT) by IUCN. Five other species listed in appendix II of CITES include *Agalychnis callidryas, Andinobates claudiae, Dendrobates auratus, Oophaga pumilio,* and *Phyllobates lugubris.* This appendix also includes species that, although not currently threatened by collection for the pet trade, may become so in the imminent future.

Eight species of amphibians found across the archipelago are considered nationally threatened (MiAmbiente 2016), two species are deemed Critically Endangered (CE) and six species are considered Vulnerable (VU). Initially, in order to assign a category of threat, the Ministry of Environment of Panama considers each species' current IUCN category, population status, endemism, distribution within the country, and whether the species has been identified as a priority for conservation according to ongoing *ex-situ* conservation programs in Panama (Gagliardo *et al.* 2008; MiAmbiente 2011). Based on the Ministry's criteria, *Strabomantis bufoniformis* is categorized as CE due to population declines caused by the chytrid fungus in other areas of its distribution (Lips *et al.* 2006; Crawford *et al.* 2010). Gratwicke *et al.* (2016) showed that *S. bufoniformis* is highly susceptible to chytridiomycosis, but can probably avoid extinction through captive breeding. *Andinobates claudiae,* the only species endemic to the region, also is categorized as CE due to its small distributional range.

Six species of anurans are categorized as VU: *Pristimantis altae, O. pumilio, D. auratus, P. lugubris, Colostethus pratti,* and *S. nubicola.* Of these, the poison frogs (*O. pumilio, D. auratus,* and *P. lugubris*) are highly sought after as pets due to their attractive variations and patterning (see section III above). *Pristimantis altae, C. pratti,* and *S. nubicola* are listed as VU because of reductions in their populations by *Bd* in other areas of their distribution (Lips *et al.* 2003, 2006; Crawford *et al.* 2010), although some individuals can still be found in *Bd*-positive areas or sites that have not experienced severe chytridiomycosis (Gratwicke *et al.* 2016). The remaining species found across the Bocas del Toro Archipelago (Table 13.1) are designated as Least Concern (LC) or Data Deficient (DD) because their population sizes are stable or because there is not enough information to provide an evaluation, respectively.

IV. References

Alfaro, E. J., 2002. Some characteristics of the precipitation annual cycle in Central America and their relationships with its surrounding tropical oceans. *Tópicos Meteorológicos y Oceanográficos* 9: 88–103.

Anderson, R. P. and Handley, C. O., Jr., 2002. Dwarfism in insular sloths: biogeography, selection, and evolutionary rate. *Evolution* 56: 1045–1058. https://doi.org/10.1111/j.0014-3820.2002.tb01415.x

Auliya, M., García Moreno, J., Schmidt, B. R., Schmeller, D. S., Hoogmoed, M. S., Fisher, M. C., Pasmans, F., Henle, K., Bickford, D., and Martel, A., 2016. The global amphibian trade flows through Europe: the need for enforcing and improving legislation. *Biodiversity and Conservation* 25: 2581–2595. https://doi.org/10.1007/s10531-016-1193-8

Bourque, O. R., 2016. Analyzing the recent, rapid tourism development in Panama's Bocas del Toro Archipelago: is socioenvironmental justice attainable? Master's thesis, Department of Environmental Science and Policy, Clark University, Worcester, Massachusetts.

Brust, D. G., 1993. Maternal brood care by *Dendrobates pumilio*: a frog that feeds its young. *Journal of Herpetology* 27: 96–98. https://doi.org/10.2307/1564914

Camargo, M., Díaz, A. A., and Imbach, A. C., 2013. Artículo I. Percepción de la comunidad del Archipiélago de Bocas del Toro, Panamá sobre el impacto de la industria turística y la variabilidad climática. Tesis de Maestría, CATIE, Costa Rica.

Carpenter, A. I., Andreone, F., Moore, R. D., and Griffiths, R. A., 2014. A review of the international trade in amphibians: the types, levels and dynamics of trade in CITES-listed species. *Oryx* 48: 565–574. https://doi.org/10.1017/S0030605312001627

CITES. The CITES Species, https://cites.org/eng/disc/species.php/, accessed August 2017.

Crawford, A. J., 2003. Huge populations and old species of Costa Rican and Panamanian dirt frogs inferred from mitochondrial and nuclear gene sequences. *Molecular Ecology* 12: 2525–2540. https://doi.org/10.1046/j.1365-294X.2003.01910.x

Crawford, A. J., Bermingham, E., and Polanía, C., 2007. The role of tropical dry forest as a long-term barrier to dispersal: a comparative phylogeographic analysis of dry forest tolerant and intolerant frogs. *Molecular Ecology* 16: 4789–4807. https://doi.org/10.1111/j.1365-294X.2007.03524.x

Crawford, A. J., Lips, K. R., and Bermingham, E., 2010. Epidemic disease decimates amphibian abundance, species diversity, and evolutionary history in the highlands of central Panama. *Proceedings of the National Academy of Sciences* 107: 13777–13782. https://doi.org/10.1073/pnas.0914115107

Daly, J. W. and Myers, C. W., 1967. Toxicity of Panamanian poison frogs (*Dendrobates*): some biological and chemical aspects. *Science* 156: 970–973. https://doi.org/10.1126/science.156.3777.970

Duellman, W. E. and Trueb, L., 1986. *Biology of Amphibians*. McGraw-Hill, New York. https://doi.org/10.2307/1445022

Dugas, M. B., Halbrook, S. R., Killius, A. M., del Sol, J. F., and Richards-Zawacki, C. L., 2015. Colour and escape behavior in polymorphic populations of an aposematic poison frog. *Ethology* 121: 1–10. https://doi.org/10.1111/eth.12396

EPYPSA, 2008. Revisión y actualización de la estrategia de desarrollo sostenible de la Provincia de Bocas del Toro: Plan indicativo de ordenamiento territorial funcional. Panamá.

Gagliardo, R., Crump, P., Griffith, E., Mendelson, J., Ross, H., and Zippel, K., 2008. The principles of rapid response for amphibian conservation, using the programmes in Panama as an example. *International Zoo Yearbook* 42: 125–135. https://doi.org/10.1111/j.1748-1090.2008.00043.x

Galeano, S. P., 2016. Ecological assembly of leaf-litter anuran communities across a Neotropical land-bridge archipelago. PhD Dissertation, Department of Biological Sciences, Louisiana State University, Baton Rouge, Louisiana.

Galeano, S. P. and Harms, K. E., 2016. Coloration in the polymorphic frog *Oophaga pumilio* associates with level of aggressiveness in intraspecific and interspecific behavioral interactions. *Behavioral Ecology and Sociobiology* 70: 83–97. https://doi.org/10.1007/s00265-015-2027-5

Gehara, M., Summers, K., and Brown, J. L., 2013. Population expansion, isolation and selection: novel insights on the evolution of color diversity in the strawberry poison frog. *Evolutionary Ecology* 27: 797–834. https://doi.org/10.1007/s10682-013-9652-0

Gratwicke, B., Ross, H., Batista, A., Chaves, G., Crawford, A. J., Elizondo, L. Estrada, A., Evans, M., Garelle, D., and Guerrel, J., 2016. Evaluating the probability of avoiding disease related extinctions of Panamanian amphibians through captive breeding programs. *Animal Conservation* 19: 324–336. https://doi.org/10.1111/acv.12249

Hertz, A., Lotzkat, S., and Köhler, G., 2013. New distribution records and variation of the two common lowland salamanders *Bolitoglossa colonnea* (Dunn, 1924) and *B. lignicolor* (Peters, 1873) in Panama (Amphibia: Caudata: Plethodontidae). *Check List* 9: 83–91. https://doi.org/10.15560/9.1.83

Hopkins, G. R. and Lahanas, P. N., 2011. Aggregation behaviour in a neotropical dendrobatid frog (*Allobates talamancae*) in western Panama. *Behaviour* 148: 359–372. https://doi.org/10.1163/000579511X559607

Jungfer, K. H., Lötters, S., and Jörgens, D., 2000. Der kleinste Pfeilgiftfrosche—eine neue *Dendrobates*-Art aus West-Panama. Herpetofauna. *Weinstadt* 22: 11–18.

Kaufmann, K. W. and Thompson, R. C., 2005. Water temperature variation and the meteorological and hydrographic environment of Bocas del Toro, Panama. *Caribbean Journal of Science* 41: 392–413.

Lawlor, T. E., 1982. The evolution of body size in mammals: evidence from insular populations in Mexico. *American Naturalist* 119: 54–72. https://doi.org/10.1086/283890

Institute for Tropical Ecology and Conservation (ITEC), http://itec-edu.org/, accessed August 2017.

IUCN, The IUCN Red List of Threatened Species, http://iucnredlist.org/, accessed August 2017.

Lips, K. R., Brem, F., Brenes, R., Reeve, J. D., Alford, R. A., Voyles, J., Cry, C., Livo, L., Pessier, A. P., and Collins, J. P., 2006. Emerging infectious disease and the loss of biodiversity in a Neotropical amphibian community. *Proceedings of the National Academy of Sciences* 103: 3165–3170. https://doi.org/10.1073/pnas.0506889103

Lips, K. R., Reeve, J. D., and Witters, L. R., 2003. Ecological traits predicting amphibian population declines in Central America. *Conservation Biology* 17: 1078–1088. https://doi.org/10.1046/j.1523-1739.2003.01623.x

Maan, M. E. and Cummings, M. E., 2012. Poison frog colors are honest signals of toxicity, particularly for bird predators. *American Naturalist* 179: E1–E14. https://doi.org/10.1086/663197

McKone, M. J., Moore, J. W., Harbison, C. W., Holmen, I. C., Lyons, H. C., Nachbor, K. M., Michalak, J. L., Neiman, M., Nicol, J. L., and Wheeler, G. R., 2014. Rapid collapse of a population of *Dieffenbachia* spp., plants used for tadpole-rearing by a poison-dart frog

(*Oophaga pumilio*) in a Costa Rican rain forest. *Journal of Tropical Ecology* 30: 615–619. https://doi.org/10.1017/S0266467414000467

MiAmbiente. 2011. Plan de acción para la conservación de los anfibios de Panamá. Panamá.

MiAmbiente. 2016. Resolución N° DM-0657-2016, por la cual se establece el proceso para la elaboración y revisión periódica del listado de las especies de fauna y flora amenazadas de Panamá, y se dictan otras disposiciones. Panamá.

Nijman, V. and Shepherd, C. R., 2010. The role of Asia in the global trade in CITES II-listed poison arrow frogs: hopping from Kazakhstan to Lebanon to Thailand and beyond. *Biodiversity and Conservation* 19: 1963–1970. https://doi.org/10.1007/s10531-010-9814-0

Patterson, B. D. and Atmar, W., 1986. Nested subsets and the structure of insular mammalian faunas and archipelagos. *Biological Journal of the Linnean Society* 28: 65–82. https://doi.org/10.1111/j.1095-8312.1986.tb01749.x

PNUD, 2009. Apoyo a la implementación del programa de desarrollo sostenible de Bocas del Toro fase II. Panamá.

Pröhl, H., 2002. Population differences in female resource abundance, adult sex ratio, and male mating success in *Dendrobates pumilio*. *Behavioral Ecology* 13: 175–181. https://doi.org/10.1093/beheco/13.2.175

Richards, C. L., Zellmer, A. J., and Martens, L., 2008. *Batrachochytrium dendrobatidis* not detected in *Oophaga pumilio* on Bastimentos Island, Panama. *Herpetological Review* 39: 200–202.

Richards-Zawacki, C. L., Wang, I. J., and Summers, K., 2012. Mate choice and the genetic basis for colour variation in a polymorphic dart frog: inferences from a wild pedigree. *Molecular Ecology* 21: 3879–3892. https://doi.org/10.1111/j.1365-294X.2012.05644.x

Schlaepfer, M. A., Hoover, C., and Dodd, C. K., Jr., 2005. Challenges in evaluating the impact of the trade in amphibians and reptiles on wild populations. *BioSciences* 55: 256–264. https://doi.org/10.1641/0006-3568(2005)055[0256:CIETIO]2.0.CO;2

Savage, J. M., 1968. The dendrobatid frogs of Central America. *Copeia* 1968: 745–776. https://doi.org/10.2307/1441845

Savage, J. M. and Wake, M. H., 2001. Reevaluation of the status of taxa of Central American caecilians (Amphibia: Gymnophiona) with comments on their origin and evolution. *Copeia* 2001: 52–64. https://doi.org/10.1643/0045-8511(2001)001[0052:ROTSOT]2.0.CO;2

Soto, R., Guzmán, H., Pinzón, Z., Moreno, J., Gamboa, C., Montoya, M., and Vargas, R., 1998. Evaluación Ecológica Rápida del Parque Nacional Marino Isla Bastimentos y áreas adyacentes, provincia de Bocas del Toro. Tomo 3: Recursos costero-marinos. Panama. Asociación Nacional para la Conservación de la Naturaleza. Panamá.

Spalding, A. K., 2013a. Lifestyle migration to Bocas del Toro, Panama: exploring migration strategies and introducing local implications of search for paradise. *International Review of Social Research* 3: 67–86. https://doi.org/10.1515/irsr-2013-0005

Spalding, A. K., 2013b. Environmental outcomes of lifestyle migration: land cover change and land use transitions in the Bocas del Toro Archipelago in Panama. *Journal of Latin American Geography* 12: 179–202. https://doi.org/10.1353/lag.2013.0038

Summers, K., Bermingham, E., Weigt, L., McCafferty, S., and Dahistrom, L., 1997. Phernotypic and genetic divergence in three species of dart-poison frogs with contrasting parental behavior. *Journal of Heredity* 88: 8–13. https://doi.org/10.1093/oxfordjournals.jhered.a023065

Tejera, V. H. and Dupuy, O. A., 2003. Anfibios del Museo de Vertebrados de la Universidad de Panamá: Catálogo. Panamá.

Valdés, V. V., 2014. Lista de especies de vertebrados en categoría de amenaza y su uso en el Archipiélago de Bocas del Toro (Panamá). *Revista Científica Centros* 3: 119–137.

Valdespino, I. A. and Santamaría, D., 1997. Evaluación ecológica rápida del Parque Nacional Marino Isla Bastimentos y áreas de Influencia: Isla Solarte, Swan Vay, Mimitimbi (Isla Colón) y el Humedal San San-Pond Sak, Provincia de Bocas del Toro. Asociación Nacional para la Conservación de la Naturaleza. Panamá.

Vertnet. http://portal.vertnet.org/, accessed August 2017.

Wang, I. J. and Shaffer, H. B., 2008. Rapid color evolution in an aposematic species: a phylogenetic analysis of color variation in the strikingly polymorphic strawberry poison-dart frog. *Evolution* 63: 2742–2759. https://doi.org/10.1111/j.1558-5646.2008.00507.x

Wang, I. J., Crawford, A. J., and Bermingham, E., 2008. Phylogeography of the pygmy rain frog (*Pristimantis ridens*) across the lowland wet forest of isthmian Central America *Molecular Phylogenetics and Evolution* 47: 992–1004. https://doi.org/10.1016/j.ympev.2008.02.021

Glossary

Allopatric: Refers to two or more populations or species that occur in separate, non-overlapping geographical areas.
Anagenesis: The evolutionary process whereby one species gradually changes until it becomes sufficiently different from being considered a new species rather than two forms evolving by splitting an ancestral species into two contemporaneous forms.
Aposematic: Refers to organisms that have colouration associated with harmful or distasteful characteristics that serve as a warning to would-be predators.
Arbustive: Containing copses of trees or shrubs.
Batrachofauna: The component of a fauna consisting of frogs.
Boid: Refers to members of the snake family Boidae.
Corneous: Consisting of a horn or having the appearance or texture of a horn.
Diel: Refers to the 24-hour period of day and night.
Ecomorphology: The study of the relationship between the ecological role of an individual and its morphological adaptations.
Econe: A species or component of species (such as a life-history stage, age class, morph, or sex) whose members share common patterns of resource-utilisation and niche characteristics but differ from other such components or species. Short definition: An intraspecific ecological niche.
Ectotherm: Refers to an animal that depends on the external environment for its source of body heat rather than producing its heat by metabolic means.
Ectrodactyly: A congenital condition characterised by the absence or malformation of one or more of the fingers or toes.
Ectromelia: A congenital absence of one or more limbs (fins, arms, or legs).
Edaphic: Pertains to the physical, chemical, and biological properties of the soil.
Edificarian: Refers to moral or intellectual instruction leading to the improvement of someone.
Electrophoresis: the movement of charged particles in a fluid or gel under the influence of an electric field
Elfin Woodland: A type of vegetation characterised by stunted trees, often forming a zone at the upper limit of tree-growth in the mountains.
Entomophagous: that feeds on insects.
Epigeal: Refers to an organism that lives on the surface of the ground.
Euryecious: Occupying a wide range of habitats.
Eurytropic: Refers to organisms with wide ecological tolerances.
Eurytrophic: Refers to organisms that consume a wide range of foods. Synonym: euryphagous.
Eustasy: a uniform change in sea level globally.
Exotrophic: having a free-living feeding larva. Not to be confused with exotropic.
Exotropic: A type of eye misalignment where one eye deviates outward.
Haplotype: A set of genetic determinants located on a single chromosome.
Holometabolism: in some insects, a complete metamorphosis that includes four stages of development, namely, egg, larva, pupa, and adult.
Hydroperiod: the number of days per year that an area of land is wet or the length of time there is standing water at a location.
Hygroscopic: Refers to a substance that absorbs moisture from the air.
Hyolaryngeal: pertains to a complex that includes the hyoid bone, thyrohyoid membrane, and laryngeal cartilages; it serves as an attachment site for the cricopharyngeus that forms the upper sphincter of the oesophagus.

Hyperxanthism: Refers to the quality of being unusually strongly coloured by a yellow pigment (usually caused by an accumulation of carotenoids).

Karyotype: the number and visual appearance of the chromosomes in the cellular nucleus of an organism or species.

Latifoliate forest: A forest of trees with broad leaves (as opposed to needle-like leaves).

Lentic: Refers to bodies of still water, such as pools, lakes, ponds, or swamps.

Lotic: Refers to bodies of flowing water, such as streams or rivers.

Monophyly: an ancestral population and all of its lineal descendants. Monophyletic groups are typically identified by shared derived characteristics.

Orographic: Refers to effects arising from the position or form of mountains, e.g. orographic precipitation; rainfall caused by mountains deflecting moisture-laden air upward into cooler atmospheres.

Palaeoisland: A part of a present-day landmass that had been separated as an island in past geological times.

Parapatric: A type of speciation that occurs when a smaller population is isolated, usually at the periphery of a larger group, and becomes differentiated to the point of becoming a new species.

Paraphyletic: Refers to groups that are descended from a common evolutionary ancestor or ancestral group, but not including all the descendant groups.

Philopatry: the tendency of an organism to stay in or to habitually return to a particular area.

Polytypic: a taxonomic group having many subdivisions. For example, a family with many genera or a genus with many species.

Prograding: Seaward advance of land as a result of the accumulation of waterborne sediment.

Saxicolous: Refers to organisms that live on or among rocks.

Scansorial: Refers to organisms that are adapted to living or climbing on vertical surfaces.

Semifossorial: Refers to animals that have a tendency to live underground.

Stentopic: Refers to organisms with narrow ecological tolerances.

Stenotrophic: Refers to organisms that consume a narrow range of foods. Synonym: stenophagous.

Stochasticism: The theory that specified kinds of events occur as a result of random variables.

Sympatric: Refers to species, or other taxa, that live in the same geographical area.

Syntopic: Refers to species, or other taxa, that live and occupy the same habitat or microhabitat.

Syntype: Any of two or more specimens listed in the original description of a taxon when a holotype was not designated.

Tubercle: A nodule on a bone, mucous membrane, or skin.

Vitellogenesis: the formation of yolk via nutrients deposited in an egg developing in the ovary.

Vomerine odontophores: Short, conical teeth-bearing processes of the anterior roof of the mouth of amphibians and located near the internal nasal openings.

Subject Index

Taxonomic Index

Numbers in **boldface** indicate illustration of taxon. Numbers in *italic* indicate information in a table, figure or appendix.

www.ingramcontent.com/pod-product-compliance
Ingram Content Group UK Ltd.
Pitfield, Milton Keynes, MK11 3LW, UK
UKHW051500020326
468520UK00002B/2

9 781784 272678